International Congress of Mathematicians (ICM 2022)

AA005058

Prize Lectures
Volume 1 of 7

Online
6-14 July 2022

Editors:

Dmitry Beliaev
Stanislav Smirnov

ISBN: 978-1-7138-9868-9

ICM INTERNATIONAL CONGRESS OF MATHEMATICIANS 2022 JULY 6–14

PRIZE LECTURES

EDITED BY D. BELIAEV AND S. SMIRNOV

INTERNATIONAL MATHEMATICAL UNION IMU

EMS PRESS

Editors

Dmitry Beliaev
Mathematical Institute
University of Oxford
Andrew Wiles Building
Radcliffe Observatory Quarter
Woodstock Road
Oxford OX2 6GG, UK

Email: belyaev@maths.ox.ac.uk

Stanislav Smirnov
Section de mathématiques
Université de Genève
rue du Conseil-Général 7–9
1205 Genève, Switzerland

Email: stanislav.smirnov@unige.ch

2020 Mathematics Subject Classification: 00B25

ISBN 978-3-98547-058-7, eISBN 978-3-98547-558-2, DOI 10.4171/ICM2022

⟶ Volume 1. Prize Lectures
ISBN 978-3-98547-059-4, eISBN 978-3-98547-559-9, DOI 10.4171/ICM2022-1
Volume 2. Plenary Lectures
ISBN 978-3-98547-060-0, eISBN 978-3-98547-560-5, DOI 10.4171/ICM2022-2
Volume 3. Sections 1–4
ISBN 978-3-98547-061-7, eISBN 978-3-98547-561-2, DOI 10.4171/ICM2022-3
Volume 4. Sections 5–8
ISBN 978-3-98547-062-4, eISBN 978-3-98547-562-9, DOI 10.4171/ICM2022-4
Volume 5. Sections 9–11
ISBN 978-3-98547-063-1, eISBN 978-3-98547-563-6, DOI 10.4171/ICM2022-5
Volume 6. Sections 12–14
ISBN 978-3-98547-064-8, eISBN 978-3-98547-564-3, DOI 10.4171/ICM2022-6
Volume 7. Sections 15–20
ISBN 978-3-98547-065-5, eISBN 978-3-98547-565-0, DOI 10.4171/ICM2022-7

This edition printed by Curran Associates, Inc.
ISBN 978-1-7138-9868-9

e-Book available: https://ems.press/books/standalone/273

No changes have been made to the content of these proceedings. There may be changes to pagination and minor adjustments for aesthetics.

The content of this volume is licensed under the CC BY 4.0 license, with the exception of the logos and branding of the International Mathematical Union and EMS Press, and where otherwise noted.

Bibliographic information published by the Deutsche Nationalbibliothek
The Deutsche Nationalbibliothek lists this publication in the Deutsche Nationalbibliografie; detailed bibliographic data are available on the Internet at http://dnb.dnb.de.

Published by EMS Press, an imprint of the

European Mathematical Society – EMS – Publishing House GmbH
Institut für Mathematik
Technische Universität Berlin
Straße des 17. Juni 136
10623 Berlin, Germany

https://ems.press

© 2023 International Mathematical Union

FOREWORD

It is often debated whether we should print proceedings in the Internet era. Printed books have been with us for centuries, they are great for archival purposes, and there is a significant number of users who prefer to read paper copies rather than stare at electronic screens. We believe that there is great value in the availability of printed volumes, so we are very happy that the Proceedings of the ICM 2022 are now available in print. At the same time, the Proceedings are available electronically and freely at the website of the International Mathematical Union and individual contributions are available through the website of the EMS Press, the publishing house of the European Mathematical Society.

These Proceedings consist of seven volumes. The first volume contains addresses from the IMU leadership, IMU Awards and Prizes (Fields Medals, Abacus Medal, Gauss Prize, Chern Medal and Leelavati Prize) presentations and laudations, the special Emmy Noether Lecture, and popular scientific expositions by Andrei Okounkov about the work of all Fields medalists as well as brief texts by Allyn Jackson about other prize recipients. The second volume contains all plenary lectures. The remaining five volumes contain sectional and special lectures.

We would like to thank Apostolos Damialis as well as Gabriella Böhm, Theresa Haney, Simon Winter, and the rest of the EMS Press team for their tireless work preparing these volumes, and Natalia Agapova for designing the covers and the style. We are grateful to the IMU for funding the production and printing of the Proceedings. We are particularly thankful to Andrei Okounkov who has written accessible notes about the work of the Fields medalists. Finally, we would like to thank all authors for preparing their contributions – without your work these volumes would not exist!

Oxford and Geneva, December 2023 — Dmitry Beliaev and Stanislav Smirnov

CONTENTS

VOLUME 1

PRIZE LECTURES

POPULAR SCIENTIFIC EXPOSITIONS

SUMMARIES OF PRIZE WINNERS' WORK

VOLUME 2

SPECIAL PLENARY LECTURES

PLENARY LECTURES

VOLUME 3

1. LOGIC

2. ALGEBRA

3. NUMBER THEORY – SPECIAL LECTURE

3. NUMBER THEORY

4. ALGEBRAIC AND COMPLEX GEOMETRY – SPECIAL LECTURE

4. ALGEBRAIC AND COMPLEX GEOMETRY

VOLUME 4

5. GEOMETRY – SPECIAL LECTURES

5. GEOMETRY

6. TOPOLOGY

7. LIE THEORY AND GENERALIZATIONS

8. ANALYSIS – SPECIAL LECTURE

8. ANALYSIS

VOLUME 5

9. DYNAMICS

10. PARTIAL DIFFERENTIAL EQUATIONS

11. MATHEMATICAL PHYSICS – SPECIAL LECTURE

11. MATHEMATICAL PHYSICS

VOLUME 6

12. PROBABILITY – SPECIAL LECTURE

12. PROBABILITY

13. COMBINATORICS – SPECIAL LECTURE

13. COMBINATORICS

14. MATHEMATICS OF COMPUTER SCIENCE – SPECIAL LECTURES

14. MATHEMATICS OF COMPUTER SCIENCE

VOLUME 7

15. NUMERICAL ANALYSIS AND SCIENTIFIC COMPUTING

18. STOCHASTIC AND DIFFERENTIAL MODELLING

19. MATHEMATICAL EDUCATION AND POPULARIZATION OF MATHEMATICS

20. HISTORY OF MATHEMATICS

INTERNATIONAL CONGRESSES OF MATHEMATICIANS

1897	Zürich	**1966**	Moscow
1900	Paris	**1970**	Nice
1904	Heidelberg	**1974**	Vancouver
1908	Rome	**1978**	Helsinki
1912	Cambridge, UK	**1983**	Warsaw
1920	Strasbourg	**1986**	Berkeley
1924	Toronto	**1990**	Kyoto
1928	Bologna	**1994**	Zürich
1932	Zürich	**1998**	Berlin
1936	Oslo	**2002**	Beijing
1950	Cambridge, USA	**2006**	Madrid
1954	Amsterdam	**2010**	Hyderabad
1958	Edinburgh	**2014**	Seoul
1962	Stockholm	**2018**	Rio de Janeiro

FIELDS MEDALISTS AND IMU AWARD WINNERS

FIELDS MEDAL

2018 Caucher Birkar, Alessio Figalli, Peter Scholze, Akshay Venkatesh
2014 Artur Avila, Manjul Bhargava, Martin Hairer, Maryam Mirzakhani
2010 Elon Lindenstrauss, Ngô Bảo Châu, Stanislav Smirnov, Cédric Villani
2006 Andrei Okounkov, Grigori Perelman,* Terence Tao, Wendelin Werner
2002 Laurent Lafforgue, Vladimir Voevodsky
1998 Richard E. Borcherds, W. Timothy Gowers, Maxim Kontsevich, Curtis T. McMullen
1994 Jean Bourgain, Pierre-Louis Lions, Jean-Christophe Yoccoz, Efim Zelmanov
1990 Vladimir Drinfeld, Vaughan F. R. Jones, Shigefumi Mori, Edward Witten
1986 Simon K. Donaldson, Gerd Faltings, Michael H. Freedman
1983 Alain Connes, William P. Thurston, Shing-Tung Yau
1978 Pierre R. Deligne, Charles L. Fefferman, Gregori A. Margulis, Daniel G. Quillen
1974 Enrico Bombieri, David B. Mumford
1970 Alan Baker, Heisuke Hironaka, Sergei Novikov, John G. Thompson
1966 Michael F. Atiyah, Paul J. Cohen, Alexander Grothendieck, Stephen Smale
1962 Lars Hörmander, John W. Milnor
1958 Klaus F. Roth, René Thom
1954 Kunihiko Kodaira, Jean-Pierre Serre
1950 Laurent Schwartz, Atle Selberg
1936 Lars V. Ahlfors, Jesse Douglas

ROLF NEVANLINNA PRIZE

2018 Constantinos Daskalakis
2014 Subhash Khot

---

*Grigori Perelman declined to accept the Fields Medal.

2010	Daniel Spielman
2006	Jon Kleinberg
2002	Madhu Sudan
1998	Peter W. Shor
1994	Avi Wigderson
1990	Alexander A. Razborov
1986	Leslie Valiant
1982	Robert Tarjan

CARL FRIEDRICH GAUSS PRIZE

2018	David L. Donoho
2014	Stanley Osher
2010	Yves Meyer
2006	Kiyosi Itô

CHERN MEDAL AWARD

2018	Masaki Kashiwara
2014	Phillip Griffiths
2010	Louis Nirenberg

LEELAVATI PRIZE

2018	Ali Nesin
2014	Adrián Paenza
2010	Simon Singh

OPENING GREETINGS BY THE IMU PRESIDENT

I am Carlos E. Kenig, the President of the IMU, and I would like to welcome those of you here in person in Helsinki, and the great majority of you who are participating online, to the 2022 virtual ICM.

As many of you know, the 19th General Assembly (GA) of the IMU and the 2022 ICM were scheduled to take place this July in Saint Petersburg, Russia. This did not happen. Due to the brutal Russian invasion of Ukraine, the Executive Committee (EC) of the IMU decided, at is virtual meeting in February 2022, that it would be impossible to host the scheduled GA and ICM in Russia. The EC expressed, at the time – and we reiterate it here today – its deepest sympathy to the Ukrainian people and to our Ukrainian colleagues in these grave circumstances.

The EC decided at this meeting to hold the GA as a traditional in-person event outside of Russia. After receiving several offers from our members, for which we are truly grateful, we decided to accept the extremely generous offer from Finland to host the GA. The GA was successfully held in Helsinki two days ago and many important decisions for the future of our Union were taken.

The ICM is a much more complex and extensive event than the GA. At the same EC meeting, we also decided on how to proceed with the ICM. Since ICM speakers had already been selected and notified, and the EC had a strong desire to follow the tradition of reporting the most exciting mathematical developments of the last four years, and given the infeasibility of alternative options, we decided to opt for a fully virtual ICM, with its organization taken over by the EC. Indeed, modern technology and the experience of the pandemic made this a realizable option. Moreover, the ongoing barbaric war that Russia continues to wage against Ukraine shows that there was no feasible alternative. The organization of the virtual ICM has been an enormously challenging and labor-intensive task. It had to be carried out under very difficult and constrained financial circumstances, and it has put a severe strain on IMU personnel, since we cut-off entirely the financial and organizational ties that Russia had been providing. Thus we are immensely grateful to the Heidelberg Laureate Forum Foundation for its very generous financial support. We are also indebted to the London Mathematical Society, the Mathematical Society of Japan, NTNU (the Norwegian University of Science and Technology), the Commission for Developing Countries (CDC), the International Commission on Mathematical Instruction (ICMI) and the Friends of the International Mathematical Union (FIMU) for their generous financial support. The IMU is also extremely grateful to the IMU Secretary General, Helge Holden, and to the IMU staff at the Secretariat in Berlin, especially to Scott Jung, the IMU Manager and Vanessa Chung, the IMU Assistant, for their remarkable efforts in dealing with the many administrative and logistical challenges that arose. We are also especially thankful to EC members Nalini Joshi and Paolo Piccione, and to Martin Hairer, chair of the Program Committee, for their very valuable help in the preparation of the virtual ICM. The self-organized ICM Satellite Coordination Group, led by Alexei Borodin, Martin Hairer and Terence Tao, has also worked independently of the IMU to coordinate the organization of various community based, overlay events in conjunction with the virtual ICM. We are extremely grateful for all these efforts.

We are all now greatly looking forward to this wonderful opportunity – unfolding over the next nine days – to learn about the most important mathematical discoveries of the last four years!

The scientific organization of an ICM is a daunting task. There are two very important committees of the IMU dedicated to this. The first one is the Structure Committee, whose remit it is to decide the scientific structure of the ICM. This is the inaugural congress for the Structure Committee, which is to remain a standing committee (with rotating membership) to enable the continuous evaluation and revision of the scientific structure of the Congress. The Structure Committee is chaired by Terence Tao (UCLA).

The second committee is the Program Committee (PC), whose charge is to select the speakers for the Congress. It is chaired by Martin Hairer (Imperial College London). The PC carried out its work with the advice of 20 sectional panels.

We are very thankful to all the self-less volunteers who did a wonderful job in developing the outstanding scientific organization of the Congress.

Traditionally, at the opening ceremony of an ICM there is an award ceremony, at which all the recipients of the IMU prizes are announced. At the virtual February 2022 meeting of the EC, it was decided that on July 5, the day after the GA, there would be a live award ceremony, at which all the IMU prizes would be announced. This took place yesterday, in Helsinki, with the generous help of our Finnish colleagues. At this ceremony, Hugo Duminil-Copin, June Huh, James Maynard and Maryna Viazovska were awarded the 2022 Fields Medal, Mark Braverman was awarded the inaugural IMU Abacus Medal, Barry Mazur was awarded the Chern Medal, Elliott H. Lieb was awarded the Gauss Prize and Nikolai Andreev was awarded the Leelavati Prize.

Today, July 6, is the opening day of the virtual ICM. We will take advantage of the fact that the Fields Medalists and the IMU Abacus Medalist are all present in Helsinki, so that the recipients can give their ICM plenary talks in front of a live audience in Helsinki.

Their lectures will, of course, also be streamed via the virtual ICM platform, for those who are not fortunate enough to be in Helsinki today.

Without further ado I declare the 2022 virtual ICM open!

Carlos E. Kenig
President of the IMU

ACKNOWLEDGEMENTS. I would like to thank Scott Jung, the IMU Manager, for his careful reading of the text, and numerous improvements.

CLOSING REMARKS BY THE IMU PRESIDENT

The virtual ICM 2022 has come to an end. We have been very fortunate to be able to carry out in full this fantastic event despite the many obstacles that we have encountered along the way. After the IMU Executive Committee (EC) decided – in the wake of the brutal Russian invasion of Ukraine – to hold this ICM as a virtual event, we worked tirelessly to set up the virtual platform for the ICM, working with the K.I.T. Group based in Berlin. The opportunity to do so was only afforded to us thanks to the financial assistance we received from several sources. We are particularly indebted to the Heidelberg Laureate Forum Foundation for their extremely generous contribution here. As was also planned by the EC at roughly the same time, the General Assembly (GA) of the IMU and the first ever IMU Award Ceremony, were carried out live and very successfully in Helsinki, Finland, on July 3–4 and July 5 respectively. We are extremely grateful to our Finnish colleagues and to the Council of Finnish Academies for their generous support.

The GA made very important decisions for the future of the IMU at its meeting. At the IMU Award Ceremony, we presented all the IMU awards (four Fields Medals, the IMU Abacus Medal, the Chern Medal Award, the Gauss Prize and the Leelavati Prize) and heard the laudations for the prizewinners. This beautiful event was attended in person by most of the award winners, and was opened by Mr Sauli Niinistö, the President of Finland. On July 6 we held the opening ceremony for the virtual ICM. The Fields Medal recipients, Hugo Duminil-Copin, June Huh, James Maynard and Maryna Viazovska, as well as the IMU Abacus Medal recipient, Mark Braverman, delivered their ICM plenary lectures in person in Helsinki. This was streamed live to the virtual ICM platform and also (fortunately) to the YouTube channel of the IMU. However, in the last few days before the virtual ICM launched, we experienced serious technical problems with the functionality of the virtual ICM platform that we had set up over several months with the K.I.T. Group. The K.I.T. Group, with the support of some members of the IMU EC, worked through the night of July 6 so that an alternative format was available by the morning of July 7, namely streaming via the IMU YouTube channel. This allowed the entire program of the virtual ICM to proceed as originally scheduled.

The virtual ICM was a great success. The newest results in many areas of mathematics and its applications were presented by the world's leading experts to a global audience. The videos of the lectures and panels are now permanently available via the IMU's YouTube channel, with free access for all. The Proceedings, edited by Dmitry Beliaev and Stanislav Smirnov, will be finalized soon, and will be published by the European Mathematical Society – EMS – Publishing House GmbH, with open access in perpetuity.

I would like to warmly thank all the speakers and panelists of the virtual ICM for their great effort in providing excellent talks and panel discussions under very challenging circumstances. I would also like to warmly thank the IMU Manager, Scott Jung, the IMU Assistant, Vanessa Chung, our colleague Martin Hairer and the K.I.T. Group for their tremendous help with the organization of the virtual ICM. In addition, I also wish to thank Nalini Joshi and Paolo Piccione, from the IMU EC, for their very valuable contributions to the organization of this virtual ICM. Something else contributed greatly to the success of this virtual ICM, namely the many community-based initiatives (carried out independently of the IMU), which organized satellite overlay events, allowing for personal interactions. We are very grateful for that, and especially for the efforts of Alexei Borodin, Martin Hairer and Terence Tao, who organized and led the ICM Satellite Coordination Group. Finally, I would like to give my warmest thanks to Helge Holden, the Secretary General of the IMU, for his indefatigable efforts and his tireless dedication, without which this event would not have been possible.

Let me conclude by mentioning two of the many important decisions taken by the GA.

Firstly, the incoming President and Secretary General of the IMU will be Hiraku Nakajima (Japan) and Christoph Sorger (France), respectively. We wish them all the best in carrying on the work of the IMU.

Secondly, the next General Assembly and ICM will take place in 2026, in New York City and Philadelphia respectively.

See you all in Philadelphia in July 2026!

Carlos E. Kenig
President of the IMU

ACKNOWLEDGEMENTS. I would like to thank Scott Jung, the IMU Manager, for his careful reading of the text, and numerous improvements.

STATUS REPORT FOR THE IMU

1. INTRODUCTION

As Secretary General of the IMU, I would like to take this opportunity to report on some of the activities of the International Mathematical Union (IMU) since we met in Rio de Janeiro, Brazil, in 2018.

The ultimate governing body of the IMU is the General Assembly (GA), consisting of Delegates appointed by the Adhering Organizations of the IMU, together with the members of the Executive Committee, and of the Representatives of Associate and Affiliate IMU Members. The GA normally meets once every four years, usually at a place and date close to an International Congress of Mathematicians (ICM). In the intervening period between GA meetings, the daily activities of the IMU are carried out by its Executive Committee (EC), which – in accordance with the IMU Statutes and subject to the direction and review of the members – conducts the business of the Union. The EC meets once a year (before the COVID-19 pandemic always in person), and in 2020, just before the onset of the pandemic, the EC met for the first time in Africa, in Cape Town. In addition, the focused activities of the IMU are organized into three Commissions and two Committees – the International Commission on Mathematical Instruction (ICMI), the Commission for Developing Countries (CDC), the International Commission on the History of Mathematics (ICHM), the Committee on Electronic Information and Communication (CEIC), and the Committee for Women in Mathematics (CWM). The members of these Commissions and Committees (C&Cs) all do excellent work for the global mathematical community in their respective areas – and they do so on a voluntary basis, for which we are extremely grateful. I refer to the respective websites of the C&Cs for a more detailed description of their ongoing activities.

Over the past term, there has been a considerable increase of attention devoted worldwide to the issues of diversity and inclusion. The IMU EC thus decided to create the

Committee on Diversity (CoD) as an ad hoc committee that can offer us advice on these issues. The purpose of CoD is threefold:

(1) Assess how the IMU has performed to date;

(2) Offer advice on how we can improve our performance;

(3) Offer advice on how the Adhering Organizations can improve their performance nationally.

The IMU is a founding and funding partner in the International Year of Basic Sciences for Sustainable Development (IYBSSD2022). In addition, the IMU recently signed the Declaration on Research Assessment (DORA) and the International Science Council's (ISC) declaration on Science in Exile.

In addition to the gratitude owed to those who serve on the IMU's C&Cs, I would also like to thank multiple organizations and individuals – too numerous to name individually – for their generous financial support of our activities. The IMU does not charge any overheads, and all IMU officers are true volunteers without any remuneration for their work. It is only because of this fact that we are able to utilize generous donations to the full effect and employ them wholly for their intended purpose.

2. MEMBERSHIP

Members of the IMU are countries, and only countries. In practice, for each country there is an Adhering Organization and a (National) Committee of mathematics that the IMU communicates with. It is not easy to describe in detail what the IMU really "is", except that the ICMs are organized under our auspices and that we award some of the most coveted prizes in mathematics. If one had to summarize it, one might say that the IMU is the "United Nations of Mathematics."

The IMU's regular membership has experienced positive developments in the past period, which were threefold in nature – we acquired new members, several of the present members upped their group of adherence, and, finally, but no less important, there are currently fewer countries experiencing serious problems in covering their membership dues.

At the 18th GA in 2018, Kyrgyzstan was admitted as a group I member, while both Chile and Portugal progressed from group II to group III. In the period 2019–2022, Mongolia was accepted as a new Associate Member, while Cyprus and Belarus were admitted as regular members in group I. We also had the following changes in group of adherence: Uzbekistan from Associate Member to a (regular) member in group I, Indonesia from group I to II, Denmark from group II to group III, and finally, Republic of Korea from group IV to group V.

Unfortunately, the associate membership of Cambodia, Madagascar, Moldova, and Nepal came to an end during the period 2019–2022 following the maximum period of two terms. The next stage would be an application for regular membership, which we hope will be forthcoming in due course. However, it appears to be difficult for many countries to secure the stable funding required to cover membership dues and thus take the step from associate to regular member.

Currently, the IMU has 87 regular and Associate Members and five Affiliate Members.

3. CONSEQUENCES FOR THE IMU OF THE COVID-19 PANDEMIC AND THE RUSSIAN INVASION OF UKRAINE

This term was characterized by two big events with global consequences, including for the IMU, namely the COVID-19 pandemic and the Russian invasion of Ukraine on 24 February 2022.

The COVID-19 pandemic has changed the life of billions of people across the globe. In addition to the tragic loss of millions of lives, the global economic consequences have been severe. We express our deepest sympathies with all affected. The pandemic has also had dramatic consequences for the way we work – and the IMU itself has not been immune to these effects. We have all had to practice remote teaching, participate in video and hybrid meetings, and attend fully virtual conferences. Furthermore, we have yet again seen the absolute necessity of fast and reliable internet to facilitate our daily work. While modern technology is truly amazing and makes rapid progress every day, we have also been keenly reminded that humans are social beings who need to interact in person. I am happy to be able to report below on a number of occasions on which the global mathematics community was able to come together and connect.

In passing, I can mention that mathematicians have been able to profit from pandemics in the past. In 1665, the University of Cambridge had to close due to the great plague, and the young Isaac Newton had to go home to Woolsthorpe where he remained for two years. During this period, he developed the theories of calculus, optics, and gravitation! While we have yet to see similar developments again this time round, we have been reminded of the importance of fundamental, curiosity driven research.

In response to the situation, the IMU established a COVID-19 resource page on the IMU website in May 2020, allowing for the submission of mathematically related papers on the emerging pandemic and also providing links to international online seminars, such as the One World series.

In terms of the specific ramifications of the pandemic for the work of the IMU, these were profound and touched every aspect of our activities.

ICME-14, the quadrennial congress organized under the auspices of ICMI, scheduled for July 2020 in Shanghai, had to be postponed due to the pandemic. It was however successfully carried out as a hybrid congress one year later, in July 2021 in Shanghai. The General Assembly of ICMI had already been carried out in 2020 as a fully virtual event, including the use of electronic voting for its elections.

Many of the programs of CDC involve exchange of scholars, and with travel and other restrictions in place for much of the last period, many of these programs came to a halt. However, alternative ways of carrying out scholarly exchange using modern technology have been investigated and developed. This may be one of the positive outcomes to result from the pandemic period, from which we will hopefully continue to benefit in the future.

The 2018 GA awarded the ICM 2022 and 19th GA to Saint Petersburg, Russia, scheduled for July 2022. The preparations for the ICM 2022 started at the first meeting of the IMU EC in 2019 and have been discussed at every meeting of the IMU EC since. The collaboration with the Russian organizers went well and ICM 2022 was shaping up to be an exciting and impressive congress. However, the brutal invasion by Russia of Ukraine on 24 February 2022 made holding the ICM and GA in Saint Petersburg impossible. The IMU EC had its annual meeting for 2022 scheduled for 24–27 February, and so it coincidentally had the opportunity to discuss how to proceed while the tragic events unfolded in real time.

Following the deliberations during its meeting, the IMU EC issued several statements in this connection (the full statements can be found on the IMU website), communicating its decision that:

- The General Assembly would be organized outside Russia on 3–4 July;
- The ICM would be organized as a fully virtual congress over 6–14 July;
- There would be an IMU Award Ceremony held outside Russia on 5 July;
- The ICM would be open to all participants;
- The GA and the ICM would be conducted without any financial contribution from the Russian Government;
- No official or representative of the Russian Government would be part of the organization or activities of the ICM.

This left the IMU EC with little time and no additional financial or human resources to organize a virtual ICM – which had never been done before – and an in-person GA at an unconfirmed location.

The IMU EC was approached by several member countries, offering to host the GA. In addition, we solicited bids from a few countries. We are extremely grateful to all those countries. Ultimately, we accepted an exceptionally generous offer from the Council of Finnish Academies to host the GA in Helsinki, Finland.

For the organization of the virtual ICM, our initial priority was to source a suitable platform for the event. We solicited bids from several companies, and we eventually accepted the bid from K.I.T. Group, a subsidiary of Messe Berlin GmbH.

These changes to the format of the ICM and GA prompted a further novelty, with the IMU EC deciding to hold a separate award ceremony as a live event on the day between the GA and the opening of the ICM. We were again supported by colleagues in Finland in hosting and organizing this unique event.

On 5 July we held the first ever IMU Award Ceremony, hosted in the Aula of Aalto University. We were honored that Mr Sauli Niinistö, the President of Finland, opened the ceremony. Fortunately, all four Fields Medalists – Hugo Duminil-Copin, June Huh, James

Maynard, and Maryna Viazovska – the inaugural IMU Abacus Medalist Mark Braverman, as well as the recipient of the Leelavati Prize, Nikolai Andreev, were able to be present in Helsinki. The recipients of the Chern Medal and the Gauss Prize, Barry Mazur and Elliott H. Lieb, participated in the ceremony remotely. For each winner, the audience heard the brief citation of the prize committee and a laudatio by an expert in the field, and also watched the superb videos produced by the Simons Foundation. The fully packed Aula delighted in the celebrations, with an electric atmosphere pervading the entire occasion.

Regarding the financial support, the IMU had promised to offer travel support for one delegate from each member country and offer full accommodation support for all delegates at the GA prior to the changes occasioned by the war in Ukraine. Despite the significant financial ramifications of having to plan the events anew after the cancellation of the ICM in Saint Petersburg, the IMU stood by this offer. While the costs were not insubstantial, it proved vital in facilitating member representation at the GA, especially in view of the particularly challenging times experienced in the last years.

Fortunately, the IMU received generous support towards the organization of the General Assembly, the Award Ceremony, and the virtual ICM from the following institutions and organizations:

- Heidelberg Laureate Forum Foundation
- Council of Finnish Academies
- London Mathematical Society
- Mathematical Society of Japan
- Norwegian University of Science and Technology
- CDC/ICMI/FIMU (Friends of the IMU)

The work in preparing both the GA and the virtual ICM put a severe strain on the leadership of the IMU and the IMU Secretariat. However, by the concerted effort of all involved, we were able to host a virtual ICM as well as a mostly in-person GA.

The 19th GA was the first hybrid GA in IMU history. The overwhelming majority of delegates wanted to participate in person – and indeed the quality of the meeting improves considerably with the informal discussions that only an in-person meeting allows for – but the travel restrictions due to the pandemic, air traffic problems in Europe, and many last-minute positive tests, resulted in about 30 remote participants, while 165 were present in Helsinki. For the first time all balloting was carried out electronically, in part due to the element of remote participation but also as a means of modernizing the voting procedure. The live voting system worked extremely smoothly and efficiently.

4. ICM 2022

As per the decision of the 2018 GA, the scientific program of the ICM is to be analyzed by the ICM Structure Committee. This includes novel ways to organize the congress, as well as the structure and relative size of the sections of the ICM. Based on recommendations from the Structure Committee, the IMU EC determined the scientific structure for the ICM. The selection of speakers for the ICM is the responsibility of the Program Committee, whose chair is appointed by the IMU President. The Program Committee is assisted by sectional committees for each section. All in all, there is approximately an equal number of people involved in the selection of speakers as there are speakers. The final selection of speakers is taken by the Program Committee, whose members, apart from the chair, are confidential until the opening of the ICM. On the other hand, membership in the Structure Committee is public, and the committee actively invites proposals and suggestions from the community.

The IMU Structure Committee for ICM 2022 was composed as follows: Terence Tao (US, chair), Carlos E. Kenig (IMU President, ex officio, US), Nalini Anantharaman (France), Alexei Borodin (US), Annalisa Buffa (Switzerland/Italy), Hélène Esnault (France/Germany), Irene Fonseca (US), Terry Lyons (UK), Stephane Mallat (France), Hiraku Nakajima (Japan), Éva Tardos (US), Peter Teichner (Germany), Akshay Venkatesh (Australia/US), Anna Wienhard (Germany).

The ICM 2022 Program Committee was composed as follows: Martin Hairer (UK, chair), Sanjeev Arora (US), Anna-Karin Tornberg (Sweden), Shafi Goldwasser (US), Ursula Hamenstaedt (Germany), Sergei Ivanov (Russia), Mikhail Kapranov (Japan), Yiming Long (China), Felix Otto (Germany), Jonathan Pila (UK), T. N. Venkataramana (India), Geordie Williamson (Australia).

On 6 July, we took advantage of the fact that the Fields Medalists and the IMU Abacus Medalist were all present in Helsinki, and arranged for them to give their plenary ICM prize lectures in front of a live audience in the Auditorium of Aalto University. This was a great success, with outstanding lectures delivered by the laureates.

Over 7–14 July we had a fully virtual ICM for the first time in IMU history. In the lead up to the ICM, we worked closely with K.I.T. Group to develop a platform that would not only deliver the lectures, but also allowed for a Q&A with the speakers. However, in the last few days before the virtual ICM launched, we were facing serious technical problems, which necessitated the restructuring of the format for the virtual ICM at very short notice. This was an exceptionally stressful period for all involved. In short, we ended up with a simplified platform that posted all talks to the IMU YouTube channel but eliminated the possibility of a Q&A with the lecturers. This was of course disappointing, but the upside was that, in this format, no registration was necessary, and thus the ICM was truly open to all.

Since a fully virtual ICM had not been tried before, all speakers were given the opportunity to submit a prerecorded lecture, or to give their talk live via zoom (with the possibility to submit a prerecorded lecture as a back-up). The program of lectures was scheduled according to the CEST time zone. Many lectures were given in person in front of a live audience, and we felt that the speakers had gone the extra mile to make their lectures as engaging and accessible as possible. The experience we have all gained in giving video lectures over the two years of the pandemic was certainly evident. Many of these efforts were coordinated by the independently-organized ICM Satellite Coordination Group led by Martin Hairer, Alexei Borodin, and Terence Tao. As satellite events, the IMU does not provide financial or organizational support to these activities, but we wholly commend the organizers for their efforts and are grateful for their engagement.

All lectures of the virtual ICM 2022 are now published on the IMU YouTube channel and it is positive to see that many are taking the opportunity to watch the lectures after they were given at the ICM. We hope that this will continue to serve as a repository of the most exciting and cutting-edge work currently being undertaken in mathematics, and one which is freely and openly accessible to the world. We are also certain that this feature of recording all ICM lectures and posting them on the IMU YouTube channel will become a standard for future ICMs.

5. INTERNATIONAL DAY OF MATHEMATICS – MARCH 14

As anticipated at the GA in 2018, UNESCO secured enough support for the proposal, and on 26 November 2019, declared March 14 ("π day") as the official International Day of Mathematics (IDM). The IMU EC set up a Governing Board for the IDM (IDMGB) and put out a call for both a logo and a host for the IDM website. Ultimately, the proposal from IMAGINARY, based in Germany, was chosen, with the dedicated IDM website idm314.org. The grand, official launch in 2020 of the new IDM at the UNESCO Paris Headquarters unfortunately had to be cancelled due to the pandemic, but the online events organized for the first IDM in 2020 were a great success. Each year a specific theme for the IDM is decided. The themes in the first three years of the IDM were:

Mathematics is Everywhere (2020)

Mathematics for a Better World (2021)

Mathematics Unites (2022)

As part of the events, the IDMGB coordinates activities worldwide, and material is developed in several languages. The IDM allows the IMU to reach out to new groups outside research mathematicians and, in a very short time, has managed to establish itself with a truly global presence.

INTERNATIONAL DAY OF MATHEMATICS MARCH 14

6. THE IMU CENTENNIAL

The IMU was founded in Strasbourg, France, on 20 September 1920. Together with our colleagues in France, the IMU organized the conference "Mathematics without Borders" at Université de Strasbourg, the same venue at which the IMU was founded. The conference – originally planned for 2020 – had to be postponed to September 2021 due to the pandemic. Although some of the talks still had to be given by video link due to the pandemic, many were delivered in person in front of a live audience. For many it was the first in-person event since the beginning of the pandemic and allowed many from the international community to reconnect. This, together with the celebratory occasion, helped make the conference a huge success.

To mark the occasion of the IMU's centennial, the IMU solicited the mathematician and historian Norbert Schappacher to write a book on the history of the IMU in May 2019. The book has just been published under the title "Framing Global Mathematics: The International Mathematical Union between Theorems and Politics" by Springer. Schappacher's book provides the most comprehensive and exciting take on the history of the IMU since Olli Lehto's "Mathematics without borders, a history of the International Mathematical Union", and more broadly analyzes the historical context of today's mathematics and its place in world culture. It is available in open access and can be freely downloaded. The project is generously funded by the Klaus Tschira Foundation.

7. A CHANGE IN THE STATUTES OF THE IMU

As a member of the ISC, the IMU already subscribes to the ISC's mission to "Defend the free and responsible practice of science," as described in Article 7 of the ISC Statutes. The GA wanted to make the IMU's commitment to this principle more explicit and thus adopted an amendment to the IMU Statutes to underscore its importance to the mathematical community. Specifically, the Article

> *"The Union adheres to the International Science Council's principle of embodying the free and responsible practice of science, freedom of movement, association, expression and communication for scientists, as well as equitable opportunities for access to science, its production and benefits, access to data, information and research material; and actively upholds this principle, by opposing any discrimination on the basis of such factors as ethnic origin, religion, citizenship, language, political or other opinion, gender, gender identity and sexual orientation, disability or age."*

was added to the IMU Statutes as a new Article 3.

The GA also passed a resolution expressing support for all mathematicians affected by the war in Ukraine, and in particular the IMU calls upon its members and other scientific organizations to do everything they can to assist our Ukrainian colleagues in these difficult times.

Some of the IMU's Adhering Organizations face temporary adverse circumstances that result in problems paying their IMU membership dues. Thus, the GA decided to establish a "reserve fund," based on earmarked donations to assist members in financial difficulties. The IMU EC will administer this fund and consider requests for assistance from members on a case-by-case basis.

8. ELECTIONS

The Executive Committee for the term 2023–2026 comprises President Hiraku Nakajima (Japan), Secretary General Christoph Sorger (France), Vice Presidents Ulrike Tillmann (UK) and Tatiana Toro (US/Colombia). Members-at-Large are Mouhamed Moustapha Fall (Senegal), Nalini Joshi (Australia), JongHae Keum (Republic of Korea), Paolo Piccione (Brazil), Günter Ziegler (Germany), Tamar Ziegler (Israel). Carlos E. Kenig (US) will serve as past President.

The majority of the members of the CDC are also elected by the GA. For the term 2023–2026 the President will be Andrea Solotar (Argentina), Secretary for Policy Ludovic Rifford (France), and Secretary of Grants Jose Maria P. Balmaceda (Philippines). Members-at-Large: Norbert Hounkonnou (Benin, representing Africa). Le Tuan Hoa (Vietnam, representing Asia). Mariel Saez (Chile, representing Latin America). Further members will be appointed by the IMU EC.

The IMU representatives on the ICHM Executive Committee for the term 2023–2026 are Guillermo Curbera (Spain), and Isobel Falconer (UK).

9. MEMBERS OF AWARD COMMITTEES

The IMU depends on a high number of volunteers that decide on the recipients of the many prizes that the IMU awards. The reputation of the prizes relies on the quality of the work carried out by the committees, and the IMU is grateful to committee members for their commitment and engagement. As is the tradition, all committee members except the chair are confidential until the prize recipients have been announced. For the 2022 IMU awards – including the first awarding of the IMU Abacus Medal, a continuation of the Rolf Nevanlinna Prize – the selection committees were composed as follows:

Fields Medal: Carlos E. Kenig (IMU President, chair, US), Artur Avila (Switzerland/Brazil), Camillo de Lellis (US/Switzerland), Michael Hopkins (US), Antti Kupiainen (Finland), Rahul Pandharipande (Switzerland), Alfio Quarteroni (Italy/Switzerland), Vera Serganova (US), Laure Saint-Raymond (France), Richard Taylor (US), Weiping Zhang (China), Tamar Ziegler (Israel).

IMU Abacus Medal: James Demmel (US, chair), Annalisa Buffa (Switzerland/Italy), Meena Mahajan (India), Amit Singer (US), Daniel Spielman (US).

Carl Friedrich Gauss Prize: Éva Tardos (US, chair), Liliana Borcea (US), Albert Cohen (France), Bernd Sturmfels (US/Germany), Tao Tang (Hong Kong, SAR).

Chern Medal Award: Yakov Eliashberg (US, chair) Sun-Yung Alice Chang (US), Henri Darmon (Canada), Alice Guionnet (France), Hiraku Nakajima (Japan).

Leelavati Prize: Pavel Etingof (US, chair), Kannan Soundararajan (US), Tadashi Tokieda (US), Chandrashekhar Khare (US), Michèle Artigue (France).

ICM Emmy Noether Lecture: Sylvia Serfaty (US, chair), Manindra Agrawal (India), Jesper Grodal (Denmark), Hinke Osinga (New Zealand), Anna Wienhard (Germany).

10. THE ICM IN 2026

It is the prerogative of the GA to decide on the venue for the next ICM. The IMU had received one bid for ICM 2026, namely from the Adhering Organization of the US. The GA gratefully accepted the bid. Thus, the 20th GA will take place in New York City on 19–20 July 2026, while ICM 2026 will be hosted in Philadelphia over 22–29 July 2026.

Helge Holden
Secretary General of the IMU

ACKNOWLEDGEMENTS. I would like to thank Scott Jung, the IMU Manager, for his careful reading of the text, and numerous improvements.

FUNDING. The IMU is proudly funded by its members, Germany for the IMU Secretariat, and its generous sponsors.

PHOTOGRAPHS

These photographs were taken by Jussi Rekiaro/Unigrafia and are courtesy of the International Mathematical Union.

James Maynard
Counting Primes

THE WORK OF THE FIELDS MEDALISTS AND THE IMU PRIZE WINNERS

THE WORK OF HUGO DUMINIL-COPIN

MARTIN HAIRER

ABSTRACT

The past decade has seen tremendous progress in our understanding of the behavior of many probabilistic models at or near their "critical point." On the 5th of July 2022, Hugo Duminil-Copin was awarded the Fields medal for the crucial role he played in many of these developments. In this short review article, we will try to put his work into context and present a small selection of his results.

MATHEMATICS SUBJECT CLASSIFICATION 2020

Primary 82B20; Secondary 82B26, 82B43

KEYWORDS

Ising model, Potts model, percolation

ICM INTERNATIONAL CONGRESS OF MATHEMATICIANS 2022 JULY 6–14

© 2022 International Mathematical Union
Proc. Int. Cong. Math. 2022, Vol. 1, pp. 26–49
DOI 10.4171/ICM2022/210

Published by EMS Press
and licensed under
a CC BY 4.0 license

1. INTRODUCTION

Hugo Duminil-Copin was awarded the Fields medal in Helsinki during the opening ceremony of the 2022 virtual ICM. In this short note, I will try to put his work into context and to give the reader a glimpse of why the questions it addresses are not only very interesting from a purely mathematical perspective, but also contribute to further our understanding of nature at a fundamental level. I should start first of all with a disclaimer. Hugo Duminil-Copin is an astounding problem solver and, while his interest falls squarely into the general area of probability theory and in particular the type of probabilistic problems that arise when studying microscopic models for statistical mechanics, I will not be able to do justice to the breadth of his contributions. Furthermore, my own area of expertise is somewhat tangential to that of Duminil-Copin, so this note should be taken as the point of view of an interested outsider. In particular, any misrepresentations of his results and/or techniques will be entirely due to my own ignorance.

In its broadest form, classical statistical mechanics can be thought of as the study of the global behavior of "large" systems (of "size" $N \gg 1$) that comprise many identical "small" subsystems interacting with each other. One typically indexes the subsystems by a discrete set Λ_N with $\lim_{N\to\infty} |\Lambda_N| = \infty$ and one is interested in quantities that are stable as $N \to \infty$. In many cases of interest, one has $\Lambda_N \subset \Lambda$ for Λ a discrete subset of Euclidean space (typically, a regular lattice) and its elements are interpreted as a physical location of the corresponding subsystem; the interaction between subsystems may then depend on their locations. (In most models they actually depend only on their relative positions, a notion that generalizes very well to locations taking values in more general symmetric spaces.)

Let us write S for the state space of one single such subsystem, so that the state space for the full system is $\mathcal{S}_N \stackrel{\text{def}}{=} S^{\Lambda_N}$. In *equilibrium statistical mechanics*, we furthermore assume that S is equipped with a "reference" probability measure μ (think of μ as being normalized counting measure if S is a finite set, normalized volume measure if it is a compact manifold, etc.) and that our system is described by an *energy function* $H^{(N)}: \mathcal{S}_N \to \mathbf{R}$, which typically comprises a contribution for each subsystem, as well as additional interaction terms. In full generality, one would have something like

$$H^{(N)}(\sigma) = \sum_{A \subset \mathcal{S}_N} H_A(\sigma_A), \tag{1.1}$$

where σ_A denotes the restriction of $\sigma \in S^{\Lambda_N}$ to S^A and the function H_A typically only depends on the "shape" of the subset A, so satisfies natural invariance properties under translations and possibly reflections and/or discrete rotations. In many classical models, the only nonvanishing terms in (1.1) are those with $|A| \leq 2$.

Given such an energy function, we obtain a probability measure $\mu_{\beta,N}$ on $\mathcal{S}_N$ by setting

$$\mu_{\beta,N}(d\sigma) = Z_{\beta,N}^{-1} \exp\bigl(-\beta H^{(N)}(\sigma)\bigr) \prod_{u \in \Lambda_N} \mu(d\sigma_u), \tag{1.2}$$

where $Z_{\beta,N}$ is chosen in such a way that $\mu_{\beta,N}(\mathcal{S}_N) = 1$. Physically, the parameter $\beta > 0$ appearing in this expression is the inverse of the temperature of the system. To a large extent, (equilibrium) statistical mechanics is the study of $\mu_{\beta,N}$ as $N \to \infty$ with a particular emphasis on the behavior under $\mu_{\beta,N}$ of observables that take a "macroscopic" (of the order of the size of the domain Λ_N) or "mesoscopic" (tending to infinity as $N \to \infty$ but much smaller than $|\Lambda_N|$) number of components of σ into account.

1.1. Bernoulli percolation

The simplest such example is that of $S = \{-1, 1\}$, $H_N = 0$, and $\mu(\{-1\}) = \mu(\{1\}) = \frac{1}{2}$. Regarding the index set Λ_N, we consider the case of the even elements of a large box in $\mathbf{Z}^2$, namely $\Lambda_N = \{u \in \{-N, \dots, N\}^2 : u_1 + u_2 \text{ even}\}$. (The reason why we make this strange choice rather than simply taking all elements of $\{-N, \dots, N\}^2$ will soon become clear.)

One of the simplest kind of "global" observables for this system is given by the following kind of linear statistics. Given a smooth function $\phi\colon [-1, 1]^2 \to \mathbf{R}$, we define $I_\phi^N \colon \mathcal{S}_N \to \mathbf{R}$ by

$$I_\phi^N(\sigma) = N^{-\alpha} \sum_{u \in \Lambda_N} \sigma_u \phi(u/N). \tag{1.3}$$

Note that this is exhaustive: for any fixed N, if we know $I_\phi^N(\sigma)$ for every smooth function ϕ, then we can in principle recover the argument σ itself. A version of the central limit theorem then immediately yields the following result:

Theorem 1.1. *Setting $\alpha = 1$, the joint distribution of $I_\phi^N(\sigma)$ for any finite collection of test functions ϕ as above converges as $N \to \infty$ to the law of a collection of jointly centered Gaussian random variables I_ϕ such that*

$$\mathbf{E} I_\phi I_\psi = \frac{1}{2} \int_{[-1,1]^2} \phi(x)\psi(x)\, dx.$$

(The factor $\frac{1}{2}$ appearing here comes from the fact that the local density of Λ_N in $\mathbf{Z}^2$ is $\frac{1}{2}$.)

A much more interesting kind of global observables is given by the connectivity properties of σ, which were first studied by Broadbent and Hammersley [12]. These are however *much* harder to analyze and, even though the model just described appears at first sight to be somewhat trivial, most of its results already lead us squarely into the 21st century mathematics. In order to describe what we mean by "connectivity" in this context, instead of interpreting elements $u \in \Lambda_N$ as points in $\mathbf{Z}^2$, we interpret them as nearest-neighbor edges of a suitable sublattice of $\mathbf{Z}^2$ by associating to u the unique edge e_u of $\mathbf{Z}_{\text{even}} \times \mathbf{Z}_{\text{odd}}$ with midpoint u. We will also write e_u^* for the edge of $\mathbf{Z}_{\text{odd}} \times \mathbf{Z}_{\text{even}}$ with midpoint u. In other words, we set

$$e_u = \begin{cases} (u_\downarrow, u_\uparrow) & \text{if } u_1 \text{ is even,} \\ (u_\leftarrow, u_\rightarrow) & \text{if } u_1 \text{ is odd,} \end{cases} \qquad e_u^* = \begin{cases} (u_\leftarrow, u_\rightarrow) & \text{if } u_1 \text{ is even,} \\ (u_\downarrow, u_\uparrow) & \text{if } u_1 \text{ is odd.} \end{cases}$$

Here, given $u = (u_1, u_2) \in \mathbf{Z}^2$, we write $u_\leftarrow = (u_1 - 1, u_2)$, etc. The endpoints of these edges do belong to the stated sublattices of $\mathbf{Z}^2$ since $u_1 + u_2$ is even, so either both u_1 and u_2 are even or both are odd.

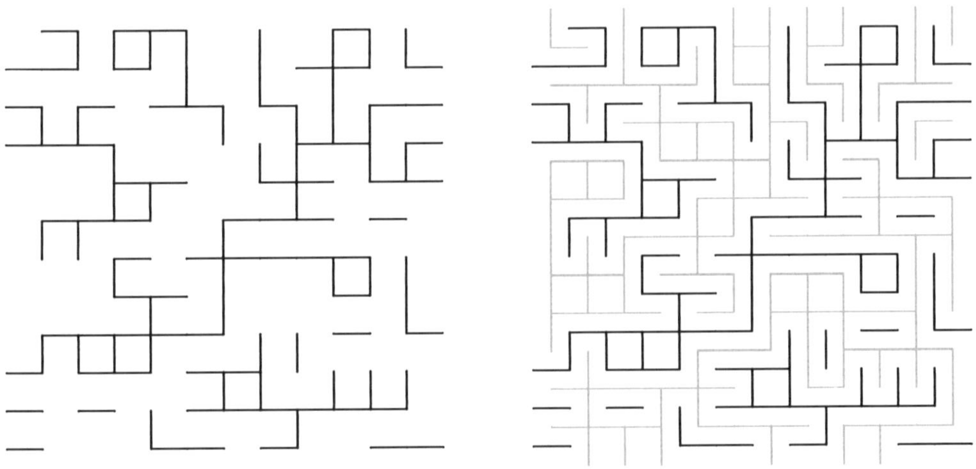

FIGURE 1

On the left, we draw a typical percolation configuration for $N = 11$. On the right, the same configuration is drawn together with its dual configuration in light blue.

Given a configuration $\sigma \in \mathcal{S}_N$, we interpret edges e_u with $\sigma_u = -1$ as "open" and draw them in black, while the remaining edges are considered "closed" and are drawn in light grey. This yields a picture like shown on the left in Figure 1. We can then ask, for example, what is the probability p_N that it is possible to go from the left boundary of the light gray graph to the right boundary (the "boundary" here consists of the ends of the dangling edges) while only traversing black edges. It turns out that this probability does take nontrivial values even for large values for N. As a matter of fact, it is independent of N as the following classical result (see, for example, [36, LEMMA 11.21]) shows.

Theorem 1.2. *One has* $p_N = \frac{1}{2}$ *for every* N.

Proof. The trick is to observe that given a configuration $\sigma \in \mathcal{S}_N$, if we draw the dual configuration $\sigma^* \in \mathcal{S}_N$ defined by $\sigma_u^* = -\sigma_u$ by coloring (in blue, say) the edges e_u^* with $\sigma_u^* = -1$, then we obtain a drawing with the property that blue edges never intersect black edges. As a consequence, it is possible to cross the square from left to right by traversing only black edges if and only if it is *not* possible to cross it from top to bottom by traversing only blue edges. (See Figure 1.) On the other hand, the law of the collection of blue edges is the same as that of the collection of black edges, only rotated by 90°, so that we must have $p_N = 1 - p_N$ as claimed. ■

Remark 1.3. If, instead of choosing edges to be open with probability $\frac{1}{2}$, we choose them to be open with some probability p, then we have $p_N \to 1$ for $p > \frac{1}{2}$ and $p_N \to 0$ for $p < \frac{1}{2}$. This is an example of *phase transition*: an abrupt change in the behavior of some global observables as a parameter of the model is varied continuously. In this specific example, we were able to determine the *critical value* $p_c = \frac{1}{2}$ explicitly by exploiting an exact duality.

It is similarly possible to obtain a large collection of interesting global observables by taking a shape $\mathcal{U} \subset [-1, 1]^2$ diffeomorphic to a square and considering the analogous event $A_{\mathcal{U}}^{(N)} \subset \mathcal{S}_N$ asking whether it is possible to connect the left and right edges of $N\mathcal{U}$ (without ever leaving $N\mathcal{U}$) by a path following only open edges of a given configuration $\sigma \in \mathcal{S}_N$. Again, the knowledge of these events is an exhaustive statistics for any given fixed value of N. It is furthermore known that for any finite number of such shapes $\{\mathcal{U}_i\}_{i \in I}$ (for I some finite index set) the random variables $\{[A_{\mathcal{U}_i}^{(N)}]\}_{i \in I}$ converge in law to a nondegenerate limit $\{[A_{\mathcal{U}_i}]\}_{i \in I}$ as $N \to \infty$ [56]. (Here, we write $[A]$ for the indicator function of an event A.) An amazing fact is that this scaling limit is conformally invariant: if $\phi: D \to D'$ is a conformal map between two smooth simply connected domains $D, D' \subset \mathbf{C}$ such that $[-1, 1]^2 \subset D$ and such that $\mathcal{V}_i \stackrel{\text{def}}{=} \phi(\mathcal{U}_i) \subset [-1, 1]^2$, then the joint law of the random variables $\{[A_{\mathcal{V}_i}]\}_{i \in I}$ is *the same* as that of $\{[A_{\mathcal{U}_i}]\}_{i \in I}$.

This conformal invariance turns out to be a crucial feature of the scaling limits of many equilibrium statistical mechanics models in two dimensions. It provides a link to conformal field theory which, at a purely mathematical level, can be thought of as the study of irreducible representations of the Virasoro algebra. In particular, it strongly suggests that the possible large-scale behaviors one can see for two-dimensional equilibrium models come in a one-parameter family of "universality classes" parametrized by the central charge of the corresponding conformal field theory. (In the case of percolation, it turns out that this central charge is given by $c = 0$.)

1.2. The Ising model

The next-"simplest" model of statistical mechanics falling into the category of equilibrium models described above is the Ising model [41, 43]. (See also the review article [16] in these proceedings which contains a more detailed account of the various developments spawned by this model.) In this case, the index set is given by $\Lambda_N = \{-N, \dots, N\}^d$ for some $d \geq 1$, the reference measure μ and local state space S are as above, but this time one has $H_A = 0$ unless $A = \{u, v\}$ with $u, v \in \mathbf{Z}^d$ such that $|u - v| = 1$, in which case one sets $H_A(\sigma) = -\sigma_u \sigma_v$. This time, the model has a nontrivial dependence on the parameter β appearing in (1.2), which plays a role somewhat similar to the parameter p that appeared in Remark 1.3.

At a very qualitative level, the situation is somewhat similar to what happened in the case for percolation: in every dimension $d \geq 2$, there exists a critical (dimension-dependent) value β_c which delineates two different regimes. At "high temperature," namely for $\beta < \beta_c$, the *spontaneous magnetization*, namely the random quantity $N^{-d} \sum_{i \in \Lambda_N} \sigma_i$, converges to 0 in probability as $N \to \infty$. For $\beta > \beta_c$, on the other hand, it converges in probability to a limiting random variable that can take exactly two possible values $\pm h_\beta \neq 0$ with equal probabilities. The actual value of β_c is only known in dimension 2 where it equals $\beta_c = \log \sqrt{1 + \sqrt{2}}$ [50]. (There is no phase transition at all in dimension 1 and the spontaneous magnetization always vanishes, so in some sense $\beta_c = +\infty$ there.)

It is again possible to ask the same questions as in the case of Bernoulli percolation. This time, however, even the analogue of Theorem 1.1, which was an essentially trivial consequence of the central limit theorem (or at least a version thereof), is already highly nontrivial. It was shown in a recent series of works [13,14] that if one chooses $\beta = \beta_c$ and $\alpha = 15/8$ in the expression (1.3) in dimension $d = 2$, then it converges in law to nontrivial limiting random variables, jointly for any fixed number of test functions. This time, however, the limiting distributions are not Gaussian (they actually exhibit an even faster decaying tail behavior). Note that the exponent α is closely related to the behavior of $\mathbf{E}_c\sigma_u\sigma_v$ (where $\mathbf{E}_c$ denotes the expectation under the Gibbs measure (1.2) for the critical value of the inverse temperature β) since, assuming that $\mathbf{E}_c\sigma_u\sigma_v \approx |u-v|^{-\delta}$, one finds that

$$\mathbf{E}_c\big(I_\phi^N(\sigma)\big)^2 = N^{-2\alpha}\sum_{u,v}\phi(u/N)\phi(v/N)\mathbf{E}_c\sigma_u\sigma_v \lesssim N^{-2\alpha}\sum_{u,v}|u-v|^{-\delta} \approx N^{2d-(\delta\wedge d)-2\alpha},$$

so that one expects the relation $\alpha = d - (\delta \wedge d)/2$, which (correctly) leads to the prediction $\delta = \frac{1}{4}$. This and a number of other properties of the Ising model at criticality allow associating it to the conformal field theory with central charge $c = \frac{1}{2}$.

The picture in higher dimensions is much less clear, however. For $d \geq 5$, it was shown in [1,2,29] that the correct scaling exponent to use in (1.3) at $\beta = \beta_c$ is $\alpha = 1 + \frac{d}{2}$ and that the limit is a Gaussian Free Field, namely the Gaussian random distribution with correlation function given by the Green's function of the Laplacian (with Neuman boundary conditions on the square). In dimension $d = 3$, virtually nothing is known rigorously about the critical Ising model, not even the value of its scaling exponents, although much progress has been made at a nonrigorous level with the development of the "conformal bootstrap" [23,24]. Regarding the case $d = 4$, it was somewhat unclear until very recently whether the Ising model at criticality should be "trivial" (i.e., described by Gaussian distributions) or not. This was eventually settled by Aizenman and Duminil-Copin in the work [3] where they show that any subsequential limit for expressions of the form (1.3) as $N \to \infty$ (and $\beta \to \beta_c$) must necessarily be Gaussian.

In fact, some of the results just mentioned are shown for the "lattice Φ^4 model" which is the equilibrium model with $S = \mathbf{R}$, as well as

$$H_{\{u\}}(\sigma) = V(\sigma_u) \stackrel{\text{def}}{=} \sigma_u^4 - \alpha\sigma_u^2, \quad H_{\{u,v\}}(\sigma) = \frac{1}{2}(\sigma_u - \sigma_v)^2,$$

again provided that u and v are nearest-neighbors, and with c an additional parameter. While this appears to be very different from the Ising model at first sight, we can see that it is actually a generalization of it: if the constant α is large, then the potential V has two very deep wells with minima located at $\pm\sqrt{\alpha}$, so its effect is to impose that $\sigma_u \approx \pm\sqrt{\alpha}$ with high probability. The main contribution then comes from the cross-term of the square in the two-body term which is the same as for the Ising model. These kind of considerations lead one to expect that, since these models exhibit long-range correlations at the critical temperature (in the sense that the correlation $\mathbf{E}\sigma_x\sigma_y$ decays slowly in $|x-y|$ as already pointed out earlier) which should furthermore lead to some form of self-averaging, the Ising model and the Φ^4 model exhibit the same behavior at criticality.

1.3. A general picture

The general picture that should by now be emerging from our discussion can be summarized as follows:

(1) Many of the simplest local equilibrium systems do exhibit a phase transition, namely there exists a critical value β_c at which the qualitative large scale behavior of the system changes abruptly. In general, a system may depend on additional parameters in which case one may see a more complicated *phase diagram* with several regions in parameter space where the global behavior of the system displays qualitatively different behavior. In any case, the "high temperature / small β phase" is expected to behave in such a way that what happens in well separated regions of space is very close to independent.

(2) In dimension 2, many of these systems appear to exhibit a form of conformal invariance at criticality, even though no rotation symmetry is built a priori into their description. When this happens, the link to $2d$ conformal field theories (and the associated probabilistic objects like SLE **[55]**, QLE **[45]**, etc.) provides a hugely powerful machinery to predict – and in a number of cases also rigorously prove – their behavior.

(3) The universe of local statistical mechanics models can be subdivided into broad classes of models that exhibit a shared large-scale behavior at criticality. These are called "universality classes" and, in the $2d$ equilibrium case, they are expected to come in families parametrized by a real parameter, the central charge. (For certain values of the central charge, one expects to have several "subclasses," but we will not discuss this kind of subtlety here.)

(4) Although one still expects conformal invariance at criticality in higher dimensions, this is a much smaller symmetry there and therefore appears to provide somewhat less insight.[1] One also expects the situation there to be more rigid than in two dimensions, with fewer universality classes. (Possibly only a discrete family.)

(5) Models that have "obvious" variants in every dimension typically have a critical dimension above which their behavior at criticality is "trivial" in the sense that it exhibits Gaussian behavior. (Typically, with correlation function given by the Green's function of the Laplacian.) In the case of the Ising universality class, this critical dimension is 4, while in the case of Bernoulli percolation it is 6.

One important branch of modern probability theory aims to put this general picture onto rigorous mathematical footing. The remainder of this article is devoted to a short overview of some of Hugo Duminil-Copin's many contributions to this vast programme.

---

1 See, however, the recent breakthrough made in the approximation of the critical exponents of the $3d$ Ising model using the "conformal bootstrap" **[23, 24]** already mentioned above.

This represents, of course, a mere sliver of his work and completely ignores very substantial chunks of it. By presenting not just a long laundry list of results that he proved and conjectures that he settled but instead an overview of the strategy of proof for a few select results, I hope to be able to convey one of the features of Duminil-Copin's body of work, namely that he has a knack for finding just the right way of looking at a problem that had hitherto been overlooked. In many cases, this only provides small cracks in the problem's armor that still require tremendous technical skill to be wedged open, but in some cases it results in surprisingly simple but ingenious proofs. Either way, I am very much looking forward to learning more from Duminil-Copin's insights for many years to come.

2. (DIS)CONTINUITY OF PHASE TRANSITIONS

One very natural question in this area is whether one can take the limit $N \to \infty$ in (1.2). At this stage, we note that the definition of $H^{(N)}$ given in (1.1) is not necessarily the most natural one since it restricts the sum over those clusters A that are constrained to *entirely* lie in S_N. Another possibility that appears just as natural would be to restrict the sum over clusters that merely intersect S_N, but to specify some fixed "boundary condition" $\bar{\sigma} \in S^\Lambda$ that is used to compute the values of the H_A with A intersecting both Λ_N and $\Lambda \setminus \Lambda_N$ in the sense that we interpret σ_A in (1.1) as $\sigma_{A,x} = \sigma_x$ for $x \in A \cap \Lambda_N$ and $\sigma_{A,x} = \bar{\sigma}_x$ otherwise.

In many examples of interest (including the case of the Ising model, but *not* the case of percolation), the measure $\mu_\beta = \lim_{N\to\infty} \mu_{\beta,N}$ is well-defined (i.e., independent of the choice of boundary condition) for $\beta < \beta_c$ while one can obtain several distinct limits in the case $\beta > \beta_c$. Figure 2 shows typical samples drawn from μ_β for the Ising model with $\bar{\sigma} \equiv 1$. In the case $\beta > \beta_c$, the resulting sample clearly "remembers" the bias introduced by $\bar{\sigma}$ in the sense that a typical configuration consists of a "sea" of spins taking the dominant value $+1$ (brown) with small "islands" of spins taking the value -1 (yellow). Had we set $\bar{\sigma} \equiv -1$, we

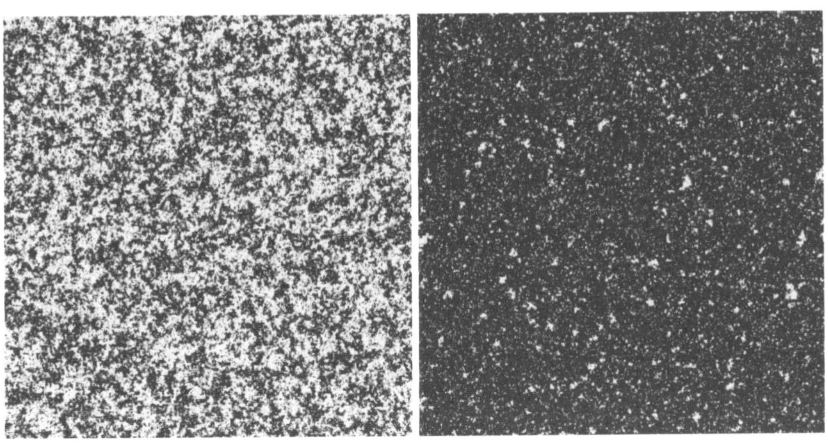

FIGURE 2
Typical Ising configurations for $\beta < \beta_c$ (left) and $\beta > \beta_c$ (right).

would have obtained a sample with the opposite behavior, which illustrates the nonuniqueness of the infinite-volume measure μ_β in this case. In the case $\beta < \beta_c$, on the other hand, each one of the two possible spin values is about equally represented and the measure is symmetric under the substitution $1 \leftrightarrow -1$, which illustrates the uniqueness of μ_β. It is in fact a theorem in the case of the Ising model that for $\beta > \beta_c$ there exist exactly two translation invariant infinite volume measures $\mu_\beta^\pm$ corresponding to boundary conditions $\bar{\sigma} \equiv \pm 1$ and that every accumulation point of $\mu_{\beta,N}$ for any sufficiently homogeneous boundary condition as $N \to \infty$ is a convex combination of μ_β^+ and μ_β^-.

This raises the question of the uniqueness of μ_β at $\beta = \beta_c$. If it is, then we say that the phase transition is *continuous*, otherwise it is said to be *discontinuous*. The reason for this terminology is that continuity in this sense turns out to be equivalent to the continuity of the maps $\beta \mapsto \mu_\beta^\pm$ at $\beta = \beta_c$. It has been known for quite some time [5, 60] that the phase transition for the Ising model is continuous in dimensions $d = 1, 2$ as well as $d \geq 4$. The reason why dimensions 1 and 2 are typically much better understood is that the Ising model is "solvable" in these dimensions in the sense that explicit expressions can be derived for the expectation of a large number of observables under $\mu_{\beta,N}$ (this solution is straightforward in $d = 1$ [41] where no phase transition is present, but it was a major breakthrough when Onsager obtained his exact solution for $d = 2$ [50]). Dimension $d = 4$, on the other hand, is the "upper critical dimension" beyond which the model is expected to be "trivial" (i.e., described by Gaussian random variables in the scaling limit) which allows using a number of powerful techniques, including, for example, the *lace expansion* [39, 54].

This leaves the case $d = 3$ which is, of course, the physically most interesting one since the Ising model is a toy model of ferromagnetism and its dimensions represent the usual spatial dimensions. Heuristic considerations suggest that the phase transition is also continuous there, and this is consistent with physical experiments, assuming that the Ising model belongs to the same universality class as that of a genuine physical magnet. In the article [4], Duminil-Copin et al. gave the first rigorous proof that this is indeed the case. The proof relies on the introduction of the quantity

$$M(\beta) = \inf_{B \subset \mathbf{Z}^3} \frac{1}{|B|^2} \sum_{x,y \in B} \int \sigma_x \sigma_y \, \mu_\beta^0(d\sigma),$$

where μ_β^0 denotes the infinite volume limit obtained from using "free" conditions, as well as three main steps. First, they rely on results of [30, 31] to argue that the Fourier transform of $x \mapsto \int \sigma_0 \sigma_x \, \mu_\beta^0(d\sigma)$ belongs to L^1 at $\beta = \beta_c$, which implies that $M(\beta_c) = 0$. Then, and this is the main step, they show that having $M(\beta) = 0$ implies that a certain percolation model with long-range correlations constructed from the Ising model admits no infinite clusters. Finally, they use a variant of the "switching lemma" [35] to show that the quantity $\int \sigma_0 \sigma_x \, \mu_\beta^+(d\sigma) - \int \sigma_0 \sigma_x \, \mu_\beta^0(d\sigma)$ is dominated by an explicit function times the probability of the origin belonging to an infinite cluster in the above mentioned model and therefore has to vanish at $\beta = \beta_c$. Once this is known, it is not too difficult to show that the spontaneous magnetization of the Ising model at criticality must vanish (namely, one has $\int \sigma_0 \, \mu_{\beta_c}^+(d\sigma) = 0$), which in turn yields the desired uniqueness statement.

To illustrate the fact that continuity of the phase transition, whatever the dimension, is a rather nontrivial property that is not necessarily expected in general, a good example is that of the Potts model [53]. This is defined similarly to the Ising model, but this time the local state space S is given by $S = \{1, \dots, q\}$ for some $q \geq 2$ endowed again with the normalized counting measure as its reference measure. As in the Ising model, one sets $H_A = 0$ unless $A = \{u, v\}$ with $u, v \in \mathbf{Z}^d$ such that $|u - v| = 1$, in which case one sets $H_A(\sigma) = \mathbf{1}_{\sigma_u = \sigma_v}$. For $q = 2$, this is equivalent to the Ising model since their energy functionals only differ by a constant. Let us also remark that there is an essentially equivalent model called the random cluster model (or sometimes the FK model after Fortuin and Kasteleyn who introduced it in [28]) in which one directly considers partitions of $\mathbf{Z}^d$ into connected "clusters" (which one should think of as the edge-connected components of the sets $\{u : \sigma_u = i\}$ for $i \in S$ and a given configuration σ of the Potts model) and which makes sense also for noninteger values of $q \geq 1$. (In the case $q = 1$, the FK model actually reduces to regular Bernoulli percolation.) See (4.1) below for a more precise definition of this model.

It was conjectured by Baxter in the 1970s [8,9] that the Potts model on $\mathbf{Z}^2$ exhibits a continuous phase transition if and only if $q \leq 4$. The pair of articles [17,21] by Duminil-Copin et al. provides proofs of both directions of this conjecture. For the sake of brevity, we will not comment on the proofs in any detail, but we note that the proof of continuity of the phase transition for $q \leq 4$ is almost completely disjoint from that in the case of the $3d$ Ising model. A milestone is again to show that the model at criticality with boundary condition set to one fixed element of S admits no infinite cluster. However, both the proof of this fact (exploiting a form of discrete holomorphicity of certain cleverly chosen observables) and the proof of its equivalence with the uniqueness of the infinite-volume measure at criticality (actually they show equivalence of a list of 5 quite distinct properties which are of independent interest for the study of the critical Potts model) are completely different.

Regarding the proof of *dis*continuity when $q > 4$, the main tool is a close relation, first discovered by Temperley–Lieb [59] in a restricted context and then by Baxter et al. [10] in more generality, between the FK model on $\mathbf{Z}^2$ and the so-called six-vertex model Configurations of the latter can be visualized as jigsaws where one assigns to each vertex of $\mathbf{Z}^2$ (or a subset thereof) one of the six (oriented) tiles

and one enforces the admissibility constraint that the tiles fit together seamlessly. One further postulates that the probability of seeing a given admissible configuration is proportional to $c^{\#p}$, where $\#p$ denotes the number of purple tiles in the configuration and c is some fixed constant. The relation between the six-vertex model and the critical FK model holds for the specific choice $c = \sqrt{2 + \sqrt{q}}$. The advantage gained from this relation is that the six-vertex model is "solvable" in a certain sense using the transfer matrix formalism. This does not get one out of the woods since the transfer matrices V_N involved are very large: they act on a vector space of dimension 2^N, but are block diagonal with each block $V_N^{[n]}$ acting on a subspace of dimension $\binom{n}{N}$. Each of these blocks is irreducible with positive entries and therefore admits a Perron–Frobenius vector. The main technical result of [21] is a very sharp asymptotic for the Perron–Frobenius eigenvalues of $V_N^{[N/2-r]}$ for fixed r as $N \to \infty$. Interestingly, the authors are able to prove that the ratios between these values converge to finite (and explicit, at least as explicit convergent series) limits as $N \to \infty$ and that the values themselves diverge exponentially in N with known exponent, but the common lower-order behavior of that divergence is not known. This asymptotic is, however, sufficient to obtain good control over the partition function of the six vertex model and to exploit it to compute an explicit expression for the inverse correlation length of the critical Potts model with free boundary conditions when $q > 4$. The finiteness of that expression finally allows deducing the discontinuity of the phase transition.

To conclude this section, I would like to mention the beautiful article [20] which, although not quite dealing with the question of continuity of the phase transition, does have a related flavor. The question there is that of the "sharpness" of the phase transition which in this particular case is couched as the question whether it is really true that the measure μ_β has exponentially decaying correlations (in the sense that the covariance between $f(\sigma_0)$ and $f(\sigma_x)$ decays exponentially fast as $|x| \to \infty$ for any "nice enough" function $f: S \to \mathbf{R}$) for *every* $\beta < \beta_c$ and not just for small enough values where a perturbation argument around $\beta = 0$ (where $f(\sigma_0)$ and $f(\sigma_x)$ are independent under μ_0 as soon as $x \neq 0$) may apply. One difficulty with this type of statement is that one will in general not know any closed-form expression for β_c: in the case of the FK model on the square lattice, such an expression can be derived by a duality argument [11], but it is not known for more general situations. The main result of [20] is that the phase transition of the FK model on *any* vertex-transitive infinite graph is sharp.

The main tool in their proof is a novel and far-reaching generalization of the OSSS inequality [49]. The context here is that of increasing random variables $f: \{0,1\}^E \to [0,1]$ (for a finite set E and for the natural coordinate-wise partial order on $\{0,1\}^E$) where $\{0,1\}^E$ is furthermore equipped with a probability measure $\mathbf{P}$ that is itself *monotonic* in the sense that for every $F \subset E$ and every $e \in E \setminus F$, the conditional probabilities $\mathbf{P}(w_e = 1 \mid \mathcal{F}_F)$ are increasing functions. (Here $\mathcal{F}_F$ denotes the σ-algebra generated by the evaluations $w \mapsto w_e$ for $e \in F$.) One then considers *any* algorithm that reveals one by one the values of an input $w \in \{0,1\}^E$ in such a way that the coordinate to be revealed next depends in a deterministic way on the information gleaned from the revealment up to that point. (In particular, the first coordinate to be revealed is always the same since no information has been obtained yet at that point.) The algorithm stops once the revealed values are sufficient to determine the value of $f(w)$, thus yielding a random set $\hat{E} \subset E$ of revealed values. The result of [20] is then that one has the inequality

$$\operatorname{Var}(f) \leq \sum_{e \in E} \mathbf{P}(e \in \hat{E}) \operatorname{Cov}(f, w_e), \tag{2.1}$$

which looks formally the same as the result of [49], but the assumption there was that the measure $\mathbf{P}$ is simply the uniform measure. Since the latter is clearly monotonic (it is such that $\mathbf{P}(w_e = 1 \mid \mathcal{F}_F)$ is constant), the results of [49] follow as a special case.

Using this result, the authors of [20] then obtain the following dichotomy which yields the desired sharpness statement.

Theorem 2.1. *Let G be any transitive graph and let $\mathbf{P}_{\beta,n}$ be the FK measure on the ball Λ_n of radius n in G. Then, there exists $\beta_c \in \mathbf{R}$ such that, for every $\beta < \beta_c$ there exists $c_\beta > 0$ such that $\mathbf{P}_{\beta,n}(0 \leftrightarrow \partial\Lambda_n) \lesssim e^{-c_\beta n}$, uniformly in n. For $\beta > \beta_c$, on the other hand, there exists $c > 0$ such that $\mathbf{P}_{\beta,n}(0 \leftrightarrow \partial\Lambda_n) \geq c \min\{1, \beta - \beta_c\}$.*

Once (2.1) is known, the proof is surprisingly simple and relies on two ingredients. First, one can show that the measures $\mathbf{P}_{\beta,n}$ and the function $\mathbf{1}_{0\leftrightarrow\partial\Lambda_n}$ satisfy the assumptions of (2.1). Setting $\theta_n(\beta) = \mathbf{P}_{\beta,n}(0 \leftrightarrow \partial\Lambda_n)$, a clever choice of search algorithm for the (potential) cluster connecting the origin 0 to $\partial\Lambda_n$ then allows showing that one has the bound

$$\theta_n'(\beta) \gtrsim \sum_{e\in E} \mathrm{Cov}_\beta(\mathbf{1}_{0\leftrightarrow\partial\Lambda_n}, w_e) \geq \frac{n}{8\Sigma_n(\beta)}\theta_n(\beta)\big(1-\theta_n(\beta)\big), \tag{2.2}$$

where $\Sigma_n = \sum_{k=0}^{n-1}\theta_n$. The fact that the first inequality holds is known and can be checked in an elementary way. The second fact is that *any* sequence of functions $\beta \mapsto \theta_n(\beta)$ satisfying a differential inequality of the form (2.2) necessarily satisfies a dichotomy of the type appearing in the statement of Theorem 2.1. Since we are not interested in the regime where θ_n is large, we can rewrite (2.2) as $\theta_n' \geq \frac{cn}{\Sigma_n}\theta_n$. The fact that the θ_n then should satisfy such a dichotomy is quite clear: if β is such that they converge to a nonvanishing limit θ, then $\Sigma_n/n \sim \theta$ and one must have $\theta' \geq c$. If, on the other hand, they converge to 0 on a whole interval $[a, b]$, then that convergence must take place sufficiently fast so that $\Sigma_n/n \gg \theta_n$ (since otherwise the previous argument applies). Since $\Sigma_n/n \sim \theta_n$ for $\theta_n \sim n^{-\alpha}$ as soon as $\alpha < 1$, it is then plausible that for any $c < b$ one has $\theta_n \ll n^{-1/2}$ (say), implying $\theta_n' \gtrsim \sqrt{n}\theta_n$ and therefore $\theta_n \lesssim e^{-\sqrt{n}(c-\beta)}$ for $\beta < c$. This shows that Σ_n is bounded for $\beta < c$, leading to $\theta_n' \gtrsim n\theta_n$ and therefore an exponentially (in n) small bound as claimed.

3. TRIVIALITY OF Φ_4^4

It has been known since the groundbreaking work of Osterwalder and Schrader [51, 52] that, at least in some cases, the construction of a (bosonic) quantum field theory satisfying the Wightman axioms is equivalent to the construction of a probability measure on the space of distributions satisfying a number of natural properties. One of the pinnacles of that line of enquiry was the construction in the seventies of the Φ_2^4 and Φ_3^4 measures [22, 25, 27, 33, 34, 47, 48, 57], which corresponds to the simplest case of an interacting theory in two or three space-time dimensions with one type of boson.

At a heuristic level, the Φ_d^4 measure is the measure $\mu^{(d)}$ on the space of Schwartz distributions $\mathcal{S}'(\mathbf{R}^d)$ (or on the d-dimensional torus) given by

$$\mu^{(d)}(d\Phi) = Z^{-1}\exp\Big(-\frac{1}{2}\int\big(|\nabla\Phi(x)|^2 - C\Phi^2(x) + \Phi^4(x)\big)\,dx\Big)\,d\Phi,$$

where "$d\Phi$" denotes the infinite-dimensional Lebesgue measure on $\mathcal{S}'(\mathbf{R}^d)$. This expression is, of course, problematic at many levels: infinite-dimensional Lebesgue measure does not exist, distributions cannot be squared, etc. If it were only for the term $|\nabla\Phi|^2$, one could reasonably interpret this expression as the Gaussian measure μ_0 with covariance operator given by the Green's function of the Laplacian, which is a well-defined probability measure (modulo technicalities arising from the constant mode which can easily be fixed). The measure μ_0 is called the Gaussian Free Field (GFF) since it corresponds to a quantum field theory in which particles are free, i.e., do not interact with each other at all.

This suggests that a more refined interpretation of the Φ^4_d measure could be given by

$$\mu^{(d)}(d\Phi) = Z^{-1} \exp\left(-\frac{1}{2}\int \Phi^4(x)\,dx\right)\mu_0(d\Phi). \tag{3.1}$$

This is still ill-defined since the GFF is supported on distributions rather than functions for any dimension $d \geq 2$. However, setting $\Phi_\varepsilon = \rho_\varepsilon \star \Phi$, the *Wick power*

$$:\!\Phi^4\!: = \lim_{\varepsilon\to 0}\left(\Phi^4_\varepsilon - 3\Phi^2_\varepsilon \mathbf{E}\Phi^2_\varepsilon\right), \tag{3.2}$$

turns out to be a well-defined random Schwartz distribution (i.e., the limit exists and is independent of the choice of ρ_ε) in dimensions $d < 4$. In dimension 2, Nelson showed in [47] that the Radon–Nikodym factor appearing in (3.1) with Φ^4 replaced by $:\!\Phi^4\!:$ yields an integrable random variable, thus leading to a definition of $\mu^{(2)}$. In particular, the Φ^4_2 measure is equivalent to the GFF. In dimension 3, this turns out not to be the case, but it is still possible to show that the measure

$$\mu^{(3)}(d\Phi) = \lim_{\varepsilon\to 0} Z_\varepsilon^{-1} \exp\left(-\frac{1}{2}\int \Phi^4_\varepsilon(x) - C_\varepsilon \Phi^2_\varepsilon(x)\,dx\right)\mu_0(d\Phi) \tag{3.3}$$

is well-defined for a suitable choice of the constant C_ε which differs from the choice $3\mathbf{E}\Phi^2_\varepsilon \sim \varepsilon^{-1}$ suggested by (3.2) by a logarithmically divergent term. (An alternative construction of this measure was recently obtained by completely different techniques in [37, 38, 46].)

This discussion begs the question of what happens for $d \geq 4$ and especially when $d = 4$ which is the physically most interesting case from the QFT perspective (remember that dimension here corresponds to space-time). Regarding the case $d > 4$, it was already shown in the eighties by Aizenman and Fröhlich [1, 2, 29] that pretty much any "reasonable" definition of the Φ^4_d measure actually coincides with the GFF. This still left the case $d = 4$ which has always been expected to be the hard case since it is "critical" in the sense that, at least at a formal level, the terms Φ^4 and $|\nabla\Phi|^2$ scale in the same way in the following sense. Writing $\mathcal{S}_\lambda$ for the transformation $(\mathcal{S}_\lambda F)(x) = F(\lambda x)$, the GFF has the self-similarity property $\mathcal{S}_\lambda \Psi \stackrel{\text{law}}{=} \lambda^{\frac{2-d}{2}}\Psi$ for Ψ drawn from μ_0. Pretending that Ψ behaves like a function (even though it really is a random distribution), we deduce that

$$\mathcal{S}_\lambda|\nabla\Psi|^2 = \lambda^{-2}|\nabla\mathcal{S}_\lambda\Psi|^2 \stackrel{\text{law}}{=} \lambda^{-d}|\nabla\Psi|^2, \quad \mathcal{S}_\lambda(\Psi^4) = (\mathcal{S}_\lambda\Psi)^4 \stackrel{\text{law}}{=} \lambda^{4-2d}\Psi^4.$$

These exponents are indeed equal if and only if $d = 4$. A heuristic calculation actually suggests that, at higher order, the term $|\nabla\Psi|^2$ dominates the term Ψ^4 at large scales. Variants of this observation have been made rigorous in a number of works [26, 32, 48], including most recently in an impressive series of works by Bauerschmidt–Brydges–Slade (see [6,7] and the references therein).

One way of formulating one of their main results is the framework given in our introduction with $S = \mathbf{R}$, μ being Lebesgue measure, $H_{\{x\}}(\phi) = \frac{g}{4}\phi_x^4 + \frac{\nu}{2}\phi_x^2$, $H_{\{x,y\}}(\phi) = |\phi_x - \phi_y|^2$ when x and y are neighboring lattice sites in $\mathbf{Z}^4$, and $H_A = 0$ otherwise. This model behaves in a way that is very similar to the Ising model, to which it degenerates in the regime $g \to \infty$ and $\nu = -g$. Traditionally, one considers the Φ^4 model with $\beta = 1$, since one can always reduce oneself to this case by adjusting g and ν, and possibly rescaling the ϕ_x's by a factor. One typically also considers g fixed, it is therefore the parameter ν that is tuneable and plays the role of a "temperature" in this model. Just like the Ising model, it exhibits a phase transition at some value $\nu_c \in \mathbf{R}$: for $\nu > \nu_c$, there exists a unique infinite volume measure which is symmetric under $\phi \mapsto -\phi$. For $\nu < \nu_c$, on the other hand, one finds two distinct infinite-volume measures (as well as their convex combinations) depending on the boundary conditions one chooses.

A state $\phi \in S^{\Lambda_N}$ with $\Lambda_N = \{-N, \dots, N\}^4$ is viewed as a distribution $\iota\phi$ on the torus (of size 2) by setting, for every smooth test function $f: \mathbf{T}^4 \to \mathbf{R}$,

$$(\iota\phi)(f) = \sum_{x \in \Lambda_N} \sigma_N \phi_x f(x/N),$$

for a sequence of values σ_N chosen in such a way that $\mathbf{E}((\iota\phi)(1)^2) = 1$. It is then shown in [6] that if g is sufficiently small and ν is chosen in a suitable way (close but not quite equal to the critical value ν_c), then $\iota\phi$ converges to a massive GFF, namely the Gaussian field with covariance given by $(m^2 - \Delta)^{-1}$ for some $m \in \mathbf{R}$ (which depends on the specific way in which ν is being tuned to approach ν_c as $N \to \infty$).

While this result strongly suggests that there exists no nontrivial Φ_4^4 measure, it does not rule out the possibility of having a nontrivial scaling limit for the discrete field we just described at (or near) criticality when the constant g is sufficiently large (in other words, "at strong coupling"). The technique of proof of [6] was to implement a rigorous version of the "renormalization group technique," which relies on a subtle analysis of the behavior of the renormalization map near the fixed point given by the GFF. This is unfortunately perturbative in nature and so has little hope of being able to deal with arbitrary g. In the recent work [3], however, Aizenman and Duminil-Copin finally succeeded in showing the following result.

Theorem 3.1. *For every way of adjusting $g = g_N$ and $\nu = \nu_N$ as $N \to \infty$ such that $\nu_N \geq \nu_{c,N}$, every $M_N \to \infty$ with $1 \ll M_N \ll N$, and every smooth compactly supported test function f, the law of $\xi_N^f = \sum_{x \in \Lambda_N} \phi_x f(x/M_N)$, normalized so that its variance is one, converges to a normal distribution.*

Remark 3.2. The condition $\nu_N \geq \nu_{c,N}$ can actually be slightly relaxed but not too much. This is because, in the "low temperature" regime and with free (or periodic) boundary conditions, one would expect the law of ξ_N^f to converge to a Bernoulli random variable rather than a Gaussian.

At a high level, the main reason why [3] can deal with arbitrary couplings is that one can think of their setting as being more akin to "perturbing around $g = \infty$" rather than around $g = 0$. In the setting of the introduction, they start by considering the Ising model as described there (i.e., with $\mu = \frac{1}{2}(\delta_1 + \delta_{-1})$), but then expand their class of models to allow for each site to represent a collection of spins with arbitrary ferromagnetic interactions within a site, instead of a single spin. This has the effect of replacing μ by any measure that can be obtained as the law of $\delta \sum_{i=1}^{K} s_i$ for some $\delta > 0$ and $K \in \mathbf{N}$, and where the $s_i \in \{-1, 1\}$ are random variables with a joint distribution proportional to $\exp(-\sum_{ij} a_{ij} s_i s_j)$ for some arbitrary but *positive* coefficients a_{ij}. As was shown already in the 1970s [58, THEOREM 1], all probability measures on $\mathbf{R}$ of the type $Z^{-1} \exp(cx^2 - gx^4)\,dx$ can be obtained as limits of such measures, so that the discrete Φ^4_4 model can be viewed as a limit of block-spin models.

Recall that to show that a collection $\{X_a\}_{a\in A}$ of real-valued random variables is jointly Gaussian it suffices to show that all joint fourth cumulants $\mathbf{E}_c\{X_{a_1}, \dots, X_{a_4}\}$ with $a_i \in A$ vanish. It is therefore not surprising that fourth cumulants of the spin variables play an important role in any proof of Gaussianity for Ising-type models. In dimension $d \geq 5$, the proof in [2] relied on two very important facts. First, writing $C(x, y) = \mathbf{E}(\sigma_x \sigma_y)$ for the spin correlation function, one shows that for any temperature any any ferromagnetic interaction, one has the bound

$$\left|\mathbf{E}_c\{\sigma_{x_1}, \dots, \sigma_{x_4}\}\right| \leq 2 \sum_{y\in\mathbf{Z}^d} C(x_1, y) \cdots C(x_4, y). \tag{3.4}$$

One then observes that *at the critical temperature*, the function C is bounded by

$$C(x, y) \lesssim |x - y|^{2-d}. \tag{3.5}$$

Consider now four smooth compactly supported test functions f_i and define

$$X_i = \sum_{x\in\mathbf{Z}^d} \sigma_x f_i(x/M).$$

In particular, the sum ranges over $\mathcal{O}(M^d)$ sites. If one assumes that (3.5) is sharp, then one expects to have $\mathbf{E}X_i^2 \approx M^{d+2}$, so that the "correct" normalization for the X_i's to have unit variance is expected to be $\xi_i = M^{-\frac{d+2}{2}} X_i$. On the other hand, combining the covariance bound with the bound on the fourth cumulant, a powercounting argument shows that $\mathbf{E}_c\{\xi_1, \dots, \xi_4\} \lesssim M^{-2(d+2)} M^{d+8} = M^{4-d}$, which does indeed converge to 0 as $M \to \infty$ when $d > 4$, thus showing that the ξ_i's are jointly Gaussian in the limit.

Clearly, this calculation does not allow us to conclude anything when $d = 4$. The main contribution of [3] is to show that (3.4) can actually be improved to a bound of the type

$$\left|\mathbf{E}_c\{\sigma_{x_1}, \dots, \sigma_{x_4}\}\right| \lesssim \frac{\sum_{y\in\mathbf{Z}^d} C(x_1, y) \cdots C(x_4, y)}{(\sum_{|x|\leq M} C(0, x)^2)^c}, \tag{3.6}$$

for some (possibly very small) $c > 0$. Here, one assumes that the x_i's are all at distances at least M of each other.

Remark 3.3. If one believes that the bound (3.5) represents the correct behavior of C at criticality, then the denominator appearing in (3.6) is of order $(\log M)^c$ in dimension 4.

This, however, is not known and is also not used by [3], whether for deriving (3.6) or for deducing Theorem 3.1 from it.

The proof of (3.6) relies on the "random current" representation of the Ising model in which the configuration space consists of "currents", namely maps $\mathbf{n}\colon E \to \mathbf{N}$ where E denotes the set of (unoriented) nearest-neighbor pairs in our lattice. The Ising measure then naturally leads to a weight w on currents defined by $w(\mathbf{n}) = \prod_{e \in E} \frac{\beta^{\mathbf{n}(e)}}{\mathbf{n}(e)!}$ as well as the notion of "source" of a current given by

$$\partial \mathbf{n} \stackrel{\text{def}}{=} \Big\{x : \sum_{e \ni x} \mathbf{n}(e) \text{ is odd}\Big\}.$$

The link between currents and the Ising model is the following formula. Given any finite set $A \subset \mathbf{Z}^d$, one has

$$\mathbf{E} \prod_{a \in A} \sigma_a = \frac{\sum_{\mathbf{n} : \partial \mathbf{n} = A} w(\mathbf{n})}{\sum_{\mathbf{n} : \partial \mathbf{n} = \emptyset} w(\mathbf{n})}.$$

A natural notion then is that of a "random current with source A" for which the probability of seeing a given current $\mathbf{n}$ is nonvanishing only when $\partial \mathbf{n} = A$ in which case it is proportional to $w(\mathbf{n})$. When $A = \{x, y\}$, a current $\mathbf{n}$ with source A can be interpreted (not uniquely!) as the occupation measure of a collection of loops in $\mathbf{Z}^d$, together with a non-self-intersecting path joining x and y. In particular, the restriction of $\mathbf{n}$ to the collection of loops connected (either directly or indirectly through other loops) to the path joining x and y can be thought of as the occupation measure of one single random path joining x to y.

The bound (3.6) can then be reformulated in terms of intersection properties of such random paths. From a heuristic perspective, one gets a lot of mileage from thinking of these random paths as simple random walk trajectories. Note that dimension 4 is critical for the question whether the traces of two random walk trajectories intersect or not: in $d < 4$, the trajectories of two independent random walks with any two starting points will intersect almost surely. In $d > 4$, on the other hand, they only intersect with positive probability (going to 0 as the two starting points are taken far from each other) and, if they do, they only have a finite number of intersection points. In dimension $d = 4$, the probability that two random walks starting at distance of order M from each other do intersect decays like $1/\log M$, but the *expected* number of intersection times remains of order one as $M \to \infty$. This shows that if they do intersect, then the number of intersection points is typically quite large, of order $\log M$.

The bulk of the hard work performed in [3] is to show that the random paths arising in the random cluster representation of the Ising model at criticality exhibit a similar behavior, but with $\log M$ replaced by some quantity of size at least $(\log M)^c$ for some $c > 0$. The argument is a masterpiece combining a delicate multiscale analysis, topological arguments, and probabilistic reasoning. One of the main problem the authors have to overcome is the fact that these random paths are *very* far from being simple random walks and only satisfy some spatial version of the Markov property.

4. ROTATIONAL INVARIANCE FOR THE CRITICAL FK MODELS

As already mentioned a number of times, a crucial feature of $2d$ equilibrium statistical mechanics is the fact that most models are expected to obey a form of conformal invariance (or equivariance) when considering large-scale observables at the critical temperature. This expectation and the resulting link to the well understood world of $2d$ conformal field theories allows to generate a plethora of conjectures regarding the large-scale behavior of these models, but these are in many cases extremely hard to prove. Consider for example the N-step $2d$ self-avoiding random walk which is simply the uniform measure on all functions $h\colon \{0,\ldots,N\} \to \mathbf{Z}^2$ such that $h(0) = 0$ and such that $|h(i+1) - h(i)| = 1$ for all $i < N$. Exploiting the expected conformal invariance of its suitably rescaled large-N limit, one expects the size of $h(N)$ to be of order $N^{3/4}$ and its rescaling by $N^{3/4}$ to converge to a specific continuous random curve, namely $\mathrm{SLE}_{8/3}$ [42]. Rigorously, almost *nothing* nontrivial is known: although the diameter of the range of h trivially has to be at least $\sqrt{N/\pi}$, the current best lower bound on the endpoint does not even match that! Instead, one only knows the bound $(\mathbf{E}|h(N)|^p)^{1/p} \geq \frac{1}{6} N^{p/(2p+2)}$ that was recently obtained by Madras [44]. Similarly, while one trivially has $|h(N)| \leq N$, the best nontrivial upper bound is pretty much the weakest possible improvement, namely that for every $p \geq 1$ one has $\lim_{N\to\infty} N^{-1}(\mathbf{E}|h(N)|^p)^{1/p} = 0$, obtained around the same time by Duminil-Copin and Hammond [18]. One main obstruction is that there is at the moment no proof showing that the self-avoiding random walk is conformally invariant at large scales.

While this illustrates the importance of showing that statistical models are conformally invariant (or at least rotationally invariant as a crucial first step) at criticality, the strategy of proof for such claims has so far mostly relied on finding a large enough collection of observables that already satisfy a discrete analogue of conformal invariance, typically by solving a discrete analogue of the Cauchy–Riemann equations. See, for example, Chelkak and Smirnov's proof of conformal invariance for the Ising model on isoradial graphs [15] and Smirnov's proof of conformal invariance for critical percolation [56]. The two-dimensional FK model with $q \leq 4$ already mentioned in Section 2 is one of the simplest models where conformal invariance at criticality is expected, but where it is not known how to obtain this from a suitable discrete conformal invariance. In the recent work [19], Duminil-Copin et al. show that the large-scale behavior of these models is indeed rotationally invariant.

To define the notion of "large-scale behavior," we recall that the configuration space of the FK model is the same as that for regular percolation, see Figure 1. Such a configuration can alternatively be described as a collection of non-self-intersecting loops separating the percolation clusters from the clusters of the dual configuration. (Actually, it naturally yields *two* collections of loops, depending on whether the loop encloses a percolation cluster of the primary or of the dual configuration, but we will ignore this detail for the sake of our exposition.) Given two collections $\mathcal{F}$ and $\bar{\mathcal{F}}$ of non-self-intersecting loops in the plane, one then defines a distance between them in the following way. Given (small) $\eta > 0$, write $\mathcal{B}_\eta \subset \mathbf{R}^2$ for a large chunk of a fine lattice in $\mathbf{R}^2$, for example, $\mathcal{B}_\eta = \eta\mathbf{Z}^2 \cap [-\eta^{-1}, \eta^{-1}]^2$. Given a loop γ and assuming that its image does not intersect the set $\mathcal{B}_\eta$, one then denotes by $[\eta]_\gamma$ its homotopy class in $\mathbf{R}^2 \setminus \mathcal{B}_\eta$. One then postulates that $d_H(\mathcal{F}, \bar{\mathcal{F}}) \le \eta$ if and only if, for every $\gamma \in \mathcal{F}$ that encloses at least two elements of $\mathcal{B}_\eta$ but not all of it, there exists $\bar{\gamma} \in \bar{\mathcal{F}}$ such that $[\gamma]_\eta = [\bar{\gamma}]_\eta$ and vice versa. (The H here stands for "homotopy.")

Given a metric space (M, d), the metric d lifts naturally to a metric on the space of probability measures on M which metrizes the topology of weak convergence (at least when M is "nice," for example, Polish). This is done by considering the Wasserstein (also sometimes called Kantorovich–Rubinstein or Monge–Kantorovich) distance

$$d(\mu, \nu) = \inf_{\mathbf{P} \in \mathcal{C}(\mu,\nu)} \int d(x, y)\, \mathbf{P}(dx, dy),$$

where $\mathcal{C}(\mu_1, \mu_2)$ denotes the set of all couplings between μ_1 and μ_2, that is, probability measures on M^2 with the ith marginal equal to μ_i. Note that with this definition, the map that assigns to x the probability measure δ_x concentrated at x is an isometry.

Fix now once and for all $q \in [1, 4]$ and consider a smooth bounded simply connected domain $\Omega \subset \mathbf{R}^2$. For $\varepsilon > 0$, write $\mathbf{P}_{\varepsilon,\Omega}$ for the critical FK measure (viewed as a measure on collections of loops) on $\varepsilon\mathbf{Z}^2 \cap \Omega$ with free boundary conditions. We also write $\mathbf{P}_\varepsilon$ for the limit of $\mathbf{P}_{\varepsilon,\Omega}$ as $\Omega \to \mathbf{R}^2$. Given an angle $\theta \in \mathbf{R}$, we also write R_θ for the rotation by θ, which naturally acts on loops in $\mathbf{R}^2$. The large-scale rotational invariance of the critical FK model can then be formulated as follows.

Theorem 4.1. *For every domain $\Omega \subset \mathbf{R}^2$ as above and every angle θ, one has*

$$\lim_{\varepsilon \to 0} d_H\big(R_\theta^* \mathbf{P}_{\varepsilon,\Omega}, \mathbf{P}_{\varepsilon, R_\theta \Omega}\big) = 0.$$

Furthermore, one has $\lim_{\varepsilon \to 0} d_H(R_\theta^* \mathbf{P}_\varepsilon, \mathbf{P}_\varepsilon) = 0$.

We only focus on the second statement since it turns out that the first can be deduced from it without too much effort. In fact, the authors of [19] show a type of universality statement for the FK model on rectangular lattices, but its formulation requires some preparation. We start by defining a specific class of isoradial embeddings of the two-dimensional square lattice into the plane. Recall that a planar graph embedded in the plane is isoradial if, for each face f, there exists a circle of radius 1 containing all the vertices of f. (For example, the canonical embedding of the square lattice is isoradial.)

Given a biinfinite sequence $\alpha: \mathbf{Z} \to (-\frac{\pi}{2}, \frac{\pi}{2})$, we consider the map $\iota_\alpha: \mathbf{Z}^2 \to \mathbf{R}^2$ given by

$$\iota_\alpha: (x, y) \mapsto (x + s_y, c_y), \quad s_y = \sum_{k \in (0,y]} \sin(\alpha_k), \quad c_y = \sum_{k \in (0,y]} \cos(\alpha_k),$$

with the convention that for $y < 0$, $\sum_{(0,y]} = -\sum_{(y,0]}$. This defines an isoradial graph $L(\alpha)$ by considering the embedding of $\{(x, y) : x + y \text{ even}\}$ (joined by diagonal edges) under ι_α (see Figure 3). The dual graph $L^*(\alpha)$ of $L(\alpha)$ is then given by the embedding of $\{(x, y) : x + y \text{ odd}\}$. The associated "diamond graph" has as its vertices both the vertices of $L(\alpha)$ and the centers of its faces, and its edges are given by all pairs (v, f) with v a vertex and f a face such that $v \in f$. The diamond graph is simply given by the embedding of the usual lattice $\mathbf{Z}^2$ with nearest-neighbor edges under ι_α.

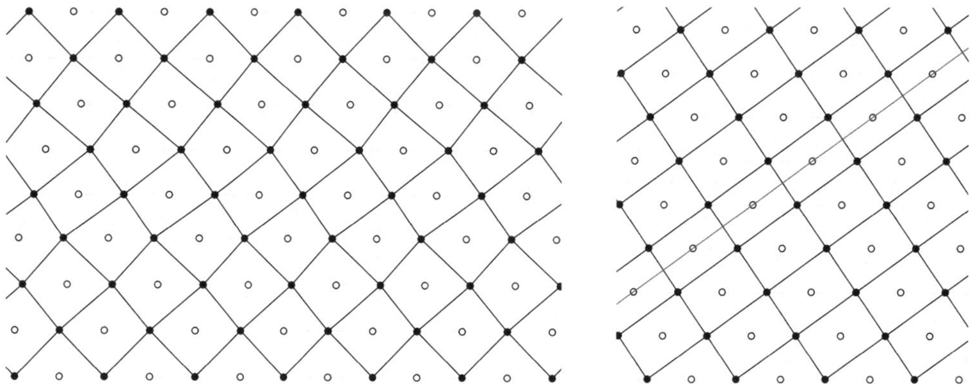

FIGURE 3
Examples of graphs $L(\alpha)$. On the left is a generic α while on the right α is constant but nonzero. The graph itself is drawn in black, the vertices of its dual graph are drawn in white, and the associated diamond graph is light gray. In red, we draw one of the symmetry axes of the second graph.

It is crucial at this stage to note that the critical FK model on $L(\alpha)$ is *not* given by simply pushing forward the critical FK model on $\mathbf{Z}^2$ under the map ι_α. Instead, one reweighs each edge of the graph in a very specific way that depends on the length of the edge. More specifically, viewing a configuration of the FK model as a subset $\omega \subset E$ of the set of edges of the (finite) graph on which the model is considered, the probability of seeing a given configuration ω is proportional to

$$\Big(\prod_{e\in\omega} p_e\Big)\Big(\prod_{e\in E\setminus\omega} (1-p_e)\Big)q^{k(\omega)}, \tag{4.1}$$

where $k(\omega)$ denotes the number of connected components of the subgraph ω. The formula for p_e as a function of q and the length of the edge e is explicit but not relevant for the sake of this discussion.

The most important step in the proof is to show that the large-scale connectivity properties of the critical FK model on $L(\alpha)$ are very close to those of the model on $L(T_j\alpha)$, where T_j swaps the jth and $(j+1)$th component:

$$(T_j\alpha)_k = \begin{cases} \alpha_{j+1} & \text{if } k = j, \\ \alpha_j & \text{if } k = j+1, \\ \alpha_k & \text{otherwise.} \end{cases}$$

Furthermore, there exists a natural coupling between the FK measures on the two lattices which implements this "closedness." This part of the proof exploits the link to the six vertex model and its "solvability" using the transfer matrix formalism. One then deduces from this that the model on the standard lattice $L(0)$ is very close to that on a rotated rectangular lattice $L(\alpha)$ with $k \mapsto \alpha_k$ constant (see the right half of Figure 3). This works by fixing some large $N > 0$ (which is then eventually sent to infinity) and starting from $\alpha_k^{(i)} = \alpha\mathbf{1}_{k\geq N}$ and then swapping components in such a way as to move some of the nonzero components down until one ends up with $\alpha_k^{(f)} = \alpha(\mathbf{1}_{|k|\le N} + \mathbf{1}_{k>3N})$. Since one has $L(0) \approx L(\alpha^{(i)})$ and $L(\alpha) \approx L(\alpha^{(f)})$, the desired statement follows if one can control the error made at each step of the argument. This turns out to be extremely delicate and one has to exploit subtle stochastic cancellations along the way. One trick is to allow the vertices of the set $\mathcal{B}_\eta$ around which the homotopy classes are computed to move a little bit with each application of a swapping operator T_j and to show that this motion ends up being diffusive (and therefore "slow") rather than ballistic.

Once one knows that $\lim_{\varepsilon\to 0} d_H(\mathbf{P}_{\varepsilon,L(0)}, \mathbf{P}_{\varepsilon,L(\alpha)}) = 0$, the second part of Theorem 4.1 follows at once. The idea is simply to note that $L(\alpha)$ is invariant under reflection along a line with angle $\frac{\pi}{4} - \frac{\alpha}{2}$, but that the effect of this reflection on $L(0)$ is the same as that of a rotation by angle α (since it is itself invariant under reflection along a line with angle $\frac{\pi}{4}$), so that

$$d_H\big(\mathbf{P}_\varepsilon, R_\alpha^*\mathbf{P}_\varepsilon\big) \le d_H(\mathbf{P}_{\varepsilon,L(0)}, \mathbf{P}_{\varepsilon,L(\alpha)}) + d_H\big(\mathbf{P}_{\varepsilon,L(\alpha)}, R_\alpha^*\mathbf{P}_{\varepsilon,L(0)}\big) = 2d_H(\mathbf{P}_{\varepsilon,L(0)}, \mathbf{P}_{\varepsilon,L(\alpha)}),$$

and the claim follows.

FUNDING

This work was partially supported by the Royal Society through a research professorship.

REFERENCES

[1] M. Aizenman, Proof of the triviality of φ_d^4 field theory and some mean-field features of Ising models for $d > 4$. *Phys. Rev. Lett.* **47** (1981), no. 1, 1–4.

[2] M. Aizenman, Geometric analysis of φ^4 fields and Ising models. I, II. *Comm. Math. Phys.* **86** (1982), no. 1, 1–48.

[3] M. Aizenman and H. Duminil-Copin, Marginal triviality of the scaling limits of critical 4D Ising and ϕ_4^4 models. *Ann. of Math. (2)* **194** (2021), no. 1, 163–235.

[4] M. Aizenman, H. Duminil-Copin, and V. Sidoravicius, Random currents and continuity of Ising model's spontaneous magnetization. *Comm. Math. Phys.* **334** (2015), no. 2, 719–742.

[5] M. Aizenman and R. Fernández, On the critical behavior of the magnetization in high-dimensional Ising models. *J. Stat. Phys.* **44** (1986), no. 3–4, 393–454.

[6] R. Bauerschmidt, D. C. Brydges, and G. Slade, Scaling limits and critical behaviour of the 4-dimensional n-component $|\phi|^4$ spin model. *J. Stat. Phys.* **157** (2014), no. 4–5, 692–742.

[7] R. Bauerschmidt, D. C. Brydges, and G. Slade, A renormalisation group method. III. Perturbative analysis. *J. Stat. Phys.* **159** (2015), no. 3, 492–529.

[8] R. J. Baxter, Generalized ferroelectric model on a square lattice. *Stud. Appl. Math.* **50** (1971), 51–69.

[9] R. J. Baxter, Potts model at the critical temperature. *J. Phys. C, Solid State Phys.* **6** (1973), no. 23, L445–L448.

[10] R. J. Baxter, S. B. Kelland, and F. Y. Wu, Equivalence of the Potts model or Whitney polynomial with an ice-type model. *J. Phys. A: Math. Gen.* **9** (1976), no. 3, 397–406.

[11] V. Beffara and H. Duminil-Copin, The self-dual point of the two-dimensional random-cluster model is critical for $q \geq 1$. *Probab. Theory Related Fields* **153** (2012), no. 3–4, 511–542.

[12] S. R. Broadbent and J. M. Hammersley, Percolation processes: I. Crystals and mazes. *Math. Proc. Cambridge Philos. Soc.* **53** (1957), no. 3, 629–641.

[13] F. Camia, C. Garban, and C. M. Newman, Planar Ising magnetization field I. Uniqueness of the critical scaling limit. *Ann. Probab.* **43** (2015), no. 2, 528–571.

[14] F. Camia, C. Garban, and C. M. Newman, Planar Ising magnetization field II. Properties of the critical and near-critical scaling limits. *Ann. Inst. Henri Poincaré Probab. Stat.* **52** (2016), no. 1, 146–161.

[15] D. Chelkak and S. Smirnov, Universality in the 2D Ising model and conformal invariance of fermionic observables. *Invent. Math.* **189** (2012), no. 3, 515–580.

[16] H. Duminil-Copin, 100 years of the (critical) Ising model on the hypercubic lattice. In *Proceedings of the International Congress of Mathematicians, Vol. 1*, pp. 164–210, EMS Press, 2022.

[17] H. Duminil-Copin, M. Gagnebin, M. Harel, I. Manolescu, and V. Tassion, Discontinuity of the phase transition for the planar random-cluster and Potts models with $q > 4$. *Ann. Sci. Éc. Norm. Supér. (4)* **54** (2021), no. 6, 1363–1413.

[18] H. Duminil-Copin and A. Hammond, Self-avoiding walk is sub-ballistic. *Comm. Math. Phys.* **324** (2013), no. 2, 401–423.

[19] H. Duminil-Copin, K. K. Kozlowski, D. Krachun, I. Manolescu, and M. Oulamara, Rotational invariance in critical planar lattice models. 2020, arXiv:2012.11672.

[20] H. Duminil-Copin, A. Raoufi, and V. Tassion, Sharp phase transition for the random-cluster and Potts models via decision trees. *Ann. of Math. (2)* **189** (2019), no. 1, 75–99.

[21] H. Duminil-Copin, V. Sidoravicius, and V. Tassion, Continuity of the phase transition for planar random-cluster and Potts models with $1 \leq q \leq 4$. *Comm. Math. Phys.* **349** (2017), no. 1, 47–107.

[22] J.-P. Eckmann and K. Osterwalder, On the uniqueness of the Hamiltionian and of the representation of the CCR for the quartic boson interaction in three dimensions. *Helv. Phys. Acta* **44** (1971), 884–909.

[23] S. El-Showk, M. F. Paulos, D. Poland, S. Rychkov, D. Simmons-Duffin, and A. Vichi, Solving the 3d Ising model with the conformal bootstrap. *Phys. Rev. D* **86** (2012), 025022.

[24] S. El-Showk, M. F. Paulos, D. Poland, S. Rychkov, D. Simmons-Duffin, and A. Vichi, Solving the 3d Ising model with the conformal bootstrap II. c-minimization and precise critical exponents. *J. Stat. Phys.* **157** (2014), no. 4–5, 869–914.

[25] J. Feldman, The $\lambda\varphi_3^4$ field theory in a finite volume. *Comm. Math. Phys.* **37** (1974), 93–120.

[26] J. Feldman, J. Magnen, V. Rivasseau, and R. Sénéor, Construction and Borel summability of infrared Φ_4^4 by a phase space expansion. *Comm. Math. Phys.* **109** (1987), no. 3, 437–480.

[27] J. S. Feldman and K. Osterwalder, The Wightman axioms and the mass gap for weakly coupled $(\Phi^4)_3$ quantum field theories. *Ann. Physics* **97** (1976), no. 1, 80–135.

[28] C. M. Fortuin and P. W. Kasteleyn, On the random-cluster model. I. Introduction and relation to other models. *Physica* **57** (1972), 536–564.

[29] J. Fröhlich, On the triviality of $\lambda\varphi_d^4$ theories and the approach to the critical point in $d > 4$ dimensions. *Nuclear Phys. B* **200** (1982), no. 2, 281–296.

[30] J. Fröhlich, R. Israel, E. H. Lieb, and B. Simon, Phase transitions and reflection positivity. I. General theory and long range lattice models. *Comm. Math. Phys.* **62** (1978), no. 1, 1–34.

[31] J. Fröhlich, B. Simon, and T. Spencer, Infrared bounds, phase transitions and continuous symmetry breaking. *Comm. Math. Phys.* **50** (1976), no. 1, 79–95.

[32] K. Gawędzki and A. Kupiainen, Massless lattice φ_4^4 theory: a nonperturbative control of a renormalizable model. *Phys. Rev. Lett.* **54** (1985), no. 2, 92–94.

[33] J. Glimm, Boson fields with the $:\Phi^4:$ interaction in three dimensions. *Comm. Math. Phys.* **10** (1968), 1–47.

[34] J. Glimm and A. Jaffe, Positivity of the ϕ_3^4 Hamiltonian. *Fortschr. Phys.* **21** (1973), 327–376.

[35] R. B. Griffiths, C. A. Hurst, and S. Sherman, Concavity of magnetization of an Ising ferromagnet in a positive external field. *J. Math. Phys.* **11** (1970), 790–795.

[36] G. Grimmett, *Percolation*. 2nd edn., Grundlehren Math. Wiss. 321, Springer, Berlin, 1999.

[37] M. Hairer, A theory of regularity structures. *Invent. Math.* **198** (2014), no. 2, 269–504.

[38] M. Hairer and J. Mattingly, The strong Feller property for singular stochastic PDEs. *Ann. Inst. Henri Poincaré Probab. Stat.* **54** (2018), no. 3, 1314–1340.

[39] T. Hara and G. Slade, Mean-field behaviour and the lace expansion. In *Probability and phase transition (Cambridge, 1993)*, pp. 87–122, NATO Adv. Sci. Inst. Ser. C: Math. Phys. Sci. 420, Kluwer Acad. Publ., Dordrecht, 1994.

[40] T. Hara and H. Tasaki, A rigorous control of logarithmic corrections in four-dimensional ϕ^4 spin systems. II. Critical behavior of susceptibility and correlation length. *J. Stat. Phys.* **47** (1987), no. 1–2, 99–121.

[41] E. Ising, Beitrag zur Theorie des Ferromagnetismus. *Z. Phys.* **31** (1925), no. 1, 253–258.

[42] G. F. Lawler, O. Schramm, and W. Werner, On the scaling limit of planar self-avoiding walk. In *Fractal geometry and applications: a jubilee of Benoît Mandelbrot, Part 2*, pp. 339–364, Proc. Sympos. Pure Math. 72, Amer. Math. Soc., Providence, RI, 2004.

[43] W. Lenz, Beitrag zum Verständnis der magnetischen Erscheinungen in festen Körpern. *Z. Phys.* **21** (1920), 613–615.

[44] N. Madras, A lower bound for the end-to-end distance of the self-avoiding walk. *Canad. Math. Bull.* **57** (2014), no. 1, 113–118.

[45] J. Miller and S. Sheffield, Quantum Loewner evolution. *Duke Math. J.* **165** (2016), no. 17, 3241–3378.

[46] A. Moinat and H. Weber, Space-time localisation for the dynamic Φ^4_3 model. *Comm. Pure Appl. Math.* **73** (2020), no. 12, 2519–2555.

[47] E. Nelson, A quartic interaction in two dimensions. In *Mathematical theory of elementary particles (Proc. Conf., Dedham, Mass., 1965)*, pp. 69–73, MIT Press, Cambridge, MA, 1966.

[48] E. Nelson, Construction of quantum fields from Markoff fields. *J. Funct. Anal.* **12** (1973), 97–112.

[49] R. O'Donnell, M. Saks, O. Schramm, and R. Servedio, Every decision tree has an influential variable. In *46th annual IEEE symposium on foundations of computer science (FOCS'05)*, pp. 31–39, IEEE, Pittsburg, PA, 2005.

[50] L. Onsager, Crystal statistics. I. A two-dimensional model with an order-disorder transition. *Phys. Rev. (2)* **65** (1944), 117–149.

[51] K. Osterwalder and R. Schrader, Axioms for Euclidean Green's functions. *Comm. Math. Phys.* **31** (1973), 83–112.

[52] K. Osterwalder and R. Schrader, Axioms for Euclidean Green's functions II. *Comm. Math. Phys.* **42** (1973), 281–305.

[53] R. B. Potts, Some generalized order-disorder transformations. *Proc. Camb. Philos. Soc.* **48** (1952), 106–109.

[54] A. Sakai, Lace expansion for the Ising model. *Comm. Math. Phys.* **272** (2007), no. 2, 283–344.

[55] O. Schramm, Scaling limits of loop-erased random walks and uniform spanning trees. *Israel J. Math.* **118** (2000), 221–288.

[56] O. Schramm and S. Smirnov, On the scaling limits of planar percolation. *Ann. Probab.* **39** (2011), no. 5, 1768–1814.

[57] B. Simon, *The $P(\phi)_2$ Euclidean (quantum) field theory*. Princeton Ser. Phys., Princeton University Press, Princeton, NJ, 1974.

[58] B. Simon and R. B. Griffiths, The $(\phi^4)_2$ field theory as a classical Ising model. *Comm. Math. Phys.* **33** (1973), 145–164.

[59] H. N. V. Temperley and E. H. Lieb, Relations between the "percolation" and "colouring" problem and other graph-theoretical problems associated with regular planar lattices: Some exact results for the "percolation" problem. *Proc. R. Soc. Lond. Ser. A* **322** (1971), no. 1549, 251–280.

[60] C. N. Yang, The spontaneous magnetization of a two-dimensional Ising model. *Phys. Rev. (2)* **85** (1952), 808–816.

MARTIN HAIRER

Imperial College London, UK, m.hairer@imperial.ac.uk

THE WORK OF JUNE HUH

GIL KALAI

ABSTRACT

June Huh found striking connections between algebraic geometry and combinatorics, solved central problems in combinatorics that had remained open for decades, and developed a theory of great importance for both fields. June Huh has been awarded the 2022 Fields Medal "for bringing the ideas of Hodge theory to combinatorics, the proof of the Dowling–Wilson conjecture for geometric lattices, the proof of the Heron–Rota–Welsh conjecture for matroids, the development of the theory of Lorentzian polynomials, and the proof of the strong Mason conjecture." In this paper I will review some of June Huh's contributions.

MATHEMATICS SUBJECT CLASSIFICATION 2020

Primary 05E14; Secondary 52C35, 05B35, 05C31, 14T15, 52B40

KEYWORDS

Matroids, log-concavity, hard Lefschetz theorems, Hodge–Riemann relations

INTERNATIONAL CONGRESS OF MATHEMATICIANS 2022 JULY 6–14

© 2022 International Mathematical Union

Proc. Int. Cong. Math. 2022, Vol. 1, pp. 50–65
DOI 10.4171/ICM2022/211

Published by EMS Press and licensed under a CC BY 4.0 license

June Huh has made groundbreaking contributions in combinatorics and algebraic geometry and his work established profound connections between these two areas. This paper describes some of Huh's main achievements and gives some background, primarily on the combinatorial aspects of his work.

The Heron–Rota–Welsh unimodality conjecture ([33, 54, 63]) asserts that the coefficients of the characteristic polynomial of a matroid form a log-concave sequence. This implies that the coefficients are unimodal. A special case of the conjecture is an earlier conjecture by Read, asserting that the coefficients of the chromatic polynomial of a graph are unimodal. In 2009 June Huh used algebraic geometry to prove Read's unimodality conjecture [34] for graphs, and the more general Heron–Rota–Welsh conjecture for matroids represented over a field of characteristic 0. The case of matroids representable over a field of a nonzero characteristic and the case of general matroids remained open. In 2010 June Huh and Eric Katz [37] found a different algebraic-geometric approach and proved the case of matroids representable over a field of an arbitrary characteristic. Finally, in 2015 the Heron–Rota–Welsh conjecture was proved in full generality by Karim Adiprasito, June Huh, and Eric Katz [2]. For this purpose, it was necessary to extend theorems from algebraic geometry (primarily the Hodge–Riemann relations and the hard Lefschetz theorem) to cases well beyond the scope of algebraic geometry. Huh and his coauthors developed an entirely novel theory of great interest and importance.

June Huh and Botong Wang [39] used connections with algebraic geometry to prove the Dowling–Wilson conjecture. Consider a configuration $\mathcal{P}$ of n points spanning a d-dimensional space. Let w_i be the number of linear spaces of dimension i spanned by the points.

Motzkin conjectured in his 1936 PhD thesis, and proved over the reals in 1951 [49], that $w_1 \leq w_{d-1}$. The case of $d = 3$ (in a planar affine formulation) was proved in 1948 by de Bruijn and Erdős, and their abstract combinatorial proof applies to every characteristic.

The Dowling–Wilson "top heavy" conjecture [22] asserts that

$$w_i \leq w_{d-i}, \quad i \leq [d/2]. \tag{1}$$

An extension of the Dowling–Wilson conjecture for arbitrary matroids (of rank d) was proved by Tom Braden, June Huh, Jacob Matherne, Nicholas Proudfoot, and Botong Wang [13].

The Mason conjecture (on independence numbers) asserts [44] that the sequence of numbers of independent sets of size k of general matroids is log-concave and it comes in several strengths. Following Huh's first result the conjecture was proved by Mathias Lenz [42] for representable real matroids and it was proved for general matroids in [2]. The strong Mason conjecture for arbitrary matroids was proved by June Huh, Benjamin Schröter, and Botong Wang [38] who relied on [2].

These works have led to further advances by several groups of researchers, and I would especially like to mention the solution of the Mihail–Vazirani conjecture on the expansion constant and rapid mixing for random walks on matroids, by Anari, Oveis Gharan, and

Vinzant [4], as well as the works by Brändén and Huh [14] on correlation inequalities for the Potts model.

The structure of this paper is as follows. In Section 1 we discuss chromatic polynomials and Read's conjecture. In Section 2 we discuss matroids and the Heron–Rota–Welsh conjecture. In Section 3 we discuss the Dowling–Wilson conjecture. Section 4 is devoted to algebraic geometry, Hodge theory, and the Grothendieck standard conjectures. In Section 5 we discuss the Mason conjectures and some recent applications and connections. A recent review paper aimed for a general audience on Huh's work and mathematical background was written by Andrei Okounkov [50].

1. GRAPHS, CHROMATIC POLYNOMIALS, AND READ'S CONJECTURE

1.1. The four-color conjecture and chromatic polynomials

A proper coloring of a graph G is a coloring of the vertices of G such that every two adjacent vertices are colored with different colors. Graph coloring is of central importance in graph theory and in graph algorithms.

Theorem 1 (The four-color theorem (Appel and Haken 1976)). *Every planar graph can be properly colored with 4 colors.*

The four-color conjecture was proposed (in a dual form, for planar maps) by Francis Guthrie in 1852 and proved by Kenneth Appel and Wolfgang Haken [6] in 1976.

For a graph G, let $\chi_G(k)$ be the number of proper colorings of G with k colors. $\chi_G(k)$ is called the chromatic polynomial of the graph G. Chromatic polynomials were introduced by George Birkhoff [11] for planar maps as a possible tool for the study of the four-color conjecture. Later Hassler Whitney extended the definition to general graphs. William Tutte found a far-reaching generalization, now called the Tutte polynomial and also introduced the related Tutte–Grothendieck invariants for graphs, which can be seen as an early bridge between graph theory and algebraic geometry. A starting point of Tutte's work is the deletion–contraction operations. For a graph G and an edge e of G, let $G\backslash e$ denote the graph obtained by deleting the edge e, and G/e denote the graph obtained by contracting the edge e, that is, by merging its two vertices to a single vertex adjacent to neighbors of both. A fundamental relation for chromatic polynomials is

$$\chi_G(k) + \chi_{G/e}(k) = \chi_{G\backslash e}(k). \tag{2}$$

This relation gives an easy inductive proof of the fact that the chromatic polynomial is indeed a polynomial. A graph H is called a minor of a graph G if it can be obtained from G by a sequence of deletions and contractions. Richard Stanley proved [58] that $\chi_G(-1)$ equals the number of acyclic orientations of G.

1.2. Read's conjecture

In 1968 Ronald Read [53] proposed the following conjecture. Suppose that

$$\chi_G(x) = a_n x^n - a_{n-1}x^{n-1} + \cdots + (-1)^i a_{n-i}x^{n-i} + \cdots; \tag{3}$$

then, the sequence $a_0, a_1, \dots, a_n$ is unimodal.

A much more general conjecture was posed a short time later by Andrew Heron, Gian-Carlo Rota, and Dominic Welsh. They also conjectured a stronger statement, namely, that the sequence of coefficients is actually log-concave:

$$a_k^2 \geq a_{k-1}a_{k+1}, \quad k = 1, 2, \dots, n-1. \tag{4}$$

Theorem 2 (June Huh [34]). *The coefficients of the chromatic polynomial $\chi_G(x)$ of every graph G are log-concave.*

The unimodality and log-concavity of sequences arising in combinatorics and algebra have been studied by many researchers and in this context I would like to refer the reader to the survey articles [16, 17, 61]. A stronger property than log-concavity of the coefficients of real polynomials is that of having only real roots. This is not the case for chromatic polynomials of graphs in general (but the location of the roots is still a fascinating topic). Unimodality of the numbers of elements according to their heights in general graded posets is also related to the important "Sperner property" of posets. We note that there are cases where unimodality was expected but failed, e.g., unimodality of face numbers of polytopes [12], and of Young lattices [62].

I first heard about Huh's startling proof of the Read conjecture from a 2011 paper by Jiří Matoušek [45] who regarded this result, among a few other results, as the beginning of a new era in discrete geometry and wrote:

> *"To me, 2010 looks as annus mirabilis, a miraculous year, in several areas of my mathematical interests. Below I list seven highlights and breakthroughs, mostly in discrete geometry, hoping to share some of my wonder and pleasure with the readers."*

Huh's proof relied on connections of the problem to singularities of local analytic functions and ultimately to mixed multiplicities of certain ideals. In his proof Huh related the coefficients of the chromatic polynomial to the Milnor numbers of a complex hyperplane arrangement associated with the graph G and, as we discuss in the next section, his proof extends to arbitrary complex hyperplane arrangements. Huh's connection between chromatic polynomials of graphs and algebraic geometry was, on the one hand, a complete surprise but, on the other hand, it tied in with several developments in and around algebraic combinatorics dating to the mid-1970s. Huh's subsequent discoveries where he further applied algebraic geometry and especially Hodge theory to combinatorics, beautifully combined new and old ideas.

2. MATROIDS AND THE HERON–ROTA–WELSH CONJECTURE

2.1. Matroids

Let $X = \{x_1, x_2, \ldots, x_n\}$ be a set of points in some vector space. We can associate with X:

- The set of linearly independent subsets of X.
- The set of bases of X (a base is a maximal independent set).
- The set of circuits of X (a circuit is a minimal dependent set).
- The set of flats of X (a flat is a subset that is closed under linear combination).
- The rank function that associates to a subset Y of X the dimension of the vector space spanned by Y.

Matroids were introduced by Hassler Whitney [65] as a generalization of configurations of points in linear spaces or as an abstraction of the notion of linear dependence. Matroid theory is an example of both a highly successful abstraction and a source of very useful and explicit examples. Matroid theory has various connections to the theory of algorithms and mathematical optimization, and also to mathematical logic.

Each of the five notions we mentioned above, independent sets, bases, circuits, flats, and rank functions, give rise to an axiomatic definition of matroids (and all these axiomatic definitions are equivalent). The definition of matroids based on independent sets is given by the following properties:

(1) Subsets of independent sets are themselves independent.

(2) For every subset Y of X, all maximal independent subsets of Y have the same cardinality.

The first property means that the set of independent sets is an abstract simplicial complex while the second property asserts that for every subset Y of the ground set X, the induced complex on Y is pure.

For an abstract simplicial complex K on a ground set X, we can define its dual (also called its blocker) by

$$K^* = \{S \subset X : X \backslash S \not\subset M\}.$$

If M is a matroid, we can define its dual as the matroid whose independent set complex is the dual of the independent set complex of M.

2.2. From graphs to matroids

Let G be a (connected) graph on n vertices $\{v_1, v_2, \ldots, v_n\}$, and suppose that $e_1, e_2, \ldots, e_n$ is the standard basis in an n-dimensional vector space over a field F. We associate to every edge $e = \{v_i, v_j\}, i < j$ the vector $e_i - e_j$. Remarkably, we get the same matroid for every field we start with. This matroid is called the graphic matroid associated

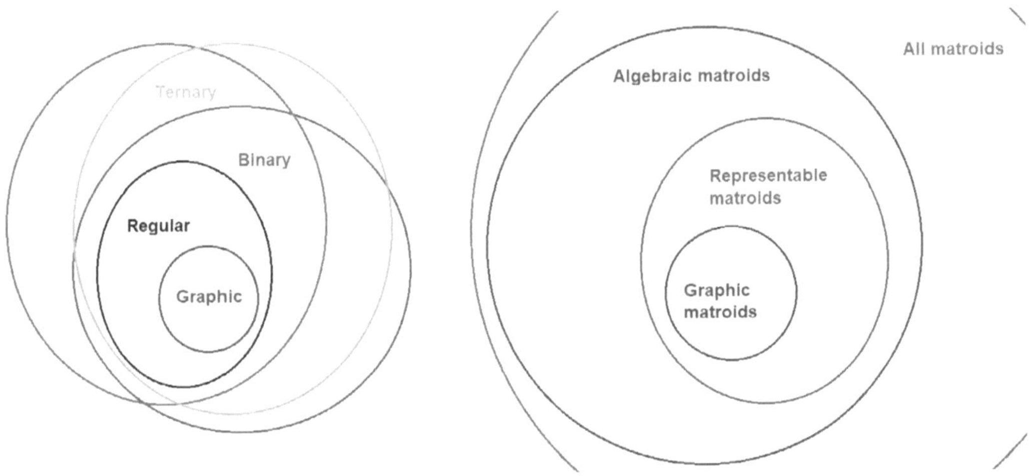

FIGURE 1

Important classes of matroids. (Right) Matroids also provide an abstraction of the notion of algebraic dependence. The large and mysterious class of algebraic matroids consists of matroids that can be represented by algebraic dependence relations over some field. (Left) Tutte characterized graphic matroids in terms of forbidden minors. Regular matroids are those matroids that can be represented over every field, and Paul Seymour **[56]** developed a structure theory for this class. Jim Geelen, Bert Gerards, and Geoff Whittle (see **[28]**) have recently proved that matroids represented over every field are characterized by a finite list of forbidden minors.

with G. It is easy to see that in this case, bases correspond to spanning trees, circuits correspond to simple cycles, independent sets correspond to spanning forests, and the rank function for subgraph H that corresponds to a set of edges is n minus the number of the connected components of H.

If M is a graphic matroid, the dual matroid need not be graphic. However, for planar graphs the dual matroid is the matroid associated to the dual graph. The notion of deletion and contraction extend from graph theory to matroid theory. (Indeed, these two operations are dual under matroid duality.)

2.3. Rank functions, characteristic polynomials, and the Heron–Rota–Welsh conjecture

The rank function of a matroid associates a nonnegative integer $r(Y)$ to every subset $Y \subset X$, with the following properties:

(i) $r(\emptyset) = 0$,

(ii) $r(A \cup B) \leq r(A) + r(B) - r(A \cap B)$,

(iii) $r(A) \leq r(A \cup \{b\}) \leq r(A) + 1$.

The characteristic function of a matroid M with ground set X is defined as follows:

$$\chi_M(\lambda) := \sum_{S \subseteq E} (-1)^{|S|} \lambda^{r(M)-r(S)}. \tag{5}$$

If M is a graphic matroid for the graph G, then $\chi_M(\lambda)$ is the chromatic polynomial of G.

Theorem 3 (Adiprasito, Huh, and Katz [2]). *The coefficients of the characteristic polynomial of a matroid M are log-concave.*

June Huh [34] proved the results for matroids (regarded as hyperplane arrangements) representable over a field of characteristic 0 and, as we mentioned above, the proof uses the Milnor numbers of the arrangement. The proof by Huh and Katz [37] for the case of an arbitrary characteristic relied on the intersection theory of "wonderful compactification" defined by Corrado De Concini and Claudio Procesi [21] for complements of hyperplane arrangements combined with an inequality of Askold Khovanskii and Bernard Teissier.

Adiprasito, Huh, and Katz [2] proved the full result. This requires far-reaching extensions of results from algebraic geometry to cohomology rings of algebraic varieties that do not exist. Here is the description of one of the early steps in the argument: the original definition of De Concini and Procesi of the "wonderful compactification" applied to realizable matroids, but Feichtner and Yuzvinsky defined in 2004 [25] a commutative ring associated to an arbitrary matroid that specializes to the cohomology ring of a wonderful compactification in the realizable case.

Let me quote from [2]: "After the completion of [37], it was gradually realized that the validity of the Hodge–Riemann relations for the Chow ring of M is a vital ingredient for the proof of the log-concavity conjectures. While the Chow ring of M could be defined for arbitrary [matroid] M, it was unclear how to formulate and prove the Hodge–Riemann relations. From the point of view of [25], the ring $A^*(M)_{\mathbb{R}}$ is the Chow ring of a smooth, but noncompact toric variety $X(\Sigma_M)$, and there is no obvious way to reduce to the classical case of projective varieties."

We will discuss some of the algebraic geometry aspects in Section 4. We note that the algebraic results of [2] actually apply to more general geometric objects well beyond matroids. For more on matroid theory see [7,32,43,51,64].

3. THE DOWLING–WILSON CONJECTURE

3.1. Background: Theorems by de Bruijn–Erdős, Motzkin, Greene, and Ryser's linear algebraic proof

Theorem 4. *A set of n points in the plane not all on the same line determines at least n lines.*

Here we say that a configuration of points determines a line ℓ if the line contains two (distinct) points from the configuration.

Proof. The Gallai–Sylvester theorem asserts that there exists a line that contains precisely two points of the configuration. The theorem now follows by induction when you delete one of these two points from the configuration. ■

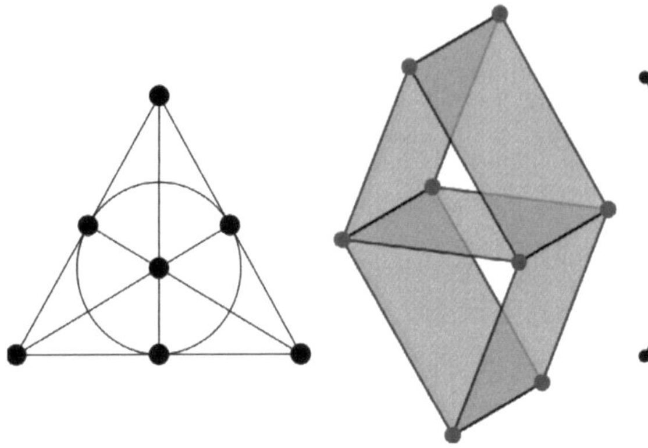

FIGURE 2

Important examples of matroids. From left to right: The Fano matroid, the Vámos matroid, and the non-Pappus matroid. The points of the Fano plane violate the Gallai–Sylvester theorem, hence it is not representable over the reals. As a matter of fact, the Fano matroid is representable over a field F if and only the characteristic of F is 2. The Vámos matroid is not algebraic. Pappus ancient theorem implies that the non-Pappus matroid is not representable over any field. Bernt Lindström proved that it is algebraic. Picture credit: Wikipedia and the "matroid union" blog.

The assertion of the Gallai–Sylvester theorem does not apply over characteristic two as seen by the Fano plane, nor does it apply for the complex plane. By contrast, the proof by Nicolaas de Bruijn and Paul Erdős uses an abstract combinatorial reasoning that is based only on the very first axiom of Euclid: "Every two points span a unique line." An algebraic proof of the theorem was given by Herbert Ryser [55].

Proof. Ryser's proof Consider the 0–1 *incidence matrix* with rows corresponding to points in the configuration and columns to lines determined by these points. Suppose that the columns of the incidence matrix are $c_1, c_2, \dots, c_m$. Note that the inner product of every two distinct rows is one. Write $b_i = \langle c_i, c_i \rangle$ for the number of points on the ith line ($b_i > 1$). Suppose that

$$\sum \alpha_i c_i = 0.$$

We write

$$0 = \left\langle \sum \alpha_i c_i, \sum \alpha_i c_i \right\rangle = \sum \alpha_i^2 (b_i - 1) + \left(\sum \alpha_i\right)^2.$$

It follows that the rows are linearly independent and therefore we must have $m \leq n$. ■

Ryser's proof was a starting point for many algebraic proofs in combinatorics. We leave it as an exercise to show that it implies that there is bijection $\psi(p)$ from points to lines such that $p \in \psi(p)$.

Theodore Motzkin considered the theorem in higher dimensions. He conjectured (already in his 1936 thesis) that n points in a d-dimensional space that affinely span the space span at least n hyperplanes. Motzkin himself proved the result, as well as an extension of the Gallai–Sylvester theorem for configurations in higher-dimensional real vector spaces [49]. Curtis Greene [30] proved a stronger theorem: there is a one-to-one map ψ from every point p to a hyperplane containing p.

Let us now move to matroids of rank d. (Note that affine dependence of points in a d-dimensional vector space describe a matroid of rank $d+1$.) In 1974 Thomas Dowling and Richard Wilson conjectured that

$$w_i \leq w_{d-i}, \quad \text{whenever } i < d - i. \tag{6}$$

This conjecture is referred to as the *top-heavy conjecture.*

3.2. The proof of the Dowling–Wilson conjecture

Theorem 5 (Braden, Huh, Matherne, Proudfoot, and Wang 2020 [13]). *Let M be a matroid, and let $\mathcal{L}^k(M)$ denote the set of k-flats of M; then, for any k, j, $k \leq j \leq \operatorname{rank}(M) - k$:*

(1) *The cardinality of $\mathcal{L}^k(M)$ is at most the cardinality of $\mathcal{L}^j(M)$.*

(2) *There is an injective map ψ from $\mathcal{L}^k(M)$ to $\mathcal{L}^j(M)$, satisfying $F \subset \psi(F)$.*

An additional result from the same paper asserts that if Γ is any group acting on M, then

(3) There is an injective map ψ from $\mathbb{Q}\mathcal{L}^k(M)$ to $\mathcal{L}^j(M)$, of permutation representation of Γ.

The case of representable matroids was proved earlier by Huh and Wang 2017 [39]. The paper [13] also gives consequences for Kazhdan–Lusztig polynomials of matroids (introduced by Elias and Proudfoot).

We note that it is still an outstanding open question (even for representable matroids) that the sequence $w_1, w_2, \ldots, w_n$ is log-concave. It is not even known for rank-3 matroids that

$$w_2^2 \geq w_1 w_3, \tag{7}$$

and this is referred to as the "point–lines–planes" conjecture. A stronger form of this conjecture (due to Mason) asserts that

$$w_2^2 \geq \frac{3}{2}\frac{w_1 - 1}{w_1 - 2} w_1 w_3.$$

In 1982 Paul Seymour [57] proved this conjecture for matroids having no five points on a line.

4. THE CONNECTION WITH HODGE THEORY AND ALGEBRAIC GEOMETRY

4.1. Three fundamental ideas and other ingredients from the proof of the Heron–Rota–Welsh conjecture

"I like the solution even more than the problem."

June Huh at a lecture at ICERM, 2015.

In his 2015 lecture at ICERM (see also [36]), June Huh explained three fundamental ideas that were used in the proof of the general Heron–Rota–Welsh conjecture:

(1) The idea of Bernd Sturmfels that a matroid can be viewed as a tropical linear space.

Indeed, tropical geometry provided both a necessary framework and insights into the solution. Briefly, tropical mathematics replaces traditional addition with the operation of "taking the minimum," and multiplication with ordinary addition. This idea arose in several areas of mathematics and in physics and it played an important role in enumerative algebraic geometry. (For more details, see [36] and Section 5.4 and appendix C of [50].)

(2) The idea of Richard Stanley [60] that a polarized Hodge structure on the cohomology of projective toric varieties produces important combinatorial inequalities.

Here the main example was the g-theorem for convex polytopes ([10, 46, 59]) where Stanley used the hard Lefschetz theorem for the cohomology ring. Another notable example was Stanley's proof of the Erdős–Moser conjecture.

(3) The idea of Peter McMullen [47, 48] that the g-conjecture can be proved entirely within the realm of convex polytope theory using the "flip connectivity" of simplicial polytopes of a given dimension.

Any two simplicial polytopes are connected by a sequence of "flips" (also known as "Pachner moves") and McMullen proved that the validity of the hard Lefschetz theorem and the Hodge–Riemann relations are preserved under flips.

In the same lecture, June Huh mentioned quite a few more ideas by many people working in algebraic combinatorics and in algebraic geometry that play a role in the proof. We already mentioned Tessier, Khovanskii, De Concini and Procesi, and Fleischer and Yuzvinsky, and Huh mentioned also Federico Ardila and Caroline Klivans [8], Angela Gibney and Diane Maclagan [29], Kalle Karu [40], and William Fulton and Robert MacPherson [26, 27]. Of course, the proof involved a large number of additional original (at times crazy) ideas by Adiprasito, Huh, and Katz themselves.

4.2. Poincaré duality, the hard Lefschetz theorem, and the Hodge–Riemann relations

Hodge theory gives rise to three conjectures (PD), (HL), and (HR), referred to as the standard conjectures, for certain algebras associated with geometric and combinatorial objects:

(PD) stands for the Poincaré duality, and it asserts that certain vector spaces A_i and A_{d-i} are dual (and thus have the same dimension).

(HD) stands for hard Lefschetz theorem and it asserts that certain linear maps ϕ_k from A_k to $A_k + 1$ have the property that their composition from A_i all the way to A_{d-i} is an injection.

(HR) stands for the Hodge–Riemann relations. (PD) and (HD) imply that a certain bilinear form is nondegenerate and (HR) is a stronger statement that this form is definite.

For the case of smooth projective algebraic variety M, we can consider its cohomology ring $A_i = H^{2i}(M)$. (For the case of singular algebraic varieties, that come into play in the strongest versions of the Dowling–Wilson conjecture, we need to use intersection cohomology.)

In [35] June Huh considered five examples (we are somewhat imprecise here): the cohomology of a compact Kähler manifold, the ring of algebraic cycles modulo homological equivalence on a smooth projective variety, McMullen's algebra generated by the Minkowski summands of a simple convex polytope, the combinatorial intersection cohomology of a convex polytope, the reduced Soergel bimodule of a Coxeter group element, and the Chow ring of a matroid. The only case among these examples where the standard conjectures are not known is in their original appearance in Grothendieck's work [31] toward the Weil conjectures. The example of Soergel bimodules is related to the celebrated 2014 solution of the Kazhdan–Lusztig conjecture for general Coxeter groups by Ben Elias and Geordie Williamson [23]. While it may be premature to expect it, it is not premature to hope that some connections will be found between the combinatorial appearances of the standard conjectures and their appearances in representation theory and number theory.

Remarks. (1) The proof of the Heron–Rota–Welsh conjecture by June Huh and his collaborators largely exploits "positivity," namely the Hodge–Riemann relations. For another central problem in algebraic combinatorics, the "g-conjecture for spheres," positivity is no longer available, and remarkable techniques to replace it and thus prove the conjecture were recently developed first by Adiprasito [1] (the "Hall–Laman property"), subsequently by Stavros Argyrios Papadakis and Vasiliki Petrotou [52] (the "anisotropy property"), and ultimately by Adiprasito, Papadakis, and Petrotou [3]. (See also Kalle and Elizabeth Xiao [41] for a simplified proof.)

(2) The work of Karu ([40]) on a hard Lefshetz theorem for general polytopes (see also [9,18]), of Elias and Williamson [23] on the Kazhdan–Lusztig conjecture, and of Braden, Huh, Matherne, Proudfoot, and Wang [13] on the Dowling–Wilson conjecture rely on (HL) and (HR), not for (combinatorial extensions of) the ordinary homology but for (combinatorial extensions of) Goresky and MacPherson's intersection homology.

5. THE STRONG MASON CONJECTURE (ON INDEPENDENCE NUMBERS), AND RELATED DEVELOPMENTS AND APPLICATIONS

5.1. Mason conjecture, regular strength, strong, and ultra-strong

Let M be an n-element matroid and let $i_k(M)$ denote the number of independent sets of M of size k. The Mason conjecture [44] comes in several strengths.

The Mason conjecture:

$$i_k^2(M) \geq i_{k-1}(M) i_{k+1}(M).$$

The strong Mason conjecture:

$$i_k^2(M) \geq \left(1 + \frac{1}{k}\right) i_{k-1}(M) i_{k+1}(M).$$

The ultra-strong Mason conjecture:

$$i_k^2(M) \geq \left(1 + \frac{1}{k}\right)\left(1 + \frac{1}{n-k}\right) i_{k-1}(M) i_{k+1}(M).$$

Mathias Lenz showed [42] how to derive the Mason conjecture for representable matroids, based on the work of Huh and Katz. Adiprasito, Huh, and Katz showed how to derive the Mason conjecture from their Hodge theory techniques and Huh, Schröter, and Wang extended these techniques to prove the strong Mason conjecture. The ultra-strong conjecture was proved in parallel by direct combinatorial reasoning by Nima Anari, Kuikui Liu, Shayan Oveis Gharan, and Cynthia Vinzant [5] and based on Hodge theory by Brändén and Huh [14,15].

June Huh's results of the past decade have led to much further research on the unimodality and log-concavity of various sequences arising in combinatorics. In some cases new combinatorial proofs were found. Let me refer the reader to recent papers by Swee Hong Chan and Igor Pak [19,20].

5.2. The Mihail–Vazirani conjecture

For a matroid M on a ground set X, consider a graph whose vertices are all bases of the matroids and two bases are adjacent if their symmetric difference has two elements. Milena Mihail and Umesh Vazirani conjectured that for every set Y of vertices in this graph, the number of edges between Y to its complement $\bar{Y}$ is at least $\min(|Y||, \bar{Y}|)$.

If M consists of the elements of the standard basis in $\mathbb{R}^d$ and their negatives, then the graph we obtain is the graph of the discrete n-dimensional discrete cube and the assertion of the Mihail–Vazirani conjecture is a well-known isoperimetric inequality of the discrete cube.

In a pioneering 1992 paper, Tomás Feder and Milena Mihail [24] proved the conjecture for balanced matroids. In 2018 Nima Anari, Shayan Oveis Gharan, and Cynthia Vinzant [4] proved the Mihail–Vazirani conjecture. Their proof relied on the Adiprasito–Huh–Katz paper, although gradually they were able to find elementary proofs not depending on Hodge theory of crucial inequalities they needed. Their result leads to a polynomial-time algorithm to approximate the number of bases in a matroid.

CONCLUSION

June Huh found striking connections between algebraic geometry and combinatorics, solved central problems in combinatorics that had remained open for decades, and developed a theory of great importance for both fields. In my review, I naturally concentrated on the combinatorial side of the story. I did not describe in this review the connection with tropical geometry, a major area both in algebraic combinatorics and algebraic geometry. The reader is also referred to Huh's papers to learn about the theory of Lorentzian polynomials developed by June Huh and his coauthors.

It is a great pleasure to congratulate June Huh for his spectacular achievements.

FUNDING

Supported by ERC grant 834735 and by an ISF grant 2669/21.

REFERENCES

[1] K. Adiprasito, Combinatorial Lefschetz theorems beyond positivity. 2019, arXiv:1812.10454v4.

[2] K. Adiprasito, J. Huh, and E. Katz, Hodge theory for combinatorial geometries. *Ann. of Math.* **188** (2018), 381–452.

[3] K. Adiprasito, S. A. Papadakis, and V. Petrotou, Anisotropy, biased pairings, and the Lefschetz property for pseudomanifolds and cycles. 2021, arXiv:2101.07245v2.

[4] N. Anari, S. O. Gharan, and C. Vinzant, Log-concave polynomials, entropy, and a deterministic approximation algorithm for counting bases of matroids. *Duke Math. J.* **170** (2021), 3459–3504. (Preliminary version, FOCS 2018.)

[5] N. Anari, S. Oveis Gharan, K. Liu, and C. Vinzant, Log-concave polynomials III: Mason's ultra-log-concavity conjecture for independent sets of matroids. 2018, arXiv:1811.01600.

[6] K. Appel and W. Haken, *Every planar map is four-colorable*, with the collaboration of John Koch. Contemp. Math. 98, Providence, RI, American Mathematical Society, 1985.

[7] F. Ardila, G. Denham, and J. Huh, Lagrangian geometry of matroids. *J. Amer. Math. Soc.* (2022).

[8] F. Ardila and C. Klivans, The Bergman complex of a matroid and phylogenetic trees. *J. Combin. Theory Ser. B* **96** (2006), 38–49.

[9] G. Barthel, J.-P. Brasselet, K.-H. Fieseler, and L. Kaup, Combinatorial duality and intersection product: a direct approach. *Tohoku Math. J.* **57** (2005), 273–292.

[10] L. J. Billera and C. W. Lee, A proof of the sufficiency of McMullen's conditions for f-vectors of simplicial convex polytopes. *J. Combin. Theory Ser. A* **31** (1981), 237–255.

[11] G. Birkhoff, A determinant formula for the number of ways of coloring a map. *Ann. of Math.* **14** (1912), 42–46.

[12] A. Björner, The unimodality conjecture for convex polytopes. *Bull. Amer. Math. Soc.* **4** (1981), 187–188.

[13] T. Braden, J. Huh, J. P. Matherne, N. Proudfoot, and B. Wang, Singular Hodge theory for combinatorial geometries. 2022, arXiv:2010.06088v3.

[14] P. Brändén and J. Huh, Hodge–Riemann relations for Potts model partition functions. 2018, arXiv:1811.01696.

[15] P. Brändén and J. Huh, Lorentzian polynomials. 2019, arXiv:1902.03719.

[16] F. Brenti, *Unimodal, log-concave and Pólya frequency sequences in combinatorics*. Mem. Amer. Math. Soc. 413, 1989.

[17] F. Brenti, Log-concave and unimodal sequences in algebra, combinatorics, and geometry: an update. In *Jerusalem Combinatorics '93*, edited by H. Barcelo and G. Kalai, pp. 71–89, Contemp. Math. 178, American Mathematical Society, Providence, RI, 1994.

[18] P. Bressler and V. A. Lunts, Hard Lefschetz theorem and Hodge–Riemann relations for intersection cohomology of nonrational polytopes. *Indiana Univ. Math. J.* **54** (2005), 263–307.

[19] S. H. Chan and I. Pak, Log-concave poset inequalities. 2021, arXiv:2110.10740v2.

[20] S. H. Chan and I. Pak, Introduction to the combinatorial atlas. 2022, arXiv:2203.01533.

[21] C. De Concini and C. Procesi, Wonderful models of subspace arrangements. *Selecta Math. (N.S.)* **12** (1995), 459–494, 36–70.

[22] T. Dowling and R. Wilson, Whitney number inequalities for geometric lattices. *Proc. Amer. Math. Soc.* **47** (1975), 504–512.

[23] B. Elias and G. Williamson, The Hodge theory of Soergel bimodules. *Ann. of Math.* **180** (2014), 1089–1136.

[24] T. Feder and M. Mihail, Balanced matroids. In *Proc. 24th Annual ACM Symposium on Theory of Computing*, pp. 26–38, ACM Press, 1992.

[25] E. M. Feichtner and S. Yuzvinsky, Chow rings of toric varieties defined by atomic lattices. *Invent. Math.* **155** (2004), 515–536.

[26] W. Fulton, R. MacPherson, F. Sottile, and B. Sturmfels, Intersection theory on spherical varieties. *J. Algebraic Geom.* **4** (1995), 181–193.

[27] W. Fulton and B. Sturmfels, Intersection theory on toric varieties. *Topology* **36** (1997), 335–353.

[28] J. Geelen, B. Gerards, and G. Whittle, Solving Rota's conjecture. *Notices Amer. Math. Soc.* **61** (2014), 736–743.

[29] A. Gibney and D. Maclagan, Lower and upper bounds on nef cones. *Int. Math. Res. Not.* **14** (2012), 3224–3255.

[30] C. Greene, A rank inequality for finite geometric lattices. *J. Combin. Theory* **9** (1970), 357–364.

[31] A. Grothendieck, Standard conjectures on algebraic cycles. In *Algebraic Geometry*, pp. 193–199, Oxford University Press, 1969.

[32] Y. O. Hamidoune and I. Salaün, On the independence numbers of a matroid. *J. Combin. Theory Ser. B* **47** (1989), 146–152.

[33] A. Heron, Matroid polynomials. In *Combinatorics (Proceedings of the Conference on Combinatorial Mathematics, Mathematics Institute)*, pp. 164–202, Oxford, 1972.

[34] J. Huh, Milnor numbers of projective hypersurfaces and the chromatic polynomial of graphs. *J. Amer. Math. Soc.* **25** (2012), 907–927.

[35] J. Huh, Combinatorial applications of the Hodge–Riemann relations. In *Proceedings of the International Congress of Mathematicians, Vol. IV. Invited lectures*, pp. 3093–3111, World Scientific Publishers, Hackensack, NJ, 2018.

[36] J. Huh, Tropical geometry of matroids. In *Current Developments in Mathematics 2016*, pp. 1–46, International Press, 2018.

[37] J. Huh and E. Katz, Log-concavity of characteristic polynomials and the Bergman fan of matroids. *Math. Ann.* **354** (2012), 1103–1116.

[38] J. Huh, B. Schröter, and B. Wang, Correlation bounds for fields and matroids. *J. Eur. Math. Soc. (JEMS)* **24** (2022), 1335–1351.

[39] J. Huh and B. Wang, Enumeration of points, lines, planes, etc. *Acta Math.* **218** (2017), 297–317.

[40] K. Karu, Hard Lefschetz theorem for nonrational polytopes. *Invent. Math.* **157** (2004), 419–447.

[41] K. Karu and E. Xiao, On the anisotropy theorem of Papadakis and Petrotou. 2022, arXiv:2204.07758v2.

[42] M. Lenz, The f-vector of a representable-matroid complex is log-concave. *Adv. in Appl. Math.* **51** (2013), 543–545.

[43] L. Lovász, Matroid matching and some applications. *J. Combin. Theory Ser. B* **28** (1980), 208–236.

[44] J. Mason, Matroids: Unimodal conjectures and Motzkin's theorem. In *Combinatorics (Proceedings of the Conference on Combinatorial Mathematics)*, pp. 207–220, Mathematics Institute, Oxford, 1972.

[45] J. Matoušek, The dawn of an algebraic era in discrete geometry? In *Proceedings of the 27th European Workshop on Computational Geometry (EuroCG'11)*, 2011.

[46] P. McMullen, The numbers of faces of simplicial polytopes. *Israel J. Math.* **9** (1971), 559–570.

[47] P. McMullen, On simple polytopes. *Invent. Math.* **113** (1993), 419–444.

[48] P. McMullen, Weights on polytopes. *Discrete Comput. Geom.* **15** (1996), 363–388.

[49] T. Motzkin, The lines and planes connecting the points of a finite set. *Trans. Amer. Math. Soc.* **70** (1951), 451–464.

[50] A. Okounkov, Combinatorial geometry takes the lead. In *Proceedings of the International Congress of Mathematicians, Vol. 1*, pp. 414–458, EMS Press, 2022.

[51] J. Oxley, *Matroid theory*. Oxf. Grad. Texts Math., Oxford University Press, Oxford, 2011.

[52] S. A. Papadakis and V. Petrotou, The characteristic 2 anisotropicity of simplicial spheres. 2020, arXiv:2012.09815.

[53] R. Read, An introduction to chromatic polynomials. *J. Combin. Theory* **4** (1968), 52–71.

[54] G.-C. Rota, Combinatorial theory, old and new. In *Actes du Congress International des Mathématicians (Nice, 1970)*, Tome 3, pp. 229–233, Gauthier-Villars, Paris, 1971.

[55] H. J. Ryser, An extension of a theorem of de Bruijn and Erdős on combinatorial designs. *J. Algebra* **10** (1968), 246–261.

[56] P. D. Seymour, Decomposition of regular matroids. *J. Combin. Theory Ser. B* **28** (1980), 305–359.

[57] P. D. Seymour, On the points-lines-planes conjecture. *J. Combin. Theory Ser. B* **33** (1982), 17–26.

[58] R. P. Stanley, Acyclic orientations of graphs. *Discrete Math.* **5** (1973), 171–178.

[59] R. P. Stanley, The number of faces of a simplicial convex polytope. *Adv. Math.* **35** (1980), 236–238.

[60] R. P. Stanley, Combinatorial applications of the hard Lefschetz theorem. In *Proceedings of the International Congress of Mathematicians, Vols. 1, 2 (Warsaw, 1983)*, pp. 447–453. PWN, Warsaw, 1984.

[61] R. Stanley, Log-concave and unimodal sequences in algebra, combinatorics, and geometry. In *Graph Theory and its Applications: East and West (Jinan, 1986)*, Ann. New York Acad. Sci., vol. 576, pp. 500–535, New York, 1989.

[62] D. Stanton, Unimodality and Young's lattice. *J. Combin. Theory Ser. A* **54** (1990), 41–53.

[63] D. Welsh, Combinatorial problems in matroid theory. In *Combinatorial Mathematics and its Applications (Oxford, 1969)*, pp. 291–306, Academic Press, London, 1971.

[64] D. Welsh, *Matroid Theory*. London Math. Soc. Monogr. Ser. 8, Academic Press, London, 1976.

[65] H. Whitney, On the abstract properties of linear dependence. *Amer. J. Math.* **57** (1935), 509–533.

GIL KALAI

Hebrew University of Jerusalem, Jerusalem, Israel, and Reichman University, Herzliya, Israel, kalai@math.huji.ac.il

THE WORK OF JAMES MAYNARD

KANNAN SOUNDARARAJAN

ABSTRACT

We give a brief account of some of the most spectacular results established by James Maynard, for which he has been awarded the Fields Medal.

MATHEMATICS SUBJECT CLASSIFICATION 2020

Primary 11N05; Secondary 11N32, 11N35, 11J83

KEYWORDS

Distribution of primes, sieve methods, metric Diophantine approximation

INTERNATIONAL CONGRESS OF MATHEMATICIANS 2022 JULY 6–14

© 2022 International Mathematical Union

Proc. Int. Cong. Math. 2022, Vol. 1, pp. 66–80

DOI 10.4171/ICM2022/212

Published by EMS Press

and licensed under a CC BY 4.0 license

James Maynard has established several spectacular results in analytic number theory. While the proofs of these results involve many deep ideas, their statements are remarkable for their simplicity and elegance. To illustrate, we state two such striking results of Maynard concerning prime numbers, before setting them in context.

Theorem 1 (Maynard [32]). *For each natural number $m \geq 2$, there exists a positive integer $C(m)$ with the following property: There are infinitely many natural numbers n such that the interval $[n, n + C(m)]$ contains at least m prime numbers.*

Theorem 2 (Maynard [36]). *There are infinitely many prime numbers p whose decimal representation does not contain the digit 7.*

Background. To place these results in context, recall that the prime number theorem gives an asymptotic for $\pi(x)$, the number of primes below x; namely,

$$\pi(x) \sim \mathrm{li}(x) = \int_0^x \frac{dt}{\log t}.$$

We may think of this asymptotic as roughly saying that the "chance" of a number n being prime is about $1/\log n$. One overarching theme in analytic number theory may be formulated as asking in what ways does the sequence of primes resemble, or differ from, a random sequence of integers with each integer $n \geq 3$ chosen independently to be in the random sequence with probability $1/\log n$ (this is also known as the Cramér model). One obvious difference is that all primes larger than 2 must be odd, whereas a random sequence would surely contain many even numbers. But if we could account for divisibility by small primes (such as 2 in our example), would a modified random model describe accurately the behavior of prime numbers?

There are many ways in which we could try to make this theme precise. For instance, the Riemann hypothesis predicts that $|\pi(x) - \mathrm{li}(x)|$ is bounded by $C(\varepsilon)x^{\frac{1}{2}+\varepsilon}$ for any $\varepsilon > 0$ and some constant $C(\varepsilon)$. Fluctuations of size about $\sqrt{x}$ are indeed what one would expect if we select random sets of integers with $n \geq 3$ included in the set with probability $1/\log n$. Thus the Riemann hypothesis is, at a crude level, consistent with a random model of primes, although if we inspect the error term $\pi(x) - \mathrm{li}(x)$ in finer detail then the influence of zeros of $\zeta(s)$ would be visible, and such features would deviate (in small but significant ways) from the random model.

At the 1912 ICM, Landau posed four "unattackable" problems on primes: (i) the Goldbach problem that every even integer larger than 2 is the sum of two primes, (ii) the twin prime problem that there are infinitely many prime pairs n and $n + 2$, (iii) there is always a prime between two consecutive squares, and (iv) there are infinitely many primes of the form $n^2 + 1$. All four problems remain open today, and all statements are exactly what one would expect for random sequences. For example, the Cramér model would suggest that the chance that n and $n + 2$ are both "prime" is about $1/\log n \times 1/\log(n + 2)$, which would predict about $x/(\log x)^2$ twin primes up to x. Of course, some care is needed, since the same prediction could be made for n and $n + 1$ being prime, and we will address this soon. Similarly, we may expect that an even number N may have about $N/(\log N)^2$ representations as a sum of two primes, making the Goldbach conjecture very plausible, and related arguments suggest the last two Landau problems as well.

For the third Landau problem on the number of primes between n^2 and $(n+1)^2$, the random model already predicts what we believe to be the right answer – namely, there should be about $(2n+1)/\log(n^2) \approx n/\log n$ primes in this interval. For the other three problems, some modification must be made to the Cramér model, to take into account the deterministic features of these problems with respect to divisibility by small primes. Precise conjectures for these problems were first made by Hardy and Littlewood motivated by their work on the circle method. These conjectures are widely believed to be true, and are supported by extensive heuristic and numerical evidence. For instance, Hardy and Littlewood formulated the following conjecture for the number of twin primes below x:

$$\#\{n \le x : n, n+2 \text{ both prime}\} \sim \mathfrak{S}\big(\{0,2\}\big) \int_2^x \frac{dt}{(\log t)^2}.$$

Here $\int_2^x dt/(\log t)^2$ is asymptotically $x/(\log x)^2$, and corresponds to the prediction of the Cramér model, while $\mathfrak{S}(\{0,2\})$, known as the *singular series*, is a correction factor

$$\mathfrak{S}(\{0,2\}) = 2 \prod_{p \ge 3} \Big(1 - \frac{2}{p}\Big)\Big(1 - \frac{1}{p}\Big)^{-2} = 1.32\ldots$$

The constant $\mathfrak{S}(\{0,2\})$ has a compelling probabilistic interpretation: it is a product over all primes p (the first factor 2 corresponds to the prime $p = 2$), with the factor at p keeping track of the ratio between the chance that n and $n+2$ are not divisible by p, and the chance that two random numbers are not divisible by p. Thus, for $p = 2$, the chance that n and $n+2$ are both not divisible by 2 is $(1 - 1/2)$ (n must be odd), while the chance that two random numbers are both not divisible by 2 is $(1-1/2)^2 = 1/4$; the ratio of these chances gives the correction factor 2. For primes $p \ge 3$, the chance that n and $n+2$ are both not divisible by p is $(1 - 2/p)$ whereas the chance that two random numbers are both not divisible by p is $(1-1/p)^2$, and we see the corresponding correction factor in the definition of $\mathfrak{S}(\{0,2\})$.

Similar conjectures can be made for the binary Goldbach problem, or for the number of primes of the form n^2+1, modifying and correcting the naive predictions of the Cramér model. To illustrate, we give a generalization of the conjecture for twin primes for counting prime k-tuples: given distinct integers $h_1, h_2, \ldots, h_k$, for large x how many integers $n \le x$ are there with $n+h_1, \ldots, n+h_k$ all being prime. Here the Hardy–Littlewood conjecture predicts that

$$\#\{n \le x : n+h_1, \ldots, n+h_k \text{ all prime}\} \sim \mathfrak{S}\big(\{h_1, \ldots, h_k\}\big) \int_2^x \frac{dt}{(\log t)^k} \tag{1}$$

where, with $\mathcal{H} = \{h_1, \ldots, h_k\}$,

$$\mathfrak{S}(\mathcal{H}) = \prod_p \Big(1 - \frac{\nu(\mathcal{H}, p)}{p}\Big)\Big(1 - \frac{1}{p}\Big)^{-k}, \tag{2}$$

and $\nu(\mathcal{H}, p)$ denotes the number of distinct residue classes occupied by the set $\mathcal{H}$ viewed mod p. Since $\nu(\mathcal{H}, p) = k$ if p is larger than $\max |h_i - h_j|$, the product defining $\mathfrak{S}(\mathcal{H})$

converges absolutely to a nonnegative real number, and it equals zero only if $\nu(\mathcal{H}, p) = p$ for some prime p. If $\nu(\mathcal{H}, p) = p$, then for any integer n at least one of the numbers $n + h_1, \ldots, n + h_k$ would be a multiple of p, and therefore there can be only finitely many integers n with $n + h_1, \ldots, n + h_k$ all being prime; for example, this is what happens if we ask for n and $n + 1$ to be prime, or n, $n + 2$, $n + 4$ all to be prime. When there is no such divisibility obstruction to $n + h_1, \ldots, n + h_k$ all being prime, the Hardy–Littlewood conjecture predicts a rich supply of such prime k-tuples. This is perhaps the most central question in prime number theory, and remains open in any situation where $\mathfrak{S}(\mathcal{H})$ is nonzero.

Sieve theory. We have described quickly some of the main motivating questions in the theory of primes. One main source of progress towards these questions is *sieve theory*, and a large part of Maynard's work lies broadly in this area. A typical problem in sieve theory is to bound the size of sets of integers $\mathcal{A}$ whose elements are constrained to omit $\nu(p)$ given residue classes mod p for primes p. For instance, the twin prime problem is of this form, as we seek to find integers n that are neither 0 nor -2 mod p for all primes $p \le \sqrt{n+2}$ (so that n and $n + 2$ would both be prime). In great generality sieve methods can produce upper bounds of the conjectured order of magnitude; for example, one can show that the number of twin primes up to x is no more that 4 times the conjectured Hardy–Littlewood asymptotic. Producing corresponding lower bounds has proved to be a much harder problem, but sieve methods have led to striking partial results such as Chen's theorem that there are many primes p for which $p + 2$ has at most two prime factors, or Iwaniec's theorem that there are many n for which $n^2 + 1$ has at most two prime factors. For a comprehensive treatment of the subject, see **[14]**.

Chen's theorem and Iwaniec's theorem exhibit a limitation of traditional sieve methods, known as the *parity problem*, which often prevents us from knowing the parity of elements left unsieved, and thus from producing prime numbers. But in some special cases, sieve methods in conjunction with other analytic input have produced prime numbers. For instance, for large x, Baker, Harman, and Pintz **[2]** showed that the interval $[x, x + x^\theta]$ contains at least $cx^\theta / \log x$ primes, where $c > 0$ is a constant and $\theta = 0.525$; the Landau problem of producing primes between consecutive squares corresponds to intervals with $\theta = \frac{1}{2}$. Another spectacular example is due to Friedlander and Iwaniec **[13]** who established an asymptotic formula for the number of primes up to x that may be written as $n^2 + m^4$, an approximation to the Landau problem of primes of the form $n^2 + 1$. A closely related result of Heath-Brown and Li **[25]** produces an asymptotic formula for primes of the form $n^2 + p^4$, where p is prime. Yet another beautiful result due to Heath-Brown **[24]** establishes an asymptotic formula for the number of primes below x of the form $n^3 + 2m^3$ with $m, n \in \mathbb{N}$. Heath-Brown's result may be viewed as an approximation to the problem of producing primes of the form $n^3 + 2$, but before his work it was not even known if there are infinitely many primes that are the sum of three cubes of natural numbers! A crucial feature of these results is that they deal with primes represented by specializations of *norm forms*. The Friedlander–Iwaniec result is concerned with the norm form $x^2 + y^2 = N(x + iy)$ associated to the field $\mathbb{Q}(i)$, and specializing y to be a square; Heath-Brown's result is concerned with the norm form $N(x + y\alpha + z\alpha^2)$ taking the norm over the field $\mathbb{Q}(\alpha)$ with $\alpha = 2^{\frac{1}{3}}$, and specializing z to be 0. The results of Friedlander–Iwaniec and Heath-Brown gave the first examples of thin sequences (in the sense that the number of integers below X in the sequence is $\leq X^{1-\delta}$ for some $\delta > 0$) of polynomial values in two or more variables that represent infinitely many primes; no example is known of a polynomial in 1 variable of degree more than 1 that represents infinitely many primes.

Maynard's work [40] gives a substantial generalization of Heath-Brown's approach, and produces many further examples of thin sequences of polynomial values in many variables that represent primes. Consider an algebraic number $\omega \in \mathbb{C}$ of degree n, and let K denote the field $\mathbb{Q}(\omega)$. We can associate to this the norm form $N(\sum_{i=1}^{n} x_i \omega^{i-1})$, which is a homogeneous polynomial of degree n in the variables $x_1, \ldots, x_n$. A thin polynomial in many variables would be obtained by specializing some of the variables in this norm form to be zero; say, we set $x_{n-k+1}, \ldots, x_n = 0$, and the number integers below x represented by such an incomplete norm form would be about $x^{1-k/n}$. In the range $n \geq 4k$, Maynard establishes an asymptotic formula for the number of primes represented by such an incomplete norm form, when the variables $x_1, \ldots, x_{n-k}$ take integer values in the range $[1, X]$.

The circle method. Apart from sieve theory, another important source of progress towards problems on primes is the *circle method*, which as we already mentioned formed the original motivation for Hardy and Littlewood in formulating their conjectures. To illustrate, consider the Goldbach problem of representing an even integer N as a sum of two primes. Using Fourier analysis, the number of such representations of N may be written as

$$r(N) = \int_0^1 S(\alpha)^2 e^{-2\pi i N\alpha} \, d\alpha, \quad \text{where } S(\alpha) = \sum_{p \leq N} e^{2\pi i p\alpha}. \tag{3}$$

The idea in the circle method is that generating functions such as $S(\alpha)$ above tend to be large near rational numbers with small denominator (the *major arcs*) and small away from them (the *minor arcs*).

While the circle method has not been able to tackle the binary Goldbach problem or the problem of twin primes, it has been extremely effective in problems where there is a bit more freedom. For instance, the ternary Goldbach problem asks to represent odd numbers as a sum of three primes, and there is one extra variable to play with here. Vinogradov famously used the circle method to show that all large odd numbers are the sum of three primes, and Helfgott [26] has extended this to show that all odd numbers larger than 5 may be so represented. Here we may mention an impressive result of Matomäki, Maynard, and Shao [38] which shows that large odd numbers n may be expressed as $p_1 + p_2 + p_3$, where all three primes p_i lie in a short interval $[n/3 - n^{\theta}, n/3 + n^{\theta}]$ for any $\theta > 11/20$. We mentioned earlier the work of Baker, Harman, and Pintz [2] showing the existence of primes in short intervals $[x, x + x^{0.525}]$, and the work of [38] is remarkable in solving the ternary Goldbach problem using primes in only slightly longer intervals.

A second example of what it might mean to have an extra degree of freedom is the Green–Tao theorem that the primes contain arbitrarily long arithmetic progressions $n, n +$

$d, \dots, n+(k-1)d$. The Hardy–Littlewood conjecture would predict a stronger "one-dimensional" version of such a result with specified choices for the common difference d; for instance, there should be infinitely many k-tuples primes of the form $n, n+k!, n+2\cdot k!, \dots, n+(k-1)\cdot k!$. The work of Green, Tao, and Ziegler [19–21] may be thought of as a far-reaching generalization of the circle method, obtaining asymptotic formulae for the number of prime solutions to linear systems with "at least two degrees of freedom."

Maynard's beautiful result on primes with missing digits (Theorem 2 stated above) is a rare occasion where the circle method can be used to solve a binary problem. Let $\mathcal{M}$ denote the set of natural numbers with no 7 in their decimal expansion (naturally one could omit any other digit instead of 7). The number of integers in $\mathcal{M}$ up to N is about $N^{\log 9/\log 10} = N^{1-\delta}$ with $\delta = 0.046\dots$, so that $\mathcal{M}$ is a thin set making the problem of finding primes in it a challenge. Before Maynard's work, Dartyge and Mauduit [7,8] had used sieve theory to show that $\mathcal{M}$ contains integers with at most two prime factors. To count the number of primes in $\mathcal{M}$ up to N, we use Fourier analysis writing this as

$$\sum_{\substack{p\le N\\ p\in\mathcal{M}}} 1 = \int_0^1 S(\alpha)M(-\alpha)d\alpha,$$

where $S(\alpha)$ is the exponential sum over primes defined in (3), and $M(\alpha) = \sum_{m\le N, m\in\mathcal{M}} e^{2\pi i\alpha}$ is the corresponding exponential sum over the set $\mathcal{M}$. Usually, such a binary problem is hopeless to attack via the circle method – the reason being that even most optimistically we may only expect "square-root cancellation" in the exponential sums $S(\alpha)$ and $M(-\alpha)$ for generic α, and even that would produce an integrand of size $N^{\frac{1}{2}} \times N^{\frac{1}{2}(1-\delta)}$, which is bigger than the expected main term of size about $N^{1-\delta}/\log N$. A crucial feature in this problem is that the set $\mathcal{M}$ has a very convenient structure which results in the exponential sum $M(\alpha)$ often being unusually small. For instance, Maynard shows that its L^1-norm satisfies

$$\int_0^1 |M(\alpha)|d\alpha \ll N^{0.32},$$

with the key point being that the exponent 0.32 is smaller even than $(1-\delta)/2$, which is the optimistic square-root cancellation that we mentioned. Such estimates raise the hope of being able to attack Theorem 2, and the main idea can be seen transparently in Maynard's expository article [35], where he proves an easier version of Theorem 2 treating primes missing a digit in base b with b sufficiently large. The set of integers up to N missing a digit in base b has size about $N^{\log(b-1)/\log b}$, and so the problem becomes easier as the base b gets larger. Getting the base down to 10 turns out to be a fiendishly difficult problem, and is arguably more significant psychologically than for any mathematical reason. Maynard [36] tackles this brilliantly by introducing a number of new ideas, including ideas from the geometry of numbers, different aspects of sieve theory, and comparisons with a Markov process. We may expect that even in base 3 there should be infinitely many primes with a given digit missing; in base 2, the only digit that might be omitted is 0, and we find the problem of whether there are infinitely many Mersenne primes, which lies beyond reasonable mathematics. We close this discussion by pointing out two other beautiful results on the digits of prime numbers which have elements in common with Maynard's work: namely, work of Mauduit and Rivat **[31]** which shows (in particular) that the sum of the decimal digits of primes is equally likely to be odd or even, and work of Bourgain **[6]** which allows one to specify a small proportion of the binary digits of primes.

Gaps between primes. We now turn to a discussion of Maynard's most spectacular result – the sun amidst small stars – namely, Theorem 1 above on finding many primes in bounded intervals. To describe the recent history of this problem, let us first discuss how primes are spaced typically. The prime number theorem tells us that the nth prime p_n is about $n \log n$, so that the average spacing between two consecutive primes, $p_{n+1} - p_n$, is about $\log p_n$. What is the distribution of the normalized spacings $(p_{n+1} - p_n)/\log p_n$? The Cramér random model for primes would predict that these normalized spacings should behave like a Poisson process, and that for any fixed interval $[\alpha, \beta] \in \mathbb{R}_{\geq 0}$,

$$\lim_{N\to\infty} \frac{1}{N}\#\left\{n \leq N : \frac{p_{n+1} - p_n}{\log p_n} \in [\alpha, \beta]\right\} = \int_\alpha^\beta e^{-t}\,dt = e^{-\alpha} - e^{-\beta}. \tag{4}$$

Gallagher **[16]** showed that this prediction is also implied by the more refined Hardy–Littlewood conjectures, the key point being that the singular series constants $\mathfrak{S}(\mathcal{H})$ (see (2)) are approximately 1 (matching the naive Cramér model) on average over k-element sets $\mathcal{H}$.

This conjecture on the normalized spacings between primes is wide open. Indeed, if we denote by $\mathcal{L}$ the set of limit points of $(p_{n+1} - p_n)/\log p_n$, then even the qualitative statement that $\mathcal{L} = [0, \infty]$ (which follows at once from (4)) is currently unknown. By creating long strings of composite numbers, Westzynthius established that $\mathcal{L}$ contains ∞, but for a long time no other limit point was known (although Erdős and Ricci had established that $\mathcal{L}$ has positive Lebesgue measure). Dramatic progress was made in the 2005 with the path-breaking work of Goldston, Pintz, and Yıldırım **[17]**, who showed that for any $\varepsilon > 0$ there are infinitely many n with $p_{n+1} - p_n \leq \varepsilon \log p_n$. Thus there are small gaps between primes in comparison to the average, and 0 is now known to be in $\mathcal{L}$. Before the work of Goldston, Pintz, and Yıldırım, it was only known that the difference between consecutive primes became smaller than about $\frac{1}{4}$ of the average spacing, and their work opened the door to later advances including Maynard's Theorem 1.

Suppose $h_1, \ldots, h_k$ are distinct integers with $\mathfrak{S}(\{h_1, \ldots, h_k\}) > 0$; such tuples are called *admissible*, and for example $\{k!, 2 \cdot k!, \ldots, k \cdot k!\}$ is admissible. The Hardy–Littlewood conjecture predicts that there are infinitely many n with $n + h_1, \ldots, n + h_k$ all being prime. Instead of wanting all k of these numbers to be prime, what if we only ask for at least two of them to be prime? This would already show that infinitely often there are bounded gaps between consecutive prime numbers. Suppose we could find nonnegative weights $w(n)$ with the property that for large x and each $j = 1, \ldots, k$,

$$\sum_{\substack{x \leq n \leq 2x \\ n+h_j \text{ prime}}} w(n) > \frac{1}{k} \sum_{x \leq n \leq 2x} w(n). \tag{5}$$

Then summing (5) over all $j = 1, \ldots, k$, we would obtain

$$\sum_{x \le n \le 2x} \#\{1 \le j \le k : n + h_j \text{ prime}\} w(n) > \sum_{x \le n \le 2x} w(n), \tag{6}$$

from which it would follow that there must be some n with at least 2 primes among $n + h_1$, ..., $n + h_k$. Thinking of the weights as giving a probability measure on $x \le n \le 2x$, we may interpret (6) as saying that the expected number of primes among the $n + h_j$ is greater than 1, so that there must be n with at least 2 primes in this k-tuple.

The difficult problem is to construct weights satisfying (5), and natural choices for such weights are suggested by sieve theory, in particular the theory of the Selberg sieve. The standard choice of Selberg sieve weights (which are used to give an upper bound for the number of prime k-tuples $n + h_1, \ldots, n + h_k$) takes the shape

$$w(n) = \Bigg(\sum_{\substack{d \mid (n+h_1)\cdots(n+h_k) \\ d \le R}} \mu(d) \Big(\frac{\log R/d}{\log R} \Big)^k \Bigg)^2 .$$

Clearly, $w(n) \ge 0$ always. Expanding out the sum, the right-hand side of (5) (the sum over all $n \in [x, 2x]$) may be evaluated asymptotically so long as $R^2 \le x^{1-\varepsilon}$. The left-hand side of (5) is more involved, and relies on understanding the distribution of primes in arithmetic progressions with the modulus of the progression going up to R^2. The Bombieri–Vinogradov theorem permits such an understanding (at a level comparable to what the Generalized Riemann Hypothesis would give) so long as $R^2 \le x^{\frac{1}{2}-\varepsilon}$, so that R is now constrained to be $\le x^{\frac{1}{4}-\varepsilon}$. For this choice of weights, the expected number of primes among the $n + h_j$ turns out to be about $(2k/(k+1)) \log R/\log x$, so that with $R \le x^{1/4-\varepsilon}$ one only expects to find $\frac{1}{2}$ a prime in the k-tuple.

Although the Selberg sieve weights described above had been optimized for upper bounds in the prime k-tuple problem, Goldston, Pintz, and Yıldırım made the surprising discovery that there are better choices of weights for optimizing the ratio of the sums in (5). They considered weights of the form

$$w(n) = \Bigg(\sum_{\substack{d \mid (n+h_1)\cdots(n+h_k) \\ d \le R}} \mu(d) \Big(\frac{\log R/d}{\log R} \Big)^{k+\ell} \Bigg)^2 ,$$

for a suitable parameter ℓ, which turns out in the optimal case to be around $\sqrt{k}$. With this choice of weights, they found that the expected number of primes among $n + h_j$ is about twice as large as previously, being $(4 + O(1/k^{\frac{1}{2}})) \log R/\log x$. With $R = x^{\frac{1}{4}-\varepsilon}$, this barely fails to give the desired relation (5), and thus barely falls short of proving bounded gaps between primes. By considering an additional possible prime value $n + h$ for $1 \le h \le \varepsilon \log x$, Goldston, Pintz, and Yıldırım were able to deduce from this argument that there are infinitely many n with $p_{n+1} - p_n \le \varepsilon \log n$. For a more detailed discussion of these ideas, see **[47]**.

If one could take R to be $x^{\frac{1}{4}+\delta}$ for any $\delta > 0$, then the argument of Goldston, Pintz, and Yıldırım would give bounded gaps between primes. To take such a value for R, one would need to understand the distribution of primes up to x in arithmetic progressions, when the modulus of the progression is as large as $x^{\frac{1}{2}+2\delta}$. The Elliott–Halberstam conjectures predict that such results should hold (on average) when the modulus is as large as $x^{1-\varepsilon}$. Partial progress towards such extensions of the Bombieri–Vinogradov theorem was made by Fouvry and Iwaniec [12], and Bombieri, Friedlander, and Iwaniec [5], but these results did not apply immediately to the problem of showing bounded gaps between primes. In April 2013, Yitang Zhang [49] made a spectacular breakthrough by establishing a version of the Bombieri–Vinogradov theorem in an extended range which was sufficient for the method of Goldston, Pintz, and Yıldırım. Zhang established that if $k > 3.5 \times 10^6$ then for any admissible k-tuple $h_1, \ldots, h_k$ there are infinitely many n with at least two of the $n + h_j$ being prime. This implied that infinitely often the gaps between consecutive primes is less than 70 million. Refinements of Zhang's work on the equidistribution of primes in arithmetic progressions were made by the Polymath project [46], and still further qualitative and quantitative refinements of such results may be found in the recent papers of Maynard [37–39].

Zhang's work established the case $m = 2$ of Theorem 1. However, even if one could take the largest possible range for R, namely $R = x^{\frac{1}{2}-\varepsilon}$ (which would be permitted by the Elliott–Halberstam conjecture), the Goldston–Pintz–Yıldırım weights would only yield that the expected number of primes in an admissible k-tuple is $\geq 2 - \varepsilon$. In other words, even under the Elliott–Halberstam conjecture, one would fall short of establishing the existence of three primes in bounded intervals.

The proof of Theorem 1 is based on a different choice of the weights $w(n)$, discovered just months after Zhang's work by Maynard (who announced the results in a memorable talk at Oberwolfach in October 2013) and independently by Tao (in unpublished work). The Maynard–Tao weights are a multidimensional extension of the weights considered earlier, and take (roughly speaking) the shape

$$w(n) = \Bigg(\sum_{\substack{d_1,\ldots,d_k \\ d_i | n+h_i \\ \prod d_i \leq R}} \prod_{i=1}^{k} \mu(d_i) F\left(\frac{\log d_1}{\log R}, \ldots, \frac{\log d_k}{\log R}\right)\Bigg)^2,$$

for suitable smooth functions $F : [0,1]^k \to \mathbb{R}$. Astonishingly, it turns out that for an appropriate choice for F, the expected number of primes in the tuple $n + h_1, \ldots, n + h_k$ (recall (6) above) is $\geq c \log k \frac{\log R}{\log x}$, for a positive constant c; in fact c may be taken close to 1 if k is large enough. The key point is that this expected number of primes in k-tuples tends to infinity with k, and in fact we only need R to grow like any power of x for the method to succeed, so that Bombieri–Vinogradov which permits $R = x^{\frac{1}{4}-\varepsilon}$ is already sufficient! Thus the following more precise version of Theorem 1 holds, which may be viewed as a partial result towards the Hardy–Littlewood prime k-tuples conjecture.

Theorem 3 (Maynard [32]). *Let $m \geq 2$ be a natural number. Let k be sufficiently large in terms of m, and let $\mathcal{H} = \{h_1, \ldots, h_k\}$ be any set of k integers with $\mathfrak{S}(\mathcal{H}) > 0$. Then there exist infinitely many n such that the k-tuple $n + h_1, \ldots, n + h_k$ contains at least m primes.*

Maynard showed that k may be taken smaller than Cm^2e^{4m} for a suitable constant C, and further refinements of this (incorporating also the work of Zhang) have been made in the work of Baker and Irving [3] who showed that k may be taken as $Ce^{3.815m}$. Of special interest is the case $m = 2$ where the Polymath project [45] optimized these arguments to establish that any admissible 50-tuple contains 2 primes infinitely often. In particular, they showed that $p_{n+1} - p_n \le 246$ infinitely often, and conditional on the Elliott–Halberstam conjecture that infinitely often there are at least two primes in the triple $n, n+2, n+6$. Let us mention one other uniform variant of these results: Maynard [33] shows, for instance, that there are at least $cX \exp(-\sqrt{\log X})$ values of $x \in [X, 2X]$ such that the interval $[x, x + \log X]$ contains at least $c \log \log X$ primes (here c is a positive constant). For detailed expositions on these results of Zhang, Maynard, and Tao, see [18, 29].

The Maynard–Tao weights offer a flexible new method to study many problems on primes and related sequences, and have found a number of applications. We describe two other results using these weights, both still concerned with spacings between consecutive primes. We referred earlier to the result of Westzynthius on large gaps between consecutive primes, which showed that ∞ lies in the set $\mathcal{L}$ of limit points of the normalized spacings $(p_{n+1} - p_n)/\log p_n$. This was quantified in the 1930s by Erdős and Rankin who showed that, for a positive constant C,

$$\max_{p_n \le X} (p_{n+1} - p_n) \ge C \log X \frac{(\log\log X) \log\log\log\log X}{(\log\log\log X)^2}. \tag{7}$$

The random model would suggest that the maximal gap between primes up to X should be about $(\log X)^2$. This is known as Cramér's conjecture, and while this is very delicate, it is widely believed that the maximal gap is no more than $(\log X)^{2+\varepsilon}$, although even this is far beyond Landau's unattackable problem of the existence of a prime between consecutive squares. Erdős drew attention to the problem of finding larger gaps between consecutive primes, offering \$10 000 for a bound that would replace C in (7) with a function tending to ∞ with X. For more than 75 years, this problem resisted attack, with only improvements of the constant C being known. Then, by a remarkable coincidence, in 2014 *two* different techniques emerged, both establishing (7) with C replaced by a function tending to infinity with X. One approach, by Ford, Green, Konyagin, and Tao [11], built upon the work of Green–Tao on arithmetic progressions in the primes, while the other approach, by Maynard [34], found a way to adapt the Maynard–Tao sieve weights. The second approach was better suited for quantifying the large gaps that are produced, and, joining forces, Ford, Green, Konyagin, Maynard, and Tao [10] established that for some constant $C > 0$,

$$\max_{p_n \le X} (p_{n+1} - p_n) \ge C \log X \frac{(\log\log X) \log\log\log\log X}{\log\log\log X}, \tag{8}$$

improving the bound in (7) by a factor of $\log\log\log X$.

The results on small gaps and large gaps between consecutive primes show that 0 and ∞ lie in the set $\mathcal{L}$ of limit points of the normalized prime spacings. No other explicit numbers are known to lie in $\mathcal{L}$, although we expect $\mathcal{L}$ to include all nonnegative real numbers. Following Zhang's breakthrough, Pintz [42] showed that $\mathcal{L}$ contains an interval $[0, c]$ for some $c > 0$, which, however, is ineffective and cannot be computed explicitly. Using the Maynard–Tao sieve weights, Banks, Freiberg, and Maynard [4] established the following beautiful result: If $\beta_1 \le \beta_2 \le \cdots \le \beta_9$ are any nine real numbers, then at least one of their differences $\beta_j - \beta_i$ (with $i < j$) must be an element of $\mathcal{L}$. Their result has been refined by Pintz [43], and Merikoski [41], and Merikoski shows that the same result holds if we start with just four real numbers $\beta_1 \le \beta_2 \le \beta_3 \le \beta_4$. Moreover, Merikoski has also shown that for any $T > 0$, the set $\mathcal{L} \cap [0, T]$ has measure at least $T/3$.

The Duffin–Schaeffer conjecture. So far we have focussed entirely on Maynard's work concerned with prime numbers. In a very different direction, Maynard in joint work with Koukoulopoulos [28], resolved one of the central problems in the metric theory of Diophantine approximation, known as the Duffin–Schaeffer conjecture.

Diophantine approximation is concerned with finding rational approximations a/q to a given irrational number α, with an emphasis on making $|\alpha - a/q|$ small in terms of q. The most basic result is Dirichlet's theorem that for every irrational number α, there are infinitely many rational approximations a/q, with $a \in \mathbb{Z}$, $q \in \mathbb{N}$ and $(a, q) = 1$ (so that the fraction is in reduced form) such that $|\alpha - a/q| \le 1/q^2$. For quadratic irrationals (like $\sqrt{2}$ or the golden ratio), Dirichlet's theorem is essentially the best possible, and for every such α there exists a positive constant $C(\alpha)$ such that $|\alpha - a/q| \ge C(\alpha)/q^2$ for any rational approximation a/q. A celebrated result of Roth establishes that for any algebraic irrational α and any $\varepsilon > 0$ one has $|\alpha - a/q| \ge C(\alpha, \varepsilon)/q^{2+\varepsilon}$, for a suitable positive constant $C(\alpha, \varepsilon)$. For particular interesting transcendental numbers, such as π, it remains an outstanding open problem to determine how well they can be approximated by rational numbers.

Metric Diophantine approximation is concerned with such approximation problems that hold for *almost all* irrational numbers α, with *almost all* interpreted in the sense of Lebesgue measure. Since the problem of approximating α by rationals is identical to that of approximating $\alpha + 1$, we may restrict attention to irrational numbers $\alpha \in [0, 1)$. The most basic problem is the following: suppose $\psi : \mathbb{N} \to \mathbb{R}_{\ge 0}$ is a given function, what can be said about the measure of $\alpha \in [0, 1)$ for which there exist infinitely many rational numbers a/q in reduced form (that is, $(a, q) = 1$) with $|\alpha - a/q| \le \psi(q)$. For instance, Dirichlet's theorem tells us that if $\psi(q) = 1/q^2$, then all irrational $\alpha \in [0, 1)$ admit infinitely many such rational approximations.

Let $\mathcal{A}_q = \mathcal{A}_q(\psi)$ denote the set of $\alpha \in [0, 1)$ for which there exists some reduced fraction a/q with $|\alpha - a/q| \le \psi(q)$, and let $\mathcal{A}$ denote the set of $\alpha \in [0, 1)$ lying in infinitely many of the sets $\mathcal{A}_q$. Thus

$$\mathcal{A} = \bigcap_{Q=1}^{\infty} \tilde{\mathcal{A}}(Q), \quad \text{with } \tilde{\mathcal{A}}(Q) = \bigcup_{q=Q}^{\infty} \mathcal{A}_q.$$

Now the measure of $\mathcal{A}_q$ is $\le 2\phi(q)\psi(q)$, since there are $\phi(q)$ possible choices for the numerator a, and if $\psi(q) \le 1/(2q)$ so that the intervals for different a do not overlap then equality holds here. If $\sum_{q=1}^{\infty} \phi(q)\psi(q)$ converges, then the measure of $\tilde{\mathcal{A}}(Q)$ is bounded by $2\sum_{q=Q}^{\infty} \phi(q)\psi(q)$, which is the tail of a convergent series and thus tends to 0 as $Q \to \infty$. It follows that $\mathcal{A}$ has measure 0. This argument is identical to the easy part of the Borel–Cantelli Lemma.

In 1941, Duffin and Schaeffer made the remarkable conjecture that in the complementary case when $\sum_{q=1}^{\infty} \phi(q)\psi(q)$ diverges, the measure of $\mathcal{A}$ is 1. Since then the Duffin–Schaeffer conjecture has remained one of the central motivating questions in the theory of metric Diophantine approximations. A number of partial results towards this conjecture were established: for example, a beautiful result of Gallagher [15] showed that the measure of the set $\mathcal{A}(\psi)$ is always either 0 or 1, work of Erdős [9] and Vaaler [48] established the conjecture when $\psi(q)$ is $O(1/q^2)$ for all q, higher dimensional analogues of the conjecture were proved by Pollington and Vaughan [44], and weaker versions of the conjecture with extra divergence conditions were established in [1, 22, 23]. But the full problem resisted until the recent work of Koukoulopoulos and Maynard [28]:

Theorem 4 (Koukoulopoulos and Maynard [28]). *Let $\psi : \mathbb{N} \to \mathbb{R}_{\geq 0}$ be such that $\sum_{q=1}^{\infty} \phi(q)\psi(q)$ diverges. Then the set of $\alpha \in [0, 1)$ that have infinitely many rational approximations $|\alpha - a/q| \leq \psi(q)$ with $(a, q) = 1$ has Lebesgue measure 1. In other words, the Duffin–Schaeffer conjecture holds.*

We refer to Koukoulopoulos's talk at this ICM [27] for a more detailed exposition of this result, and the ideas behind its proof.

We have given an overview of some of Maynard's most spectacular achievements in analytic number theory. Maynard's work is characterized by ingenious but simple ideas, which are carried very far with his powerful technical ability. As impressive as his work so far has been, it may only mark a beginning.

FUNDING

This work was partially supported by grants from the National Science Foundation, and a Simons Investigator Award from the Simons Foundation.

REFERENCES

[1] C. Aistleitner, T. Lachmann, M. Munsch, N. Technau, and A. Zafeiropoulos, The Duffin–Schaeffer conjecture with extra divergence. *Adv. Math.* **356** (2019), 106808, 11.

[2] R. C. Baker, G. Harman, and J. Pintz, The difference between consecutive primes. II. *Proc. Lond. Math. Soc. (3)* **83** (2001), no. 3, 532–562.

[3] R. C. Baker and A. J. Irving, Bounded intervals containing many primes. *Math. Z.* **286** (2017), no. 3–4, 821–841.

[4] W. D. Banks, T. Freiberg, and J. Maynard, On limit points of the sequence of normalized prime gaps. *Proc. Lond. Math. Soc. (3)* **113** (2016), no. 4, 515–539.

[5] E. Bombieri, J. B. Friedlander, and H. Iwaniec, Primes in arithmetic progressions to large moduli. *Acta Math.* **156** (1986), no. 3–4, 203–251.

[6] J. Bourgain, Prescribing the binary digits of primes, II. *Israel J. Math.* **206** (2015), no. 1, 165–182.

[7] C. Dartyge and C. Mauduit, Nombres presque premiers dont l'écriture en base r ne comporte pas certains chiffres. *J. Number Theory* **81** (2000), no. 2, 270–291.

[8] C. Dartyge and C. Mauduit, Ensembles de densité nulle contenant des entiers possédant au plus deux facteurs premiers. *J. Number Theory* **91** (2001), no. 2, 230–255.

[9] P. Erdős, On the distribution of the convergents of almost all real numbers. *J. Number Theory* **2** (1970), 425–441.

[10] K. Ford, B. Green, S. Konyagin, J. Maynard, and T. Tao, Long gaps between primes. *J. Amer. Math. Soc.* **31** (2018), no. 1, 65–105.

[11] K. Ford, B. Green, S. Konyagin, and T. Tao, Large gaps between consecutive prime numbers. *Ann. of Math. (2)* **183** (2016), no. 3, 935–974.

[12] E. Fouvry and H. Iwaniec, On a theorem of Bombieri–Vinogradov type. *Mathematika* **27** (1980), no. 2, 135–152 (1981).

[13] J. Friedlander and H. Iwaniec, The polynomial $X^2 + Y^4$ captures its primes. *Ann. of Math. (2)* **148** (1998), no. 3, 945–1040.

[14] J. Friedlander and H. Iwaniec, *Opera de cribro*. Amer. Math. Soc. Colloq. Publ. 57, American Mathematical Society, Providence, RI, 2010.

[15] P. Gallagher, Approximation by reduced fractions. *J. Math. Soc. Japan* **13** (1961), 342–345.

[16] P. X. Gallagher, On the distribution of primes in short intervals. *Mathematika* **23** (1976), no. 1, 4–9.

[17] D. A. Goldston, J. Pintz, and C. Y. Yıldırım, Primes in tuples. I. *Ann. of Math. (2)* **170** (2009), no. 2, 819–862.

[18] A. Granville, Primes in intervals of bounded length. *Bull. Amer. Math. Soc. (N.S.)* **52** (2015), no. 2, 171–222.

[19] B. Green and T. Tao, Linear equations in primes. *Ann. of Math. (2)* **171** (2010), no. 3, 1753–1850.

[20] B. Green and T. Tao, The Möbius function is strongly orthogonal to nilsequences. *Ann. of Math. (2)* **175** (2012), no. 2, 541–566.

[21] B. Green, T. Tao, and T. Ziegler, An inverse theorem for the Gowers $U^{s+1}[N]$-norm. *Ann. of Math. (2)* **176** (2012), no. 2, 1231–1372.

[22] G. Harman, *Metric number theory*. London Math. Soc. Monogr. Ser. 18, The Clarendon Press, Oxford University Press, New York, 1998.

[23] A. K. Haynes, A. D. Pollington, and S. L. Velani, The Duffin–Schaeffer conjecture with extra divergence. *Math. Ann.* **353** (2012), no. 2, 259–273.

[24] D. R. Heath-Brown, Primes represented by $x^3 + 2y^3$. *Acta Math.* **186** (2001), no. 1, 1–84.

[25] D. R. Heath-Brown and X. Li, Prime values of $a^2 + p^4$. *Invent. Math.* **208** (2017), no. 2, 441–499.

[26] H. A. Helfgott, The ternary Goldbach problem. In *Proceedings of the International Congress of Mathematicians—Seoul 2014. Vol. II*, pp. 391–418, Kyung Moon Sa, Seoul, 2014.

[27] D. Koukoulopoulos, Rational approximations of irrational numbers. 2021, arXiv:2109.11003.

[28] D. Koukoulopoulos and J. Maynard, On the Duffin–Schaeffer conjecture. *Ann. of Math. (2)* **192** (2020), no. 1, 251–307.

[29] E. Kowalski, Gaps between prime numbers and primes in arithmetic progressions [after Y. Zhang and J. Maynard]. Astérisque (2015), no. 367-368, Exp. No. 1084, ix, 327–366.

[30] K. Matomäki, J. Maynard, and X. Shao, Vinogradov's theorem with almost equal summands. *Proc. Lond. Math. Soc. (3)* **115** (2017), no. 2, 323–347.

[31] C. Mauduit and J. Rivat, Sur un problème de Gelfond: la somme des chiffres des nombres premiers. *Ann. of Math. (2)* **171** (2010), no. 3, 1591–1646.

[32] J. Maynard, Small gaps between primes. *Ann. of Math. (2)* **181** (2015), no. 1, 383–413.

[33] J. Maynard, Dense clusters of primes in subsets. *Compos. Math.* **152** (2016), no. 7, 1517–1554.

[34] J. Maynard, Large gaps between primes. *Ann. of Math. (2)* **183** (2016), no. 3, 915–933.

[35] J. Maynard, Digits of primes. In *European Congress of Mathematics*, pp. 641–661, Eur. Math. Soc., Zürich, 2018.

[36] J. Maynard, Primes with restricted digits. *Invent. Math.* **217** (2019), no. 1, 127–218.

[37] J. Maynard, Primes in arithmetic progressions to large moduli I: Fixed residue classes. 2020, arXiv:2006.06572.

[38] J. Maynard, Primes in arithmetic progressions to large moduli II: Well-factorable estimates. 2020, arXiv:2006.07088.

[39] J. Maynard, Primes in arithmetic progressions to large moduli III: Uniform residue classes. 2020, arXiv:2006.08250.

[40] J. Maynard, Primes represented by incomplete norm forms. *Forum Math. Pi* **8** (2020), e3, 128.

[41] J. Merikoski, Limit points of normalized prime gaps. *J. Lond. Math. Soc. (2)* **102** (2020), no. 1, 99–124.

[42] J. Pintz, Polignac numbers, conjectures of Erdős on gaps between primes, arithmetic progressions in primes, and the bounded gap conjecture. In *From arithmetic to zeta-functions*, pp. 367–384, Springer, Cham, 2016.

[43] J. Pintz, A note on the distribution of normalized prime gaps. *Acta Arith.* **184** (2018), no. 4, 413–418.

[44] A. D. Pollington and R. C. Vaughan, The k-dimensional Duffin and Schaeffer conjecture. *Mathematika* **37** (1990), no. 2, 190–200.

[45] D. H. J. Polymath, New equidistribution estimates of Zhang type. *Algebra Number Theory* **8** (2014), no. 9, 2067–2199.

[46] D. H. J. Polymath, Variants of the Selberg sieve, and bounded intervals containing many primes. *Res. Math. Sci.* **1** (2014), Art. 12, 83 .

[47] K. Soundararajan, Small gaps between prime numbers: the work of Goldston–Pintz–Yıldırım. *Bull. Amer. Math. Soc. (N.S.)* **44** (2007), no. 1, 1–18.

[48] J. D. Vaaler, On the metric theory of Diophantine approximation. *Pacific J. Math.* **76** (1978), no. 2, 527–539.

[49] Y. Zhang, Bounded gaps between primes. *Ann. of Math. (2)* **179** (2014), no. 3, 1121–1174.

KANNAN SOUNDARARAJAN

Department of Mathematics, Stanford University, Stanford, CA, 94305, USA, ksound@stanford.edu

This Page Intentionally Left Blank

THE WORK OF MARYNA VIAZOVSKA

HENRY COHN

ABSTRACT

On July 5th, 2022, Maryna Viazovska was awarded a Fields Medal for her solution of the sphere packing problem in eight dimensions, as well as further contributions to related extremal problems and interpolation problems in Fourier analysis. This article explains some of the ideas behind her work to a broad mathematical audience.

MATHEMATICS SUBJECT CLASSIFICATION 2020

Primary 52C17; Secondary 11F03, 11H31

KEYWORDS

Sphere packing, modular forms

INTERNATIONAL CONGRESS OF MATHEMATICIANS 2022 JULY 6–14

© 2022 International Mathematical Union

Proc. Int. Cong. Math. 2022, Vol. 1, pp. 82–105
DOI 10.4171/ICM2022/213

Published by EMS Press and licensed under a CC BY 4.0 license

1. INTRODUCTION

The sphere packing problem asks how we can fill as large a fraction of space as possible with congruent balls, if they are not allowed to overlap except tangentially.[1] This problem sits at the interface between many branches of mathematics, and of science more generally, with connections ranging from materials science to information theory. Sphere packing is a natural problem in Euclidean geometry, with a simple statement, and one might expect an equally elementary and self-contained solution. Instead, the topic is dominated by unexpected connections.

Before Viazovska's breakthrough work, the optimal sphere packing density was known only in one, two, and three dimensions. One dimension is trivial, because intervals can tile the real line with density 1. The two-dimensional case is not trivial, but Thue **[26]** showed that arranging six neighbors around each disk is optimal, with density $\pi/\sqrt{12} = 0.9068\ldots$. The three-dimensional case was solved by Hales **[16]** via an ingenious and elaborate computer-assisted proof, which has since been formally verified **[17]**. The unsurprising answer is shown in Figure 1: optimal two-dimensional layers are nestled together as densely as possible, to achieve density $\pi/\sqrt{18} = 0.7404\ldots$.

FIGURE 1
An optimal packing of cannonballs.

These prior results paint a misleading picture of what happens in higher dimensions. Stacking optimal layers from the previous dimension generally produces suboptimal packings, and nobody has any idea what the densest sphere packings might be in most dimensions. We do not even know whether they should be crystalline or disordered.

---

1 To state the problem precisely, "as large a fraction as possible" must be made precise. One way to do so is by taking a limit of the packing problem in a bounded region as its size grows relative to the sphere radius. The sphere packing problem turns out to be very robust, in the sense that just about all reasonable formulations are equivalent.

High-dimensional packings are not merely of pure mathematical interest, but also important for practical applications, because sphere packings are error-correcting codes for a continuous communication channel (such as radio). In this model, the packing is in an abstract signal space, whose dimension is the number of measurements used to characterize the signal and is generally much larger than three.

There does not seem to be any simple pattern in the optimal packings that persists across many dimensions, and the best upper and lower bounds known for the packing density in $\mathbb{R}^d$ remain exponentially far apart as d grows. However, a handful of dimensions stand out as special, most notably 8 and 24 dimensions. These dimensions feature exceptional packings, namely the E_8 root lattice and the Leech lattice Λ_{24}, with remarkable symmetries and numerous connections to different branches of mathematics. Thanks to Viazovska's work [10,27], we now know that they are truly optimal. The jump from 3 dimensions to 8 and 24 in the known solutions is remarkable, and it illustrates the exceptional nature of these packings.

The E_8 and Leech lattices had long been viewed as the most compelling candidates for further solutions of the sphere packing problem. However, a direct geometric proof seems infeasible: it is natural to try to work with a decomposition of space into cells, but the curse of dimensionality means we are faced with an unmanageable number of potential cell shapes and ways they could adjoin each other. Perhaps there exists a proof along these lines, but nobody has found a workable approach.

Instead, Viazovska proved the optimality of E_8 via a dramatic new connection to the theory of modular forms, following which she and several collaborators extended her ideas to the case of the Leech lattice:

Theorem 1.1 (Viazovska [27]). *The E_8 root lattice achieves the optimal sphere packing density in $\mathbb{R}^8$, namely $\pi^4/384$.*

Theorem 1.2 (Cohn, Kumar, Miller, Radchenko, and Viazovska [10]). *The Leech lattice Λ_{24} achieves the optimal sphere packing density in $\mathbb{R}^{24}$, namely $\pi^{12}/12!$.*

As Peter Sarnak said at the time [19], her paper [27] is "stunningly simple, as all great things are." This simplicity is characteristic of Viazovska's work: she has a gift for linking concepts and posing bold conjectures, and these insights lead her to striking arguments. Her proofs engage directly with the heart of the matter, without any extraneous complications. Of course, simple is very much not the same thing as easy. What makes her work extraordinary is how different her ideas are from what came before.

In the remainder of this article, we will examine Viazovska's proof of the optimality of E_8, as well as its motivation and place in mathematics more broadly. In particular, this article can serve as an introduction and guide to Viazovska's techniques, alongside other expositions [6,20]. For background on sphere packing and lattices, see [12,15,25].

Of course, we should keep in mind that this topic represents only one strand of Viazovska's research. For example, [3] is a beautiful and decisive paper on a quite different topic. What will she be known for in 20 or 30 years? I look forward to finding out.

2. THE PAST

Before we turn to Viazovska's proof, we will need some background. In this section, we will construct the E_8 lattice and explain a method for proving upper bounds for the sphere packing density.

Sphere packings can be constructed in many ways, among which lattice packings are the simplest possibility. A *lattice packing* of spheres centers the spheres at the points of a *lattice* Λ in $\mathbb{R}^d$, i.e., a discrete subgroup of $\mathbb{R}^d$ of rank d, or equivalently the integral span of a basis of $\mathbb{R}^d$. There is no reason why an optimal sphere packing should have this algebraic structure, and, for example, the best sphere packing known in $\mathbb{R}^{10}$ does not. However, many of the best sphere packings known in low dimensions are lattice packings.

To form a packing from a lattice Λ, we must choose the sphere radius r so that neighboring spheres do not overlap. Specifically, we should take

$$r = \frac{1}{2} \min_{x \in \Lambda \setminus \{0\}} |x|.$$

The volume of a sphere of radius r in $\mathbb{R}^d$ is $\pi^{d/2} r^n / (d/2)!$, where $(d/2)!$ means $\Gamma(d/2+1)$ when d is odd, and the *density* of the overall packing (i.e., the fraction of space covered by the balls) is the sphere volume times the number of spheres per unit volume in space. Let $\operatorname{vol}(\mathbb{R}^d/\Lambda)$ denote the *covolume* of the lattice, i.e., the volume of the quotient torus, or equivalently the absolute value of the determinant of a lattice basis. Then the number of spheres per unit volume in space is $1/\operatorname{vol}(\mathbb{R}^d/\Lambda)$, and so the lattice packing density is

$$\frac{\pi^{d/2} r^n}{(d/2)! \operatorname{vol}(\mathbb{R}^d/\Lambda)}.$$

One of the most remarkable lattices is the E_8 *root lattice*, which originated in Lie theory but has since become widespread across mathematics. We will see below how to obtain E_8 as a modification of the D_d lattice, the checkerboard lattice in d dimensions, which is defined by

$$D_d = \{(x_1, \dots, x_d) \in \mathbb{Z}^d : x_1 + \dots + x_d \text{ is even}\}.$$

In other words, D_d simply omits every other point in the cubic lattice $\mathbb{Z}^d$. As a special case, D_3 is the face-centered cubic lattice in three dimensions, which Hales showed achieves the optimal sphere packing density [16], and D_4 and D_5 are the best packings known in their dimensions. However, D_d is not optimal beyond five dimensions.

The problem with D_d in higher dimensions is that its holes are too large. A *hole* is a point in space that is a local maximum for distance from the lattice. There are two types of holes in D_d, shallow holes at distance 1 from the lattice, such as $(1, 0, \dots, 0)$, and deep holes at distance $\sqrt{d/4}$ from the lattice, such as $(\frac{1}{2}, \frac{1}{2}, \dots, \frac{1}{2})$. As $d \to \infty$, so does $\sqrt{d/4}$, and so the deep holes become large enough to fit enormous numbers of additional spheres. In particular, D_d cannot be optimal when d is large.

When $d = 8$, something beautiful happens. The distance $\sqrt{8/4}$ from a deep hole to the lattice exactly equals the distance $\sqrt{2}$ between lattice points in D_8, and that means the deep holes are just large enough to be filled with additional spheres. If we plug these

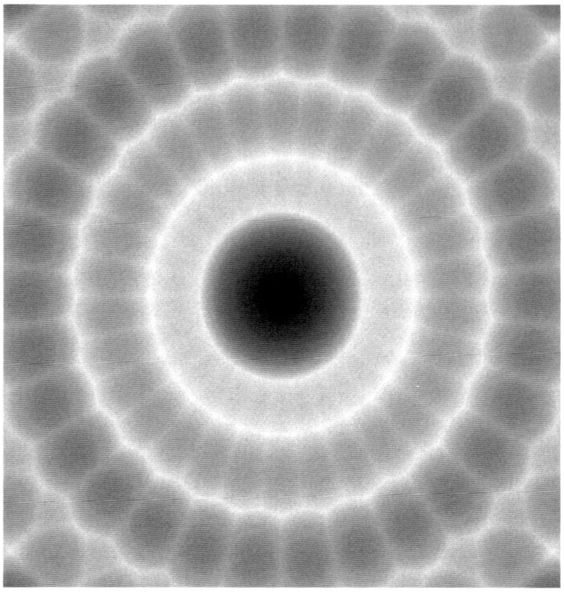

FIGURE 2
A two-dimensional cross-section of $\mathbb{R}^8$ through a Coxeter plane of E_8, colored according to the squared distance to the nearest point in E_8 (dark is close) and inspired by **[22]**.

holes with spheres, then the resulting packing is the union of D_8 with its translate $D_8 + (\frac{1}{2}, \frac{1}{2}, \dots, \frac{1}{2})$. It is not hard to check that this packing is a lattice (it amounts to the fact that $2 \cdot (\frac{1}{2}, \frac{1}{2}, \dots, \frac{1}{2}) \in D_8$), which is called the E_8 *root lattice*.

The E_8 lattice packing has packing radius $r = \sqrt{2}/2$ and covolume $\mathrm{vol}(\mathbb{R}^8/E_8) = \mathrm{vol}(\mathbb{R}^8/D_8)/2 = 1$, and so it has a packing density of $\pi^4/384 = 0.2536\dots$ It is by no means obvious that this construction is optimal. In fact, the construction feels a little ad hoc. However, the E_8 lattice turns out to be far more beautiful and symmetric than its construction indicates. For example, see Figure 2 for a view of E_8 with 30-fold symmetry. This is a common pattern with exceptional structures in mathematics: they are typically obtained by piecing together several substructures that each have less symmetry individually.

Now that we have the E_8 lattice, the next question is how we could try to obtain a matching upper bound for the sphere packing density in eight dimensions. Obtaining a matching bound seems completely infeasible in most dimensions, but in a few special dimensions bounds based on harmonic analysis work remarkably well. This idea, called the *linear programming bound*, goes back to a fundamental paper by Delsarte **[13]** on error-correcting codes, and the corresponding bound for sphere packings was developed by Cohn and Elkies **[7]**.

The linear programming bound is formulated in terms of the *Fourier transform* $\hat{f}$ of an integrable function $f: \mathbb{R}^d \to \mathbb{C}$, which we will normalize as

$$\hat{f}(y) = \int_{\mathbb{R}^d} f(x) e^{-2\pi i \langle x, y \rangle}\, dx,$$

where $\langle \cdot, \cdot \rangle$ is the usual inner product on $\mathbb{R}^d$. Recall that the Fourier transform decomposes f into complex exponentials; in signal processing terms, it amounts to identifying the frequencies that occur in a signal and their relative magnitudes. This decomposition amounts to the *Fourier inversion theorem*: if $\hat{f}$ is integrable as well, then

$$f(x) = \int_{\mathbb{R}^d} \hat{f}(y) e^{2\pi i \langle x, y \rangle} \, dy.$$

In other words, the Fourier transform is very nearly its own inverse, with a single sign change being the only difference. Note that $\hat{f}$ is generally complex-valued, even if f is real-valued, but $\hat{f}$ is real-valued if f is real-valued and an even function.

We will also need a few types of well-behaved functions. A function $f: \mathbb{R}^d \to \mathbb{R}$ is called *rapidly decreasing* if $f(x) = O(|x|^{-c})$ as $|x| \to \infty$ for every constant $c > 0$, and a *Schwartz function* is a smooth function such that it and all its iterated partial derivatives (of every order) are rapidly decreasing. Schwartz functions are arguably the best-behaved functions in harmonic analysis. Much of what we will discuss can be generalized somewhat beyond Schwartz functions, but they are all Viazovska needed to solve the sphere packing problem.

We can now state the linear programming bound for sphere packing:

Theorem 2.1 (Cohn and Elkies [7]). *Let $f: \mathbb{R}^d \to \mathbb{R}$ be an even Schwartz function and r a positive real number. If*

(1) $f(x) \leq 0$ *for all* $x \in \mathbb{R}^d$ *satisfying* $|x| \geq r$,

(2) $\hat{f}(y) \geq 0$ *for all* $y \in \mathbb{R}^d$, *and*

(3) $f(0) = \hat{f}(0) = 1$,

then the optimal sphere packing density in $\mathbb{R}^d$ is at most $\operatorname{vol}(B^d_{r/2}) = \pi^{d/2}(r/2)^d/(d/2)!$.

This theorem produces an upper bound for the packing density from a function f satisfying certain inequalities, but it says nothing about how to choose f to optimize the bound. Numerical optimization can produce good choices for f, which yield the bounds shown in Figure 3. These bounds are rigorous, but it is possible that other functions may produce even better bounds.

As one can see in Figure 3, the bounds in 8 and 24 dimensions appear sharp. Numerical optimization will not yield an exactly sharp bound, but it seems to come as close as desired. Based on data of this sort as well as analogies with other problems in coding theory, Cohn and Elkies conjectured the existence of *magic functions* f that would solve the sphere packing problem exactly in $\mathbb{R}^8$ and $\mathbb{R}^{24}$, by achieving $r = \sqrt{2}$ and $r = 2$, respectively. Note that this is not because the bound dips lower in these dimensions, but rather because the optimal packings rise up to meet it. No other dimensions greater than 2 seem to have a sharp linear programming bound, and it seems unlikely that others exist, but no proof is known, and the bound has been exactly optimized only for $d = 1$, 8, and 24.

The heart of Viazovska's breakthrough lies in the construction of the magic functions. What should f look like if we are to obtain a sharp bound? There are some simple

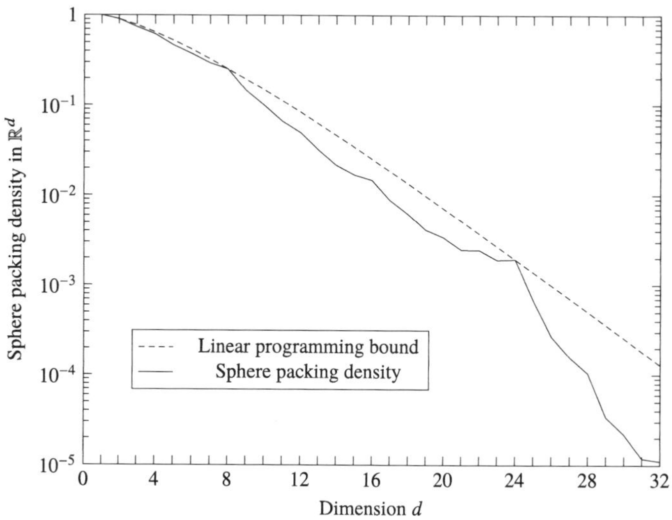

FIGURE 3
A plot of the numerically computed linear programming bound **[1]** and the best sphere packing density currently known **[12]**.

criteria, which we can obtain from the proof of Theorem 2.1. In this article we will examine a proof for just the special case of lattices, but the theorem can be proved in full generality by combining the same technique with a little additional algebra. The argument is based on the *Poisson summation formula*, which says that if $f: \mathbb{R}^d \to \mathbb{C}$ is a Schwartz function, Λ is a lattice in $\mathbb{R}^d$, and Λ^* is its *dual lattice* (i.e., the lattice generated by the dual basis of any basis of Λ with respect to the inner product $\langle \cdot, \cdot \rangle$), then

$$\sum_{x \in \Lambda} f(x) = \frac{1}{\operatorname{vol}(\mathbb{R}^d/\Lambda)} \sum_{y \in \Lambda^*} \hat{f}(y).$$

Proof of Theorem 2.1 *for lattice packings.* The sphere packing problem is scaling-invariant, and so we can use spheres of radius $r/2$. Let Λ be any lattice packing with packing radius $r/2$, which means $|x| \geq r$ for $x \in \Lambda \setminus \{0\}$. If f satisfies the hypotheses of Theorem 2.1, then $f(x) \leq 0$ for $x \in \Lambda \setminus \{0\}$ and $\hat{f}(y) \geq 0$ for all y, from which it follows that

$$1 = f(0) \geq \sum_{x \in \Lambda} f(x) = \frac{1}{\operatorname{vol}(\mathbb{R}^d/\Lambda)} \sum_{y \in \Lambda^*} \hat{f}(y) \geq \frac{\hat{f}(0)}{\operatorname{vol}(\mathbb{R}^d/\Lambda)} = \frac{1}{\operatorname{vol}(\mathbb{R}^d/\Lambda)}.$$

Therefore the packing density $\operatorname{vol}(B^d_{r/2})/\operatorname{vol}(\mathbb{R}^d/\Lambda)$ is bounded above by $\operatorname{vol}(B^d_{r/2})$, as desired. ■

A first observation is that we can assume without loss of generality that f is radial, i.e., $f(x)$ depends only on $|x|$. This reason is that we can replace f with the average of its rotations about the origin, because all the constraints are linear and rotation-invariant. One might wonder whether nonradial functions could be helpful conceptually even if they are not needed, but so far the answer appears to be no. Instead, Viazovska's work turns out to lead to a wonderful new theory of interpolation for radial functions. We will henceforth assume f is radial, and when $t \in [0, \infty)$ we will write $f(t)$ for the common value $f(x)$ with $|x| = t$, as well as $f'(t)$ for the radial derivative.

Now if we examine the central inequality in the proof of Theorem 2.1 for lattices, we can see when it could be sharp. To obtain a sharp bound, all of the discarded terms in the inequality must vanish: we must have $f(x) = 0$ for $x \in \Lambda \setminus \{0\}$ and $\hat{f}(y) = 0$ for $y \in \Lambda^* \setminus \{0\}$. In other words, f must vanish on the nonzero distances between lattice points, and $\hat{f}$ must vanish on the nonzero distances between dual lattice points.

One can check directly from the construction of E_8 given above that $E_8^* = E_8$ and that the vector lengths in E_8 are all square roots of even integers. Furthermore, it turns out that each distance $\sqrt{2n}$ with $n \geq 0$ actually occurs in E_8. We should therefore have $r = \sqrt{2}$ in Theorem 2.1, and the magic function f should have a sign change at radius $\sqrt{2}$, followed by double roots at $\sqrt{2n}$ for $n \geq 2$, as indicated in Figure 4. In other words, we wish to control the behavior of f and $\hat{f}$ to second order at these points, i.e., control both the values $f(\sqrt{2n})$ and $\hat{f}(\sqrt{2n})$ and the radial derivatives $f'(\sqrt{2n})$ and $\hat{f}'(\sqrt{2n})$.

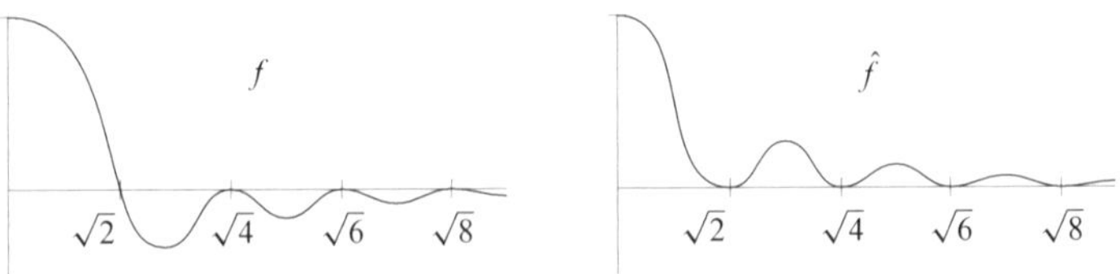

FIGURE 4

This schematic diagram, which is taken from **[6]**, shows the roots of the magic function f and its Fourier transform $\hat{f}$ in eight dimensions. It is not a plot of the actual function, which decreases very rapidly. See Figure 5 for an actual plot.

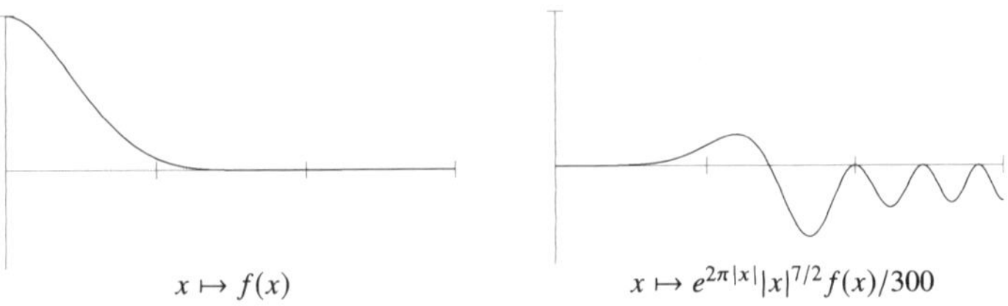

FIGURE 5

Two plots of Viazovska's magic function in eight dimensions. The first plot is scaled correctly, but it decreases so rapidly that the roots become invisible. The second plot introduces a rescaling to make them visible, based on the asymptotic decay rate.

How can one construct such a function f? The reason this task is difficult is that it involves controlling both f and $\hat{f}$ simultaneously. Either one is of course easy on its own, but handling both at once introduces profound difficulties. The underlying issue here is Heisenberg's uncertainty principle: in loose terms, whenever you try to pin down f, you lose control over $\hat{f}$, and vice versa. More precisely, we run into Bourgain, Clozel, and Kahane's uncertainty principle for controlling the signs of functions [4, 8]. These seemingly simple inequalities on f and $\hat{f}$ therefore turn out to be far more subtle than they initially appear.

When Elkies and I proposed this method in 1999, Viazovska was still in secondary school. Without realizing how profoundly difficult the remaining step was, I imagined that we had almost solved the sphere packing problem in 8 and 24 dimensions, and our inability to find the magic functions was extremely frustrating. At first, I worried that someone else would find an easy solution and leave me feeling foolish for not doing it myself. Over time I became convinced that obtaining these functions was in fact difficult, and others also reached the same conclusion. For example, Thomas Hales has said that "I felt that it would take a Ramanujan to find it" [19]. Eventually, instead of worrying that someone else would solve it, I began to fear that nobody would solve it, and that I would someday die without knowing the outcome. I am grateful that Viazovska found such a satisfying and beautiful solution, and that she introduced wonderful new ideas for the mathematical community to explore.

3. MODULAR FORMS

Viazovska's magic function is constructed using modular forms, certain special functions that play an important role in number theory. The theory of modular forms has a reputation for being somewhat forbidding, but the basics are not so difficult, and that is all that is needed for Viazovska's proof. We will outline the needed theory here. For a down to earth introduction to the case of $\mathrm{SL}_2(\mathbb{Z})$, see Chapter VII in [24], and for more detailed and general treatments, see [5, 14, 28].

We begin with an example of a modular form, namely Eisenstein series. Recall that the Riemann zeta function is defined by

$$\zeta(s) = \sum_{n=1}^{\infty} \frac{1}{n^s}$$

when this sum converges, i.e., when $\mathrm{Re}(s) > 1$. Here we are summing inverse powers of the arithmetic progression $1, 2, \dots$, and Euler obtained an exact formula when s is an even integer. What if we instead wanted to sum inverse powers of a lattice in the complex plane? Setting aside the question of why we would want to do this (the result has deeper significance than one might guess), we could write the result as the *Eisenstein series*

$$E_k(z) = \frac{1}{2\zeta(k)} \sum_{(m,n)\in\mathbb{Z}^2\setminus\{(0,0)\}} \frac{1}{(mz+n)^k} \tag{3.1}$$

for $\mathrm{Im}\, z > 0$, where we are summing over the lattice $\{mz + n : m, n \in \mathbb{Z}\}$, with the exception of the point $(0, 0)$ at which the summand blows up. Up to scaling by a complex factor, all two-dimensional lattices are of this form.

The factor of $1/(2\zeta(k))$ in the definition is merely a convenient normalizing factor, which plays no essential role in the study of E_k. Unfortunately, the notation E_k conflicts with our name for the E_8 root lattice, but that will not cause any ambiguity in practice.

We will restrict our attention to positive integers k, so that $(mz + n)^k$ is single-valued. The series (3.1) converges absolutely when $k \geq 3$, but just conditionally when $k = 2$.

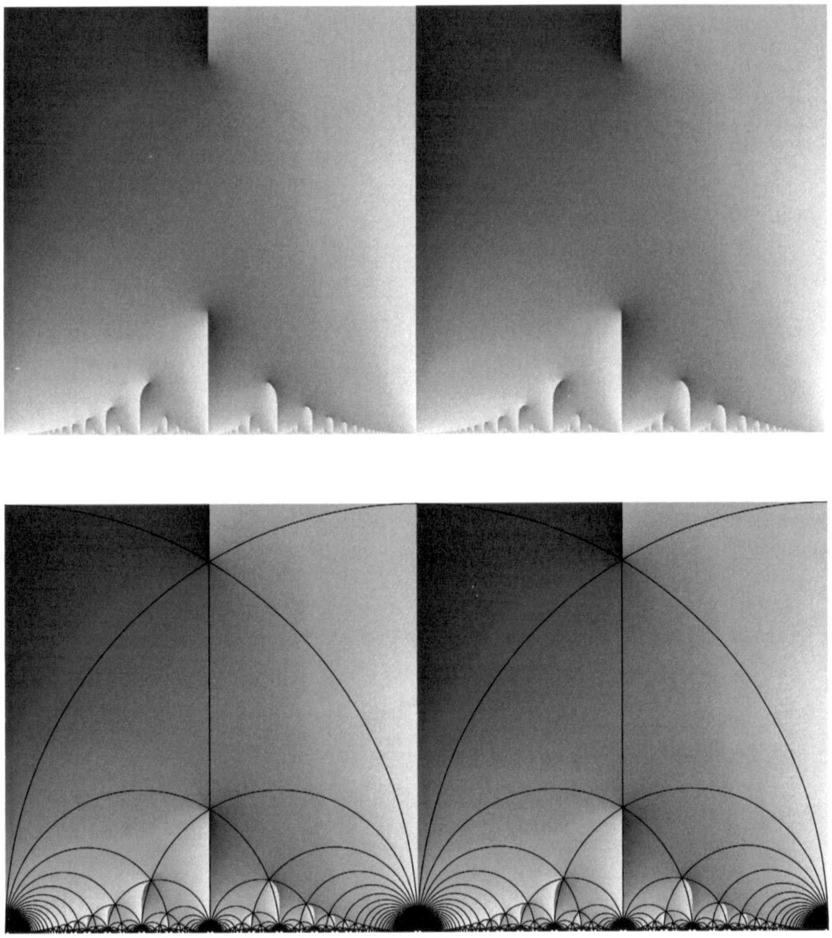

FIGURE 6

A plot of the Eisenstein series $E_4(z)$ for $-1 \le \mathrm{Re}\, z \le 1$ and $0 < \mathrm{Im}\, z \le 1$ (above) and the same plot overlaid with a tiling of $\mathcal{H}$ using fundamental domains for the action of $\mathrm{SL}_2(\mathbb{Z})$ (below).

For odd k, the (m, n) and $(-m, -n)$ terms cancel and we obtain $E_k(z) = 0$, and so only the even cases are interesting.[2] Thus, we will focus on E_k for k even and at least 4.

What does an Eisenstein series look like? Figure 6 is a plot of E_4, in which black is zero, white is infinity, and color indicates complex phase **[21]**, with the sharp transitions in color occurring at positive real values. The fractal structure visible in this plot can be explained using two functional equations:

$$E_k(z+1) = E_k(z) \quad \text{and} \quad E_k(-1/z) = z^k E_k(z).$$

These symmetries follow from rearranging the defining series (3.1) when $k > 2$, and they are the central equations in the theory of modular forms.

2 This parity phenomenon is essentially the same as in Euler's formula for the zeta function at even integers, which can be viewed as computing $\sum_{n \in \mathbb{Z} \setminus \{0\}} n^{-k}$ explicitly for all integers $k > 1$.

The mappings $z \mapsto z+1$ and $z \mapsto -1/z$ that occur in these functional equations generate a discrete group of linear fractional transforms of the *upper half-plane* $\mathcal{H} = \{z \in \mathbb{C} : \operatorname{Im} z > 0\}$. To put it into a broader context of matrix groups, we can let the matrix $\left(\begin{smallmatrix} a & b \\ c & d \end{smallmatrix}\right)$ act on $\mathcal{H}$ via

$$\begin{pmatrix} a & b \\ c & d \end{pmatrix} \cdot z = \frac{az+b}{cz+d}.$$

Then the matrices $T = \left(\begin{smallmatrix} 1 & 1 \\ 0 & 1 \end{smallmatrix}\right)$ and $S = \left(\begin{smallmatrix} 0 & -1 \\ 1 & 0 \end{smallmatrix}\right)$ satisfy $T \cdot z = z+1$ and $S \cdot z = -1/z$, and they turn out to generate the group $\mathrm{SL}_2(\mathbb{Z})$.

The *weight k action* of $\mathrm{SL}_2(\mathbb{Z})$ on functions $f : \mathcal{H} \to \mathbb{C}$ is defined by

$$(f|_k\gamma)(z) = (cz+d)^{-k} f\left(\frac{az+b}{cz+d}\right)$$

for $\gamma = \left(\begin{smallmatrix} a & b \\ c & d \end{smallmatrix}\right)$. In this notation, the functional equations $E_k(z+1) = E_k(z)$ and $E_k(-1/z) = z^k E_k(z)$ imply that the Eisenstein series E_k satisfies $E_k|_k\gamma = E_k$ for all $\gamma \in \mathrm{SL}_2(\mathbb{Z})$ when $k > 2$.

A *modular form of weight k for* $\mathrm{SL}_2(\mathbb{Z})$ is a holomorphic function $f : \mathcal{H} \to \mathbb{C}$ such that $f|_k\gamma = f$ for all $\gamma \in \mathrm{SL}_2(\mathbb{Z})$ and one additional condition holds, called being holomorphic at infinity. To state this condition, note that taking $\gamma = T$ shows that $f(z+1) = f(z)$, and thus we can expand f as a Fourier series

$$f(z) = \sum_{n \in \mathbb{Z}} a_n e^{2\pi i n z}.$$

We say f is *meromorphic at infinity* if there are only finitely many nonzero coefficients a_n with $n < 0$, and *holomorphic at infinity* if $a_n = 0$ for all $n < 0$. The name reflects the fact that this Fourier series governs the behavior of $f(z)$ as $\operatorname{Im} z$ grows, because $e^{2\pi i z} \to 0$ as $\operatorname{Im} z \to \infty$. The Fourier series of a modular form is often known as its *q-series*, with $q = e^{2\pi i z}$.

The normalization factor $1/(2\zeta(k))$ in (3.1) ensures that the q-series of E_k has rational coefficients, and even integral coefficients when k is small. For example, one can show that $E_4(z) = 1 + 240 \sum_{n \geq 1} \sigma_3(n) q^n$ and $E_6(z) = 1 - 504 \sum_{n \geq 1} \sigma_5(n) q^n$, where $\sigma_k(n)$ denotes the sum of the kth powers of the divisors of n.

The product of modular forms of weights k and ℓ is a modular form of weight $k + \ell$, and modular forms therefore form a graded ring. For $\mathrm{SL}_2(\mathbb{Z})$, one can show that this ring is generated by E_4 and E_6. In other words, the vector space of modular forms of weight k for $\mathrm{SL}_2(Z)$ is spanned by the modular forms $E_4^j E_6^\ell$ with $4j + 6\ell = k$.

In addition to using Eisenstein series directly, Viazovska also uses the *modular discriminant* Δ, which is given by

$$\Delta(z) = \frac{E_4(z)^3 - E_6(z)^2}{1728} = q \prod_{n=1}^{\infty} (1 - q^n)^{24}. \tag{3.2}$$

Its key property is that it vanishes nowhere in the upper half plane, while it vanishes at infinity (in the sense that its q-series has no constant term).

Turán said that special functions should instead be called useful functions, and modular forms are no exception to this principle. The reason we study modular forms is not that we have a special love for Eisenstein series, but rather that the functional equations $f(z+1) = f(z)$ and $f(-1/z) = z^k f(z)$ arise far more often than one might expect. For example, the E_8 lattice has an important modular form associated with it, namely its *theta series*

$$\Theta_{E_8}(z) = \sum_{n=0}^{\infty} N_n e^{2\pi i n z},$$

where $N_n = \#\{x \in E_8 : |x|^2 = 2n\}$. In other words, the theta series is a generating function that counts the number of vectors of each length in E_8.

This theta series satisfies both functional equations: $\Theta_{E_8}(z+1) = \Theta_{E_8}(z)$ follows from the definition of Θ_{E_8} as a Fourier series, while $\Theta_{E_8}(-1/z) = z^4 \Theta_{E_8}$ amounts to Poisson summation over E_8 for the complex Gaussian $x \mapsto e^{\pi i z |x|^2}$, which has eight-dimensional Fourier transform $y \mapsto z^{-4} e^{\pi i(-1/z)|y|^2}$. These functional equations tell us that Θ_{E_8} is a modular form for $\mathrm{SL}_2(\mathbb{Z})$ of weight 4, and it must therefore be proportional to E_4. In fact, $\Theta_{E_8} = E_4$, because $N_0 = 1$. Thus, we obtain the beautiful formula $240\sigma_3(n)$ for the number of vectors in E_8 of squared norm $2n$.

The theory of modular forms extends to other discrete groups, if one carefully defines what being holomorphic at infinity means.[3] Viazovska's proof makes use of one more group, namely

$$\Gamma(2) = \left\{ \gamma \in \mathrm{SL}_2(\mathbb{Z}) : \gamma \equiv \begin{pmatrix} 1 & 0 \\ 0 & 1 \end{pmatrix} \pmod 2 \right\},$$

which has index 6 in $\mathrm{SL}_2(\mathbb{Z})$. If we let

$$U(z) = \left(\sum_{n \in \mathbb{Z}} e^{\pi i n^2 z} \right)^4,$$

$W = U|_2 T$, and $V = U - W$, then U, V, and W are modular forms of weight 2 for $\Gamma(2)$ that satisfy $U = V + W$ and

$$\begin{aligned} U|_2 T &= W, & V|_2 T &= -V, & W|_2 T &= U, \\ U|_2 S &= -U, & V|_2 S &= -W, & W|_2 S &= -V. \end{aligned} \tag{3.3}$$

These identities will play a key role in the construction of Viazovska's magic function. It turns out that U and W generate the ring of modular forms for $\Gamma(2)$, and therefore every modular form of weight $2k$ for $\Gamma(2)$ is a linear combination of $U^k, U^{k-1}W, U^{k-2}W^2, \dots, W^k$.

Because modular forms are so closely connected with lattices, it is natural to turn to modular forms when attempting to construct the magic functions. However, it is entirely unclear where we should even start, because modular forms are completely different sorts

3 If Γ is a subgroup of finite index in $\mathrm{SL}_2(\mathbb{Z})$, then the condition is that for each $\gamma \in \mathrm{SL}_2(\mathbb{Z})$, $f|_k \gamma$ should be holomorphic at infinity. Note that $f|_k \gamma$ need not satisfy $(f|_k \gamma)(z+1) = (f|_k \gamma)(z)$, but one can check that it always satisfies $(f|_k \gamma)(z+n) = (f|_k \gamma)(z)$ for some positive integer n and thus has a Fourier expansion in $e^{2\pi i z/n} = q^{1/n}$.

of objects from radial Schwartz functions. Figure 6 looks nothing whatsoever like Figures 4 or 5, and there is no familiar transformation that makes it look any more similar.

4. VIAZOVSKA'S CONSTRUCTION FOR SINGLE ROOTS

The first step in Viazovska's construction of the magic function f is to split f into eigenfunctions of the Fourier transform. Radial functions satisfy $\hat{\hat{f}} = f$, and so we can write f as $f = f_+ + f_-$, where $f_+ := (f + \hat{f})/2$ satisfies $\widehat{f_+} = f_+$ and $f_- := (f - \hat{f})/2$ satisfies $\widehat{f_-} = -f_-$. If f is the magic function in eight dimensions, then f and $\hat{f}$ both have roots at $\sqrt{2n}$ for integers $n \geq 1$, and therefore f_+ and f_- do as well. Thus, we are looking for radial Fourier eigenfunctions with specified roots. Specifically, each of $f_\pm$ should have a single root at $\sqrt{2}$ and double roots at $\sqrt{2n}$ for $n \geq 2$. These roots turn out to provide enough information to determine $f_\pm$ up to scaling, and they can then be combined to obtain f.

Before we construct the actual magic function, it is worth examining a simpler variant as a warm-up exercise. Instead of trying to control the behavior of f to second order at $\sqrt{2n}$, we will instead control the behavior of a function g to first order at $\sqrt{n}$. This construction has no known applications to sphere packing, but it is nevertheless of intrinsic interest in Fourier analysis. We will also focus on the -1 eigenfunction (i.e., the case $\hat{g} = -g$) in the single-root case, for the sake of specificity.

Viazovska found a remarkable integral transform that can construct such functions. We will write a radial function $g\colon \mathbb{R}^8 \to \mathbb{C}$ as a continuous linear combination of complex Gaussians $x \mapsto e^{\pi i z |x|^2}$ with $z \in \mathcal{H}$ via the contour integral

$$g(x) = \frac{1}{2} \int_{-1}^{1} \psi(z) e^{\pi i z |x|^2}\, dz, \tag{4.1}$$

where ψ is a holomorphic function on $\mathcal{H}$ and the contour is a semicircle centered at the origin. Under which conditions on ψ will g be a Fourier eigenfunction, and how can we control its values at $\sqrt{n}$?

We can obtain the values $g(\sqrt{n})$ by imposing periodicity on ψ as follows. Suppose $\psi(z+2) = \psi(z)$ for all $z \in \mathcal{H}$, so that ψ has a Fourier series of the form

$$\psi(z) = \sum_{n \in \mathbb{Z}} a_n e^{\pi i n z}. \tag{4.2}$$

Then for integers $n \geq 0$,

$$g(\sqrt{n}) = \frac{1}{2} \int_{-1}^{1} \psi(z) e^{\pi i n z}\, dz = a_{-n}$$

by orthogonality, provided that we can interchange the sum and integral. If the Fourier expansion (4.2) has only finitely many negative terms, then $g(\sqrt{n})$ will vanish for all but finitely many n.

To compute the Fourier transform of g, we can interchange the contour integral and Fourier transform, again assuming the integral is sufficiently well behaved. Then

$$\hat{g}(y) = \frac{1}{2} \int_{-1}^{1} \psi(z) z^{-4} e^{\pi i (-1/z) |y|^2}\, dz,$$

because the d-dimensional Fourier transform of the complex Gaussian $x \mapsto e^{\pi i z |x|^2}$ with $z \in \mathcal{H}$ is given by $y \mapsto (i/z)^{d/2} e^{\pi i (-1/z)|y|^2}$, and $d = 8$ here. Changing variables to $u = -1/z$ shows that

$$\hat{g}(y) = -\frac{1}{2} \int_{-1}^{1} \psi(-1/u) u^2 e^{\pi i u |y|^2} \, du.$$

In other words, taking the Fourier transform of g amounts to replacing ψ with $-\psi|_{-2}S$, and we obtain $\hat{g} = -g$ if $\psi|_{-2}S = \psi$.

Let Γ be the subgroup of $\mathrm{SL}_2(\mathbb{Z})$ generated by S and T^2, which has index 3 in $\mathrm{SL}_2(\mathbb{Z})$. Then the conditions that $\psi|_{-2}T^2 = \psi$ (i.e., $\psi(z+2) = \psi(z)$) and $\psi|_{-2}S = \psi$ mean that ψ is *weakly modular of weight* -2 for Γ. The reason why ψ is less than a full-fledged modular form is that it is only meromorphic at infinity (this is unavoidable, since the weight is negative). We furthermore require ψ to vanish at ± 1, which will be enough to justify our integral manipulations and show that g is a Schwartz function. In terms of Fourier series, this vanishing says that $\psi|_{-2}TS$ has no negative terms in its q-series, because TS maps the cusp $i\infty$ to 1.

We will construct an example of the form $\psi = \psi_0/\Delta$ using the Δ function from (3.2), where ψ_0 is a genuine modular form of weight 10 for Γ. Note that the denominator of Δ causes no difficulties in $\mathcal{H}$, since $\Delta(z) \neq 0$ for all $z \in \mathcal{H}$, and the zero of Δ at infinity will lead to a pole of ψ.

The function ψ_0 is modular of weight 10 for Γ, and thus also for $\Gamma(2)$ because $\Gamma(2)$ is a subgroup of Γ. In particular, ψ_0 must be a linear combination of $U^5, U^4W, U^3W^2, \ldots, W^5$, because U and W generate the ring of modular forms for $\Gamma(2)$. The relations (3.3) specify the action of S and T, and they imply that the subspace invariant under S is spanned by

$$\begin{aligned} \alpha &:= U^5 - 6U^3W^2 + 4U^2W^3, \\ \beta &:= U^4W - 3U^3W^2 + 2U^2W^3, \quad \text{and} \\ \gamma &:= -U^3W^2 + 4U^2W^3 - 5UW^4 + 2W^5, \end{aligned}$$

with q-expansions

$$\begin{aligned} \frac{\alpha}{\Delta} &= -q^{-1} - 40q^{-1/2} + 752 + \cdots, & \frac{\alpha}{\Delta}\Big|_{-2}TS &= -1024 + 90112q + \cdots, \\ \frac{\beta}{\Delta} &= -16q^{-1/2} + 256 + \cdots, & \frac{\beta}{\Delta}\Big|_{-2}TS &= -512 - 20480q + \cdots, \\ \frac{\gamma}{\Delta} &= 256 - 10240q^{1/2} + \cdots, & \frac{\gamma}{\Delta}\Big|_{-2}TS &= -2q^{-1} - 32 + \cdots \end{aligned}$$

in terms of $q^{1/2} = e^{\pi i z}$. Now requiring ψ to vanish at ± 1 determines it up to scaling as

$$\psi = \frac{2\beta - \alpha}{\Delta} = q^{-1} + 8q^{-1/2} - 240 - 6176q^{1/2} - \cdots, \tag{4.3}$$

which yields a radial Schwartz function $g\colon \mathbb{R}^8 \to \mathbb{R}$ such that $\hat{g} = -g$ and

$$g(\sqrt{n}) = \begin{cases} -240 & \text{if } n = 0, \\ 8 & \text{if } n = 1, \\ 1 & \text{if } n = 2, \text{ and} \\ 0 & \text{if } n \geq 3. \end{cases}$$

Note that we do not have much flexibility here: the values $g(0)$, $g(1)$, and $g(\sqrt{2})$ are uniquely determined by Poisson summation over $\mathbb{Z}^8$ and E_8, up to scaling.

We can rewrite the definition of f in another useful form as follows. If $|x|$ is large enough (in fact, $|x|^2 > 2$ will suffice), then

$$\begin{aligned} g(x) &= \frac{1}{2} \int_{-1}^{1} \psi(z) e^{\pi i z |x|^2} \, dz \\ &= \frac{1}{2} \int_{-1}^{i} \psi(z) e^{\pi i z |x|^2} \, dz - \frac{1}{2} \int_{1}^{i} \psi(z) e^{\pi i z |x|^2} \, dz \\ &= \frac{1}{2} \int_{-1}^{-1+i\infty} \psi(z) e^{\pi i z |x|^2} \, dz - \frac{1}{2} \int_{1}^{1+i\infty} \psi(z) e^{\pi i z |x|^2} \, dz \\ &= \frac{e^{-\pi i |x|^2} - e^{\pi i |x|^2}}{2} \int_{0}^{i\infty} \psi(u+1) e^{\pi i u |x|^2} \, du. \end{aligned}$$

In these manipulations, the second line merely breaks the integral in two, the third line uses the fact that

$$\int_{-1+iR}^{1+iR} \psi(z) e^{\pi i z |x|^2} \, dz \to 0$$

as $R \to \infty$ (which holds if $|x|^2$ is large enough), and the fourth line uses $\psi(u-1) = \psi(u+1)$.

In other words, $g(x)$ is given by $\sin(\pi |x|^2)$ times the Laplace transform of $t \mapsto \psi(it+1)$ evaluated at $\pi |x|^2$:

$$g(x) = \sin\bigl(\pi |x|^2\bigr) \int_{0}^{\infty} \psi(it+1) e^{-\pi t |x|^2} \, dt. \tag{4.4}$$

While the original integral (4.1) converges for all x, this integral converges only when $|x|^2$ is large enough for the Gaussian factor $e^{-\pi t |x|^2}$ to counteract the growth of $\psi(it+1)$ as $t \to \infty$. In particular, (4.3) implies that

$$\psi(it+1) = e^{2\pi t} - 8e^{\pi t} - 240 + 6176 e^{-\pi t} - \cdots$$

as $t \to \infty$, which means we need $|x|^2 > 2$. We can use this expansion to analytically continue g by removing the divergent terms:

$$\begin{aligned} g(x) &= \sin\bigl(\pi |x|^2\bigr) \int_{0}^{\infty} (e^{2\pi t} - 8e^{\pi t} - 240) e^{-\pi t |x|^2} \, dt \\ &\quad + \sin\bigl(\pi |x|^2\bigr) \int_{0}^{\infty} \bigl(\psi(it+1) - e^{2\pi t} + 8e^{\pi t} + 240\bigr) e^{-\pi t |x|^2} \, dt \\ &= \frac{\sin(\pi |x|^2)}{\pi(|x|^2 - 2)} - \frac{8 \sin(\pi |x|^2)}{\pi(|x|^2 - 1)} - \frac{240 \sin(\pi |x|^2)}{\pi |x|^2} \\ &\quad + \sin\bigl(\pi |x|^2\bigr) \int_{0}^{\infty} \bigl(\psi(it+1) - e^{2\pi t} + 8e^{\pi t} + 240\bigr) e^{-\pi t |x|^2} \, dt, \end{aligned}$$

and this last formula holds regardless of $|x|$, with removable singularities at $|x| = 0$, 1, and $\sqrt{2}$.

5. VIAZOVSKA'S CONSTRUCTION FOR DOUBLE ROOTS

We are now in a position to obtain the magic function in eight dimensions. First, we will obtain the -1 eigenfunction f_-. It is not immediately clear how to generalize the contour integral (4.1) from single to double roots, but the Laplace transform formula (4.4) generalizes elegantly. To obtain f_-, we will look for a special function ψ such that

$$f_-(x) = -4i \sin\big(\pi |x|^2/2\big)^2 \int_0^{i\infty} \psi(z) e^{\pi i z |x|^2}\, dz$$

when $|x|$ is large enough. If we write $-4\sin(\pi|x|^2/2)^2 = e^{-\pi i |x|^2} + e^{\pi i |x|^2} - 2$, we find that

$$\begin{aligned} f_-(x) = &\int_{-1}^{-1+i\infty} \psi(z+1) e^{\pi i |x|^2 z}\, dz + \int_{1}^{1+i\infty} \psi(z-1) e^{\pi i |x|^2 z}\, dz \\ &- 2\int_0^{i\infty} \psi(z) e^{\pi i |x|^2 z}\, dz. \end{aligned}$$

We will construct a function ψ such that ψ is holomorphic on $\mathcal{H}$ and $\psi(z)$ is exponentially bounded as $\operatorname{Im} z \to \infty$. Under these conditions, when $|x|$ is sufficiently large we can shift the contours and combine the integrals to obtain

$$\begin{aligned} f_-(x) = &\int_{-1}^{i} \psi(z+1) e^{\pi i |x|^2 z}\, dz + \int_{1}^{i} \psi(z-1) e^{\pi i |x|^2 z}\, dz \\ &- 2\int_0^{i} \psi(z) e^{\pi i |x|^2 z}\, dz + \int_i^{i\infty} \big(\psi(z+1) + \psi(z-1) - 2\psi(z)\big) e^{\pi i |x|^2 z}\, dz, \end{aligned}$$

with the contours shown in Figure 7. This formula will be the analogue of (4.1), and it will define $f_-(x)$ for all x.

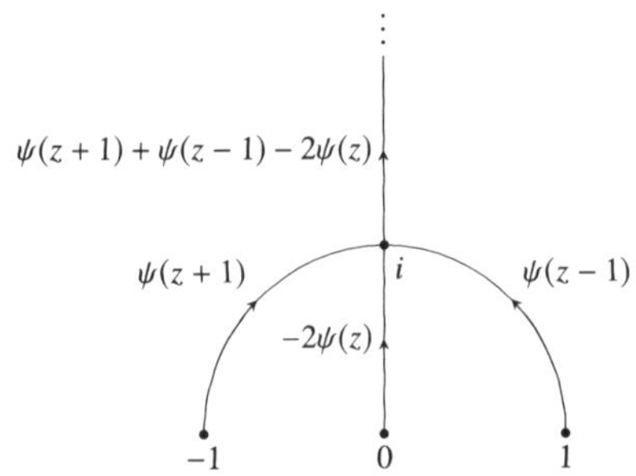

FIGURE 7
The contours used to obtain $f_-(x)$, labeled with their integrands (omitting $e^{\pi i |x|^2 z}\, dz$).

Taking the Fourier transform amounts to replacing $e^{\pi i |x|^2 z}$ with $z^{-4} e^{\pi i |y|^2 (-1/z)}$ in the formula defining f_-:

$$\widehat{f_-}(y) = \int_{-1}^{i} \psi(z+1) z^{-4} e^{\pi i |y|^2(-1/z)}\, dz + \int_{1}^{i} \psi(z-1) z^{-4} e^{\pi i |y|^2(-1/z)}\, dz$$

$$- 2\int_0^i \psi(z) z^{-4} e^{\pi i |y|^2(-1/z)}\,dz$$

$$+ \int_i^{i\infty} \big(\psi(z+1) + \psi(z-1) - 2\psi(z)\big) z^{-4} e^{\pi i |y|^2(-1/z)}\,dz.$$

We can now set $u = -1/z$, which exchanges the four contours in pairs. The simplest way to obtain $\widehat{f_-} = -f_-$ would be if the resulting formula is exactly the negative of the formula with which we began. That amounts to the functional equations

$$\psi|_{-2}TS = -\psi|_{-2}T^{-1}$$

and

$$2\psi|_{-2}S = 2\psi - \psi|_{-2}T - \psi|_{-2}T^{-1}.$$

Note that the structure of these equations reflects the integrands.

Now the question is which sorts of functions ψ satisfy these functional equations. The simplest possibility would be some sort of modular form. The functional equations are not consistent with invariance under S and T, and so ψ cannot be modular for the full group $\mathrm{SL}_2(\mathbb{Z})$. Let us suppose instead that ψ is weakly modular of weight -2 for $\Gamma(2)$ (i.e., invariant under $\Gamma(2)$ but only meromorphic at infinity). Then $\psi|_{-2}T = \psi|_{-2}T^{-1}$, because $T^2 \in \Gamma(2)$, and our functional equations become $\psi|_{-2}TS = -\psi|_{-2}T$ and $\psi = \psi|_{-2}T + \psi|_{-2}S$. Furthermore, the second equation implies the first, because $S^2 = I$. We will therefore obtain the eigenfunction equation $\widehat{f_-} = -f_-$ as long as ψ is weakly modular of weight -2 for $\Gamma(2)$ and satisfies $\psi = \psi|_{-2}T + \psi|_{-2}S$.

As in the single-root case, it is natural to multiply ψ by Δ to try to eliminate a pole at infinity. Then $\psi\Delta$ will be a genuine modular form of weight 10 for $\Gamma(2)$, and thus a linear combination of $U^5, U^4W, U^3W^2, \dots, W^5$. One can check that the solutions of the remaining functional equation form a two-dimensional subspace, spanned by

$$\alpha := 2U^4W - 4U^3W^2 + U^2W^3 + UW^4 \quad \text{and}$$
$$\beta := 5U^4W - 10U^3W^2 + 5U^2W^3 + W^5,$$

with

$$\frac{\alpha}{\Delta} = -16q^{-1/2} + 768 + \cdots \quad \text{and} \quad \frac{\beta}{\Delta} = q^{-1} - 40q^{-1/2} + 2064 + \cdots.$$

We will take

$$\psi = \frac{-5\alpha + 2\beta}{\Delta} = 2q^{-1} + 288 + \cdots,$$

so that we eliminate the $q^{-1/2}$ term in the q-series. The motivation for eliminating that term is that it prevents f_- from having a pole at radius 1. To see why, let us analytically continue

$$f_-(x) = 4\sin\!\big(\pi|x|^2/2\big)^2 \int_0^\infty \psi(it) e^{-\pi t|x|^2}\,dt$$

as in the single-root case. If $\psi(it) = a_2e^{2\pi t} + a_1e^{\pi t} + a_0 + \cdots$ as $t \to \infty$, then

$$f_-(x) = \frac{4a_2\sin(\pi|x|^2/2)^2}{\pi(|x|^2-2)} - \frac{4a_1\sin(\pi|x|^2/2)^2}{\pi(|x|^2-1)} - \frac{4a_0\sin(\pi|x|^2/2)^2}{\pi|x|^2}$$
$$+ 4\sin\!\big(\pi|x|^2/2\big)^2 \int_0^\infty \big(\psi(it) - a_2e^{2\pi t} - a_1e^{\pi t} - a_0\big)e^{-\pi t|x|^2}\,dt.$$

Here the a_1 term has a pole unless $a_1 = 0$. For our choice of ψ, $(a_2, a_1, a_0) = (2, 0, 288)$, and thus f_- has a single root at $\sqrt{2}$ and double roots at $\sqrt{2n}$ for $n \geq 2$. One can also check that $\psi(it)$ vanishes as $t \to 0+$ (equivalently, $\psi|_{-2}S$ vanishes at infinity), which is enough for f_- to be a Schwartz function and to justify all our integral manipulations.

We have therefore obtained a magic eigenfunction f_- as

$$f_-(x) = 4\sin\big(\pi|x|^2/2\big)^2 \int_0^\infty \psi(it)e^{-\pi t|x|^2}\, dt$$

for $|x|^2 > 2$, where

$$\psi = \frac{W^3(5U^2 - 5UW + 2W^2)}{\Delta}. \tag{5.1}$$

Our scaling here does not yet match the magic function for sphere packing, but aside from that we have exactly what we need.

Equation (5.1) implies that $\psi(it) > 0$ for all $t \in (0, \infty)$. (Specifically, $\Delta(it) > 0$ thanks to its product formula, $W(it) > 0$ since it is the fourth power of a real quantity, and $5U(it)^2 - 5U(it)W(it) + 2W(it)^2 > 0$ since it is a positive-definite quadratic form.) It follows that f_- never changes sign beyond radius $\sqrt{2}$, in accordance with our expectations. However, note that our eigenfunction is positive beyond radius $\sqrt{2}$, and so we will have to correct its sign later to match the magic function.

All that remains is to construct a magic eigenfunction f_+ and take a suitable linear combination of f_+ and f_- to obtain f. Constructing f_+ is very much like constructing f_-. If we define f_+ for $|x|$ sufficiently large by

$$f_+(x) = -4i\,\sin\big(\pi|x|^2/2\big)^2 \int_0^{i\infty} \phi(z)e^{\pi i z|x|^2}\, dz$$

for some holomorphic function $\phi\colon \mathcal{H} \to \mathbb{C}$, then the eigenfunction equation $\widehat{f_+} = f_+$ will follow from the functional equations

$$\phi|_{-2}TS = \phi|_{-2}T^{-1}$$

and

$$2\phi|_{-2}S = -2\phi + \phi|_{-2}T + \phi|_{-2}T^{-1}.$$

These are the same functional equations as we required for ψ, except for a factor of -1.

A little manipulation using $(ST)^3 = I$ shows that the first functional equation is equivalent to $\phi|_{-2}ST = \phi|_{-2}S$. Thus, if we set $\chi := \phi|_{-2}S$, then χ must be invariant under T. However, the second functional equation is more subtle. A short calculation shows that if $\chi|_0 S = \chi$ (equivalently, $(\chi|_{-2}S)(z) = z^2\chi(z)$), then the second functional equation holds. In other words, it is enough for χ to be weakly modular of weight 0 for $\mathrm{SL}_2(\mathbb{Z})$. However, such functions turn out not to be sufficient to obtain f_+. If one tries to solve for undetermined coefficients to construct f_+, as in the f_- case, one finds that there is no solution with the needed properties.

Instead, we can use *quasimodular forms*, not just modular forms. Recall that the Eisenstein series E_2 was not a modular form of weight 2, because conditional convergence interfered with the series manipulations needed to prove modularity. If we let

$$E_2(z) = 1 - 24 \sum_{n \geq 1} \sigma_1(n) q^n,$$

then E_2 turns out to satisfy

$$z^{-2} E_2(-1/z) = E_2(z) - \frac{6i}{\pi z},$$

with the $6i/(\pi z)$ term amounting to the deviation from modularity. A *quasimodular form of weight k and depth ℓ* for $\mathrm{SL}_2(\mathbb{Z})$ is a sum $f_k + f_{k-2}E_2 + \cdots + f_{k-\ell}E_2^{\ell}$, where each f_j is a modular form of weight $k - 2j$.

Instead of just a weakly modular form of weight 0, one can check that the function χ can be a weakly quasimodular form of weight 0 and depth 2 for $\mathrm{SL}_2(\mathbb{Z})$. Now we have enough flexibility to construct f_+, and calculations much like those in the f_- case lead to

$$\chi = \frac{(E_2E_4 - E_6)^2}{\Delta},$$

up to scaling. See Figure 8 for plots of the quasimodular forms that yield f_- and f_+.

FIGURE 8

Plots of $\psi(z)\Delta(z)$ (above) and $(\phi|_{-2}S)(z)\Delta(z)$ (below) for $-1 \leq \operatorname{Re} z \leq 1$ and $0 < \operatorname{Im} z \leq 1$.

Now that we have obtained both magic eigenfunctions, we can construct the magic function f as a linear combination of them. First, we rescale ϕ so that $f_+(0) = 1$, and then we rescale ψ so that $f'_-(\sqrt{2}) = f'_+(\sqrt{2})$, to obtain a double root at $\sqrt{2}$ for $\hat{f}$. Using these scalings, the eight-dimensional magic function is given by

$$f(x) = 4\sin\big(\pi|x|^2/2\big)^2 \int_0^\infty \big(\phi(it) + \psi(it)\big)e^{-\pi t|x|^2}\,dt$$

for $|x|^2 > 2$, and the eigenfunction property implies that

$$\hat{f}(y) = 4\sin\big(\pi|y|^2/2\big)^2 \int_0^\infty \big(\phi(it) - \psi(it)\big)e^{-\pi t|y|^2}\,dt$$

for all $y \neq 0$ (this integral turns out to converge whenever $|y| > 0$, because the exponential growth in $\phi(it)$ and $\psi(it)$ as $t \to \infty$ cancels).

The final step in the proof of Theorem 1.1 is to check the inequalities that are needed for Theorem 2.1, namely $f(x) \leq 0$ for $|x| \geq 2$ and $\hat{f}(y) \geq 0$ for all y, to make sure there are no unexpected sign changes between the roots $\sqrt{2n}$. In principle, that might seem difficult, because integral transforms of quasimodular forms could be complicated. However, these inequalities hold for the simplest reason one could hope for:

$$\phi(it) + \psi(it) < 0 \quad \text{and} \quad \phi(it) - \psi(it) > 0$$

for all $t > 0$. In other words, the desired inequalities hold directly at the level of the quasimodular forms themselves. This can be checked rigorously in any of several ways. For example, one can use asymptotics to check the inequalities as $t \to 0$ or $t \to \infty$, and then use interval arithmetic to verify them on the remaining bounded interval.

Overall, this proof feels like a miracle. Everything falls beautifully into place, with Viazovska's constructions having just enough flexibility to complete the proof in a unique way. What I find most impressive is the number of ingenious ideas required for the full proof. The single-root construction is itself remarkable, generalizing it to f_- is even more so, and still more ideas are required for f_+. Viazovska is a master of special functions, whose work would surely have excited Jacobi and Ramanujan.

6. INTERPOLATION AND CONSEQUENCES

Along the way to proving the optimality of E_8, Viazovska made the bold conjecture that the magic function is uniquely determined by its required roots, and that more generally a radial Schwartz function on $\mathbb{R}^8$ is uniquely determined by its values and radial derivatives at the radii $\sqrt{2n}$ and those of its Fourier transform. It is far from obvious that it is possible in principle to reconstruct a radial Schwartz function from discrete data of this sort.

Radchenko and Viazovska took a major step in this direction by proving a one-dimensional analogue for first-order interpolation, and the second-order theorem was proved by Cohn, Kumar, Miller, Radchenko, and Viazovska.

Theorem 6.1 (Radchenko and Viazovska [23]). *There exist even Schwartz functions $a_n: \mathbb{R} \to \mathbb{R}$ for integers $n \geq 0$ such that for every even Schwartz function $f: \mathbb{R} \to \mathbb{R}$ and $x \in \mathbb{R}$,*

$$f(x) = \sum_{n \geq 0} f(\sqrt{n}) a_n(x) + \sum_{n \geq 0} \hat{f}(\sqrt{n}) \hat{a}_n(x).$$

Theorem 6.2 (Cohn, Kumar, Miller, Radchenko, and Viazovska [11]). *Let (d, n_0) be $(8, 1)$ or $(24, 2)$. Then every radial Schwartz function $f: \mathbb{R}^d \to \mathbb{R}$ is uniquely determined by the values $f(\sqrt{2n})$, $f'(\sqrt{2n})$, $\hat{f}(\sqrt{2n})$, and $\hat{f}'(\sqrt{2n})$ for integers $n \geq n_0$. Specifically, there exists an interpolation basis a_n, b_n for $n \geq n_0$ such that for every radial Schwartz function f and $x \in \mathbb{R}^d$,*

$$\begin{aligned} f(x) &= \sum_{n=n_0}^{\infty} f(\sqrt{2n}) a_n(x) + \sum_{n=n_0}^{\infty} f'(\sqrt{2n}) b_n(x) \\ &\quad + \sum_{n=n_0}^{\infty} \hat{f}(\sqrt{2n}) \hat{a}_n(x) + \sum_{n=n_0}^{\infty} \hat{f}'(\sqrt{2n}) \hat{b}_n(x). \end{aligned}$$

The proofs construct the interpolation bases explicitly, by combining Viazovska's integral transform techniques with broader classes of special functions.

One consequence of radial Fourier interpolation is a stronger optimality theorem for E_8 and the Leech lattice. Instead of just taking into account local interactions between particles, as in the sphere packing problem, one can study optimization problems with long-range interactions. For example, one could ask for the ground state of particles interacting via an inverse power law. Cohn and Kumar [9] formulated a broad notion of optimality, called *universal optimality*, and radial Fourier interpolation yields corresponding magic functions:

Theorem 6.3 (Cohn, Kumar, Miller, Radchenko, and Viazovska [11]). *The E_8 root lattice and the Leech lattice are universally optimal in $\mathbb{R}^8$ and $\mathbb{R}^{24}$, respectively.*

7. THE FUTURE

Although Viazovska's work has settled several major questions, much remains to be understood. For example, the theory of interpolation for radial Schwartz functions is rapidly developing, with noteworthy connections to uniqueness theory for the Klein–Gordon equation [2].

One puzzling issue is two dimensions. While the two-dimensional sphere packing problem can be settled by elementary geometry, universal optimality remains a tantalizing conjecture. There seems to be a magic function for $d = 2$ in Theorem 2.1, with $r = (4/3)^{1/4}$; no proof is known, but numerical computations agree with the optimal packing density in $\mathbb{R}^2$ to over 1000 decimal places. Furthermore, analogous magic functions seem to exist for universal optimality in $\mathbb{R}^2$. However, it is unclear what sort of function space might allow a suitable interpolation theory (see Section 7 in [11]).

There are also remarkable connections with conformal field theory and quantum gravity [18]. When d is even, the linear programming bound for the sphere packing density in $\mathbb{R}^d$ turns out to be equivalent to the spinless modular bootstrap bound for the spectral gap in a theory of $d/2$ free bosons, and the conformal bootstrap program generalizes it to a family of related bounds. How these more general bounds might relate to discrete geometry remains a mystery.

REFERENCES

[1] N. Afkhami-Jeddi, H. Cohn, T. Hartman, D. de Laat, and A. Tajdini, High-dimensional sphere packing and the modular bootstrap. *J. High Energy Phys.* **2020** (2020), no. 12, Paper No. 066, 44 pp. arXiv:2006.02560 DOI 10.1007/jhep12(2020)066

[2] A. Bakan, H. Hedenmalm, A. Montes-Rodríguez, D. Radchenko, and M. Viazovska, Fourier uniqueness in even dimensions. *Proc. Natl. Acad. Sci. USA* **118** (2021), no. 15, Paper No. 2023227118, 4 pp. DOI 10.1073/pnas.2023227118

[3] A. Bondarenko, D. Radchenko, and M. Viazovska, Optimal asymptotic bounds for spherical designs. *Ann. of Math. (2)* **178** (2013), no. 2, 443–452. arXiv:1009.4407 DOI 10.4007/annals.2013.178.2.2

[4] J. Bourgain, L. Clozel, and J.-P. Kahane, Principe d'Heisenberg et fonctions positives. *Ann. Inst. Fourier (Grenoble)* **60** (2010), no. 4, 1215–1232. arXiv:0811.4360 DOI 10.5802/aif.2552

[5] H. Cohen and F. Strömberg, *Modular forms: a classical approach*. Grad. Stud. Math. 179, American Mathematical Society, Providence, RI, 2017.

[6] H. Cohn, A conceptual breakthrough in sphere packing. *Notices Amer. Math. Soc.* **64** (2017), no. 2, 102–115. arXiv:1611.01685 DOI 10.1090/noti1474

[7] H. Cohn and N. Elkies, New upper bounds on sphere packings I. *Ann. of Math. (2)* **157** (2003), no. 2, 689–714. arXiv:math/0110009 DOI 10.4007/annals.2003.157.689

[8] H. Cohn and F. Gonçalves, An optimal uncertainty principle in twelve dimensions via modular forms. *Invent. Math.* **217** (2019), no. 3, 799–831. arXiv:1712.04438 DOI 10.1007/s00222-019-00875-4

[9] H. Cohn and A. Kumar, Universally optimal distribution of points on spheres. *J. Amer. Math. Soc.* **20** (2007), no. 1, 99–148. arXiv:math/0607446 DOI 10.1090/S0894-0347-06-00546-7

[10] H. Cohn, A. Kumar, S. D. Miller, D. Radchenko, and M. Viazovska, The sphere packing problem in dimension 24. *Ann. of Math. (2)* **185** (2017), no. 3, 1017–1033. arXiv:1603.06518 DOI 10.4007/annals.2017.185.3.8

[11] H. Cohn, A. Kumar, S. D. Miller, D. Radchenko, and M. Viazovska, Universal optimality of the E_8 and Leech lattices and interpolation formulas. *Ann. of Math. (2)* **196** (2022), no. 3, 983–1082. arXiv:1902.05438 DOI 10.4007/annals.2022.196.3.3

[12] J. H. Conway and N. J. A. Sloane, *Sphere packings, lattices and groups*. 3rd edn., Grundlehren Math. Wiss. 290, Springer, New York, 1999. DOI 10.1007/978-1-4757-6568-7

[13] P. Delsarte, Bounds for unrestricted codes, by linear programming. *Philips Res. Rep.* **27** (1972), 272–289.

[14] F. Diamond and J. Shurman, *A first course in modular forms*. Grad. Texts in Math. 228, Springer, New York, 2005. DOI 10.1007/978-0-387-27226-9

[15] W. Ebeling, *Lattices and codes: a course partially based on lectures by Friedrich Hirzebruch*. 3rd edn., Adv. Lectures Math., Springer, Wiesbaden, 2013. DOI 10.1007/978-3-658-00360-9

[16] T. C. Hales, A proof of the Kepler conjecture. *Ann. of Math. (2)* **162** (2005), no. 3, 1065–1185. DOI 10.4007/annals.2005.162.1065

[17] T. Hales, M. Adams, G. Bauer, T. D. Dang, J. Harrison, L. T. Hoang, C. Kaliszyk, V. Magron, S. McLaughlin, T. T. Nguyen, Q. T. Nguyen, T. Nipkow, S. Obua, J. Pleso, J. Rute, A. Solovyev, T. H. A. Ta, N. T. Tran, T. D. Trieu, J. Urban, K. Vu, and R. Zumkeller, A formal proof of the Kepler conjecture. *Forum Math. Pi* **5** (2017), e2, 29 pp. arXiv:1501.02155 DOI 10.1017/fmp.2017.1

[18] T. Hartman, D. Mazáč, and L. Rastelli, Sphere packing and quantum gravity. *J. High Energy Phys.* **2019** (2019), no. 12, 048, 66 pp. arXiv:1905.01319 DOI 10.1007/jhep12(2019)048

[19] E. Klarreich, Sphere packing solved in higher dimensions. *Quanta Mag.* (March 30, 2016). https://www.quantamagazine.org/sphere-packing-solved-in-higher-dimensions-20160330/

[20] D. de Laat and F. Vallentin, A breakthrough in sphere packing: the search for magic functions. *Nieuw Arch. Wiskd. (5)* **17** (2016), no. 3, 184–192. arXiv:1607.02111

[21] D. Lowry-Duda, Visualizing modular forms. In *Arithmetic geometry, number theory, and computation*, edited by J. S. Balakrishnan, N. Elkies, B. Hassett, B. Poonen, A. V. Sutherland, and J. Voight, Simons Symposia, Springer, 2021. arXiv:2002.05234 DOI 10.1007/978-3-030-80914-0_19

[22] D. Madore, Sections du diagramme de Voronoï du réseau E_8. *David Madore's WebLog* (April 9, 2017). http://www.madore.org/~david/weblog/d.2017-04-09.2433.html

[23] D. Radchenko and M. Viazovska, Fourier interpolation on the real line. *Publ. Math. Inst. Hautes Études Sci.* **129** (2019), 51–81. arXiv:1701.00265 DOI 10.1007/s10240-018-0101-z

[24] J.-P. Serre, *A course in arithmetic*. Grad. Texts in Math. 7, Springer, New York–Heidelberg, 1973. DOI 10.1007/978-1-4684-9884-4

[25] T. M. Thompson, *From error-correcting codes through sphere packings to simple groups*. Carus Math. Monogr. 21, Mathematical Association of America, Washington, DC, 1983.

[26] A. Thue, Om nogle geometrisk-taltheoretiske Theoremer. *Forhandlingerne ved de Skand. Naturforskeres* **14** (1892), 352–353.

[27] M. S. Viazovska, The sphere packing problem in dimension 8. *Ann. of Math. (2)* **185** (2017), no. 3, 991–1015. arXiv:1603.04246 DOI 10.4007/annals.2017.185.3.7

[28] D. Zagier, Elliptic modular forms and their applications. In *The 1–2–3 of modular forms*, edited by K. Ranestad, pp. 1–103, Universitext, Springer, Berlin, 2008. DOI 10.1007/978-3-540-74119-0_1

HENRY COHN
Microsoft Research New England, One Memorial Drive, Cambridge, MA 02140, USA, cohn@microsoft.com

THE WORK OF MARK BRAVERMAN

RAN RAZ

ABSTRACT

Mark Braverman was awarded the 2022 IMU Abacus medal for his work on Information Complexity and additional work. Mark is a world leader of the research area of information complexity and his works are among the most influential in this research area. Mark has a broad research interest and key works in several other research areas, that in some cases solved central long-standing open problems. We describe some of his work, focusing mainly on contribution to information complexity and related topics at the interface of computational complexity and information theory.

MATHEMATICS SUBJECT CLASSIFICATION 2020

Primary 68Q11; Secondary 68P30, 94A15, 68Q01, 68Q17

KEYWORDS

Computational complexity, communication complexity, information theory, information complexity

INTERNATIONAL CONGRESS OF MATHEMATICIANS 2022 JULY 6–14

© 2022 International Mathematical Union
Proc. Int. Cong. Math. 2022, Vol. 1, pp. 106–117
DOI 10.4171/ICM2022/214
Published by EMS Press
and licensed under a CC BY 4.0 license

1. COMMUNICATION COMPLEXITY

Communication complexity, first introduced by Yao [54], is a central model in complexity theory that studies the amount of communication needed to solve a problem, when the input to the problem is distributed among two or more parties.

In the two-player distributional model, each of two players gets an input, where the two inputs X, Y are random variables sampled from some joint distribution (known to both players). The players' goal is to solve a communication task that depends on both inputs, such as computing a function $f(X, Y)$, where $f : \{0, 1\}^n \times \{0, 1\}^n \to \{0, 1\}$ is known to both players and X, Y are inputs of length n bits. The players communicate in rounds, where in each round one of the players sends a message to the other player. At the end of the protocol, in the example given above, both players need to know the value of $f(X, Y)$. The players are allowed to use both public and private random strings and are allowed to err with some fixed small probability.

The communication complexity of a protocol is the maximal number of bits communicated by the players in the protocol, where the maximum is taken over all possible inputs (in the support of the input distribution). The communication complexity of a communication task is the minimal communication complexity of a protocol that solves the task with high probability (say, probability larger than $\frac{2}{3}$).

2. INFORMATION COMPLEXITY

Information complexity, introduced by [1, 2, 25], studies the amount of information that two players need to reveal about their inputs in order to solve a communication task. The model was motivated by fundamental information-theoretical questions of compressing communication, as well as by fascinating relations to communication complexity, and in particular to proving lower bounds for communication complexity and to the direct-sum problem in communication complexity, a problem that has a rich history and has been studied in many works and various settings.

The paper by Barak, Braverman, Chen, and Rao distinguishes between internal and external information complexity of a communication protocol [2]. Roughly speaking, the external information complexity of a protocol, first defined in [25], is the amount of information that an external observer, who watches the execution of the protocol, learns about the players' inputs, while the internal information complexity of a protocol, implicit in [1] and explicitly defined in [2], is the amount of information that the players learn about each other's input, when running the protocol.

Formally, if M is the transcript of the protocol and R is the public random string, external and internal information complexity are defined by

$$\mathrm{Ext} = I\big((X, Y); M \mid R\big),$$
$$\mathrm{Int} = I(X; M \mid Y, R) + I(Y; M \mid X, R),$$

where I is the conditional mutual information function. (It is known that the private random strings of the protocol can be ignored here.)

The (internal or external) information complexity of a communication task is the infimum of the (internal or external) information complexity of a protocol that solves the task with high probability (say, probability larger than $\frac{2}{3}$).

It is not hard to prove that for any protocol (and thus also for any communication task), the internal information complexity of the protocol is at most its external information complexity, which, in turn, is at most its communication complexity. This motivated the study of information complexity as a tool for proving lower bounds for communication complexity.

A beautiful and useful property of internal information complexity, that motivated its definition, is the additivity property, or direct-sum property. Roughly speaking, the internal information complexity of performing two communication tasks, on two independent pairs of inputs, is equal to the sum of the internal information complexities of the two tasks. Consequently, the internal information complexity of performing k copies of a communication task, on k independent pairs of inputs, is equal to k times the internal information complexity of the communication task ([1, 2, 20], using techniques from [43, 45]). The direct-sum property also relates information complexity to the direct-sum problem in communication complexity.

Finally, we note that in the case where the inputs X, Y for the two players are sampled independently, the internal and external information complexity of any protocol are equal.

Much of the work on information complexity was consolidated into a theory in Braverman's works [2, 6, 7, 20]. Braverman also defined a variant of information complexity that does not depend on the prior distribution of the input, proving that several possible definitions are essentially equivalent [7].

A priori, it was not clear whether information complexity is computable, in the sense that there is an algorithm that approximates it, but this was proved by Braverman and Schneider (for the zero-error case) [23].

3. INTERACTIVE COMPRESSION

The classical works of Shannon, Fano, and Huffman show that if a player wants to send a message X to another player, it is sufficient for her to send $\lceil H(X) \rceil$ bits, in expectation, where H denotes Shannon's entropy function [29, 35, 50]. That is, the length of the message can be compressed to roughly $H(X)$, the information content of the message. Are there analogous results in the interactive setting, where two players engage in an interactive communication protocol?

Barak, Braverman, Chen, and Rao initiated a study of the interactive compression problem [2]. Given a communication protocol with small information complexity, can the protocol be compressed so that the total number of bits communicated by the protocol is also small? More formally, given a communication protocol Π with communication complexity C and (internal or external) information complexity $I \ll C$, is there always an equivalent protocol Π' (possibly with slightly higher error probability), with communication complexity significantly smaller than C (and arbitrary information complexity)?

Barak, Braverman, Chen, and Rao gave two different compression protocols, one for internal and one for external information complexity. For internal information complexity, they proved that any communication protocol Π with communication complexity C and internal information complexity I can be compressed to an equivalent protocol Π', with communication complexity $O(\sqrt{C \cdot I} \cdot \log C)$. For external information complexity, they proved that any communication protocol Π with communication complexity C and external information complexity I can be compressed to an equivalent protocol Π', with communication complexity $O(I \cdot \log C)$ [2]. Recall that internal information complexity is always smaller or equal to external information complexity, and hence compressing the communication complexity to an expression close to the external information complexity of the original protocol is easier.

These results were followed by many additional works that further studied the interactive compression problem. Braverman and Rao proved that any one-round (or small number of rounds) communication protocol with internal information complexity I can be compressed to an equivalent protocol, with communication complexity $O(I)$ [20]. Braverman proved that any communication protocol with internal information complexity I can be compressed to an equivalent protocol, with communication complexity $2^{O(I)}$ [7].

Kol's breakthrough work proved that in the important special case where the two inputs X, Y are independent, any communication protocol with internal/external information complexity I can be compressed to an equivalent protocol, with communication complexity $O(I^2 \cdot \text{polylog}(I))$ [37]. This was culminated by Sherstov who improved the last communication complexity to $O(I \cdot \text{polylog}(I))$ [51]. Note that this last expression does not depend on the communication complexity of the original protocol at all and almost matches the lower bound of $\Omega(n)$. Recall that in the case where X, Y are independent, the internal and external communication complexity are equal. Building on these works, Braverman and Kol proved that any communication protocol with communication complexity C and external information complexity I can be compressed to an equivalent protocol, with communication complexity $\text{poly}(I) \cdot \text{loglog}(C)$ [15].

As for lower bounds, Braverman suggested a candidate for a communication task with communication complexity exponentially larger than (internal or external) information complexity [5]. This task and other communication tasks were analyzed in subsequent works, establishing exponential gaps between communication complexity and information complexity [30–32, 42], namely, examples for communication tasks with (internal or external) information complexity I and communication complexity $2^{\Omega(I)}$. In particular, these works show that Braverman's compression of the communication complexity of a protocol to $2^{O(I)}$ [7] is the best possible, and one cannot hope for compression to $\text{poly}(I)$ in the general case (as obtained by Kol and Shesrtov for the special case of independent inputs X, Y [37, 51]). Building on this line of works, Braverman and Minzer established exponential gaps between internal and external information complexity [17]. An important open problem asks whether compression to $\text{poly}(I) \cdot \text{polylog}(C)$, where I is the internal information complexity, is possible in the general case [11]. (As described above, the best known today are compressions to $O(\sqrt{C \cdot I} \cdot \log C)$ [2] and $2^{O(I)}$ [7].)

In each of the above mentioned compression protocols, the two players manage to sample together, with low communication, a transcript of the original protocol, such that the transcript is sampled (approximately) from the correct distribution on transcripts and both players agree on the same transcript with high probability. One of the challenges is that none of the players knows the correct distribution of transcripts.

As an illustration of the flavor of techniques used in these results, we state a brilliant theorem from the work of Braverman and Rao [20]:

Theorem. *Assume that player 1 knows a distribution P and player 2 knows a distribution Q over the same finite set U. For every $\varepsilon > 0$, there is a public coin communication protocol that uses an expected number of $D(P \| Q) + 2\log(1/\varepsilon) + O(\sqrt{D(P \| Q)} + 1)$ bits of communication (where $D(P \| Q) = \sum_x P(x)\log(P(x)/Q(x))$ is the Kullback–Leibler informational divergence), such that at the end of the protocol Player 1 outputs an element a distributed according to P and Player 2 outputs an element b such that for every $x \in U$, $\Pr[b = x | a = x] > 1 - \varepsilon$.*

4. DIRECT SUM

One of the first motivations for studying information complexity came from relations to the direct-sum problem in communication complexity. The direct-sum problem asks what are the relations between the communication complexity of a communication task and the communication complexity of performing k copies of the same task on k independently chosen inputs.

Let T be a communication task. For every k, let T^k be the task of performing k copies of the task T, on k inputs that are independently chosen according to the input distribution of T, and allowing to err on each copy with the same probability of error that is allowed for the task T. The amortized communication complexity of a task T is defined by

$$\lim_{k\to\infty} \frac{\mathrm{CC}(T^k)}{k}$$

where CC denotes communication complexity.

Braverman and Rao proved that the amortized communication complexity of any task T exactly equals to its internal information complexity [20] (see also [41]). This surprising result relates the direct-sum problem in communication complexity to the interactive compression problem.

A priori, one could think that the amortized communication complexity of a task should always be close to its communication complexity. However, using Braverman and Rao's equivalence between amortized communication complexity and internal information complexity, the above mentioned exponential gaps between communication complexity and internal information complexity also imply exponential gaps between communication complexity and amortized communication complexity, showing that there are communication tasks with communication complexity C and amortized communication complexity $O(\log C)$ [30,31,42]. This shows that a strong direct-sum property does not hold for communication complexity.

Conversely, each of the above mentioned compression protocols, in terms of internal information complexity, implies a lower bound on amortized communication complexity. For example, the compression protocols of Kol and Sherstov [37,51] imply that for the special case of independent X, Y, communication complexity and amortized communication complexity are essentially equal (up to polylogarithmic factors), and the compression protocol of Braverman [7] implies that amortized communication complexity is at least logarithmic in the communication complexity.

Additional works by Braverman, Rao, Weinstein and Yehudayof [22] and Braverman and Weinstein [24] show that if a protocol tries to solve T^k with communication complexity significantly smaller than k times the amortized communication complexity of T, then the success probability of the protocol is exponentially small.

5. COMMUNICATION COMPLEXITY OF SET-INTERSECTION

Set-Intersection, or Set-Disjointness, is a central problem in communication complexity. In this problem, each of two (or more) players gets a vector in $\{0,1\}^n$ and their goal is to determine whether there exists a coordinate $i \in [n]$ where they both (or all) have 1. This simple problem inspired a lot of progress in both communication complexity and information complexity.

It has been known since 1987 that the probabilistic communication complexity of Set-Intersection is at least $\Omega(n)$ [36,45]. The main result of the paper by Bar-Yossef, Jayram, Kumar, and Sivakumar, one of the papers that started the research area of information complexity, was a new proof for the lower bound of $\Omega(n)$ for Set-Intersection, using information complexity [1]. This proof was one of their main motivations for studying information complexity.

Braverman used information complexity to study many additional aspects of the communication complexity of Set-Intersection.

While it was known that the probabilistic communication complexity of Set-Intersection is $\Theta(n)$ [1,36,45], Braverman, Garg, Pankratov, and Weinstein studied the information complexity of the Boolean AND function and from that analysis they figured out the exact constant in the $\Theta(n)$ expression, that is, they computed the probabilistic communication complexity of Set-Intersection exactly, up to second-order terms [14].

Braverman and Moitra studied communication protocols for Set-Intersection that get advantage of at least ε over a random guess. They proved a tight lower bound of $\Omega(\varepsilon n)$ for the communication complexity of any such protocol [18], while previous proofs only implied a lower bound of $\Omega(\varepsilon^2 n)$. From their improved lower bound, they obtained as an application lower bounds for the size of linear programs.

Braverman, Ellen, Oshman, Pitassi, and Vaikuntanathan [10] and Braverman and Oshman [19] used information complexity to prove tight lower bounds for the communication complexity of Set-Intersection with more than two players. Braverman, Garg, Kun-Ko, Mao, and Touchette used a quantum variant of information complexity to prove lower bounds for the quantum communication complexity of Set-Intersection with bounded number of rounds [13].

6. PARALLEL REPETITION OF TWO-PROVER GAMES

Information complexity is closely related to the study of parallel repetition of two-prover games. Both areas make substantial use of information theory, but the connection is deeper; the two areas use many similar ideas, intuitions, definitions, tools, and techniques (such as, subadditivity of entropy, correlation-breaking events, and correlated sampling).

In a two-prover (two-player) game, a referee samples questions (x, y) from some (publicly known) distribution, and sends x to the first player and y to the second player. The first player responds by $a = a(x)$ and the second by $b = b(y)$ (without communicating with each other). The players jointly win if a (publicly known) predicate $V(x, y, a, b)$ is satisfied. The value of the game is the maximal probability of success that the players can achieve, where the maximum is taken over all protocols $a = a(x)$, $b = b(y)$.

Roughly speaking, a parallel repetition of a two-prover game is a game where the players try to win n copies of the original game simultaneously. More precisely, the referee generates questions $x = (x_1, \ldots, x_n)$, $y = (y_1, \ldots, y_n)$, where each pair (x_i, y_i) is chosen independently according to the original distribution. The players respond by $a = (a_1, \ldots, a_n) = a(x)$ and $b = (b_1, \ldots, b_n) = b(y)$. The players win if they win simultaneously on all the coordinates, that is, if for every i, $V(x_i, y_i, a_i, b_i)$ holds.

The parallel repetition theorem states that for any two-prover game, with value smaller than 1, the value of the game repeated in parallel n times decreases exponentially fast in n [43]. The parallel repetition theorem, and other results about parallel repetition of two-prover games, have many applications in computational complexity and other research areas.

While it was known for a long time that parallel repetition reduces the value of two-prover games exponentially fast, the exact rate of exponential decrease was not known when the value of the game was already small, to begin with. (A tight analysis for games with small value was only known for the special case of projection games [27]).

Braverman and Garg solved this problem. They proved that if the value of the game is $v < 1/2$ and the length of answers is s then the value of the game repeated in parallel n times is at most $v^{\Omega(n \log(1/v)/s)}$ [12]. Only a bound of $2^{-\Omega(n/s)}$ was previously known [43].

7. INTERACTIVE CODING THEORY

Shannon's celebrated 1948 paper, "A Mathematical Theory of Communication," initiated (among many other famous contributions) the field of error correcting codes. Suppose that a player wants to send a message of length n bits to another player, but the only available channel is noisy and changes every bit that is sent with some constant probability (smaller than $1/2$). Shannon proved that the player can send a message of length $O(n)$ bits, over the noisy channel, such that from that message the original message can be retrieved with high probability and with no errors [50]. Are there analogous results in the interactive setting, where two players engage in an interactive communication protocol?

This question was first asked and answered by Schulman in 1992. Schulman showed how to translate any interactive communication protocol to an equivalent noise-resilient protocol that runs over a noisy channel, with only a constant overhead in the communication complexity (even when the noise is adversarially chosen) [47–49]. These results initiated interactive coding theory, the study of how to perform an interactive communication protocol reliably in the presence of noise.

In 2011, Braverman and Rao initiated a study of the question of what is the maximal fraction of errors that can be recovered in an interactive protocol. While Schulman's work only recovered a fraction of errors that is bounded by $1/240$, Braverman and Rao showed how to recover $1/4 - \varepsilon$ fraction of errors, when the encoding alphabet size is some constant, and $1/8 - \varepsilon$ fraction of errors, when the encoding alphabet size is just 2. The result holds even in the adversarial case, and at a cost of increasing the communication complexity of the protocol by only a constant factor [21]. (The fraction of errors of $1/8 - \varepsilon$ for an encoding alphabet of size 2 was recently improved to an optimal fraction of $1/6 - \varepsilon$ [28, 34].)

This work by Braverman and Rao initiated a renewed interest in interactive coding theory and inspired many follow-up works. Braverman studied additional aspects of interactive coding theory in many subsequent works. For example, Braverman and Efremenko studied list decoding for interactive communication [8], and Braverman, Efremenko, Gelles, and Haeupler proved that constant-rate coding for multiparty interactive communication is impossible [9].

8. LOWER BOUNDS FOR BOUNDED-DEPTH CIRCUITS

Bounded-depth Boolean circuits are among the most important subclasses of Boolean circuits and have been extensively studied in numerous works. They are central in many subareas of complexity theory, as well as in analysis of Boolean functions. Roughly speaking, a Boolean circuit computes a Boolean function of n binary input variables using AND, OR, and NOT gates, where the fan-in of the AND and OR gates is unbounded. The size of the circuit is the number of wires in it and the depth of the circuit is the length of the longest directed path from an input variable to the output (not counting NOT gates).

In 1990, Linial and Nisan conjectured that circuits of size m and depth d cannot distinguish between the uniform distribution over the inputs and any k-wise independent distribution over the inputs, with $k \geq (\log m)^{d-1}$ [40]. The conjecture means that a bounded-depth circuit cannot recognize global structure, as long as it does not come with some local structure. This was an important conjecture but there was very little progress for many years and the conjecture was only proved for DNFs (that is, for circuits of depth 2) [3, 46].

In 2009, Braverman proved that circuits of size m and depth d cannot distinguish between the uniform distribution over the inputs and any k-wise independent distribution over the inputs, with $k \geq (\log m)^{O(d^2)}$ [4]. This result qualitatively proves the conjecture, with somewhat weaker parameters.

To prove this result, Braverman brilliantly combines two different types of approximation of bounded-depth circuits by low-degree polynomials. The first, by Razborov and Smolensky [44, 52], gives a polynomial that is equal to the function computed by the circuit almost everywhere but may be very different from it on a small fraction of inputs. Braverman observes that the difference between the function computed by the circuit and the approximating polynomial can itself be computed by a bounded-depth circuit. He then approximates that difference by a low-degree polynomial using a different type of approximation, the approximation given by Linial, Mansour, and Nisan [39], that approximates a bounded-depth circuit by a low-degree polynomial that is close to the function computed by the circuit on average. The final result is a low-degree polynomial that approximates the original circuit so well that the trivial proof that low-degree polynomials cannot distinguish between the uniform distribution and k-wise independent distributions (with k larger or equal to the degree of the polynomial) works [4].

Since Braverman published his work, it was improved and became more important in two ways. First, Tal's breakthrough work [53] improved the approximation given by Linial, Mansour, and Nisan [39] and by plugging in the new parameters into Braverman's proof he obtained an improved result: Circuits of size m and depth d cannot distinguish between the uniform distribution over the inputs and any k-wise independent distribution over the inputs, with $k \geq (\log m)^{O(d)}$ [53]. This comes even closer to proving the original conjecture. Additionally, Chattopadhyay and Zuckerman used these results in their breakthrough construction of explicit two-source extractors [26].

9. GROTHENDIECK'S CONSTANT VS. KRIVINE'S BOUND

In 1953, Grothendieck proved that there is a positive constant $K \in \mathbb{R}$, such that, for any $m \times n$ real matrix $(a_{ij})_{i \in [m], j \in [n]}$,

$$\max_{\{X_i\},\{Y_j\}} \sum_{i,j} a_{ij} \langle X_i, Y_j \rangle \leq K \cdot \max_{\{x_i\},\{y_j\}} \sum_{i,j} a_{ij} x_i y_j,$$

where X_i, Y_j (on the left-hand side) are unit vectors in $\mathbb{R}^{m+n}$ and x_i, y_j (on the right-hand side) are in $\{-1, 1\}$ [33]. The smallest value of K that satisfies this inequality is called Grothendieck's constant.

This is an important theorem, with applications in several areas. In computer science, Grothendieck's constant can be viewed as the integrality gap between a maximum obtained over values in $\{-1, 1\}$, on the right-hand side, that is many times desirable but is often hard to compute, and the maximum obtained over unit vectors, on the left-hand side, that can be computed in polynomial time.

The exact value of Grothendieck's constant is still not known. In 1979, Krivine proved that Grothendieck's constant is at most $\frac{\pi}{2\ln(1+\sqrt{2})}$ and conjectured that this is an equality [38]. The conjecture was disproved by Braverman, Makarychev, Makarychev, and Naor [16].

FUNDING

Supported by the Simons Collaboration on Algorithms and Geometry, by a Simons Investigator Award and by the National Science Foundation grants No. CCF-1714779, CCF-2007462.

REFERENCES

[1] Z. Bar-Yossef, T. S. Jayram, R. Kumar, and D. Sivakumar, An information statistics approach to data stream and communication complexity. *J. Comput. System Sci.* **68** (2004), no. 4, 702–732.

[2] B. Barak, M. Braverman, X. Chen, and A. Rao, How to compress interactive communication. *SIAM J. Comput.* **42** (2013), no. 3, 1327–1363.

[3] L. M. J. Bazzi, Polylogarithmic independence can fool DNF formulas. *SIAM J. Comput.* **38** (2009), no. 6, 2220–2272.

[4] M. Braverman, Polylogarithmic independence fools AC0 circuits. *J. ACM* **57** (2010), no. 5, 28:1–28:10.

[5] M. Braverman, A hard-to-compress interactive task? In *51th Annual Allerton Conference on Communication, Control, and Computing*, pp. 8–12, IEEE, 2013.

[6] M. Braverman, Interactive information and coding theory. In *Proceeding of the international congress of mathematicians, ICM 2014*, pp. 535–559, KYUNG MOON SA Co. Ltd, 2014.

[7] M. Braverman, Interactive information complexity. *SIAM J. Comput.* **44** (2015), no. 6, 1698–1739.

[8] M. Braverman and K. Efremenko, List and unique coding for interactive communication in the presence of adversarial noise. *SIAM J. Comput.* **46** (2017), no. 1, 388–428.

[9] M. Braverman, K. Efremenko, R. Gelles, and B. Haeupler, Constant-rate coding for multiparty interactive communication is impossible. *J. ACM* **65** (2018), no. 1, 4:1–4:41.

[10] M. Braverman, F. Ellen, R. Oshman, T. Pitassi, and V. Vaikuntanathan, A tight bound for set disjointness in the message-passing model. In *FOCS*, pp. 668–677, IEEE, 2013.

[11] M. Braverman, A. Ganor, G. Kol, and R. Raz, A candidate for a strong separation of information and communication. In *ITCS*, pp. 11:1–11:13, Schloss Dagstuhl - Leibniz-Zentrum für Informatik, 2018.

[12] M. Braverman and A. Garg, Small value parallel repetition for general games. In *STOC*, pp. 335–340, ACM, 2015.

[13] M. Braverman, A. Garg, Y. Kun-Ko, J. Mao, and D. Touchette, Near-optimal bounds on the bounded-round quantum communication complexity of disjointness. *SIAM J. Comput.* **47** (2018), no. 6, 2277–2314.

[14] M. Braverman, A. Garg, D. Pankratov, and O. Weinstein, From information to exact communication. In *STOC*, pp. 151–160, ACM, 2013.

[15] M. Braverman and G. Kol, Interactive compression to external information. In *STOC*, pp. 964–977, ACM, 2018.

[16] M. Braverman, K. Makarychev, Y. Makarychev, and A. Naor, The Grothendieck constant is strictly smaller than Krivine's bound. In *FOCS*, pp. 453–462, IEEE, 2011.

[17] M. Braverman and D. Minzer, New separations results for external information. In *STOC*, pp. 248–258, ACM, 2021.

[18] M. Braverman and A. Moitra, An information complexity approach to extended formulations. In *STOC*, pp. 161–170, ACM, 2013.

[19] M. Braverman and R. Oshman, A rounds vs. communication tradeoff for multi-party set disjointness. In *FOCS*, pp. 144–155, IEEE, 2017.

[20] M. Braverman and A. Rao, Information equals amortized communication. *IEEE Trans. Inf. Theory* **60** (2014), no. 10, 6058–6069.

[21] M. Braverman and A. Rao, Toward coding for maximum errors in interactive communication. *IEEE Trans. Inf. Theory* **60** (2014), no. 11, 7248–7255.

[22] M. Braverman, A. Rao, O. Weinstein, and A. Yehudayoff, Direct products in communication complexity. In *FOCS*, pp. 746–755, IEEE, 2013.

[23] M. Braverman and J. Schneider, Information complexity is computable. In *ICALP*, pp. 87:1–87:10, Schloss Dagstuhl - Leibniz-Zentrum für Informatik, 2016.

[24] M. Braverman and O. Weinstein, An interactive information odometer and applications. In *STOC*, pp. 341–350, ACM, 2015.

[25] A. Chakrabarti, Y. Shi, A. Wirth, and A. C. C. Yao, Informational complexity and the direct sum problem for simultaneous message complexity. In *FOCS*, pp. 270–278, IEEE, 2001.

[26] E. Chattopadhyay and D. Zuckerman, Explicit two-source extractors and resilient functions. In *STOC*, pp. 670–683, ACM, 2016.

[27] I. Dinur and D. Steurer, Analytical approach to parallel repetition. In *STOC*, pp. 624–633, ACM, 2014.

[28] K. Efremenko, G. Kol, and R. Saxena, Binary interactive error resilience beyond $1/8$ (or why $(1/2)^3 > 1/8$). In *FOCS*, pp. 470–481, IEEE, 2020.

[29] R. M. Fano, *The transmission of information*. Massachusetts Institute of Technology, Research Laboratory of Electronics, 1949.

[30] A. Ganor, G. Kol, and R. Raz, Exponential separation of information and communication. In *FOCS*, pp. 176–185, IEEE, 2014.

[31] A. Ganor, G. Kol, and R. Raz, Exponential separation of information and communication for boolean functions. *J. ACM* **63** (2016), no. 5, 46:1–46:31.

[32] A. Ganor, G. Kol, and R. Raz, Exponential separation of communication and external information. *SIAM J. Comput.* **50** (2021), no. 3.

[33] A. Grothendieck, Résumé de la Théorie Métrique des Produits Tensoriels Topologiques. *Bol. Soc. Mat. São Paulo* **8** (1953), 1–79.

[34] M. Gupta and R. Y. Zhang, The optimal error resilience of interactive communication over binary channels. In *STOC*, pp. 948–961, ACM, 2022.

[35] D. A. Hufman, A method for the construction of minimum redundancy codes. *Proc. IRE* **40** (1952), no. 9, 1098–1101.

[36] B. Kalyanasundaram and G. Schnitger, The probabilistic communication complexity of set intersection. *SIAM J. Discrete Math.* **5** (1992), no. 4, 545–557.
[37] G. Kol, Interactive compression for product distributions. In *STOC*, pp. 987–998, ACM, 2016.
[38] J. L. Krivine, Constantes de Grothendieck et Fonctions de Type Positif sur les Spheres. *Adv. Math.* **31** (1979), no. 1, 16–30.
[39] N. Linial, Y. Mansour, and N. Nisan, Constant depth circuits, Fourier transform, and learnability. *J. ACM* **40** (1993), no. 3, 607–620.
[40] N. Linial and N. Nisan. Approximate inclusion-exclusion. *Combinatorica* **10** (1990), no. 4, 349–365.
[41] N. Ma and P. Ishwar, The infinite-message limit of two-terminal interactive source coding. *IEEE Trans. Inf. Theory* **59** (2013), no. 7, 4071–4094.
[42] A. Rao and M. Sinha, Simplified separation of information and communication. *Theory Comput.* **14** (2018), no. 1, 1–29.
[43] R. Raz, A parallel repetition theorem. *SIAM J. Comput.* **27** (1998), no. 3, 763–803.
[44] A. A. Razborov, Lower bounds on the size of bounded-depth networks over a complete basis with logical addition. *Math. Notes Acad. Sci. USSR* **41** (1987), no. 4, 333–338.
[45] A. A. Razborov, On the distributional complexity of disjointness. *Theoret. Comput. Sci.* **106** (1992), no. 2, 385–390.
[46] A. A. Razborov, A simple proof of Bazzi's theorem. *ACM Trans. Comput. Theory* **1** (2009), no. 1, 3:1–3:5.
[47] L. J. Schulman, Communication on noisy channels: a coding theorem for computation. In *FOCS*, pp. 724–733, IEEE, 1992.
[48] L. J. Schulman, Deterministic coding for interactive communication. In *STOC*, pp. 747–756, ACM, 1993.
[49] L. J. Schulman, Coding for interactive communication. *IEEE Trans. Inf. Theory* **42** (1996), no. 6, 1745–1756.
[50] C. E. Shannon, A mathematical theory of communication. *Bell Syst. Tech. J.* 1948.
[51] A. A. Sherstov, Compressing interactive communication under product distributions. *SIAM J. Comput.* **47** (2018), no. 2, 367–419.
[52] R. Smolensky, Algebraic methods in the theory of lower bounds for boolean circuit complexity. In *STOC*, pp. 77–82, ACM, 1987.
[53] A. Tal, Tight bounds on the Fourier spectrum of AC0. In *Computational Complexity Conference*, pp. 15:1–15:31, Schloss Dagstuhl - Leibniz-Zentrum für Informatik, 2017.
[54] A. C. C. Yao, Some complexity questions related to distributive computing (preliminary report). In *STOC*, pp. 209–213, ACM, 1979.

RAN RAZ

Department of Computer Science, Princeton University, Princeton, NJ 08540, USA, ranr@princeton.edu

THE WORK OF BARRY MAZUR

HENRI DARMON

ABSTRACT

Barry Mazur is awarded the 2022 Chern Medal "for his profound discoveries in topology, arithmetic geometry and number theory, and his leadership and generosity in forming the next generation." This *laudatio* surveys some of the highlights of Mazur's remarkable mathematical career.

MATHEMATICS SUBJECT CLASSIFICATION 2020

01A70

KEYWORDS

Schoenflies conjecture, primes, knots, elliptic curves, modular curves, modular forms, Eisenstein ideal, Galois representations, deformation theory, eigencurves, Fermat's last theorem, Iwasawa theory, p-adic L-functions, Euler systems, rational points

INTERNATIONAL CONGRESS OF MATHEMATICIANS 2022 JULY 6–14

© 2022 International Mathematical Union

Proc. Int. Cong. Math. 2022, Vol. 1, pp. 118–141
DOI 10.4171/ICM2022/217

Published by EMS Press and licensed under a CC BY 4.0 license

Barry Mazur was born in 1937 in New York City. After graduating from the Bronx High School of Science in 1954, he completed his undergraduate studies at MIT in just two years, and his PhD at Princeton University in a further two years, during which he also spent a semester in Paris, attending, among others, the seminars of Cartan and Chevalley. After a one-year stint at the Institute for Advanced Study, he joined the faculty of the Mathematics Department at Harvard in 1959, first as a member of Harvard's Society of Fellows, and currently, as the Gerhard Gade University Professor.

Through a remarkable career spanning over six decades at Harvard alone, Barry Mazur has profoundly influenced the scientific outlooks of generations of graduate students, postdoctoral fellows, and colleagues. He has shaped the modern landscape of number theory by successfully tackling the most difficult problems in the area, laying the groundwork for important theories, and initiating legions of disciples to fertile new perspectives. His scientific achievements place him squarely among the greatest mathematicians of the 20th century. The following report touches on a few of the topics, in roughly chronological order, where Barry Mazur has had a transformative impact.

1. GEOMETRIC AND DIFFERENTIAL TOPOLOGY

(References: **[1–19]**.)

Barry Mazur's earliest contributions were to the field of geometric topology and differential geometry. His 1959 PhD thesis at Princeton **[4]** caused a sensation by proving the *generalised Schoenflies conjecture*, a higher-dimensional generalization of the Jordan curve theorem. It asserts that an $(n-1)$-sphere S embedded in the n-sphere S^n in a way that extends to an embedding of a small thickening of S can be mapped to the standard $(n-1)$-sphere by a homeomorphism of S^n **[4–6]**. The necessity of some regularity hypotheses on the embedding is illustrated by well-known counterexamples like the Alexander horned sphere. One of Mazur's ingenious ideas in the proof is the eponymous "swindle," which demonstrates that the connected sum of two nontrivial knots or manifolds is necessarily nontrivial. The seductively simple argument is based on the fact that infinite connected sums make rigorous sense in the setting of "wild knots"; if K_1 and K_2 are knots or manifolds for which $K_1 + K_2$ is trivial, then

$$K_1 = K_1 + (K_2 + K_1) + (K_2 + K_1) + \cdots = (K_1 + K_2) + (K_1 + K_2) + \cdots = 0,$$

and likewise for K_2. Mazur was awarded the Oswald Veblen Prize of the AMS with Morton Brown in 1966 for his work on the generalized Schoenflies conjecture.

Among other key notions, Mazur also discovered, independently and at roughly the same time as Valentin Poenaru, what are now commonly referred to in the literature as "Mazur manifolds" or "Poenaru–Mazur manifolds" **[7]**: compact, contractible, smooth four-manifolds with boundary which are not diffeomorphic to the standard four-ball.

Mazur's article **[15]** on dynamical systems, in collaboration with Michael Artin, studies the space $\mathcal{F}$ of k-differentiable self-maps on a compact differentiable manifold M, equipped with the suitable (C^k) topology, and proves that there is a dense subset of $\mathcal{F}$

consisting of maps whose number of isolated periodic points of period n grows at most exponentially with n. The proof is obtained by invoking an approximation theorem of Nash to reduce to an analogous statement for real algebraic varieties, which can then be tackled with the methods of intersection theory of algebraic cycles.

2. ALGEBRAIC GEOMETRY

(References: [15, 20–29].)

With its appealing blend of differential and algebraic methods, [15] marked a gradual widening of Mazur's mathematical interests to encompass algebraic geometry at a time when the subject was experiencing a profound renewal under the impetus of the Grothendieck school. It is during this period, in the 1960s and early 1970s, that Mazur produced a number of seminal works in algebraic geometry, nourished by regular visits to the IHES.

His articles [20] and [21] study the interplay between the Frobenius operator and the Hodge filtration on the de Rham cohomology of a variety V over $\mathbb{Q}_p$ admitting a smooth model over $\mathbb{Z}_p$. It establishes the fundamental "Mazur inequality," originally conjectured by Nick Katz [132], asserting that "the Newton polygon lies above the Hodge polygon." The Newton polygon measures the slopes, or valuations at p, of the eigenvalues the Frobenius endomorphism acting on the ith cristalline cohomology of V, or equivalently, of the canonical lift of Frobenius to the de Rham cohomology of V over $\mathbb{Q}_p$. The Hodge polygon encodes the dimensions of the successive quotients of this de Rham cohomology relative to the Hodge filtration. The latter invariants were classically calculated via complex transcendental methods, by studying the Hodge decomposition on the de Rham cohomology of varieties over $\mathbb{C}$. Mazur's inequality captures a fundamental feature of the behavior of the algebraic de Rham cohomology of a variety under "mod p reduction," and provides subtle p-adic information about the zeta-functions of varieties over finite fields of characteristic p.

Mazur's treatise [22] with Messing on cristalline cohomology represents a foundational contribution to the study of p-adic cohomology theories. This subject has gradually emerged as a powerful tool for understanding the p-adic representations of the Galois groups of p-adic fields that arise from the étale cohomology of algebraic varieties. It has been vigorously developed in the past decades and acquired a growing importance in number theory, notably in the theory of motives and in the Langlands program.

Some of Mazur's later contributions incorporating perspectives from p-adic Hodge theory shall be evoked in greater detail below, most notably, in §9, his celebrated conjecture with Jean-Marc Fontaine characterising the global p-adic Galois representations realized in the p-adic étale cohomology of varieties over number fields. The theory of p-adic periods also plays a key role in extending to higher weight modular forms the definition of the $\mathcal{L}$-*invariant* of Mazur, John Tate, and Jeremy Teitelbaum arising in the leading terms of certain p-adic L-functions in the presence of an "exceptional zero" (cf. §8).

Another notable achievement from roughly this period is the article [28] with M. Artin laying the foundations for a homotopy theory for schemes, based on the étale topology which had been introduced less than a decade earlier and has since come to play a central role in arithmetic geometry.

3. ARITHMETIC TOPOLOGY

(Reference: [30].)

In his gradual transition from topology and geometry to number theory, Mazur seems to have drawn guidance and inspiration from a suggestive analogy between knots and primes.

A knot is a copy of the circle S^1 embedded in a three-sphere S^3. Many invariants of knots arise from studying the fundamental group of the knot complement. There is a beautiful and tantalizing parallel between this knot complement and the complement of a prime in the scheme $\mathrm{Spec}(\mathbb{Z})$. Namely, the latter space shares some of the same homological properties as S^3 insofar as its interesting cohomology is concentrated in degree 3, whereas $\mathrm{Spec}(\mathbb{F}_p)$ behaves like a circle since its fundamental group is (topologically pro-) cyclic.

The pursuit of this analogy leads to a beguiling dictionary between number theory and knot theory, in which quadratic reciprocity resonates with the symmetry of the linking number of two knots, and the higher quadratic residue symbols of Redei can be envisaged as analogues of the higher linking of knot configurations like the famous Borromean rings, both notions being manifestations of higher Massey products.

Mazur's unpublished but widely influential manuscript [30] enriches the number theory-knot theory lexicon by explicating the parallel between the Alexander polynomial of a knot and Iwasawa's conjectural algebraic description of the Kubota–Leopoldt p-adic zeta-function as the characteristic power series of a certain Iwasawa module constructed out of ideal class groups of p-power cyclotomic fields. The Iwasawa module in question can be identified via global class field theory with the maximal abelian (pro-p) extension of the maximal abelian extension of $\mathbb{Q}$ ramified only at p. It can then be understood as the second graded piece relative to a natural filtration on (the pro-solvable completion of) the fundamental group of the complement of $\mathrm{Spec}(\mathbb{F}_p)$ in $\mathrm{Spec}(\mathbb{Z})$. Iwasawa's interpretation of the p-adic zeta-function resembles the Alexander polynomial of a knot K, which encodes the characteristic polynomial of a generator of the homology of the knot complement acting on the next graded piece in the filtration of $\pi_1(S^3 - K)$ given by its derived central series.

The rich analogy between knots and primes which guided Mazur in his transition from topology to number theory has subsequently spawned an entire new field, known as *arithmetic topology*, which is elegantly described in the recent textbook of Masanori Morishita [138]. (See also [146] for further striking manifestations of the analogy.)

4. TORSION SUBGROUPS OF ELLIPTIC CURVES

(References: [32–38, 122].)

The deep and systematic study of rational torsion points on elliptic curves carried out in roughly the decade from 1975 to 1985 stands among Mazur's landmark contributions to number theory.

An elliptic curve over a field F is a smooth projective curve E of genus one over F equipped with a distinguished rational point $O \in E(F)$. What makes these curves particularly rich arithmetically is that they are endowed with the structure of a projective *algebraic group*. In particular, the set $E(\mathbb{Q})$ of rational points on an elliptic curve over $\mathbb{Q}$ is an abelian group, known to be finitely generated by the Mordell–Weil theorem, and thus is isomorphic to

$$E(\mathbb{Q}) = \mathbb{Z}^r \times T,$$

where T is a finite group, called the *torsion subgroup* of E over $\mathbb{Q}$. Mazur's celebrated theorem [36] lists all the possibilities for the groups T that can arise in this way:

Theorem 4.1. *The torsion subgroup T of an elliptic curve over $\mathbb{Q}$ can only be isomorphic to one of the following* 15 *groups:*

$$\mathbb{Z}/n\mathbb{Z}, \ \textit{with}\ 1 \le n \le 10 \ \textit{or}\ n = 12, \quad \mathbb{Z}/2\mathbb{Z} \times \mathbb{Z}/n\mathbb{Z} \ \textit{with}\ n = 2, 4, 6, \ \textit{or}\ 8.$$

This striking result was apparently anticipated by the Italian geometer Beppo Levi [142] in 1908. It became more widely known as a precise conjecture formulated by Andrew Ogg [139] and provides the backdrop for an active area of investigation to which mathematicians like Kamienny [130], Merel [135], and many others, have made important subsequent contributions. Indeed, the study of rational points on modular curves remains a lively terrain of investigation to which a variety of approaches grounded in the pioneering insights of [36] have been applied (cf. [127, 128, 133, 134, 136, 137], ...).

Beyond the appealing nature of the final statement "for its own sake," the perspectives that Mazur introduced into the subject in order to prove Theorem 4.1 also had a tremendous impact on other related developments. Both the statement and the proof of Theorem 4.1 are indispensable ingredients in the proof of the modularity of elliptic curves and of Fermat's Last Theorem, as will be explained further in Sections 5, 6, and 10.

In a subsequent article [37], Mazur also classifies the primes N for which there are elliptic curves over $\mathbb{Q}$ possessing a rational subgroup of order N, i.e., a nontrivial isogeny of degree N defined over $\mathbb{Q}$, simplifying his earlier proof of Theorem 4.1 at the same time:

Theorem 4.2. *Let N be a prime number such that some elliptic curve admits an isogeny of degree N defined over $\mathbb{Q}$. Then $N = 2, 3, 5, 7, 13$ (with infinitely many possible E for each N) or $N = 11, 17, 19, 37, 43, 67,$ or* 163.

The values $N = 11, 17, 19, \ldots, 163$ are primes for which the imaginary quadratic field $\mathbb{Q}(\sqrt{-N})$ has class number one. Elliptic curves with complex multiplication by the maximal orders of these fields admit models over $\mathbb{Q}$ and the kernel of multiplication by $\sqrt{-N}$ gives a cyclic subgroup of order N in E, defined over $\mathbb{Q}$. It is a measure of the delicacy of Mazur's argument that it accounts for these arithmetically nontrivial exceptions while ruling out all other eventual occurrences.

Theorem 4.1 has been extended by Sheldon Kamienny, leading to the classification of possible torsion subgroups for elliptic curves defined over number fields of small degree over $\mathbb{Q}$ (cf. [130] and [38]). The most definitive result in this direction was then obtained by Loïc Merel [135], who showed that the torsion subgroups of elliptic curves defined over a number field K are bounded by a constant B_K depending only on K, and indeed, only on the degree of K over $\mathbb{Q}$.

5. RATIONAL POINTS ON MODULAR CURVES

Theorems 4.1 and 4.2 can be recast in terms of rational points on *modular curves*, which arise naturally as *moduli spaces* parametrizing isomorphism classes of elliptic curves with auxiliary "level structures."

If E is an elliptic curve over a field F in which 6 is invertible, there are two rational functions x and y which are regular on $E - \{O\}$, have poles of order 2 and 3, respectively, at O, and satisfy an equation of the form

$$y^2 = x^3 + ax + b, \quad \text{with } a, b \in F.$$

The functions x and y are uniquely determined by these properties up to replacing (x, y) by (t^2x, t^3y) for some $t \in F^\times$, which has the effect of replacing the coefficients (a, b) by (t^4a, t^6b). In particular, the expression

$$j(E) := 1728\frac{4a^3}{4a^3 + 27b^2},$$

known as the j-invariant of E, depends only on (the $\bar{F}$-isomorphism class of) E and not on the choice of x and y. It is in fact a complete isomorphism invariant: two elliptic curves over F are isomorphic (over the algebraic closure of F) precisely when they have the same j-invariant. The affine j-line, viewed as an algebraic curve over $\mathbb{Q}$, is thus a (coarse) moduli space of elliptic curves: its points over any field F of characteristic zero are in bijection with the $\bar{F}$-isomorphism classes of elliptic curves over F. This affine j-line is the simplest instance of a *modular curve*.

More interesting examples can be obtained by classifying elliptic curves *with extra level structure*. A typical level N structure on E amounts to the datum of a subgroup or a point of order N on E, or eventually a basis for the full N-torsion of E. The curves that classify solutions of these problems are commonly denoted $Y_0(N)$, $Y_1(N)$, and $Y(N)$, respectively. They are affine curves over $\mathbb{Q}$, which can be completed to smooth projective curves by adjoining to them a finite set of cusps: the resulting projective curves are called $X_0(N)$, $X_1(N)$, and $X(N)$.

For example, any elliptic curve admits a degree 2 map π to $\mathbb{P}_1$ which is ramified precisely at the set $E[2]$ of its two-torsion points. The extra datum of a basis (P_1, P_2) for $E[2]$ over F can be used to rigidify the choice of π by requiring that

$$\pi(P_1) = 0, \quad \pi(P_2) = 1, \quad \pi(O) = \infty.$$

The invariant $\lambda := \pi(P_1 + P_2) \in F - \{0, 1\}$ determines the triple (E, P_1, P_2) uniquely up to isomorphism over $\bar{F}$, and the assignment $(E, P_1, P_2) \mapsto \lambda$ gives an identification

$$Y(2) = \mathbb{P}_1 - \{0, 1, \infty\}, \tag{5.1}$$

in which $\lambda \in \mathbb{P}_1$ corresponds to the Legendre elliptic curve $y^2 = x(x-1)(x-\lambda)$ with basis $((0,0),(1,0))$ for its 2-division points.

The following is merely a reformulation of Theorem 4.2 from the perspective of rational points on modular curves:

Theorem 5.1. *Let N be a prime number for which $Y_0(N)(\mathbb{Q})$ is nonempty. Then $N = 2, 3, 5, 7, 13$ (when $X_0(N)$ is isomorphic to the projective line, and has infinitely many rational points) or $N = 11, 17, 19, 37, 43, 67,$ or 163 (when $Y_0(N)$ contains a finite set of "sporadic" rational points).*

Concrete (but ultimately not very useful) equations for modular curves can be written down. If E and E' are related by a cyclic isogeny of degree N, then their j-invariants j and j' give rise to a root (j, j') of the so-called *modular polynomial* $\Phi_N(x, y)$, which is a rational polynomial of bidegree $N+1$ when N is a prime number. The curve $Y_0(N)$ is birationally equivalent to the plane curve defined by this polynomial. These defining equations tend to be quite complicated. For instance,

$$\begin{aligned}\Phi_2(x,y) = {} & x^3 - x^2y^2 + 1488x^2y - 162000x^2 + 1488xy^2 + 40773375xy \\ & + 8748000000x + y^3 - 162000y^2 + 8748000000y - 157464000000000,\end{aligned}$$

and tackling the associated diophantine equations through a direct elementary approach seems decidedly unpromising.

Mazur's opening gambit is to embed the modular curve – $X_0(N)$, say – in its *Jacobian* $J_0(N)$, an abelian variety whose rational points can then be studied through Fermat's method of infinite descent, in the conceptual modern framework given for it by André Weil, in which the consideration of explicit equations can largely be avoided.

Mazur is able to show that if N is a prime for which $J_0(N)$ is nontrivial (i.e., if $N = 11$ or $N > 13$) then this Jacobian admits nontrivial quotients with finite Mordell–Weil group over $\mathbb{Q}$, which he calls *Eisenstein quotients*. This immediately implies, a decade before Faltings' proof of the Mordell conjecture, that $X_0(N)$ has finitely many rational points whenever it has genus ≥ 1, and, with more care, can be used to derive bounds on the set of rational points sufficiently precise to deduce Theorem 4.1, and, with even greater care, Theorem 4.2.

The Eisenstein quotients of $J_0(N)$ are attached to the different primes p dividing the numerator of $(N-1)/12$, and denoted $J_{\text{eis}}^{(p)}(N)$. The Mordell–Weil group $J_{\text{eis}}^{(p)}(N)(\mathbb{Q})$ contains an element of order p, and it becomes natural to calculate this Mordell–Weil group by a p-descent argument involving the Selmer group for a p-torsion module on which the Galois group of $\mathbb{Q}$ acts through an abelian quotient. The "Eisenstein descent" which Mazur developed for this purpose thus places the study of $J_{\mathrm{eis}}^{(p)}(\mathbb{Q})$ in proximity with more classical questions surrounding the class groups of cyclotomic fields.

In constructing $J_{\mathrm{eis}}^{(p)}$ and establishing the finiteness of its Mordell–Weil group, Mazur is able to marshal several special features of modular curves that make their diophantine properties more amenable to analysis. Most critically, modular curves are endowed with a plentiful supply of algebraic correspondences over $\mathbb{Q}$, which emerge naturally from their moduli description and are geometric incarnations of *Hecke operators*. The resulting endomorphisms break up $J_0(N)$ into arithmetically simpler pieces with a large endomorphism algebra, whose Tate modules give rise to (compatible systems of) two-dimensional ℓ-adic representations of $\mathrm{Gal}(\bar{\mathbb{Q}}/\mathbb{Q})$. These abelian variety quotients "of $\mathbf{GL}(2)$ type" offer a fertile testing ground for the general program of understanding linear representations of the Galois groups of number fields, a cornerstone of the Langlands program. The two-dimensional representations of $\mathrm{Gal}(\bar{\mathbb{Q}}/\mathbb{Q})$ represent a prototypical first step in this program, going beyond the abelian setting of global class field theory. It is partly for this reason that Mazur's Eisenstein descent has largely transcended in importance the diophantine application for which it was originally designed. The ideas Mazur introduced into the subject have played a key role, notably, in Andrew Wiles' proof **[148]** almost 20 years later of the Taniyama–Weil conjecture on the modularity of elliptic curves over $\mathbb{Q}$, as will be explained further below.

The nontrivial point of order p on $J_0(N)$ which Mazur so spectacularly exploits in his proofs of Theorems 4.1 and 4.2 arises from the image of a divisor supported on the cusps of $X_0(N)$. In addition to the cusps, the modular curve $X_0(N)$ is also endowed with a plentiful supply of points defined over various ring class fields of imaginary quadratic fields – the *Heegner points* arising from the moduli of suitable elliptic curves with complex multiplication. A formula of Benedict Gross and Don Zagier connects the heights of these points to the first derivatives of the Hasse–Weil L-series of abelian variety quotients of $J_0(N)$. In the late 1980s, Victor Kolyvagin parlayed this connection into a proof of the finiteness of the Mordell–Weil group of any quotient of $J_0(N)$ whose Hasse–Weil L-series does not vanish at the center, consistent with the Birch and Swinnerton-Dyer conjecture for these quotients. The somewhat larger quotient of $J_0(N)$ with finite Mordell–Weil group that emerges from Kolyvagin's theorem is called the *winding quotient* (a terminology that can be traced back to Mazur's "winding element" **[42]**). The winding quotient was later exploited to great effect by Merel in his extension of Theorem 4.1 to number fields of arbitrary degree **[135]**.

6. FERMAT'S LAST THEOREM

Mazur's Theorem 5.1 on rational points on modular curves asserts that an infinite collection of curves, of increasing genus and arithmetic complexity – the modular curves $X_0(N)$ indexed by the parameter N – have no rational points except the trivial ones when N is large enough. This statement is reminiscent of Fermat's Last Theorem, which makes the same assertion for the Fermat curves F_N with equation

$$F_N : x^N + y^N = z^N.$$

The relation between the two statements goes far beyond a superficial analogy. Theorem 5.1 turns out to be a critical ingredient – indeed, the key diophantine ingredient – in the proof of Fermat's Last Theorem.

The tight connection between the diophantine properties of modular curves and Fermat curves can seem surprising at first, since only rarely are there explicit maps between the two types of curves. A charming exception to this statement is the modular curve $X(7)$ with full level 7 structure, a genus 3 curve having a maximal size automorphism group for its genus, the group $\mathbf{PSL}(2,7)$ of order 168. This property determines it uniquely up to isomorphism over $\bar{\mathbb{Q}}$, and a model for it is provided by the famous *Klein quartic* with equation

$$X(7) : u^3 v + v^3 w + w^3 u = 0.$$

It turns out that $X(7)$ is the image of the Fermat curve

$$F_7 : x^7 + y^7 + z^7 = 0$$

under the degree 7 map $\pi : F_7 \to X(7)$ sending $(x, y, z) \in F_7$ to

$$(u, v, w) = \pi(x, y, z) := (x^3 z, y^3 x, z^3 y).$$

A nontrivial solution to Fermat's Last Theorem would thus give rise to a nontrivial rational point on $X(7)$, and the assertion that this modular curve has no nontrivial rational points (satisfying $uvw \neq 0$) therefore implies Fermat's last theorem for exponent 7. More interesting is the converse implication that was first proved by Hurwitz, namely, that $X(7)$ has no nontrivial rational points because the same is true for F_7. (At the time, Fermat's Last Theorem for exponent 7 was already known through the work of Lamé.) Hurwitz notes that if (u, v, w) is a point on the Klein quartic with integer coordinates, satisfying $\gcd(u, v, w) = 1$, then these coordinates need not be pairwise coprime. Setting

$$x = \gcd(u, v), \quad y = \gcd(v, w), \quad z = \gcd(w, u),$$

a direct reasoning involving the fundamental theorem of arithmetic shows (after changing the signs of x, y, and/or z if necessary) that (x, y, z) lies on the Fermat curve F_7 and that $\pi(x, y, z) = (u, v, w)$. Through this argument, Hurwitz shows that the map $\pi : F_7 \to X(7)$ is surjective on rational points. Unlike the purely algebraic implication

$$F_7 \text{ has a non-trivial rational point} \quad \Rightarrow \quad X(7) \text{ has a nontrivial rational point,} \tag{6.1}$$

the reverse implication is more genuinely arithmetic, resting on ingredients like unique factorization. Essential for this implication is the fact that the degree 7 map π (viewed as a map of Riemann surfaces, on the complex points of the curves, say) is *everywhere unramified*.

The proof of Fermat's Last Theorem for the general (prime) exponent p rests on an analogous but substantially more general geometric relation between the modular curve $X(2p)$ and the pth Fermat curve F_p. Namely, both are equipped with natural surjective maps

$$F_p \xrightarrow{\pi_1} \mathbb{P}_1 \xleftarrow{\pi_2} X(2p)$$

to the projective line $\mathbb{P}_1$ with "common local features." The map π_1 sends the Fermat triple (x, y, z) to x^p / y^p, and the map π_2 is simply the one that "forgets about the level p structure," sending a point on $X(2p)$ to its natural image in $X(2)$, identified with the projective line via the identification in (5.1).

Although they have different degrees and are defined on different curves, the maps π_1 and π_2 exhibit the following striking affinity: they are both ramified only at 0, 1, and ∞, and their ramification degrees at these three points are equal to p. This suggests that, if $(a, b, c) \in F_p(\mathbb{Q})$ is a nontrivial solution to Fermat's Last Theorem, then the image $\pi_1(a, b, c) = a^p/b^p \in \mathbb{P}_1(\mathbb{Q})$ ought to lift to a point of $X(2p)$ whose field of definition exhibits a *limited amount of ramification*, bounded independently of the solution (a, b, c). One is led to study the field generated by the p-division points of the "Frey elliptic curve"

$$E_{a,b,c} : y^2 = x(x - a^p)(x + b^p),$$

which is indeed (after eventually re-ordering a, b and c appropriately, and modifying their signs) unramified outside of 2 and p.

The ultimate proof of Fermat's Last Theorem rests on a supremely delicate analysis of this field, or, better yet, of the $\mathbb{Z}/p\mathbb{Z}$-linear representation

$$\varrho_{a,b,c} : G_{\mathbb{Q}} := \mathrm{Gal}(\bar{\mathbb{Q}}/\mathbb{Q}) \longrightarrow \mathrm{Aut}\big(E_{a,b,c}[p]\big) \simeq \mathbf{GL}_2(\mathbb{Z}/p\mathbb{Z})$$

of the absolute Galois group of $\mathbb{Q}$ acting on the p-torsion points of $E_{a,b,c}$. The startling insight that emerged from the work of Gerhard Frey, Jean-Pierre Serre **[143]**, and Kenneth Ribet **[141]** is that the modularity of $E_{a,b,c}$, which was ultimately proved by Wiles **[148]**, can be parlayed into the conclusion that $\varrho_{a,b,c}$ is necessarily *reducible*. Because of this, any nontrivial solution $(a, b, c) \in F_p(\mathbb{Q})$ to Fermat's Last Theorem can be transferred to a nontrivial rational point on $X_0(p)$, by chasing it through the following diagram of maps of curves:

$$\begin{array}{ccccc} F_p & & X(2p) & \longrightarrow & X_0(p) \\ & \searrow^{\pi_1} & \swarrow_{\pi_2} & & \\ & \mathbb{P}_1 & & & \end{array}$$

Thanks to the implication

$$F_p \text{ has a nontrivial rational point} \quad \Rightarrow \quad X_0(p) \text{ has a nontrivial rational point} \qquad (6.2)$$

(which is reminiscent of (6.1), is even closer in spirit to its converse, and is considerably deeper than either statement), a Diophantine question about the Fermat curves F_p is reduced to the same question about the modular curves $X_0(p)$: precisely the question that is answered in Mazur's Theorem 5.1.

As will be further explained in Section 10, the ideas that Mazur introduced to prove Theorem 5.1 are also instrumental in the the proof of (6.2): they are thus woven into the very fabric of Wiles' extraordinary proof of the Taniyama–Weil conjecture and of Fermat's Last Theorem.

7. IWASAWA MAIN CONJECTURES

(References: [45, 47, 51].)

The proof of the main conjecture of Iwasawa theory by Mazur and Wiles [47] is another milestone of number theory, occurring roughly a decade before the proof of Fermat's Last Theorem. Iwasawa theory starts with the fact that the p-parts of the ideal class groups of the p-power cyclotomic fields, obtained by adjoining to $\mathbb{Q}$ the p^nth roots of unity, exhibit a remarkably regular growth as a function of n. The main conjecture of Iwasawa theory ties this behavior to the zeroes of the Kubota–Leopoldt p-adic zeta-function. It grew out of an analogy with Weil's formulation of the Riemann hypothesis for varieties over finite fields, and can be envisaged as its counterpart in a p-adic setting, insofar as it assigns to the mysterious zeroes of the p-adic zeta function a *spectral interpretation*. Namely these zeroes are the eigenvalues of a certain operator – a topological generator of the Galois group of the cyclotomic $\mathbb{Z}_p$-extension generated by all p-power roots of unity – acting on an Iwasawa module formed by piecing together the ideal class groups of the finite layers of this $\mathbb{Z}_p$-extension. A remarkable feature of the proof of Mazur and Wiles is that it rests on a careful study of the two-dimensional Galois representations arising from the quotients of the Jacobians of modular curves, particularly those that are *reducible*, to prove a statement that is ostensibly part of the more classical abelian theory of class groups of cyclotomic fields. Global class field theory is used to convert questions about class groups into ones about constructing unramified abelian extensions of cyclotomic fields, and the extensions that are predicted to arise from the zeroes of the p-adic zeta function are ultimately shown to be cut out by the Galois representations arising from the p-power torsion points of these modular Jacobians.

The proof of the Iwasawa Main conjecture – justifying the analogy between the p-adic zeta function of Kubota–Leopoldt and the Alexander polynomial of a knot which Mazur had perceived decades earlier – stands as one of the notable achievements in number theory in the latter half of the 20th century. Its method has been vastly generalized, notably by Wiles for totally real fields [147], and by Chris Skinner and Eric Urban [144] in the setting of elliptic curves, a framework which also owes much to Mazur's vision and will be discussed in the following section.

8. ELLIPTIC CURVES AND THE BIRCH AND SWINNERTON-DYER CONJECTURE

(References: [40–44, 46, 48, 50, 52, 56, 58, 101].)

Throughout the 1970s and 1980s, Mazur reflected extensively on the arithmetic of elliptic curves, focusing on the most notoriously difficult and central open problem in the area, namely the Birch and Swinnerton-Dyer conjecture. Rather than tackling the problem head-on, he initiated a parallel study in the p-adic setting, opening up a new terrain of investigation which has been remarkably fruitful and witnessed decades of sustained progress.

The article [41] champions the introduction of Iwasawa-theoretic ideas in the arithmetic study of elliptic curves and abelian varieties. The relevant Iwasawa modules are obtained by replacing the p-parts of ideal class groups with relevant p-Selmer groups over the finite layers of a $\mathbb{Z}_p$-extension, appropriately pieced together. The importance of this new perspective can hardly be overstated: entire mathematical careers (the author's among them) have been enjoyably spent fleshing out Mazur's vision for the Iwasawa theory of abelian varieties over towers of number fields.

Mazur's article [42] with Peter Swinnerton-Dyer introduces what has since come to be known as the Mazur–Swinnerton-Dyer p-adic L-function of an elliptic curve over $\mathbb{Q}$, the direct counterpart of the Hasse–Weil L-function in the p-adic world. Relating analytically defined p-adic L-functions like this one to the characteristic power series of Mazur's Iwasawa modules leads to a rich variety of "Iwasawa main conjectures" for elliptic curves.

The foundations that are laid in [41] and [42] lead naturally to a *p-adic analogue* of the Birch and Swinnerton-Dyer conjecture, which was formulated roughly ten years later in a profoundly influential article [50] by Mazur, Tate, and Teitelbaum.

The p-adic Birch and Swinnerton-Dyer conjecture is more tractable than its archimedean precursor, because of the tight connection one can hope to establish between p-adic L-functions and Mazur's Iwasawa modules, as expressed in the main conjecture. The main conjecture explains why elliptic curves of large rank, for example, ought to exhibit high order zeroes in their associated p-adic L-functions: it is because the Mordell–Weil group provides a subspace of the relevant Iwasawa module that is fixed by Galois and thus contributes to the multiplicity of the trivial character as a zero of the p-adic L-function.

Such a spectral interpretation is sorely lacking for the zeroes of the Hasse–Weil L-function in the archimedean setting, and indeed there is not a single elliptic curve over $\mathbb{Q}$ whose L-series can be shown to vanish to order > 3 at $s = 1$, although elliptic curves of rank > 3 (and even > 23) are known to exist in relative abundance.

In the non-archimedean framework that Mazur pioneered, the situation is better understood. The requisite divisibility in the main conjecture for elliptic curves over $\mathbb{Q}$ was shown by Kazuya Kato in the early 1990s by exploiting, much as Kolyvagin with Heegner points, special elements in the K-theory of modular curves arising from pairs of Siegel units and (crucially) their p-adic deformations [131]. Thanks to Kato's result, the Mazur–Swinnerton-Dyer p-adic L-function of an elliptic curve is known to vanish to order at least the rank of the Mordell–Weil group.

The opposite divisibility in the main conjecture for elliptic curves was established by Skinner and Urban [144], by building on the very different circle of ideas that arose in the proof of the original Iwasawa main conjecture by Mazur and Wiles. Significant mysteries relating to the finiteness of the Shafarevich-Tate group and nondegeneracies in p-adic heights (and the eventual nonsemisimplicity of the relevant Iwasawa modules) still prevent this divisibility in the main conjecture from leading to the correct upper bound on the order of vanishing of the p-adic L-function. So the p-adic Birch and Swinnerton-Dyer conjecture of Mazur, Tate, and Teitelbaum still offers alluring mysteries in spite of its relative accessibility compared to the original archimedean conjecture.

Another appealing feature of the p-adic Birch and Swinnerton-Dyer conjecture is the appearance of new phenomena that seem to have no immediate counterpart in the archimedean setting, most notably, the phenomenon of *exceptional zeroes* of p-adic L-functions that can arise, for instance, from the vanishing of an Euler factor at p that needs to be inserted to ensure the interpolation of the special values. This phenomenon was first observed and explored in [50]. While they may appear somewhat specialized to the uninitiated, leading terms of p-adic L-functions at points where there is an exceptional zero encode rich arithmetic information, and their careful examination is often rewarded with fruitful new insights. The original "exceptional zero conjecture" of Mazur, Tate, and Teitelbaum involved the Tate p-adic period of an elliptic curve with multiplicative reduction. A series of suggestive proposals have been formulated to extend this conjecture to modular forms of higher weight, notably by Jeremy Teitelbaum [145] in terms of the Cerednik–Drinfeld theory of p-adic uniformization of Shimura curves, and by Fontaine and Mazur [67], exploiting the filtered Frobenius monodromy module which p-adic Hodge theory attaches to the local p-adic Galois representation of a modular form of higher weight. As a further instance of the importance of exceptional zeros, let us mention that they also sometimes arise in p-adic L-series attached to totally odd characters of totally real fields at $s = 0$, where they are central to Gross's p-adic variant of the Stark conjectures.

Towards the end of the 1980s, Mazur also introduced, in collaboration with John Tate, a *tame refinement* of the p-adic Birch and Swinnerton-Dyer conjecture which consists, roughly speaking, in replacing the Iwasawa algebra – the completed group ring of the Galois group of a $\mathbb{Z}_p$ extension – by the group ring of the Galois group of a finite abelian extension [52]. The more refined conjectures that emerge from the tame framework turn out to offer a congenial setting in which to study and organize the behavior of Euler systems, and these ideas have undergone something of a recent revival, notably through their connections with conjectures of Harris and Venkatesh concerning Venkatesh's "derived Hecke operators" acting on the cohomology of coherent sheaves on modular curves attached to modular forms of weight one [129].

9. THE FONTAINE–MAZUR CONJECTURE

Like many of the great number theorists of the 20th century, Mazur has contributed significantly to the study of Galois representations and their connection with automorphic forms. These ideas are central to a number of the achievements of Mazur that have already been recounted.

One of Mazur's important contributions in this direction is the deep conjecture, formulated in [71] with Jean-Marc Fontaine, which has widely come to be known as the *Fontaine–Mazur conjecture*. It aims to characterize the global p-adic Galois representations that arise from the p-adic étale cohomology of varieties over number fields. The characterization is via their restriction to the decomposition group at p (one demands that these p-adic representations of the Galois groups of p-adic fields be *potentially semistable*, a notion based on comparison functors between p-adic étale cohomology over p-adic fields and the p-adic cohomologies studied by Mazur in earlier decades) combined with a natural requirement of otherwise being ramified at finitely many primes other than p. This conjecture provides an elegant framework in which much of the recent progress on the Langlands program can be understood and conceptualized.

10. DEFORMATIONS OF GALOIS REPRESENTATIONS

(References: [53,54,61,73,75,76,78,108].)

The p-adic variation of modular forms and Galois representations is a theme that underlies much of Mazur's work in number theory, starting with his early work on the Eisenstein ideal. His fundamental article [53] formalizes this notion on the Galois theory side by introducing the *universal deformation ring* attached to a Galois representation with coefficients in a complete local ring. With this idea, Mazur launched the new field of Galois deformation theory, which almost immediately after its inception found a spectacular application in Wiles' proof of the Taniyama–Weil conjecture. This proof proceeds by constructing a natural map from one of Mazur's universal Galois deformation rings to a suitably completed ring of Hecke operators, and showing this map is an isomorphism. The deep study of the ring theoretic structure of completed Hecke algebras had already been initiated, more than a decade earlier, in Mazur's work on the Eisenstein ideal. With the introduction of universal deformation rings, Mazur can be credited for a substantial part of the theoretical infrastructure that enabled the proof of the Taniyama–Weil conjecture. Mazur's ideas are thus present in the very foundations of the remarkably successful strategy for establishing the modularity of Galois representations that has been extensively developed and generalized in the wake of Wiles' breakthrough on the modularity of elliptic curves.

Mazur's subsequent work [73,78] with Robert Coleman represents an attempt to partially globalize the study of deformation spaces of Galois representations, leading to the fundamental notion of Coleman–Mazur "eigencurves" and "eigenvarieties." The framework initiated by Coleman and Mazur in these foundational papers has been extensively developed in the past decades, spawning a fruitful area that underlies much of the recent progress in the Langlands program via p-adic methods.

11. DIOPHANTINE GEOMETRY

(References: [49,62,68,70,72,79,86,95,100].)

Mazur's work on diophantine geometry distinguishes itself by insights that are often stunning in their audacity. The article [62] ventures the striking conjecture that if the rational points of a variety V are Zariski dense, then their topological closure in $V(\mathbb{R})$ for the real topology is a union of connected components of $V(\mathbb{R})$.

Just as far reaching are the celebrated conjectures Mazur formulated with Lucia Caporaso and Joe Harris [70,72], asserting that the number of rational points on a curve of genus g over a number field K is *uniformly bounded* by a constant that depends only on g and K, and even just on g if one tolerates a finite number of exceptions. In [70] it is shown that this conjecture, which is both remarkably strong and pleasingly concrete, follows from the earlier, and at the time more widely accepted, conjecture of Lang that the set of rational points on a variety of general type can never be Zariski dense.

Such fearless conjectures, applying to all varieties at once or to the number of points on all curves of given genus, shine an unexpected light on venerable questions about rational points and have guided a lot of subsequent efforts by other researchers.

Many of Mazur's articles devoted to diophantine topics reveal unexpected connections to other mathematical themes. This is the case, notably, for [109] and [114], which study the variation in 2-Selmer ranks of elliptic curves over number fields, revealing a surprising connection between the notion of "Diophantine stability" and Hilbert's tenth problem concerning the undecidability of diophantine questions over certain number fields.

12. EULER SYSTEMS AND RELATED AREAS

(References: [92, 99, 104, 106, 110, 115, 117–120].)

The method of *Euler systems* is a powerful technique that emerged in the late 1980s from the works of mathematicians like Francisco Thaine, Karl Rubin, Victor Kolyvagin, and Kazuya Kato. It parlays the presence of special elements in the global Galois cohomology of (a compatible system of) p-adic Galois representations into a proof of at least one inequality in the associated main conjecture. The existence of the global elements making up an Euler system is still poorly understood, and their construction remains as much an art as a science.

The articles [92, 99, 104, 110, 119], and [120], all joint with Karl Rubin, are part of a systematic attempt to formalize (via the notion of what the authors call a "Kolyvagin system") the procedure whereby such norm-compatible collections of global classes with ties to L-function behavior can be exploited to obtain results in the direction of a main conjecture, or possibly a tame counterpart in the spirit of [52]. The perspectives introduced by Mazur and Rubin have had a decisive influence on an entire generation of researchers who are currently exploring the ramifications of the Euler system method.

13. EXPOSITION

(References: [59, 63, 75, 83, 87, 89, 93, 97, 105, 107, 112, 113, 116, 121, 124, 126].)

Mazur is a master expositor who revels in the joy of mathematical and philosophical ideas. He is the author of a fascinating, eclectic collection of essays in which his erudition and intellectual curiosity range far and wide. Some of these essays are devoted to broad mathematical topics like local-global principles in number theory [63], the deformation theory of Galois representations [75], diophantine questions related to perfect powers [83], the general idea of deformation in various parts of mathematics [93], the notion of a motive [97], the Sato–Tate conjecture [105], and the Riemann hypothesis [121]. Others examine ideas through the lens of their historical development, treating complex numbers as they were envisioned in the 16th century [87], or Hermann Weyl's foundational article on spectral theory [112]. Mazur also ventures into more philosophical topics like dreams in mathematics told through an evocation of Kronecker's *Jugendtraum* [113], the concept of number and mathematical abstraction [89], the subtle and elusive concept of equality in mathematics [107], the notion of plausibility [116], the overarching unity of mathematics [124], and thoughts on doing mathematics during the pandemic [126]. Mazur's infectious enthusiasm easily transmits itself to the reader, and his reflections on a diverse range of mathematical, historical and philosophical subjects never fail to delight, uplift, and enlighten. (The range and depth of Mazur's intellectual interests are vividly evoked in the engaging documentary movie "Barry Mazur and the infinite cheese of knowledge" directed by Oliver Ralfe [140].)

14. MENTORSHIP

According to the Mathematics genealogy website, Mazur has had (at least) 57 students and 325 descendants, figures that are bound to be obsolete by the time this *laudatio* goes to press. Beyond the direct impact he has had on his students, Mazur has shaped the views of an entire generation of number theorists who have been enriched by his ideas and enjoyed the privilege of pursuing his capacious intellectual legacy. This legacy, which is now being recognized through the awarding of the Chern medal, is a central and integral part of modern number theory and its influence will be felt for a very long time.

REFERENCES

[1] B. Mazur, The definition of equivalence of combinatorial imbeddings. *Inst. Hautes Études Sci. Publ. Math.* **1959** (1959), 97–109.

[2] B. Mazur, On the structure of certain semi-groups of spherical knot classes. *Inst. Hautes Études Sci. Publ. Math.* **1959** (1959), 111–119.

[3] B. Mazur, Orthotopy and spherical knots. *Inst. Hautes Études Sci. Publ. Math.* **1959** (1959), 121–140.

[4] B. Mazur, *On embeddings of spheres*. PhD thesis, Princeton University, 1959.

[5] B. Mazur, On embeddings of spheres. *Bull. Amer. Math. Soc.* **65** (1959), 59–65.

[6] B. Mazur, On embeddings of spheres. *Acta Math.* **105** (1961), 1–17.

[7] B. Mazur, A note on some contractible 4-manifolds. *Ann. of Math. (2)* **73** (1961), 221–228.

[8] B. Mazur, Stable equivalence of differentiable manifolds. *Bull. Amer. Math. Soc.* **67** (1961), 377–384.

[9] B. Mazur, Simple neighborhoods. *Bull. Amer. Math. Soc.* **68** (1962), 87–92.

[10] B. Mazur, Symmetric homology spheres. *Illinois J. Math.* **6** (1962), 245–250.

[11] B. Mazur, Relative neighborhoods and the theorems of Smale. *Ann. of Math. (2)* **77** (1963), 232–249.

[12] B. Mazur, Differential topology from the point of view of simple homotopy theory. *Inst. Hautes Études Sci. Publ. Math.* **15** (1963), 93 pp.

[13] B. Mazur, The method of infinite repetition in pure topology. I. *Ann. of Math. (2)* **80** (1964), 201–226.

[14] B. Mazur, Combinatorial equivalence versus topological equivalence. *Trans. Amer. Math. Soc.* **111** (1964), 288–316.

[15] M. Artin and B. Mazur, On periodic points. *Ann. of Math. (2)* **81** (1965), 82–99.

[16] B. Mazur, Morse theory. In *1965 Differential and combinatorial topology (A Symposium in Honor of Marston Morse)*, pp. 145–165, Princeton University Press, Princeton, NJ, 1965.

[17] M. Artin and B. Mazur, On the van Kampen theorem. *Topology* **5** (1966), 179–189.

[18] B. Mazur, The method of infinite repetition in pure topology. II. Stable applications. *Ann. of Math. (2)* **83** (1966), 387–401.

[19] M. W. Hirsch and B. Mazur, *Smoothings of piecewise linear manifolds*. Ann. of Math. Stud. 80, Princeton University Press, Princeton, NJ; University of Tokyo Press, Tokyo, 1974.

[20] B. Mazur, Frobenius and the Hodge filtration. *Bull. Amer. Math. Soc.* **78** (1972), 653–667.

[21] B. Mazur, Frobenius and the Hodge filtration (estimates). *Ann. of Math. (2)* **98** (1973), 58–95.

[22] B. Mazur and W. Messing, *Universal extensions and one dimensional crystalline cohomology*. Lecture Notes in Math. 370, Springer, Berlin–New York, 1974.

[23] B. Mazur, Eigenvalues of Frobenius acting on algebraic varieties over finite fields. In *Algebraic geometry (Humboldt State Univ., Arcata, Calif., 1974)*, pp. 231–261, Proc. Sympos. Pure Math., 29, Amer. Math. Soc., Providence, RI, 1975.

[24] M. Artin and B. Mazur, Formal groups arising from algebraic varieties. *Ann. Sci. Éc. Norm. Supér. (4)* **10** (1977), no. 1, 87–131.

[25] M. Artin and B. Mazur, Homotopy of varieties in the etale topology. In *1967 Proc. Conf. Local Fields (Driebergen, 1966)*, pp. 1–15, Springer, Berlin, 1967.

[26] B. Mazur and L. Roberts, Local Euler characteristics. *Invent. Math.* **9** (1969/70), 201–234.

[27] B. Mazur, Finite flat structures. In *1970 Applications of Categorical Algebra (New York, 1968)*, pp. 219–225, Proc. Sympos. Pure Math. XVII, Amer. Math. Soc., New York, 1970.

[28] M. Artin and B. Mazur, *Etale homotopy*. Reprint of the 1969 original. Lecture Notes in Math. 100, Springer, Berlin, 1986.

[29] B. Mazur, Local flat duality. *Amer. J. Math.* **92** (1970), 343–361.

[30] B. Mazur, *Remarks on the Alexander polynomial*. Unpublished. Available at https://people.math.harvard.edu/~mazur/papers/alexander_polynomial.pdf.

[31] B. Mazur, Notes on étale cohomology of number fields. *Ann. Sci. Éc. Norm. Supér. (4)* **6** (1973), 521–552 (1974).

[32] B. Mazur and J. Vélu, Courbes de Weil de conducteur 26. *C. R. Acad. Sci. Paris Sér. A-B* **275** (1972), A743–A745.

[33] B. Mazur and J. Tate, Points of order 13 on elliptic curves. *Invent. Math.* **22** (1973/74), 41–49.

[34] B. Mazur and J.-P. Serre, Points rationnels des courbes modulaires $X_0(N)$ (d'après A. Ogg). In *Séminaire Bourbaki (1974/1975)*, Exp. No. 469, pp. 238–255, Lecture Notes in Math. 514, Springer, Berlin, 1976.

[35] B. Mazur, Rational points on modular curves. In *Modular functions of one variable, V (Proc. Second Internat. Conf., Univ. Bonn, Bonn, 1976)*, pp. 107–148, Lecture Notes in Math. 601, Springer, Berlin, 1977.

[36] B. Mazur, Modular curves and the Eisenstein ideal. *Inst. Hautes Études Sci. Publ. Math.* **47** (1977), 33–186 (1978).

[37] B. Mazur, Rational isogenies of prime degree. *Invent. Math.* **44** (1978), no. 2, 129–162.

[38] S. Kamienny and B. Mazur, Rational torsion of prime order in elliptic curves over number fields. Columbia University Number Theory Seminar (New York, 1992). *Astérisque* No. **228** (1995), 3, 81–100.

[39] N. M. Katz and B. Mazur, *Arithmetic moduli of elliptic curves*. Ann. of Math. Stud. 108, Princeton University Press, Princeton, NJ, 1985.

[40] B. Mazur, Courbes elliptiques et symboles modulaires. In *Séminaire Bourbaki, 24ème année (1971/1972)*, Exp. No. 414, pp. 277–294, Lecture Notes in Math. 317, Springer, Berlin, 1973.

[41] B. Mazur, Rational points of abelian varieties with values in towers of number fields. *Invent. Math.* **18** (1972), 183–266.

[42] B. Mazur and P. Swinnerton-Dyer, Arithmetic of Weil curves. *Invent. Math.* **25** (1974), 1–61.

[43] B. Mazur, p-adic analytic number theory of elliptic curves and Abelian varieties over $\mathbb{Q}$. In *Proceedings of the International Congress of Mathematicians (Vancouver, BC, 1974)*, Vol. 1, pp. 369–377, Academic Press, 1970.

[44] B. Mazur, On the arithmetic of special values of L-functions. *Invent. Math.* **55** (1979), no. 3, 207–240.

[45] B. Mazur and A. Wiles, Analogies between function fields and number fields. *Amer. J. Math.* **105** (1983), no. 2, 507–521.

[46] B. Mazur and J. Tate, Canonical height pairings via biextensions. In *Arithmetic and geometry, Vol. I*, pp. 195–237, Progr. Math. 35, Birkhäuser, Boston, MA, 1983.

[47] B. Mazur and A. Wiles, Class fields of abelian extensions of $\mathbb{Q}$. *Invent. Math.* **76** (1984), no. 2, 179–330.

[48] B. Mazur, Modular curves and arithmetic. In *Proceedings of the International Congress of Mathematicians, Vol. 1, 2 (Warsaw, 1983)*, pp. 185–211, PWN, Warsaw, 1984.

[49] B. Mazur, Arithmetic on curves. *Bull. Amer. Math. Soc. (N.S.)* **14** (1986), no. 2, 207–259.

[50] B. Mazur, J. Tate, and J. Teitelbaum, On p-adic analogues of the conjectures of Birch and Swinnerton-Dyer. *Invent. Math.* **84** (1986), no. 1, 1–48.

[51] B. Mazur and A. Wiles, On p-adic analytic families of Galois representations. *Compos. Math.* **59** (1986), no. 2, 231–264.

[52] B. Mazur and J. Tate, Refined conjectures of the "Birch and Swinnerton-Dyer type". *Duke Math. J.* **54** (1987), no. 2, 711–750.

[53] B. Mazur, Deforming Galois representations. In *Galois groups over* $\mathbb{Q}$, pp. 385–437, Math. Sci. Res. Inst. Publ. 16, Springer, New York, 1989.

[54] N. Boston and B. Mazur, Explicit universal deformations of Galois representations. In *Algebraic number theory*, pp. 1–21, Adv. Stud. Pure Math. 17, Academic Press, Boston, MA, 1989.

[55] B. Mazur, Two-dimensional p-adic Galois representations unramified away from p. *Compos. Math.* **74** (1990), no. 2, 115–133.

[56] B. Mazur and J. Tilouine, Représentations galoisiennes, différentielles de Kähler et "conjectures principales". *Inst. Hautes Études Sci. Publ. Math.* **71** (1990), 65–103.

[57] F. Gouvêa and B. Mazur, The square-free sieve and the rank of elliptic curves. *J. Amer. Math. Soc.* **4** (1991), no. 1, 1–23.

[58] B. Mazur and J. Tate, The p-adic sigma function. *Duke Math. J.* **62** (1991), no. 3, 663–688.

[59] B. Mazur, Number theory as gadfly. *Amer. Math. Monthly* **98** (1991), no. 7, 593–610.

[60] B. Mazur and K. Ribet, Two-dimensional representations in the arithmetic of modular curves. In *Courbes modulaires et courbes de Shimura (Orsay, 1987/1988)*, Astérisque No. 196–197 (1991), 6, 215–255 (1992).

[61] F. Gouvêa and B. Mazur, Families of modular eigenforms. *Math. Comp.* **58** (1992), no. 198, 793–805.

[62] B. Mazur, The topology of rational points. *Exp. Math.* **1** (1992), no. 1, 35–45.

[63] B. Mazur, On the passage from local to global in number theory. *Bull. Amer. Math. Soc. (N.S.)* **29** (1993), no. 1, 14–50.

[64] F. Q. Gouvêa and B. Mazur, On the characteristic power series of the U operator. *Ann. Inst. Fourier (Grenoble)* **43** (1993), no. 2, 301–312.

[65] E. M. Friedlander and B. Mazur, Filtrations on the homology of algebraic varieties. With an appendix by Daniel Quillen. *Mem. Amer. Math. Soc.* **110** (1994), no. 529.

[66] B. Mazur, Questions of decidability and undecidability in number theory. *J. Symbolic Logic* **59** (1994), no. 2, 353–371.

[67] B. Mazur, On monodromy invariants occurring in global arithmetic, and Fontaine's theory. In *p-adic monodromy and the Birch and Swinnerton-Dyer conjecture (Boston, MA, 1991)*, pp. 1–20, Contemp. Math. 165, Amer. Math. Soc., Providence, RI, 1994.

[68] B. Mazur, Speculations about the topology of rational points: an update. *Columbia University Number Theory Seminar (New York, 1992). Astérisque* **228** (1995), 4, 165–182.

[69] F. Q. Gouvêa and B. Mazur, Searching for p-adic eigenfunctions. *Math. Res. Lett.* **2** (1995), no. 5, 515–536.

[70] L. Caporaso, J. Harris, and B. Mazur, How many rational points can a curve have? In *The moduli space of curves (Texel Island, 1994)*, pp. 13–31, Progr. Math. 129, Birkhäuser, Boston, MA, 1995.

[71] J.-M. Fontaine and B. Mazur, Geometric Galois representations. In *Elliptic curves, modular forms, & Fermat's last theorem (Hong Kong, 1993)*, pp. 41–78, Ser. Number Theory, I, Int. Press, Cambridge, MA, 1995.

[72] L. Caporaso, J. Harris, and B. Mazur, Uniformity of rational points. *J. Amer. Math. Soc.* **10** (1997), no. 1, 1–35.

[73] B. Mazur, An "infinite fern" in the universal deformation space of Galois representations. *Journées Arithmétiques* (Barcelona, 1995). *Collect. Math.* **48** (1997), no. 1–2, 151–193.

[74] D. Eisenbud and B. Mazur, Evolutions, symbolic squares, and Fitting ideals. *J. Reine Angew. Math.* **488** (1997), 189–201.

[75] B. Mazur, An introduction to the deformation theory of Galois representations. In *Modular forms and Fermat's last theorem (Boston, MA, 1995)*, pp. 243–311, Springer, New York, 1997.

[76] F. Gouvêa and B. Mazur, On the density of modular representations. In *Computational perspectives on number theory (Chicago, IL, 1995)*, pp. 127–142, AMS/IP Stud. Adv. Math. 7, Amer. Math. Soc., Providence, RI, 1998.

[77] J. Harris, B. Mazur, and R. Pandharipande, Hypersurfaces of low degree. *Duke Math. J.* **95** (1998), no. 1, 125–160.

[78] R. Coleman and B. Mazur, The eigencurve. In *Galois representations in arithmetic algebraic geometry (Durham, 1996)*, pp. 1–113, London Math. Soc. Lecture Note Ser. 254, Cambridge Univ. Press, Cambridge, 1998.

[79] B. Mazur, Open problems regarding rational points on curves and varieties. In *Galois representations in arithmetic algebraic geometry (Durham, 1996)*, pp. 239–265, London Math. Soc. Lecture Note Ser. 254, Cambridge University Press, Cambridge, 1998.

[80] B. Mazur, Open problems in number theory. In *Current developments in mathematics, 1997 (Cambridge, MA)*, pp. 199–203, Int. Press, Boston, MA, 1999.

[81] B. Mazur, Visualizing elements of order three in the Shafarevich–Tate group. In *Sir Michael Atiyah: A great mathematician of the twentieth century. Asian J. Math.* **3** (1999), no. 1, 221–232.

[82] D. Kazhdan, B. Mazur, and C.-G. Schmidt, Relative modular symbols and Rankin–Selberg convolutions. *J. Reine Angew. Math.* **519** (2000), 97–141.

[83] B. Mazur, Questions about powers of numbers. *Notices Amer. Math. Soc.* **47** (2000), no. 2, 195–202.

[84] B. Mazur, The theme of p-adic variation. In *Mathematics: Frontiers and perspectives*, pp. 433–459, Amer. Math. Soc., Providence, RI, 2000.

[85] J. E. Cremona and B. Mazur, Visualizing elements in the Shafarevich–Tate group. *Exp. Math.* **9** (2000), no. 1, 13–28.

[86] B. Mazur, Abelian varieties and the Mordell–Lang conjecture. In *Model theory, algebra, and geometry*, pp. 199–227, Math. Sci. Res. Inst. Publ. 39, Cambridge University Press, Cambridge, 2000.

[87] F. La Nave and B. Mazur. Reading Bombelli. *Math. Intelligencer* **24** (2002), no. 1, 12–21.

[88] B. Mazur and K. Rubin, Elliptic curves and class field theory. In *Proceedings of the International Congress of Mathematicians, Vol. II (Beijing, 2002)*, pp. 185–195, Higher Ed. Press, Beijing, 2002.

[89] B. Mazur, *Imagining numbers. Particularly the square root of minus fifteen*. Farrar, Straus and Giroux, New York, 2003.

[90] B. Mazur and K. Rubin, Studying the growth of Mordell–Weil. In *Kazuya Kato's fiftieth birthday*, pp. 585–607, Doc. Math., Extra Vol., 2003.

[91] T. Graber, J. Harris, B. Mazur, and J. Starr, Jumps in Mordell–Weil rank and arithmetic surjectivity. In *Arithmetic of higher-dimensional algebraic varieties (Palo Alto, CA, 2002)*, pp. 141–147, Progr. Math. 226, Birkhäuser, Boston, MA, 2004.

[92] B. Mazur and K. Rubin, Kolyvagin systems. *Mem. Amer. Math. Soc.* **168** (2004).

[93] B. Mazur, Perturbations, deformations, and variations (and "near-misses") in geometry, physics, and number theory. *Bull. Amer. Math. Soc. (N.S.)* **41** (2004), no. 3, 307–336.

[94] B. Mazur and K. Rubin, Pairings in the arithmetic of elliptic curves. In *Modular curves and abelian varieties*, pp. 151–163, Progr. Math. 224, Birkhäuser, Basel, 2004.

[95] T. Graber, J. Harris, B. Mazur, and J. Starr, Arithmetic questions related to rationally connected varieties. In *The legacy of Niels Henrik Abel*, pp. 531–542, Springer, Berlin, 2004.

[96] B. Mazur and K. Rubin, Introduction to Kolyvagin systems. In *Stark's conjectures: recent work and new directions*, pp. 207–221, Contemp. Math. 358, Amer. Math. Soc., Providence, RI, 2004.

[97] B. Mazur, What is … a motive? *Notices Amer. Math. Soc.* **51** (2004), no. 10, 1214–1216.

[98] B. Mazur and K. Rubin, Organizing the arithmetic of elliptic curves. *Adv. Math.* **198** (2005), no. 2, 504–546.

[99] B. Mazur and K. Rubin, Finding large Selmer groups. *J. Differential Geom.* **70** (2005), no. 1, 1–22.

[100] T. Graber, J. Harris, B. Mazur, and J. Starr, Rational connectivity and sections of families over curves. *Ann. Sci. Éc. Norm. Supér. (4)* **38** (2005), no. 5, 671–692.

[101] B. Mazur, W. Stein, and J. Tate, Computation of p-adic heights and log convergence. Doc. Math., Extra Vol. (2006), 577–614.

[102] B. Bektemirov, B. Mazur, W. Stein, and M. Watkins, Average ranks of elliptic curves: Tension between data and conjecture. *Bull. Amer. Math. Soc. (N.S.)* **44** (2007), no. 2, 233–254.

[103] B. Mazur, K. Rubin K, and A. Silverberg, Twisting commutative algebraic groups. *J. Algebra* **314** (2007), no. 1, 419–438.

[104] B. Mazur and K. Rubin, Finding large Selmer rank via an arithmetic theory of local constants. *Ann. of Math. (2)* **166** (2007), no. 2, 579–612.

[105] B. Mazur, Finding meaning in error terms. *Bull. Amer. Math. Soc. (N.S.)* **45** (2008), no. 2, 185–228.

[106] B. Mazur and K. Rubin, Growth of Selmer rank in nonabelian extensions of number fields. *Duke Math. J.* **143** (2008), no. 3, 437–461.

[107] B. Mazur, When is one thing equal to some other thing? In *Proof and other dilemmas*, pp. 221–241, MAA Spectrum, Math. Assoc. America, Washington, DC, 2008.

[108] F. Calegari and B. Mazur, Nearly ordinary Galois deformations over arbitrary number fields. *J. Inst. Math. Jussieu* **8** (2009), no. 1, 99–177.

[109] B. Mazur and K. Rubin, Ranks of twists of elliptic curves and Hilbert's tenth problem. *Invent. Math.* **181** (2010), no. 3, 541–575.

[110] B. Mazur and K. Rubin, Refined class number formulas and Kolyvagin systems. *Compos. Math.* **147** (2011), no. 1, 56–74.

[111] B. Mazur, How can we construct abelian Galois extensions of basic number fields? *Bull. Amer. Math. Soc. (N.S.)* **48** (2011), no. 2, 155–209.

[112] B. Mazur, About Hermann Weyl's "Ramifications, old and new, of the eigenvalue problem". *Bull. Amer. Math. Soc. (N.S.)* **49** (2012), no. 2, 325–326.

[113] B. Mazur, Visions, dreams, and mathematics. In *Circles disturbed*, pp. 183–210, Princeton University Press, Princeton, NJ, 2012.

[114] Z. Klagsbrun, B. Mazur, and K. Rubin, Disparity in Selmer ranks of quadratic twists of elliptic curves. *Ann. of Math. (2)* **178** (2013), no. 1, 287–320.

[115] J. B. Friedlander, H. Iwaniec, B. Mazur, and K. Rubin, The spin of prime ideals. *Invent. Math.* **193** (2013), no. 3, 697–749.

[116] B. Mazur, Is it plausible? *Math. Intelligencer* **36** (2014), no. 1, 24–33.

[117] Z. Klagsbrun, B. Mazur, and K. Rubin, A Markov model for Selmer ranks in families of twists. *Compos. Math.* **150** (2014), no. 7, 1077–1106.

[118] B. Mazur and K. Rubin, Selmer companion curves. *Trans. Amer. Math. Soc.* **367** (2015), no. 1, 401–421.

[119] B. Mazur and K. Rubin, Controlling Selmer groups in the higher core rank case. *J. Théor. Nombres Bordeaux* **28** (2016), no. 1, 145–183.

[120] B. Mazur and K. Rubin, Refined class number formulas for G_m. *J. Théor. Nombres Bordeaux* **28** (2016), no. 1, 185–211.

[121] B. Mazur and W. Stein, *Prime numbers and the Riemann hypothesis*. Cambridge University Press, Cambridge, 2016. xi+142 pp.

[122] M. Derickx, B. Mazur, and S. Kamienny, Rational families of 17-torsion points of elliptic curves over number fields. In *Number theory related to modular curves—Momose memorial volume*, pp. 81–104, Contemp. Math. 701, Amer. Math. Soc., 2018.

[123] B. Mazur and K. Rubin, Diophantine stability. *Amer. J. Math.* **140** (2018), no. 3, 571–616.

[124] B. Mazur, Grand unity. *ICCM Not.* **7** (2019), no. 1, 76.

[125] B. Mazur and K. Rubin, Big fields that are not large. *Proc. Amer. Math. Soc. Ser. B* **7** (2020), 159–169.

[126] B. Mazur, Math in the time of plague. *Math. Intelligencer* **42** (2020), no. 4, 1–6.

[127] J. Balakrishnan, N. Dogra, J. S. Müller, J. Tuitman, and J. Vonk, Explicit Chabauty–Kim for the split Cartan modular curve of level 13. *Ann. of Math. (2)* **189** (2019), no. 3, 885–944.

[128] Y. Bilu, P. Parent, and M. Rebolledo, Rational points on $X_0^+(p^r)$. *Ann. Inst. Fourier (Grenoble)* **63** (2013), no. 3, 957–984.

[129] M. Harris and A. Venkatesh, Derived Hecke algebra for weight one forms. *Exp. Math.* **28** (2019), no. 3, 342–361.

[130] S. Kamienny, Torsion points on elliptic curves and q-coefficients of modular forms. *Invent. Math.* **109** (1992), no. 2, 221–229.

[131] K. Kato, p-adic Hodge theory and values of zeta functions of modular forms. Cohomologies p-adiques et applications arithmétiques III.. *Astérisque* No. **295** (2004), ix, 117–290.

[132] N. M. Katz, Slope filtration of F-crystals. *Journées de Géométrie Algébrique de Rennes* Vol. I, 113–163. *Astérisque* **63**, Soc. Math. France, Paris, 1979.

[133] M. A. Kenku, Rational torsion points on elliptic curves defined over quadratic fields. *J. Nigerian Math. Soc.* **2** (1983), 1–16.

[134] M. A. Kenku and F. Momose, Torsion points on elliptic curves defined over quadratic fields. *Nagoya Math. J.* **109** (1988), 125–149.

[135] L. Merel, Bornes pour la torsion des courbes elliptiques sur les corps de nombres. *Invent. Math.* **124** (1996), no. 1–3, 437–449.

[136] F. Momose, Rational points on the modular curves $X_{\mathrm{split}}(p)$. *Compos. Math.* **52** (1984), no. 1, 115–137.

[137] F. Momose, Rational points on the modular curves $X_0^+(p^r)$. *J. Fac. Sci., Univ. Tokyo, Sect. 1A, Math.* **33** (1986), no. 3, 441–466.

[138] M. Morishita, *Knots and primes. An introduction to arithmetic topology*. Universitext, Springer, London, 2012. 191 pp.

[139] A. P. Ogg, Rational points of finite order on elliptic curves. *Invent. Math.* **12** (1971), 105–111.

[140] O. Ralfe, *Barry Mazur and the infinite cheese of knowledge*, Sheepstreet films.

[141] K. A. Ribet, On modular representations of $\mathrm{Gal}(\bar{\mathbb{Q}}/\mathbb{Q})$ arising from modular forms. *Invent. Math.* **100** (1990), no. 2, 431–476.

[142] N. Schappacher and R. Schoof, Beppo Levi and the arithmetic of elliptic curves. *Math. Intelligencer* **18** (1996), no. 1, 57–69.

[143] J.-P. Serre, Sur les représentations modulaires de degré 2 de Gal$(\bar{\mathbb{Q}}/\mathbb{Q})$. *Duke Math. J.* **54** (1987), no. 1, 179–230.

[144] C. Skinner and E. Urban, The Iwasawa main conjectures for GL_2. *Invent. Math.* **195** (2014), no. 1, 1–277.

[145] J. Teitelbaum, Values of p-adic L-functions and a p-adic Poisson kernel. *Invent. Math.* **101** (1990), no. 2, 395–410.

[146] A. Venkatesh, Primes and knots, Public Lecture, IAS, Princeton, video online at https://www.youtube.com/watch?v=jvoYgNYKyk0.

[147] A. Wiles, The Iwasawa conjecture for totally real fields. *Ann. of Math. (2)* **131** (1990), no. 3, 493–540.

[148] A. Wiles, Modular elliptic curves and Fermat's last theorem. *Ann. of Math. (2)* **141** (1995), no. 3, 443–551.

HENRI DARMON

McGill University, Montreal, Canada, henri.darmon@mcgill.ca

THE WORK OF ELLIOTT LIEB

RUPERT L. FRANK

ABSTRACT

On the occasion of Elliott Lieb being awarded the Gauss Prize 2022, we give a nontechnical overview over some of his seminal works in mathematical physics. We emphasize, in particular, his work on Coulomb many-body systems and functional inequalities.

MATHEMATICS SUBJECT CLASSIFICATION 2020

Primary 01-01; Secondary 46-02, 49-02, 81-02, 82-02

KEYWORDS

Quantum many body systems, functional inequalities, sharp constants, mathematical condensed matter physics, statistical mechanics

ICM INTERNATIONAL CONGRESS OF MATHEMATICIANS 2022 JULY 6–14

© 2022 International Mathematical Union

Proc. Int. Cong. Math. 2022, Vol. 1, pp. 142–158
DOI 10.4171/ICM2022/216

Published by EMS Press

and licensed under
a CC BY 4.0 license

Elliott Lieb is awarded the Gauss Prize 2022 "for deep mathematical contributions of exceptional breadth which have shaped the fields of quantum mechanics, statistical mechanics, computational chemistry, and quantum information theory."

It is my great pleasure to congratulate Elliott on this honor. In the following pages I will try to give a nontechnical overview over some of his seminal works.

Lieb is a mathematical physicist. This is a field that lies at the boundary between physics and mathematics and that Freeman Dyson [From Eros to Gaia, pp. 164–165] has described as follows:

> *"Mathematical physics is the discipline of people who try to reach a deep understanding of physical phenomena by following the rigorous style and method of mathematics."*

As mentioned in the citation, Lieb has made ground-breaking contributions to both mathematics and physics. In this connection it is worthwhile to mention that about half a year ago, Lieb was awarded the 2022 APS Medal for Exceptional Achievement in Research, the highest honor of the American Physical Society.

A distinctive feature of Lieb's work is its timelessness. Independently of fashions and trends, he has worked, and continues to work, on problems that he considers deep and fundamental. Sometimes it is decades later that the full potential of his ideas is understood. Prime examples are the Lieb–Robinson bounds, proved in 1972, and the strong subadditivity of entropy, proved in 1973, that have both come to play a key role in quantum information theory in the 21st century.

Lieb's publication list contains, at the time of this writing, 404 items with the first one dating back to 1955 and the most recent one just about to appear. Four volumes of his selected works have been published so far [26–29] and, on the occasion of his 90th birthday, a 1300+ page collection of articles was edited [8], where the contributors explain the content and the ramifications of Lieb's work.

It is impossible to give a complete overview over this monumental body of work. I have chosen here two main areas of Lieb's research, namely quantum many-body Coulomb systems and functional inequalities, and omitted all others, except for a brief mention of some in the last section. This selection is influenced by my predilections and my ignorance, and I ask the readers' indulgence for all the omissions.

1. QUANTUM COULOMB SYSTEMS

One of the recurring themes in Lieb's research since the early 1970s is the analysis of continuous quantum many-body systems and, in particular, of systems of particles interacting through Coulomb forces. Here one neglects other forces, but this description is appropriate for ordinary matter and much of the world relevant to everyday life. As a very readable account of Lieb's research program in this area we recommend his Gibbs lecture in 1989 [25], which contains a much more detailed description than what we provide here.

We consider a system consisting of N quantum electrons and K classical nuclei in $\mathbb{R}^3$. The latter are fixed at positions $R_1, \dots, R_K \in \mathbb{R}^3$ and have charges $Z_1, \dots, Z_K \in (0, \infty)$ (in units where the electron charge is -1). The properties of such a system are described by the Hamiltonian

$$H := \sum_{n=1}^{N}(-\Delta_n) - \sum_{n=1}^{N}\sum_{k=1}^{K}\frac{Z_k}{|x_n - R_k|} + \sum_{1\leq n<m\leq N}\frac{1}{|x_n - x_m|} + \sum_{1\leq k<\ell\leq K}\frac{Z_k Z_\ell}{|R_k - R_\ell|}, \tag{1.1}$$

acting on functions in $\mathbb{R}^{3N}$. Here we use coordinates $x = (x_1, \dots, x_N) \in \mathbb{R}^{3N}$. The four sums in the definition of H correspond, respectively, to the kinetic energy of the electrons, the electron–nucleus attraction, the electron–electron repulsion, and the nucleus–nucleus repulsion.

This Hamiltonian can be realized as an unbounded, self-adjoint operator in the Hilbert space consisting of all antisymmetric functions in $L^2(\mathbb{R}^{3N})$, that is, square-integrable functions ψ satisfying $\psi(\dots, x_n, \dots, x_m, \dots) = -\psi(\dots, x_m, \dots, x_n, \dots)$ for all $n \neq m$. The antisymmetry reflects the Pauli exclusion principle. One should also take the electron spin into account, but mathematically this only leads to minor changes and will be ignored in what follows.

The ground state energy of the system is, by definition,

$$\inf \operatorname{spec} H.$$

It is the infimum of $\langle \psi, H\psi \rangle$ with respect to all normalized, antisymmetric ψ.

The fundamental feature of the problem of analyzing the ground state energy of H (and of quantum many-body systems in general) is the huge number of dimensions of the underlying Hilbert space. This makes a numerical computation virtually impossible and for quantitative results one typically has to rely on approximate theories that are numerically more tractable. This situation underlines the importance of analytical studies about the full Schrödinger problem and also about its relation to approximate theories. Lieb has made fundamental contributions to this problem, as we will review in the remainder of this section.

Stability of matter

The problem of *stability of matter* consists in showing that for every $z > 0$ there is a constant C such that, for all $N, K \in \mathbb{N}$, $R_1, \dots, R_K \in \mathbb{R}^3$ and $Z_1, \dots, Z_K \in [0, z]$, one has

$$\inf \operatorname{spec} H \geq -C(N + K). \tag{1.2}$$

This inequality says that, despite the fact that the number of interactions grows *quadratically* in the total number of particles, the ground state energy only behaves *linearly*. This is fundamental for the existence of matter as we know it.

We emphasize that the stability of matter depends on the fact that electrons are fermions, that is, on the fact that H is considered only on the subspace of antisymmetric functions in $L^2(\mathbb{R}^{3N})$ and not on the full space $L^2(\mathbb{R}^{3N})$. It is, in fact, wrong on the latter, bigger space, as we shall see.

The first proof of stability of matter was achieved in 1967 by Dyson and Lenard. Lieb and Thirring gave a new proof in 1975 **[44]**. The latter proof gives a much more realistic value for the constant C in (1.2), namely about $C \approx 5$ instead of Dyson and Lenard's $C \approx 10^{14}$. It also helps to clarify the reason for the validity of stability of matter on a conceptual level. Here is what Dyson writes in the preface of the Selecta of Elliott Lieb **[29]**:

> *"Our proof was so complicated and so unilluminating that it stimulated Lieb and Thirring to find the first decent proof [...]. Why was our proof so bad and why was theirs so good? The reason is simple. Lenard and I began with mathematical tricks and hacked our way through a forest of inequalities without any physical understanding. Lieb and Thirring began with physical understanding and went on to find the appropriate mathematical language to make their understanding rigorous. Our proof was a dead end. Theirs was a gateway to the new world of ideas [...]."*

The fundamental mechanism of the Lieb–Thirring proof of stability of matter is the fact that atoms do not bind in an approximate model of a Coulomb system known as Thomas–Fermi theory. (Here "no binding" means, mathematically, that the ground state energy of a molecule is simply the sum of the energies of the individual atoms.) Therefore the goal of Lieb and Thirring was to show that this approximate theory provides, after an appropriate adjustment of constants, a rigorous lower bound to the Schrödinger ground state energy. Later in this section we will discuss both this Thomas–Fermi theory and a functional inequality (known as Lieb–Thirring inequality) that leads to the claimed lower bound.

The above mentioned work by Lieb and Thirring from 1975 was the starting point of Lieb's thorough investigation of the problem of stability of matter, a topic to which he would return repeatedly for several decades and with many coworkers. Notable are, among other things, a proof of stability of matter in the presence of magnetic fields that covers the physical value of the fine structure constant **[33]**. An introduction to this area is provided in his book with Seiringer "The Stability of Matter in Quantum Mechanics" **[36]**.

Existence of the thermodynamic limit for real matter with Coulomb forces

Lieb's first result on Coulomb many-body systems, even before his work on stability of matter, settled an open problem in the foundations of statistical mechanics. Namely, in 1969 he and Lebowitz proved the existence of the thermodynamic limit for real matter with Coulomb forces **[30]**.

Here one considers a large number of particles (electrons and nuclei, for instance) confined to an open set Ω in $\mathbb{R}^3$. One considers the limit where Ω tends to $\mathbb{R}^3$ (in a sense to be made precise) and where the densities of the different particles (that is, the number of particles divided by the volume of Ω) tend to given constants. The particles interact through Coulomb forces, as in (1.1). Moreover, one typically discusses this question at a given positive temperature. The theorem of Lieb and Lebowitz states that the limit of the corresponding free energy exists, is independent of the shape of the approximating domains Ω and has appropriate convexity and concavity properties.

The Lieb–Lebowitz theorem uses the stability of matter theorem as an ingredient. In the proof of the existence of the thermodynamic limit the main concern is the slow decay of $|x|^{-1}$ as $|x| \to \infty$. Indeed, the existence of the thermodynamic limit had been known in the case of short-range interactions. In the long range case charge neutrality is an essential input. Lieb and Lebowitz exploit the electrostatic screening very originally via Newton's theorem. In this way they are led to the geometric problem of how to efficiently pack large balls by smaller balls, which they solve by their "Cheese Theorem."

Thomas–Fermi theory and density functional theory

In our discussion of the Lieb–Thirring proof of stability of matter we have already mentioned the Thomas–Fermi model of a Coulomb system. This was proposed independently by Thomas and Fermi in 1927, remarkably soon after Schrödinger had introduced his theory. In the variational approach to Thomas–Fermi theory one starts from the energy functional, defined for nonnegative function ρ on $\mathbb{R}^3$ by

$$\mathcal{E}[\rho] = \gamma^{\mathrm{TF}} \int_{\mathbb{R}^3} \rho(x)^{\frac{5}{3}}\, dx - \sum_{k=1}^{K} \int_{\mathbb{R}^3} \frac{Z_k \rho(x)}{|x - R_k|}\, dx + D[\rho] + \sum_{1 \leq k < \ell \leq K} \frac{Z_k Z_\ell}{|R_k - R_\ell|}$$

with

$$D[\rho] := \frac{1}{2} \iint_{\mathbb{R}^3 \times \mathbb{R}^3} \frac{\rho(x)\rho(y)}{|x - y|}\, dx\, dy.$$

The function ρ describes the distribution of electrons and its integral has the meaning of the total number of electrons. The four terms in the definition of $\mathcal{E}$ correspond, respectively, to the kinetic energy of the electrons, the electron–nucleus attraction, the electron–electron repulsion, and the nucleus–nucleus repulsion. In the next subsection we will briefly describe how one can arrive at the $\rho^{\frac{5}{3}}$ approximation to the kinetic energy. This approximation also leads to a certain value for the constant $\gamma^{\mathrm{TF}} > 0$.

The ground state energy in Thomas–Fermi theory is

$$\inf\left\{ \mathcal{E}[\rho] : \rho \geq 0,\ \int_{\mathbb{R}^3} \rho\, dx = N \right\}.$$

We emphasize that, in contrast to Schrödinger theory, Thomas–Fermi theory is a nonlinear theory. The important simplifying feature is that in Thomas–Fermi theory one optimizes over functions of only three variables, as opposed to functions of $3N$ variables in Schrödinger theory.

While Thomas–Fermi theory had been around and had been used for a long time, it was only in the 1970s that Lieb and Simon rigorously established its mathematical foundations [40]. They answered, among other things, questions about the existence and uniqueness of solutions, their regularity and decay. Later, Lieb and collaborators investigated systematically density functional theories that are refinements of Thomas–Fermi theory. These findings are summarized in the review [17].

Another fundamental result that Lieb and Simon prove in their Thomas–Fermi paper is a rigorous relation between the ground state energy of the Hamiltonian H in (1.1) and the minimal Thomas–Fermi energy in the joint limit $N \to \infty$ and $Z \to \infty$ with $N/Z \to \lambda \in (0, \infty)$. (Here we consider, for simplicity, the atomic case $K = 1$ and set $Z = Z_1$. The most important case is $N = Z$, that is, the case of a neutral atom.) It is shown that

$$\lim \frac{\inf \operatorname{spec} H}{\inf \mathcal{E}} = 1.$$

This convergence of energies is supplemented by convergence results of the one-particle densities of (approximate) ground states of the Schrödinger Hamiltonian. Technically, this result is related to semiclassical analysis, but outside of the typical regularity assumptions in this theory. The Lieb–Simon result has become a blueprint for other derivations of effective theories in scaling limits.

As an aside, we also mention the first proof, given by Lieb and Simon, of the existence of solutions to the Hartree–Fock equations for atoms and molecules [39]. Hartree–Fock theory is a more precise approximation to Schrödinger theory than Thomas–Fermi theory and used in the computation of atomic and molecular energies. In contrast to Thomas–Fermi theory, which is a density functional theory, where the unknown is a scalar function, Hartree–Fock theory is a density matrix theory, where the unknown is an operator. The Lieb–Simon paper is a foundational paper in the noncommutative calculus of variations.

In Thomas–Fermi theory, the Coulomb repulsion between the electrons is approximated by the quantity $D[\rho]$. In 1981, Lieb and Oxford [34] proved a lower bound on the difference between these two quantities, which became known as the Lieb–Oxford bound for the exchange energy. This bound is well known to quantum chemists and guides their thinking about the exchange correlation energy in molecules.

In 1983, Lieb published the paper "Density Functionals for Coulomb Systems" [19], which laid out the theoretic foundations of density functional theory and is widely cited. It introduced a universal functional, known as the Levy–Lieb functional, which gives the lowest energy that can be reached with all possible quantum states having a given density function. This functional yields, by definition, the ground state energy of interacting quantum Coulomb systems (even if it is not known explicitly). This point of view has played a very important role. Density functional theory has exploded in the 1990s and is widely used in industry. It is now the method of choice to compute the quantum state of molecules and solids.

Among Lieb's more recent works in this direction are a mathematically rigorous justification of the Local Density Approximation in density functional theory, and a proof of the equivalence in the thermodynamic limit of three different definitions of the minimum energy of a homogeneous electron gas. These are joint works with Lewin and Seiringer [12,13].

Lieb–Thirring inequalities

Arguably the least obvious part in the approximation of the Schrödinger energy functional by the Thomas–Fermi functional is that the correct kinetic energy

$$\int_{\mathbb{R}^3} \cdots \int_{\mathbb{R}^3} \sum_{n=1}^{N} |\nabla_n \psi|^2 \, dx_1 \cdots dx_N$$

is replaced by the term

$$\gamma^{\mathrm{TF}} \int_{\mathbb{R}^3} \rho(x)^{\frac{5}{3}} \, dx$$

for a certain explicit constant $\gamma^{\mathrm{TF}} > 0$. It is in this step (and only in this step) of the approximation that the fermionic nature of the wave function ψ enters. Behind this approximation is the observation that the kinetic energy per unit volume of a noninteracting Fermi gas in its ground state with constant density ρ is $\gamma^{\mathrm{TF}} \rho^{\frac{5}{3}}$. Imagining that particles described by a wave function ψ with low energy are locally essentially in the ground state of the Fermi gas with the corresponding local density one arrives at the Thomas–Fermi expression for the kinetic energy.

The question arises whether this approximation can be substantiated by rigorous bounds. This is accomplished by the famous Lieb–Thirring inequality, which was a crucial ingredient in their proof of stability of matter **[43]**. This inequality states that for any antisymmetric, normalized ψ on $\mathbb{R}^{3N}$, one has

$$\int_{\mathbb{R}^3} \cdots \int_{\mathbb{R}^3} \sum_{n=1}^{N} |\nabla_n \psi|^2 \, dx_1 \cdots dx_N \geq K \int_{\mathbb{R}^3} \rho_\psi(x)^{\frac{5}{3}} \, dx \tag{1.3}$$

with

$$\rho_\psi(x) = \sum_{n=1}^{N} \int_{\mathbb{R}^3} \cdots \int_{\mathbb{R}^3} \left|\psi(x_1, \ldots, x_{n-1}, x, x_{n+1}, \ldots, x_N)\right|^2 dx_1 \cdots dx_{n-1} \, dx_{n+1} \cdots dx_N .$$

The important point here is that the constant K is independent of N.

The Lieb–Thirring inequality (1.3) can be viewed as a mathematical expression of both the exclusion and the uncertainty principles in quantum mechanics. The connection with the exclusion principle is that the constant K is independent of N – if symmetric functions ψ (describing a bosonic system) would be allowed, the constant K in (1.3) would have to deteriorate with N. (As an example, take ψ as a product function $\varphi(x_1) \cdots \varphi(x_N)$.) For $N = 1$, the Lieb–Thirring inequality reduces to a certain Sobolev interpolation inequality, and for $N \geq 2$ it can be considered as a generalization thereof. The conclusion of Sobolev-type inequalities, namely that an L^p with a "large" p can be controlled, is a nonconcentration result and therefore quantifies the uncertainty principle in quantum mechanics.

The constant K in (1.3) that Lieb and Thirring obtained was smaller than γ^{TF}, but it retained the important feature of being independent of N (and of ψ, of course). The famous Lieb–Thirring conjecture states that the inequality should be valid with constant equal to γ^{TF}. This would mean that the Thomas–Fermi approximation for the kinetic energy is a universal lower bound to its Schrödinger expression. There has been a lot of work on this constant, leading to the currently best bound of $(0.7785)\,\gamma^{\mathrm{TF}}$.

Lieb and Thirring did not prove inequality (1.3) directly, but first showed that it is equivalent to a certain inequality about sums of negative eigenvalues of one-body Schrödinger operators, and then verified the latter inequality. In their follow-up paper **[44]** they extended this latter inequality to arbitrary dimensions and arbitrary powers of eigenvalues. They proved that the negative eigenvalues (E_j) of the Schrödinger operator $-\Delta + V$ in $L^2(\mathbb{R}^d)$ satisfy

$$\sum_j |E_j|^\gamma \leq L_{\gamma,d} \int_{\mathbb{R}^d} V(x)_-^{\gamma+d/2}\, dx$$

for all $\gamma > 1/2$ if $d = 1$ and $\gamma > 0$ if $d \geq 2$. For $\gamma = 1$ and $d = 3$, this inequality is equivalent to (1.3). Soon afterwards, Lieb **[16]** proved the corresponding inequality in the endpoint case, known as Cwikel–Lieb–Rozenblum inequality, namely

$$\#\{j : E_j < 0\} \leq L_{0,d} \int_{\mathbb{R}^d} V(x)_-^{d/2}\, dx$$

for $d \geq 3$.

The general form of the famous Lieb–Thirring conjecture concerns the optimal values of the constants $L_{\gamma,d}$. Apart from the obvious relevance of the values of these constants in applications, the conjecture addresses on a conceptual level the strength of the exclusion principle in different dimensions. Lieb, together with Hundertmark and Thomas, have proved the only known case of an optimal inequality where the constant is not given by that arising from a semiclassical (or Thomas–Fermi-like) approximation. Also, after more than four decades, Lieb's value of the constant $L_{0,3}$ is still the smallest one that is known.

The Lieb–Thirring and Cwikel–Lieb–Rozenblum inequalities and their generalizations are of great importance in the study of large fermionic systems. Apart from that, they have found applications in the context of nonlinear evolution equations like the Navier–Stokes equation **[23]**. Moreover, they have attracted great interest from a purely mathematical point of view and the fact that orthogonality of functions leads to an improved dependence on the number of functions has been verified in a number of other functional inequalities as well, including **[9, 18]**.

The ionization problem

Let us return to the quantum many-body Hamiltonian H in (1.1). The infimum of its spectrum may or may not be an eigenvalue. If it is, we interpret the corresponding eigenfunction as describing the ground state of the system and we think of the N electrons as bound to the nuclei. Physical intuition suggests that given nuclei with given charges can only bound a finite number of electrons, but even this is not quite obvious mathematically. The quantitative version of this question, namely how many electrons an atom (or a molecule) can bind, is still not settled, despite serious efforts.

For simplicity, let us restrict our attention to the atomic case, that is, $K = 1$ in (1.1). Experimental data and numerical estimates suggest that a nucleus of charge Z can bind at most $Z + 1$, or possibly $Z + 2$ electrons. To prove (or disprove) this rigorously in the above Schrödinger model is a famous open problem.

One of the few nonasymptotic results in this direction is due to Lieb [22] and states that an atomic nucleus cannot bind $2Z + 1$ or more electrons. The factor 2 in front of Z looks "too large" for large Z, but, for instance, for $Z = 1$, the bound is optimal. Only three decades later was Lieb's result improved for large Z.

A striking discovery by Benguria and Lieb [2] is that the purported bound on the excess charge would not be true if the electrons were bosons, that is, if the antisymmetry requirement on admissible functions was replaced by the symmetry requirement $\psi(\dots, x_n, \dots, x_m, \dots) = \psi(\dots, x_m, \dots, x_n, \dots)$ for $n \neq m$. They showed that there is a number $\lambda > 1$ (numerically, $\lambda \approx 1.21$) such that a "bosonic atom" can bind at least $(\lambda + o(1))Z$ electrons as $Z \to \infty$. As a consequence of the Benguria–Lieb result, the Pauli principle (i.e., the antisymmetry requirement) needs to enter any possible proof. This excludes, in particular, any naive, purely electrostatic argument.

Returning to the original, fermionic case, one can ask whether there is at least asymptotic neutrality in the sense that, as $Z \to \infty$, the number of electrons that can be bound is $Z + o(Z)$. That this is indeed the case was proved by Lieb together with Sigal, Simon, and Thirring [38]. Other researchers obtained later quantitative bounds on the $o(Z)$ remainder, but showing that it is bounded seems to be out of reach of current techniques.

Bosonic systems

While our focus in this section was mostly on fermionic systems, Lieb has also made many fundamental contributions to the study of bosonic systems. Among those are the following:

(a) The construction, with Liniger, of a model of a one-dimensional interacting Bose gas [31]. This model has served as a prototype for later theoretical developments, and it has also been realized experimentally.

(b) Together with Yngvason, Lieb proved an asymptotic formula, conjectured 58 years earlier, for the ground state energy of a dilute Bose gas [45]. Subsequently, together with Seiringer and Yngvason, he rigorously derived the Gross–Pitaevskii equation for the ground state energy of dilute bosons in a trap, starting with many-body quantum mechanics [37]. This result has had a tremendous impact on the development of mathematical physics in the past two decades.

(c) Together with Conlon and Yau [7] and later with Solovej [41, 42], Lieb proved the $N^{7/5}$ law for charged bosons. This means, roughly, that the energy does not obey a linear bound as in (1.2), but rather decreases like $-CN^{7/5}$. This was the first rigorous validation of Bogolubov's pairing theory of the Bose gas and paved the way for many current developments.

2. FUNCTIONAL INEQUALITIES

Lieb's name is inseparably connected with the subject of inequalities and a whole 700 page volume of his Selecta is dedicated to this topic [26]. In the previous section, in the discussion of the Lieb–Thirring inequality, we have already seen one instance of a functional inequality. In this section we will review three more examples, namely entropy inequalities in matrix analysis, the Brascamp–Lieb inequalities, and the sharp form of the Hardy–Littlewood–Sobolev inequality.

Lieb's concavity theorem and the strong subadditivity

In 1973, Lieb and Ruskai proved the *strong subadditivity of the quantum-mechanical entropy* [35]. This theorem has many equivalent formulations, such as the concavity of the conditional entropy, the joint convexity of the relative entropy or the monotonicity of the relative entropy. Strong subadditivity, or one of its equivalents, has come to play an essential role in the modern and very active area of quantum information and quantum computation.

Let us state this theorem in its monotonicity formulation, originally derived by Lindblad in 1974 from a result of Lieb. A density matrix is a Hermitian, positive semidefinite matrix of trace one. The relative entropy (or Kullbach–Leibler divergence) of two density matrices ρ and σ is defined to be

$$D(\rho\|\sigma) = \operatorname{Tr} \rho \ln \rho - \operatorname{Tr} \rho \ln \sigma.$$

This quantity is nonnegative and vanishes if and only if $\rho = \sigma$. Roughly speaking, it measures how distinguishable ρ and σ are, even though the quantity is not symmetric in ρ and σ. Quantum operations are described by completely positive, trace-preserving maps, and the theorem of monotonicity of the relative entropy (also known as the *data processing inequality*) states that for any such operation $\mathcal{E}$, one has

$$D(\mathcal{E}\rho\|\mathcal{E}\sigma) \le D(\rho\|\sigma).$$

Thus, applying a quantum operation can only make the states harder to distinguish. This makes it clear that the monotonicity of the relative entropy lies at the very foundation of quantum information theory.

Both the Lieb–Ruskai proof of the strong subadditivity of the entropy and the Lindblad proof of monotonicity of the relative entropy rest on a deep theorem that Lieb proved in his 1973 paper "Convex trace functions and the Wigner–Yanase–Dyson conjecture" [14]. This theorem states that, for numbers $p, q \ge 0$ with $p + q \le 1$, the map $(A, B) \mapsto \operatorname{Tr} A^p B^q$, defined on nonnegative Hermitian matrices, is jointly concave.

Lieb's paper has generated considerable further work containing alternative proofs, generalizations, and applications. He himself, often jointly with Carlen, has also returned several times to this circle of ideas and deepend our understanding of matrix analysis; see, e.g., [1, 6].

The Brascamp–Lieb inequalities

In 1976 Brascamp and Lieb published the paper "Best Constants in Young's Inequality, Its Converse and Its Generalization to More Than Three Functions" [3]. As the title suggests, this paper treats three related, but different topics and it is famous for all three.

The first one concerns a basic inequality in real analysis, namely the Young inequality

$$\left| \iint_{\mathbb{R}^d \times \mathbb{R}^d} f(x) g(x-y) h(y)\, dx\, dy \right| \leq C_{p,q,r,d} \|f\|_{L^p(\mathbb{R}^d)} \|g\|_{L^q(\mathbb{R}^d)} \|h\|_{L^r(\mathbb{R}^d)} \quad (2.1)$$

for three functions f, g, h on $\mathbb{R}^d$ and parameters $1 \leq p, q, r \leq \infty$ satisfying $\frac{1}{p} + \frac{1}{q} + \frac{1}{r} = 2$. This inequality appears frequently in theory and applications, since convolution is a basic operation in mathematical analysis and the Young inequality bounds its effect in Lebesgue spaces.

The question is to find the optimal (that is, smallest possible) constant $C_{p,q,r,d}$ in (2.1) and to characterize all cases of equality. The first task was solved by Beckner around the same time as [3]. Brascamp and Lieb give an alternative proof and solve the second task. Their proof combines in an original way the technique of symmetric decreasing rearrangement with the operation of taking tensor products, leading to a convergence result in the spirit of the central limit theorem. It follows that the optimal constant is determined by Gaussian functions and attained only for those.

The second topic of the Brascamp–Lieb paper is that Young's inequality holds, with the inequality sign reversed, if $0 < p, q \leq 1$ and f, g, h are nonnegative. Again the authors are able to compute the optimal constant. This reverse Young's inequality contains, as a limiting case, the Prékopa–Leindler theorem.

The third topic of the Brascamp–Lieb paper is a far-reaching generalization of Young's inequality to an arbitrary number of functions, together with a replacement of the linear functions x, y and $x - y$ on $\mathbb{R}^{2d}$, appearing in (2.1), by a much larger class of linear functions. The resulting family of inequalities is now known as Brascamp–Lieb inequalities. It contains not only Young's inequality, but also those of Hölder and of Loomis and Whitney as special cases. Brascamp and Lieb show that for given linear maps the inequality holds with a finite constant if and only if it holds for Gaussian functions, and in this case the optimal constant can be computed using the latter class of functions. Their argument is again based on a central limit theorem.

Besides the original application to statistical mechanics, the Brascamp–Lieb inequalities have come to play an important role in convexity theory, harmonic analysis and other parts of mathematics.

In his 1990 paper with the title "Gaussian Kernels Have Only Gaussian Maximizers" [24], Lieb revisits the topic of Young and Brascamp–Lieb inequalities. He proves a general theorem about the norm of a large class of integral operators. Among other things, he characterizes the cases of equality in the Hausdorff–Young inequality with optimal constant. Again, they are given by Gaussian functions.

We finally mention that there is yet another celebrated inequality known as Brascamp–Lieb inequality. This is a Poincaré (or spectral gap) inequality for log-concave probability distributions **[4]**.

The sharp Hardy–Littlewood–Sobolev inequality

A fundamental result of real analysis, known as the Hardy–Littlewood–Sobolev inequality, the weak Young inequality or the theorem of fractional integration, states that, for $0 < \lambda < d$ and $1 < p, r < \infty$ with $\frac{1}{p} + \frac{\lambda}{d} + \frac{1}{r} = 2$, one has

$$\left| \iint_{\mathbb{R}^d \times \mathbb{R}^d} \frac{f(x)h(y)}{|x-y|^{\lambda}} \, dx \, dy \right| \leq C_{\lambda,p,r} \|f\|_{L^p(\mathbb{R}^d)} \|h\|_{L^r(\mathbb{R}^d)}. \tag{2.2}$$

This inequality has many applications in pure and applied mathematics. For instance, in the case $\lambda = d-2$, $d \geq 3$, the kernel in this inequality is the Coulomb kernel and the quantity on the left-hand side represents the Coulomb interaction of two charge distributions f and h. For $\lambda = d-2$ and $p = r$, inequality (2.2) is equivalent to the Sobolev inequality.

Lieb's 1983 paper "Sharp constants in the Hardy–Littlewood–Sobolev and related inequalities" **[21]** had a profound impact on the field of Calculus of Variations. Indeed, this paper and his related 1983 paper "A relation between pointwise convergence of functions and convergence of functionals" **[5]** with Brezis are Lieb's most-cited pure mathematics papers.

The "Sharp constants" paper is remarkable for at least two distinct aspects, namely for the development of a rather general compactness argument and for an ingenious observation concerning the problem at hand. We briefly discuss these two points.

The compactness aspect of Lieb's HLS paper concerns the question whether there is a nontrivial pair (f, h) such that equality holds in (2.2) with $C_{\lambda,p,r}$ taken to be the minimal constant. Lieb had worked earlier on questions about existence of optimizers for certain variational problems. In contrast to classical variational problems, these problems were often translation-invariant and so Lieb had to find methods to deal with the corresponding loss of compactness (for instance, by symmetric decreasing rearrangement **[15]** or by an original "running-after argument" **[20]**). The optimization problem for the HLS inequality, however, features another potential loss of compactness, namely through dilations. In order to deal with these problems, Lieb found a strengthening of Fatou's lemma, which says that, if functions f_j on a measure space X are bounded in L^p and converge pointwise a.e. to a function f, then

$$\lim_{j\to\infty} \int_X \left| |f_j|^p - |f|^p - |f_j - f|^p \right| dx = 0. \tag{2.3}$$

In particular,

$$\int_X |f_j|^p \, dx = \int_X |f|^p \, dx + \int_X |f_j - f|^p \, dx + o(1).$$

For comparison, in Fatou's lemma, the second term on the right-hand side is omitted, thus leading to an inequality rather than an equality. The statement (2.3), with $|\cdot|^p$ replaced by more general functions, is known as the Brezis–Lieb lemma and constitutes a fundamental tool in functional analysis and the calculus of variations.

Lieb's compactness method, sometimes called the method of the missing mass, tracks carefully the remainder term $f_j - f$ in (2.3). This technique is quite robust and has found various applications to different settings, including [11]. Also other compactness methods use the Brezis–Lieb lemma as a fundamental ingredient.

The second remarkable aspect of Lieb's HLS paper concerns the special case $p = r$ (but $0 < \lambda < d$ is arbitrary). Lieb managed to compute the smallest possible constant $C_{\lambda,p,r}$ explicitly and to characterize all pairs (f, h) of functions for which equality is attained. The crucial observation in Lieb's proof was a "hidden" symmetry, namely the conformal invariance of (2.2) for $p = r$. Lieb's paper has spawned a field of variational problems with conformal invariance. Among these developments is also the sharp form of the HLS inequality on the Heisenberg group [10].

3. TOPICS NOT COVERED

In the previous sections we have discussed two areas of Lieb's work, namely quantum Coulomb systems and functional inequalities. Those make up just a fraction of the body of Lieb's work and it is not unlikely that other writers would have picked completely different topics. We would be remiss if we would not at least briefly mention a few more.

We have completely ignored the important chapter of exactly soluble models in Lieb's work in the 1960s. This field was revolutionized by Lieb's solutions of the ice problem, the Fierz–Rys F-model and Slater's KDP model, as well as the already mentioned Lieb–Liniger model. In connection with ice type models, Lieb and Temperley constructed what became known as the Temperley–Lieb algebra, which has applications in several areas of mathematics, e.g., knot theory, and in physics. Lieb's work lies at the foundations of modern statistical mechanics and has a long-lasting influence on integrable probability and combinatorics, to name just a few.

Concerning Lieb's contributions to condensed matter physics Nachtergaele, Solovej, and Yngvason write in the preface of the third volume of Lieb's selected works [27]:

> *"The impact of Lieb's work in mathematical condensed matter physics is unrivaled. It is fair to say that if one were to name a founding father of the field, Elliott Lieb would be the only candidate to claim this singular position."*

This area includes, in particular, Lieb's seminal work on the Hubbard model, including its solution in the one-dimensional case with Wu and the discovery of Lieb ferromagnetism and the Lieb lattice, as well as the highly cited joint works with Schultz and Mattis on two soluble models of an antiferromagnetic chain and with Mattis on the Luttinger model and the discovery of bosonization.

Already in the introduction we mentioned the Lieb–Robinson bounds, which establish the deep fact that there is a finite group velocity for information propagation in quantum spin systems. This turned out to be quite important for ideas about quantum computation and condensed matter physics.

Moreover, together with Affleck, Kennedy, and Tasaki, Lieb developed what came to be known as the AKLT model of a spin-one spin chain. This has proved to be an important prototype of a class of models and eigenstates known as matrix product states.

From the area of statistical mechanics we mention Lieb's proof of the existence of phase transitions in classical and quantum spin systems, obtained jointly with Fröhlich, Israel, and Simon and with Dyson and Simon. The proofs are based on the method of reflection positivity, which Lieb has masterfully used in several situations.

Another famous result is Lieb's work with Heilman on the zeros of the partition function of the monomer–dimer problem, which is related to the matching problem in combinatorics. This is also of importance in computer science.

A notable mention is a new approach to the physics and mathematics of the second law of thermodynamics, developed jointly with Yngvason.

The list of topics in this section has so far focused on the more physics-oriented side of Lieb's work. From the more mathematics-oriented side, let us just mention his fundamental contributions to the theory of symmetric decreasing rearrangement. This includes, among other things, a general rearrangement inequality for many functions, obtained with Brascamp and Luttinger, as well as the proof that symmetric decreasing rearrangement can be discontinuous in Sobolev spaces, obtained with Almgren.

Finally, we would like to highlight the book by Lieb and Loss [32], which has become a standard textbook for graduate courses in mathematical analysis and which eloquently and concisely promotes the philosophy of rigorous mathematics with a view towards applications.

We hope that these pages may serve as an invitation to look at Lieb's original research papers. Only in this way can the reader feel the clarity, vitality and beauty of Lieb's work, a work that has, and continues to, inspire generations.

Congratulations on receiving the Gauss Prize, Elliott!

FUNDING

Partial support through US National Science Foundation grant DMS-1954995, as well as through the Deutsche Forschungsgemeinschaft (DFG, German Research Foundation) through Germany's Excellence Strategy EXC-2111-390814868 is acknowledged. The author is grateful to M. Lewin for helpful remarks.

REFERENCES

[1] K. Ball, E. A. Carlen, and E. H. Lieb, Sharp uniform convexity and smoothness inequalities for trace norms. *Invent. Math.* **115** (1994), 463–482.

[2] R. Benguria and E. H. Lieb, Proof of the stability of highly negative ions in the absence of the Pauli principle. *Phys. Rev. Lett.* **50** (1983), 1771–1774.

[3] H. J. Brascamp and E. H. Lieb, Best constants in Young's inequality, its converse and its generalization to more than three functions. *Adv. Math.* **20** (1976), 151–172.

[4] H. J. Brascamp and E. H. Lieb, On extensions of the Brunn–Minkowski and Prékopa–Leindler theorems, including inequalities for log concave functions and with an application to the diffusion equation. *J. Funct. Anal.* **22** (1976), 366–389.

[5] H. Brezis and E. H. Lieb, A relation between pointwise convergence of functions and convergence of functionals. *Proc. Amer. Math. Soc.* **88** (1983), 486–490.

[6] E. A. Carlen and E. H. Lieb, A Minkowski type trace inequality and strong subadditivity of quantum entropy. In *Differential operators and spectral theory*, pp. 59–69, Amer. Math. Soc. Transl. Ser. 2 189, American Mathematical Society, Providence, 1999.

[7] J. Conlon, E. H. Lieb, and H.-T. Yau, The $N^{7/5}$ law for charged bosons. *Comm. Math. Phys.* **116** (1988), 417–448.

[8] R. L. Frank, A. Laptev, M. Lewin, and R. Seiringer (eds.), *The physics and mathematics of Elliott Lieb: The 90th Anniversary Volume I and II*, EMS Press, 2022, to appear, DOI 10.4171/90.

[9] R. L. Frank, M. Lewin, E. H. Lieb, and R. Seiringer, Strichartz inequality for orthonormal functions. *J. Eur. Math. Soc. (JEMS)* **16** (2014), 1507–1526.

[10] R. Frank and E. H. Lieb, Sharp constants in several inequalities on the Heisenberg group. *Ann. of Math. (2)* **176** (2012), 349–381.

[11] R. Frank, E. H. Lieb, and J. Sabin, Maximizers for the Stein–Tomas inequality. *Geom. Funct. Anal.* **26** (2016), 1095–1134.

[12] M. Lewin, E. H. Lieb, and R. Seiringer, Statistical mechanics of the uniform electron gas. *J. Éc. Polytech. Math.* **5** (2018), 79–116.

[13] M. Lewin, E. H. Lieb, and R. Seiringer, The local density approximation in density functional theory. *Pure Appl. Anal.* **2** (2020), 35–73.

[14] E. H. Lieb, Convex trace functions and the Wigner–Yanase–Dyson conjecture. *Adv. Math.* **11** (1973), 267–288.

[15] E. H. Lieb, Existence and uniqueness of the minimizing solution of Choquard's non-linear equation. *Stud. Appl. Math.* **57** (1977), 93–105.

[16] E. H. Lieb, The number of bound states of one-body Schrödinger operators and the Weyl problem. *Proc. Amer. Math. Soc. Sympos. Pure Math.* **36** (1980), 241–252.

[17] E. H. Lieb, Thomas–Fermi and related theories of atoms and molecules. *Rev. Modern Phys.* **53** (1981), 603–641.

[18] E. H. Lieb, An L^p bound for the Riesz and Bessel potentials of orthonormal functions. *J. Funct. Anal.* **51** (1983), 159–165.

[19] E. H. Lieb, Density functionals for Coulomb systems. *Int. J. Quantum Chem.* **24** (1983), 243–277.

[20] E. H. Lieb, On the lowest eigenvalue of the Laplacian for the intersection of two domains. *Invent. Math.* **74** (1983), 441–448.

[21] E. H. Lieb, Sharp constants in the Hardy–Littlewood–Sobolev and related inequalities. *Ann. of Math. (2)* **118** (1983), 349–374.

[22] E. H. Lieb, Bound on the maximum negative ionization of atoms and molecules. *Phys. Rev. A* **29** (1984), 3018–3028.

[23] E. H. Lieb, On characteristic exponents in turbulence. *Comm. Math. Phys.* **92** (1984), 473–480.

[24] E. H. Lieb, Gaussian kernels have only Gaussian maximizers. *Invent. Math.* **102** (1990), 179–208.

[25] E. H. Lieb, The stability of matter: from atoms to stars. *Bull. Amer. Math. Soc. (N.S.)* **22** (1990), 1–49.

[26] E. H. Lieb, *Inequalities. Selecta of Elliott H. Lieb.* Edited, with a preface and commentaries, by M. Loss and M. B. Ruskai, Springer, Berlin, 2002.

[27] E. H. Lieb, *Condensed matter physics and exactly soluble models. Selecta of Elliott H. Lieb.* Edited by B. Nachtergaele, J. P. Solovej and J. Yngvason, Springer, Berlin, 2004.

[28] E. H. Lieb, *Statistical mechanics. Selecta of Elliott H. Lieb.* Edited, with a preface and commentaries, by B. Nachtergaele, J. P. Solovej and J. Yngvason, Springer, Berlin, 2004.

[29] E. H. Lieb, *The stability of matter: from atoms to stars. Selecta of Elliott H. Lieb.* Edited by W. Thirring, and with a preface by F. Dyson, 4th edn., Springer, Berlin, 2005.

[30] E. H. Lieb and J. L. Lebowitz, The constitution of matter: existence of thermodynamics for systems composed of electrons and nuclei. *Adv. Math.* **9** (1972), 316–398.

[31] E. H. Lieb and W. Liniger, Exact analysis of an interacting Bose gas. I. The general solution and the ground state. *Phys. Rev.* **130** (1963), 1605–1616.

[32] E. H. Lieb and M. Loss, *Analysis*. 2nd edn., Grad. Stud. Math. 14, American Mathematical Society, Providence, RI, 2001.

[33] E. H. Lieb, M. Loss, and J. P. Solovej, Stability of matter in magnetic fields. *Phys. Rev. Lett.* **75** (1995), no. 6, 985–989.

[34] E. H. Lieb and S. Oxford, An improved lower bound on the indirect Coulomb energy. *Int. J. Quantum Chem.* **19** (1981), 427–439.

[35] E. H. Lieb and M. B. Ruskai, Proof of the strong subadditivity of quantum-mechanical entropy. *J. Math. Phys.* **14** (1973), 1938–1941.

[36] E. H. Lieb and R. Seiringer, *The stability of matter in quantum mechanics*. Cambridge University Press, Cambridge, 2010.

[37] E. H. Lieb, R. Seiringer, and J. Yngvason, Bosons in a trap: a rigorous derivation of the Gross–Pitaevskii energy functional. *Phys. Rev. A* **61** (2000), art. 043602.

[38] E. H. Lieb, I. M. Sigal, B. Simon, and W. Thirring, Approximate neutrality of large-Z ions. *Comm. Math. Phys.* **116** (1988), 635–644.

[39] E. H. Lieb and B. Simon, The Hartree–Fock theory for Coulomb systems. *Comm. Math. Phys.* **53** (1977), 185–194.

[40] E. H. Lieb and B. Simon, Thomas–Fermi theory of atoms, molecules and solids. *Adv. Math.* **23** (1977), 22–116.

[41] E. H. Lieb and J. P. Solovej, Ground state energy of the one-component charged Bose gas. *Comm. Math. Phys.* **217** (2001), 127–163; Erratum ibid. **225** (2002), 219–221.

[42] E. H. Lieb and J. P. Solovej, Ground tate energy of the two-component charged Bose gas. *Comm. Math. Phys.* **252** (2004), 485–534.

[43] E. H. Lieb and W. Thirring, Bound for the kinetic energy of fermions which proves the stability of matter. *Phys. Rev. Lett.* **35** (1975), 687–689; Errata ibid. **1116** (1975).

[44] E. H. Lieb and W. Thirring, Inequalities for the moments of the eigenvalues of the Schrödinger Hamiltonian and their relation to Sobolev inequalities. In *Studies in Mathematical Physics*, edited by E. Lieb, B. Simon, and A. Wightman, pp. 269–303, Princeton University Press, 1976.

[45] E. H. Lieb and J. Yngvason, Ground state energy of the low density Bose gas. *Phys. Rev. Lett.* **80** (1998), 2504–2507.

RUPERT L. FRANK

Mathematisches Institut, Ludwig-Maximilans Universität München, Theresienstr. 39, 80333 München, Germany, and Munich Center for Quantum Science and Technology, Schellingstr. 4, 80799 München, Germany, and Mathematics 253-37, Caltech, Pasadena, CA 91125, USA, r.frank@lmu.de

This Page Intentionally Left Blank

NIKOLAI ANDREEV AND THE ART OF MATHEMATICAL ANIMATION AND MODEL-BUILDING

TADASHI TOKIEDA

ABSTRACT

The Leelavati Prize of the IMU was awarded at the ICM 2022 to Nikolai Andreev, for his original development of mathematical animation and of mathematical model-building, in a style which inspires the young and the old alike, and which mathematicians around the world can adapt to their varied uses—as well as for his indefatigable efforts to popularize genuine mathematics among the public.

MATHEMATICS SUBJECT CLASSIFICATION 2020

Primary 97A80; Secondary 97U80, 97U60

KEYWORDS

Leelavati, Andreev, animation, model, popularization

INTERNATIONAL CONGRESS OF MATHEMATICIANS 2022 JULY 6–14

© 2022 International Mathematical Union
Proc. Int. Cong. Math. 2022, Vol. 1, pp. 160–162
DOI 10.4171/ICM2022/215

Published by EMS Press and licensed under a CC BY 4.0 license

In the realm of visual and tactile mathematics, Nikolai Nikolayevich Andreev is a master of a wondrous art of animation and of model-building. Animation differs from simulation: it is a cartoonish video of a story unfolding in front of our eyes and precipitating some mathematical surprise, delightful to watch. Models differ from 3D-printing: they are made by hand, of wood or paper, entertaining to touch and manipulate. His animation and models are minimalist, yet executed with consummate craftsmanship. Andreev's signature style of this art captures the imagination of both the young and the old, and offers potential for a variety of uses in the popularization of mathematics. In parallel, he is recognized for tremendous resilience in often overcoming hardship to continue kindling enthusiasm for mathematics among a large number of people of all life-circumstances, via web resources, lectures, and a book. For these he is being awarded the 2022 Leelavati Prize.

Eastern Europe has a grassroots tradition, harking back to the era of Tchebyshev, of organizing nationwide ladders, so to speak, of mathematical activities from small children to college students; some of it has been exported abroad, as witnessed by 'math circles' that flourish in hubs of initiative around the globe today. Out of this tradition came *Kvant*, arguably the highest-quality magazine of popularization in mathematics and theoretical physics that the world has seen, commanding in its heyday a circulation of 2×10^5. (As of 2012 a younger sibling *Kvantik* [1] came into action.) It is to this tradition that Andreev, or Kolya to his many friends, was born in 1975 in Saratov, and to this tradition that he has claim to be a leading successor in our 21st century.

His early training was as a researcher. Andreev graduated from the Faculty of Mechanics and Mathematics of Moscow State University, completing a candidate's degree (PhD) in 2000 in the area of extremal problems and approximation theory, codes, and designs. In the same year he began working at Steklov Mathematical Institute, where he has been based ever since.

The year 2002 marked a watershed: Andreev gathered a team of R. A. Koksharov (senior developer, web design), M. A. Kalinichenko (graphics, video producer), N. M. Panyunin (mathematics), and created the project 'Mathematical Études' [2]. The project is a treasure trove of animation videos, available to everybody free of charge. Each video gives a brief but genuine mathematical experience of an interesting point that is elementary, but little known; as such, it is orthogonal to the common practice of presenting journalistically a fashionable topic. He also recruited A. D. Leshinskii, a wright of stunning skill who realizes curious mathematical phenomena in beautiful wooden models. In 2010 Andreev was appointed Head of the Laboratory of Popularization of Steklov Institute. The productions of his Lab include, alongside the ongoing growth of Études plus models, the collection 'mechanisms by Tchebyshev' [3], moveable gadgets which make intriguing uses of classical mathematics, and the book *Matematicheskaia͡ sostavlia͡iu͡shchaia͡* (which we may translate freely as *Mathematical take on things*) [4], an anthology of about 30 mathematicians on a luxuriant diversity of material reminiscent of *Kvant*. The book, first published 2015 by him, S. P. Konovalov, N. M. Panyunin, R. A. Koksharov, earned a gold medal for scientific writing 2017; the second edition, more than double in content, followed in 2019.

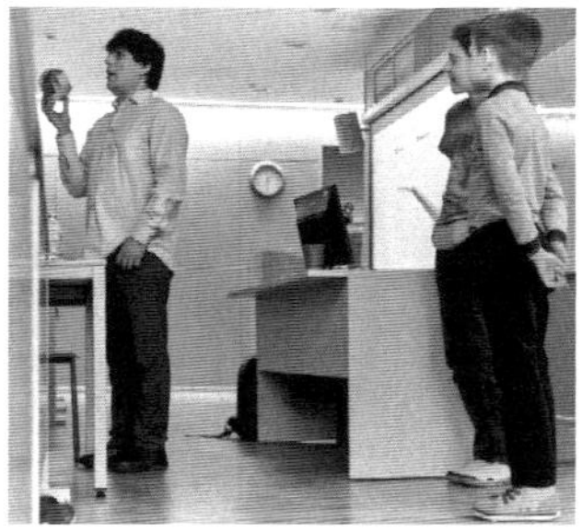

Andreev travels the length and breadth of a vast continent to deliver lectures, well over 1000 in 20 years, reaching out to by now countless members of the public, especially the adolescents. Time and again extraordinary dedication and perseverance saw his cause through: chronic administrative and financial trammels, endless negotiations and set-backs. Whenever his team's funding dried up, he divided his own salary in equal parts among himself and the other staff of the team, in order to keep the work alive.

For all his accomplishments, much of Andreev's career is still ahead of him. We salute Kolya, as one representative of the community of mathematicians through the centuries who gave of themselves selflessly to doing mathematics with each rising generation. We look forward to being raised by his future work for decades to come.

REFERENCES

[1] https://kvantik.com/en/

[2] https://etudes.ru

[3] https://tcheb.ru

[4] https://book.etudes.ru

TADASHI TOKIEDA

Department of Mathematics, Stanford University, Stanford CA 94305-2125, USA

tokieda@stanford.edu

PRIZE LECTURES

100 YEARS OF THE (CRITICAL) ISING MODEL ON THE HYPERCUBIC LATTICE

HUGO DUMINIL-COPIN

Dedicated to the memory of colleagues and friends, Dmitry Ioffe and Vladas Sidoravicius.

ABSTRACT

We take the occasion of this article to review 100 years of the physical and mathematical study of the Ising model. The model, introduced by Lenz in 1920, has been at the cornerstone of many major revolutions in statistical mechanics. We wish, through its history, to outline some of these amazing developments. We restrict our attention to the ferromagnetic nearest-neighbor model on the hypercubic lattice, and essentially focus on what happens at or near the so-called critical point.

MATHEMATICS SUBJECT CLASSIFICATION 2020

Primary 82B20; Secondary 60K35

KEYWORDS

Ising model, percolation theory, phase transition, statistical mechanics

ICM INTERNATIONAL CONGRESS OF MATHEMATICIANS 2022 JULY 6–14

© 2022 International Mathematical Union
Proc. Int. Cong. Math. 2022, Vol. 1, pp. 164–210
DOI 10.4171/ICM2022/204

Published by EMS Press and licensed under a CC BY 4.0 license

1. SHORT MOTIVATION

How to provide an introduction to (part of) statistical physics aimed at a wide audience of mathematicians? The question is not easy, especially since the domain is positioned halfway between theoretical physics and mathematics, and that, contrarily to (some) other fields of mathematics, it is hard to identify a theory that would embrace most of statistical physics. A partial answer may be to follow the standard approach of teaching by examples, and to pick what is maybe the most classical model of statistical physics, namely the Ising model. Through its history, one may trace many of the revolutions, both on the theoretical physics and mathematical sides, that statistical physics underwent in the last century.

We therefore chose to streamline this history, from the emergence of the model to explain experimental results, to its modern applications in mathematics, physics, and beyond. Obviously, the story will be tainted by the expertise of the author since thousands of papers have been published on the subject, which ranges over many subfields of mathematics and theoretical physics. A subjective selection of papers has therefore been made, and the attention has been restricted to the model on the hypercubic lattice, at equilibrium, (for most of the review) at criticality, and always with nearest neighbor ferromagnetic interactions (we are well aware that noncritical and dynamical aspects, as well as long-range, random, or antiferromagnetic interactions, also are of prime importance).

We tried to respect the timing of the appearance of the different notions pertaining to the model, and avoided as much as possible some tempting anachronisms. As a result, certain readers may be surprised by some statements, knowing that simpler and more natural versions exist nowadays. Also, the large number of breakthroughs in the Ising model's history—Peierls' argument, Onsager's solution on the square lattice and the exact integrability results that followed, Kadanoff's scaling and universality hypotheses, the Lee–Yang theorem, correlation inequalities, the Fortuin–Kasteleyn representation, reflection positivity, Aizenman's treatment of the random current representation and use of differential inequalities, conformal field theory, rigorous renormalization group, Chelkak–Smirnov's conformal invariance, 3D conformal bootstrap, to cite but a few—forced us to be very quick on some of these developments. References are added for the avid reader. We also refer to [20, 85] for historical introductions, and [48] for a book on statistical physics including a study of the Ising model.

2. THE FIRST 20 YEARS: A LABORIOUS START

2.1. Ising model's prehistory

In 1895, the French physicist Pierre Curie [30] noticed that magnets lose their magnetic attraction when they are heated above a certain critical temperature, now called the *Curie temperature* (the phenomenon seemed to have been discovered before by the French physicist Pouillet in 1832, see [69] for precise references and a discussion). While the Curie temperature varies from slightly over 100 degrees Celsius for certain alloys, to 769.85 degrees Celsius for magnets made of iron, the underlying phenomenon is always the same:

at a certain temperature, a magnet ceases to be able to keep a *spontaneous magnetization* and exhibits magnetization *only* when an external field is applied to it. This phenomenon is called a *phase transition* between a paramagnetic phase above the Curie temperature, and a ferromagnetic phase below it.

Curie also identified a law, now called *Curie's law*, relating the magnetic susceptibility of the system to the temperature and the external magnetic field applied to the magnet. He noticed the similarity between the ferromagnetic and paramagnetic phases of a magnet in terms of the temperature and the external magnetic field applied to it, and respectively the liquid and gas phases of a fluid in terms of the pressure and temperature. Pierre Weiss [98] tried to produce an efficient physical explanation of this phenomenon by introducing an assumption, referred to today as the *mean-field approximation*. This mean-field model, called the *Curie–Weiss model*, gave rise to an interesting, yet not fully accurate, description of the phase transition.

The German physicist Wilhelm Lenz got interested in Curie's law. Lenz agreed with one of Weiss' suggestions that magnets are made of elementary pieces that behave themselves as small magnets. Yet, he was at the same time in line with his contemporary physicists, thinking that one of Weiss' assumption, namely that elementary magnets can rotate freely within a solid, was wrong. Taking this into account, he challenged the rotational freeness. Observing that a crystal selects certain directions corresponding to its symmetries, he made the assumption that elementary magnets also behave in this way. By analogy, he then suggested that a crystal-like mechanism for magnets should favor that neighboring elementary magnets are aligned, therefore corresponding to either pointing in the same or opposite directions. At the end, the reasoning of Lenz led to the assumption that elementary magnets were taking only two possible directions that are opposite of each other. He formalized this reasoning in [80].

At this stage, Lenz did not propose an explicit form for the interaction between elementary magnets. Also, the paper approximately explained the typical behavior of a paramagnet having respectively zero magnetization when no magnetic field is applied, and a magnetization when such a magnetic field is applied, but Lenz made no mention of what will later be referred to as the ferromagnetic behavior.

Ernst Ising was a German physicist born in 1900, who was a PhD student of Lenz in Hamburg. He graduated in 1924 and published a paper [68] on Lenz's model in 1925. So, what did Ising actually achieve in his famous paper from 1925?

First of all, he went one step further than Lenz by specifying the interaction between elementary magnets. He first made the assumption that interactions "*decay rapidly with distance, so that we, in general, to a first approximation, only have to take the influence on neighboring elements into account*." He also assumed that "*of all the possible positions that the neighboring atoms can assume in relation to each other, the one that requires the minimum energy is when they are both acting in the same direction*." These two assumptions led to the mathematical model that we will define formally in the next section. In order to treat this model, Ising made a further assumption: he assumed that the elementary magnets are positioned on a linear chain.

From all of this, Ising could deduce Curie's law in the paramagnetic phase. While this was a source for optimism, the latter was severely challenged by the observation that the magnetization was tending to 0 as the magnetic field vanished, irrespective of the temperature. In other words, no explanation for ferromagnetism was in sight. Even worse, despite a few attempts at generalizing the model (Ising considered nonnearest neighbor interactions, more possible directions for the elementary magnets, and a hybrid three-dimensional model that would correspond to a limit in which only pairs of neighboring elementary magnets in one direction truly interact), the ferromagnetism did not seem to be explainable. This led Ising to conjecture that the model was not a good explanation for ferromagnetism (even when considering higher dimensional base graph for the spins), a thought that he gathered in a letter to American historian Stephen Brush years later: "*I discussed the result of my paper widely with Professor Lenz and with Dr. Wolfgang Pauli, who at that time was teaching in Hamburg. There was some disappointment that the linear model did not show the expected ferromagnetic properties.*"

After his PhD, Ising left academia to become a teacher in Germany before being forced to step down due to his Jewish origins. He fled Nazi Germany and emigrated to the United States, where he became a Professor in Physics at Bradley University. He never published after his first original paper, and only later became aware of how famous the model had grown into.

2.2. Formal definition

Let us turn to the formal definition of the model for our magnet. Consider a finite nonoriented subgraph $G = (V, E)$ of the hypercubic lattice $\mathbb{Z}^d$ with vertex-set V corresponding to the position of its elementary magnet constituents, and edge-set E modeling the links between neighboring ones. An edge $e \in E$ is often written $e = \{x, y\}$, where x and y are its endpoints. The elementary magnet at $x \in V$ will be a quantity $\sigma_x \in \{-1, +1\}$, where -1 and $+1$ correspond to the two opposite directions that it may take. The value σ_x is called the *spin at* x, and the collection $(\sigma_x : x \in V) \in \{-1, 1\}^V$ of all spins at vertices in V is called the *spin configuration*, and should be understood as the state of our magnet.

The *energy*—or *Hamiltonian*—of a configuration σ on G is given by

$$H_{G,h}(\sigma) := -\sum_{\{x,y\}\in E} \sigma_x\sigma_y - h\sum_{x\in V} \sigma_x, \tag{2.1}$$

where $h \in \mathbb{R}$ is called the *magnetic field*. Sometimes, one may want to generalize the model to accommodate nonnearest neighbor and nonferromagnetic interactions by setting

$$H_{G,h,(J_{x,y})}(\sigma) := -\sum_{x,y\in V} J_{x,y}\sigma_x\sigma_y - h\sum_{x\in V} \sigma_x, \tag{2.2}$$

where the $(J_{x,y} : x, y \in V)$ are called the *coupling constants* of the model. Except when otherwise stated, we focus here on the Hamiltonian $H_{G,h}$ corresponding to what is called the *nearest neighbor ferromagnetic* (n.n.f.) Ising model on G.

Following Boltzmann, one considers the *(grand) partition function* of the Ising model on G at inverse-temperature β and magnetic field h defined by

$$Z(G,\beta,h) := \sum_{\sigma\in\{-1,1\}^V} \exp\bigl[-\beta H_{G,h}(\sigma)\bigr]. \tag{2.3}$$

The quantity β is interpreted as the inverse of the temperature, as the latter corresponds to the thermal excitation of elementary magnets, for which it is natural to predict that the larger their excitation, the less relevant their interaction.

Physicists then consider the linear form defined for every function $X : \{-1,1\}^V \to \mathbb{R}$ by the formula

$$\langle X\rangle_{G,\beta,h} := \frac{1}{Z(G,\beta,h)} \sum_{\sigma\in\{-1,1\}^V} X(\sigma)\exp\bigl[-\beta H_{G,\beta,h}(\sigma)\bigr]. \tag{2.4}$$

At this stage, we do not consider $\langle\cdot\rangle_{G,\beta,h}$ itself, and instead focus on a thermodynamical quantity of the system called the free energy. Consider a box $\Lambda_n := [-n,n]^d \cap \mathbb{Z}^d$ and define the *free energy* of the d-dimensional Ising model by the formula

$$f(\beta,h) := -\frac{1}{\beta}\lim_{n\to\infty}\frac{1}{|\Lambda_n|}\ln Z(\Lambda_n,\beta,h) \tag{2.5}$$

(the existence of the limit is justified by a subadditivity argument left to the reader).

Originally, Lenz and Ising were interested in a quantity

$$m(\beta,h) := -\frac{\partial}{\partial h} f(\beta,h), \tag{2.6}$$

which is interpreted as the *magnetization* of the system in the presence of a magnetic field of strength h. One may then define the *spontaneous magnetization*, which corresponds to the remaining magnetization when removing the magnet from the ambient magnetic field,

$$m^*(\beta) := \lim_{h\searrow 0} m(\beta,h) \tag{2.7}$$

(to justify the limit, one may prove that $m(\beta,h)$ decreases as h decreases). The cases $m^*(\beta) = 0$ and $m^*(\beta) > 0$ are respectively called the paramagnetic and ferromagnetic cases as they correspond to the cases where the magnet respectively loses or keeps its magnetization even without external magnetic field.

2.3. What does the Ising model truly model?

The Ising model did not develop quickly after its introduction. The original paper was cited very sporadically in the ten years that followed. In fact, Ising himself was aware of one citation to his paper only, and this lack of interest was one of the reasons that pushed him to abandon academia.

There are several explanations why the paper received little attention. The first is that the negative result of the paper, stating that the model did not explain ferromagnetism, was a pretty disappointing one. The second is a timing problem. A few years after Ising's paper, Heisenberg introduced another model of ferromagnetism [61] based on quantum mechanics, in which the "classical" spins of the Ising model are replaced by the quantum spins of electrons. In other words, Heisenberg's model tries to explain ferromagnetism via the interaction of the spin angular momentum of the electrons in the atoms, while the Ising model was relying on their magnetic moments. In a certain sense, the Ising model was a semiclassical version of Heisenberg model, and as such was violating the latest developments of quantum mechanics. The discrepancy between the great predictive successes of the Heisenberg model, and the impossibility to reconcile the Ising model with the recent advancements in modern physics almost entirely disqualified the model as a good description of ferromagnetic materials.

At this point, one may wonder why this model, initially introduced in theoretical physics to explain ferromagnetism but seemingly unable to do so, did not simply fall into darkness after this rocky start. An element of answer can be found in the developments of other fields of physics, which we now review.

In 1919, the Russian–German chemical physicist Gustav Tamman presented an interesting experiment in which atoms in alloys of copper and gold tend to be surrounded by atoms of the other kind (to picture this, think of a chessboard coloring of the square lattice). In Tamman's experiment, the thermal agitation has a direct impact on how much the atoms tend to be in the right places. In 1935, Bragg and Williams [19] explained this phenomenon by a statistical mechanics's argument involving the energy cost of having an atom in the wrong place. Hans Bethe simplified the model by assuming that only nearest atoms interact.

In 1936, Ralph Fowler and his team in Cambridge introduced another theoretical model to understand the adsorption of metal vapor on a glass. Fowler more generally identified a class of experiments exhibiting similar behaviors, that he named *cooperative phenomena*.

The German theoretical physicists Rudolf Peierls later noticed the similarity between Bethe's approximation of the Bragg–Williams model, Fowler's theory of adsorption, and the Ising model. While the original physical problems are different, the mathematical treatment is in fact similar. In retrospect, Peierls was perhaps the first person to identify that the Ising model could treat a number of different phenomena, even though the model was a coarse caricature for each phenomenon in question.

This observation was maybe what kept the Ising model alive for some years, but it is mathematics that truly changed the nature of the model and made it what it is today. We now turn to the first mathematical breakthrough in the model.

2.4. Peierls' argument

While Peierls agreed with the majority of the physics community that the Ising model was not a good model for ferromagnetism, he certainly recognized that the model was of mathematical interest. Furthermore, he totally disagreed with the naive generalization, based on the few attempts of Ising, of the absence of a ferromagnetic phase to higher dimensional lattices. This led him to reconsider the problem of the Ising model in two and three dimensions. As a result, he produced what is probably one of the most important papers in the early Ising history [89], in which he developed a technique which is now widely known in statistical physics as *Peierls' argument*.

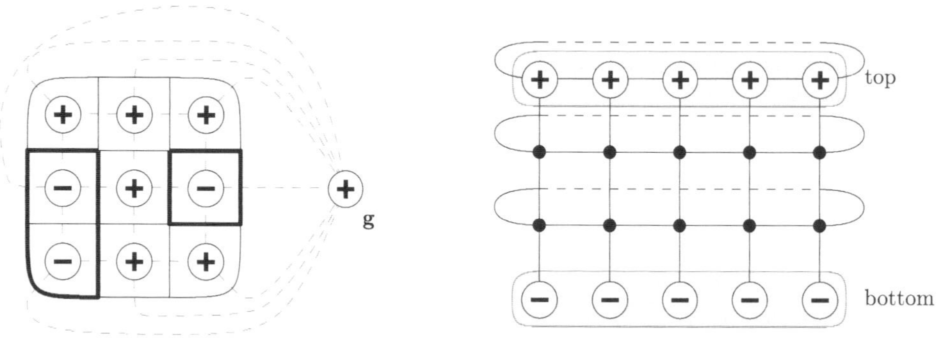

FIGURE 1

(Left) A picture of the low-temperature expansion on Λ_1^+. The set Λ_1^+ is depicted in dashed gray, and $(\Lambda_1^+)^*$ in plain black. The edges in the dual configuration are depicted in bold. (Right) The set $\mathbb{S}(5,3)$ with the bottom and top sets depicted. In this case $\underline{\tau}$ and $\overline{\tau}$ are respectively constant equal to -1 and $+1$.

Roughly speaking, the argument runs as follows. When considering a configuration σ of the Ising model on $\mathbb{Z}^2$, or a finite subgraph of it, one may associate a subset $E(\sigma)$ composed of the edges $\{x, y\}$ of the graph with $\sigma_x \neq \sigma_y$. In a planar context, one may draw these edges $e \in E(\sigma)$ by considering the dual edges $(e^* : e \in E(\sigma))$ on the dual graph[1]; see Figure 1. These dual edges and their endpoints form an even subgraph of the dual graph (call Even(G^*) the set of such even subgraphs) which can be interpreted as a collection of loops on the dual graph. The representation of configurations σ in terms of even subgraphs is called the *low-temperature expansion*. Using the mapping between σ and $E(\sigma)$, one may rewrite the partition function as

$$Z(G,\beta,0) = \sum_{\sigma\in\{-1,1\}^V} e^{-\beta H_{G,h}(\sigma)} = e^{\beta|E(G)|} \sum_{F\in\mathrm{Even}(G^*)} e^{-2\beta|F|}. \tag{2.8}$$

This formula immediately highlights the fact that a large β renders configurations $F \in \mathrm{Even}(G^*)$ with large loops unlikely. Building on this observation, Peierls was able to obtain that $m^*(\beta) > 0$ for large values of β, see Frame 1 for more details.

The idea to introduce a model of "domain walls" separating the different phases (here pure $+1$ and pure -1) from each other is not restricted to the Ising model: it has been very fruitful to prove the existence of phase transitions, and Peierls' argument is now one of the most famous and robust arguments in statistical physics.

1 The dual graph $G^* = (V^*, E^*)$ of a planar graph $G = (V, E)$ is the planar graph with vertex-set given by the faces of G (including the exterior one) and edge-set E^* given by unordered pairs $\{u, v\}$, where u and v are two faces that are bordered by the same edge. When this edge is e, we denote the dual edge $\{u, v\}$ by e^*. The map $e \mapsto e^*$ is therefore a bijection between E and E^*. On the square lattice, the dual graph is nothing but the translate by $(\frac{1}{2}, \frac{1}{2})$ of the square lattice itself.

Frame 1: A quick version of Peierls' argument

We do not consider the magnetization $m^*(\beta)$ but rather the correlation $\langle\sigma_0\sigma_{\mathbf{g}}\rangle_{\Lambda_n^+,\beta,0}$, where Λ_n^+ is the graph Λ_n plus a vertex $\mathbf{g}$, sometimes referred to as Griffiths' "ghost" vertex, connected to all the vertices on the boundary of Λ_n; see Figure 1 on the left. The limit as n tends to infinity can be shown to be $m^*(\beta)$, so it is sufficient to prove that the quantity is bounded away from 0 uniformly in n.

If one denotes by $\mathbf{C} = \mathbf{C}(\sigma)$ the connected component of 0 in $\mathbb{R}^2 \setminus \{e^* : e \in E(\sigma)\}$, one may decompose the magnetization depending on the value of $\mathbf{C}$ to get

$$\langle\sigma_0\sigma_{\mathbf{g}}\rangle_{\Lambda_n^+,\beta,0} = 1 - 2 \sum_{\mathbf{g}\notin C \in \mathrm{Even}((\Lambda_n^+)^*)} \langle\mathbb{I}(\mathbf{C} = C)\rangle_{\Lambda_n^+,\beta,0}. \tag{2.9}$$

Now things become interesting. For every $C \not\ni \mathbf{g}$, consider the configuration $\mathrm{Flip}_C(\sigma)$ obtained from σ by flipping the values of the spins inside C. This effectively corresponds to removing the set $\partial_e C$ of edges in $E(\sigma)$ with *exactly one* endpoint in C. Taking into account the cost of this operation leads to

$$\langle\mathbb{I}(\mathbf{C} = C)\rangle_{\Lambda_n^+,\beta,0} \leq e^{-2\beta|\partial_e C|}$$

for every $C \not\ni \mathbf{g}$. At this stage, the fact that $\partial_e C$ is a loop and that there are at most $(k+1)4^k$ possible loops of length k surrounding the origin gives

$$\langle\sigma_0\sigma_{\mathbf{g}}\rangle_{\Lambda_n^+,\beta,0} \geq 1 - 2\sum_{k\geq 1} k4^k e^{-2\beta k} > 1 - \frac{8e^{-2\beta}}{(1-4e^{-2\beta})^2}. \tag{2.10}$$

3. ONSAGER'S 1944 REVOLUTION AND THE INTEGRABILITY OF THE ISING MODEL

3.1. Kramers–Wannier treatment of the Ising model and duality

While Peierls' result is certainly one of the first key rigorous steps in the understanding of the Ising model, the work [78] of Hans Kramers and Gregory Wannier in 1941 propelled the Ising model in another dimension in terms of mathematical interest. Indeed, the two physicists agreed that the Ising model was not necessarily an accurate description of ferromagnetism, but they were precursors in strongly believing that having mathematical models that can be rigorously analyzed was of crucial interest for the understanding of physical phenomena, even if only approximate.

Kramers and Wannier's goal was to understand what happens for the Ising model at *arbitrary* inverse-temperature. Peierls' argument shows that the model behaves like a ferromagnet when β is large. A fairly simple argument, see Frame 4, shows that it behaves like a paramagnet when β is small. It is therefore tempting to think that there is an intermediate inverse-temperature, playing the theoretical role of the inverse of Curie's temperature, that separates a paramagnetic phase from a ferromagnetic phase, i.e., a *critical inverse-*

temperature β_c defined by the formula

$$\beta_c = \beta_c(\mathbb{Z}^d) := \inf\{\beta : m^*(\beta) > 0\}. \tag{3.1}$$

Of course, the notion of critical inverse-temperature immediately leads to the following question: Can one compute the value of the critical point β_c?

The work [78] represented an important historical step towards this computation. It involved a number of ideas that deeply influenced the way mathematicians and physicists approach the Ising model. The first key observation is that Kramers and Wannier did not work with the Ising model in the presence of a magnetic field (in other words, they set h to be 0). Instead, they proposed to look at the *specific heat* defined by

$$C(\beta) := -\beta^2 \frac{\partial^2}{(\partial\beta)^2}(\beta f)(\beta, 0). \tag{3.2}$$

Kramers and Wannier argued that the critical point of the model on $\mathbb{Z}^2$ should correspond to a value of β at which $C(\beta)$ blows up. The next step is maybe the most interesting one. By *assuming* that there exists a unique point at which $C(\beta)$ blows up, they were able to predict the value of β_c. The reason behind this prediction is the following *duality relation* for the free energy on $\mathbb{Z}^2$:

$$\beta f(\beta, 0) = \beta^* f(\beta^*, 0) - 2\beta + \ln 2 + 2 \ln \cosh(\beta^*), \tag{3.3}$$

where β and β^* are related via the formula $\tanh(\beta^*) = e^{-2\beta}$. The uniqueness implies that β_c must be the self-dual point satisfying $\beta^* = \beta$, i.e., β_c must be equal to $\frac{1}{2}\ln(1+\sqrt{2})$. Of course, this reasoning is not a formal proof as it is not a priori obvious that the singular point is unique.

The proof of Kramers and Wannier of the duality relation is also of great interest. Originally, they used so-called *transfer matrices* to do it; see Frame 2 for details. While they did not invent those matrices (they already appeared in the work of Montroll [84]), they probably made the first important use of them. Today, the derivation of this relation is straightforward and does not rely on transfer matrices. It involves relating the partition functions $Z(G, \beta, 0)$ and $Z(G^*, \beta^*, 0)$ using, for the first one, the expression given by the low-temperature expansion (2.8), and for the second, an alternative representation called the *high-temperature expansion*, obtained by van der Waerden [96] and briefly described in Frame 4. When observing that the dual of a box in the square lattice is (except on the boundary) a box of the square lattice, one obtains the identity above by considering larger and larger boxes.

Frame 2: Transfer matrices of the Ising model

To lighten the presentation, we restrict our attention to the case $h = 0$. Consider the slices $\mathbb{S}(N, M) := (\mathbb{Z}/N\mathbb{Z})^{d-1} \times [\![0, M]\!]$ with no edges between the vertices of the bottom $(\mathbb{Z}/N\mathbb{Z})^{d-1} \times \{0\}$ (we call $(\mathbb{Z}/N\mathbb{Z})^{d-1} \times \{M\}$ the top of the slice); see

Figure 1 on the right. Let $\sigma_{|\text{bottom}}$ and $\sigma_{|\text{top}}$ be the restrictions of σ to the top and bottom of $\mathbb{S}(N, M)$, considered as two elements of $\{-1, 1\}^{(\mathbb{Z}/N\mathbb{Z})^{d-1}}$. Introduce the quantity

$$Z(N, M, \underline{\tau}, \overline{\tau}) := \sum_{\sigma \in \{-1,1\}^{\mathbb{S}(N,M)}} \exp[-\beta H_{\mathbb{S}(N,M),h}(\sigma)] \mathbb{I}(\sigma_{|\text{bottom}} = \underline{\tau}, \sigma_{|\text{top}} = \overline{\tau}), \quad (3.4)$$

where $\underline{\tau}, \overline{\tau} \in \{-1, 1\}^{(\mathbb{Z}/N\mathbb{Z})^{d-1}}$, as well as the so-called *transfer matrix*

$$V_N(\underline{\tau}, \overline{\tau}) := Z(N, 1, \underline{\tau}, \overline{\tau}) = \exp\Big[-\beta\Big(\sum_{x \in (\mathbb{Z}/N\mathbb{Z})^{d-1}} \underline{\tau}_x \overline{\tau}_x + \sum_{\{x,y\} \in E((\mathbb{Z}/N\mathbb{Z})^{d-1})} \overline{\tau}_x \overline{\tau}_y\Big)\Big]. \quad (3.5)$$

One immediately finds that $Z(N, M, \underline{\tau}, \overline{\tau}) = V_N^M(\underline{\tau}, \overline{\tau})$. Other quantities of the model may be written in terms of transfer matrices, for instance, the partition function of the model on the d-dimensional torus $(\mathbb{Z}/N\mathbb{Z})^d$ becomes the trace of V_N^N.

One important aspect of those transfer matrices V_N is that certain questions on the behavior of the model are rephrased as spectral questions on the transfer matrix. For instance, by letting M and then N go to infinity, one observes that the asymptotic behavior of the partition function on $(\mathbb{Z}/N\mathbb{Z})^{d-1} \times (\mathbb{Z}/M\mathbb{Z})$, and therefore the value of the free energy, are connected to the asymptotic behavior of the leading eigenvalue of V_N as N tends to infinity. This can very well be an intractable problem, but in some cases it is not.

3.2. Onsager's result

Kramers and Wannier's results unraveled the potential mathematical interest of the Ising model, but the real revolution came only a few years after with one of the most impressive achievements in mathematical physics. Lars Onsager, Nobel prize winner in 1968, was a Norwegian specialist in theoretical chemistry. He was particularly interested in mathematical problems and focused his attention on the Ising model for the formidable challenge that its exact solution represented more than for his physical relevance.[2]

To everyone's surprise, Onsager announced at a conference of the New York Academy of Sciences in 1942 that he obtained the following *exact* expression for the free energy (at zero magnetic field) of the Ising model on the square lattice $\mathbb{Z}^2$:

$$-\beta f(\beta, 0) = \ln 2 + \frac{1}{8\pi^2} \int_0^{2\pi} \int_0^{2\pi} \ln\big[\cosh(2\beta)^2 - \sinh(2\beta)(\cos\theta_1 + \cos\theta_2)\big]\, d\theta_1\, d\theta_2. \quad (3.6)$$

This implies, in physics jargon, that the model is *exactly solvable*. This solvability is itself linked to a deep property of the model called *integrability*. Onsager's underlying idea was

2 His opinion on this fact seemingly evolved: in his first 1944 paper **[86]**, he presents it as a poor model of ferromagnetism, but a fairly good model for binary alloys, while in his paper with Kaufman in 1949 **[75]** he describes it as a model for ferromagnetism.

that the transfer matrices of the 2D Ising model are a product of two matrices that generate (by taking successive brackets) a finite-dimensional Lie algebra. He used this observation to derive the asymptotic behavior of the leading eigenvalue of these matrices in his famous 1944 paper [86]. In 1949, Bruria Kaufman [75] provided an alternative and simpler derivation.

A few years later, Onsager surprised the world of theoretical physicists again by claiming an exact expression for the spontaneous magnetization on $\mathbb{Z}^2$: for $\beta \geq \beta_c$,

$$m^*(\beta) = \big(1 - \sinh(2\beta)^{-4}\big)^{1/8}. \tag{3.7}$$

While the result was announced by Onsager first, it was a young physicist, that would later become one of the most influential theoretical physicists of the second half of the twentieth century, Chen-Ning Yang (from Yang–Baxter's equation, Yang–Mills's theory, Lee–Yang's theory, etc.), who provided a mathematical proof [102] of this statement by achieving a mathematical tour de force involving Toeplitz determinants. The proof relies on a computation, again using transfer matrices but much more evolved than for the free energy, of the two-point function $\langle\sigma_{(0,0)}\sigma_{(n,0)}\rangle_{(\mathbb{Z}/N\mathbb{Z})^2,\beta,0}$, and the observation that

$$m^*(\beta)^2 = \lim_{n\to\infty}\lim_{N\to\infty}\langle\sigma_{(0,0)}\sigma_{(n,0)}\rangle_{(\mathbb{Z}/N\mathbb{Z})^2,\beta,0} \tag{3.8}$$

(at the time, such an identity was not obviously true, but nowadays this can be proved easily using, for instance, the FK percolation, see Section 7.2).

In the 1940s and 1950s, these successes were considered by physicists as a mathematical curiosity rather than a truly crucial advance. Yet, they had a revolutionary impact on theoretical physics for multiple reasons: First, the *level of sophistication* of the mathematical tools used in the proofs is without any common measure with what was previously used in such kinds of problems, and these techniques created whole new types of mathematical physics. Second, the behavior of the model did not correspond to previous *mean-field approximations*, thus invalidating rigorously the Curie–Weiss or Landau theories and opening a new era in statistical mechanics. Third, the results had *many direct applications* for the Ising model itself, for instance, the specific heat $C(\beta)$ can easily be shown to blow up logarithmically as β approaches $\frac{1}{2}\log(1+\sqrt{2})$, thus confirming rigorously that this value is the critical point of the system (the logarithmic blow-up is one example of non-mean-field behavior).

Numerous alternatives have been proposed to the approach of Onsager–Kaufman–Yang, often referred to as the *algebraic method*. As a joke, Baxter and Enting named their 1978 paper [11], introducing a solution to the 2D Ising model involving the notion of star-triangle transformation, the "*399th solution of the Ising model*." This count is, of course, overestimated, but one can list a large number of alternative strategies.

The first such strategy is called the *combinatorial approach* and is referring to an original argument of Kac and Ward [70] rewriting the partition function of the model in terms of the square root of the determinants of so-called Kac–Ward's matrices using a combinatorial expansion of the partition function generalizing the van der Waerden high-temperature expansion [96]. The advantage of such an approach is that it does not rely on transfer matrices, and therefore is applicable to *every* finite planar graph, even with arbitrary nearest-neighbor coupling constants. Unfortunately, the original argument was not entirely rigorous and one had to wait until 1999 [33] to finally obtain a mathematical derivation of this approach. Nowadays, the method is very well understood and especially useful in relation to discrete holomorphicity and higher genus graphs, see [29] and the references therein for a more complete account.

The (nowadays) most classical method is probably the *Pfaffian method*. It came as an attempt to go around the substantial difficulties to make the combinatorial approach rigorous. Due to Hurst and Green [64], Kasteleyn [74], and Fisher [45], the strategy consists in writing the Ising partition function on a finite planar graph G in terms of the dimer (a dimer configuration is a subset of edges which covers every vertex exactly once) partition function on a related graph $K(G)$ (the precise definition of the graph depends on the implementation of the Pfaffian method). It is then possible to relate the partition function to a skew-symmetric adjacency matrix and express the partition as a Pfaffian, hence the name of the method. This strategy has been the basis of a number of more refined results about the model, in particular the computation of the spin–spin correlations of the model at and away from criticality. For the deepest and most impressive results, we recommend that the reader takes a look at the two books of McCoy–Wu [82] and Palmer [88].

Another approach of importance was proposed by Schultz–Mattis–Lieb in [93] to tackle the cases for which a transfer matrix can be used. In this paper, they connected the transfer matrix with the exponential of a quantum Hamiltonian. This connection to 1D quantum spin chains has been very fruitful and understood in a number of alternative ways since then. As a byproduct, the authors were able to express the partition function as a Grassmann "Gaussian" integral. The advantage of this way of writing the partition function is that the Pfaffians emerge naturally. This approach is at the basis of renormalization schemes in two dimensions; see Section 8.3.

Yet another approach dealing with the context in which transfer matrices can be applied is worth mentioning, as it is by far the most generalizable to other models. It is based on the commutation of the transfer matrices attached to the model with different critical parameters. Pioneered by Rodney Baxter, this approach consists in using the so-called Yang–Baxter equation. The advantage is that the same strategy can be applied to a very large variety of *integrable systems*, such as the six-vertex model, to cite only one example. We refer to [10] and references therein for more details.

4. THE 1950S AND 1960S: THE ISING MODEL BECOMES A LABORATORY FOR UNDERSTANDING CRITICAL PHENOMENA

The 1950s and 1960s were probably the decades during which the Ising model became an "unavoidable" model. The realization that having a tractable model of statistical physics could be a useful explanatory but also predicting tool became more and more obvious. The Ising model, with Onsager's solution, was a prime example of a model with such qualities.

The model therefore developed tremendously in the postwar era in theoretical physics as well as in a rapidly growing field called *mathematical physics*. The latter gathered more and more physicists that were interested in rigorous aspects of the objects they studied, and mathematicians willing to study problems that were naturally emerging from physical modeling. The Ising model offered a wonderful playground for such scientists, and the number of papers mentioning the model started to be counted in the hundreds.

4.1. Progress in mathematical physics: From perturbative regions of the phase diagram to the vicinity of the critical point

During this period, the newly developing community of mathematical physicists recognized that the study of phase transitions, and in particular of the critical phase (when β is equal to β_c), was a vast field of its own. While the previous developments mostly concerned the values of β and h that were far from the critical regime (Peierls' argument [89] or Baker's use of Padé approximant [8], for instance), the situation changed drastically around the 1950s. The interest in the intermediate values of β became stronger and stronger. Onsager's solution offers a precise understanding of the critical behavior of the 2D Ising model, yet it has clear downsides related to the relative fragility of the integrability of the system. As a consequence, mathematical physicists started using the Ising model not only as a solvable system, but more generally as a good mathematical model that one should not reduce to its integrability aspects. New rigorous techniques emerged during this period to try to understand the vicinity of the critical point for nonintegrable cases, for instance in higher dimensions.

4.1.1. Correlation inequalities

It is natural to ask which monotonicity properties are satisfied by the system, in particular by the *spin–spin correlations* $\langle\sigma_A\rangle_{G,\beta,h}$ where $\sigma_A := \prod_{x\in A}\sigma_x$, when the parameters vary (for instance, G, β, or h).

To tackle such questions, mathematical physicists started proving what we now call *correlation inequalities* using combinatorial arguments. Among the first such examples are Griffiths' inequalities [56]: for every $\beta, h \geq 0$ and every $A, B \subset V$,

$$\langle\sigma_A\rangle_{G,\beta,h} \geq 0 \quad \text{and} \quad \langle\sigma_A\sigma_B\rangle_{G,\beta,h} \geq \langle\sigma_A\rangle_{G,\beta,h}\langle\sigma_B\rangle_{G,\beta,h}. \tag{4.1}$$

A byproduct of the second inequality, when applied to $B = \{x, y\}$ and summed over all edges $\{x, y\}$, is that correlations $\langle\sigma_A\rangle_{G,\beta,h}$ are increasing in β (and also in G with a little bit of additional work). One may derive the same for the spontaneous magnetization $m^*(\beta)$, so that the definition of β_c can now be rephrased as

$$\beta_c = \inf\{\beta \geq 0 : m^*(\beta) > 0\} = \sup\{\beta \geq 0 : m^*(\beta) = 0\}. \tag{4.2}$$

This implies in particular that there is indeed a unique transition between paramagnetic and ferromagnetic phases.

Other interesting correlation inequalities were obtained in subsequent years. Let us contemplate a few examples (we do not write them in full generality, and we drop the subscript after $\langle\cdot\rangle$):

- GHS's inequality [58]: for $h \geq 0$ and $x \in G$,

$$\frac{\partial^2}{(\partial h)^2}\langle\sigma_x\rangle \leq 0. \tag{4.3}$$

- Simon–Lieb's inequality [81]: for $S \ni 0$ and $x \notin S$, when $\langle\cdot\rangle_S$ refers to the model in S,

$$\langle\sigma_0\sigma_x\rangle \leq \sum_{y\in\partial S}\langle\sigma_0\sigma_y\rangle_S\langle\sigma_y\sigma_x\rangle. \tag{4.4}$$

- Messager–Miracle–Solé's inequality [83]: for $x, y \in \mathbb{Z}_+^d$ (below $\langle\cdot\rangle$ is defined on $\mathbb{Z}^d$)

$$\langle\sigma_0\sigma_{x+y}\rangle \leq \langle\sigma_0\sigma_x\rangle. \tag{4.5}$$

- FKG's inequality [47]: for any increasing functions $f, g : \{-1, 1\}^V \to \mathbb{R}$,

$$\langle fg\rangle \geq \langle f\rangle\langle g\rangle. \tag{4.6}$$

This far from exhaustive list, which we did not discuss in detail, is intended to show the variety of possible correlation inequalities. Clever use of these inequalities provided the embryo of what would be considered later as the theory of noncritical statistical physics systems at equilibrium, as the correlation inequalities and their consequences often generalize in the same (or slightly altered) form to a wider class of lattice spin models.

4.1.2. The Ising model with a magnetic field: The Lee–Yang theory

While studying the whole phase diagram is a Herculean task that was far beyond the techniques developed at the time, a beautiful development enabled mathematical physicists to understand the case $h \neq 0$.

The twin papers [79], referred to as the Lee–Yang theory, relate the regularity properties of the free energy (and therefore the location of singular points corresponding to places where a phase transition occurs) to the locus of the complex zeroes of the partition function $Z(G, \beta, h)$ when seen as a function of $h \in \mathbb{C}$. Beyond the result itself, the philosophy consisting in studying the complex zeroes of the partition function had a resounding effect on the field of mathematical physics. This can be put in parallel with the analysis of zeroes of the Riemann zeta function: one learns something about prime numbers by studying the zeroes of a generating-type function associated with them.

The result of Lee and Yang is not restricted to the n.n.f. Ising model on $G \subset \mathbb{Z}^d$, but the latter gives an important application of it. In our context, let $Z(G, \beta, \mathbf{h}) \in \mathbb{C}$ (for $\mathbf{h} = (\mathbf{h}_x : x \in V) \in \mathbb{C}^V$) be the partition function defined as in (2.3) with the difference that the magnetic field is allowed to vary with the vertex, i.e., that the $\sum_{x\in V} h\sigma_x$ term of the Hamiltonian in (2.2) is replaced by $\sum_{x\in V}\mathbf{h}_x\sigma_x$. The result states that for this model, the zeroes of the function $\mathbf{h} \mapsto Z(G, \beta, \mathbf{h})$ are satisfying $\mathrm{Re}(\mathbf{h}_x) = 0$ for every $x \in V$.

As a consequence of this theorem, the free energy $f(\beta, h)$ (which we recall from (2.5) is expressed in terms of the limit of the logarithm of partition functions) is analytic as soon as $h \neq 0$. Other consequences follow, such as exponential decay of so-called truncated correlations of the system, as well as analyticity of the other thermodynamical quantities when the magnetic field is nonzero. Roughly put, the Lee–Yang theory enables understanding in full detail the part of the phase diagram corresponding to a nonzero magnetic field.

4.2. Revolutionary progress on the physics front

In parallel to these first successes in mathematical physics, revolutionary progress was made during this period on the physical understanding of phase transitions. Among other things, the scaling and universality hypotheses were formulated, and the pillars of the renormalization group were cast, in both cases using the Ising model as an important source of inspiration.

4.2.1. Critical exponents and the success of scaling theory

A fundamental notion of physics is the assumption that thermodynamical quantities of physical systems near criticality tend to take simple forms when expressed in terms of the parameters of the system. A major advance was achieved in the 1960s by American chemist Benjamin Widom who proposed in [99] that these quantities are powers in each parameter. For the Ising model, the parameters are β and h, and this *scaling hypothesis* translates into the existence of so-called *critical exponents*. To give a few examples related to already defined quantities, one may, for instance, predict that

$$m^*(\beta) = (\beta - \beta_c)_+^{\boldsymbol{\beta}+o(1)}, \quad m(\beta_c, h) = h^{1/\boldsymbol{\delta}+o(1)}, \quad \langle \sigma_0 \sigma_x \rangle_{\beta_c,0} = \frac{1}{|x|^{d-2+\boldsymbol{\delta}}} \tag{4.7}$$

(notice that $\boldsymbol{\beta}$ and β have nothing to do with each other), where $o(1)$ is a quantity tending to 0 as β tends to β_c, h tends to 0, or $|x|$ tends to infinity, respectively. In fact, the whole family of such exponents, denoted by $\boldsymbol{\alpha}, \boldsymbol{\beta}, \boldsymbol{\gamma}, \boldsymbol{\delta}, \boldsymbol{\eta}, \boldsymbol{\nu}$ (for the most classical ones), can be defined for each model. Understanding the phase transition boils down to, among other things, deriving those exponents.

Dealing with such exponents, one may naturally wonder how many degrees of freedom truly exist in statistical physics models. For instance, could some of these critical exponents be connected via direct relations that would transcend the precise definition of each model? In the 1960s, physicists such as Essam, Fisher, and Widom himself, to cite only those three (see [42, 46, 99] for some early works on the subject), started unraveling systematic connections between the exponents, thus hinting towards the fact that only two degrees of freedom exist and that exponents are related by so-called *scaling relations*

$$\boldsymbol{\nu} d = 2 - \boldsymbol{\alpha} = 2\boldsymbol{\beta} + \boldsymbol{\gamma} = \boldsymbol{\beta}(\boldsymbol{\delta} + 1) = \boldsymbol{\gamma}\frac{\boldsymbol{\delta} + 1}{\boldsymbol{\delta} - 1}, \quad 2 - \boldsymbol{\eta} = \frac{\boldsymbol{\gamma}}{\boldsymbol{\nu}} = d\frac{\boldsymbol{\delta} - 1}{\boldsymbol{\delta} + 1}. \tag{4.8}$$

The scaling relations apply in a context which is far more general than just the Ising model (see, for instance, [38] for a proof in the case of a large family of two-dimensional percolation models). In the course of discovering these different scaling relations, the Ising model in two and three dimensions played the important role of a sanity check. While other experimental systems were used as testing grounds, the Ising model was the only example of a theoretical system which did not exhibit mean-field behavior (and therefore was not too "trivial") and for which such exponents were available, either rigorously thanks to the exact solution in 2D, or approximately thanks to Baker's use of Padé approximant [8] in 3D.

To conclude this section, let us mention an important quantity, called the *correlation length* $\xi(\beta)$ of the system, that plays an important role in the scaling hypothesis (it corresponds to the exponent $\boldsymbol{\nu}$). We consider the case $\beta < \beta_c$ but a similar notion can be introduced for $\beta > \beta_c$, with analogous interpretations.

When considering, say, spin–spin correlations at criticality, one expects an algebraic decay as mentioned in (4.7). Yet, when $\beta < \beta_c$, the scaling hypothesis cannot hold uniformly in $|x|$ and such a decay does not occur. In fact, it was found in many systems that spin–spin correlations decay exponentially fast (see Section 7.1 for more details) and the inverse-rate of decay is the correlation length $\xi(\beta)$. This correlation length has an interesting interpretation: it is the smallest scale at which the system with $\beta < \beta_c$ is off-critical, meaning that when looking at a system with a size which is much smaller than $\xi(\beta)$, the difference between the system and a critical system will be invisible to the physicist's eye, while on the contrary when the size is much larger than $\xi(\beta)$, the model looks similar to the case of $\beta \ll \beta_c$. In other words, when approaching the critical point, a system becomes more and more "critical." By how much this is true depends on the size of the system, and the correlation length separates between the sizes at which the system looks critical, and the sizes at which it looks clearly non critical.

4.2.2. Kadanoff's block-spin renormalization and universality

While Widom's scaling hypothesis provides compelling evidence that critical exponents exist, the underlying justification of the hypothesis itself remained slightly superficial until Russian physicist Leo Kadanoff provided an illuminating argument for it. In his famous 1966 paper [71], Kadanoff suggested that the block-spin renormalisation transformation—i.e., replacing a block of neighboring sites by one site having a spin equal to the dominant spin in the block—corresponds to appropriately changing the scale and the parameters β and h of the model. Assuming that iterating this procedure somehow converges suggests that the asymptotic properties of the system are described by a fixed point of a renormalization map. As a result, one ends up with the *scale invariance* of the model. This argument, inspired by the study of the Ising model, turned out to be the basis of the monumental theory of the renormalization group (RG) that was put in a general framework a few years later by Kenneth Wilson [101].

The block-spin argument of Kadanoff achieved much more than a physical justification of the scaling hypothesis. Assuming *uniqueness of the fixed point* also implies that the renormalization of Ising models defined on different d-dimensional lattices should converge to the same fixed point, and therefore share the same critical exponents. This was already partially realized in 2D by observing the Ising model on the square, hexagonal, and triangular lattices (they are all exactly solvable), as well as in 3D by approximations using series expansions [34], but the renormalization argument suggests that the few examples of equalities between exponents are, in fact, the illustration of a much more general phenomenon.

What is now known as the *universality hypothesis* was explicitly formulated in parallel by Robert B. Griffiths and Kadanoff in 1971 [57,72]. Roughly speaking, it states that the critical properties of a physical system only depend on

- the lattice dimension d;
- the symmetry of the space of possible spins ($\mathbb{Z}/2\mathbb{Z}$ symmetry for Ising);
- the speed of decay of coupling constants (this is only relevant when the $J_{x,y}$ are allowed to decay polynomially with $\|x - y\|$, which is not the case in this text).

This realization of universality is fundamental to the relevance of statistical physics as a whole. To borrow from Kadanoff's wording: "*Why study a simplified model like the Ising model? The strategy of studying physical questions by using highly simplified models is made rewarding by a characteristic of physical systems called "universality," in that many systems may show the very same qualitative features, and sometimes even the same quantitative ones. To study a given qualitative feature, it often pays to look for the simplest possible example.*"

To summarize Section 4, by the end of the 1960s it became clear to mathematical physicists and theoretical physicists that the Ising model was one of the most striking examples of a simple physical system which was rich enough to grasp a large variety of phenomena falling in the range of statistical physics. Results on the Ising model started to play a role similar to experimental results in the sense that they could corroborate or, on the contrary, invalidate the embryo of a theory. It is fair to say that the importance of the model was never argued upon later on and that it was finally recognized as one of the centerpieces of modern statistical physics.

5. THE 1960S AND 1970S: EMERGENCE OF THE PROBABILISTIC INTERPRETATION

Physicists and mathematical physicists think of the quantity $\langle\cdot\rangle_{G,\beta,h}$ as a form attributing to each function $X : \{-1,1\}^V \to \mathbb{R}$ (resp. $\mathbb{C}$) a value in $\mathbb{R}$ (resp. $\mathbb{C}$). In the late 1960s and 1970s, the rise of probabilistic methods led to an alternative interpretation of the Ising model in which $\langle\cdot\rangle_{G,\beta,h}$ is now understood as (dual to) a probability measure $\mu_{G,\beta,h}$. As a consequence of this reinterpretation, it becomes natural to ask what the properties of a randomly chosen spin configuration are, and what the possible measures on the infinite lattice that can be obtained as limits of measures in finite volume are.

5.1. The random geometry of the spin configuration

As mentioned above, $\langle\cdot\rangle_{G,\beta,h}$ is the linear form associated with the probability measure $\mu_{G,\beta,h}$ on $\{-1,1\}^V$ defined for every configuration σ by the formula

$$\mu_{G,\beta,h}\big[\{\sigma\}\big] := \frac{1}{Z(G,\beta,h)} \exp\big[-\beta H_{G,h}(\sigma)\big]. \tag{5.1}$$

Then, quantities like $\langle\sigma_A\rangle_{G,\beta,h}$ can be interpreted as the *correlations* between the random variables σ_x with $x \in A$. Note that in this interpretation the partition function is a normalizing factor making the measure at hand a probability measure.

Let us assume for a moment that $h = 0$ and interpret the phase transition in terms of probability. The structure of the probability measure is such that configurations have greater probability if they have more pairs of neighbors with a similar spin. In this interpretation, the larger β is the more important it is that neighbors have the same spins. In particular, in the limit as β tends to infinity, one ends up with one of the two configurations where all spins are the same. It becomes then natural to expect that for β large, typical configurations have an excess of one spin compared to the other. On the other hand, when β is very small, how much the measure takes the agreements into account is fairly limited, and one may expect that spins behave roughly independently, at least at large distance of each other.

The interpretation in terms of random variables opens new uncharted territories: one can interpret probabilistically natural thermodynamical quantities such as magnetization (which corresponds to the expectation of the spin at a vertex) or surface tension. It also opens a way to new problems, such as dynamics on the space of spin configurations or large deviations (for instance, for an Ising model at an inverse-temperature β, but with an excess of $+1$ spins in a region and of -1 spins in another); see Frame 3.

Frame 3: Sampling the Ising model—Glauber dynamics

The probabilistic interpretation naturally raises the question of sampling random configurations according to $\mu_{G,\beta,0}$ (set $h = 0$ for simplicity). A classical method consists in expressing the measure as the invariant measure of a Markovian dynamics $(\sigma(t) : t \geq 0) \in (\{-1, 1\}^V)^{\mathbb{R}_+}$, called the Glauber dynamics and defined as follows: attach an exponential clock to each vertex of G. Each time a clock rings, say at time t at $x \in V$,

- If $\sigma_x(t) \sum_{y:\{x,y\}\in E} \sigma_y(t) < 0$, switch the value of the spin at x,
- Otherwise, switch the value of the spin at x with a probability equal to $\exp[-2\beta \sum_{y:\{x,y\}\in E} \sigma_y(t)]$, and do not switch otherwise.

Since $\mu_{G,\beta,0}$ is the only invariant measure for this dynamics, the limit as t tends to infinity, irrespectively of the initial value $\sigma(0)$, is sampled according to $\mu_{G,\beta,h}$.

This dynamics was named after the American physicist Roy J. Glauber. Alternative choices of dynamics are obtained by changing the jump probabilities. In Figure 2, three simulations of the Ising model are shown respectively below (on the left), at (in the middle) and above (on the right) β_c.

5.2. Boundary conditions and the Gibbs formalism

An important output of the probabilistic interpretation of the model is that it becomes natural to *condition* on spins in a subset of V. More precisely, let $W \subset V$ and let H be the graph with vertex-set W and edge-set induced by the edges of the graph G. Let $\tau \in \{-1, 1\}^V$ be a spin configuration on G. One may ask what is the law of the spins in W when conditioning σ outside W to be equal to τ, i.e., what is $\mu_{G,\beta,h}[\cdot\,|\sigma_x = \tau_x, \forall x \notin W]$?

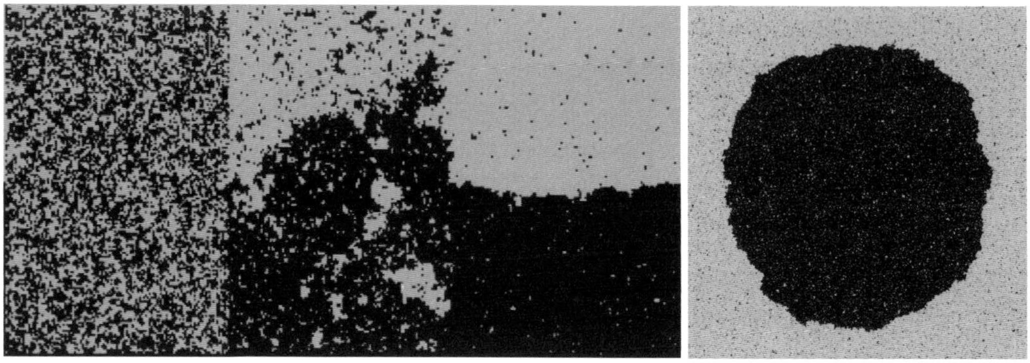

FIGURE 2

(Left) Simulations at three different temperatures ($\beta < \beta_c$, $\beta = \beta_c$, and $\beta > \beta_c$) of the Ising model with plus boundary conditions on the top and minus boundary conditions on the bottom. Pluses are in gray and minuses in black. *Credit: S. Smirnov.* (Right) An example of a bubble of minuses in an environment of pluses at $\beta > \beta_c$. *Credit: Y. Velenik.*

The answer to this question is best cast when introducing the notion of boundary conditions. For a subgraph G of $\mathbb{Z}^d$ and a configuration $\tau \in \{-1, 1\}^{\mathbb{Z}^d}$, introduce the measure $\mu^\tau_{G,\beta,h}$ with τ *boundary conditions* defined like $\mu_{G,\beta,h}$ except that $H_{G,h}$ is replaced by

$$H^\tau_{G,h}(\sigma) := H_{G,h}(\sigma) - \sum_{\{x,y\}\in E(\mathbb{Z}^d):x\in V,y\notin V} \sigma_x \tau_y. \tag{5.2}$$

Note that the only values of τ that matter are on the exterior boundary of G, i.e., on the vertices that are connected by an edge of $\mathbb{Z}^d$ to a vertex in V.

With this definition, we obtain the following important property of the Ising model, called the *spatial Markov property*: for every finite subgraph G of $\mathbb{Z}^d$, every $W \subset V$, and every configuration $\tau \in \{-1, 1\}^{\mathbb{Z}^d}$, if H denotes the graph induced by the set W,

$$\mu_{G,\beta,h}[\,\cdot\,|\sigma_x = \tau_x, \forall x \notin W] = \mu^\tau_{H,\beta,h}[\,\cdot\,]. \tag{5.3}$$

In words, when conditioning the Ising model on G to coincide with a given configuration outside W, one gets the measure in H with the corresponding boundary condition.

This property offers a natural consistency relation between measures $\mu^\tau_{G,\beta,h}$ for varying τ and G. As a byproduct, one is naturally led to postulate that any reasonable infinite-volume version of Ising measures should satisfy the same consistency relation. One therefore ends up with the following notion: a measure μ on $(\{-1, 1\}^{\mathbb{Z}^d}, \mathfrak{F}_{\mathbb{Z}^d})$ is called a *Gibbs measure* of the Ising model with parameters β and h if it satisfies the *Dobrushin–Lanford–Ruelle (DLR) property*: for every finite $V \subset \mathbb{Z}^d$ and $\tau \in \{-1, 1\}^{\mathbb{Z}^d}$,

$$\mu[\,\cdot\,|\mathfrak{F}_{\mathbb{Z}^d\setminus V}] = \mu^\tau_{G,\beta,h}[\,\cdot\,] \quad \text{on } E_\tau \ \mu\text{-almost surely}, \tag{5.4}$$

where

- G is the graph induced by the vertex-set V;

- E_τ is the event that σ and τ agree on the exterior boundary of G;

- $\mathfrak{F}_{\mathbb{Z}^d \setminus V}$ is the σ-algebra generated by the random variables $(\sigma_x : x \notin V)$.

The notion of Gibbs measure is not restricted to the Ising model (see [51] for a book on the subject), but the classification of such Gibbs measures has been the object of intense study in the specific case of the Ising model, with a very successful outcome.

The first question that one may ask is the existence of Gibbs measures. At least three such measures can be defined in a fairly straightforward way. By taking limits as G tends to $\mathbb{Z}^d$ of the measures $\mu_{G,\beta,h}$, $\mu^+_{G,\beta,h}$, and $\mu^-_{G,\beta,h}$ (where + and – refer, with a slight abuse of notation, to τ equal to all $+1$ or all -1), one ends up with three (possibly equal) Gibbs measures $\mu_{\beta,h}$, $\mu^+_{\beta,h}$, and $\mu^-_{\beta,h}$. More generally, one may construct measures by taking all possible subsequential limits of measures of the form $\mu^\tau_{G,\beta,h}$, where one may even consider τ as a random variable.

In general, the set of possible Gibbs measures on $\mathbb{Z}^d$ is a nonempty simplex whose extremal measures are called *extremal states*. One can therefore try to classify such extremal Gibbs measures.

Some cases are quite simple to treat: for $h \neq 0$ or $h = 0$ and $\beta < \beta_c$, the simplex is reduced to a singleton, i.e., there exists a *unique* Gibbs measure. When $h = 0$ and $\beta = \beta_c$, it was recently proved that this is also the case [5]. On the contrary, when $h = 0$ and $\beta > \beta_c$, things are more interesting. It was realized very early on that there may be more extremal states than the two obvious $\mu^+_{\beta,0}$ and $\mu^-_{\beta,0}$, but examples that were found did not exhibit translation invariance. The most important such specimen was provided by Russian mathematical physicist Roland Dobrushin [31], who explained that in three dimensions the measure $\mu^{\mathrm{dobr}}_{\beta,0}$ obtained by taking the limit of measures $\mu^\tau_{[-n,n]^3,\beta,0}$, where τ is all plus on the upper half-space, and all minus on the lower half-space, was not translation invariant in the vertical direction at high values of β. The existence of these Dobrushin states is related to a very deep and still mysterious (at least on a mathematical level) phenomenon in 3D statistical physics often referred to as the *roughening phase transition*.

Leaving non-translation invariant measures aside, many efforts were made to prove that every *translation invariant* Gibbs state is a convex combination of $\mu^+_{\beta,0}$ and $\mu^-_{\beta,0}$. The first result in this direction proved a stronger statement that draws a direct link with the previous paragraph. In two dimensions, Aizenman [1] and Higuchi [62] proved in the 1980s that *every* Gibbs state, not only translation invariant ones, is a mixture of $\mu^+_{\beta,0}$ and $\mu^-_{\beta,0}$. In particular, $\mu_{\beta,0} = \frac{1}{2}\mu^+_{\beta,0} + \frac{1}{2}\mu^-_{\beta,0}$. In higher dimensions, it took 20 more years to obtain the result for every *translation invariant* Gibbs measure. We refer to the historical proof of Bodineau [17] and to the recent generalization of Raoufi [91].

5.3. Phase coexistence and Wulff shape

The classification of Gibbs states naturally raises the question of the coexistence of different so-called *phases*. When $h = 0$ and $\beta > \beta_c$, $\mu^+_{\beta,0}$ and $\mu^-_{\beta,0}$ are not equal: they correspond to two extremal states, sometimes referred to as the plus and minus phases. Now, what happens when one tries to "mix" the two states? For instance, how does it look if one asks that part of the space is in one state, and the other part is in the other?

In 2D, an *interface* is created between the two phases (see Figure 2 for simulations at different temperatures). While it is not obvious to define such an object in general, let us consider the simple example of the Ising model on a finite box $[-n, n]^2$ of the triangular lattice with plus spins on the part of the boundary above the x-axis, and minus spins on the rest of the boundary. In this case, one can draw a unique interface going from $(-n, 0)$ to $(n, 0)$ winding between pluses and minuses. It was understood heuristically early on that above criticality this interface should have the same fluctuations as Brownian motions, but it took decades to turn this intuition into a rigorous proof, first in the large β regime and then in the whole $\beta > \beta_c$ regime; see [55] and references therein. The techniques involved also enabled mathematicians to understand precise asymptotics of spin–spin correlations in the noncritical regimes. The theory, known under the coined name of Ornstein–Zernike theory, is now an area of intense research and spans over a large variety of statistical physics models. We refer to [23] for details on the Ising case.

When conditioning on the neighborhood of the origin to be in a plus phase inside a minus phase, one ends up with a "bubble" (see Figure 2 on the right) converging when taking larger and larger volume to the so-called *Wulff shape*. In 2D, this bubble was analyzed in detail, see the book [32] and the article [65]. In 3D, the story is even more complex. The boundary between the plus and minus phases is a kind of two-dimensional surface. The study of this object is quite intricate, and the fluctuations of the surface are still widely open. We refer to [16, 18, 24] and references therein.

6. THE 1970S AND 1980S: THE ISING MODEL AND FIELD THEORY

6.1. Constructive quantum field theory

Quantum field theories with local interaction are central in most subfields of theoretical physics, from high energy to condensed matter physics. The mathematical challenge of the proper formulation of this concept led to the program of constructive quantum field theory (CQFT). A path towards that goal was charted through the proposal to define quantum fields satisfying Wightman axioms [100] using the Osterwalder–Schrader theorem [87], in which case the construction boils down to producing relevant random distributions defined over the corresponding Euclidean space that meet a number of conditions such as suitable analyticity, permutation symmetry, Euclidean covariance, and reflection-positivity.

Finding these *Euclidean fields* boils down to constructing probability averages over random distributions $\Phi(x)$ of the form

$$\langle F(\Phi) \rangle \approx \frac{1}{\text{norm}} \int F(\Phi) \exp[-H(\Phi)] \prod_{x \in \mathbb{R}^d} d\Phi(x), \tag{6.1}$$

where

- $F(\Phi)$ is a smeared average of the form $T_f(\Phi) := \int_{\mathbb{R}^d} f(x)\Phi(x)\,dx$ associated with continuous functions of compact support f.

- $H(\Phi)$ is a Hamiltonian $H(\Phi) :\approx (\Phi, A\Phi) + \int_{\mathbb{R}^d} P(\Phi(x))\, dx$ with $(\Phi, A\Phi)$ a positive definite and reflection-positive (see Section 6.2) quadratic form, and $P(\Phi(x))$ an even polynomial whose terms of order $\Phi(x)^{2k}$ are interpreted heuristically as representing k-particle interactions.

By linearity, the expectation values of products of such variables can be rewritten as

$$\left\langle \prod_{j=1}^{n} T_{f_j}(\Phi) \right\rangle := \int_{(\mathbb{R}^d)^n} S_n(x_1, \ldots, x_n) \prod_{j=1}^{n} f(x_j)\, dx_1 \cdots dx_n, \tag{6.2}$$

where the $S_n(x_1, \ldots, x_n)$ are the *Schwinger functions* of the corresponding Euclidean field theory which can be interpreted heuristically as pointwise correlations $\langle \prod_{j=1}^{n} \Phi(x_j) \rangle$. Interpreting (6.1) raises a number of problems of varying difficulty.

The simplest example of Euclidean fields are the reflection-positive (see Section 6.2 again) Gaussian fields, for which $H(\Phi)$ contains only quadratic terms. Gaussian fields are alternatively characterized by $2n$-point Schwinger functions satisfying Wick's law:

$$S_{2n}(x_1, \ldots, x_{2n}) = \sum_{\pi \text{ pairings}} \prod_{j=1}^{n} S_2(x_{\pi(2j-1)}, x_{\pi(2j)}). \tag{6.3}$$

The field theoretical interpretation of (6.3) is the absence of interaction. Due to that and to their algebraically simple structure, such fields are referred to as *trivial*.

The next level of difficulty is to add the next lowest order even term, i.e., $\lambda\Phi^4$ for $\lambda > 0$. Note that, if it exists at all, the corresponding field is a random distribution so making sense of this fourth power is not straightforward. The heuristic RG approach to the problem by Wilson **[101]** indicates that in low enough dimensions, the problem could be tackled through a renormalization procedure. The CQFT program has successfully yielded nontrivial scalar field theories over $\mathbb{R}^2$ **[54]** and $\mathbb{R}^3$ **[44,53]**, and is still a lively field of mathematical physics.

A natural example aimed at constructing a Φ_d^4 functional integral is to regularize it with a pair of cutoffs: at a short distance (*ultraviolet*) scale and a large distance (*infrared*) scale. A lattice version of that is the restriction of $\Phi(\cdot)$ to the vertices of a finite graph $\Lambda_R^{(a)} := (a\mathbb{Z})^d \cap [-R, R]^d$, where a and R play respectively the roles of the ultraviolet and infrared cutoffs. For the corresponding finite collection of variables $(\phi_x : x \in \Lambda_R^{(a)})$, the Hamiltonian is then interpreted in terms of a Riemann-sum style discrete analog of the integral expressions, leading to the following statistical-mechanics Gibbs equilibrium state average

$$\langle F(\phi) \rangle = \frac{1}{\text{norm}} \int_{\mathbb{R}^{\Lambda_R^{(a)}}} F(\phi) \exp\left[-H(\phi)\right] \prod_{x \in \Lambda_R^{(a)}} d\rho(\phi_x), \tag{6.4}$$

with a Hamiltonian $H(\phi)$ and an a-priori measure ρ of the form

$$H(\phi) = - \sum_{\{x,y\} \subset E(\Lambda_R^{(a)})} \phi_x \phi_y, \quad d\rho(\phi_x) = e^{-\lambda \phi_x^4 - b\phi_x^2} d\phi_x, \tag{6.5}$$

where $d\phi_x$ is the Lebesgue measure on $\mathbb{R}$. This is called the ϕ^4 *lattice model*.

The cutoffs are removed through the limit $R \nearrow \infty$ followed by $a \searrow 0$. Parameters may be added to adjust in the process the spin-spin correlations $\langle \phi_{x_1} \dots \phi_{x_n} \rangle$ in such a way that they stabilize to the Schwinger functions $S_n(x_1, \dots, x_n)$ in the continuum limit scale.

The Ising model can be thought of as a limiting case of a ϕ^4 lattice model as it is obtained by letting $\lambda = b/2$ tend to infinity (the limit of the measures ρ then forces the spins ϕ_x to take the values ± 1). Actually, the discrete approximations of the ϕ^4 functional integral and the Gibbs states of an Ising model are *always* connected. This relation is based on a construction which was initiated by Griffiths to obtain the Lee–Yang theorem for the ϕ^4 lattice models, and was advanced further by Griffiths and Simon [59]. A probability measure on $\rho(d\phi)$ on $\mathbb{R}$ is said to belong to the *Griffiths–Simon class* if the expectation values with respect to ρ can be represented as an Ising model on the complete graph with well-chosen coupling constants, or as a limit of such models (satisfying some mild tail conditions). The ϕ^4 lattice model belongs to the Griffiths–Simon class. For this reason, most techniques that are at our disposal for the Ising model apply to the Griffiths–Simon class. This makes the Ising model an object of major interest when working on CQFT. The developments of the model have therefore been deeply connected to CQFT in the 1980s, and we now discuss some examples of such interactions.

6.2. Reflection positivity

The notion of reflection positivity was introduced in Quantum Field Theory in the work of Osterwalder-0Schrader [87], and we refer to [15] for a review. While reflection positivity did not emerge initially as a property of the Ising model, the model remains one of the most natural instances of a reflection-positive model, and some of the most striking applications of reflection positivity are indeed dealing with the Ising model.

Consider the Ising model on a d-dimensional torus $\mathbb{T}_L := (\mathbb{Z}/L\mathbb{Z})^d$ with L even and split equally the torus into two pieces $\mathbb{T}_L^+$ and $\mathbb{T}_L^-$ using hyperplanes (the two pieces are isomorphic to $[0, L/2] \times (\mathbb{Z}/L\mathbb{Z})^{d-1}$) and consider a reflection ϑ with respect to one of these hyperplanes mapping $\mathbb{T}_L^+$ to $\mathbb{T}_L^-$. We say that $\langle \cdot \rangle$ is *reflection positive* if for all $f, g : \mathbb{T}_L^+ \to \mathbb{R}$,

$$\langle f \vartheta g \rangle = \langle g \vartheta f \rangle \quad \text{and} \quad \langle f \vartheta f \rangle \geq 0, \tag{6.6}$$

or, in other words, that $f, g \mapsto \langle f \vartheta g \rangle$ is a positive semidefinite symmetric bilinear form. The archetypical examples of reflection positive measures are the Ising n.n.f. measures $\langle \cdot \rangle_{\mathbb{T}_L,\beta,0}$, but many other examples exist, including some Ising models with long-range interactions.

Reflection positivity has two important implications, namely *Gaussian domination* leading to the *infrared bound*, and the *chessboard estimate*. Due to lack of space, and since most of the applications of reflection positivity to the specific example of the Ising model rely on the infrared bound, let us focus on it and Gaussian domination.

Gaussian domination is a statement linking the partition function of the Ising model with magnetic field to the partition function of the model without it. Formally, it states that for every function $\mathbf{h} : V \to \mathbb{R}$, $Z_L(\mathbf{h}) \le Z_L(0)$, where

$$Z_L(\mathbf{h}) := \sum_{\sigma \in \{-1,1\}^V} \exp\Big[-\beta \sum_{\{x,y\} \in E(\mathbb{T}_L)} (\sigma_x - \sigma_y + \mathbf{h}_x - \mathbf{h}_y)^2\Big]. \tag{6.7}$$

Gaussian domination can be proved via reflection positivity through the two hyperplanes mentioned above to show that, for each $\mathbf{h}$, a symmetric version of $\mathbf{h}$ with respect to a hyperplane has a larger value of $Z_L(\cdot)$. Gaussian domination immediately implies a Fourier version of the infrared bound by using a second-order expansion of $Z_L(\mathbf{h})$ near 0: for $d > 2$ and every $(a_x) \in \mathbb{C}^{\mathbb{T}_L}$ summing to zero,

$$\sum_{x,y \in \mathbb{T}_L} a_x \overline{a}_y \langle \sigma_x \sigma_y \rangle_{\mathbb{T}_L,\beta,0} \le \frac{2}{\beta} \sum_{x,y \in \mathbb{T}_L} a_x \overline{a}_y G(x,y), \tag{6.8}$$

where $G(x,y)$ is the Green function of the simple random walk on $\mathbb{Z}^d$.

In the specific case of the Ising model, the Messager–Miracle-Solé inequality enables to turn this Fourier estimate into a pointwise estimate on the two-point function: there exist $C, C' > 0$ such that for every $\beta > 0$ and every $x, y \in \mathbb{Z}^d$,

$$\langle \sigma_x \sigma_y \rangle_{\beta,0} - m^*(\beta)^2 \le \frac{C}{\beta} G(x,y) \le \frac{C'}{\|x-y\|_2^{d-2}}. \tag{6.9}$$

This is particularly interesting when β approaches β_c from below, as it implies that the spin–spin correlations decay algebraically fast at β_c, with an exponent at least $d-2$.

6.3. The random current revolution

The context of CQFT was also at the origin of one of the most important revolutions in our understanding of the Ising model that we will describe in Section 6.4. The technique, called the *random current*, was introduced by Griffiths and greatly developed by Aizenman who realized that it provides a graphical representation of the Ising model. It became one of the most powerful and robust tools available to mathematicians to study the Ising model. We describe it now (see **[35]** for a review).

The whole story starts with the observation that the component $\exp[\beta \sigma_x \sigma_y]$ of the Hamiltonian term attached to each edge can be rewritten using Taylor's expansion to get

$$Z(G,\beta,0) = \sum_{\sigma \in \{-1,1\}^V} \prod_{\{x,y\} \in E} \sum_{\mathbf{n}_{\{x,y\}}=0}^{\infty} \frac{(\beta \sigma_x \sigma_y)^{\mathbf{n}_{\{x,y\}}}}{\mathbf{n}_{\{x,y\}}!} = \sum_{\mathbf{n} \in \mathbb{Z}_+^E} w_\beta(\mathbf{n}) \sum_{\sigma \in \{-1,1\}^V} \prod_{x \in V} \sigma_x^{\Delta_x(\mathbf{n})}, \tag{6.10}$$

where

$$w_\beta(\mathbf{n}) := \prod_{\{x,y\} \in E} \frac{\beta^{\mathbf{n}_{\{x,y\}}}}{\mathbf{n}_{\{x,y\}}!} \quad \text{and} \quad \Delta_x(\mathbf{n}) := \sum_{y \in V : \{x,y\} \in E} \mathbf{n}_{\{x,y\}}. \tag{6.11}$$

Now, the involutions on spin configurations switching the spins at a vertex immediately imply that the sum on σ on the right-hand side is either equal to $2^{|V|}$ if $\Delta_x(\mathbf{n})$ is even for all $x \in V$, or 0 otherwise (this seems like a very elementary observation, but it bears at the heart of it the $+/-$ symmetry of the space of possible spins).

Call a function from E to $\mathbb{Z}_+$ a *current*. A *source* of the current will be a vertex x with $\Delta_x(\mathbf{n})$ odd. The set of sources will be denoted by $\partial\mathbf{n}$. The previous discussion and the notation lead to the identity

$$Z(G,\beta,0) = 2^{|V|} \sum_{\partial\mathbf{n}=\emptyset} w_\beta(\mathbf{n}), \tag{6.12}$$

where from now on we omit to specify that we consider currents when using the notation $\mathbf{n}$.

A current $\mathbf{n}$ with $\partial\mathbf{n} = A$ can be interpreted as the occupation time of a collection of paths pairing vertices of A and loops or, equivalently, the number of times the collection of paths and loops goes through an edge. The decomposition into loops and paths is not unique; nonetheless, it remains interesting to interpret currents in terms of them.

Proceeding in a similar fashion with the numerator of the spin–spin correlations, we get that

$$\langle\sigma_A\rangle_{G,\beta,0} = \frac{\sum_{\partial\mathbf{n}=A} w_\beta(\mathbf{n})}{\sum_{\partial\mathbf{n}=\emptyset} w_\beta(\mathbf{n})}. \tag{6.13}$$

In words, one may write spin–spin correlations in terms of weighted sums of currents with specific source constraints $\partial\mathbf{n} = A$ and $\partial\mathbf{n} = \emptyset$. Note that the source constraint is not the same for the numerator and denominator.

Frame 4: The high-temperature expansion and $\beta_c > 0$

The *high-temperature expansion* of the Ising model, due to van der Waerden [96], can be neatly defined here as the set of edges with an *odd* current (it can also be obtained by a direct expansion using that $\exp[\beta\sigma_x\sigma_y] = \cosh(\beta) + \sinh(\beta)\sigma_x\sigma_y$). One ends up with another expression of the partition function in terms of even subgraphs

$$Z(G,\beta,0) = \cosh(\beta)^{|E|} \sum_{F\in\mathrm{Even}(G)} \tanh(\beta)^{|F|}, \tag{6.14}$$

which resembles the low-temperature expansion, except that it is on G instead of G^* and that it is valid for arbitrary graphs and not only planar ones. In particular, one may easily deduce the Kramers–Wannier duality between the low and high temperature expansions at temperatures β and β^* satisfying $\tanh(\beta) = e^{-2\beta^*}$ in the case of the square lattice.

One application of currents (or alternatively high-temperature expansion) is obtained by considering a mapping from currents with $\partial\mathbf{n} = \{x,y\}$ to currents with $\partial\mathbf{n} = \emptyset$ setting the current on a path from x to y of odd current (such a path necessarily exists) to 0. This many-to-one mapping (one has to keep track of the path and the value of the current on it to reconstruct the preimage) increases drastically the weight of the current as soon as $\beta \ll 1$, which shows that the spin-spin correlations $\langle\sigma_x\sigma_y\rangle_{G,\beta,0}$ are decaying exponentially fast in this regime. This implies in particular that $\beta_c > 0$.

A key observation of Aizenman is that the so-called *switching lemma*, see Frame 5, pertaining to combinatorial properties of the random current model, could be used to reinterpret spin–spin correlations as well as many other properties in terms of *probabilities* involving multiple independent currents. This lemma completely changed the point of view on currents, as it transforms them from a combinatorial type object into a probabilistic one. In particular, intuitions coming from probabilistic models such as random walks and percolation was later used to prove new theorems on the Ising model; see Sections 6.4, 6.6, and 7.1.

Frame 5: The switching lemma for random currents

Write $\mathbf{n} \in \mathcal{F}_A$ if there exists $\mathbf{k} \le \mathbf{n}$ with $\partial\mathbf{k} = A$. Note that if $A = \{x, y\}$, this is equivalent to the existence of a path from x to y which is made of edges with a positive current. Recall that $A \Delta B$ denotes the symmetric difference of the sets A and B. With this notation, the *switching lemma* [58] states that for every $F : \mathbb{Z}_+^E \to \mathbb{R}$ and every two sets of vertices $A, B \subset V$,

$$\sum_{\substack{\partial\mathbf{n}_1 = A \\ \partial\mathbf{n}_2 = B}} w(\mathbf{n}_1)w(\mathbf{n}_2)F(\mathbf{n}_1 + \mathbf{n}_2) = \sum_{\substack{\partial\mathbf{n}_1 = A\Delta B \\ \partial\mathbf{n}_2 = \emptyset}} w(\mathbf{n}_1)w(\mathbf{n}_2)F(\mathbf{n}_1 + \mathbf{n}_2)\mathbb{I}(\mathbf{n}_1 + \mathbf{n}_2 \in \mathcal{F}_B). \tag{6.15}$$

The name of the lemma is fairly self-explanatory, as it consists, when considering sums of two currents, of a recipe to switch the sources from the second to the first. The proof is a very entertaining combinatorial problem that is left to the reader.

A direct application (to illustrate the strength of the lemma) is the case $A = B$, which gives immediately that

$$\langle \sigma_A \rangle_{G,\beta,0}^2 = \mathbb{P}_G^\emptyset \otimes \mathbb{P}_G^\emptyset[\mathbf{n}_1 + \mathbf{n}_2 \in \mathcal{F}_A], \tag{6.16}$$

where $\mathbb{P}_G^B$ is the measure on currents $\mathbf{n}$ on G with $\partial\mathbf{n} = B$ attributing to each such $\mathbf{n}$ a probability that is proportional to $w(\mathbf{n})$, and $\otimes$ denotes the product for probability measures. In words, one may interpret the *square of* spin–spin correlations $\langle \sigma_A \rangle_{G,\beta,0}$ as the probability, for the sum of two *independent* random currents, of pairing the elements of A by paths of positive current. One may also try as an exercise to recover Griffiths' inequalities from the switching lemma.

6.4. Triviality in dimension $d > 4$

In 1982, Michael Aizenman and Juerg Fröhlich [2,49] independently proved that the scaling limit of the Ising model is trivial in dimension five and more in the following sense. Consider discrete smeared averages defined by

$$T_{f,L}(\sigma) := \frac{1}{\sqrt{\Sigma_L}} \sum_{x \in \mathbb{Z}^d} f(x/L)\sigma_x, \tag{6.17}$$

where f ranges over compactly supported continuous functions, and $\Sigma_L := \langle (\sum_{x\in\Lambda_L} \sigma_x)^2 \rangle$ denotes the variance of the sum of spins over the box of size L. The theorem states that when $d > 4$, these smeared averages $T_{f,L}(\sigma)$ are approximately Gaussian of variance $\langle T_{f,L}(\sigma)^2 \rangle_\beta$

in the sense that there exists an explicit constant $C_f > 0$ such that for every $\beta \le \beta_c$, every $L \le \xi(\beta)$, and every $z > 0$,

$$\left|\left\langle \exp\left[zT_{f,L}(\sigma) - \frac{z^2}{2}\langle T_{f,L}(\sigma)^2\rangle_\beta \right]\right\rangle_\beta - 1\right| \le \frac{C_f z^4}{L^{d-4}}. \tag{6.18}$$

In words, the previous statement claims that the characteristic function of $T_{f,L}(\sigma)$ is close to the one of a Gaussian random variables.

As a direct consequence of this result, one obtains that any well-defined scaling limit of the Ising model, and in fact more generally of the ϕ^4 lattice model, is inevitably Gaussian. The result marked a brutal stop in the CQFT program outlined in Section 6.1 as the proofs suggested, while not proving, that the model should also be trivial in four dimensions.

As mentioned above, one of the most striking applications of the random current representation is related to CQFT. Indeed, Aizenman's proof of this theorem relies on a beautiful parallel between random walks and the paths joining sources in currents. We do not resist discussing this link below. But before doing so, let us mention that the approach of Fröhlich in [49], based on the Brydges–Fröhlich–Spencer (BFS) walk representation of spin–spin correlations [21], is deeply connected to the random current as well. The walks in the BFS representation play the roles of the paths between sources in the random current. The advantage of this alternative approach is that it works for more general models, at the cost of losing the switching lemma and its benefits.

Let us focus on the four-point function and define the corresponding Ursell function given, for $x_1, \dots, x_4 \in \mathbb{Z}^d$, by

$$U_4^\beta(x_1,\dots,x_4) := \langle \sigma_{x_1}\cdots\sigma_{x_4}\rangle_\beta - \sum_{\pi \text{ pairing}} \prod_{i=1}^{2} \langle \sigma_{x_{\pi(2i-1)}}\sigma_{x_{\pi(2i)}}\rangle_\beta. \tag{6.19}$$

A simple exercise involving the switching lemma shows that

$$\begin{aligned} U_4^\beta(x_1,\dots,x_4) = & -2\langle\sigma_{x_1}\sigma_{x_2}\rangle\langle\sigma_{x_3}\sigma_{x_4}\rangle \\ & \mathbb{P}^{\{x_1,x_2\}} \otimes \mathbb{P}^{\{x_3,x_4\}}[x_1,\dots,x_4 \text{ all connected in } \mathbf{n}_1 + \mathbf{n}_2], \end{aligned} \tag{6.20}$$

where connected in $\mathbf{n}_1 + \mathbf{n}_2$ means being connected by a path of edges with $\mathbf{n}_1 + \mathbf{n}_2$ not equal to zero. If one remembers that one can think of a current with sources x_1 and x_2 as a path connecting the two vertices together with a collection of loops, one can reinterpret the right-hand side of the previous identity at the light of so-called random walks (a random walker traces his way through the vertices of a graph by picking its next steps at random among neighbors of where it currently stands—this Markov process is one of the most fundamental objects of probability theory). It is a classical result that two random walks connecting two pairs of points that are at a mutual distance of order L intersect with a probability bounded away from 0 as L tends to infinity in dimensions $d < 4$, and tending to zero in dimension $d \ge 4$.

At this stage, it is totally unclear why the paths linking the points x_1 and x_2 in $\mathbf{n}_1$, and x_3 and x_4 in $\mathbf{n}_2$, would behave as random walks. It is also unclear what would be the impact of the additional loops. Still, it is tempting to think that if an analogy with random walks was valid, then it would single out dimensions $d \geq 4$ as being dimensions for which U_4^β becomes much smaller than products of two-point correlations or, in other words, for which Wick's law would become asymptotically valid, thus hinting at triviality.

When the dimension is strictly larger than 4, the story for random walks becomes even simpler, as the *expected number of intersections* is also tending to zero with L. Using the infrared bound to estimate the spin–spin correlations of the Ising model, one may go around the difficulty of proving a random walk type behavior for currents to show that the intersection probability is tending to 0.

Making the argument work for currents in dimension 4 is more subtle because, contrarily to larger dimensions, the expected number of intersections does not tend to 0 when L tends to infinity. Hence, in order to prove that the intersection probability goes to 0, one inevitably has to go deeper in the understanding of the analogy between currents and random walks.

6.5. Rigorous renormalization group in 4D Ising

The triviality of the Ising model in dimension $d > 4$ naturally raises the question of its triviality in dimension $d = 4$, which is not only the pertinent physical dimension for CQFT, but also for the so-called $4 - \varepsilon$ expansions providing information on dimension 3. In the 1980s, Wilson's renormalization group method was already in every physicists' toolbox, yet the challenges to overcome to cast the general theory in a mathematical framework seemed out of reach. Interestingly, a very relevant case became an important exception.

Consider the lattice version of the ϕ^4 model discussed in Section 6.1. The case $b = \lambda = 0$ corresponds to a Gaussian field known under the name of discrete Gaussian Free Field (GFF), which enjoys a number of striking features. One of them is that the model converges, when rescaling the lattice, to the continuum GFF. In a series of impressive papers [43, 50, 60], mathematical physicists proved in the 1980s that, when starting from a weakly coupled ϕ^4 lattice model (meaning that λ is small), one may apply a multiscale analysis to prove convergence of the model to the continuum GFF.

Several methods were used at the time, but let us mention that the method of Gawedski and Kupiainen [50] can be thought of as a rigorous version of Kadanoff block-spin renormalization procedure. It consists of writing the model in terms of averages of spins over large blocks of size L^k, and to average them out scale by scale. At leading order, each step of the procedure boils down to modifying the parameters of the model. Of course, the reality is much more complicated than the first-order analysis suggests, and the renormalization scheme is quite complex.

An alternative to this block-spin renormalization was later developed by Bauerschmidt, Brydges, and Slade [9] in order to obtain refined results, as well as to treat more general models. In these alternative approaches, the block-spin analysis is replaced by the following strategy: one thinks of quantities in the ϕ^4 lattice model as being expressed in terms of the discrete GFF itself. In order to control the asymptotic behavior of such quantities, one decomposes the covariance of the discrete GFF into a sum of finite-range covariances that one integrates out one by one. At each step a change of the parameters of the system is required to keep things converging towards a limit. Doing so enables the authors to focus their attention on how the parameters evolve under this procedure. This evolution can be thought of as the renormalization map in the renormalization group.

The level of sophistication of these techniques is quite astonishing, and the precision of the results outstanding. As one may guess, this comes at a price. At the bottom of both strategies lies the fact that the original ϕ^4 lattice model is in the "vicinity of a model," the Gaussian Free Field, that enjoys a number of nice properties. As a result, the technique is (as for today) *perturbative* in nature, which is somehow its main limitation. We will see another instance of such a renormalization scheme, this time near another fixed point, when discussing the 2D Ising model.

6.6. Forty years later: The random current strikes back

While renormalization techniques provided impressive rigorous results in dimension 4, they remained as we mentioned perturbative, meaning that they required that the lattice ϕ^4 model one starts from has a small ϕ^4 term. Yet, if one would like to construct a nontrivial 4D quantum field theory, one would definitely try to start with a strongly coupled ϕ^4 lattice model (meaning with a ϕ^4 terms which is not a priori small), for instance, working with the Ising model which in some sense can be thought of as the model with the strongest possible coupling, thus excluding existing renormalization group techniques.

This asks for another approach, and this is probably why one had to wait for 40 years to finally obtain a proof of the triviality of the 4D Ising and ϕ^4 lattice models, which states [4] that there exists $c > 0$ such that for the n.n.f. ϕ^4 lattice model on $\mathbb{Z}^4$ with parameters b, λ, and a compactly supported continuous function f, there exists $C_f > 0$ such that for every $\beta \leq \beta_c = \beta_c(b, \lambda)$, every $L \leq \xi(\beta)$, and every $z > 0$,

$$\left|\left\langle \exp\left[zT_{f,L}(\varphi) - \frac{z^2}{2}\langle T_{f,L}(\varphi)^2\rangle_\beta \right]\right\rangle_\beta - 1\right| \leq \frac{C_f z^4}{(\log L)^c}. \tag{6.21}$$

The strategy of the proof uses a more delicate probabilistic perspective on the random current than in [2], still keeping in mind the interpretation in terms of random walks of the paths joining the sources of the current. Indeed, it can be proved that two random walkers in four dimensions going from points to points that are all at a mutual distance of order L intersect with probability of order $(\log L)^{-c}$ for some universal constant $c > 0$. The reason is that while the expected number of intersections is of order 1, the number of intersections, when such intersections exist, is with high probability quite large in L (and is growing with L). The core of the paper is to apply a similar argument to the paths in the random current. Of course, challenges emerge when trying to handle the highly non-Markovian paths obtained by considering the paths joining the sources in currents. Nevertheless, guided by the random walk intuition, one can build a multiscale analysis to prove that conditioned on intersecting, random currents intersect a large number of times, and ultimately deduce from this the triviality result.

7. THE LAST 50 YEARS: ISING MODEL AND PERCOLATION

Percolation theory gathers under its umbrella a variety of random graph systems. A configuration on $G = (V, E)$ is an element $\omega = (\omega_e : e \in E) \in \{0,1\}^E$ which is interpreted as a subgraph with vertex-set V and edge-set $\{e \in E : \omega_e = 1\}$. Then, different percolation models can be defined by considering different measures on $\{0,1\}^E$. Historically, the original model, called Bernoulli percolation, is defined in such a way that the ω_e are independent Bernoulli random variables. It was introduced to understand the behavior of liquid in a porous medium. Nevertheless, the theory of non-Bernoulli models has been found to be related to a variety of other models of statistical physics explaining various physical phenomena.

As often, the Ising model has played an essential role in the development of percolation theory, and conversely certain advances in percolation theory have been fundamental to our understanding of the Ising model. Sometimes, the link between the two models is simply an analogy between their behaviors, but sometimes the connection is much more direct. For instance, spin–spin correlations can be rewritten in terms of a percolation model, in which case we speak of the percolation model as being a *graphical representation* of the Ising model. We now propose to discuss some examples of these links between the Ising model and percolation.

7.1. Percolation interpretation of random currents

We have seen one example of a graphical representation in Frame 5 where the squares of spin–spin correlations get rephrased as connectivity properties of the sum of two currents. One may easily define a percolation model out of the pair of currents above by saying that for an edge $\{x, y\}$, $\omega_{\{x,y\}} = 1$ if $(\mathbf{n}_1 + \mathbf{n}_2)_{\{x,y\}} > 0$. Then, the square of the spin–spin correlations between two points becomes the probability, for this percolation model, that x and y are connected in ω.

The best illustration of how intuition from percolation or the Ising model can drive developments on the other model is provided by an important result on the Ising model in the regime $\beta < \beta_c$. This result from 1987, due to Aizenman, Barsky, and Fernandez **[3]** (see **[39]** for an alternative argument), states that correlations of the n.n.f. Ising model decay exponentially fast as soon as $\beta < \beta_c$ in the sense that for each such β, there exists $\tau > 0$ such that for every $x, y \in \mathbb{Z}^d$,

$$\langle \sigma_x \sigma_y \rangle_{\beta,0} \le \exp\bigl(-\tau \|x - y\|\bigr). \tag{7.1}$$

We say that the phase transition is *sharp*: there is no intermediate phase $(\beta_{\exp}, \beta_c)$ in the Ising model in which spin–spin correlations would decay polynomially. Let us mention that a similar exponential decay was obtained recently for truncated correlations $\langle \sigma_x \sigma_y \rangle_{\beta,0} - m^*(\beta)^2$ when $\beta > \beta_c$, see **[36]**.

This theorem is of fundamental importance for the following reason. Perturbative results, which are combinatorial in nature, are valid under the assumption that certain quantities decay exponentially fast, and in fact with a rate of decay which is sufficiently large. While this hypothesis is important to apply the techniques, it happens to be of little relevance from a physical point of view. In fact, one expects that most of the phenomenology remains unchanged as long as spin–spin correlations decay exponentially fast. As a consequence, (7.1) can be thought of as a bottleneck in the understanding of the phase $\beta < \beta_c$: as soon as it is obtained, a number of important results can be derived from it. As an example, the results on fluctuations of interfaces and Ornstein–Zernike estimates were proved to hold in the whole regime $\beta < \beta_c$. The result also provides meaning to the correlation length $\xi(\beta)$ mentioned in Section 4.1, as it proves that it is finite as soon as $\beta < \beta_c$.

Let us now comment on the proof. The argument relies on a fruitful idea consisting in deriving differential inequalities between thermodynamical quantities of the Ising model. The archetypical example of such differential inequalities are given, for the problem at hand, by (recall that the magnetization $m = m(\beta, h)$ is a function of β and h)

$$m \le \tanh(\beta h)\frac{\partial}{\partial(\beta h)}m + m^2\left(\beta\frac{\partial}{\partial\beta}m + m\right) \quad \text{and} \quad m\frac{\partial}{\partial\beta}m \ge c. \tag{7.2}$$

The interesting feature here is that similar differential inequalities appear when studying Bernoulli percolation. In fact, a number of results were obtained in parallel during the 1980s, where each result for Ising had its pendant for Bernoulli percolation, and vice versa. As an example, critical exponents for $d > 4$ were obtained by Aizenman and Fernandez [7] using differential inequalities that can be adapted to Bernoulli percolation. These techniques are useful to transform qualitative results (e.g., a quantity tends to 0) to quantitative ones (e.g., exponentially fast). We do not resist mentioning one of them: for $h = 0$ and $\beta < \beta_c$,

$$\left(1 - \frac{B}{\chi}\right)\frac{2d\chi^2}{1+B} \le \frac{\partial}{\partial\beta}\chi \le 2d\chi^2, \tag{7.3}$$

where $\chi(\beta) := \sum_x \langle\sigma_0\sigma_x\rangle_{\beta,0}$ is the *susceptibility*, and $B(\beta)$ is the *Bubble diagram* and is given by

$$B(\beta) := \sum_{x\in\mathbb{Z}^d} \langle\sigma_0\sigma_x\rangle_{\beta,0}^2. \tag{7.4}$$

Since the Infrared Bound implies that $B(\beta)$ remains bounded uniformly in $\beta < \beta_c$ as soon as $d > 4$, $\chi(\beta)$ must blow up like $1/|\beta - \beta_c|$ as β approaches β_c from below.

Another striking instance of how fruitful the connection between percolation models and the Ising model was for the development of both models is the following *continuity result* of the phase transition of the 3D Ising model, due to [5], stating that the n.n.f. Ising model satisfies $m^*(\beta_c) = 0$ for every $d \ge 3$.

The argument relies on percolation methods applied to the double random current representation of an argument of Burton and Keane proving the uniqueness of the infinite connected component of percolation. The whole argument can be improved and extended to study all translation-invariant Gibbs measures, obtaining the classification result already mentioned in Section 5.2.

7.2. Fortuin–Kasteleyn percolation

Another (and in fact older) example of a graphical representation is provided by a special case of the Fortuin–Kasteleyn (FK) percolation. In this model, introduced in [47], the

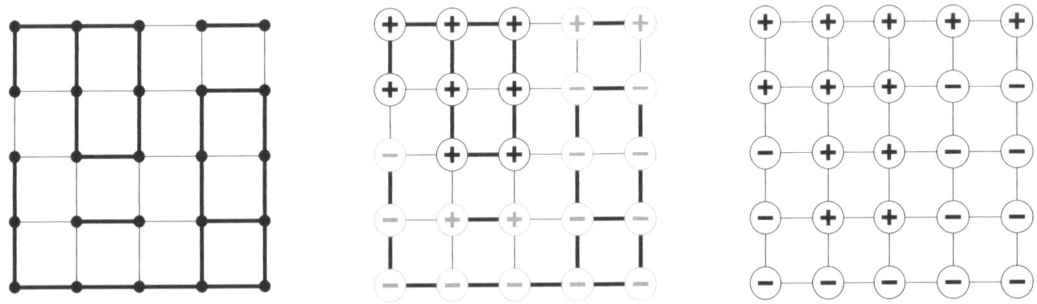

FIGURE 3
The Edwards–Sokal coupling, with a picture of the FK Ising configuration on the left (bold edges are those with $\omega_e = 1$); in the middle, spins are attached to each cluster (one example in black and others in grey); and on the right, the spins without the FK Ising configuration.

measure $\phi_{G,p,q}$ is given, for $G = (V, E)$ finite and $\omega \in \{0, 1\}^E$, by

$$\phi_{G,p,q}\big[\{\omega\}\big] := \frac{1}{Z(G, p, q)} p^{|\omega|}(1 - p)^{|E|-|\omega|} q^{k(\omega)}, \tag{7.5}$$

where $p \in [0, 1]$ and $q > 0$ are the parameters of the model, called respectively the *edge-weight* and the *cluster-weight*, $|\omega| := \sum_{e\in E} \omega_e$ is interpreted as the number of edges in ω, and $k(\omega)$ is the number of connected components of ω.

When $q = 1$, one ends up with the classical Bernoulli percolation model in which the ω_e are independent. When $q \neq 1$, the state of edges is no longer independent and one ends up with a *dependent percolation model* whose study is central in modern probability theory. From now on, we focus on the case $q = 2$, which we call the *FK Ising model*. We confine our discussion to two features of this percolation model, namely its link to the Ising model, and the FKG inequality.

Let us start with the former, which provides a recipe to obtain the Ising model configuration out of FK Ising; see Figure 3. Consider a random variable $\omega \in \{0, 1\}^E$ with the law of FK Ising with parameter $p \in [0, 1]$ and construct $\sigma \in \{-1, 1\}^V$ by

- choosing for every connected component $\mathcal{C}$ of ω a spin $\sigma_{\mathcal{C}}$ uniformly between -1 and $+1$, and independently of the other connected components;
- defining $\sigma_x = \sigma_{\mathcal{C}}$ for every $\mathcal{C}$ and every $x \in \mathcal{C}$.

Then, σ has the law of the Ising model on G with parameter $\beta = \frac{1}{2}\log[1/(1 - p)]$ and $h = 0$. This coupling, due to Fortuin and Kasteleyn and often referred to as the *Edwards–Sokal coupling* due to the paper **[40]**, enables to express correlation functions of the Ising model in terms of FK Ising. For instance, by decomposing on the events that x is connected to y or not in ω, one easily gets that

$$\langle\sigma_x\sigma_y\rangle_{G,\beta,0} = \phi_{G,1-e^{-2\beta},2}[x \text{ connected to } y \text{ in } \omega]. \tag{7.6}$$

Similarly, $\langle\sigma_A\rangle_{G,\beta,0} = \phi_{G,1-e^{-2\beta},2}[\mathcal{F}_A]$, where $\mathcal{F}_A$ is the event that each connected component of ω contains an even number (possibly equal to 0) of vertices in A. Another interesting feature of this coupling is that it is at the basis of so-called *cluster algorithms* due to Swendsen and Wang, who used it to speed up the Glauber dynamics and the simulation of the Ising model, in particular near the critical point.

The interest of FK Ising and more generally FK percolation models *with* $q \geq 1$ is that they enjoy some nice monotonicity properties (dependent percolation models satisfying these properties have been an object of intense study in the past ten years). Let us mention two such properties. The Fortuin–Kasteleyn–Ginibre (FKG) inequality states that for every increasing functions $f, g : \{0, 1\}^E \to \mathbb{R}$,

$$\phi_{G,p,q}[fg] \geq \phi_{G,p,q}[f]\phi_{G,p,q}[g]. \tag{7.7}$$

This inequality is often used for indicator functions of *increasing events* (i.e., events for which the indicator function is an increasing function), in which case the inequality states that increasing events are positively correlated. Another manifestation of the monotonicity properties is the monotonicity in p: for every increasing function $f : \{0, 1\}^E \to \mathbb{R}$ and $p' \geq p$,

$$\phi_{G,p',q}[f] \geq \phi_{G,p,q}[f]. \tag{7.8}$$

These monotonicity properties are particularly useful. The latter applied to FK Ising and the indicator function of $\mathcal{F}_A$ implies that $\langle \sigma_A \rangle_{G,\beta,0}$ is increasing in β, and the former applied to indicator functions of $\mathcal{F}_A$ and $\mathcal{F}_B$ implies the second Griffiths inequality $\langle \sigma_A \sigma_B \rangle_{G,\beta,0} \geq \langle \sigma_A \rangle_{G,\beta,0} \langle \sigma_B \rangle_{G,\beta,0}$.

7.3. The broader impact of the Ising model on dependent percolation models

In the first 50 years that followed its introduction, the theory of percolation was much more advanced for Bernoulli percolation than for other dependent percolation models. The past ten years have seen tremendous progress in bridging the gap between our understanding of the Bernoulli case and the others. The interplay between dependent percolation models and the Ising model has been fundamental for these developments.

We already saw that the Ising model is related to FK Ising and a percolation model created out of random currents. It does not come as a surprise that one of the first dependent percolation models to see significant progress in its understanding was the FK Ising. Of course, the Edwards–Sokal coupling enables to transfer immediately certain known facts about the Ising model to its percolation representation (for instance, the critical point of the FK Ising on $\mathbb{Z}^2$ is $1 - e^{-2\beta_c} = \sqrt{2}/(1 + \sqrt{2})$ thanks to Onsager's result). Also, the model enjoys some specific features that make its direct analysis simpler than for other dependent percolation models.

For all these reasons, the FK Ising became the entrance gate to a new realm of results on dependent percolation models. A perfect illustration of this is provided by the study of *crossing probabilities* for planar dependent percolation models. Let us provide slightly more detail.

One important feature of critical dependent percolation models in two dimensions is that they satisfy the box-crossing property (BCP), and its connected notion the Russo–Seymour–Welsh theory (RSW). More precisely, if for a rectangle R, the event $\text{Cross}(R)$

corresponds to the existence of a path in ω between the left- and right-hand sides of R, the properties (BCP) and (RSW) for a percolation model on $\mathbb{Z}^2$ with measure $\mathbb{P}$ are the following:

- (BCP) for all $\rho > 0$, there exists $c > 0$ such that for every $n \geq 1$,

$$c \leq \mathbb{P}\big[\mathrm{Cross}\big([0, \rho n] \times [0, n]\big) | \omega_{|[-n,(\rho+1)n]\times[-n,2n]^c}\big] \leq 1 - c \quad \text{almost surely.} \tag{7.9}$$

- (RSW) for all $\rho > 0$, there exists $C > 0$ such that for every $n \geq 1$,

$$\mathbb{P}\big[\mathrm{Cross}\big([0, \rho n] \times [0, n]\big)\big] \geq \mathbb{P}\big[\mathrm{Cross}\big([0, n] \times [0, \rho n]\big)\big]^C. \tag{7.10}$$

These two properties have been the driving force of the progress in our understanding of the 2D dependent percolation models. The FK-Ising model played an essential role in these developments, as it was the first dependent percolation model for which (BCP) could be proved [37]. This development triggered a whole new direction of research that led to substantial progress in our understanding of (BCP) and (RSW) for various percolation models.

8. OVER THE LAST TEN YEARS: CONFORMAL INVARIANCE OF THE ISING MODEL

8.1. What is conformal invariance?

As mentioned before, Kadanoff used his block-spin renormalization to predict that the large scale properties of the critical Ising model were invariant under scaling. The same argument also leads to postulate *translation* and *rotation invariance*. In 1970, Polyakov [90] suggested a much stronger invariance of the model. Since we saw that it is natural to associate a QFT with the large scale properties of the critical Ising model, and since this QFT is a local field, these properties should be invariant under any map which is *locally* a composition of translation, rotation and homothety. As a corollary one predicts full *conformal invariance*, i.e., invariance under all one-to-one holomorphic maps. This prediction was turned into a classification of possible *conformal field theories* (CFT) in 2D in seminal papers by Belavin, Polyakov, and Zamolodchikov [12] that generated an explosion of activity, allowing nonrigorous explanations of many critical phenomena.

From a mathematical perspective, the notion of conformal invariance of a model is not straightforward to define. A number of interpretations of the limit of large scale properties—called the *scaling limit*—can be taken, and we mention a few now.

For clarity of the exposition, we focus on the critical Ising model on $\mathbb{Z}^d$ and its rescaled versions $a\mathbb{Z}^d$ for $a > 0$. We drop the subscript referring to β and h as they are fixed to be equal to β_c and 0, respectively. Consider a simply connected domain $\Omega \subsetneq \mathbb{R}^d$.

(Spins) The most natural approach is to consider the spin–spin correlations defined for every $a > 0$ and $x_1, \ldots, x_n \in \Omega$ by

$$S_\Omega^{(a)}(x_1, \ldots, x_n) := \langle \sigma_{[x_1]_a} \cdots \sigma_{[x_n]_a} \rangle_{a\mathbb{Z}^d \cap \Omega}, \tag{8.1}$$

where $[x]_a$ is the vertex of $a\mathbb{Z}^d \cap \Omega$ closest to x. These Schwinger functions already appeared as the key players in CQFT. One is then interested in the limit as a tends to 0 of these properly renormalized quantities. If the limit exists, we call it $S_\Omega(x_1, \dots, x_n)$.

(Energies) Another object of interest is the *energy–energy* correlations. For $a > 0$ and $x_1, \dots, x_n \in \Omega$, one considers at the quantities

$$T_\Omega^{(a)}(x_1, \dots, x_n) := \langle \varepsilon_{(x_1)_a} \cdots \varepsilon_{(x_n)_a} \rangle_{a\mathbb{Z}^d \cap \Omega}, \tag{8.2}$$

where $\varepsilon_{\{u,v\}} := \sigma_u \sigma_v - \langle \sigma_u \sigma_v \rangle_{a\mathbb{Z}^d}$ and $(x)_a$ is the edge closest to x. The quantity ε_x is called the *energy*. One is again interested in the limit $T_\Omega(x_1, \dots, x_n)$ as a tends to 0 of these properly rescaled quantities.

(Geometry of interfaces) In two dimensions, another direction was proposed in the 1990s. It consists in considering the low-temperature representation, i.e., the interfaces between plus and minus spins. In a domain Ω, it creates a family of nonintersecting loops together with arcs from boundary to boundary. Let $\mathfrak{C}_\Omega$ be the set of such collections of loops and arcs. The set $\mathfrak{C}_\Omega$ can be turned into a metric space by attaching a distance d_Ω which, heuristically, states that two configurations are close to each other when the large loops and arcs are close to each other. Let us call $\mathcal{C}_\Omega^{(a)}$ the random variable obtained by considering the low-temperature expansion of a critical Ising model configuration in $a\mathbb{Z}^d \cap \Omega$. Here, we are interested in the limit of $\mathcal{C}_\Omega^{(a)}$ as a random object.

Now, what do we mean by conformal invariance? Roughly speaking, we mean that certain quantities of the model are conformally covariant/invariant. With the definitions above, it would for instance mean that there exists a way of renormalizing the $S_\Omega^{(a)}(x_1, \dots, x_n)$ and $T_\Omega^{(a)}(x_1, \dots, x_n)$ in such a way that they converge to quantities $S_\Omega(x_1, \dots, x_n)$ and $T_\Omega(x_1, \dots, x_n)$ that satisfy that there exist $\mathbf{\Delta}_\sigma, \mathbf{\Delta}_\varepsilon$ such that for every conformal (i.e., holomorphic and one-to-one) map $f : \Omega \to f(\Omega)$, we have

$$S_{f(\Omega)}\big(f(x_1), \dots, f(x_n)\big) = \big|f'(x_1)\big|^{-\mathbf{\Delta}_\sigma} \cdots \big|f'(x_n)\big|^{-\mathbf{\Delta}_\sigma} S_\Omega(x_1, \dots, x_n), \tag{8.3}$$

$$T_{f(\Omega)}\big(f(x_1), \dots, f(x_n)\big) = \big|f'(x_1)\big|^{-\mathbf{\Delta}_\varepsilon} \cdots \big|f'(x_n)\big|^{-\mathbf{\Delta}_\varepsilon} T_\Omega(x_1, \dots, x_n). \tag{8.4}$$

For the geometry of interfaces, the situation is even simpler as one means that the family of loops and arcs $\mathcal{C}_\Omega^{(a)}$ converges to a limit $\mathcal{C}_\Omega$ as a tends to 0 and that this limit satisfies that $\mathcal{C}_{f(\Omega)}$ and $f(\mathcal{C}_\Omega)$ have the same law for every conformal map $f : \Omega \to f(\Omega)$.

8.2. Conformal invariance of the 2D Ising model

Around 15 years ago, Smirnov [95] and Chelkak and Smirnov [28] obtained a major breakthrough towards proving conformal invariance of 2D Ising model. This fundamental proof, that we discuss below, opened the way to a very deep understanding of the scaling limit of the model.

A few years later, Chelkak–Izyurov–Hongler [27] proved conformal covariance of the spin–spin correlations (with $\mathbf{\Delta}_\sigma = 1/8$). It was later proved in [22] that the quantities $S_\Omega(x_1, \dots, x_n)$ are the Schwinger functions of a random distribution, that can be understood as the *spin-field* that physicists sometimes refer to. In the same spirit, conformal covariance of the energy–energy correlations was proved in [63] (with $\mathbf{\Delta}_\varepsilon = 1$). In this case, one may prove that the correlations are *not* the Schwinger functions of a random distribution. Turning to interfaces, the following result was the culmination of the theory: the arcs in $\mathscr{C}_\Omega$ are given by the so-called free arc ensemble of parameter 3 and the loops by conformal loop ensembles of parameter 3 in the simply connected domains obtained as the complements of the arcs (see [13, 14]). In particular, the scaling limit is conformally invariant. This body of work uses the ideas from [28, 95] together with the theory of the Schramm–Loewner evolution and its consequences.

As mentioned above, an important breakthrough came from the works [28, 95] where conformal covariance of so-called *fermionic observables* $f_\Omega^{(a)}$ is proved. Those observables are linear combinations of order–disorder operators (see Frame 6) considered by Kadanoff and Ceva in [73], see also [26] for several connections to other classical objects.

Frame 6: Fermionic observable

Consider a simply connected domain $\Omega \subset \mathbb{C}$ and for $a > 0$, let be the largest connected component of $a\mathbb{Z}^2 \cap \Omega$. Consider n vertices $x_1, \ldots, x_n$ of , and n faces $f_1, \ldots, f_n$ of such that f_i is bordered by x_i for every $1 \leq i \leq n$. Choose n disjoint *cuts* $\ell_1, \ldots, \ell_n$, i.e., families of dual edges $(e_i^*(j))$ forming self-avoiding paths in the dual from the unbounded face to the center of f_i. Define the *disorder operator* μ_ℓ for a cut ℓ as the observable that effectively switches the coupling constants of the edges $e_i(j)$ associated with the $e_i^*(j)$ in the cut (it can be written as a product of terms of the form $\exp[-2\beta\sigma_x\sigma_y]$ over edges appearing in the family of edges $\{e_i(j) : i, j\}$). Then, the *order–disorder* correlations are given by the formula

$$F_\Omega^{(a)}(x_1, f_1, \ldots, x_n, f_n) := \langle \sigma_{x_1}\mu_{\ell_1} \cdots \sigma_{x_n}\mu_{\ell_n} \rangle. \tag{8.5}$$

Let us mention that these quantities can be expressed in terms of correlations of Grassmann variables in the Schultz–Mattis–Lieb representation [93].

Smirnov introduced a *fermionic observable* $f_\Omega^{(a)}$ defined at centers of edges $\{x, y\}$ of that can be written as a linear combination (with complex coefficients) of the $F_\Omega^{(a)}$ with x_1 equal to x or y, and f_1 to one of the two faces bordered by $\{x, y\}$. The details of the definition are unimportant here and the take-home message is that Chelkak and Smirnov proved that the limit (as a tends to 0) of these fermionic observables is conformally covariant.

The conformal covariance of the fermionic observable should be understood as the first brick among the conformal covariance results of spin–spin, energy–energy correlations, and even of the conformal invariance of interfaces. Let us mention that these results require substantial additional ideas compared to [28, 95]. In fact, conformal covariance/invariance of virtually all quantities one may be interested in the 2D Ising model can be recovered today.

The proof of the theorem relies on the observation that $f_\Omega^{(a)}$ is the solution of a *discrete* version of a Riemann–Hilbert boundary value problem. More precisely, the function can be proved, via combinatorial arguments involving the van der Waerden high-temperature expansion, to be *preholomorphic* (see Frame 7), and to satisfy certain boundary conditions. These special features are connected to the integrability of the model. From general principles on preholomorphic functions, the limit as a tends to 0 of these objects must be the holomorphic solution of a continuum Riemann–Hilbert boundary value problem, which can be computed and proved to be conformally covariant. Such reasoning has been used in several existing proofs of conformal invariance, for instance for dimers or Bernoulli site percolation on the triangular lattice. It has created an explosion of results in the field as many quantities can be proved to converge using a similar strategy.

Frame 7: Preholomorphic observables

The notion of preholomorphic function on a planar graph G appeared implicitly in the work of Kirchhoff on electrical networks [76]. It was explicitly linked to holomorphicity in the work of Isaacs [66, 67], in which the author proposed to discretize the Cauchy–Riemann equation to get to the definition (on the square lattice)

$$F(\mathrm{NW}) - F(\mathrm{SE}) = i\big[F(\mathrm{NE}) - F(\mathrm{SW})\big], \tag{8.6}$$

where NW, SW, SE, and NE are the four corners found in counterclockwise order around each face, when starting from the top left vertex.

The properties of preholomorphic functions have been the object of a renewed interest with the emergence of the question of conformal invariance in connections to boundary value problems. Indeed, general theorems stating that preholomorphic functions satisfying certain boundary value conditions converge when taking finer and finer mesh size to holomorphic solutions of the continuum version of the boundary value problem took a central place in the theory.

In the case of the Ising model, the complexity of the boundary value problem (involving a condition on the argument of the fermionic observable) pushed Smirnov to introduce a stronger notion of preholomorphicity, called *s-holomorphicity*, which is also satisfied by fermionic observables. The advantage of this notion is that it enables one to define the imaginary part of the primitive of the square of the observable, which, roughly speaking, becomes the discrete solution of a Dirichlet boundary value problem, a much more tractable problem for which convergence (when a tends to 0) can be proved very elegantly.

8.3. Towards universality of the 2D Ising model

As mentioned in Section 4.2.2, the large-scale properties of the critical Ising model should not depend on the precise properties of the underlying graph. With the tremendous successes that have been achieved over the years in the case of the Ising model on $\mathbb{Z}^2$ and more generally on planar graphs, it is natural to test the validity of the universality hypothesis in this context. Several advances have been made in this direction in the last 15 years.

The first impressive progress can be found in the work of Chelkak and Smirnov themselves [28]. They observed that the preholomorphicity argument leading to conformal invariance can be articulated naturally in the setting of so-called *isoradial graphs*. An isoradial graph is an *embedding* of a graph G in the plane such that every face of the graph is inscribed in a circle of radius 1. In this context, one may define special coupling constants $J_{x,y}$ depending on the graph in such a way that $\beta_c = 1$ and that the fermionic observable is naturally preholomorphic on this graph. Then, the strategy of Chelkak and Smirnov on the square lattice applies to isoradial graphs with the same conclusions. Note that this result can be understood as a universality result on the graph (isoradial graphs are a fairly large family of planar graphs, even though not fully general), but that the choice of $J_{x,y}$ is *determined* by the embedded graph itself. Moreover, a striking feature of this theorem is that no transitivity or quasitransitivity is required for this to work.

In recent developments, Chelkak generalized the conformal invariance result to a wider class of Ising models, namely those defined on planar locally-finite doubly periodic weighted graphs (G, J), i.e., weighted graphs which are invariant under the action of some lattice $\Lambda \approx \mathbb{Z} \oplus \mathbb{Z}$ (in such case G/Λ is a finite graph embedded in the torus). For such models, Chelkak proved in [25] that there exists an embedding in the plane, called an *s-embedding*, with the property that the scaling limit of the critical model defined on this embedding is conformally invariant.

This result is a strong indication of universality for planar graphs. Now what happens beyond planar graphs? The universality conjecture asserts that the scaling limit depends on the *large scale* geometry of the graph (for instance, a planar Euclidean geometry). In particular, one may consider the graph obtained with the vertex-set $\mathbb{Z}^2$ and edge-set given by pairs of vertices at a distance at most R of each other. This model, called the *finite-range* model on $\mathbb{Z}^2$, should have a behavior that is similar to the nearest-neighbor case as it is "almost planar." The additional difficulty is that nonplanarity immediately breaks the integrability of the system. The universality of such Ising models has been investigated in two different directions.

First, one may consider finite-range models that are perturbations of the nearest-neighbor integrable case, meaning that non-nearest neighbor interactions are very weak, i.e., that $J_{x,y}$ is small when $1 < \|x - y\|_2 \leq R$. Using the Schultz–Mattis–Lieb Grassmann representation [93] of the nearest neighbor case, one may express the partition function and more generally the energy–energy and spin–spin correlations in terms of Grassmann variables, and therefore at the end in terms of the nearest-neighbor model. Using an elaborate multiscale analysis and studying the renormalization of parameters induced by this multiscale analysis, Giuliani–Greenblatt–Mastropietro derived in [52] the large-scale behavior of energy–energy correlations in the full plane. While the previously mentioned renormalization schemes in dimension 4 were enabled by the fact that the model is a small perturbation of the discrete GFF (which is a Gaussian process), the two-dimensional case relies on a similar connection, this time to the n.n.f. Ising model on $\mathbb{Z}^2$ (which has a Grassmannian structure).

As a consequence, the strategy suffers from the same limitations as the 4D case in the sense that it is restricted to small perturbations of the n.n.f. Ising model on $\mathbb{Z}^2$.

A totally different approach explaining the emergence of planarity in finite range Ising models was proposed in [6] based on the random current representation. The underlying idea relies on the fact that thanks to the switching lemma, intersection properties of random currents with sources are related to the structure of n-point correlations in the model. Yet, the intersection properties of long paths on the graph induced by $\mathbb{Z}^2$ and the edges between vertices at a distance R of each other resemble the ones that can be obtained for planar graphs. As an example of a possible application, one can obtain that spin–spin correlations on the boundary of a domain Ω have a Pfaffian structure, a result which is specific to the universality class of the 2D Ising model. More precisely, for any collection of points $x_1 = (k_1, 0), \ldots, x_{2n} = (k_{2n}, 0)$ satisfying $k_1 < k_2 < \cdots < k_{2n}$ on the boundary of the upper half-plane $\mathbb{H} := \mathbb{Z} \times \mathbb{Z}_+$,

$$\langle \sigma_{x_1} \cdots \sigma_{x_{2n}} \rangle_{\mathbb{H},\beta_c} = \mathrm{Pfaff}_n\big(\big[\langle \sigma_{x_i}\sigma_{x_j}\rangle_{\mathbb{H},\beta_c}\big]_{1\le i<j\le 2n}\big)\big[1+o(1)\big], \tag{8.7}$$

where $o(1)$ is a function of the points $x_1, \ldots, x_{2n}$ which tends to zero for configuration sequences with $\min\{|x_i - x_j| : 1 \le i < j \le 2n\}$ tending to infinity.

This is, to the author's knowledge, the first property witnessing the 2D Ising universality class that can be obtained in a level of generality that is not restricted to planar graphs and their perturbations. Also, the proof relies on the key properties of the Ising model that one would like to use: the $\pm$ spin symmetry (entering the story through the use of the random current representation) and the large scale planarity of the underlying graph (which for finite range models on $\mathbb{Z}^2$ is the reason behind the "almost" intersection properties of long paths). The trade-off is that full conformal invariance of this family of models is still out of reach.

8.4. Conformal bootstrap in 3D Ising model

At this point, we already mentioned that the 1D Ising model was trivially solved in the original paper of Ising [68], and that it took 20 more years to achieve a solution of the 2D Ising model [86]. We also saw that the model in dimensions 4 and higher is much simpler as its large-scale properties should be Gaussian. This singles out 3D as the remaining challenging dimension. To the best of our knowledge [97], it is not known whether the model is integrable or not. This is particularly problematic as the third dimension is probably the most relevant one physically (for instance, the model should be in the universality class of liquid–vapor systems, and totally anisotropic magnets).

In recent years, a striking progress has been made on the physics side using the so-called *conformal bootstrap*. A conformal field theory (CFT) is characterized by the correlation functions $\langle - - \rangle$ of an infinite number of *local operators* $\mathcal{A}(x)$, which in the case of Ising should be understood as the objects obtained by taking the limit of random variables defined in terms of spins next to a given position of space. For example, the scaling limit of spin and energy observables σ_x and $\varepsilon_{\{x,y\}} = \sigma_x\sigma_y - \langle \sigma_x\sigma_y \rangle$ give such local operators in the case of Ising, but one may think of more complicated ones, such as the scaling limit of (products of) the gradient $\sigma_{x+y} - \sigma_x$ of the spins.

Conformal invariance already forces huge constraints on the correlations of operators in the theory. Oversimplifying slightly, for scalar local operators there must exist exponents $\mathbf{\Delta}_{\mathcal{A}}$ and coefficients $f_{\mathcal{A}\mathcal{B}\mathcal{C}}$ such that

$$\langle \mathcal{A}(x)\mathcal{A}(y)\rangle = \frac{1}{\|x-y\|_2^{\mathbf{\Delta}_{\mathcal{A}}}}, \tag{8.8}$$

$$\langle \mathcal{A}(x)\mathcal{B}(y)\mathcal{C}(z)\rangle = \frac{f_{\mathcal{A}\mathcal{B}\mathcal{C}}}{\|x-y\|_2^{\mathbf{\Delta}_{\mathcal{A}}+\mathbf{\Delta}_{\mathcal{B}}-\mathbf{\Delta}_{\mathcal{C}}}\|y-z\|_2^{\mathbf{\Delta}_{\mathcal{B}}+\mathbf{\Delta}_{\mathcal{C}}-\mathbf{\Delta}_{\mathcal{A}}}\|z-x\|_2^{\mathbf{\Delta}_{\mathcal{C}}+\mathbf{\Delta}_{\mathcal{A}}-\mathbf{\Delta}_{\mathcal{B}}}} \tag{8.9}$$

(in (8.8), we adopted without loss of generality the normalization of $\mathcal{A}$ that makes the constant in the numerator equal to 1). The exponents and coefficients depend a priori on the CFT, but a striking feature is that there exists a way, called the *conformal block decomposition*, to express multipoint correlations of local operators in terms of three-point functions by gluing points together using the so-called operator product expansion. This theoretically shows that all the information in a CFT can be encoded in terms of the $\Delta_{\mathcal{A}}$ and the $f_{\mathcal{A}\mathcal{B}\mathcal{C}}$. Of course, determining these coefficients is very difficult.

While in 2D this was done in the 1980s, the analogous question remains widely open in 3D. Nevertheless, one can proceed in a slightly different way by asking which choices of these quantities can lead to a consistent CFT. This approach, called the *conformal bootstrap*, was shown to be amazingly powerful in 3D. The underlying idea is that one is facing an infinite family of consistency relations coming from different ways of applying the conformal block decomposition (which is not unique). For instance, one may start with $\langle \mathcal{A}(x_1)\mathcal{A}(x_2)\mathcal{A}(x_3)\mathcal{A}(x_4)\rangle$ and proceed by gluing first x_1 and x_2 or, on the contrary, x_3 and x_4. This leads to two decompositions of the same object as a linear combination (with positive coefficients in the Ising case) of known objects called the *conformal blocks*. Equalling these two decompositions, one ends up with constraints on the possible exponents.

There is a priori no reason to be able to determine the critical exponents as the unique values satisfying a (finite) number of constraints thus obtained. Indeed, the set of possible values may not shrink when considering more and more conditions, but it happens that in the case of the Ising model, the region of the plane for possible critical exponents $(\mathbf{\Delta}_\sigma, \mathbf{\Delta}_\varepsilon)$ for the spin and energy local operators can be reduced drastically, to a point where estimates—namely $(\mathbf{\Delta}_\sigma, \mathbf{\Delta}_\varepsilon) = (0.5181489(10), 1.412625(10))$—using this bootstrap technique become way better than Monte Carlo simulations. We refer to **[41, 77, 92]** for some of the original papers and **[94]** for a review of the most recent progress in this very exciting area of modern theoretical physics.

Let us conclude that even if one may use conformal bootstrap to *exactly* identify the critical exponents, this would leave the question of proving that the critical 3D Ising model *indeed* converges to a CFT widely open. In some sense, getting sufficient information on the possible scaling limits and proving that these scaling limits indeed exist are two almost entirely disjoint questions even though, of course, one may hope that information on the former question would help answer the latter.

9. A TAIL TO THIS STORY

The Ising model has always played the role of a locomotive in the developments of statistical physics. Its central place and incredible properties turn it into an amazing playground for both mathematicians and physicists. As a consequence, during most of its history novel techniques were developed to solve problems on it, which later led to whole independent fields of mathematical physics (integrable systems, graphical representations, rigorous renormalization methods, etc.).

Let us mention several long-standing problems remaining widely open for this model. At the top of the list, universality of the 2D behavior (see Section 8.3), critical properties of the 3D model (see Section 8.4), and the roughening phase transition (see Section 5.3) are among the most important unsolved puzzles. Solving them will probably require the development of new techniques that will again, through cross-fertilization, benefit the whole field of statistical mechanics.

ACKNOWLEDGMENTS

We thank all our coauthors for the wonderful years of joint research, both past and future. We also wish to thank D. Cimasoni, T. Gunaratnam, D. Krachun, I. Manolescu, R. Panis, V. Tassion, and Y. Velenik who, without knowing its ultimate aim, took the time to take a look at this review and to give the author feedback.

FUNDING

This work has received funding from the European Research Council (ERC) under the European Union's Horizon 2020 research and innovation programme (grant agreements No. 757296). The author acknowledges funding from the NCCR SwissMap, the Swiss FNS, and the Simons collaboration on localization of waves.

REFERENCES

[1] M. Aizenman, Translation invariance and instability of phase coexistence in the two dimensional Ising system. *Comm. Math. Phys.* **73** (1980), no. 1, 83–94.

[2] M. Aizenman, Geometric analysis of φ^4 fields and Ising models. I, II. *Comm. Math. Phys.* **86** (1982), no. 1, 1–48.

[3] M. Aizenman, D. J. Barsky, and R. Fernández, The phase transition in a general class of Ising-type models is sharp. *J. Stat. Phys.* **47** (1987), no. 3–4, 343–374.

[4] M. Aizenman and H. Duminil-Copin, Marginal triviality of the scaling limits of critical 4D Ising and ϕ_4^4 models. *Ann. of Math.* **194** (2021), no. 1, 163–235.

[5] M. Aizenman, H. Duminil-Copin, and V. Sidoravicius, Random Currents and Continuity of Ising Model's Spontaneous Magnetization. *Comm. Math. Phys.* **334** (2015), 719–742.

[6] M. Aizenman, H. Duminil-Copin, V. Tassion, and S. Warzel, Emergent planarity in two-dimensional Ising models with finite-range interactions. *Invent. Math.* **216** (2019), no. 3, 661–743.

[7] M. Aizenman and R. Fernández, On the critical behavior of the magnetization in high-dimensional Ising models. *J. Stat. Phys.* **44** (1986), no. 3–4, 393–454.

[8] J. R. Baker and A. George, Application of the Padé approximant method to the investigation of some magnetic properties of the Ising model. *Phys. Rev.* **124** (1961), no. 3, 768.

[9] R. Bauerschmidt, D. C. Brydges, and G. Slade, Scaling limits and critical behaviour of the 4-dimensional n-component $|\phi^4|$ spin model. *J. Stat. Phys.* **157** (2014), 692–742.

[10] R. J. Baxter, *Exactly solved models in statistical mechanics*. Academic Press Inc., London, 1989. Reprint of the 1982 original.

[11] R. J. Baxter and I. G. Enting, 399th solution of the Ising model. *J. Phys. A: Math. Gen.* **11** (1978), no. 12, 2463.

[12] A. A. Belavin, A. M. Polyakov, and A. B. Zamolodchikov, Infinite conformal symmetry in two-dimensional quantum field theory. *Nuclear Phys. B* **241** (1984), no. 2, 333–380.

[13] S. Benoist, H. Duminil-Copin, and C. Hongler, Conformal invariance of crossing probabilities for the Ising model with free boundary conditions. *Ann. Inst. Henri Poincaré* **52** (2016), no. 4, 1784–1798.

[14] S. Benoist and C. Hongler, The scaling limit of critical Ising interfaces is CLE(3). *Ann. Probab.* **47** (2019), no. 4, 2049–2086.

[15] M. Biskup, Reflection positivity and phase transitions in lattice spin models. In *Methods of contemporary mathematical statistical physics*, pp. 1–86, Lecture Notes in Math. 1970, Springer, Berlin, 2009.

[16] T. Bodineau, The Wulff construction in three and more dimensions. *Comm. Math. Phys.* **207** (1999), no. 1, 197–229.

[17] T. Bodineau, Translation invariant Gibbs states for the Ising model. *Probab. Theory Related Fields* **135** (2006), no. 2, 153–168.

[18] T. Bodineau, D. Ioffe, and Y. Velenik, Rigorous probabilistic analysis of equilibrium crystal shapes. *J. Math. Phys.* **41** (2000), no. 3, 1033–1098.

[19] W. Bragg and E. Williams, The effect of thermal agitation on atomic arrangement in alloys. (–II). *Proc. R. Soc. Lond.* **145** (1934), no. 855, 699–730; **151** (1935), no. 874, 540–566.

[20] S. G. Brush, History of the Lenz–Ising model. *Rev. Modern Phys.* **39** (1967), no. 4, 883.

[21] D. Brydges, J. Fröhlich, and T. Spencer, The random walk representation of classical spin systems and correlation inequalities. *Comm. Math. Phys.* **83** (1982), no. 1, 123–150.

[22] F. Camia, C. Garban, and C. M. Newman, Planar Ising magnetization field I. Uniqueness of the critical scaling limit. *Ann. Probab.* **43** (2015), no. 2, 528–571.

[23] M. Campanino, D. Ioffe, and Y. Velenik, Ornstein–Zernike theory for finite range Ising models above T_c. *Probab. Theory Related Fields* **125** (2003), no. 3, 305–349.

[24] R. Cerf and A. Pisztora, On the Wulff crystal in the Ising model. *Ann. Probab.* **28** (2000), 947–1017.

[25] D. Chelkak, Ising model and s-embeddings of planar graphs. 2020, arXiv:2006.14559.

[26] D. Chelkak, D. Cimasoni, and A. Kassel, Revisiting the combinatorics of the 2D Ising model. *Ann. Inst. Henri Poincaré D* **4** (2017), no. 3, 309–385.

[27] D. Chelkak, C. Hongler, and K. Izyurov, Conformal invariance of spin correlations in the planar Ising model. *Ann. of Math. (2)* **181** (2015), no. 3, 1087–1138.

[28] D. Chelkak and S. Smirnov, Universality in the 2D Ising model and conformal invariance of fermionic observables. *Invent. Math.* **189** (2012), no. 3, 515–580.

[29] D. Cimasoni, The Critical Ising Model via Kac–Ward Matrices. *Comm. Math. Phys.* **316** (2012), 99–126.

[30] P. Curie, Propriétés magnétiques des corps à diverses températures. 4, Gauthier-Villars et fils.

[31] R. L. Dobrushin, Gibbs state describing coexistence of phases for a three-dimensional Ising model. *Theory Probab. Appl.* **17** (1973), no. 4, 582–600.

[32] R. L. Dobrushin, R. Kotecký, and S. Shlosman, *Wulff construction: a global shape from local interaction* 104. American Math. Society, Providence, 1992, x+204 pp.

[33] N. P. Dolbilin, Y. M. Zinov'ev, A. S. Mishchenko, M. A. Shtan'ko, and M. I. Shtogrin, The two-dimensional Ising model and the Kac–Ward determinant. *Izv. Math.* **63** (1999), no. 4, 707.

[34] C. Domb and M. F. Sykes, On the susceptibility of a ferromagnetic above the Curie point. *Proc. R. Soc. Lond. Ser. A, Math. Phys. Sci.* **240** (1957), no. 1221, 214–228.

[35] H. Duminil-Copin, Random currents expansion of the Ising model. 2016, arXiv:1607.06933.

[36] H. Duminil-Copin, S. Goswami, and A. Raoufi, Exponential decay of truncated correlations for the Ising model in any dimension for all but the critical temperature. *Comm. Math. Phys.* **374** (2020), no. 2, 891–921.

[37] H. Duminil-Copin, C. Hongler, and P. Nolin, Connection probabilities and RSW-type bounds for the two-dimensional FK Ising model. *Comm. Pure Appl. Math.* **64** (2011), no. 9, 1165–1198.

[38] H. Duminil-Copin and I. Manolescu, Planar random-cluster model: Scaling relations. 2020, arXiv:2011.15090.

[39] H. Duminil-Copin and V. Tassion, A new proof of the sharpness of the phase transition for Bernoulli percolation and the Ising model. *Comm. Math. Phys.* **343** (2016), no. 2, 725–745.

[40] R. G. Edwards and A. D. Sokal, Generalization of the Fortuin–Kasteleyn–Swendsen–Wang representation and monte carlo algorithm. *Phys. Rev. D* **38** (1988), no. 6, 2009.

[41] S. El-Showk, M. F. Paulos, D. Poland, V. S. Rychkov, D. Simmons-Duffin, and A. Vichi, Solving the 3D Ising model with the conformal bootstrap. *Phys. Rev. D* **86** (2012), no. 2, 025022.

[42] J. W. Essam and M. E. Fisher, Padé approximant studies of the lattice gas and Ising ferromagnet below the critical point. *J. Chem. Phys.* **38** (1963), no. 4, 802–812.

[43] J. Feldman, J. Magnen, V. Rivasseau, and R. Sénéor, Construction and Borel Summability of Infrared Φ_4^4 by a Phase Space Expansion. *Comm. Math. Phys.* **109** (1987), 437–480.

[44] J. S. Feldman and K. Osterwalder, The Wightman axioms and the mass gap for weakly coupled ϕ_3^4 quantum field theories. *Ann. Phys.* **97** (1976), no. 1, 80–135.

[45] M. E. Fisher, On the dimer solution of planar Ising models. *J. Math. Phys.* **7** (1963), 1776.

[46] M. E. Fisher, Correlation functions and the critical region of simple fluids. *J. Math. Phys.* **5** (1964), no. 7, 944–962.

[47] C. M. Fortuin and P. W. Kasteleyn, On the random-cluster model. I. Introduction and relation to other models. *Physica* **57** (1972), 536–564.

[48] S. Friedli and Y. Velenik, *Statistical mechanics of lattice systems: A concrete mathematical introduction*. Cambridge University Press, 2017.

[49] J. Fröhlich, On the triviality of $\lambda\phi_d^4$ theories and the approach to the critical point in $d(-) > 4$ dimensions. *Nuclear Phys. B* **200** (1982), no. 2, 281–296.

[50] K. Gawedzki and A. Kupiainen, Massless Lattice Φ_4^4 Theory: Rigorous Control of a Renormalizable Asymptotically Free Model. *Comm. Math. Phys.* **99** (1985), 197–252.

[51] H. O. Georgii, *Gibbs measures and phase transitions*. 2nd edn., de Gruyter Stud. Math. 9, Walter de Gruyter and Co., Berlin, 2011.

[52] A. Giuliani, R. L. Greenblatt, and V. Mastropietro, The scaling limit of the energy correlations in non-integrable Ising models. *J. Math. Phys.* **53** (2012), no. 9, 095214.

[53] J. Glimm and A. Jaffe, Positivity of the ϕ_3^4 hamiltonian. *Fortschr. Phys.* **21** (1973), no. 7, 327–376.

[54] J. Glimm and A. Jaffe, *Quantum physics: a functional integral point of view*. Springer, 2012.

[55] L. Greenberg and D. Ioffe, On an invariance principle for phase separation lines. *Ann. Inst. Henri Poincaré B, Probab. Stat.* **41** (2005), no. 5, 871–885.

[56] R. B. Griffiths, Correlation in Ising ferromagnets I, II. *J. Math. Phys.* **8** (1967), 478–489.

[57] R. B. Griffiths, Dependence of critical indices on a parameter. *Phys. Rev. Lett.* **24** (1970), no. 26, 1479.

[58] R. B. Griffiths, C. A. Hurst, and S. Sherman, Concavity of magnetization of an Ising ferromagnet in a positive external field. *J. Math. Phys.* **11** (1970), 790–795.

[59] R. B. Griffiths and B. Simon, The $(\Phi_2)^4$ Field Theory as a Classical Ising Model. *Comm. Math. Phys.* **33** (1973), 145–164.

[60] T. Hara and H. Tasaki, A Rigorous Control of Logarithmic Corrections in Four-Dimensional $(\phi_4)^4$ Spin Systems. II. Critical Behaviour of Susceptibility and Correlation Length. *J. Stat. Phys.* **47** (1987), no. 1/2, 99–121.

[61] W. Heisenberg, Zur Theorie des Ferromagnetismus. *Z. Phys.* **49** (1928), no. 9, 619–636.

[62] Y. Higuchi, On limiting Gibbs states of the two-dimensional Ising models. *Publ. Res. Inst. Math. Sci.* **14** (1978), no. 1, 53–69.

[63] C. Hongler, *Conformal invariance of Ising model correlations*. Ph.D. thesis, Université de Genève, 2010.

[64] C. A. Hurst and H. S. Green, New solution of the Ising problem for a rectangular lattice. *J. Chem. Phys.* **33** (1960), 1059.

[65] D. Ioffe and R. H. Schonmann, Dobrushin–Kotecký–Shlosman theorem up to the critical temperature. *Comm. Math. Phys.* **199** (1998), no. 1, 117–167.

[66] R. P. Isaacs, A finite difference function theory. *Univ. Nac. Tucumán. Rev. A* **2** (1941), 177–201.

[67] R. P. Isaacs, Monodiffric functions. Construction and applications of conformal maps. In *Proc. of a symp* 18, pp. 257–266, 1941.

[68] E. Ising, Beitrag zur Theorie des Ferromagnetismus. *Z. Phys.* **31** (1925), 253–258.

[69] P. Jossang, En Bref. "En Bref" de La Recherche (1996, juillet-août), no. 289, 7 pp.

[70] M. Kac and J. C. Ward, A combinatorial solution of the two-dimensional Ising model. *Phys. Rev.* **88** (1952), 1332.

[71] L. P. Kadanoff, Scaling laws for Ising models near T_c. *Physics Physique Fizika* **2** (1966), no. 6, 263.

[72] L. P. Kadanoff, Critical behaviour. Universality and scaling. In *Critical phenomena*, edited by M. S. Green, pp. 100–107, Proceedings of the International School 'Enrico Fermi' 51. Italian Physical Society, Academic Press, New York.

[73] L. P. Kadanoff and H. Ceva, Determination of an operator algebra for the two-dimensional Ising model. *Phys. Rev. B* **3** (1971), no. 11, 3918.

[74] P. W. Kasteleyn, The statistics of dimers on a lattice. *Physica* **27** (1961), 1209; Dimer statistics and phase transitions. *J. Math. Phys.* **4** (1963), 287.

[75] B. Kaufman and L. Onsager, Crystal statistics. III. Short-range order in a binary Ising lattice. *Phys. Rev.* **76** (1949), no. 8, 1244.

[76] G. Kirchhoff, Ueber die Auflösung der Gleichungen, auf welche man bei der Untersuchung der linearen Vertheilung galvanischer Ströme geflührt wird. *Ann. Phys.* **148** (1847), no. 12, 497–508.

[77] F. Kos, D. Poland, and D. Simmons-Duffin, Bootstrapping mixed correlators in the 3D Ising model. *J. High Energy Phys.* **1411** (2014), 109.

[78] H. A. Kramers and G. H. Wannier, Statistics of the two-dimensional ferromagnet. Part I. *Phys. Rev.* **60** (1941), no. 3, 252.

[79] T. D. Lee and C. N. Yang, Statistical theory of equations of state and phase transitions. I. Theory of condensation. II. Lattice gas and Ising model. *Phys. Rev.* **87** (1952), no. 3, 404, 410.

[80] W. Lenz, Beitrag zum Verständnis der magnetischen Eigenschaften in festen Körpern. *Z. Phys.* **21** (1920), 613–615.

[81] E. H. Lieb, A refinement of Simon's correlation inequality. *Comm. Math. Phys.* **77** (1980), no. 2, 127–135.

[82] B. McCoy and T. T. Wu, *The two-dimensional Ising model*. Harvard Univ. Press, 2013.

[83] S. Miracle-Solé and A. Messager, Correlation functions and boundary conditions in the Ising ferromagnet. *J. Stat. Phys.* **17** (1977), no. 4, 245–262.

[84] E. W. Montroll, Statistical mechanics of nearest neighbour systems. *J. Chem. Phys.* **9** (1941), no. 9, 706–721.

[85] M. Niss, History of the Lenz–Ising model 1920–1950: From ferromagnetic to cooperative phenomena. 1950–1965: From irrelevance to relevance. 1965–1971: The role of a simple model in understanding critical phenomena. *Arch. Hist. Exact Sci.* **59** (2005), no. 3, 267–318; **63** (2009), no. 3, 267–318 243–287; **65** (2011), no. 6, 625–658.

[86] L. Onsager, Crystal statistics. I. A two-dimensional model with an order–disorder transition. *Phys. Rev.* **65** (1944), no. 3–4, 117.

[87] K. Osterwalder and R. Schrader, Axioms for Euclidean Green's functions I. II. *Comm. Math. Phys.* **31** (1973), 83–112; *Comm. Math. Phys.* **42** (1975), 281–305. .

[88] J. Palmer, *Planar Ising Correlations* 49. Springer, 2007.

[89] R. Peierls, On Ising's model of ferromagnetism. *Math. Proc. Cambridge Philos. Soc.* **32** (1936), no. 3, 477–481.

[90] A. M. Polyakov, Conformal symmetry of critical fluctuations. *JETP Lett.* **12** (1970), 381–383.

[91] A. Raoufi, Translation-invariant Gibbs states of the Ising model: General setting. *Ann. Probab.* **48** (2020), no. 2, 760–777.

[92] R. Rattazzi, V. S. Rychkov, E. Tonni, and A. Vichi, Bounding scalar operator dimensions in 4D CFT. *J. High Energy Phys.* **0812** (2008), 031.

[93] T. D. Schultz, D. C. Mattis, and E. H. Lieb, Two-dimensional Ising model as a soluble problem of many fermions. *Rev. Modern Phys.* **36** (1964), no. 3, 856.

[94] D. Simmons-Duffin, The conformal bootstrap. In *New Frontiers in Fields and Strings: TASI 2015 Proceedings of the 2015 Theoretical Advanced Study Institute in Elementary Particle Physics*, pp. 1–74, 2017.

[95] S. Smirnov, Conformal invariance in random cluster models. I. Holomorphic fermions in the Ising model. *Ann. of Math. (2)* **172** (2010), no. 2, 1435–1467.

[96] B. L. van der Waerden, Die lange Reichweite der regelmassigen Atomanordnung in Mischkristallen. *Z. Phys.* **118** (1941), 473–488.

[97] G. M. Viswanathan, M. A. G. Portillo, E. P. Raposo, and M. G. E. da Luz, What does it take to solve the 3D Ising model? Minimal necessary conditions for a valid solution. 2022, arXiv:2205.12357.

[98] P. Weiss, L'hypothèse du champ moléculaire et la propriété ferromagnétique. *J. Phys. Theor. Appl.* **6** (1907), no. 1, 661–690.

[99] B. Widom, Equation of state in the neighborhood of the critical point. *J. Chem. Phys.* **43** (1965), no. 11, 3898–3905.

[100] A. S. Wightman, Quantum Field Theory in Terms of Vacuum Expectation Values. *Phys. Rev.* **101** (1956), 860.

[101] K. G. Wilson, Renormalization Group and Critical Phenomena. I. Renormalization Group and the Kadanoff Scaling Picture. *Phys. Rev. B* **4** (1971).

[102] C. N. Yang, The spontaneous magnetization of a two-dimensional Ising model. *Phys. Rev.* **85** (1952), no. 5, 808.

HUGO DUMINIL-COPIN

Unige, 7-9 rue du Conseil-Général, 1205 Genève, Switzerland; IHES, 35 route de Chartres, 91440 Bures-Sur-Yvette, France, hugo.duminil@unige.ch, duminil@ihes.fr

This Page Intentionally Left Blank

COMBINATORICS AND HODGE THEORY

JUNE HUH

ABSTRACT

I will tell two interrelated stories illustrating fruitful interactions between combinatorics and Hodge theory. The first is that of Lorentzian polynomials, based on my joint work with Petter Brändén. They link continuous convex analysis and discrete convex analysis via tropical geometry, and they reveal subtle information on graphs, convex bodies, projective varieties, Potts model partition functions, log-concave polynomials, and highest weight representations of general linear groups. The second is that of intersection cohomology of matroids, based on my joint work with Tom Braden, Jacob Matherne, Nick Proudfoot, and Botong Wang. It shows a surprising parallel between the theory of convex polytopes, Coxeter groups, and matroids. After giving an overview of the similarity, I will outline proofs of two combinatorial conjectures on matroids, the nonnegativity conjecture for their Kazhdan–Lusztig coefficients and the top-heavy conjecture for the lattice of flats.

MATHEMATICS SUBJECT CLASSIFICATION 2020

Primary 05A20; Secondary 05B35, 14C17, 14N10

KEYWORDS

Lorentzian polynomials, matroids, intersection cohomology

ICM INTERNATIONAL CONGRESS OF MATHEMATICIANS 2022 JULY 6–14

© 2022 International Mathematical Union

Proc. Int. Cong. Math. 2022, Vol. 1, pp. 212–239
DOI 10.4171/ICM2022/205

Published by EMS Press and licensed under a CC BY 4.0 license

1. INTRODUCTION

One may seek unity in mathematics through the eyes of *cohomology*. Let X be a mathematical object of "dimension" d. The object may be analytic, arithmetic, geometric, or combinatorial, and the precise notion of dimension will depend on the context. Curiously, often it is possible to construct from X in a natural way a graded real vector space

$$A(X) = \bigoplus_{k=0}^{d} A^k(X).$$

The new object $A(X)$, called the cohomology of X, often encodes essential information on X. When two objects X and Y of the same kind are related in a particular way, the relationship is often reflected on their cohomologies $A(X)$ and $A(Y)$, and this property can be exploited to extend our understanding. Primary consumers of this viewpoint so far were topologists and geometers, and a great number of triumphs in topology and geometry are based on a construction of $A(X)$ from X. Interestingly, sometimes, satisfactory and equally useful cohomologies exist even when X does not have a geometric structure in the conventional sense. In particular, when X is a *matroid*, the study of $A(X)$ led to proofs of a few combinatorial conjectures that were beyond reach with traditional methods [1, 6, 13].

There are a few pieces of evidence for the unity in the above context. The list is short, but the pattern is remarkable. For example, $A(X)$ can be the ring of algebraic cycles modulo homological equivalence on a smooth projective variety [36], the combinatorial cohomology of a convex polytope [45], the Soergel bimodule of a Coxeter group element [26], the Chow ring of a matroid [1], the conormal Chow ring of a matroid [6], or the intersection cohomology of a matroid [13]. In these cases, the cohomology comes equipped with a symmetric bilinear pairing $P : A^*(X) \times A^{d-*}(X) \to \mathbb{R}$ and a graded linear map $L : A^*(X) \to A^{*+1}(X)$ that are symmetric in the sense that

$$P(x, y) = P(y, x) \quad \text{and} \quad P(x, Ly) = P(Lx, y) \quad \text{for all } x \text{ and } y.$$

The linear map L is allowed to vary in a family $K(X)$, a convex cone in the space of linear operators on $A(X)$. Here P is for Poincaré, L is for Lefschetz, and K is for Kähler, who first emphasized the importance of the respective objects in topology and geometry. In good cases, $A^0(X)$ has a distinguished generator 1, and one expects the following properties to hold for every nonnegative integer $k \leq \frac{d}{2}$:

(1) The symmetric bilinear pairing

$$A^k(X) \times A^{d-k}(X) \to \mathbb{R}, \quad (x_1, x_2) \mapsto P(x_1, x_2)$$

is nondegenerate (*Poincaré duality* for X).

(2) For any $L_1, \ldots, L_{d-2k} \in K(X)$, the linear map

$$A^k(X) \to A^{d-k}(X), \quad x \mapsto \left(\prod_{i=1}^{d-2k} L_i\right) x$$

is an isomorphism (*hard Lefschetz property* for X).

(3) For any $L_0, L_1, \ldots, L_{d-2k} \in K(X)$, the symmetric bilinear form

$$A^k(X) \times A^k(X) \to \mathbb{R}, \quad (x_1, x_2) \mapsto (-1)^k \, \mathrm{P}\left(x_1, \left(\prod_{i=1}^{d-2k} L_i\right) x_2\right)$$

is positive definite on the kernel of the linear map

$$A^k(X) \to A^{d-k+1}(X), \quad x \mapsto \left(\prod_{i=0}^{d-2k} L_i\right) x$$

(*Hodge–Riemann relations* for X).

In the classical setting, $A(X)$ is the cohomology of real (k, k)-forms on a compact Kähler manifold, and the three statements are consequences of Hodge theory **[43, CHAPTER 3]**.[1] All three statements are known to hold for $A(X)$ listed above except the first one, which is the subject of Grothendieck's standard conjectures on algebraic cycles **[36]**. In every case, the three statements for $A(X)$ reveal a fundamental property of X: Weil conjectures on the number of solutions to a system of polynomial equations over finite fields when X is a smooth projective variety **[36, 67]**, the generalized lower bound conjecture on the number of faces when X is a convex polytope **[45, 70]**, and Kazhdan–Lusztig's nonnegativity conjecture when X is a Coxeter group element **[26]**. When X is a matroid, the hard Lefschetz property and the Hodge–Riemann relations for different choices of $A(X)$ are used to settle Rota's conjecture on the characteristic polynomial **[1]**, Brylawski's and Dawson's conjectures on the h-vectors of the broken circuit complex and the independence complex **[6]**, and Dowling–Wilson's top-heavy conjecture on the number of flats **[13]**. The known proofs of the Poincaré duality, the hard Lefschetz property, and the Hodge–Riemann relations for the objects listed above have certain structural similarities, but there is no known way of deducing one from the others. Could there be a Hodge-theoretic framework general enough to explain this miraculous coincidence?

A related goal is to produce a flexible analytic theory that would reflect certain basic features of the unified theory: If one postulates the existence of the satisfactory cohomology $A(X)$, what can we say about X at an elementary and numerical level? This is a worthwhile question because, depending on X, the construction and the study of $A(X)$ might be beyond the reach of our current understanding. A step in this direction is taken in a joint work with Petter Brändén on *Lorentzian polynomials* **[17]**, where the difficult goal of finding $A(X)$ is replaced by an easier goal of producing a Lorentzian polynomial from X. Such a Lorentzian polynomial can be used to settle and generate conjectures on various X (Section 2) and, sometimes, leads to a satisfactory theory of $A(X)$ (Section 3).

---

1 In **[13, 26, 36, 43, 45]**, the hard Lefschetz property and the Hodge–Riemann relations are considered only in the "unmixed" case where $L = L_i$ for all i. According to **[18]**, this special case implies the general case stated above.

2. LORENTZIAN POLYNOMIALS

Lorentzian polynomials link continuous convex analysis and discrete convex analysis via tropical geometry, and they reveal subtle information on graphs, convex bodies, projective varieties, Potts model partition functions, log-concave polynomials, and highest weight representations of general linear groups. Let H_n^d be the space of degree d homogeneous polynomials in n variables with real coefficients. The members of H_n^d will be written

$$f = \sum_{\alpha} c_\alpha \frac{w^\alpha}{\alpha!},$$

where the sum is over the nonnegative integral vectors $\alpha \in \mathbb{Z}^n_{\geq 0}$ with $|\alpha|_1 = d$ and

$$\frac{w^\alpha}{\alpha!} := \frac{w_1^{\alpha_1}}{\alpha_1!}\frac{w_2^{\alpha_2}}{\alpha_2!}\cdots\frac{w_n^{\alpha_n}}{\alpha_n!}.$$

Note that a polynomial f can be viewed as a function in at least two different ways. The continuous f is the function given by the evaluation

$$f : \mathbb{R}^n_{\geq 0} \to \mathbb{R}, \quad w \mapsto f(w),$$

and the discrete f is the function given by the coefficients

$$f : \mathbb{Z}^n_{\geq 0} \to \mathbb{R}, \quad \alpha \mapsto c_\alpha.$$

Throughout we write $\operatorname{supp}(f)$ for the *support* of the discrete f, the set of monomials appearing in f with nonzero coefficients. The theory of Lorentzian polynomials shows that the log-concavity of the continuous f is related to the log-concavity of the discrete f in an interesting way. Before defining Lorentzian polynomials in Definition 4, we list three applications of the theory to demonstrate the usefulness and ubiquity of Lorentzian polynomials. Each item below presents an elementary statement that is difficult to prove without the Lorentzian point of view.

Example 1 (Analysis). Let f be a homogeneous polynomial of degree d in n variables with nonnegative coefficients. Such a polynomial f is said to be *strongly log-concave* if, for all $\alpha \in \mathbb{Z}^n_{\geq 0}$, we have

$\partial^\alpha f$ *is identically zero or* $\log(\partial^\alpha f)$ *is concave on the positive orthant* $\mathbb{R}^n_{>0}$.

For bivariate polynomials, one can show that $f = \sum_{k=0}^d c_k w_1^k w_2^{d-k}$ is strongly log-concave exactly when the sequence $\{c_k\}$ has *no internal zeros* and is *ultra log-concave*:

$$\frac{c_k^2}{\binom{d}{k}^2} \geq \frac{c_{k-1}}{\binom{d}{k-1}}\frac{c_{k+1}}{\binom{d}{k+1}} \quad \text{for all } 0 < k < d.$$

In [17, COROLLARY 2.32], the theory of Lorentzian polynomials is used to prove the following statement:

The product of strongly log-concave homogeneous polynomials is strongly log-concave.

This answers a question of Gurvits [37, SECTION 4.5] for homogeneous polynomials, and extends the following theorem of Liggett [50, THEOREM 2]:

> *The convolution product of two ultra log-concave sequences with no internal zeros is an ultra log-concave sequence with no internal zeros.*

The short proof in [17] is based on the following analytic characterization of Lorentzian polynomials [17, THEOREM 2.30]:

> *A homogeneous polynomial with nonnegative coefficients is Lorentzian if and only if it is strongly log-concave.*

It is interesting to compare the argument with the computational proof in [50] for bivariate polynomials.

Example 2 (Combinatorics). Let $\mathscr{A}$ be a set of n vectors in a vector space. For any k, set

$$f_k(\mathscr{A}) := \text{the number of } k \text{ element linearly independent subsets of } \mathscr{A}.$$

For example, if $\mathscr{A}$ is the set of all seven nonzero vectors in a three-dimensional vector space over the field with two elements, then there are seven dependencies among the triples shown below, and hence

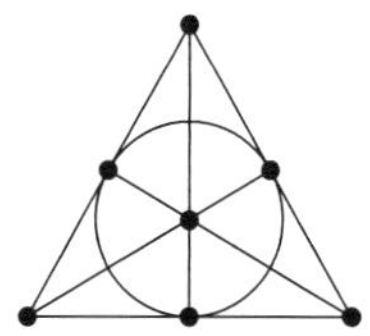

$$f_0(\mathscr{A}) = 1, \quad f_1(\mathscr{A}) = 7, \quad f_2(\mathscr{A}) = 21, \quad f_3(\mathscr{A}) = 28.$$

Mason's conjecture from [52] predicts that, for any $\mathscr{A}$ and any positive integer k,

$$\frac{f_k(\mathscr{A})^2}{\binom{n}{k}^2} \geq \frac{f_{k-1}(\mathscr{A})}{\binom{n}{k-1}} \frac{f_{k+1}(\mathscr{A})}{\binom{n}{k+1}}.$$

The same statement was conjectured more generally for all *matroids* (Definition 9), and the general statement is proved in [17, THEOREM 4.14] using the theory of Lorentzian polynomials.[2] The proof is based on the Lorentzian property of the Potts model partition function for matroids introduced in [68].

2 Nima Anari, Kuikui Liu, Shayan Oveis Gharan, and Cynthia Vinzant have independently developed methods that partly overlap with [17] in a series of papers [2–4]. They study the class of *completely log-concave polynomials*, which agrees with the class of Lorentzian polynomials in the homogeneous case. The main overlap is an independent proof of Mason's conjecture in [4].

Example 3 (Algebra). Schur polynomials are the characters of finite-dimensional irreducible polynomial representations of the general linear group $\mathrm{GL}_n(\mathbb{C})$. Combinatorially, the *Schur polynomial* of a partition λ in n variables is

$$s_\lambda(w_1, \ldots, w_n) = \sum_\alpha K_{\lambda\alpha} w^\alpha,$$

where $K_{\lambda\alpha}$ is the *Kostka number* counting Young tableaux of given shape λ and weight α. Correspondingly, the irreducible representation $V(\lambda)$ of the general linear group with the highest weight λ has the weight space decomposition

$$V(\lambda) = \bigoplus_\alpha V(\lambda)_\alpha \quad \text{with } \dim V(\lambda)_\alpha = K_{\lambda\alpha}.$$

Schur polynomials were first studied by Cauchy, who defined them as ratios of alternants. The connection to the representation theory of $\mathrm{GL}_n(\mathbb{C})$ was found by Schur. For a gentle introduction to these remarkable polynomials, and for any undefined terms, we refer to **[31]**.

In **[40, THEOREM 2]**, the authors use the Lorentzian property for normalized Schur polynomials to show that the sequence of weight multiplicities of $\mathrm{V}(\lambda)$ one encounters is always log-concave if one walks in the weight diagram along any root direction $e_i - e_j$. In other words, for any $\alpha \in \mathbb{Z}^n_{\geq 0}$ and any $i, j \in [n]$,

$$K^2_{\lambda\alpha} \geq K_{\lambda\alpha - e_i + e_j} K_{\lambda\alpha + e_i - e_j}.$$

This verifies a special case of Okounkov's conjecture from **[61, CONJECTURE 1]**.[3]

We now define Lorentzian polynomials. As before, we write H^d_n for the space of degree d homogeneous polynomials in n variables with real coefficients. Let $\mathring{L}^2_n \subseteq H^2_n$ be the open subset of quadratic forms with positive coefficients that have the *Lorentzian signature* $(+, -, \ldots, -)$. For d larger than 2, we define an open subset $\mathring{L}^d_n \subseteq H^d_n$ by setting

$$\mathring{L}^d_n = \{ f \in H^d_n \mid \partial_i f \in \mathring{L}^{d-1}_n \text{ for all } i \in [n] \},$$

where ∂_i is the partial derivative with respect to the ith variable. Thus f belongs to $\mathring{L}^d_n$ if and only if all quadratic polynomials of the form $\partial_{i_1}\partial_{i_2}\cdots\partial_{i_{d-2}} f$ belongs to $\mathring{L}^2_n$.

Definition 4 (Lorentzian polynomials). The polynomials in $\mathring{L}^d_n$ are called *strictly Lorentzian*, and the limits of strictly Lorentzian polynomials are called *Lorentzian*.

The prototypical examples of Lorentzian polynomials, which motivated Definition 4, are those obtained from the various examples of $A(X)$ in Section 1 in the following way. For any linear operators $L_1, \ldots, L_d$ on $A(X)$, we set

$$\deg\left(\prod_{i=1}^d L_i\right) := P\left(1, \prod_{i=1}^d L_i \cdot 1\right),$$

where 1 is the distinguished generator of $A^0(X)$ defining $P(1, -) : A^d(X) \simeq \mathbb{R}$.

---

3 The general conjecture is that the discrete function $(\nu, \kappa, \lambda) \mapsto \log c^\nu_{\kappa\lambda}$ is concave, where $c^\nu_{\kappa\lambda}$ are the *Littlewood–Richardson coefficients* **[61, CONJECTURE 1]**. The conjecture holds in the "classical limit" **[61, SECTION 3]**, but the general case is refuted in **[19]**.

Proposition 5. *Let $L_1, \dots, L_n$ be members of the closure $\overline{K}(X)$, and let f the polynomial*

$$f(w_1, \dots, w_n) = \frac{1}{d!} \deg(w_1 L_1 + \cdots + w_n L_n)^d.$$

If $A(X)$ satisfies the Hodge–Riemann relations in degrees ≤ 1, then f is Lorentzian.

Before deducing Proposition 5 from Theorem 12 below, we give two prominent cases.

Example 6 (Volume polynomials of convex bodies). For any collection of convex bodies $C = (C_1, \dots, C_n)$ in $\mathbb{R}^d$, consider the function

$$\mathrm{vol}_C : \mathbb{R}^n_{\geq 0} \to \mathbb{R}, \quad w \mapsto \frac{1}{d!} \mathrm{vol}(w_1 C_1 + \cdots + w_n C_n),$$

where $w_1 C_1 + \cdots + w_n C_n$ is the Minkowski sum and vol is the Euclidean volume. Minkowski showed that $\mathrm{vol}_C(w)$ is a polynomial **[66, CHAPTER 5]**. One may approximate the convex bodies with convex polytopes to prove that vol_C is Lorentzian. Using Proposition 5, where X is the Minkowski sum of the approximating convex polytopes and $A(X)$ is the combinatorial cohomology in **[45]**, we get the following statement:

The polynomial $\mathrm{vol}_C(w)$ *is Lorentzian for any convex bodies $C_1, \dots, C_n$ in $\mathbb{R}^d$.*

Alternatively, one can use Brunn–Minkowski theory to deduce the Lorentzian property of the volume polynomial **[17, SECTION 4.1]**.

Example 7 (Volume polynomials of projective varieties). Let X be a d-dimensional irreducible projective variety over an algebraically closed field. A Cartier divisor on X is said to be *nef* if it intersects every irreducible curve in X nonnegatively.[4] For any collection of nef divisors $H = (H_1, \dots, H_n)$ on X, consider the function

$$\mathrm{vol}_H : \mathbb{R}^n_{\geq 0} \to \mathbb{R}, \quad w \mapsto \frac{1}{d!} \deg(w_1 H_1 + \cdots + w_n H_n)^d,$$

where deg is the degree map on the Chow group of 0-dimensional cycles on X. When X admits a resolution of singularities Y, one can deduce the following statement from Proposition 5 and the Hodge–Riemann relations in degree ≤ 1 for the ring of algebraic cycles $A(Y)$:

The polynomial $\mathrm{vol}_H(w)$ *is Lorentzian for any nef divisors $H_1, \dots, H_n$ on X.*

In general, one can use Bertini's theorem to reduce the statement to the case of surfaces and apply Hodge's index theorem **[17, SECTION 4.2]**.

Next we formulate the main structural results on Lorentzian polynomials. A central definition is that of generalized permutohedra. Let E be a finite set, and let $\{e_i\}_{i \in E}$ be the standard basis of $\mathbb{R}^E$.

---

4 By Kleiman's theorem **[47, SECTION 1.4]**, any nef divisor on a projective variety is a limit of ample $\mathbb{R}$-divisors, which form the convex cone $K(X)$ in this setting.

Definition 8. A *generalized permutohedron* is a polytope in $\mathbb{R}^E$ all of whose edges are in the direction $e_i - e_j$ for some i and j in E.

For example, the *standard permutohedron* in $\mathbb{R}^n$, which is the convex hull of all permutations of $(1, 2, \ldots, n)$, and the *hyperoctahedron* in $\mathbb{R}^n$, which is the convex hull of all permutations of $(\pm 1, 0, \ldots, 0)$, are generalized permutohedra. The following pictures show the two polytopes in $\mathbb{R}^4$:

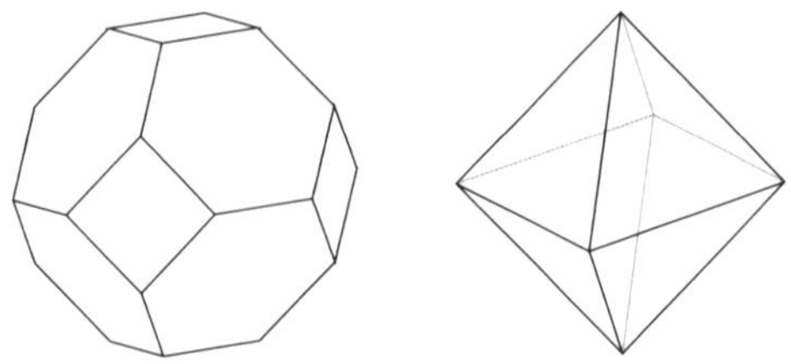

Generalized permutohedra are precisely the translates of the base polytopes of *polymatroids* [24], and they are obtained from the standard permutohedron by moving the vertices so that all the edge directions are preserved [63]. They lead to the central notion of M-*convexity* in the study of discrete convex analysis [59].

Definition 9. A subset $J \subseteq \mathbb{Z}^E_{\geq 0}$ is M-*convex* if it is the set of all lattice points of an integral generalized permutohedron. A *matroid* on E is an M-convex subset of $\mathbb{Z}^E_{\geq 0}$ consisting of zero-one vectors. The vectors in a matroid J are called *bases* of J.

A subset $J \subseteq \mathbb{Z}^E_{\geq 0}$ is M-convex exactly when it satisfies the *symmetric basis exchange property* [24, 39]: For any $\alpha, \beta \in J$ and an index i satisfying $\alpha_i > \beta_i$, there is an index j that satisfies

$$\alpha_j < \beta_j \quad \text{and} \quad \alpha - e_i + e_j \in J \quad \text{and} \quad \beta - e_j + e_i \in J.$$

In [59, CHAPTER 4], one can find several other equivalent characterizations of M-convexity. The above definition of matroids goes back to the study of moment map images of torus orbits in Grassmannians by Gelfand, Goresky, MacPherson, and Serganova in [33]. For a general introduction to matroids, and for any undefined matroid terms, we refer to [62]. Hereafter we identify the subsets of E with the zero-one vectors in $\mathbb{Z}^E_{\geq 0}$.

Example 10 (Graphic matroids). For any finite connected graph G with the edge set E, consider the set of indicator vectors

$$\mathscr{B}(G) := \{e_B \mid B \text{ is a spanning tree of } G\} \subseteq \mathbb{Z}^E_{\geq 0}.$$

The subset $\mathscr{B}(G)$ is M-convex for any G. Such matroids are said to be *graphic*.

Example 11 (Representable matroids). For any function $\varphi : E \to W$ from a finite set E to a vector space W over a field $\mathbb{F}$, consider the set of indicator vectors

$$\mathscr{B}(\varphi) := \{e_B \mid \varphi(B) \text{ is a bases of } W\} \subseteq \mathbb{Z}^E_{\geq 0}.$$

The subset $\mathscr{B}(\varphi)$ is M-convex for any $\varphi : E \to W$. Such matroids are said to be *representable over* $\mathbb{F}$, and the function φ is called a *representation over* $\mathbb{F}$. One typically requires without loss of generality that the image of φ spans W. A graphic matroid is representable over every field [62, SECTION 5.1]. In general, a matroid may or may not have a representation over $\mathbb{F}$:

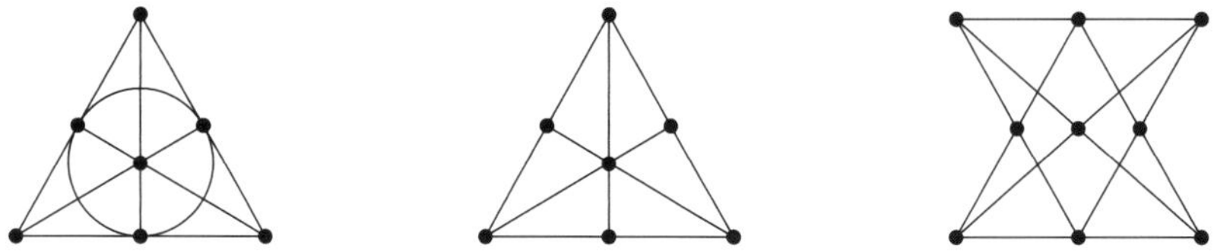

Among the three matroids pictured above, where the bases are given by all triples of points not on a line, the first is representable over $\mathbb{F}$ if and only if the characteristic of $\mathbb{F}$ is 2, the second is representable over $\mathbb{F}$ if and only if the characteristic of $\mathbb{F}$ is not 2, and the third is not representable over any field.

Let $L_n^2 \subseteq H_n^2$ be the closed subset of quadratic forms with nonnegative coefficients that have at most one positive eigenvalue. For d larger than 2, we define $L_n^d \subseteq H_n^d$ by setting

$$L_n^d = \{f \in \mathrm{M}_n^d \mid \partial_i f \in L_n^{d-1} \text{ for all } i\},$$

where $\mathrm{M}_n^d \subseteq H_n^d$ is the set of polynomials with nonnegative coefficients whose supports are M-convex. The following characterization in [17, THEOREM 2.25] is central to the theory of Lorentzian polynomials.

Theorem 12. *L_n^d is the set of Lorentzian polynomials in H_n^d.*

In other words, L_n^d is the closure of $\mathring{L}_n^d$ in H_n^d. *Theorem* 12 *makes it possible to decide whether a given polynomial is Lorentzian or not.* For example, the following polynomials are not Lorentzian because their supports are not M-convex:

$$w_1^3 + w_2^3, \quad w_1w_2^2 + w_1w_3^2 + w_2w_3^2 + w_1w_2w_3, \quad w_1^2w_3 + w_2^3.$$

One can also use Theorem 12 to show that a given polynomial is Lorentzian. For example, the elementary symmetric polynomial of degree d in n variables is Lorentzian because its support is M-convex and all its associated quadratic forms are

$$\begin{pmatrix} 0 & 1 & 1 & \dots & 1 \\ 1 & 0 & 1 & \dots & 1 \\ 1 & 1 & 0 & \dots & 1 \\ \vdots & \vdots & \vdots & \ddots & \vdots \\ 1 & 1 & 1 & \dots & 0 \end{pmatrix},$$

which have exactly one positive eigenvalue $n - d + 1$. One can also use Theorem 12 and the relevant Hodge–Riemann relations to show that the volume polynomials in Example 6 and Example 7 are Lorentzian. In particular, the supports of these volume polynomials must be M-convex for any collection of convex bodies and any collection of nef divisors.

Proof of Proposition 5. We may suppose that $L_1, \dots, L_n$ are members of $K(X)$. Under this assumption, all the coefficients of f are positive by the Hodge–Riemann relations in degree 0, so the support of f is M-convex. Choose any $d-2$ among the linear operators, say $L_1, \ldots, L_{d-2}$, and observe that

$$\partial_1 \ldots \partial_{d-2} f(w_1, \ldots, w_n) = \deg L_1 \cdots L_{d-2} \, (w_1 L_1 + \cdots + w_n L_n)^2.$$

Thus, by Theorem 12, it is enough to observe that the symmetric bilinear pairing

$$B^1(X) \times B^1(X) \to \mathbb{R}, \quad (x_1, x_2) \mapsto P(x_1, L_1 \cdots L_{d-2} \cdot x_2)$$

has the Lorentzian signature, where $B^1(X)$ is the span of $L_1 \cdot 1, \ldots, L_n \cdot 1$ in $A^1(X)$. This follows from the Hodge–Riemann relations in degrees ≤ 1: For any L in $K(X)$, the pairing is positive on $L \cdot 1$ by the Hodge–Riemann relations in degree 0, and it is negative definite on the orthogonal complement of $L \cdot 1$ by the Hodge–Riemann relations in degree 1. ■

Example 13. Not all Lorentzian polynomials are volume polynomials of convex bodies. In fact, the basis generating polynomial of a matroid on $[n]$ is the volume polynomial of n convex bodies precisely when the matroid is representable over every field [17, REMARK 4.3]. For example, the elementary symmetric polynomial

$$w_1 w_2 + w_1 w_3 + w_1 w_4 + w_2 w_3 + w_2 w_4 + w_3 w_4$$

is not the volume polynomial of four convex bodies in $\mathbb{R}^2$ because its support is not representable over the field $\mathbb{F}_2$.

Example 14. Not all Lorentzian polynomials are volume polynomials of nef divisors on a projective variety. For example, consider the cubic polynomial

$$f = 14w_1^3 + 6w_1^2 w_2 + 24w_1^2 w_3 + 12w_1 w_2 w_3 + 6w_1 w_3^2 + 3w_2 w_3^2.$$

One can use Theorem 12 to check that f is Lorentzian. To see that f is not the volume polynomial of nef divisors, one can use the *reverse Khovanskii–Teissier inequality* [49, THEOREM 5.7]: For any nef divisors L_1, L_2, L_3 on a d-dimensional projective variety and any $k \leq d$,

$$\binom{d}{k} (L_2^k \cdot L_1^{d-k})(L_1^k \cdot L_3^{d-k}) \geq (L_1^d)(L_2^k \cdot L_3^{d-k}).$$

The complex analytic proof of the inequality in [49] relies on the Calabi–Yau theorem [74]. The algebraic proof of the inequality in [44] using Okounkov bodies works over any algebraically closed field.

The theory of toric varieties shows that the volume polynomial of any set of convex bodies is the limit of a sequence of volume polynomials of nef divisors on projective varieties [30, SECTION 5.4]. Thus, the Lorentzian cubic f provides a counterexample to Gurvits' conjecture that a strongly log-concave homogeneous polynomial in three variables with nonnegative coefficients is the volume polynomial of three convex bodies [37, CONJECTURE 4.1].

The space of Lorentzian polynomials has numerous surprising properties. For example, writing $\mathbb{P}L$ for the image of $L \setminus 0 \subseteq H_n^d$ in the real projective space $\mathbb{P}H_n^d$, one can show that

$\mathbb{P}L_n^d$ is compact contractible subset with contractible interior $\mathbb{P}\mathring{L}_n^d$.

The contractibility follows from the following semigroup action [17, THEOREM 2.10]:

Any nonnegative linear change of coordinates preserves L_n^d. More generally, when $f(w) \in L_n^d$, then $f(Av) \in L_m^d$ for any $n \times m$ matrix A with nonnegative entries.

In fact, Brändén showed in [16] that $\mathbb{P}L_n^d$ is homeomorphic to a closed Euclidean ball, verifying a conjecture posed in [17, CONJECTURE 2.29]. The main feature of this *Lorentzian ball* is the following stratification labeled by M-convex sets [17, THEOREM 3.10 AND PROPOSITION 3.25]:

The set L_J of Lorentzian polynomials with support J is nonempty if and only if J is M*-convex. In this case, $\mathbb{P}L_J$ deformation retracts to the exponential generating function $\sum_{\alpha \in J} \frac{1}{\alpha!} w^\alpha$.*

This supports the opinion that *matroid theory provides the correct level of generality*. Leaving out any one matroid, say not representable over any field, will make the Lorentzian ball noncompact.[5]

The connection between discrete convex analysis and Lorentzian polynomials can be strengthened as follows. For a function $\nu : \mathbb{Z}_{\geq 0}^n \to \mathbb{R} \cup \{\infty\}$, we write $\mathrm{dom}(\nu) \subseteq \mathbb{Z}_{\geq 0}^n$ for the subset on which ν is finite, called the *effective domain* of ν. For a positive real parameter q, consider the exponential generating function

$$f_q^\nu(w) = \sum_{\alpha \in \mathrm{dom}(\nu)} \frac{q^{\nu(\alpha)}}{\alpha!} w^\alpha.$$

By [17, THEOREM 3.14], the polynomial f_q^ν is Lorentzian for all sufficiently small q if and only if the function ν is M-*convex* in the sense of discrete convex analysis [59]: For any index i and any $\alpha, \beta \in \mathrm{dom}(\nu)$ whose ith coordinates satisfy $\alpha_i > \beta_i$, there is an index j satisfying

$$\alpha_j < \beta_j \quad \text{and} \quad \nu(\alpha) + \nu(\beta) \geq \nu(\alpha - e_i + e_j) + \nu(\beta - e_j + e_i).$$

Considering the special case when ν takes values in $\{0, \infty\}$, we see that J is an M-convex set if and only if its exponential generating function $\sum_{\alpha \in J} \frac{1}{\alpha!} w^\alpha$ is a Lorentzian polynomial [17, THEOREM 3.10]. Another corollary is that a homogeneous polynomial with nonnegative coefficients is Lorentzian if the natural logarithms of its normalized coefficients form an M-concave function [17, COROLLARY 3.16]. Working over the field of real Puiseux series $\mathbb{K}$, we see that the tropicalization of any Lorentzian polynomial over $\mathbb{K}$ is an M-convex function, and that *all* M-convex functions are limits of tropicalizations of Lorentzian polynomials

5 Almost all matroids are not representable over any field. More precisely, the portion of matroids in $\mathbb{Z}_{\geq 0}^n$ that are representable over some field goes to zero as n goes to infinity [60]. For logical discussions of the "missing axiom" of matroid theory, see [53, 54, 73].

over $\mathbb{K}$ [17, COROLLARY 3.28]. This generalizes a result of Brändén [15], who showed that the tropicalization of any homogeneous stable polynomial over $\mathbb{K}$ is M-convex. In particular, for any matroid M with the set of bases $\mathscr{B}$, the *Dressian* of all valuated matroids on M can be identified with the tropicalization of the space of Lorentzian polynomials over $\mathbb{K}$ with support $\mathscr{B}$. For example, the tropicalization of the space of multiaffine Lorentzian quadrics in five variables is the tropical Grassmannian trop Gr(2, 5), a cone over the Petersen graph in $\mathbb{R}^{10}/\mathbb{R}\mathbf{1}$:

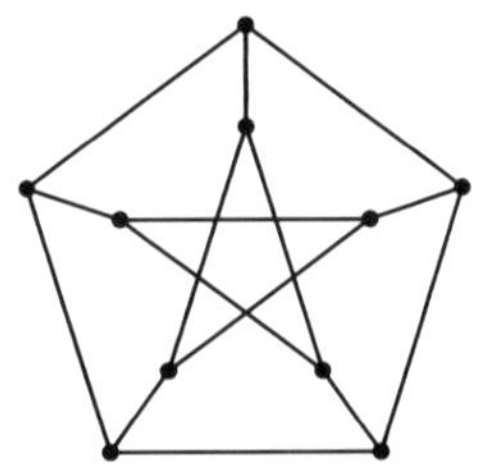

The figure shows a shadow of the Lorentzian ball $\mathbb{P}L_5^2$ over $\mathbb{K}$, highlighting its nonconvexity. We refer to [51, CHAPTER 4] for a friendly introduction to Dressians and tropical Grassmannians.

The theory of Lorentzian polynomials is not only useful for proving conjectures but also for generating them. Once one has identified a combinatorial polynomial f that is either provably or conjecturally Lorentzian, it is natural to look for an algebraic object $A(X)$ satisfying the Hodge–Riemann relations that explains the Lorentzian property of f. In good cases, one can further speculate that there is a projective variety X that produces f as a volume polynomial for some choices of nef divisors on X.

One such speculation concerns the basis generating polynomial for a morphism of matroids. Let M and N be matroids on finite sets E and F. The *rank function* of M is the function defined by

$$\mathrm{rk}_{\mathrm{M}} : 2^E \to \mathbb{Z}, \quad \mathrm{rk}_{\mathrm{M}}(S) = \max_{B \in \mathscr{B}} |B \cap S|,$$

where the maximum is taken over the set of bases of M. A *morphism* $g : \mathrm{M} \to \mathrm{N}$ is a function $E \to F$ that satisfies the rank inequalities

$$\mathrm{rk}_{\mathrm{N}}\big(g(S_2)\big) - \mathrm{rk}_{\mathrm{N}}\big(g(S_1)\big) \le \mathrm{rk}_{\mathrm{M}}(S_2) - \mathrm{rk}_{\mathrm{M}}(S_1) \quad \text{for any } S_1 \subseteq S_2 \subseteq E.$$

A function between the ground sets is a morphism if and only if the preimage of a flat is a flat (Definition 22). A subset $S \subseteq E$ is a *basis* of g if S is contained in a basis of M and $g(S)$ contains a basis of N. For a general discussion of morphisms of matroids, we refer to [46].

In [27, COROLLARY 4.6], the authors show that the *homogenous basis generating polynomial*

$$f_g(w_0, w_i)_{i \in E} := \sum_{S \in \mathscr{B}(g)} w_0^{|E|-|S|} \prod_{i \in S} w_i$$

is Lorentzian for any morphism of matroids $\varphi : \mathrm{M} \to \mathrm{N}$, where $\mathscr{B}(g)$ is the set of bases of g. When N is the rank-zero matroid on one element, one recovers the Lorentzian property of the homogenous independent set generating polynomial of M in [17, SECTION 4.3]. Setting the variables $(w_i)_{i \in E}$ equal to each other, we get a bivariate Lorentzian polynomial witnessing the validity of Mason's conjecture in Example 2. When g is the identity morphism, one recovers the Lorentzian property of the basis generating polynomial of a matroid [17, SECTION 3.2].

Example 15 (Continued from Example 10). A homomorphism from a graph G_1 to a graph G_2 is a function from the vertex set of G_1 to the vertex set of G_2 that maps adjacent vertices to adjacent vertices. The induced map from the edge set of G_1 to the edge set of G_2 is a morphism from the graphic matroid $\mathscr{B}(G_1)$ to the graphic matroid $\mathscr{B}(G_2)$. Such morphisms of matroids are said to be *graphic*.

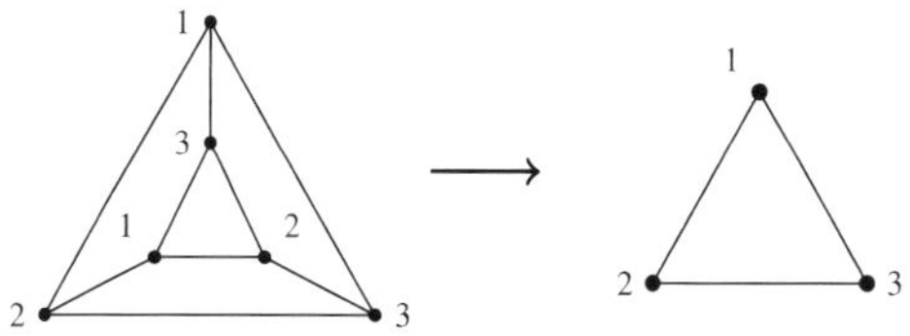

The graphic morphism of matroids depicted above has 27 bases of cardinality two, 79 bases of cardinality three, 111 bases of cardinality four, and 75 bases of cardinality five.

Example 16 (Continued from Example 11). Let M_i be matroids on E_i with representations $\varphi_i : E_i \to W_i$ over a field $\mathbb{F}$. A function g from E_1 to E_2 is a morphism from M_1 to M_2 if it fits into a commutative diagram

$$\begin{array}{ccc} E_1 & \xrightarrow{\varphi_1} & W_1 \\ \downarrow & & \downarrow \\ E_2 & \xrightarrow{\varphi_2} & W_2, \end{array}$$

where $W_1 \to W_2$ is a linear map between the vector spaces. Such morphisms of matroids are said to be *representable over* $\mathbb{F}$. A graphic morphism of matroids is representable over every field.

Continuing Example 7, we say that a degree d Lorentzian polynomial f in variables $w_1, \dots, w_n$ is a *volume polynomial over* $\mathbb{F}$ if there are nef divisors $H_1, \dots, H_n$ on a d-dimensional irreducible projective variety X over $\mathbb{F}$ that satisfy

$$f = \frac{1}{d!} \deg(w_1 H_1 + \cdots + w_n H_n)^d.$$

The following existence conjecture was made in [27, CONJECTURE 5.6]. It strengthens the Lorentzian property of the homogeneous basis generating polynomial of g when g is representable over $\mathbb{F}$.

Conjecture 17. *If g is a morphism of matroids that is representable over $\mathbb{F}$, then the homogenous basis generating polynomial of g is a volume polynomial over $\mathbb{F}$.*

Let M be a matroid on E that is representable over $\mathbb{F}$. In [5], the authors construct a collection of nef divisors $(L_i)_{i \in E}$ on an irreducible projective variety Y over $\mathbb{F}$ such that

$$\sum_{B \in \mathscr{B}} \prod_{i \in B} w_i = \frac{1}{d!} \deg\Big(\sum_{i \in E} w_i L_i\Big)^{\dim X},$$

where the first sum is over the set of bases $\mathscr{B}$ of M. This verifies Conjecture 17 when g is the identity morphism. A detailed study of this Y and its resolution of singularities in [41], in turn, was used to define the *matroid intersection cohomology* in [13]. It plays a central role in the resolution of two combinatorial conjectures on matroids, the top-heavy conjecture for the lattice of flats and the nonnegativity conjecture for the Kazhdan–Lusztig coefficients. We outline their proofs in Section 3.

Another speculation on Lorentzian polynomials is based on the Lorentzian property of the *normalized Schur polynomial*

$$\mathrm{N}\big(s_\lambda(w_1, \dots, w_n)\big) = \sum_{\alpha} K_{\lambda\alpha} \frac{w^\alpha}{\alpha!}.$$

Here, as in Example 3, λ is a partition and $K_{\lambda\alpha}$ are the Kostka coefficients.

Definition 18. The *normalization operator* is the linear operator N defined on the space of Laurent generating functions defined by

$$\mathrm{N}\Big(\sum_{\alpha \in \mathbb{Z}^n} c_\alpha w^\alpha\Big) = \sum_{\alpha \in \mathbb{Z}^n_{\geq 0}} c_\alpha \frac{w^\alpha}{\alpha!}.$$

For example, we have $\mathrm{N}(\frac{1}{z(1-z)}) = e^z$.

In [17, PROPOSITION 4.4], it was observed that the *Alexandrov–Fenchel inequality* for volume polynomials of convex bodies holds more generally for any Lorentzian polynomial in n variables:

If $\sum_\alpha c_\alpha \frac{w^\alpha}{\alpha!}$ is Lorentzian, then $c_\alpha^2 \geq c_{\alpha - e_i + e_j} c_{\alpha + e_i - e_j}$ for any α and any $i, j \in [n]$.

Since the Kostka coefficients are the weight multiplicities of the finite-dimensional irreducible representation $V(\lambda)$ of $\mathrm{GL}_n(\mathbb{C})$, the Lorentzian property of $\mathrm{N}(s_\lambda)$ thus implies

$$\big(\dim V(\lambda)_\alpha\big)^2 \geq \dim V(\lambda)_{\alpha - e_i + e_j} \dim V(\lambda)_{\alpha - e_j + e_i} \quad \text{for any } i, j \in [n].$$

Could this be a special case of a more general discrete log-concavity for weight multiplicities?

Let Λ be the integral weight lattice of the Lie algebra $\mathfrak{sl}_n(\mathbb{C})$. For $\lambda \in \Lambda$, write $\mathrm{V}(\lambda)$ for the irreducible $\mathfrak{sl}_n(\mathbb{C})$-module with the highest weight λ and consider its decomposition into finite-dimensional weight spaces

$$\mathrm{V}(\lambda) = \bigoplus_{\alpha} \mathrm{V}(\lambda)_\alpha.$$

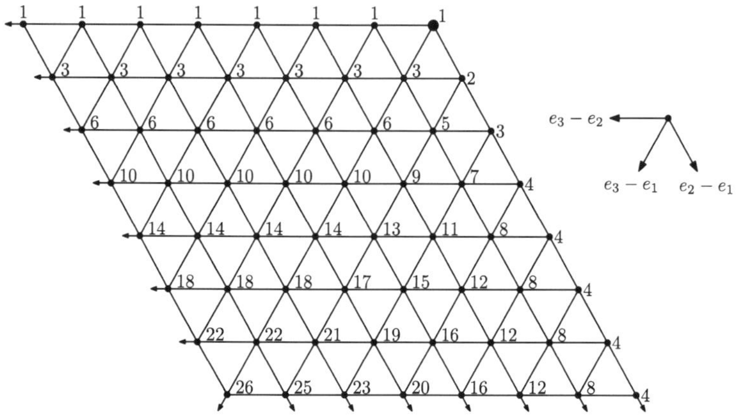

FIGURE 1

The figure shows some of the weight multiplicities of the irreducible $\mathfrak{sl}_4(\mathbb{C})$-module with the highest weight $-2\varpi_1 - 3\varpi_2$. We start from the highlighted vertex $\varpi_1 - 6\varpi_2 - 3\varpi_3$ and walk along negative root directions in the hyperplane spanned by $e_2 - e_1$ and $e_3 - e_2$. In the shown region, the sequence of weight multiplicities along any line is log-concave, as predicted by Conjecture 19.

We point to **[42]** for background on the representation theory of semisimple Lie algebras. The following conjecture was proposed in **[40, SECTION 3.1]**.

Conjecture 19. *For any $\lambda \in \Lambda$ and any $\alpha \in \Lambda$, we have*

$$\big(\dim \mathrm{V}(\lambda)_\alpha\big)^2 \geq \dim \mathrm{V}(\lambda)_{\alpha-e_i+e_j} \dim \mathrm{V}(\lambda)_{\alpha-e_j+e_i} \quad \textit{for any } i,j \in [n].$$

When λ is dominant, the dimension of the weight space $\mathrm{V}(\lambda)_\alpha$ is the Kostka number $K_{\lambda\alpha}$, and the Lorentzian property of the normalized Schur polynomial $\mathrm{N}(s_\lambda)$ implies that Conjecture 19 holds in this case. When λ is antidominant, $\mathrm{V}(\lambda)$ is the *Verma module* $\mathrm{M}(\lambda)$, the universal highest weight module of highest weight λ. Using the connection between the Kostant partition function and the volumes of flow polytopes in **[8]**, one can produce Lorentzian polynomials that witness the validity of the conjecture in this case **[40, PROPOSITION 11]**. Figure 1 illustrates some cases of Conjecture 19 when λ is neither dominant nor antidominant.

Conjecture 19 suggests the following existence statements of increasing strength.

There is a Lorentzian polynomial f that implies the discrete log-concavity in Conjecture 19 for given λ and α.

There is a cohomology A satisfying the Hodge–Riemann relations that implies the Lorentzian property of f for given λ and α.

There is a projective variety X that implies the Hodge–Riemann relations of A for given λ and α.

We give a precise formulation of the first prediction. For $\lambda \in \Lambda$, consider the Laurent generating function

$$\mathrm{ch}_\lambda(w_1, \ldots, w_n) := \sum_{\alpha \in \Lambda} \dim \mathrm{V}(\lambda)_\alpha w^{\alpha-\lambda}.$$

Note that every monomial appearing in ch_λ is a product of degree zero monomials of the form $w_i w_j^{-1}$.

Conjecture 20. $\mathrm{N}(w^\delta \mathrm{ch}_\lambda(w_1, \ldots, w_n))$ *is Lorentzian for any* $\lambda \in \Lambda$ *and* $\delta \in \mathbb{Z}^n_{\geq 0}$.

Conjecture 20 holds for any δ when λ is either dominant or antidominant. In general, the homogeneous polynomial $\mathrm{N}(w^\delta \mathrm{ch}_\lambda)$ can be computed using the Kazhdan–Lusztig theory [42, CHAPTER 8]. The authors of [40] tested Conjecture 20 for $\lambda = -\sigma\rho - \rho$ and $\delta = (1, \ldots, 1)$, where ρ is the sum of all the fundamental weights, for all permutations σ in S_n for $n \leq 6$. Conjecture 19 for λ and α follows from Conjecture 20 for λ and any sufficiently large δ.

Similar conjectures can be made for various other polynomials appearing in representation theory and symmetric function theory. For relevant definitions, we refer to [40, SECTION 3] and references therein.

Conjecture 21. *The following polynomials are Lorentzian* [40, CONJECTURES 15,19,20,22,23]*:*

(1) *The normalized Schubert polynomial* $\mathrm{N}(\mathfrak{S}_\sigma)$ *for any permutation* σ.

(2) *The normalized skew Schur polynomial* $\mathrm{N}(s_{\lambda/\nu})$ *for any skew partition* λ/ν.

(3) *The normalized Schur P-polynomial* $\mathrm{N}(P_\lambda)$ *for any strict partition* λ.

(4) *The normalized key polynomial* $\mathrm{N}(\kappa_\mu)$ *for any composition* μ.

(5) *The normalized homogeneous Grothendieck polynomial* $\mathrm{N}(\tilde{\mathfrak{G}}_\sigma)$ *for any permutation* σ.

The M-convexity of the support is known for the Schubert polynomial [29, COROLLARY 8], the skew Schur polynomial [56, PROPOSITION 2.9], the Schur P-polynomial [56, PROPOSITION 3.5], and the key polynomial [29, COROLLARY 8]. The potential validity of each of these conjectures suggests the existence of certain Hodge–Riemann relations, or perhaps more strongly, projective varieties.

3. INTERSECTION COHOMOLOGY OF MATROIDS

The set of bases of a matroid M on a finite set E is a subset $\mathscr{B} \subseteq 2^E$ that satisfies the *symmetric basis exchange property*: For any $B_1, B_2 \in \mathscr{B}$ and any $i \in B_1 \setminus B_2$, there is $j \in B_2 \setminus B_1$ such that

$$(B_1 \setminus i) \cup j \in \mathscr{B} \quad \text{and} \quad (B_2 \setminus j) \cup i \in \mathscr{B}.$$

Any two bases of M have the same cardinality $d = \mathrm{rk}\,\mathrm{M}$, called the *rank* of M. When M has a representation $\varphi : E \to W$ over a field $\mathbb{F}$, the authors of [5] construct a collection of nef divisors $(L_i)_{i \in E}$ on a d-dimensional irreducible projective variety Y over $\mathbb{F}$ whose volume polynomial is the basis generating polynomial of M:

$$\frac{1}{d!}\deg\Big(\sum_{i\in E} w_i L_i\Big)^d = \sum_{B\in\mathscr{B}}\prod_{i\in B} w_i.$$

The projective variety Y, called the *matroid Schubert variety* of φ, is the closure of the image of the dual map $\varphi^\vee : W^\vee \to \mathbb{F}^E$ in the product of projective lines $(\mathbb{P}^1)^E$. In view of Proposition 5, one can say that Y is a geometric source of the Lorentzian property of the basis generating polynomial. A detailed study of this Y and its resolution of singularities in [41] was used to define the *intersection cohomology* IH(M) of M in [13]. When M is not representable over any field, there is no known projective variety that explains the Lorentzian property of the basis generating polynomial of M. However, for any M, one can construct IH(M) as a graded $\mathbb{Q}$-vector space equipped with a symmetric pairing $P : \mathrm{IH}^*(\mathrm{M}) \times \mathrm{IH}^{d-*}(\mathrm{M}) \to \mathbb{Q}$ and graded linear operators $L_i : \mathrm{IH}^*(\mathrm{M}) \to \mathrm{IH}^{*+1}(\mathrm{M})$ for each i in E. The main result of [13] is that IH(M) satisfies the Poincaré duality, the hard Lefschetz theorem, and the Hodge–Riemann relations with respect to any positive linear combination of $(L_i)_{i\in E}$. When M is representable over the complex numbers, the intersection cohomology of M is the intersection cohomology of Y with $\mathbb{Q}$-coefficients. When M is representable over a finite field, the intersection cohomology of M is a rational form of the ℓ-adic étale intersection cohomology of Y for which the Hodge–Riemann relations hold.[6] The existence of IH(M) plays a central role in the resolution of two combinatorial conjectures on M, the top-heavy conjecture for the lattice of flats and the nonnegativity conjecture for the Kazhdan–Lusztig coefficients. Below we outline the construction of IH(M) and explain its relation to the two conjectures.

The *top-heavy conjecture* was proposed by Dowling and Wilson in [22,23]. It originates from the following theorem of de Bruijn and Erdős [20]:

> *Every finite set of points E in a projective plane determines at least $|E|$ lines, unless E is contained in a line.*

In other words, if E is not contained in a line, then the number of lines in the plane containing at least two points in E is at least $|E|$. The result is valid for any projective plane, not necessarily Desarguesian, and in this sense the statement is purely combinatorial. The figures below depict the two possibilities when $|E| = 4$.

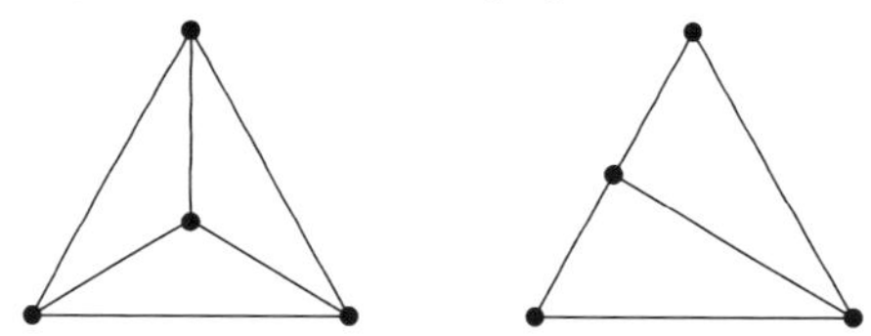

(4 points determining 6 lines) (4 points determining 4 lines)

6 Since $\mathbb{Q}_\ell$ is not ordered, there are no Hodge–Riemann relations for the ℓ-adic intersection cohomology. When M is representable over some field, we suspect that IH(M) is a Chow analogue of the intersection cohomology of X.

The following more general statement, conjectured by Motzkin in [57], was subsequently proved by many in various settings:

> *Every finite set of points E in a projective space determines at least $|E|$ hyperplanes, unless E is contained in a hyperplane.*

Motzkin proved the above for E in real projective spaces in [58]. Basterfield and Kelly [9] showed the statement in general, and Greene [35] strengthened the result by showing that there is an *order-matching* from E to the set of hyperplanes determined by E, unless E is contained in a hyperplane:

> *For every point in E one can choose a hyperplane containing the point in such a way that no hyperplane is chosen twice.*

Mason [52] and Heron [38] obtained similar results by different methods.

Based on these and other known results, Dowling and Wilson formulated the top-heavy conjecture in the generality of matroids, in terms of their flats.

Definition 22. A *flat* of a matroid M on a finite set E is a subset of E that is maximal for its rank.

In other words, a subset of E is a flat of M if the addition of any other element to the set increases its rank in M. Since the intersection of flats of M is a flat of M, the collection of all flats of M form a lattice $\mathscr{L} = \mathscr{L}(\mathrm{M})$, the *lattice of flats* of M. The lattice $\mathscr{L}$ is graded, and the rank of a subset S of E in M is the height of the smallest flat of M containing S in the graded lattice $\mathscr{L}$. Thus, one can recover the rank function of M, and hence the set of bases $\mathscr{B}$ of M, from the lattice of flats $\mathscr{L}$ of M.

We write $\mathscr{L}^k$ for the set of rank k flats of M. When M has a representation $\varphi : E \to W$ over a field $\mathbb{F}$, we have

$$\mathscr{L}^k = \{\varphi^{-1}(V) \mid V \text{ is a } k\text{-dimensional subspace of } W\}.$$

When φ injects E into the projective space $\mathbb{P}V$, there are bijections

$$\mathscr{L}^1 \simeq \text{the set of points in } E \quad \text{and} \quad \mathscr{L}^2 \simeq \text{the set of lines joining points in } E.$$

The top-heavy conjecture extends the relation between $|\mathscr{L}^1|$ and $|\mathscr{L}^2|$ in de Bruijn–Erdős theorem as follows.

Conjecture 23 (Top-heavy conjecture). *Let $\mathscr{L}$ be the lattice of flats of a rank d matroid.*

(1) *For every nonnegative integer k less than $\frac{d}{2}$,*

$$|\mathscr{L}^k| \le |\mathscr{L}^{d-k}|.$$

In fact, there is an injective map $\iota : \mathscr{L}^k \to \mathscr{L}^{d-k}$ satisfying $x \le \iota(x)$ for all x.

(2) *For every nonnegative integer k less than $\frac{d}{2}$,*

$$\left|\mathscr{L}^k\right| \le \left|\mathscr{L}^{k+1}\right|.$$

In fact, there is an injective map $\iota : \mathscr{L}^k \to \mathscr{L}^{k+1}$ satisfying $x \le \iota(x)$ for all x.

When $\mathscr{L}$ is a finite Boolean lattice or a finite projective geometry, Conjecture 23 is a classical result; see, for example, [72, COROLLARY 4.8 AND EXERCISE 4.4]. In these self-dual cases, the second statement of Conjecture 23 says that $\mathscr{L}$ admits order-matchings

$$\mathscr{L}^0 \hookrightarrow \mathscr{L}^1 \hookrightarrow \cdots \hookrightarrow \mathscr{L}^{\lfloor \frac{d}{2} \rfloor} \leftrightarrow \mathscr{L}^{\lceil \frac{d}{2} \rceil} \hookleftarrow \cdots \hookleftarrow \mathscr{L}^{d-1} \hookleftarrow \mathscr{L}^d.$$

These order-matchings partition $\mathscr{L}$ into $|\mathscr{L}^{\lfloor \frac{d}{2} \rfloor}|$ disjoint chains, and hence $\mathscr{L}$ has the *Sperner property*:

> *The maximal number of pairwise incomparable subsets of $[n]$ is the maximum among the binomial coefficients $\binom{n}{k}$. Similarly, the maximal number of pairwise incomparable subspaces of $\mathbb{F}_q^n$ is the maximum among the q-binomial coefficients $\binom{n}{k}_q$.*

Let M be a rank d matroid on a finite set E. The proof of Conjecture 23 in [13] is based on a detailed analysis of the *graded Möbius algebra*

$$\mathrm{H}(\mathrm{M}) := \bigoplus_{F \in \mathscr{L}(\mathrm{M})} \mathbb{Q} y_F.$$

The grading is defined by declaring the degree of the element y_F to be $\operatorname{rk} F$, the rank of F in M. The multiplication is defined by the formula

$$y_F y_G := \begin{cases} y_{F \vee G} & \text{if } \operatorname{rk} F + \operatorname{rk} G = \operatorname{rk}(F \vee G), \\ 0 & \text{if } \operatorname{rk} F + \operatorname{rk} G > \operatorname{rk}(F \vee G), \end{cases}$$

where $\vee$ stands for the join in the lattice of flats of M. Unlike its ungraded counterpart, which is isomorphic to the product of $\mathbb{Q}$'s as a $\mathbb{Q}$-algebra [69, THEOREM 1], the graded Möbius algebra has a nontrivial algebra structure.

There is a straightforward relation between the basis generating polynomial of M and the graded Möbius algebra of M. For each i in E, we associate a degree 1 element

$$L_i := \begin{cases} y_{\bar{i}} & \text{if the smallest flat } \bar{i} \text{ containing } i \text{ has rank } 1, \\ 0 & \text{if the smallest flat } \bar{i} \text{ containing } i \text{ has rank } 0. \end{cases}$$

Writing deg for the isomorphism $\mathrm{H}^d(\mathrm{M}) \simeq \mathbb{Q}$ with $\deg(y_E) = 1$, we have

$$\frac{1}{d!} \deg\Big(\sum_{i \in E} w_i L_i\Big)^d = \sum_{B \in \mathscr{B}} \prod_{i \in B} w_i.$$

For the top-heavy conjecture, of central importance is the element $L := \sum_{i \in E} L_i$. The following elementary statement on H(M), proposed in [41, CONJECTURE 7], is one of the main conclusions of [13]. Its analogue for Weyl groups and for general Coxeter groups can be found in [11] and [55].

Theorem 24. *For every nonnegative integer $k \le \frac{d}{2}$, the multiplication map*

$$\mathrm{H}^k(\mathrm{M}) \to \mathrm{H}^{d-k}(\mathrm{M}), \quad x \mapsto L^{d-2k}x$$

is injective (the injective hard Lefschetz property *for* M*).*

To deduce Conjecture 23 from Theorem 24, consider the matrix of the multiplication map with respect to the standard bases of the source and the target. Entries of this matrix are labeled by pairs of elements of $\mathscr{L}$, and all the entries corresponding to incomparable pairs are zero. The matrix has full rank by Theorem 24, so there is a maximal square submatrix with a nonzero determinant. In the standard expansion of this determinant, there must be a nonzero term, and the permutation corresponding to this term produces the injective map ι in Conjecture 23.

It seems difficult to prove Theorem 24 directly. One possible reason for this is the lack of Poincaré duality for H(M): Typically, for small k, a matroid has much more corank k flats than rank k flats. In known settings where the hard Lefschetz property is the main statement needed for applications [13, 26, 45], it was necessary to prove Poincaré duality, the hard Lefschetz property, and the Hodge–Riemann relations together as a single package.

The intersection cohomology IH(M) is an H(M)-module that repairs the failure of Poincaré duality of H(M) in an efficient way. The construction of IH(M) is inspired by the Kazhdan–Lusztig theory of matroids developed in [25]. For any flat F of M, we define the *localization* of M at F to be the matroid M^F on the ground set F whose flats are the flats of M contained in F. Similarly, we define the *contraction* of M at F to be the matroid M_F on the ground set $E \setminus F$ whose flats are $G \setminus F$ for flats G of M containing F.[7] According to [14, THEOREM 2.2], there is a unique way to assign a polynomial $P_{\mathrm{M}}(t)$ to each matroid M, called the *Kazhdan–Lusztig polynomial* of M, subject to the following three conditions:

(1) If $\operatorname{rk} \mathrm{M} = 0$, then $P_{\mathrm{M}}(t)$ is the constant polynomial 1.

(2) If $\operatorname{rk} \mathrm{M} > 0$, then the degree of $P_{\mathrm{M}}(t)$ is strictly less than $\operatorname{rk} \mathrm{M}/2$.

(3) We have $Z_{\mathrm{M}}(t) = t^{\operatorname{rk} \mathrm{M}} Z_{\mathrm{M}}(t^{-1})$, where $Z_{\mathrm{M}}(t) := \sum_{F \in \mathscr{L}(\mathrm{M})} t^{\operatorname{rk}_{\mathrm{M}} F} P_{\mathrm{M}_F}(t)$.

The polynomial $Z_{\mathrm{M}}(t)$, called the *Z-polynomial* of M, was introduced in [65] using a different but equivalent definition of $P_{\mathrm{M}}(t)$.

Example 25. It is straightforward to check that the Kazhdan–Lusztig polynomial is 1 for matroids of rank at most two. Thus, when the rank of M is three, we should have

$$P_{\mathrm{M}}(t) + |\mathscr{L}^1|t + |\mathscr{L}^2|t^2 + t^3 = t^3 P_{\mathrm{M}}(t^{-1}) + |\mathscr{L}^1|t^2 + |\mathscr{L}^2|t + 1.$$

Since the degree of $P_{\mathrm{M}}(t)$ is at most 1, it follows that $P_{\mathrm{M}}(t) = 1 + |\mathscr{L}^2|t - |\mathscr{L}^1|t$.

---

7 In [25], as well as several other references on Kazhdan–Lusztig polynomials of matroids, the localization is denoted M_F and the contraction is denoted M^F. Our notational choice here is consistent with [1] and [12, 13].

Example 26. When the rank of M is four, computing as in Example 25, we get $P_{\mathrm{M}}(t) = 1 + |\mathscr{L}^3|t - |\mathscr{L}^1|t$. When the rank of M is five [25, PROPOSITION 2.16], we have

$$P_{\mathrm{M}}(t) = 1 + |\mathscr{L}^4|t - |\mathscr{L}^1|t + |\mathscr{L}^3|t^2 - |\mathscr{L}^2|t^2 + |\mathscr{L}^{1,2}|t^2 - |\mathscr{L}^{1,4}|t^2 + |\mathscr{L}^{2,4}|t^2 - |\mathscr{L}^{2,3}|t^2,$$

where $|\mathscr{L}^{i,j}|$ is the number of incidences between the flats of rank i and rank j. For example, if M is the uniform matroid of rank 5 on 6 elements, $P_{\mathrm{M}}(t) = 1 + 9t + 5t^2$.

The following *nonnegativity conjecture* was proposed in [25, CONJECTURE 2.8], where it was proved for matroids representable over some field using ℓ-adic étale intersection cohomology theory of [10]. For sparse paving matroids, a combinatorial proof of the nonnegativity was given in [48]. The general case of the conjecture is proved in [13, THEOREM 1.3] using the intersection cohomology of matroids.

Conjecture 27 (Nonnegativity conjecture). *$P_{\mathrm{M}}(t)$ has nonnegative coefficients for any* M.

Kazhdan–Lusztig polynomials of matroids are special cases of Kazhdan–Lusztig–Stanley polynomials [64, 71]. Several important families of Kazhdan–Lusztig–Stanley polynomials turn out to have nonnegative coefficients, including classical Kazhdan–Lusztig polynomials associated with Bruhat intervals [26] and g-polynomials of convex polytopes [45]. Each of the known proofs of the nonnegativity of the three Kazhdan–Lusztig–Stanley polynomials involves numerous details that are unique to that specific case.

The following existence result of [13] implies Conjecture 23 and Conjecture 27. Let $K(\mathrm{M})$ be the open convex cone of degree 1 elements

$$K(\mathrm{M}) = \left\{ \sum_{F \in \mathscr{L}^1} c_F y_F \mid c_F \text{ is positive} \right\} \subseteq \mathrm{H}^1(\mathrm{M}).$$

The elements of $K(\mathrm{M})$ act as linear operators by multiplication on any H(M)-module.

Theorem 28. *There is a graded* H(M)*-module* IH(M) *and a symmetric bilinear pairing*

$$P : \mathrm{IH}^*(\mathrm{M}) \times \mathrm{IH}^{d-*}(\mathrm{M}) \to \mathbb{Q}$$

that satisfies the following properties for any nonnegative integer $k \leq \frac{d}{2}$:

(1) *The symmetric bilinear pairing*

$$\mathrm{IH}^k(\mathrm{M}) \times \mathrm{IH}^{d-k}(\mathrm{M}) \to \mathbb{Q}, \quad (x_1, x_2) \mapsto P(x_1, x_2)$$

*is nondegenerate (*Poincaré duality theorem *for* M*).*

(2) *For any $L_1, \ldots, L_{d-2k} \in K(\mathrm{M})$, the multiplication map*

$$\mathrm{IH}^k(\mathrm{M}) \to \mathrm{IH}^{d-k}(\mathrm{M}), \quad x \mapsto \left(\prod_{i=1}^{d-2k} L_i \right) x$$

*is an isomorphism (*hard Lefschetz theorem *for* M*).*

(3) *For any* $L_0, L_1, \dots, L_{d-2k} \in K(X)$, *the symmetric bilinear form*

$$\mathrm{IH}^k(\mathrm{M}) \times \mathrm{IH}^k(\mathrm{M}) \to \mathbb{Q}, \quad (x_1, x_2) \mapsto (-1)^k \mathrm{P}\left(x_1, \left(\prod_{i=1}^{d-2k} L_i\right) x_2\right)$$

is positive definite on the kernel of the linear map

$$\mathrm{IH}^k(\mathrm{M}) \to \mathrm{IH}^{d-k+1}(\mathrm{M}), \quad x \mapsto \left(\prod_{i=0}^{d-2k} L_i\right) x$$

*(*Hodge–Riemann relations *for* M*).*

(4) *Writing* $\mathrm{IH}_\varnothing$ *for the graded vector space* $\mathrm{IH}(\mathrm{M}) \otimes_{\mathrm{H}(\mathrm{M})} \mathbb{Q}$, *we have*

$$P_{\mathrm{M}}(t) = \sum_{k \geq 0} \dim(\mathrm{IH}^k_\varnothing) t^k \quad \text{and} \quad Z_{\mathrm{M}}(t) = \sum_{k \geq 0} \dim(\mathrm{IH}^k(\mathrm{M})) t^k$$

*(*Kazhdan–Lusztig identities *for* M*).*

(5) $\mathrm{IH}^0(\mathrm{M})$ *generates a submodule isomorphic to* H(M) *(*Purity *for* M*).*

Since injective maps restrict to injective maps, the injective hard Lefschetz property for M in Theorem 24, and hence the top-heavy conjecture for M, follows from the hard Lefschetz theorem and the purity for M. The nonnegativity conjecture for M follows from the Kazhdan–Lusztig identities for M. More generally, when a finite group Γ acts on M, one can define the *equivariant Kazhdan–Lusztig polynomial* $P^\Gamma_{\mathrm{M}}(t)$ as in **[32]**. This is a polynomial with coefficients in the ring of virtual representations of Γ, with the property that taking dimensions recovers the ordinary polynomial $P_{\mathrm{M}}(t)$. The authors of **[13]** show that Γ acts naturally on IH(M) and that

$$P^\Gamma_{\mathrm{M}}(t) = \sum_{k \geq 0} [\Gamma \curvearrowright \mathrm{IH}^k_\varnothing] t^k \in \mathrm{VRep}(\Gamma)[t].$$

This proves the equivariant nonnegativity conjecture proposed in **[32, CONJECTURE 2.13]**. Conjecture 27 is the special case when Γ is trivial.

The construction of IH(M) is inspired by geometry in the representable case. Consider the case when M has a representation $\varphi : E \to W$ over $\mathbb{C}$, and recall that the matroid Schubert variety Y of φ is the closure of $W^\vee$ in the product of projective lines $(\mathbb{P}^1)^E$. The additive group $W^\vee$ acts on Y with finitely many orbits, each of which is isomorphic to an affine space. The poset of cells in this stratification of Y is isomorphic to the poset of cells is isomorphic to the lattice of flats of M, and, in fact, the singular cohomology $\mathrm{H}^{2*}(Y, \mathbb{Q})$ is isomorphic to the graded Möbius algebra $\mathrm{H}^*(\mathrm{M})$ **[41, THEOREM 14]**.[8]

The Schubert variety admits a distinguished resolution of singularities $f : X \to Y$ obtained by blowing up all the strata in the order of increasing dimension. The resulting smooth projective variety X is the *augmented wonderful variety* of φ studied in **[12]**. Adopting

---

8 All the cohomology rings and intersection cohomology groups of varieties in this paper vanish in odd degrees, and our isomorphisms double degrees.

the computations in [21,28], one can show that its singular cohomology and Chow rings are isomorphic to the *augmented Chow ring*

$$\mathrm{CH}(\mathrm{M}) := \mathbb{Q}[y_i, x_F \mid i \text{ is an element of } E \text{ and } F \text{ is a proper flat of M}]/(I_\mathrm{M} + J_\mathrm{M}),$$

where I_M is the ideal generated by the linear forms

$$y_i - \sum_{i \notin F} x_F, \quad \text{for every element } i \text{ of } E,$$

and J_M is the ideal generated by the quadratic monomials

$$x_{F_1}x_{F_2}, \quad \text{for every pair of incomparable proper flats } F_1 \text{ and } F_2 \text{ of M, and}$$
$$y_i x_F, \quad \text{for every element } i \text{ of } E \text{ and every proper flat } F \text{ of M not containing } i.$$

As expected from the identification with $\mathrm{H}^{2*}(X, \mathbb{Q})$ in the representable case, for any M, the augmented Chow ring of M vanishes in degrees larger than d. Furthermore, there is a unique linear map

$$\deg\colon \mathrm{CH}^d(\mathrm{M}) \to \mathbb{Q}, \quad \prod_{F \in \mathscr{F}} x_F \mapsto 1,$$

where $\mathscr{F}$ is any complete flag of proper flats of M, defining a symmetric pairing on CH(M).

The main observation is that the pullback homomorphism in singular cohomology

$$f^* : \mathrm{H}^*(Y, \mathbb{Q}) \to \mathrm{H}^*(X, \mathbb{Q})$$

only depends on M and not on φ. In terms of the graded Möbius algebra and the augmented Chow ring of M, the pullback homomorphism is given by

$$f^* : \mathrm{H}(\mathrm{M}) \to \mathrm{CH}(\mathrm{M}), \quad L_i \mapsto y_i.$$

Applying the decomposition theorem of Beilinson–Bernstein–Deligne–Gabber [10] to f, we find that the intersection cohomology $\mathrm{IH}^*(Y)$ is isomorphic as a graded $\mathrm{H}^*(Y)$-module to a direct summand of $\mathrm{H}^*(X)$. Furthermore, a slight extension of an argument of Ginzburg [34] shows that $\mathrm{IH}^*(Y)$ is indecomposable as an $\mathrm{H}^*(Y)$-module. This motivates the following definition.

Definition 29. The intersection cohomology IH(M) of a matroid M is the unique indecomposable graded H(M)-module direct summand of CH(M) that is nonzero in degree zero.

The above defines the intersection cohomology of M up to isomorphism of graded H(M)-modules, where the uniqueness is given by the general Krull–Schmidt theorem [7, THEOREM 1]. The intersection cohomology inherits a symmetric pairing P from CH(M). In [13], the authors construct a canonical submodule IH(M) $\subseteq$ CH(M) that is preserved by all the symmetries of M. The construction of IH(M) as an explicit submodule of CH(M), or more generally the construction of the *canonical decomposition* of CH(M) as a graded H(M)-module, is essential in inductively proving Poincaré duality, the hard Lefschetz theorem, and the Hodge–Riemann relations for IH(M).

ACKNOWLEDGMENTS

I thank my past and current collaborators: Karim Adiprasito, Federico Ardila, Farhad Babaee, Tom Braden, Petter Brändén, Graham Denham, Chris Eur, Eric Katz, Matt Larson, Jacob Matherne, Karola Mészáros, Nick Proudfoot, Benjamin Schröter, Avery St. Dizier, Bernd Sturmfels, and Botong Wang. It was a privilege to have connected with your minds, and I am grateful for our mathematical adventures together.

FUNDING

This work was partially supported by Simons Investigator Grant and NSF Grant DMS-2053308.

REFERENCES

[1] K. Adiprasito, J. Huh, and E. Katz, Hodge theory for combinatorial geometries. *Ann. of Math. (2)* **188** (2018), no. 2, 381–452.

[2] N. Anari, S. O. Gharan, and C. Vinzant, Log-concave polynomials, I: entropy and a deterministic approximation algorithm for counting bases of matroids. *Duke Math. J.* **170** (2021), no. 16, 3459–3504.

[3] N. Anari, K. Liu, S. O. Gharan, and C. Vinzant, Log-concave polynomials II: high-dimensional walks and an FPRAS for counting bases of a matroid. In *STOC'19—Proceedings of the 51st Annual ACM SIGACT Symposium on Theory of Computing*, pp. 1–12, ACM, New York, 2019.

[4] N. Anari, K. Liu, S. Oveis Gharan, and C. Vinzant, Log-concave polynomials III: Mason's ultra-log-concavity conjecture for independent sets of matroids. 2018, arXiv:1811.01600.

[5] F. Ardila and A. Boocher, The closure of a linear space in a product of lines. *J. Algebraic Combin.* **43** (2016), no. 1, 199–235.

[6] F. Ardila, G. Denham, and J. Huh, Lagrangian geometry of matroids. *J. Amer. Math. Soc.*, published electronically. DOI 10.1090/jams/1009.

[7] M. Atiyah, On the Krull–Schmidt theorem with application to sheaves. *Bull. Soc. Math. France* **84** (1956), 307–317.

[8] W. Baldoni and M. Vergne, Kostant partitions functions and flow polytopes. *Transform. Groups* **13** (2008), no. 3–4, 447–469.

[9] J. G. Basterfield and L. M. Kelly, A characterization of sets of n points which determine n hyperplanes. *Proc. Camb. Philos. Soc.* **64** (1968), 585–588.

[10] A. A. Beĭlinson, J. Bernstein, and P. Deligne, Faisceaux pervers. In *Analysis and topology on singular spaces, I (Luminy, 1981)*, pp. 5–171, Astérisque 100, Soc. Math. France, Paris, 1982.

[11] A. Björner and T. Ekedahl, On the shape of Bruhat intervals. *Ann. of Math. (2)* **170** (2009), no. 2, 799–817.

[12] T. Braden, J. Huh, J. P. Matherne, N. Proudfoot, and B. Wang, A semi-small decomposition of the Chow ring of a matroid. 2020, arXiv:2002.03341.

[13] T. Braden, J. Huh, J. Matherne, N. Proudfoot, and B. Wang, Singular Hodge theory for combinatorial geometries. 2020, arXiv:2010.06088.

[14] T. Braden and A. Vysogorets, Kazhdan–Lusztig polynomials of matroids under deletion. *Electron. J. Combin.* **27** (2020), no. 1, 17.

[15] P. Brändén, Discrete concavity and the half-plane property. *SIAM J. Discrete Math.* **24** (2010), no. 3, 921–933.

[16] P. Brändén, Spaces of Lorentzian and real stable polynomials are Euclidean balls. *Forum Math. Sigma* **9** (2021), e73.

[17] P. Brändén and J. Huh, Lorentzian polynomials. *Ann. of Math. (2)* **192** (2020), no. 3, 821–891.

[18] E. Cattani, Mixed Lefschetz theorems and Hodge–Riemann bilinear relations. *Int. Math. Res. Not. IMRN* **rnn025** (2008), no. 10, 20.

[19] C. Chindris, H. Derksen, and J. Weyman, Counterexamples to Okounkov's log-concavity conjecture. *Compos. Math.* **143** (2007), no. 6, 1545–1557.

[20] N. G. de Bruijn and P. Erdös, On a combinatorial problem. *Ned. Akad. Wet., Proc.* **51** (1948), 1277–1279; *Indag. Math. (N.S.)* **10** (1948), 421–423.

[21] C. De Concini and C. Procesi, Wonderful models of subspace arrangements. *Selecta Math. (N.S.)* **1** (1995), no. 3, 459–494.

[22] T. A. Dowling and R. M. Wilson, The slimmest geometric lattices. *Trans. Amer. Math. Soc.* **196** (1974), 203–215.

[23] T. A. Dowling and R. M. Wilson, Whitney number inequalities for geometric lattices. *Proc. Amer. Math. Soc.* **47** (1975), 504–512.

[24] J. Edmonds, Submodular functions, matroids, and certain polyhedra. In *Combinatorial Structures and their Applications (Proc. Calgary Internat. Conf., Calgary, Alta., 1969)*, pp. 69–87, Gordon and Breach, New York, 1970.

[25] B. Elias, N. Proudfoot, and M. Wakefield, The Kazhdan–Lusztig polynomial of a matroid. *Adv. Math.* **299** (2016), 36–70.

[26] B. Elias and G. Williamson, The Hodge theory of Soergel bimodules. *Ann. of Math. (2)* **180** (2014), no. 3, 1089–1136.

[27] C. Eur and J. Huh, Logarithmic concavity for morphisms of matroids. *Adv. Math.* **367** (2020), 107094, 19.

[28] E. M. Feichtner and S. Yuzvinsky, Chow rings of toric varieties defined by atomic lattices. *Invent. Math.* **155** (2004), no. 3, 515–536.

[29] A. Fink, K. Mészáros, and A. St. Dizier, Schubert polynomials as integer point transforms of generalized permutahedra. *Adv. Math.* **332** (2018), 465–475.

[30] W. Fulton, *Introduction to toric varieties*. Ann. of Math. Stud. 131, Princeton University Press, Princeton, NJ, 1993.

[31] W. Fulton, *Young tableaux*. London Math. Soc. Stud. Texts 35, Cambridge University Press, Cambridge, 1997.

[32] K. Gedeon, N. Proudfoot, and B. Young, The equivariant Kazhdan–Lusztig polynomial of a matroid. *J. Combin. Theory Ser. A* **150** (2017), 267–294.

[33] I. M. Gelfand, R. M. Goresky, R. D. MacPherson, and V. V. Serganova, Combinatorial geometries, convex polyhedra, and Schubert cells. *Adv. Math.* **63** (1987), no. 3, 301–316.

[34] V. Ginsburg, Perverse sheaves and $\mathbb{C}^*$-actions. *J. Amer. Math. Soc.* **4** (1991), no. 3, 483–490.

[35] C. Greene, A rank inequality for finite geometric lattices. *J. Combin. Theory* **9** (1970), 357–364.

[36] A. Grothendieck, Standard conjectures on algebraic cycles. In *Algebraic Geometry (Internat. Colloq., Tata Inst. Fund. Res., Bombay, 1968)*, pp. 193–199, Oxford Univ. Press, London, 1969.

[37] L. Gurvits, On multivariate Newton-like inequalities. In *Advances in combinatorial mathematics*, pp. 61–78, Springer, Berlin, 2009.

[38] A. P. Heron, A property of the hyperplanes of a matroid and an extension of Dilworth's theorem. *J. Math. Anal. Appl.* **42** (1973), 119–131.

[39] J. Herzog and T. Hibi, Discrete polymatroids. *J. Algebraic Combin.* **16** (2002), no. 3, 239–268.

[40] J. Huh, J. Matherne, K. Mészáros, and A. St. Dizier, Logarithmic concavity of Schur and related polynomials. *Trans. Amer. Math. Soc.* **375** (2022), 4411–4427.

[41] J. Huh and B. Wang, Enumeration of points, lines, planes, etc. *Acta Math.* **218** (2017), no. 2, 297–317.

[42] J. E. Humphreys, *Representations of semisimple Lie algebras in the BGG category $\mathcal{O}$*. Grad. Stud. Math. 94, American Mathematical Society, Providence, RI, 2008.

[43] D. Huybrechts, *Complex geometry*. Universitext, Springer, Berlin, 2005.

[44] C. Jiang and Z. Li, Algebraic reverse Khovanskii–Teissier inequality via Okounkov bodies. 2020, arXiv:2012.02847.

[45] K. Karu, Hard Lefschetz theorem for nonrational polytopes. *Invent. Math.* **157** (2004), no. 2, 419–447.

[46] J. P. S. Kung, Strong maps. In *Theory of matroids*, pp. 224–253, Encyclopedia Math. Appl. 26, Cambridge Univ. Press, Cambridge, 1986.

[47] R. Lazarsfeld, *Positivity in algebraic geometry. I. Ergebnisse der Mathematik und ihrer Grenzgebiete. 3. Folge.* A Series of Modern Surveys in Mathematics [Results in Mathematics and Related Areas. 3rd Series. A Series of Modern Surveys in Mathematics] 48, Springer, Berlin, 2004.

[48] K. Lee, G. D. Nasr, and J. Radcliffe, A combinatorial formula for Kazhdan–Lusztig polynomials of sparse paving matroids. 2020, arXiv:2002.03341.

[49] B. Lehmann and J. Xiao, Correspondences between convex geometry and complex geometry. *Épijournal Géom. Algébrique* **1** (2017), Art. 6, 29.

[50] T. M. Liggett, Ultra logconcave sequences and negative dependence. *J. Combin. Theory Ser. A* **79** (1997), no. 2, 315–325.

[51] D. Maclagan and B. Sturmfels, *Introduction to tropical geometry*. Grad. Stud. Math. 161, American Mathematical Society, Providence, RI, 2015.

[52] J. H. Mason, Matroids: Unimodal conjectures and Motzkin's theorem. In *Combinatorics (Proc. Conf. Combinatorial Math., Math. Inst., Oxford, 1972)*, pp. 207–220, Southend-on-Sea, 1972.

[53] D. Mayhew, M. Newman, and G. Whittle, Yes, the 'missing axiom' of matroid theory is lost forever. *Trans. Amer. Math. Soc.* **370** (2018), no. 8, 5907–5929.

[54] D. Mayhew, G. Whittle, and M. Newman, Is the missing axiom of matroid theory lost forever? *Q. J. Math.* **65** (2014), no. 4, 1397–1415.

[55] G. Melvin and W. Slofstra, Soergel bimodules and the shape of Bruhat intervals, 2020, preprint.

[56] C. Monical, N. Tokcan, and A. Yong, Newton polytopes in algebraic combinatorics. *Selecta Math. (N.S.)* **25** (2019), no. 5, Paper No. 66, 37.

[57] T. Motzkin, *Beiträge zur Theorie der linearen Ungleichungen*. 1936.

[58] T. Motzkin, The lines and planes connecting the points of a finite set. *Trans. Amer. Math. Soc.* **70** (1951), 451–464.

[59] K. Murota, *Discrete convex analysis*. SIAM Monographs on Discrete Mathematics and Applications, Society for Industrial and Applied Mathematics (SIAM), Philadelphia, PA, 2003.

[60] P. Nelson, Almost all matroids are nonrepresentable. *Bull. Lond. Math. Soc.* **50** (2018), no. 2, 245–248.

[61] A. Okounkov, Why would multiplicities be log-concave? In *The orbit method in geometry and physics (Marseille, 2000)*, pp. 329–347, Progr. Math. 213, Birkhäuser Boston, Boston, MA, 2003.

[62] J. Oxley, *Matroid theory. Second edn*. Oxf. Grad. Texts Math. 21, Oxford University Press, Oxford, 2011.

[63] A. Postnikov, Permutohedra, associahedra, and beyond. *Int. Math. Res. Not. IMRN* **6** (2009), 1026–1106.

[64] N. Proudfoot, The algebraic geometry of Kazhdan–Lusztig–Stanley polynomials. *EMS Surv. Math. Sci.* **5** (2018), no. 1–2, 99–127.

[65] N. Proudfoot, Y. Xu, and B. Young, The Z-polynomial of a matroid. *Electron. J. Combin.* **25** (2018), no. 1, Paper 1.26, 21.

[66] R. Schneider, *Convex bodies: the Brunn–Minkowski theory. Expanded edn*. Encyclopedia Math. Appl. 151, Cambridge University Press, Cambridge, 2014.

[67] J.-P. Serre, Analogues kählériens de certaines conjectures de Weil. *Ann. of Math. (2)* **71** (1960), 392–394.

[68] A. D. Sokal, The multivariate Tutte polynomial (alias Potts model) for graphs and matroids. In *Surveys in combinatorics 2005*, pp. 173–226, London Math. Soc. Lecture Note Ser. 327, Cambridge Univ. Press, Cambridge, 2005.

[69] L. Solomon, The Burnside algebra of a finite group. *J. Combin. Theory* **2** (1967), 603–615.

[70] R. P. Stanley, The number of faces of a simplicial convex polytope. *Adv. Math.* **35** (1980), no. 3, 236–238.

[71] R. P. Stanley, Subdivisions and local h-vectors. *J. Amer. Math. Soc.* **5** (1992), no. 4, 805–851.

[72] R. P. Stanley, *Algebraic combinatorics*. Undergrad. Texts Math., Springer, Cham, 2018.

[73] P. Vámos, The missing axiom of matroid theory is lost forever. *J. Lond. Math. Soc. (2)* **18** (1978), no. 3, 403–408.

[74] S. T. Yau, On the Ricci curvature of a compact Kähler manifold and the complex Monge-Ampère equation. I. *Comm. Pure Appl. Math.* **31** (1978), no. 3, 339–411.

JUNE HUH

Fine Hall, Washington Road, Princeton, NJ 08544-1000, USA, huh@princeton.edu

COUNTING PRIMES

JAMES MAYNARD

ABSTRACT

We survey techniques used to detect prime numbers in sets, highlighting the strengths and limitations of current techniques.

MATHEMATICS SUBJECT CLASSIFICATION 2020

Primary 11N05; Secondary 11N35, 11M06, 11N13

KEYWORDS

Prime numbers, sieve methods, Type I/II sums

INTERNATIONAL CONGRESS OF MATHEMATICIANS 2022 JULY 6–14

© 2022 International Mathematical Union

Proc. Int. Cong. Math. 2022, Vol. 1, pp. 240–268
DOI 10.4171/ICM2022/206

Published by EMS Press

and licensed under a CC BY 4.0 license

1. INTRODUCTION

Many of the most notorious open problems about prime numbers can be phrased as variations of the following question.

Question. Given a set of integers $\mathcal{A}$, how many primes are in $\mathcal{A}$?

Depending on the context, 'how many' could be asking whether there exists at least one prime in $\mathcal{A}$, whether there are infinitely many primes in $\mathcal{A}$, or asking for a quantitative estimate for the number of primes up to some threshold.

For example, we have the following special cases:

- $\mathcal{A} = \mathbb{Z}$. That there are infinitely many primes in $\mathcal{A}$ follows from Euclid's proof of the infinitude of primes. An asymptotic formula for the primes in $\mathcal{A}$ less than x is given by the Prime Number Theorem, and asking for the smallest possible error term in such an asymptotic estimate is essentially a reformulation of the Riemann Hypothesis.

- $\mathcal{A} = \{p + 2 : p \text{ prime}\}$. Asking for infinitely many primes in $\mathcal{A}$ is the famous Twin Prime Conjecture, and an asymptotic formula for the number of primes in $\mathcal{A}$ is a conjecture of Hardy and Littlewood.

- $\mathcal{A} = \{2N - p : p \text{ prime}\}$, for some fixed integer $N \geq 2$. In this case $\mathcal{A}$ contains only a finite number of positive elements (and so a finite number of primes), but asking that it contains at least one prime for every $N \geq 2$ is Goldbach's conjecture.

The final two examples are two of Landau's influential four problems on primes listed in his 1912 ICM address; all four remain unsolved.

In general, we will focus on situations where we expect (from heuristics, numerical evidence, or other guesswork) that there should be primes in $\mathcal{A}$, and the task is to try to prove this is indeed the case.

We know of no way to *construct* prime numbers theoretically, and therefore we typically need to use an indirect method to prove the existence of primes in a given set $\mathcal{A}$. If we are unable to numerically test elements, then often the only way we know how to prove the existence of a single prime in a set $\mathcal{A}$ is to perform the a priori harder task of approximately counting the number of primes in $\mathcal{A}$ and showing there are many primes in $\mathcal{A}$ of a given size. For example, Vinogradov's three primes theorem states that every sufficiently large odd number can be written as the sum of three primes (this is now actually known for *all* $N \geq 7$ thanks to work of Helfgott [42]), but the only way we know how to prove this actually shows that there are 'many' ways to write a large odd integer N as the sum of three primes.

The ultimate goal in this area is to develop a flexible toolkit which can reduce the question of counting primes in sets $\mathcal{A}$ of interest to easier (but more technical) questions about the arithmetic structure of the set in question, and then to have a set of techniques which can investigate these questions.

2. MULTIPLICATIVE NUMBER THEORY

Multiplicative number theory rests on utilizing the following crucial property of the primes, which is essentially the Fundamental Theorem of Arithmetic.

Property. Prime numbers generate the positive integers via multiplication.

This property allows us to define suitable multiplicative generating functions (L-functions) which encode properties of the primes via the integers they generate. A reformulation of the Fundamental Theorem of Arithmetic is the identity (for $\mathrm{Re}(s) > 1$)

$$\zeta(s) = \sum_{n=1}^{\infty} \frac{1}{n^s} = \prod_p \Big(1 - \frac{1}{p^s}\Big)^{-1}.$$

Since we analytically understand the integers under addition quite well, we can obtain a good understanding (analytic continuation, controlled growth) of $\zeta(s)$ via the Dirichlet series representation on the left-hand side. This understanding can then be translated into understanding about the primes. The infinitude of the primes follows from the fact that $\zeta(s)$ has a pole at $s = 1$, the Prime Number Theorem follows from (and is essentially equivalent to) the fact that $\zeta(s)$ has no zeros on the line $\mathrm{Re}(s) = 1$, and precise estimates for the count of primes are essentially equivalent to zero-free regions for $\zeta(s)$ within the critical strip $0 < \mathrm{Re}(s) < 1$. In all these cases the partial information we are interested in about primes becomes much easier to establish via translating it to a question about partial understanding of $\zeta(s)$.

Moreover, the techniques of multiplicative number theory extend well beyond just studying primes via $\zeta(s)$, but to a whole zoo of different L-functions which encode different algebraic information about primes. Prime ideals generate all ideals of the ring of integers of a number field, and so prime ideals (and hence the splitting of rational primes) can be studied via the same techniques via Dedekind L-functions $\zeta_K(s)$ (the analogue of $\zeta(s)$ for a number field K). Moreover, one can twist the $\zeta(s)$ by a Dirichlet character or the Archimedean character n^{it}, or one can twist $\zeta_K(s)$ by a Hecke character (or more generally twist an L-function by a suitable automorphic representation), to obtain further L-functions, which can study primes in arithmetic progressions, short intervals, the locations of prime ideals in lattices or similar questions.

Essentially the *only* method we have which is capable of 'producing' primes is using multiplicative number theory. Even though there are now a few ostensibly different proofs of the Prime Number Theorem, all known proofs rely fundamentally on the Fundamental Theorem of Arithmetic, and require multiplicative structure. Virtually all other results counting primes can be thought of as extensive elaborate manoeuvres which allow one to reduce to the situation of using multiplicative number theory to count primes.

2.1. Primes and zeros

The techniques of multiplicative number theory crucially allow one to understand multiplicative questions on the distribution of primes via the zeros of the corresponding L-functions. The duality between primes and zeros of $\zeta(s)$ is best seen through Riemann's famous *Explicit Formula* for $\zeta(s)$: for $x, T \geq 2$,

$$\sum_{n<x} \Lambda(n) = x - \sum_{|\rho|<T} \frac{x^\rho}{\rho} - \log\left(2\pi\sqrt{1-x^{-2}}\right) + O\left(\frac{x(\log x)^3}{T}\right), \tag{2.1}$$

where $\Lambda(n)$ is the von Mangold function and the sum is over all nontrivial zeros ρ of $\zeta(s)$ (counted with multiplicity, although all zeros are believed to be simple). For every L-function (satisfying the expected meromorphicity and growth conditions), we get a corresponding explicit formula with one side representing primes and the other zeros of the L-function.

The explicit formula points to unexpected deep *structure* within the sequence of primes; if the Riemann Hypothesis ($\mathrm{Re}(\rho) = 1/2$ for all nontrivial ρ) holds, then treating all terms apart from x trivially, we would obtain a *smaller* size error term than we would expect based on simple random model predictions (we expect that the presence of zeros alters effects such as the law of the iterated logarithm, for example). Indeed, the zeros of $\zeta(s)$ constrain the error term in the count of primes to fluctuate relatively less than we expect for other arithmetic sequences (such as twin primes) where we expect 'random-like' behavior. Another example where this structure plays a role is the fact that the error term in the Prime Number Theorem can be self-improving; if we can show that

$$\left|\pi(x) - \int_2^x \frac{dt}{\log t}\right| \ll x^{1/2+o(1)},$$

then we know that the Riemann Hypothesis holds and the error term $x^{1/2+o(1)}$ can be upgraded to the more precise $O(x^{1/2}\log x)$. It would be interesting to see if the structure implied by zeros can be exploited meaningfully in other ways.

Similarly, since the error term in (2.1) disappears as $T \to \infty$, we see that knowing *all* zeros encodes *all* information about primes, and vice versa. This observation is useless for most practical purposes, but it means that zeros of $\zeta(s)$ must also encode the distribution of primes in arithmetic progressions, and therefore encode information about zeros of Dirichlet L-functions too. This is partial justification for the idea that L-functions should be studied in families rather than individually. A spectacular example of this is Goldfeld and Gross–Zagier's [29, 30, 34] joint resolution of the Gauss class number one problem by showing that an L-function attached to a suitable Elliptic curve had a triple zero at the central point, and this triple zero had a suitably strong influence on zeros of Dirichlet L-functions to prevent there being any particularly bad Siegel zeros.

2.2. Zero density estimates

Although the Riemann Hypothesis is the most important question for any given L-function, often it would suffice for applications to primes to show a much weaker statement that 'most' zeros lie 'close' to the line $\mathrm{Re}(s) = 1/2$, rather than requiring that all zeros lie on this line. For example, under the Riemann Hypothesis we can show an asymptotic formula for primes in $[x, x + x^{1/2}(\log x)^2]$. If we let $N(\sigma, T)$ denote the number of zeros $\rho = \beta + i\gamma$ with $|\gamma| \leq T$ and $\beta \geq \sigma$ then a bound $N(\sigma, T) \ll T^{2-2\sigma+o(1)}$ (known as the 'Density Hypothesis') would allow us to deduce an asymptotic formula for primes in

$[x, x + x^{1/2+o(1)}]$, which is almost as short as what we obtain under the Riemann Hypothesis. Unfortunately, the Density Hypothesis is open in general, but a classical result of Huxley **[44]** shows $N(\sigma, T) \ll T^{12(1-\sigma)/5+o(1)}$, which implies an asymptotic formula for the number of primes in $[x, x + x^{7/12+o(1)}]$, and is essentially the best known result (Heath-Brown **[36]** used sieve methods to remove the $o(1)$.)

In many counting problems for the primes which are directly related to zeros, the limitation in our results is due to a limitation in our understanding of zeros near the 3/4-line (such as the example above of primes in short intervals), or near the 1-line (such as issues with Siegel zeros or the least quadratic nonresidue). For example, if we knew that there were no zeros ρ with $0.74 \leq \mathrm{Re}(\rho) \leq 0.76$ (or if there were 'few' such zeros), then we would improve on our understanding of primes in short intervals. The typical way to bound such zeros is to detect them via large values of a Dirichlet polynomial (see **[47, CHAPTER 10]**). The key limitation of our zero density estimates for the past 50 years reduces to the following question.

Question 1. Can we show

$$\mathrm{meas}\left\{t \in [T, 2T] : \left|\sum_{n=T^{2/5}}^{2T^{2/5}} n^{it}\right| > T^{3/10}\right\} \ll T^{3/5-\delta}$$

for some fixed positive constant δ?

The bound $T^{3/5+o(1)}$ follows quickly from straightforward bounds for the 4th or 6th mean value of the Dirichlet polynomial. Improving on the 6th moment bound is related to bounding the 6th moment of $\zeta(1/2 + it)$, but it is not unreasonable to hope that this question might be easier to study.

Even if we cannot improve our current zero-density bounds on the *number* of zeros, an alternative approach might be to see what this might imply for the *distribution* of zeros of $\zeta(s)$. (Ultimately one might hope to obtain a putative classification which either contradicts other known properties or demonstrates that there are still primes in short intervals with this distribution of zeros.)

Question 2. Imagine that $|\pi(x + x^{7/12-\epsilon}) - \pi(x) - x^{7/12-\epsilon}/\log x| \gg x^{7/12-\epsilon}/\log x$ for some large x. What does this imply about the distribution of the zeros of $\zeta(s)$?

We know that there must be roughly $T^{3/5}$ zeros of height T with real part very close to $3/4$ for $T \approx x^{5/12}$, and, moreover, it must be the case that these zeros $\rho = \beta + i\gamma$ have the fact that the fractional part of $2\pi\gamma \log x$ is quite strongly biased modulo 1. Moreover, we speculate that there should be much more prescriptive constraints on the *vertical* distribution such zeros – roughly that they occur in small clusters whose imaginary parts are roughly in an arithmetic progression. Obtaining a precise classification of this sort seems difficult (it appears related to the inverse Littlewood problem in additive combinatorics/harmonic analysis), but a suitably strong classification would open up a new manner to potentially rule out conspiracies preventing primes in short intervals. A proof-of-concept in this direction is recent work with Pratt **[66]**.

Theorem 3 (Conditional improvement to zero density estimates). *Assume that the zeros of $\zeta(s)$ lie on finitely many vertical lines. Then*

$$\#\{p \in [x, x + x^{1/2+\epsilon}]\} = (1 + o_\epsilon(1))\frac{x^{1/2+\epsilon}}{\log x}.$$

The point here is that the hypothesis still allows for the possibility of vertical arithmetic progressions of zeros, and so one of the potential limitations is actually less of an issue. We can obtain improvements on the classical exponent 7/12 to give results almost as strong as what the Riemann Hypothesis would imply by studying the vertical patterns of zeros of $\zeta(s)$, albeit under rather strong assumptions.

In a very different direction, following work of Matomäki–Radziwiłł [54], if one is interested in the Möbius function (and is happy with weaker quantitative bounds), then we can restrict attention to Dirichlet polynomials which factor in many ways (expanding on earlier ideas of [8,10,49]). This allows one to overcome the issues raised here for primes, and obtain stronger results about the Möbius function in short intervals [56] as well as almost-all short intervals [54].

2.3. Limits to multiplicative techniques

In general the multiplicative theory for counting primes points to a rich structure encoded by the zeros and a powerful set of techniques. Unfortunately, there are some issues with this from a practical point of view:

(1) Multiplicative techniques rely on the presence of multiplicative structure in the problem. In situations which are less structured (particularly when there is addition polluting multiplicative objects like in the Twin Prime Conjecture), we do not know how to make use of multiplicative techniques. Even when they can be of use, it require a lot of work to massage problems into a suitable form that the powerful multiplicative techniques can apply to.

(2) In the absence of the conjectured strong control over zeros, our estimates are often limited in their range of applicability, particularly with uniformity of estimates with respect to underlying parameters such as conductor or degree of number field.

(3) The multiplicative methods tend to either give strong asymptotic formulae or fail to give any nontrivial bound whatsoever. The strength of the analytic approach means that it is not well-suited to answering 'soft' questions with a wide degree of flexibility.

As an example of the final two points, Hooley's [43] proof of the Artin primitive root conjecture under the Generalized Riemann Hypothesis for suitable Dedekind L-functions relied crucially on the upper bound

$$\sum_{q\sim Q} \pi^*(x;q) \ll \frac{x}{Q\log x} + Qx^{1/2}(\log x)^{O(1)},$$

where $\pi^*(x;q)$ counts primes $p < x$ with $p \equiv 1 \pmod{q}$ and for which 2 is a qth power $(\text{mod } p)$. (In fact, an upper bound of the form $o_{Q\to\infty}(x/(\log x)^2)$ for $Q < x^{1/2}(\log x)^{-A}$ would have sufficed.) The only way we know how to prove an upper bound of this type is by proving an asymptotic formula of the form $\pi^*(x;q) = \pi(x)/(q\varphi(q)) + O(x^{1/2}\log x)$ via GRH, which is a much stronger statement. Unconditional techniques based on multiplicative number theory can capture the condition of being a qth power, but only with error terms that degrade quickly with q. (By contrast, other techniques such as sieve methods can be very flexible at producing upper bounds, but appear poorly suited to capturing the more algebraic qth power condition.)

Question 4. Can one produce a nontrivial upper bound for $\sum_{q\sim x^{1/2-\epsilon}} \pi^*(x;q)$ unconditionally?

3. SIEVE METHODS

Sieve methods take a different, combinatorial approach to studying primes, based on the following simple property:

Property. Primes are integers n which have no divisors smaller than $\sqrt{n}$ other than 1.

Thus primes are examples of numbers with no small divisors, and more generally one can look at integers n with no divisors (other than 1) less than some quantity z. This formulation naturally suggests that one can count such numbers in a set $\mathcal{A}$ via inclusion–exclusion:

$$\sum_{\substack{n\in\mathcal{A}\\ p|n\Rightarrow p>z}} 1 = \sum_{\substack{d\\ p|d\Rightarrow p\le z}} \mu(d) \sum_{\substack{n\in\mathcal{A}\\ d|n}} 1.$$

Let us restrict attention from now on to sets $\mathcal{A} \subseteq [x, 2x]$ for some large value x, so that all elements have roughly the same size.

Unfortunately, even if one had very good estimates for the size of the set $\mathcal{A}_d$ of multiples of d in $\mathcal{A} \subset [x, 2x]$, there would be $2^{\pi(z)}$ different integers d in the sum and so any error terms would accumulate and dominate the hope of a main term unless z was very small (such as if $z \le \log x$). The first key insight in of sieve methods is that one can use positivity to truncate the inclusion–exclusion process and avoid the presence of d's which are too large, at the cost of a small amount of precision. The basic arithmetic information required to make this work is then a moderate understanding of inner sums above, namely the size of the sets $\mathcal{A}_d = \{n \in \mathcal{A} : d|n\}$.

Let $g(d)$ be a multiplicative function which we think of as an approximation to the density of elements of $\mathcal{A}$ which are a multiple of d. We assume that $g(p) < 1 - \epsilon$ (so that there are no prime factors which are too common) and that $g(p) \approx \kappa/p$ for some fixed constant $\kappa > 0$ on average by assuming for $2 \le w$,

$$\sum_{p\le w} g(p)\log p = \kappa \log w + O(1). \tag{3.1}$$

The key arithmetic input for sieve methods is then an estimate for every $A > 0$,

$$\sum_{d<x^{\gamma}} \left|\#\mathcal{A}_d - g(d)\#\mathcal{A}\right| \ll_A \frac{\#\mathcal{A}}{(\log x)^A} \tag{3.2}$$

for some given fixed $\gamma > 0$. The larger we are able to take γ, the better we are able to understand $\mathcal{A}$ in arithmetic progressions and the more powerful the conclusions of our sieve methods will be. In most situations of interest we expect (3.2) to hold for a suitable function g and reasonably large constant $\gamma \in (0, 1)$, so (3.2) should be thought of as a reasonably mild constraint when γ is small.

The basic point of sieve methods is that for any set which does satisfy an estimate like (3.2) we can make the inclusion-exclusion argument much more accurate. This is known as the 'fundamental lemma' (see, for example, [28, COROLLARY 6.10]).

Lemma 5 (Fundamental Lemma of Sieve Methods). *Let g be a multiplicative function as above. Then we have for any $\eta, \gamma > 0$,*

$$\sum_{\substack{n\in\mathcal{A}\\ p|n \Rightarrow p>x^{\eta}}} 1 = \left(1 + O_{\kappa}(e^{-\gamma/\eta})\right) \prod_{p\le x^{\eta}} \left(1 - g(p)\right)\#\mathcal{A} + O(E),$$

where

$$E = \sum_{d<x^{\gamma}} \left|\#\mathcal{A}_d - g(d)\#\mathcal{A}\right|.$$

One should think of the case when $\mathcal{A}$ satisfies (3.2) with some fixed $\gamma > 0$, and η is taken as a sufficiently small fixed constant. The key point of the fundamental lemma is then that one can still obtain good asymptotic estimates for the number of elements in $\mathcal{A}$ with no prime factors less than z even when z is as large as x^{η}, provided we have a relatively modest estimate for the distribution of $\mathcal{A}$ in arithmetic progressions.

An immediate consequence is that $\mathcal{A}$ contains $O(\#\mathcal{A}/(\log x)^{\kappa})$ primes, and we expect that in most situations this should be the *correct* order of magnitude for the number of primes in $\mathcal{A}$. For example, returning to some of Landau's problems mentioned in Section 1, we find that there are $O(x/(\log x)^2)$ twin primes less than x, and that there are $O(x^{1/2}/\log x)$ prime values of $n^2 + 1$ which are less than x, and both estimates are conjectured to be sharp up to the multiplicative constant. We also immediately obtain that $\mathcal{A}$ contains 'many' elements with a bounded number of prime factors as soon as it satisfies something like (3.2). The fact that sieve methods can very flexibly give upper bounds of the right order of magnitude in a wide variety of situations is a very valuable fact when used inside more complicated arguments.

The fundamental lemma essentially produces optimal bounds (with care, the $O_{\kappa}(e^{-\gamma/\eta})$ error term can usually be handled satisfactorily), and so the sieving process of 'small' primes less than x^{η} is almost perfect, and as if the small primes were behaving independently of one another. This can therefore also be used just as a preliminary sieving stage, where we first remove all 'small' prime factors $\le x^{\eta}$ perfectly via an application of the Fundamental Lemma, leaving us to be more careful in trying to handle the about the $O(1/\eta)$ 'large' prime factors (bigger than x^{η}) of elements of $\mathcal{A}$. Although the behavior of the small primes is essentially that of independence and the same for all sets $\mathcal{A}$ satisfying (3.2), the distribution of the large prime factors in general will vary according to the set. Understanding how much control we have on these large prime factors from (3.2) is still something of a poorly understood art in general, and often the best sieving procedure is tailored to the question at hand.

3.1. Arranging the large prime factors

In general, for any set $\mathcal{A}$ satisfying (3.2), and x sufficiently large we will have that

$$\sum_{\substack{n\in\mathcal{A}\\ p|n\Rightarrow p\geq x^\eta}} 1 \leq \big(F(\kappa,\eta,\gamma)+o(1)\big)\#\mathcal{A}\prod_{p\leq x^\eta}\big(1-g(p)\big),$$

$$\sum_{\substack{n\in\mathcal{A}\\ p|n\Rightarrow p\geq x^\eta}} 1 \geq \big(f(\kappa,\eta,\gamma)+o(1)\big)\#\mathcal{A}\prod_{p\leq x^\eta}\big(1-g(p)\big),$$

for some functions $0\leq f(\kappa,\eta,\gamma)\leq F(\kappa,\eta,\gamma)$ depending only on the constant γ in (3.2), the 'sieve dimension' κ from (3.1) and η from the sieving threshold of x^η.

When $\kappa=1$ (the most common sieving situation) we are in the situation of the 'linear sieve', somewhat remarkably we know the optimal values of the functions.

Lemma 6 (Optimality of the linear sieve). *Let g satisfy (3.1) with $\kappa=1$ and $g(p)<1-\epsilon$. Then there are functions $F(s)$, $f(s)$ such that we have the following:*

(1) *For any set $\mathcal{A}$ satisfying (3.2), we have*

$$\sum_{\substack{n\in\mathcal{A}\\ p|n\Rightarrow p\geq x^\eta}} 1 \leq \big(F(\gamma/\eta)+o(1)\big)\#\mathcal{A}\prod_{p\leq x^\eta}\big(1-g(p)\big),$$

$$\sum_{\substack{n\in\mathcal{A}\\ p|n\Rightarrow p\geq x^\eta}} 1 \geq \big(f(\gamma/\eta)+o(1)\big)\#\mathcal{A}\prod_{p\leq x^\eta}\big(1-g(p)\big).$$

(2) *There are sets $\mathcal{A}^+,\mathcal{A}^-\subseteq[x,2x]$ which satisfy (3.2) and $g^\pm$ which satisfy (3.1) and $g^\pm(p)<1-\epsilon$ such that*

$$\sum_{\substack{n\in\mathcal{A}^+\\ p|n\Rightarrow p\geq x^\eta}} 1 = \big(F(\gamma/\eta)+o(1)\big)\#\mathcal{A}^+\prod_{p\leq x^\eta}\big(1-g^+(p)\big),$$

$$\sum_{\substack{n\in\mathcal{A}^-\\ p|n\Rightarrow p\geq x^\eta}} 1 = \big(f(\gamma/\eta)+o(1)\big)\#\mathcal{A}^-\prod_{p\leq x^\eta}\big(1-g^-(p)\big).$$

This technical looking statement says that for any $\mathcal{A}$ satisfying a linear sieving problem, we know the optimal upper and lower bounds for the number of sieved elements of the set, based purely on the distribution of $\mathcal{A}$ in arithmetic progressions to modulus x^γ. We can take $\mathcal{A}^\pm=\{n\in[x,x+x^{\gamma+\epsilon}:\lambda(n)=\mp1\}$ and $g^\pm(p)=1/p$, where $\lambda(n)$ is the Liouville function ($\lambda(n)=-1$ if n has an odd number of prime factors, and $\lambda(n)=1$ otherwise) and this gives the functions F, f, which can be written explicitly as solutions to a delay-differential equation. Thus for the basic problem of understanding the consequences of (3.2), we have an essentially complete answer.

Although the linear sieve is essentially optimal, when the sieving dimension is greater than one we have a much poorer understanding of optimality and what we can hope to achieve. The linear sieve bounds are proven using a 'combinatorial sieve', and combinatorial sieves tend to produce the best bounds when κ is reasonably small. When κ gets larger, however, it typically turns out that Selberg's sieve performs better. However, in no circumstance do we have anything like the complete understanding of the picture that we would like.

Question 7. What are the optimal sieve functions for high-degree sieves? What do the extremal sets look like?

For example, the upper bound for the number of prime k-tuplets less than x is larger than the expected truth by $2^k k!$. Although the parity phenomenon would prevent us from obtaining a bound smaller than 2^k times the expected truth, it is very unclear what sort of bound an optimal k-dimensional sieve could hope to prove in this situation. The key innovation in [58] was a new high-dimensional variant of Selberg's sieve tailored to the application at hand, which allowed for notable progress on the sieving problem of bounded gaps between primes (see Section 4). Although this does not appear to help with the direct upper and lower bounds, it indicates that there is potentially a lot left to be understood about high-dimensional sieves.

Question 8. What other arithmetic features of sets $\mathcal{A}$ of interest can be exploited to produce improved sieving bounds?

If there is extra arithmetic information which could distinguish sets $\mathcal{A}$ from extremal sets, this could then be incorporated into the sieving assumptions to hopefully produce better bounds.

For example, Chen's twist [9] was a key innovation used by Chen to show that there are infinitely many primes p with $p+2$ having at most two prime factors, and this exploited the fact that the situation could be viewed as fixing the prime factorization of either n of $n+2$ and viewing it as a sieve problem to produce bounds which are better than what the standard linear sieve would imply. High dimensional sieves often have similar features where they can be viewed as $(k-1)$-dimensional sieving problems or k-dimensional ones, and mixing these perspectives allows one to do slightly better than typical situations [57]. In a different direction, the 'interval sieve' asks for bounds when we know that $\mathcal{A}$ is just an interval – it is known in this case [18, 31] that the optimal sieve functions are closely linked to the presence of Siegel zeros, and so in many situations this limits what we can hope to achieve.

3.2. Limitations of sieve methods and the parity phenomenon

We saw above that the extremal sets $\mathcal{A}^{\pm}$ for the linear sieve were given in terms of numbers with an odd or even number of prime factors. This is an example of a fundamental limitation of sieve methods based purely on arithmetic information of the from (3.2): the parity phenomenon. Roughly, this says that sieve methods cannot distinguish between numbers with an even number of prime factors and an odd number of prime factors.

For example, all sieve upper and lower bounds are based on using sieve weights which are short divisor sums

$$w_n = \sum_{\substack{d|n \\ d<x^\gamma}} \lambda_d.$$

Thus, recalling that $\lambda(n) = -1$ if n has an odd number of prime factors, and $\lambda(n) = 1$ otherwise, we see

$$\sum_{n\in\mathcal{A}} w_n\Big(\frac{1\pm\lambda(n)}{2}\Big) = \frac{1}{2}\sum_{n\in\mathcal{A}} w_n + O\Big(\sum_{d<x^\gamma} |\lambda_d|\Big|\sum_{\substack{n\in\mathcal{A}\\ d|n}} \lambda(n)\Big|\Big).$$

For most sets $\mathcal{A}$ of interest, it is believed that the inner sums on the right-hand side should always be very small, meaning that the same total weight is put on numbers with an even number of prime factors as those with an odd number of prime factors (although actually proving this is almost as hard as proving an asymptotic formula for primes in $\mathcal{A}$).

Because the weight is equidistributed between numbers with an even and an odd number of prime factors, it means that any upper bound sieve for primes will be off by a factor of at least 2 (the weight placed upon primes can be at most the total weight of numbers with an odd number of prime factors, which in turn is at most half the total weight). It also means that we cannot hope to obtain a nontrivial lower bound for the number of primes in a set $\mathcal{A}$ by just using pure sieve methods.

In various situations, this elementary loss of a factor of 2 from sieve methods is intimately linked to the possible presence of Siegel zeros (which would cause certain residue classes to have double the expected number of primes of a certain size.) For example, the Brun–Titchmarsh Theorem **[69]** (proven using sieve methods) states that

$$\pi(x;q,a) \le \frac{2x}{\varphi(q)\log(x/q)}.$$

When x is fairly large relative to q, this is off by a factor of roughly 2 from the expected asymptotic, but improving the constant 2 to $2-\delta$ in this regime would rule out the possibility of a Siegel zero.

4. SIDE-STEPPING LIMITATIONS OF SIEVE METHODS

Although sieve methods alone cannot directly produce primes, sometimes this apparent limitation can be sidestepped. For example, consider the following result (**[58, 59, 71, 82]** and unpublished work of Tao).

Theorem 9 (Bounded gaps between primes). *Let k be a positive integer. Then*

$$\liminf_n (p_{n+k} - p_n) < \infty.$$

In the special case when $k = 1$, we can take the finite constant to be 246; for general k, we can take the bound to be $O(e^{3.815k})$ thanks to work of Baker–Irving **[3]**.

This result manifestly says something about prime numbers, but ultimately only relies on arithmetic information of the form (3.2), in this case the Bombieri–Vinogradov

Theorem. The reason this result isn't prevented from saying something by the parity phenomenon is because it sidesteps some of the issues via the pigeonhole principle, which then avoids the need to specify exactly which quantities are taking prime values. More specifically, the proof of Theorem 9 relies on considering the quantity

$$S = \sum_{n \sim x} \left(\sum_{i=1}^{K} \mathbf{1}_{\mathbb{P}}(n + h_i) - k \right) w_n$$

for some suitable fixed constants $h_1 < \cdots < h_K$ (chosen such that $\prod_{i=1}^{K}(n + h_i)$ is not always a multiple of a fixed prime p) and some nonnegative sieve weight w_n tailored to the situation at hand. Since $w_n \geq 0$, showing that $S > 0$ implies that there is some $n \sim x$ for which at least $k + 1$ of $n + h_1, \ldots, n + h_K$ are simultaneously prime, and hence there are $k + 1$ primes all contained in an interval of length $h_K - h_1$. The fact that we do not have any control over *which* of the different $n + h_i$ are prime, merely the fact that several of them are prime is what allows us to sidestep the parity phenomenon issue.

Another example of proving the existence of primes in a set by sidestepping the usual obstacles is due to Elkies **[15]**.

Theorem 10 (Elkies' Theorem). *Let $E/\mathbb{Q}$ be an elliptic curve. Then there are infinitely many supersingular primes for E.*

The proof actually only relies on Dirichlet's theorem on primes in arithmetic progressions; all the nontrivial content of the proof is showing that there are polynomials P_ℓ encoding E_p having complex multiplication by a suitable order (this happens if p divides the numerator of $P_\ell(j(E))$ and $-\ell$ is a quadratic nonresidue $\pmod p$). Carefully choosing a sequence of ℓ's then shows that there must be infinitely many distinct such p's. Thus this is an example where we started with what seemed a difficult counting problem, but by focusing on a special subsequence we were able to reduce to a much counting problem for primes.

One result about primes which relies on sieving procedures but is not directly limited by the parity phenomenon is that of large gaps between primes. In this case it is again fruitful to focus on a special case; if we have a long string of consecutive integers $n, n + 1, \ldots, n + y$ all with a small prime factor $\leq (\log n)/2$, then certainly we have a long gap between primes. The fact we only search for factors $\leq \log n$ limits our approach (we expect we cannot find gaps between primes less than x bigger than $(\log x)(\log\log x)^{2+o(1)}$ in this way), but enables us to understand the situation by looking at n in residue classes (mod $\prod_{p \leq (\log x)/2} p$), and choosing a convenient residue class to make all the consecutive integers composite. This indirect approach therefore allows us to avoid directly counting primes. The current record is **[19, 20, 60]**

Theorem 11 (Large gaps between primes).

$$\sup_{p_n \leq X} (p_{n+1} - p_n) \gg \frac{(\log x)(\log\log x)(\log\log\log\log x)}{\log\log\log x},$$

This improves upon an old bound of Erdős–Rankin **[17, 72]**. The key input for this bound was a version of Theorem 9 showing the existence of certain residue classes containing unusually many small primes – this exploited the fact that sieve results (when successful) are often very flexible and uniform with respect to other parameters.

The parity phenomenon issue applies equally to estimating primes or estimating sums involving the Liouville function $\lambda(n)$. It is therefore somewhat remarkable that Tao [74] was able to avoid this for the 2-point Chowla conjecture.

Theorem 12 (Logarithmically average 2-point Chowla).

$$\sum_{n<x} \frac{\lambda(n)\lambda(n+1)}{n} = o(\log x).$$

The key property that is exploited here is the multiplicativity of λ; by using $\lambda(np) = -\lambda(n)$ and averaging over small primes p, the problem is turned from a binary problem (which we might expect to be limited by the parity phenomenon) to a ternary one (where we might hope to use a version of the circle method and not be limited by the parity phenomenon). Unfortunately, the subsequent steps appear only able to handle very small primes p, which appears to stop this idea applying to questions about the primes.

5. PRIMES IN ARITHMETIC PROGRESSIONS AND EXTENDING THE LEVEL OF DISTRIBUTION

Most results using sieve methods rely crucially on an estimate of the form (3.2), and the strength of the final results is determined by how large we can take the constant γ to be. Natural questions are how far we can push the constant γ for a given set $\mathcal{A}$, and whether we really need the full strength of (3.2) or whether we can produce a weaker, but more technical result which would still suffice for intended applications.

How far we can extend these estimates naturally depends on the particular set $\mathcal{A}$ in question. For simplicity, we will focus on the case when $\mathcal{A}$ is closely related to the set of primes ($\mathcal{A}$ could be shifted primes, like in the Twin Prime problem, for example) since this is a common case which appears regularly. In this situation, (3.2) is asking us to understand primes in arithmetic progressions, and typically the basic tool used is the Bombieri–Vinogradov Theorem [4,77].

Theorem 13 (Bombieri–Vinogradov Theorem). *Let $\epsilon, A > 0$. Then we have*

$$\sum_{q \le x^{1/2-\epsilon}} \sup_{(a,q)=1} \left| \pi(x;a,q) - \frac{\pi(x)}{\varphi(q)} \right| \ll_{\epsilon,A} \frac{x}{(\log x)^A}.$$

This asserts that the set of primes shifted by a constant satisfies a strong form of (3.2) for any $\gamma < 1/2$. From the point of view of sieve methods (where we typically only need estimates 'on average' over arithmetic progressions) this is typically an unconditional substitute for the Generalized Riemann Hypothesis. We expect, however, that one should be able to go much further [16].

Conjecture 1 (Elliott–Halberstam Conjecture). *Let $\epsilon, A > 0$. Then we have*

$$\sum_{q \le x^{1-\epsilon}} \sup_{(a,q)=1} \left| \pi(x;q,a) - \frac{\pi(x)}{\varphi(q)} \right| \ll_{\epsilon,A} \frac{x}{(\log x)^A}.$$

Increasing the arithmetic information available to the sieve method in question naturally produces stronger results; under the Elliott–Halberstam conjecture. For example, the bound 246 of the case $k = 1$ of Theorem 9 can be improved to 12 [58], and we can obtain an upper bound for twin primes which is a factor of only 2 larger than the expected truth.

Unfortunately in this formulation we do not know how to extend the Bombieri–Vinogradov Theorem to moduli beyond $x^{1/2}$ – this is often known as the 'square-root barrier', and the difficulty of the problem increases dramatically at this point where it goes beyond the region of the Generalized Riemann Hypothesis. However, if we ask for a slightly more technical version of these results on primes in arithmetic progressions, then one *can* do better. The pioneering work of Fouvry and Bombieri–Friedlander–Iwaniec [5–7,22] produced various results accounting for moduli as large as $x^{4/7-o(1)}$. This was recently extended [64] to larger moduli still.

Theorem 14 (Beyond $x^{1/2}$ barrier for nice coefficients). *Let $\lambda(n)$ be 'triply well factorable' and $\epsilon, A > 0$. Then we have*

$$\sum_{q \le x^{3/5-\epsilon}} \lambda(q) \left(\pi(x;q,a) - \frac{\pi(x)}{\varphi(q)} \right) \ll_{a,\epsilon,A} \frac{x}{(\log x)^A}.$$

For simplicity we will not go into the precise definition of 'triply well factorable' (it roughly means that $\lambda(q)$ can be decomposed into a triple-convolution of sequences of any predetermined sizes). The key point here is that one can take any $\gamma < 3/5$ so we can consider very large moduli, and at the same time the technical weakenings (triply well factorable sequences and a dependency on the residue class) are sufficient for various applications to sieve methods. For example, Iwaniec [45] showed that the linear sieve weights can be modified to become 'well-factorable,' which then makes linear sieve estimates amenable to such results. Working a bit harder, one can show that the linear sieve weights then cancel with the error term for primes in arithmetic progressions up to moduli of size $x^{7/12-\epsilon}$. Moreover, recent work of Lichtman [51] shows one can modify the linear sieve construction itself to exploit newer equidistribution results profitably (the linear sieve is only optimal at exploiting the information (3.2)).

The spectacular work of Zhang [82] on bounded gaps between primes was an important application of breaking the square-root barrier (even though now we do not need such strong results to prove bounded gaps between primes), and similarly the work of Adleman–Fouvry–Heath-Brown [1,23] on Fermat's last Theorem relied crucially on ideals going beyond the $x^{1/2}$ barrier (although now we know Fermat's Last Theorem in full [75,80].) Even in the absence of a headline application, it still feels a fundamental and central problem in analytic number theory to concretely go beyond the Bombieri–Vinogradov range.

Question 15. Can we show for any a, A,

$$\sum_{\substack{q \le x^{1/2+\delta} \\ (q,a)=1}} \left| \pi(x; q, a) - \frac{\pi(x)}{\varphi(q)} \right| \ll_{a,A} \frac{x}{(\log x)^A}$$

for some fixed $\delta > 0$?

The work of Bombieri–Friedlander–Iwaniec [5–7] covered most terms which occur when performing a combinatorial decomposition of the primes, leaving one only to deal with products of j integers of size roughly $x^{1/j}$ for $j \in \{4, 5, 6\}$. The recent work [63] handles the case $j = 5$, but only obtains partial results for $j = 4$ and $j = 6$, which remain to be handled. In particular, we highlight the case $j = 4$, which appears to clearly need new ideas

Question 16. Can one obtain a nontrivial estimate for

$$\sum_{q \le x^{1/2+\delta}} \left| \sum_{\substack{n_1,n_2,n_3,n_4 \in [x^{1/4}, 2x^{1/4}] \\ (n_1 n_2 n_3 n_4, q)=1}} \left(\mathbf{1}_{n_1 n_2 n_3 n_4 \equiv 1 \pmod q} - \frac{1}{\varphi(q)} \right) \right| ?$$

6. BILINEAR ESTIMATES

Although basic sieve methods relying only on information about $\mathcal{A}$ in arithmetic progressions cannot detect primes because of the parity barrier, it is known that if you incorporate extra 'bilinear' information into the method, then you can count primes; this ultimately goes back to the pioneering work of Vinogradov [78]. For example, by inclusion–exclusion on the largest prime factor, for $\mathcal{A} \subseteq [x, 2x]$, we have

$$\#\{p \in \mathcal{A}\} = S(\mathcal{A}, z) - \sum_{z < p < x^{1/2}} S(\mathcal{A}_p, p).$$

When z is a small power of x, basic sieve methods can get good upper and lower bounds for $S(\mathcal{A}, z)$. The sum over primes counts products $pm \in \mathcal{A}$ where p and m are both larger than z, and the power of bilinear sums is that they can estimate the number of such products in $\mathcal{A}$ with very little arithmetic information required beyond that both factors are of moderate size.

To state things more precisely, it is often easiest to compare the set $\mathcal{A}$ of interest with a simpler set $\mathcal{B}$ where we know how to count primes using techniques from multiplicative number theory, but which is expected to have similar distributional properties. For example, if $\mathcal{A} = [x, x + x^{\theta}]$ is a short interval, then we might take $\mathcal{B} = [x, x + x\exp(-\sqrt{\log x})]$ to be a long interval. A slight extension of (3.2) is then

$$\sum_{m \sim M} \alpha_m \sum_{n \in \mathcal{I}} \left(\mathbf{1}_{nm \in \mathcal{A}} - \frac{\#\mathcal{A}}{\#\mathcal{B}} \mathbf{1}_{nm \in \mathcal{B}} \right) \ll_A \frac{\#\mathcal{A}}{(\log x)^A} \tag{6.1}$$

for any 1-bounded sequence α_m, constant A, interval $\mathcal{I}$, and any $M < x^{\gamma}$. With this formulation, we can consider similar variants, in particular the estimate

$$\sum_{m \sim M} \sum_{n} \alpha_n \beta_m \left(\mathbf{1}_{nm \in \mathcal{A}} - \frac{\#\mathcal{A}}{\#\mathcal{B}} \mathbf{1}_{mn \in \mathcal{B}} \right) \ll_A \frac{\#\mathcal{A}}{(\log x)^A} \tag{6.2}$$

for all 1-bounded sequences α_n, β_m.

We call (6.1) a 'Type I' estimate, and (6.2) a 'Type II' or 'bilinear' estimate for $\mathcal{A}$.

One should interpret condition (6.2) as saying that we can obtain an asymptotic formula for products with some prescribed prime factorization, provided these factorizations always contain a divisor of a convenient size.

Naturally, (6.2) is typically much harder to establish, and proving a nontrivial Type II estimate is normally the key technical difficulty which needs to be overcome if wanting to prove the existence of primes in some set $\mathcal{A}$. For example, we would like to establish fairly good Type I estimates for the sets mentioned in the introduction, but we currently do not know how to estimate Type II sums for most of the outstanding open problems on primes.

Question 17 (Type II estimates for twin primes). Can one estimate a Type II sum associated to Twin Primes such as

$$\sum_{n\sim N}\sum_{m\sim M}\alpha_n\beta_m\Lambda(nm+2)$$

for arbitrary 1-bounded sequences α_n, β_m?

One might also try to reduce both prime variables to bilinear terms, but sums such as

$$\sum_{n\sim N}\sum_{\substack{m\sim M\\ nm+2=rs}}\sum_{r\sim R}\sum_{s\sim S}\alpha_n\beta_m\gamma_r\delta_s$$

also appear infeasible to handle. (The natural Cauchy–Schwarz argument leads to conditions like $n_1s_2 - s_2n_1 = d$ for some $d|2n_2 - 2n_1$, and little appears to have been gained.)

Note that (6.2) cannot be expected to hold if $\mathcal{A}$ has a lot of multiplicative structure in the sense that information about n tells us a lot about which m's can have $nm \in \mathcal{A}$. This is to be expected – if $\mathcal{A}$ contained only numbers with an even number of prime factors, for example, then we could not hope to produce primes and so we expect that we cannot produce good Type II estimates. In this case the parity of the number of prime factors of n would dictate the parity of the number of prime factors of m, and so by choosing α_n, β_m to account for these we would give a counterexample to the bound (6.2). Indeed, (6.2) can be thought of as ruling out such multiplicative conspiracies, so that the arithmetic nature of n and m over products $nm \in \mathcal{A}$ is being 'independent on average'.

Although (6.2) is ruling out a certain amount of multiplicative structure within $\mathcal{A}$, somewhat perversely we are typically only able to estimate Type II terms effectively if $\mathcal{A}$ has some different multiplicative structure which we are able to exploit to show that the factors n, m behave somewhat independently of one another. For example, after some initial massaging one typically attempts to prove a Type II estimate via Cauchy–Schwarz to eliminate one of the unknown sets of coefficients (there is typically little lost in doing this, since we cannot rule out $\alpha_n = \operatorname{sgn}(\sum_m \beta_m \mathbf{1}_{nm\in\mathcal{A}})$), leaving us to estimate a quantity like

$$\#\{n \sim x/M : m_1n \in \mathcal{A}, m_2n \in \mathcal{A}\}. \qquad (6.3)$$

If we can estimate this quantity reasonably accurately (and the diagonal terms with $m_1 = m_2$ do not dominate), then we should be optimistic of obtaining a Type II estimate. It is precisely the difficulty of estimating quantities like (6.3) which limits our ability to apply Type I/II methods.

6.1. Type I/II ranges to primes

We first introduce some general notation to talk about sets where we can estimate the bilinear Type II sums in certain ranges at least.

Definition (Type I/II ranges). Given $\mathcal{A}, \mathcal{B} \subseteq [x, 2x]$:

- We say that $\mathcal{A}$ satisfies a Type I range of $[0, \gamma]$ if (6.1) holds for all choices of $M \le x^\gamma$ (for all $A > 0$, all intervals $\mathcal{I}$ and all 1-bounded sequences α_m).
- We say that $\mathcal{A} \subseteq [x, 2x]$ satisfies a Type II range of $[\alpha, \beta]$ if (6.1) holds for all choices of $M \in [x^\alpha, x^\beta]$ (for all $A > 0$ and all 1-bounded sequences α_m, β_n).

We typically suppress mentioning $\mathcal{B}$, since we assume that $\mathcal{B}$ is a simple set like $[x, 2x]$ in which we can count primes well.

Since we think of $\mathcal{A} \subseteq [x, 2x]$, we see that by switching the roles of n, m if $\mathcal{A}$ satisfies a Type II range of $[\alpha, \beta]$ then it also has a Type II range of $[1-\beta+\epsilon, 1-\alpha-\epsilon]$ for any $\epsilon > 0$.

A key basic result, is that if we have 'enough' Type I/II arithmetic information, then we can count primes in $\mathcal{A}$.

Lemma 18 (Vaughan's identity). *Let $\mathcal{A}$ satisfy a Type I range of $[0, \gamma]$ and a Type II range of $[\alpha, \alpha+\beta]$. If $\beta + \gamma > 1$ then we have*

$$\#\{p \in \mathcal{A}\} = \frac{\#\mathcal{A}}{\#\mathcal{B}} \#\{p \in \mathcal{B}\}\big(1 + o(1)\big).$$

(This formulation is somewhat different from typical statements of Vaughan's identity. Ignoring some minor technical considerations to do with separating variables and removing log-coefficients, it follows from choosing $U = x^\alpha$, $V = x^{1-\alpha-\beta}$ in [11, CHAPTER 24], for example.)

Therefore, if the length of the Type I range plus the length of the Type II range is bigger than 1, we can obtain an asymptotic formula for primes in $\mathcal{A}$. Unfortunately, if $\mathcal{A}$ satisfies some Type I/II estimates but the combined lengths are not bigger than 1, we cannot necessarily obtain an asymptotic formula for primes and the precise Type I/II regions when we can produce primes becomes a more subtle arithmetic-combinatorial question.

Although we are only considering sets $\mathcal{B}$ which are 'simple' (and so contain many primes), essentially the same arguments allow us to show that conclusions of Lemma 18 hold even if $\mathcal{B}$ is a more complicated set. Thus in principle these techniques can show different sets $\mathcal{A}, \mathcal{B}$ contain the roughly same number of primes, even if we are unable to establish precisely how many primes there are in either set. In this way the results are 'independent' of the Prime Number Theorem, but are not 'producing' primes.

For many applications, we merely wish to prove the existence of primes in $\mathcal{A}$. Therefore even if we do not have sufficient Type I/II ranges to obtain an asymptotic formula, we might still be able to obtain a nontrivial lower bound for the number of primes in $\mathcal{A}$. Methods to do this were gradually developed [39,46] culminating in Harman's sieve [35]. This allowed one to exploit positivity to drop inconvenient terms and obtain a lower bound of the correct order of magnitude, provided one still had suitably large Type I and Type II ranges. Given this, the strategy for proving the existence of primes in $\mathcal{A}$ then becomes the following:

(1) Establish a Type I estimate in as large a range as possible;

(2) Establish a Type II estimate in as large a range as possible;

(3) Use a sieve decomposition to verify the Type I/II information established is sufficient to obtain a non-trivial lower bound for primes in $\mathcal{A}$.

With this is mind, we define the upper and lower bound functions $L(\alpha, \beta, \gamma)$ and $U(\alpha, \beta, \gamma)$, obtained from an optimal translation of this arithmetic information.

Definition (Optimal constants in Harman's sieve). For given fixed constants $\alpha, \beta, \gamma \in [0, 1]$, and $\mathcal{B} = [x, 2x]$:

- Let $L_x(\alpha, \beta, \gamma)$ denote the infimum of $\pi(\mathcal{A}) \log x / \#\mathcal{A}$ over all sets $\mathcal{A} \subseteq [x, 2x]$ satisfying a Type I range $[0, \gamma]$ and a Type II range $[\alpha, \alpha + \beta]$. Let $L(\alpha, \beta, \gamma) = \liminf_{x \to \infty} L_x(\alpha, \beta, \gamma)$.
- Let $U_x(\alpha, \beta, \gamma)$ denote the supremum of $\pi(\mathcal{A}) \log x / \#\mathcal{A}$ over all sets $\mathcal{A} \subseteq [x, 2x]$ satisfying a Type I range $[0, \gamma]$ and a Type II range $[\alpha, \alpha + \beta]$. Let $U(\alpha, \beta, \gamma) = \limsup_{x \to \infty} U_x(\alpha, \beta, \gamma)$.

Clearly, $0 \le L(\alpha, \beta, \gamma) \le U(\alpha, \beta, \gamma)$. Moreover, assuming that $\gamma > 0$, we have that $U(\alpha, \beta, \gamma) \ll 1$ from Lemma 5. If $\gamma > 1/2$ then we know that $L(\alpha, \beta, \gamma)$ and $U(\alpha, \beta, \gamma)$ will be continuous functions on $[0, 1]^3$; we expect them to be piecewise smooth and continuous everywhere.

In many problems, we are most interested in showing the existence of primes in $\mathcal{A}$, which would follow if $\mathcal{A}$ satisfied (6.1) and (6.2) for some α, β, γ such that $L(\alpha, \beta, \gamma) > 0$. Therefore a crucial open question is the following.

Question 19. For which choices of α, β, γ do we have $L(\alpha, \beta, \gamma) > 0$?

The machinery of Harman's sieve allows one to compute a numerical lower bound for $L(\alpha, \beta, \gamma)$ (or an upper bound for $U(\alpha, \beta, \gamma)$) for given constants α, β, γ in terms of various multidimensional integrals, but the lower bound is not guaranteed before time to be positive. It is slightly unsatisfying that the computations often have to rely on a moderate amount of explicit numerical calculation of integrals and the decompositions need to be done by hand, but empirically this typically works well. If one has a moderately large constant γ for the Type I range, then in practice we can often succeed in showing a positive lower bound even when β is as small as $1/20$ or $1/30$, and often (but not always) an argument which produces a nontrivial Type II range will produce one of an adequate length. It is the empirical fact that one can get a nontrivial lower bound via Harman's sieve even with quite limited Type II ranges which makes it very applicable.

That said, it would be desirable to have a much better understanding of the optimal ways to apply Harman's sieve, the optimal constants which come out, and what sort of sets we would need to distinguish ourselves from if we wanted to produce stronger results.

Question 20. Given constants α, β, γ, what are the optimal values $L(\alpha, \beta, \gamma)$ and $U(\alpha, \beta, \gamma)$? What are the sets which achieve these maxima and minima?

Work-in-progress [21] makes some first steps to understanding optimality in Harman's sieve, but the general picture appears to be arithmetically quite subtle (much more so than for the linear sieve bounds) and combinatorially quite involved.

If we have some nontrivial arithmetic information about $\mathcal{A}$, but we know that $\mathcal{A}$ does not contain the expected number of primes, then we know that this must be compensated by $\mathcal{A}$ also containing a different number of products of r primes, for some small value of r.

7. PRIMES IN THIN SETS

One particularly challenging situation which encompasses many important situations is when the set $\mathcal{A}$ in question contains $O(x^{1-\theta})$ elements in $[x, 2x]$ for some fixed $\theta > 0$. In this case $\mathcal{A}$ is a sparse subset of the integers, and there are limitations on what sort of Type I and Type II information one could hope to establish even in the most optimistic scenarios.

Trivially, $\#\mathcal{A}_d$ is an integer, and so we can only hope for the approximation $\#\mathcal{A}_d \approx g(d)\#\mathcal{A}$ to be accurate when $d < \#\mathcal{A}$, which limits our Type I range to $\gamma \leq 1 - \theta$. Similarly, for typical $n \sim N$ there should be roughly $\#\mathcal{A}/N$ choices of $m \sim x/N$ with $mn \in \mathcal{A}$, and so we can only hope to obtain a nontrivial estimate for $\sum_{m:mn\in\mathcal{A}} \beta_m$ if $N < \#\mathcal{A}$. This limits our Type II range to $\alpha \geq \theta$. Finally, if we attempt to estimate our Type II sums by following the standard Cauchy–Schwarz strategy of estimating

$$\#\{n \sim N : m_1 n \in \mathcal{A}, m_2 n \in \mathcal{A}\}, \tag{7.1}$$

then (for generic m_1, m_2) we would expect this count to be roughly $N\#\mathcal{A}^2/x^2$. For this to be typically greater than 1, this would limit us to $N > x^2/\mathcal{A}^2 = x^{2\theta}$, and so $\alpha + \beta < 1 - 2\theta$ in our Type II range. Thus if $\mathcal{A} \subseteq [x, 2x]$ with $\#\mathcal{A} = x^{1-\theta}$, in the absence of more sophisticated methods we expect to be limited to a Type I range of $[0, x^{1-\theta}]$ and a Type II range of $[x^\theta, x^{1-2\theta}]$. In particular, this range would be sufficient to obtain an asymptotic formula via Vaughan's identity if $\#\mathcal{A} > x^{3/4}$, but we would expect to fail to obtain any Type II information at all if $\#\mathcal{A} < x^{2/3}$.

In various favorable situations we can obtain Type I and Type II estimates of this strength.

(1) Let $\mathcal{A} = \{n \sim x : \|\alpha n + \beta\| < n^{-\theta}\}$ for given irrationals α, β, corresponding to the question of inhomogeneous Diophantine approximation by primes. In this

situation, $\#\mathcal{A} = x^{1-\theta+o(1)}$ and it follows from work of Vaughan [76] that one can obtain a Type I range $[0, 1-\theta]$ and a Type II range $[\theta, 1-2\theta]$, therefore covering essentially the full range.

(2) Let $N(x_1 + x_2\sqrt[n]{a} + \cdots + x_{n-k}\sqrt[n]{a^{n-k-1}})$ be the incomplete norm form associated to the Kummer extension $\mathbb{Q}(\sqrt[n]{a})$, and $\mathcal{A}$ be the value set of N on $[1, x^{1/n}]^n$. Since N is a degree n polynomial in $n-k$ variables, $\mathcal{A}$ contains roughly $x^{1-k/n}$ in $[x, 2x]$ and so is a thin set of integers. In [62] we obtain a Type I range $[0, 1-k/n]$ and a Type II range $[k/n, 1-2k/n]$, therefore corresponding to the optimistic basic estimates above.

Jia [48] showed that if $\theta < 9/28$ then Harman's sieve can produce a lower bound of the correct order of magnitude for the number of primes in a set $\mathcal{A}$ satisfying a Type I estimate $[0, 1-\theta]$ and a Type II estimate $[\theta, 1-2\theta]$.

In some situations one can exploit extra structure of the problem to obtain slightly wider Type II estimates. One might hope to obtain cancellations in the error terms $E(m_1, m_2)$ occurring in estimating $\#\{n \sim N : m_1 n \in \mathcal{A}, m_2 n \in \mathcal{A}\}$, for example, which might allow one to have a Type II range beyond $1 - 2\theta$.

(1) Let $\mathcal{A} = \{n \in [x, x + x^{1-\theta}]\}$, so we are investigating primes in short intervals. In Section 2 we saw that we can use zero-density methods to obtain an asymptotic formula for $\theta < 5/12$ (note that $5/12 > 1/3$, so this is much sparser than the examples above). By using Dirichlet polynomials, we can actually obtain nontrivial arithmetic information for this problem whenever $\theta > 1/2$ (although we can only obtain Type II style estimates for coefficients of special types corresponding to convolutions of 3 rather than 2 sequences). By combining these estimates for triple convolutions (and more) with Harman's sieve we can unconditionally show the existence of primes in intervals $[x, x + x^{0.525+o(1)}]$ [2], which is only an exponent only slightly worse than what we would obtain under the Riemann Hypothesis. The most powerful arithmetic input is Watt's mean value Theorem [79] – it would be very desirable to have some new arithmetic estimates which could apply to these short interval problems, but currently our techniques do not seem able to go beyond Watt's work.

(2) Let $\mathcal{A} = \{a^3 + 2b^3 : a, b < x^{1/3}\}$. After switching to prime ideals, Heath-Brown [38] is essentially able to classify those m_1, m_2 for which there is an n with $m_1 n, m_2 n \in \mathcal{A}$ since such n can be given explicitly in terms of m_1, m_2, and then obtain suitable cancellations over these special pairs m_1, m_2. This enables him to obtain a Type II range $[1/3, 1/2]$, which is sufficient for obtaining an asymptotic formula for primes represented by $a^3 + 2b^3$, even though this only contains roughly $x^{2/3}$ elements in $[1, x]$. Li [50] is able to generalize this to further restrict b to be small, allowing him to handle sets even sparser than this.

(3) Let $\mathcal{A} = \{n \sim x : \|\alpha n\| < x^{-1/3+\epsilon}\}$. Then $\mathcal{A}$ contains roughly $x^{2/3+\epsilon}$ integers of size x, but nevertheless Matomäki [53] (building on [40]) was able to show that $\mathcal{A}$ still contained primes by establishing nontrivial arithmetic information in wider ranges. Again, to establish these wider ranges she needed to consider trilinear sums.

In a slightly different direction in [61] Type II estimates were deduced by exploiting a very nice Fourier structure in the underlying set. This is an example where the set does not have obvious 'linear structure' (such as short intervals, or the distribution of αn modulo one), and does not lack obvious multiplicative structure which makes it more feasible to estimate (7.1), but nevertheless nontrivial arithmetic information can be established (in this case within the Hardy–Littlewood circle method). It would be interesting to add to this example.

We mention in passing the recent work of Heath-Brown–Li [41] on primes of the form $X^2 + p^4$ and Merikoski [67] on $X^2 + (Y^2 + 1)^2$ and Xiao [81] on primes of the form $f(a, b^2)$ for binary quadratic forms f all generalizing the work of Friedlander–Iwaniec on $X^2 + Y^4$ [27].

Even with these proof-of-concept results that in principle one can establish some sort of nontrivial arithmetic information with fairly general coefficient sequences in some sparse sets, all approaches seem to break down completely when considering sets containing fewer than $x^{1/2}$ elements in $[x, 2x]$.

Question 21. Is there a plausible way to adapt Type I/II machinery to apply to very sparse sets with $x^{1/2-\epsilon}$ elements in $[x, 2x]$?

Without some advance in this direction, we do seem to have any means of counting primes in intervals of length smaller than $x^{1/2}$, and thereby addressing Legendre's conjecture on the existence of a prime between consecutive squares. Of course, we expect there to be primes in much shorter intervals (as short as $(\log x)^{2+o(1)}$), but going beyond $x^{1/2}$ seems out of reach for now, even if we assume the Riemann Hypothesis and things like Montgomery's Pair Correlation Conjecture [68].

8. FURTHER ARITHMETIC INFORMATION

Even if the Type I/Type II arithmetic information in insufficient for generating primes (or asymptotic formulae for primes), we can sometimes remedy the situation by incorporating further arithmetic information into the method.

For example, we mentioned in Section 7 that for the problem of primes in short intervals or for small values of αp modulo one it was important that there was additional flexibility to consider triple convolutions of sequences, rather than just bilinear sums. Often we find that the size of factors of terms produced in a decomposition of the primes is the key feature – when terms factor in a convenient manner one can produce much stronger results.

As well as higher order convolutions (corresponding to assuming some factorization properties of the sequences α_n or β_m) we can also exploit the fact that sometimes we are able to produce stronger results if some of the sequences involved are just the constant 1. For example, we have Linnik's identity [52]

$$\frac{\Lambda(n)}{\log n} = -\sum_{j=1}^{\infty} \frac{(-1)^j}{j} \tau_j'(n)$$

where $\tau_j'(n)$ counts representations of n as the product of j integers all bigger than 1. In principle this allows us to understand primes in $\mathcal{A}$ by understanding the average of $\tau_j'(n)$ for $n \in \mathcal{A}$. Understanding $\tau_j'(n)$ is similar to understanding j-fold convolutions in $\mathcal{A}$, therefore generalizing our linear and bilinear sums. Moreover, in this formulation the coefficients of each of the j factors is just 1 rather than some unknown sequence. This additional flexibility of only needing to consider smooth coefficient sequences is difficult to exploit unless some of the variables are very long like in the case of Type I estimates (and for practical applications Heath-Brown's identity [36] is often more convenient to use), but is crucial in some situations. For example, the recent work [63–65] on primes in arithmetic progressions crucially relied on estimates for the divisor function in arithmetic progressions and for $\tau_3(n)$ in arithmetic progressions [24, 25, 37].

One further comment is that the coefficients which naturally occur from Buchstab iterations are the indicator function of products of primes, where each prime is of a roughly fixed size. This means that rather than requiring estimates like (6.2) for arbitrary sequences, we only really require this when α_n and β_m look like the indicator function of primes, or products of primes. In the ground-breaking work of Friedlander–Iwaniec on $X^2 + Y^4$ representing primes [26, 27] the fact that the coefficients satisfied a suitable Siegel–Walfisz Theorem was crucial, and so the Type II estimates were only valid for this reduced class of coefficients.

One simple observation is that $\tau_j(n)$ are the coefficients of the degree j L-function $\zeta(s)^j$. There is a general principle that often estimates which can be obtained in a direct manner for $\tau(n)$ can be also obtained in a more complicated manner for the Fourier coefficients of suitable cusp forms via the spectral theory of automorphic forms. It is therefore compelling to speculate whether this would allow for further 'higher degree' arithmetic information to be incorporated.

Question 22. Can one use coefficients of other higher degree L-functions to aid counting primes?

Work of Drappeau–Maynard [14] made crucial use of the Sato–Tate distribution of Kloosterman sums to enable an estimation of a sum over primes, where arithmetic properties of the underlying sequence essentially reduced the sieve dimension. Since Fourier coefficients have similar distributional features, one might hope that this simple example could be indicative of a wider approach.

9. CHOICE OF LIFT AND COMPARISON SETS

When attempting to count primes in $\mathcal{A}$ using the Type I/II sums strategy, one wants to understand a sum

$$\sum_{p\in\mathcal{A}} a_p$$

over *primes*, and we study this by gaining arithmetic information (such as Type I/II estimates) for a sequence a_n over *integers* $n \in \mathcal{A}$. We therefore choose a *lift* of the sequence a_p supported on primes to the sequence a_n supported on integers which hopefully is more amenable to estimation. In many contexts there is a natural choice of a_n which works well (e.g., $a_p = 1$ and $a_n = 1$), but one could imagine other choices also being worthy of consideration (or perhaps multiple different lifts). For example, if one could understand the sums with $a_n = 2/\tau(n)$, then one would have a lift of the sequence $a_p = 1$ which would remain closer to the primes, and it would be correspondingly easier to detect primes given the same basic arithmetic information (it would be reducing the sieve dimension). So far our estimates appear to have been limited to the simplest possible choices, but it is natural to ask if this is really necessary.

Question 23. Are there situations where other lifts a_n of the sequence a_p can aid estimating primes?

As a very basic proof-of-concept, in some situations it is easier to lift $a_p = 1$ to $a_n = \theta(n)$ where $\theta(n)$ is a sieve weight ensuring that a_n behaves as if it is supported only on small prime factors. But ideally we would find a nontrivial way to lift to a sequence sensitive to all prime factors of n, not just small ones.

In (6.1) and (6.2) we compare arithmetic counts in a set $\mathcal{A}$ to a simpler set $\mathcal{B}$, but the choice of $\mathcal{B}$ is left to the application at hand. In most cases $\mathcal{B}$ is a truly simple set (such as an interval) where something like the Prime Number Theorem can be applied directly. However, in some cases it is advantageous (or important) to have more complicated comparison sets (or one could generalize to a weighted sequence). For example, in looking at primes in arithmetic progressions to large moduli, it is useful to compare the indicator function of the residue class $\mathbf{1}_{n\equiv a \pmod{q_1q_2}}$ *not* with the basic choice of 1 (or $\mathbf{1}_{(n,q_1q_2)=1}$), but with the 'intermediate complexity' sequences $\mathbf{1}_{n\equiv a \pmod{q_1}}$. This allows us to use additive Fourier analysis to show that $\mathbf{1}_{n\equiv a \pmod{q_1q_2}} \approx \mathbf{1}_{n\equiv a \pmod{q_1}}$ in some average sense, and then use multiplicative Fourier analysis (Dirichlet characters) to show that $\mathbf{1}_{n\equiv a \pmod{q_1}} \approx 1$. Therefore we are going through a two-step approximation process, and exploiting in a crucial manner that $\mathbb{Z}/q_1\mathbb{Z}$ is a subgroup of $\mathbb{Z}/q_1q_2\mathbb{Z}$.

Question 24. When is it helpful to use more complicated intermediate comparison sequences $\mathcal{B}$?

It would be very interesting if we could weaken the requirement that $\mathbb{Z}/q_1q_2\mathbb{Z}$ has a suitably sized subgroup for the arguments to apply.

In various works Drappeau [12, 13] has shown that it can be valuable to retain various possible secondary main terms in applications of Linnik's dispersion method, which corresponds to it being somewhat advantageous to choose a more complicated comparison set $\mathcal{B}$. (A similar feature was used in **[62]** to help account for Siegel-zero issues.) These can be thought of as examples of intermediate sequences $\mathcal{B}$ which are taking into account the possible causes of fluctuations of the number of primes in $\mathcal{A}$.

10. ABELIAN QUADRATIC LIMITATIONS

One limitation in many methods for counting primes is that we cannot rule out zeros of L-functions very close to the line $\mathrm{Re}(s) = 1$, and so even in the simplest situations such as counting primes in $[1, x]$ we cannot obtain an error term better than some exponential log factor.

One curious feature is that often the more involved counting arguments (such as Type I/II estimates) actually come with much stronger error terms (such as giving a power saving) whenever the estimate can be achieved. For example, the classical exponential sum bound shows that

$$\sum_{n<x} \Lambda(n)e(n\alpha) \ll x^{1-\epsilon}$$

unless $\alpha \approx a/q$ for some $q < x^{2\epsilon}(\log x)^{O(1)}$, in which case the possible existence of a Siegel zero would prevent a power-saving estimate.

Similarly, the error term in the Titchmarsh divisor problem of estimating $\sum_{p<x} \tau(p-1)$ is fundamentally limited by the possible existence of Siegel-zeros (see **[13]**), but for the analogue of this problem with (normalized) Fourier coefficients of holomorphic cusp forms of $\mathrm{PSL}_2(\mathbb{Z})$, we obtain a power-saving estimate $\sum_{p<x} a(p-1) < x^{391/392+o(1)}$ due to work of Pitt **[70]**.

The 'higher order Fourier analysis' pioneered by Green and Tao **[32]** involves looking at sums over primes twisted by nilsequences. Again, it is the case that it is ultimately *easier* to obtain quantitative cancellation for nilsequences when the nilsequence is suitably far from a rational phase; the limits of the results stem from possible zeros of Dirichlet L-functions (see, for example, the discussion after **[33, THEOREM 1]**). Other examples of this occur in the more recent work **[55,73]** where the ultimately key limitations to estimates are when a nilsequence is 'close' to encoding a rational phase, reducing to the classical situation.

In a slightly different direction, for many situations involving higher degree L-functions it is known that the issue of zeros very close to $s = 1$ cannot arise; Siegel zeros are essentially only a phenomenon which could arise for quadratic Dirichlet L-functions, and so we can have better results in these more complicated scenarios (unless quadratic Dirichlet character could be lurking under the surface, such as if we consider a Dedekind L-function for a number field with an index 2 – so quadratic – subfield).

In all these cases estimates for primes which at first sight seem harder that the classical setting actually avoid the limitations from the well-known obstacles and so prove to actually be easier in some sense.

ACKNOWLEDGMENTS

The author is supported by a Royal Society Wolfson Merit Award, and this project has received funding from the European Research Council (ERC) under the European Union's Horizon 2020 research and innovation programme (grant agreement No 851318).

REFERENCES

[1] L. M. Adleman and D. R. Heath-Brown, The first case of Fermat's last theorem. *Invent. Math.* **79** (1985), no. 2, 409–416.

[2] R. C. Baker, G. Harman, and J. Pintz, The difference between consecutive primes. II. *Proc. Lond. Math. Soc. (3)* **83** (2001), no. 3, 532–562.

[3] R. C. Baker and A. J. Irving, Bounded intervals containing many primes. *Math. Z.* **286** (2017), no. 3–4, 821–841.

[4] E. Bombieri, On the large sieve. *Mathematika* **12** (1965), 201–225.

[5] E. Bombieri, J. B. Friedlander, and H. Iwaniec, Primes in arithmetic progressions to large moduli. *Acta Math.* **156** (1986), no. 3–4, 203–251.

[6] E. Bombieri, J. B. Friedlander, and H. Iwaniec, Primes in arithmetic progressions to large moduli. II. *Math. Ann.* **277** (1987), no. 3, 361–393.

[7] E. Bombieri, J. B. Friedlander, and H. Iwaniec, Primes in arithmetic progressions to large moduli. III. *J. Amer. Math. Soc.* **2** (1989), no. 2, 215–224.

[8] J. Bourgain, P. Sarnak, and T. Ziegler, Disjointness of Moebius from horocycle flows. In *From Fourier analysis and number theory to Radon transforms and geometry*, pp. 67–83, Dev. Math. 28, Springer, New York, 2013.

[9] J.-R. Chen, On the representation of a larger even integer as the sum of a prime and the product of at most two primes. *Sci. Sin.* **16** (1973), 157–176.

[10] H. Daboussi and H. Delange, On multiplicative arithmetical functions whose modulus does not exceed one. *J. Lond. Math. Soc. (2)* **26** (1982), no. 2, 245–264.

[11] H. Davenport, *Multiplicative number theory. Second edn*. Grad. Texts in Math. 74, Springer, New York–Berlin, 1980.

[12] S. Drappeau, Théorèmes de type Fouvry–Iwaniec pour les entiers friables. *Compos. Math.* **151** (2015), no. 5, 828–862.

[13] S. Drappeau, Sums of Kloosterman sums in arithmetic progressions, and the error term in the dispersion method. *Proc. Lond. Math. Soc. (3)* **114** (2017), no. 4, 684–732.

[14] S. Drappeau and J. Maynard, Sign changes of Kloosterman sums and exceptional characters. *Proc. Amer. Math. Soc.* **147** (2019), no. 1, 61–75.

[15] N. Elkies, The existence of infinitely many supersingular primes for every elliptic curve over $\mathbb{Q}$. *Invent. Math.* **89** (1987), no. 3, 561–567.

[16] P. D. T. A. Elliott and H. Halberstam, A conjecture in prime number theory. In *Symposia Mathematica, Vol. IV (INDAM, Rome, 1968/69)*, pp. 59–72, Academic Press, London, 1970.

[17] P. Erdős, The difference of consecutive primes. *Duke Math. J.* **6** (1940), 438–441.

[18] K. Ford, Large prime gaps and progressions with few primes. *Riv. Mat. Univ. Parma (N.S.)* **12** (2021), no. 1, 41–47.

[19] K. Ford, B. Green, S. Konyagin, J. Maynard, and T. Tao, Long gaps between primes. *J. Amer. Math. Soc.* **31** (2018), no. 1, 65–105.

[20] K. Ford, B. Green, S. Konyagin, and T. Tao, Large gaps between consecutive prime numbers. *Ann. of Math. (2)* **183** (2016), no. 3, 935–974.

[21] K. Ford and J. Maynard, *An optimal Harman sieve*. In preparation.

[22] E. Fouvry, Autour du théorème de Bombieri–Vinogradov. *Acta Math.* **152** (1984), no. 3–4, 219–244.

[23] E. Fouvry, Théorème de Brun–Titchmarsh: Application au théorème de Fermat. *Invent. Math.* **79** (1985), no. 2, 383–407.

[24] E. Fouvry, E. Kowalski, and P. Michel, On the exponent of distribution of the ternary divisor function. *Mathematika* **61** (2015), no. 1, 121–144.

[25] J. B. Friedlander and H. Iwaniec, Incomplete Kloosterman sums and a divisor problem. *Ann. of Math. (2)* **121** (1985), no. 2, 319–350.

[26] J. B. Friedlander and H. Iwaniec, Asymptotic sieve for primes. *Ann. of Math. (2)* **148** (1998), no. 3, 1041–1065.

[27] J. B. Friedlander and H. Iwaniec, The polynomial $X^2 + Y^4$ captures its primes. *Ann. of Math. (2)* **148** (1998), no. 3, 945–1040.

[28] J. B. Friedlander and H. Iwaniec, *Opera de cribro*. Amer. Math. Soc. Colloq. Publ. 57, American Mathematical Society, Providence, RI, 2010.

[29] D. Goldfeld, The class number of quadratic fields and the conjectures of Birch and Swinnerton-Dyer. *Ann. Sc. Norm. Super. Pisa Cl. Sci. (4)* **3** (1976), no. 4, 624–663.

[30] D. Goldfeld, Gauss's class number problem for imaginary quadratic fields. *Bull. Amer. Math. Soc. (N.S.)* **13** (1985), no. 1, 23–37.

[31] A. Granville, Sieving intervals and Siegel zeros. 2020, arXiv:2010.01211.

[32] B. Green and T. Tao, Linear equations in primes. *Ann. of Math. (2)* **171** (2010), no. 3, 1753–1850.

[33] B. Green and T. Tao, The Möbius function is strongly orthogonal to nilsequences. *Ann. of Math. (2)* **175** (2012), no. 2, 541–566.

[34] B. Gross and D. Zagier, Heegner points and derivatives of L-series. *Invent. Math.* **84** (1986), no. 2, 225–320.

[35] G. Harman, *Prime-detecting sieves*. London Math. Soc. Monogr. Ser. 33, Princeton University Press, Princeton, NJ, 2007.

[36] D. R. Heath-Brown, Prime numbers in short intervals and a generalized Vaughan identity. *Canad. J. Math.* **34** (1982), no. 6, 1365–1377.

[37] D. R. Heath-Brown, The divisor function $d_3(n)$ in arithmetic progressions. *Acta Arith.* **47** (1986), no. 1, 29–56.

[38] D. R. Heath-Brown, Primes represented by $x^3 + 2y^3$. *Acta Math.* **186** (2001), no. 1, 1–84.

[39] D. R. Heath-Brown and H. Iwaniec, On the difference between consecutive primes. *Invent. Math.* **55** (1979), no. 1, 49–69.

[40] D. R. Heath-Brown and C. Jia, The distribution of αp modulo one. *Proc. Lond. Math. Soc. (3)* **84** (2002), no. 1, 79–104.

[41] D. R. Heath-Brown and X. Li, Prime values of $a^2 + p^4$. *Invent. Math.* **208** (2017), no. 2, 441–499.

[42] H. Helfgott, The ternary Goldbach problem. *Ann. of Math. Stud.*. To appear. 2015, arXiv:1501.05438.

[43] C. Hooley, On Artin's conjecture. *J. Reine Angew. Math.* **225** (1967), 209–220.

[44] M. Huxley, On the difference between consecutive primes. *Invent. Math.* **15** (1972), 164–170.

[45] H. Iwaniec, A new form of the error term in the linear sieve. *Acta Arith.* **37** (1980), 307–320.

[46] H. Iwaniec and M. Jutila, Primes in short intervals. *Ark. Mat.* **17** (1979), no. 1, 167–176.

[47] H. Iwaniec and E. Kowalski, *Analytic number theory*. Amer. Math. Soc. Colloq. Publ. 53, American Mathematical Society, Providence, RI, 2004.

[48] C. Jia, On the distribution of αp modulo one. II. *Sci. China Ser. A* **43** (2000), no. 7, 703–721.

[49] I. Kátai, A remark on a theorem of H. Daboussi. *Acta Math. Hungar.* **47** (1986), no. 1–2, 223–225.

[50] X. Li, Prime values of a sparse polynomial sequence. *Duke Math. J.* **171** (2022) no. 1, 101–208.

[51] J. Lichtman, A modification of the linear sieve, and the count of twin primes. 2021, arXiv:2109.02851.

[52] Y. Linnik, *The dispersion method in binary additive problems*. American Mathematical Society, Providence, RI, 1963.

[53] K. Matomäki, The distribution of αp modulo one. *Math. Proc. Cambridge Philos. Soc.* **147** (2009), no. 2, 267–283.

[54] K. Matomäki and M. Radziwiłł, Multiplicative functions in short intervals. *Ann. of Math. (2)* **183** (2016), no. 3, 1015–1056.

[55] K. Matomäki, X. Shao, T. Tao, and J. Teräväinen, Higher uniformity of arithmetic functions in short intervals I. All intervals. 2022, arXiv:2204.03754.

[56] K. Matomäki and J. Teräväinen, On the Möbius function in all short intervals. *J. Eur. Math. Soc.* To appear. 2019, arXiv:1911.09076.

[57] J. Maynard, 3-tuples have at most 7 prime factors infinitely often. *Math. Proc. Cambridge Philos. Soc.* **155** (2013), no. 3, 443–457.

[58] J. Maynard, Small gaps between primes. *Ann. of Math. (2)* **181** (2015), no. 1, 383–413.

[59] J. Maynard, Dense clusters of primes in subsets. *Compos. Math.* **152** (2016), no. 7, 1517–1554.

[60] J. Maynard, Large gaps between primes. *Ann. of Math. (2)* **183** (2016), no. 3, 915–933.

[61] J. Maynard, Primes with restricted digits. *Invent. Math.* **217** (2019), no. 1, 127–218.

[62] J. Maynard, Primes represented by incomplete norm forms. *Forum Math. Pi* **8** (2020), e3, 128.

[63] J. Maynard, Primes in arithmetic progressions to large moduli I: fixed residue classes. *Mem. Amer. Math. Soc.* To appear. 2019, arXiv:2006.06572.

[64] J. Maynard, Primes in arithmetic progressions to large moduli II: well-factorable estimates. *Mem. Amer. Math. Soc.* To appear. 2019, arXiv:2006.07088.

[65] J. Maynard, Primes in arithmetic progressions to large moduli III: uniform residue classes. *Mem. Amer. Math. Soc.* To appear. 2019, arXiv:2006.08250.

[66] J. Maynard and K. Pratt, *Half-isolated zeros and zero density estimates*. 2022, arXiv:2206.11729.

[67] J. Merikoski, The polynomials $X^2 + (Y^2 + 1)^2$ and $X^2 + (Y^3 + Z^3)^2$ also capture their primes. 2021, arXiv:2112.03617.

[68] H. L. Montgomery, The pair correlation of zeros of the zeta function. In *Analytic number theory (Proc. Sympos. Pure Math., Vol. XXIV, St. Louis Univ., St. Louis, Mo., 1972)*, pp. 181–193, 1973. Amer. Math. Soc., Providence, R.I.

[69] H. L. Montgomery and R. C. Vaughan, The large sieve. *Mathematika* **20** (1973), 119–134.

[70] N. Pitt, On an analogue of Titchmarsh's divisor problem for holomorphic cusp forms. *J. Amer. Math. Soc.* **26** (2013), no. 3, 735–776.

[71] D. H. J. Polymath, Variants of the Selberg sieve, and bounded intervals containing many primes. *Res. Math. Sci.* **1** (2014), Art. 12, 83.

[72] R. A. Rankin, The Difference between Consecutive Prime Numbers. *J. Lond. Math. Soc.* **11** (1936), no. 4, 242–245.

[73] X. Shao and J. Teräväinen, The Bombieri–Vinogradov theorem for nilsequences. *Discrete Anal.* **21** (2021), 55.

[74] T. Tao, The logarithmically averaged Chowla and Elliott conjectures for two-point correlations. *Forum Math. Pi* **4** (2016), e8, 36.

[75] R. Taylor and A. Wiles, Ring-theoretic properties of certain Hecke algebras. *Ann. of Math. (2)* **141** (1995), no. 3, 553–572.

[76] R. C. Vaughan, On the distribution of αp modulo 1. *Mathematika* **24** (1977), no. 2, 135–141.

[77] A. I. Vinogradov, The density hypothesis for Dirichet L-series. *Izv. Ross. Akad. Nauk Ser. Mat.* **29** (1965), 903–934.

[78] I. M. Vinogradov, *The method of trigonometrical sums in the theory of numbers*. Dover Publications, Inc., Mineola, NY, 2004.

[79] N. Watt, Kloosterman sums and a mean value for Dirichlet polynomials. *J. Number Theory* **53** (1995), no. 1, 179–210.

[80] A. Wiles, Modular elliptic curves and Fermat's last theorem. *Ann. of Math. (2)* **141** (1995), no. 3, 443–551.

[81] S. Y. Xiao, Prime values of $f(a, b^2)$ and $f(a, p^2)$, f quadratic. 2021, arXiv:2111.04136.

[82] Y. Zhang, Bounded gaps between primes. *Ann. of Math. (2)* **179** (2014), no. 3, 1121–1174.

JAMES MAYNARD

Mathematical Institute, Oxford, OX1 4AU, England,
james.alexander.maynard@gmail.com

This Page Intentionally Left Blank

ON DISCRETE FOURIER UNIQUENESS SETS IN EUCLIDEAN SPACE

MARYNA VIAZOVSKA

ABSTRACT

This paper presents a new construction of a discrete Fourier uniqueness set in Euclidean space.

MATHEMATICS SUBJECT CLASSIFICATION 2020

Primary 11F67; Secondary 11F41, 11G40

KEYWORDS

Fourier uniqueness, harmonic analysis

INTERNATIONAL CONGRESS OF MATHEMATICIANS 2022 JULY 6–14

© 2022 International Mathematical Union

Proc. Int. Cong. Math. 2022, Vol. 1, pp. 270–283
DOI 10.4171/ICM2022/207

Published by EMS Press

and licensed under a CC BY 4.0 license

1. INTRODUCTION

This paper gives a new construction of a closed discrete Fourier uniqueness set in $\mathbb{R}^d$. Let us start with a definition of Fourier uniqueness. For a Schwartz function $f : \mathbb{R}^d \to \mathbb{C}$, its Fourier transform is defined as

$$\hat{f}(y) := \int_{\mathbb{R}^d} f(x)\, e^{-2\pi i x y}\, dx, \quad y \in \mathbb{R}^d.$$

Definition 1.1. A set $X \subset \mathbb{R}^d$ is a *Fourier uniqueness set* if for any Schwartz function f the conditions

$$f|_X \equiv 0 \quad \text{and} \quad \hat{f}|_X \equiv 0$$

imply $f \equiv 0$.

In [4] we have shown that the set $X = \{\operatorname{sign}(n)\sqrt{|n|}\}_{n\in\mathbb{Z}}$ is essentially a uniqueness set in $\mathbb{R}$. More precisely, we have proven that the conditions $f|_X \equiv 0$, $\hat{f}|_X \equiv 0$, together with one more linear constraint $f'(0) = 0$, imply the vanishing of f on the whole real line. M. Stoller [5] has extended this result to $\mathbb{R}^d$ in the following way. For a positive real number r, let $S(r)$ denote the sphere in $\mathbb{R}^d$ with center at the origin and radius r. Stoller has proven that the set $X := \bigcup_{n=1}^{\infty} S(\sqrt{n})$ is a Fourier uniqueness set in $\mathbb{R}^d$ for $d \geq 5$. The following theorem is proven in [5].

Theorem 1.2. *Let $d \geq 5$ be an integer. Suppose that $f : \mathbb{R}^d \to \mathbb{C}$ is a Schwartz function such that $f|_{S(\sqrt{n})} \equiv 0$ and $\hat{f}|_{S(\sqrt{n})} \equiv 0$ for all $n \in \mathbb{Z}_{\geq 1}$. Then, f is identically zero.*

Moreover, Stoller and J. P. G. Ramos have recently shown the existence of a closed discrete Fourier uniqueness set in $\mathbb{R}^d$ [6, THEOREM 2, REMARK 1.1].

A natural question is: How "big" is this discrete Fourier uniqueness set? More precisely, for a closed discrete subset $X \subset \mathbb{R}^d$ we would like to analyze the function $M_X(r)$, $r \in \mathbb{R}_{>0}$, that counts the number of elements of X inside of the ball of radius r about the origin. For the Fourier uniqueness set X constructed in [6, THEOREM 2, REMARK 1.1], the function $M_X(r)$ grows superexponentially in r.

This paper aims to construct a closed discrete Fourier uniqueness set X such that the function $M_X(r)$ grows at most polynomially in r.

1.1. Construction of a discrete Fourier uniqueness set

In this paper we will show that for a family of sufficiently uniformly distributed finite subsets $X_n \subset S(1)$, $n \in \mathbb{Z}_{\geq 1}$, the union

$$X := \bigcup_{n\geq 1} \sqrt{n}\, X_n \tag{1.1}$$

is a Fourier uniqueness set. Let us give one possible quantitative description of the term "uniformly distributed."

Definition 1.3. A finite subset $X \subset S(1)$ is a *spherical design of strength* s if, for all polynomials p in d variables and total degree at most s, the following holds:

$$\int_{S(1)} p(\zeta)\, d\zeta = \frac{1}{|X|} \sum_{x \in X} p(x).$$

Here $d\zeta$ denotes the Lebesgue measure on $S(1)$ normalized so that $\int_{S(1)} 1\, d\zeta = 1$.

The main result of this paper is

Theorem 1.4. *For each dimension d, there exist positive constants $\tilde{A} = \tilde{A}(d)$ and $\tilde{B} = \tilde{B}(d)$ with the following property. If $(X_n)_{n=1}^{\infty}$ is a collection of finite subsets of $S(1)$ such that each set X_n is a spherical design of strength $\tilde{B} n^{\tilde{A}}$ then the set*

$$X := \bigcup_{n \geq 1} \sqrt{n}\, X_n$$

is a Fourier uniqueness set.

It is known [1] that for a dimension d, there exists a constant c_d such that for all nonnegative integers s, there exists a spherical design of strength s with at most $c_d\, s^d$ points. Therefore, the above theorem implies the existence of a closed discrete Fourier uniqueness set X with a polynomially bounded function $M_X(r)$.

2. AUXILIARY RESULTS FROM FOURIER ANALYSIS

Our proof of Theorem 1.4 relies on several facts from Fourier analysis and the theory of modular forms. First, we will use the following statements about the decomposition of a Schwartz function in $\mathbb{R}^d$. Let $\mathcal{H}_m = \mathcal{H}_m(\mathbb{R}^d)$ be the space of homogenous harmonic polynomials of total degree m on $\mathbb{R}^d$. Let $\mathcal{B}_m$ be an orthonormal basis of $\mathcal{H}_m$ with respect to the standard L_2 product on the unit sphere $S(1)$. Set $\mathcal{B} := \bigcup_{m \geq 0} \mathcal{B}_m$. Each Schwartz function $f : \mathbb{R}^d \to \mathbb{C}$ has the unique decomposition

$$f(x) = \sum_{p \in \mathcal{B}} p(x)\, g_p(\|x\|),$$

where g_p are radial Schwartz functions. For $p \in \mathcal{B}$, we denote

$$f_p(x) := p(x)\, g_p(\|x\|). \tag{2.1}$$

Theorem 2.1. *Let $f : \mathbb{R}^d \to \mathbb{C}$ be a Schwartz function. For $p \in \mathcal{B}$ and $n \in \mathbb{Z}_{\geq 1}$, we set*

$$\phi_{p,n} = \phi_{p,n}(f) := \sup_{x \in S(\sqrt{n})} |f_p(x)|.$$

For all $\alpha, \beta > 0$ we have

$$\sup_{p \in \mathcal{B}, n \in \mathbb{Z}_{\geq 1}} \big(\deg(p)^{\alpha}\, n^{\beta}\, \phi_{p,n}\big) < \infty.$$

Proof. We have

$$\phi_{p,n} = \sup_{x \in S(\sqrt{n})} |f_p(x)| = n^{\frac{\deg(p)}{2}} |g_p(\sqrt{n})| \sup_{\zeta \in S(1)} |p(\zeta)|.$$

The number $g_p(\sqrt{n})$ can be computed as follows:

$$\int_{S(1)} f(\sqrt{n}\zeta)\,\overline{p(\zeta)}\,d\zeta$$
$$= \int_{S(1)} g_p(\sqrt{n})\,p(\sqrt{n}\zeta)\,\overline{p(\zeta)}\,d\zeta$$
$$= n^{\frac{\deg(p)}{2}}\,g_p(\sqrt{n}). \tag{2.2}$$

Therefore

$$\phi_{p,n} = \left|\int_{S(1)} f(\sqrt{n}\zeta)\big\|p(\zeta)\big\|\,d\zeta\right| \cdot \sup_{\zeta\in S(1)} \big|p(\zeta)\big|$$
$$\le \sup_{x\in S(\sqrt{n})} \big|f(x)\big| \sup_{\zeta\in S(1)} \big|p(\zeta)\big|^2. \tag{2.3}$$

Note that there exist positive constants C_1 and C_2 depending only on dimension d such that $\sup_{\zeta\in S(1)} |p(\zeta)| \le C_1 \deg(p)^{C_2 d}$ for all $p \in \mathcal{B}$. This gives us the estimate

$$\phi_{p,n} \le C_1 \deg(p)^{2C_2} \sup_{x\in S(\sqrt{n})} \big|f(x)\big|. \tag{2.4}$$

Let β be a fixed positive number. Since f is a Schwartz function, we have

$$\sup_{x\in\mathbb{R}^d} \|x\|^\beta \,\big|f(x)\big| < \infty. \tag{2.5}$$

Estimates (2.4) and (2.5) imply

$$\sup_{p\in\mathcal{B},n\in\mathbb{Z}_{\ge 1}} \left(\deg(p)^{-2C_2}\, n^\beta\, \phi_{p,n}\right) < \infty. \tag{2.6}$$

Our next goal is to replace $-2C_2$ with an arbitrary positive constant α. Let $\Delta = \frac{\partial^2}{\partial x_1^2} + \cdots + \frac{\partial^2}{\partial x_1^2}$ be the Laplace operator on $\mathbb{R}^d$. For a point $x \in \mathbb{R}^d \setminus \{0\}$, we define its polar coordinates $r = \|x\|$ and $\zeta = \frac{x}{\|x\|}$. Consider the following differential operator:

$$\Delta_{S^{d-1}} f := r^2\,\Delta f - (d-1)\,r\,\frac{\partial}{\partial r}\,f - r^2\,\frac{\partial^2}{\partial r^2}\,f.$$

An important property of this operator is that it maps Schwartz functions to Schwartz functions. Indeed, we compute in polar coordinates $x = r\,\zeta$ that

$$r\frac{\partial}{\partial r} f(r\zeta_1,\ldots,r\zeta_d) = r\,\zeta_1\,\frac{\partial}{\partial x_1} f + \cdots + r\,\zeta_d\,\frac{\partial}{\partial x_d} f$$
$$= x_1\,\frac{\partial}{\partial x_1} f + \cdots + x_d\,\frac{\partial}{\partial x_d} f$$

and, analogously,

$$r^2\frac{\partial^2}{\partial r^2} f(r\zeta_1,\ldots,r\zeta_d) = r^2\,\zeta_1^2\,\frac{\partial^2}{\partial x_1^2} f + \cdots + r^2\,\zeta_d^2\,\frac{\partial^2}{\partial x_d^2} f$$
$$= x_1^2\,\frac{\partial^2}{\partial x_1^2} f + \cdots + x_d^2\,\frac{\partial^2}{\partial x_d^2} f.$$

Thus, if f is a Schwartz function, so is $\Delta_{S^{d-1}} f$. Suppose that g is a radial Schwartz function and p is a homogenous harmonic polynomial on $\mathbb{R}^d$ of total degree $\deg(p)$. Then a straightforward computation shows that

$$\Delta_{S_d}\big(g(r)\,p(x)\big) = -\deg(p)\,\big(\deg(p)+d-2\big)\,g(r)\,p(x). \tag{2.7}$$

We define $\lambda_m := -m(m+d-2)$. Clearly, $|\lambda_m| \sim m^2$ as m goes to infinity.

Now let α be a positive integer. Given a Schwartz function f, we define a new Schwartz function $\tilde{f} := \Delta^{\alpha}_{S^{d-1}} f$. Suppose that f has decomposition $f = \sum_{p\in\mathcal{B}} f_p$, then by equation (2.7) the new function $\tilde{f}$ has decomposition $\tilde{f} = \sum_{p\in\mathcal{B}} \tilde{f}_p$ where $\tilde{f}_p = \lambda^{\alpha}_{\deg(p)} f_p$. Also the numbers $\widetilde{\phi}_{p,n} := \max_{x\in S(\sqrt{n})} |\tilde{f}_p(x)|$ satisfy

$$\widetilde{\phi}_{p,n} = |\lambda_{\deg(p)}|^{\alpha}\,\phi_{p,n}.$$

Finally, we apply estimate (2.6) to the function $\tilde{f}$ and derive

$$\sup_{p\in\mathcal{B},n\in\mathbb{Z}_{\geq 1}} \big(\deg(p)^{2\alpha-2C_2}\,n^{\beta}\,\phi_{p,n}\big) < \infty$$

for arbitrary positive α and β. This finishes the proof of the theorem. ■

3. AUXILIARY RESULTS FROM THE THEORY OF MODULAR FORMS

Let k be a half-integer. We denote by $S_k(\Gamma(2),\chi_k)$ the space of holomorphic cusp forms h satisfying the transformation rule

$$\begin{cases} h(\tau+2) = h(\tau), \\ \tilde{h}(\tau) := (-i\tau)^k\,h(\tau), \\ \tilde{h}(\tau+2) = \tilde{h}(\tau). \end{cases}$$

The following statement is known as the Voronoi summation formula.

Theorem 3.1. *Let h be a cusp form in $S_{d/2}(\Gamma(2),\chi_d)$ and let $\tilde{h}(\tau) := (-i\tau)^{-d/2}\,h(\frac{-1}{\tau})$. Then, for a radial Schwartz function $f:\mathbb{R}^d\to\mathbb{C}$, the following summation formula holds:*

$$\sum_{n=1}^{\infty} f(\sqrt{n})\,c_h(n) = \sum_{n=1}^{\infty} \hat{f}(\sqrt{n})\,c_{\tilde{h}}(n).$$

For a half-integer k and a positive number ϵ, we define

$$N(k,\epsilon) := \left(\frac{\epsilon\,\Gamma(k-1/2)}{(2\pi)^{k-1}\,\zeta(k-2)\,4\pi}\right)^{1/k}.$$

A straightforward consequence of the Stirling formula is that

$$N(k,\epsilon) \sim \frac{k}{2\pi e} \quad \text{as } k\to\infty.$$

The main technical tool in our proof of Theorem 1.4 is the following statement about the space of modular forms $S_k(\Gamma(2),\chi_k)$.

Theorem 3.2. *Fix a number $\epsilon \in (0, 1/2)$ and for a half-integer k set $N(k) := \lfloor N(k,\epsilon) \rfloor$. For each half-integral weight $k \geq 5/2$, there exist elements $(h_m)_{m=1}^{N(k)-1}$ in the space $S_k(\Gamma(2), \chi_d)$ such that:*

(1) *the function h_m has the Fourier expansion*

$$h_m(\tau) = e^{\pi i m \tau} + \sum_{\substack{n \in \mathbb{Z} \\ n \geq N(k)}} c_{h_m}(n)\, e^{\pi i n \tau};$$

(2) *the function $\tilde{h}_m := (-i\tau)^{-k}\, h_m(\frac{-1}{\tau})$ has the Fourier expansion*

$$\tilde{h}_m(\tau) = \sum_{\substack{n \in \mathbb{Z} \\ n \geq N(k)}} c_{\tilde{h}_m}(n)\, e^{\pi i n \tau};$$

(3) *the Fourier coefficients $c_{h_m}(n)$ and $c_{\tilde{h}_m}(n)$ satisfy the following estimates:*

$$\left|c_{h_m}(n)\right| \leq C\, m^{-k/2+\alpha}\, n^{k/2+\alpha},$$
$$\left|c_{\tilde{h}_m}(n)\right| \leq C\, m^{-k/2+\alpha}\, n^{k/2+\alpha}.$$

Here C and α are positive constants independent of k, m, and n, and depending on ϵ.

4. PROOF OF THEOREM 3.2

Let $P_{k,\chi,m}$ be the Poincaré series for the group $\Gamma(2)$ and multiplier system χ_k (see [3, p. 47, equation (3.2)]). The Fourier coefficients of the Poincaré series can be explicitly computed by the Petersson formula,

$$c_{P_{k,\chi,m}}(n) = \delta_{m,n} + \sum_{c>0} S(m,n,c)\, \mathcal{J}_c(m,n). \tag{4.1}$$

Here $\mathcal{J}_c(m,n)$ is the following sum:

$$\mathcal{J}_c(m,n) = \frac{2\pi}{i^k c} \left(\frac{n}{m}\right)^{\frac{k-1}{2}} J_{k-1}\left(\frac{4\pi\sqrt{mn}}{c}\right),$$

the function J_ν is the Bessel J-function given by the power series

$$J_\nu(x) = \sum_{\ell=0}^{\infty} \frac{(-1)^\ell}{\ell!\,\Gamma(\ell+1+\nu)} \left(\frac{x}{2}\right)^{\nu+2\ell}.$$

And $S(m,n,c)$ is the Kloosterman sum defined in [3, p. 51, equation (3.13)]. The following estimate can be found in [5].

Lemma 4.1. *For a half-integer weight $k \geq 5/2$ and positive integers m, n, the Fourier coefficients of Poincaré series satisfy:*

$$\left|c_{P_{k,n}}(m) - \delta_{m,n}\right| \leq \left(\frac{m}{n}\right)^{\frac{k-1}{2}} \varepsilon^{-2}\, n^{1+\varepsilon}\, m^{1+\varepsilon}\, C,$$
$$\left|c_{\tilde{P}_{k,n}}(m)\right| \leq \left(\frac{m}{n}\right)^{\frac{k-1}{2}} \varepsilon^{-2}\, n^{1+\varepsilon}\, m^{1+\varepsilon}\, C.$$

Here C is an absolute constant and ε is any number in the interval $(0, \frac{1}{8}]$.

Lemma 4.2. *For a half-integer weight $k \geq 5/2$ and positive integers m, n lying in the interval $[1, N(k,\epsilon)]$, the Fourier coefficients of Poincaré series satisfy:*

(1) $|c_{P_{k,n}}(m) - \delta_{m,n}|\,(\frac{n}{m})^{\frac{k-1}{2}} \leq \frac{\epsilon}{N(k,\epsilon)}$;

(2) $|c_{\tilde{P}_{k,n}}(m)|\,(\frac{n}{m})^{\frac{k-1}{2}} \leq \frac{\epsilon}{N(k,\epsilon)}$.

Proof. Part (1) of the lemma is an immediate consequence of Stirling's formula.

The Mehler–Sonine formula [2] gives the following integral representation of the Bessel J-function:

$$J_\nu(z) = \frac{(z/2)^\nu}{\Gamma(\nu+1/2)\,\sqrt{\pi}} \int_{-1}^{1} e^{izs}\,(1-s^2)^{\nu-\frac{1}{2}}\,ds, \quad \nu > \frac{-1}{2},\ z \in \mathbb{C}.$$

This integral representation implies an estimate

$$\big|J_\nu(z)\big| \leq \frac{(z/2)^\nu\, 2}{\Gamma(\nu+1/2)\,\sqrt{\pi}}.$$

Also, we use the trivial estimate for the Klostermann sums (see [3, EQUATION (3.13)])

$$\big|S(m,n,c)\big| < c^2.$$

We combine these two estimates with the Petersson formula (4.1) for the Fourier coefficients of the Poincaré series and obtain

$$\begin{aligned}
\big|c_{P_{k,n}}(m) - \delta_{m,n}\big| \left(\frac{n}{m}\right)^{\frac{k-1}{2}} &\leq 2\pi \sum_{c>0} c \left| J_{k-1}\left(\frac{4\pi\sqrt{mn}}{c}\right)\right| \\
&\leq 4\pi \sum_{c>0} c \left| \frac{(2\pi/c)^{k-1}}{\Gamma(\nu+1/2)} \right| (mn)^{\frac{k-1}{2}} \\
&\leq 4\pi\, \frac{\zeta(k-2)\,(2\pi)^{k-1}}{\Gamma(k-1/2)}\,(mn)^{\frac{k-1}{2}}.
\end{aligned} \tag{4.2}$$

Note that $\sqrt{mn} \leq N(k,\epsilon)$, therefore inequality (4.2) and our choice of the function $N(k,\epsilon)$ imply part (2) of the lemma. Proof of part (3) is analogous. ■

Proof of Theorem 3.2. Fix a half-integral weight k and $\epsilon \in (0, 1/2)$ and set $N := \lfloor N(k,\epsilon) \rfloor$. Consider a matrix $A = (a_{m,n})_{m,n=1}^{2N}$ with entries defined by the coefficients of the Poincaré series $\mathcal{P}_m := \mathcal{P}_{k,m}$ as

$$a_{m,n} = \begin{cases} c_{\mathcal{P}_m}(n)\,(\frac{m}{n})^{\frac{k-1}{2}} & \text{if } m,n \in [1,N], \\ c_{\tilde{\mathcal{P}}_m}(n-N)(\frac{m}{n-N})^{\frac{k-1}{2}} & \text{if } m \in [1,N],\ n \in [N+1, 2N], \\ c_{\tilde{\mathcal{P}}_{m-N}}(n)(\frac{m-N}{n})^{\frac{k-1}{2}} & \text{if } m \in [N+1,2N],\ n \in [1,N], \\ c_{\tilde{\mathcal{P}}_{m-N}}(n-N)(\frac{m-N}{n-N})^{\frac{k-1}{2}} & \text{if } m,n \in [N+1, 2N]. \end{cases}$$

From Lemma 4.2, we know that A is diagonally dominated and therefore invertible. Moreover, the inverse matrix $B = (b_{m,n})_{m,n=1}^{2N} := A^{-1}$ satisfies

$$|b_{m,n} - \delta_{m,n}| < \sum_{k=1}^{\infty} (2\epsilon)^k = \frac{2\epsilon}{1-2\epsilon}. \tag{4.3}$$

Consider modular forms

$$h_\ell := \ell^{\frac{1-k}{2}} \sum_{n=1}^{N} (b_{\ell,n}\, P_n + b_{\ell,n+N}\, \tilde{P}_n)\, n^{\frac{k-1}{2}}, \quad \ell = 1, \ldots, N. \tag{4.4}$$

From the definition of coefficients $b_{\ell,n}$, we see

$$c_{h_\ell}(m) = \delta_{\ell,m} \quad \text{for } \ell, m = 1, \ldots N.$$

For the functions $\tilde{h}_\ell(\tau) := (-i\tau)^{-k}\, h_\ell(-1/\tau)$, we find

$$\tilde{h}_\ell := \ell^{\frac{1-k}{2}} \sum_{n=1}^{N} (b_{\ell,n}\, \tilde{P}_n + b_{\ell,n+N}\, P_n)\, n^{\frac{k-1}{2}}, \quad \ell = 1, \ldots, N. \tag{4.5}$$

The matrix A has symmetries $a_{m,n} = a_{m+N,n+N}$ and $a_{m+N,n} = a_{m,n+N}$ for $m, n = 1, \ldots, N$. Same symmetries are inherited by B, namely $b_{m,n} = b_{m+N,n+N}$, $b_{m+N,n} = b_{m,n+N}$ under same assumptions on indices m and n. Hence, we can rewrite (4.5) as

$$\tilde{h}_\ell := \ell^{\frac{1-k}{2}} \sum_{n=1}^{N} (b_{\ell+N,n}\, P_n + b_{\ell+N,n+N}\, \tilde{P}_n)\, n^{\frac{k-1}{2}}, \quad \ell = 1, \ldots, N.$$

Thus, we see that

$$c_{\tilde{h}_\ell}(m) = \delta_{\ell,m+N} = 0 \quad \text{for } \ell, m = 1, \ldots, N.$$

Finally, we prove part (3) of the theorem. Let ℓ and m be integers such that $\ell \in [1, N]$ and $m \in (N, \infty)$. We apply definition (4.4) and estimate the mth Fourier coefficient of h_ℓ as

$$\left|c_{h_\ell}(m)\right| \ell^{\frac{k-1}{2}}\, m^{\frac{1-k}{2}} \leq \sum_{n=1}^{N} \left(|b_{\ell,n}|\, \left|c_{P_n}(m)\right| n^{\frac{k-1}{2}}\, m^{\frac{1-k}{2}} + |b_{\ell,n+N}|\, \left|c_{\tilde{P}_n}(m)\right| n^{\frac{k-1}{2}}\, m^{\frac{1-k}{2}} \right).$$

Now we apply Lemma 4.1 and estimate (4.3) in order to obtain

$$\left|c_{h_\ell}(m)\right| \ell^{\frac{k-1}{2}}\, m^{\frac{1-k}{2}} \leq \frac{2}{1-2\epsilon} \sum_{n=1}^{N} \varepsilon^{-2}\, n^{1+\varepsilon}\, m^{1+\varepsilon}\, C \leq \frac{2C}{(1-2\epsilon)\varepsilon^2}\, N^{2+\varepsilon}\, m^{1+\varepsilon}.$$

Analogously, we show that

$$\left|c_{\tilde{h}_\ell}(m)\right| \ell^{\frac{k-1}{2}}\, m^{\frac{1-k}{2}} \leq \frac{2C}{(1-2\epsilon)\varepsilon^2}\, N^{2+\varepsilon}\, m^{1+\varepsilon}.$$

This finishes the proof of Theorem 3.2. ■

5. PROOF OF THEOREM 1.4

Lemma 5.1. *Let $(X_n)_{n=1}^{\infty}$ be a sequence of subsets of $S(1)$ such that X_n is a spherical design of strength $D(n)$ and let $X := \bigcup_{n=1}^{\infty} \sqrt{n} X_n$. Suppose that f is a Schwartz function such that $f|_X = 0$. There exist an absolute positive constant C independent of f and X and a positive number β, which depends linearly on dimension d, such that for all $p \in \mathcal{B}$ and $n \in \mathbb{Z}_{\geq 1}$,*

$$\phi_{p,n} \leq C\, \deg(p)^{\beta} \sum_{\substack{q \in \mathcal{B} \\ \deg(q) > D(n) - \deg(p)}} \phi_{q,n}.$$

Proof. By (2.3), we have

$$\phi_{p,n} = \left| \int_{S(1)} f(\sqrt{n}\zeta)\, p(\zeta)\, d\zeta \right| \cdot \sup_{\zeta \in S(1)} \left| p(\zeta) \right|.$$

For $M \in \mathbb{Z}_{\geq 0}$, we define the "head" of f as

$$h_M := \sum_{\substack{p \in \mathcal{B} \\ \deg(p) \leq M}} f_p$$

and the "tail" as

$$t_M := \sum_{\substack{p \in \mathcal{B} \\ \deg(p) > M}} f_p.$$

The integral in (2.3) can be written as

$$\int_{S(1)} f(\sqrt{n}\zeta) \left\| p(\zeta) \right\| d\zeta = \int_{S(1)} \left(h_M(\sqrt{n}\zeta) + t_M(\sqrt{n}\zeta) \right) \left\| p(\zeta) \right\| d\zeta.$$

For a finite set $Y \subset S(1)$ and a function $g : S(1) \to \mathbb{C}$, we will use the notation

$$\int_Y g(\zeta)\, d\zeta := \frac{1}{|Y|} \sum_{y \in Y} g(y).$$

Suppose the integer M is chosen so that $M + \deg(p) \leq D(n)$. Then, our assumption that the set X_n is a spherical design of strength $D(n)$ implies that

$$\int_{S(1)} h_M(\sqrt{n}\zeta)) \left\| p(\zeta) \right\| d\zeta = \int_{X_n} h_M(\sqrt{n}\zeta) \left\| p(\zeta) \right\| d\zeta.$$

Thus, we can write the integral (2.2) as

$$\begin{aligned} &\int_{X_n} h_M(\sqrt{n}\zeta) \left\| p(\zeta) \right\| d\zeta + \int_{S(1)} t_M(\sqrt{n}\zeta) \left\| p(\zeta) \right\| d\zeta \\ &= \int_{X_n} (f - t_M)(\sqrt{n}\zeta) \left\| p(\zeta) \right\| d\zeta + \int_{S(1)} t_M(\sqrt{n}\zeta) \left\| p(\zeta) \right\| d\zeta \\ &= \int_{X_n} f(\sqrt{n}\zeta) \left\| p(\zeta) \right\| d\zeta + \left(\int_{S(1)} - \int_{X_n} \right) t_M(\sqrt{n}\zeta) \left\| p(\zeta) \right\| d\zeta. \end{aligned} \tag{5.1}$$

The first summand in the above line vanishes by the assumption that $f|_{X_n} = 0$. Therefore, we can estimate the integral (2.2) in the following way:

$$\left| \int_{S(1)} f(\sqrt{n}\zeta)\, p(\zeta)\, d\zeta \right| \leq 2 \sup_{\zeta \in S(1)} \left| p(\zeta) \right| \sup_{x \in S(\sqrt{n})} \left| t_M(x) \right|. \tag{5.2}$$

We observe that

$$\sup_{x \in S(\sqrt{n})} \left| t_M(x) \right| \leq \sum_{\substack{q \in \mathcal{B} \\ \deg(q) > M}} \phi_{q,n}.$$

This finishes the proof of Lemma 5.1. ■

Theorems 3.1 and 3.2 give us other inequalities for the numbers $(\phi_{p,n})_{p \in \mathcal{B}, n \in \mathbb{Z}_{\geq 1}}$.

Lemma 5.2. *Fix $\epsilon \in (0, 1/2)$ and set $N(k) := \lfloor N(k, \epsilon) \rfloor$. Suppose that a Schwartz function f is an eigenfunction of the Fourier transform. There exists an absolute positive constant C big enough such that for all $p \in \mathcal{B}$ and all positive integers $m \leq N(\deg(p) + d/2)$, we have*

$$\phi_{p,m} \leq C\, m^{\alpha - \frac{d}{4}} \sum_{\substack{n \in \mathbb{Z} \\ n > N(\deg(p)+d/2)}} n^{\alpha + \frac{d}{4}}\, \phi_{p,n}.$$

Proof. Let f be a Schwartz function in $\mathbb{R}^d$. As described in Section 2, this function has a decomposition

$$f(x) = \sum_{p \in \mathcal{B}} f_p(x), \quad f_p(x) = p(x)\, g_p\big(|x|\big).$$

Here for each homogenous harmonic polynomial $p \in \mathcal{B}$, the function $g_p : \mathbb{R}_{\geq 0} \to \mathbb{C}$ is such that the function $x \mapsto g_p(|x|)$ on $\mathbb{R}^d$ is a radial Schwartz function. A known result in analysis implies that $x \mapsto g_p(|x|)$ is a Schwartz function on any Euclidean space $\mathbb{R}^s$. We denote by $\mathcal{F}_s$ the s-dimensional Fourier transform and have

$$\mathcal{F}_d(f_p)(x) = \mathcal{F}_d\big(p(x)\, g_p\big(|x|\big)\big) = (-i)^{\deg(p)}\, p(y)\, \mathcal{F}_{d+2\deg(p)}(g_p)\big(|y|\big).$$

Let $\{h_m\}_{m=1}^{N(d/2+\deg(p))} \subset S_{d/2+\deg(p)}(\Gamma(2), \chi)$ be the modular forms constructed in Theorem 3.2. By Theorem 3.1, for each integer m on the interval $[1, \dots, N(d/2 + \deg(p))]$, we have the following linear relation between values of g_p:

$$\sum_{n=1}^{\infty} g_p(\sqrt{n})\, c_{h_m}(n) = \sum_{n=1}^{\infty} \mathcal{F}_{d+2\deg(p)}(g_p)(\sqrt{n})\, c_{\tilde{h}_m}(n).$$

Therefore for each point ζ on the sphere $S(1)$, we have

$$\sum_{n=1}^{\infty} g_p(\sqrt{n})\, p(\sqrt{n}\,\zeta)\, n^{\frac{-\deg(p)}{2}}\, c_{h_m}(n)$$
$$= (-i)^{\deg(p)} \sum_{n=1}^{\infty} \mathcal{F}_{d+2\deg(p)}(g_p)(\sqrt{n})\, p(\sqrt{n}\,\zeta)\, n^{\frac{-\deg(p)}{2}}\, c_{\tilde{h}_m}(n).$$

This is equivalent to

$$\sum_{n=1}^{\infty} f_p(\sqrt{n}\zeta)\, n^{\frac{-\deg(p)}{2}}\, c_{h_m}(n) = (-i)^{\deg(p)} \sum_{n=1}^{\infty} \widehat{f}_p(\sqrt{n}\zeta)\, n^{\frac{-\deg(p)}{2}}\, c_{\tilde{h}_m}(n).$$

Conditions (1) and (2) of Theorem 3.2 imply that for an integer m in the interval $[1, N(d/2 + \deg(p))]$ and a point ζ on the sphere $S(1)$,

$$f_p(\sqrt{m}\,\zeta)\, m^{\frac{-\deg(p)}{2}} = \sum_{n=1}^{\infty} \big(f_p(\sqrt{n}\,\zeta)\, c_{h_m}(n) + (-i)^{\deg(p)}\, \widehat{f}_p(\sqrt{n}\,\zeta)\, c_{\tilde{h}_m}(n)\big)\, n^{\frac{-\deg(p)}{2}}.$$

Now condition (3) of Theorem 3.2 and the assumption that f is an eigenfunction of the Fourier transform imply that

$$\big|f_p(\sqrt{m}\,\zeta)\, m^{\frac{-\deg(p)}{2}}\big| \leq C \sum_{n=N(d/2+\deg(p))+1}^{\infty} \big|f_p(\sqrt{n}\,\zeta)\big|\, n^{\frac{-\deg(p)}{2}}\, n^{\frac{d}{4} + \frac{\deg(p)}{2} + \alpha}\, m^{-\frac{d}{4} - \frac{\deg(p)}{2} + \alpha}.$$

■

We set $\tilde{\alpha} := \alpha + d/4$. For all $p \in \mathcal{B}$ and all positive integers $m \leq N(\deg(p) + d/2)$, we have

$$\phi_{p,m} \leq C\, m^{\tilde{\alpha}} \sum_{\substack{n \in \mathbb{Z} \\ n > N(\deg(p)+d/2)}} n^{\tilde{\alpha}}\, \phi_{p,n}.$$

Now, we are ready for the final step in the proof of Theorem 1.4. In particular, we will define the positive constants $\tilde{A}(d)$ and $\tilde{B}(d)$. We will show that for a suitable choice of $\tilde{A}(d)$ and $\tilde{B}(d)$ the growth condition of Theorem 2.1, combined with the inequalities of Lemmas 5.1 and 5.2, implies the vanishing of the numbers $(\phi_{p,n})_{p \in \mathcal{B}, n \in \mathbb{Z}_{\geq 0}}$.

For each $\epsilon \in (0, 1/2)$, there exists a sufficiently small positive number b such that

$$N(k, \epsilon) \geq bk, \quad k \in \frac{1}{2}\mathbb{Z}_{\geq 1}.$$

For a polynomial $p \in \mathcal{B}$, we set

$$\mathcal{N}(p) := b \deg(p).$$

Note that

$$\mathcal{N}(p) \leq N\big(\deg(p) + d/2\big).$$

Let C' and γ be positive numbers (depending on dimension d) such that $\dim \mathcal{H}_m \leq C' m^{\gamma}$. Note that $\gamma = d - 2$ is admissible. We will need the following technical statement.

Lemma 5.3. *For each dimension d, we consider $D(n) := \tilde{B}\, n^{\tilde{A}}$, where*

$$\tilde{B} > 2 \max\left(b + \frac{1}{b}, \frac{C\,C'}{b^{\beta+\gamma+1}}\right), \qquad \tilde{A} = 2\tilde{\alpha} + \beta + \gamma + 3.$$

Then

(1) *for $p, q \in \mathcal{B}$ and $n \in \mathbb{Z}_{\geq 1}$, the conditions $n \geq \mathcal{N}(p)$ and $\deg(q) \geq D(n) - \deg(p)$ imply $n \leq \mathcal{N}(q)$.*

(2) *for all positive integers m and all $q \in \mathcal{B}$ with $m \geq \mathcal{N}(q)$, we have*

$$\sum_{\substack{n \in \mathbb{Z}_{\geq 1}, p \in \mathcal{B}: \\ n \geq \mathcal{N}(p) \\ D(n) - \deg(p) \leq \deg(q)}} C \cdot \deg(p)^{\beta} \cdot n^{2\tilde{\alpha}+1} < m.$$

Proof. Part (1) of the lemma follows immediately from our choice of $\tilde{A}$ and $\tilde{B}$. Indeed, we observe that $\tilde{A} > 1$ and $\tilde{B} > \frac{1}{b}$. Therefore we have

$$\mathcal{N}(q) = b \deg(q) \geq b\big(2\tilde{B} n^{\tilde{A}} - \deg(p)\big) > b\left(\frac{2n}{b} - \frac{n}{b}\right) = n.$$

We rewrite the sum in part (2) in the following way:

$$\sum_{\substack{n \in \mathbb{Z}_{\geq 1},\, p \in \mathcal{B}: \\ n \geq \mathcal{N}(p) \\ D(n) - \deg(p) \leq \deg(q)}} C \cdot \deg(p)^{\beta} \cdot n^{2\tilde{\alpha}+1}$$

$$= \sum_{\substack{n \in \mathbb{Z}_{\geq 1} \\ D(n) - \frac{n}{b} \leq \deg(q)}} \sum_{\substack{p \in \mathcal{B}: \\ \deg(p) \leq \frac{n}{b} \\ \deg(p) \geq D(n) - \deg(q)}} C \cdot \deg(p)^{\beta} \cdot n^{2\tilde{\alpha}+1}.$$

Now we use that $D(n) - \frac{n}{b} \geq \frac{1}{2}\tilde{B}\, n^{\tilde{A}}$ and estimate the above expression by

$$\leq \sum_{\substack{n\in\mathbb{Z}_{\geq 1} \\ \frac{1}{2}\tilde{B}\, n^{\tilde{A}} \leq \deg(q)}} \sum_{\substack{p\in\mathcal{B}: \\ \deg(p)\leq \frac{n}{b} \\ \deg(p)\geq D(n)-\deg(q)}} C \cdot \deg(p)^{\beta} \cdot n^{2\tilde{\alpha}+1}.$$

Next we use the fact that the dimension of $\mathcal{H}_{\deg(p)}$ is bounded by $C' \deg(p)^{\gamma}$ and bound the sum in part (2) by

$$\leq \sum_{\substack{n\in\mathbb{Z}_{\geq 1} \\ \frac{1}{2}\tilde{B}\, n^{\tilde{A}} \leq \frac{m}{b}}} \sum_{\substack{s\in\mathbb{Z}_{\geq 1}: \\ D(n)-\frac{m}{b}\leq s\leq \frac{n}{b}}} C\, C'\, s^{\beta+\gamma} \cdot n^{2\tilde{\alpha}+1}.$$

This sum does not exceed

$$\sum_{\substack{n\in\mathbb{Z}_{\geq 1} \\ n\leq (\frac{2m}{b\,\tilde{B}})^{1/\tilde{A}}}} C\, C' \left(\frac{n}{b}\right)\left(\frac{n}{b}\right)^{\beta+\gamma} n^{2\tilde{\alpha}+1}.$$

Finally, we crudely estimate each term of this sum by substituting $n \mapsto (\frac{2m}{b\,\tilde{B}})^{1/\tilde{A}}$ and bounding the number of terms by $(\frac{2m}{b\,\tilde{B}})^{1/\tilde{A}}$. This gives us an upper bound

$$\frac{C\, C'}{b^{\beta+\gamma}} \left(\frac{2m}{b\,\tilde{B}}\right)^{\frac{2\tilde{\alpha}+\beta+\gamma+3}{\tilde{A}}}.$$

Now, our choice of $\tilde{A}$ and $\tilde{B}$ guarantees that the sum in part (2) of the lemma is less than m. ■

Proof. We are ready to complete the proof of Theorem 1.4. Let $(X_n)_{n=1}^{\infty}$ be a collection of spherical designs on the sphere $S(1)$. We suppose that for each n the design X_n has strength $D(n) = \tilde{B}\, n^{\tilde{A}}$, where $\tilde{A}$ and $\tilde{B}$ are defined in the Lemma 5.3. We will show that $X = \bigcup_n \sqrt{n} X_n$ is a Fourier uniqueness set. Suppose that $f : \mathbb{R}^d \to \mathbb{C}$ is a Schwartz function that satisfies

$$f|_X \equiv 0 \quad \text{and} \quad \hat{f}|_X \equiv 0. \tag{5.3}$$

Then for each $n \in \mathbb{Z}_{\geq 1}$, we have

$$f|_{\sqrt{n}X_n} = \hat{f}|_{\sqrt{n}X_n} = 0.$$

Without loss of generality, we assume that f is an eigenfunction of the Fourier transform.

Consider the sum

$$\sum_{\substack{p\in\mathcal{B},\, n\in\mathbb{Z}: \\ n\geq \mathcal{N}(p)}} \phi_{p,n}\, n^{\tilde{\alpha}+1}. \tag{5.4}$$

By Theorem 2.1, this sum of nonnegative numbers converges to a finite limit.

By Lemma 5.1, we can estimate the sum (5.4) as

$$\sum_{\substack{p\in\mathcal{B}, n\in\mathbb{Z}: \\ n\geq \mathcal{N}(p)}} \phi_{p,n}\, n^{\tilde{\alpha}+1} \leq \sum_{\substack{p\in\mathcal{B}, n\in\mathbb{Z}: \\ n\geq \mathcal{N}(p)}} n^{\tilde{\alpha}+1}\, C \deg(p)^{\beta} \cdot \sum_{\substack{q\in\mathcal{B}: \\ \deg(q)> D(n)-\deg(p)}} \phi_{q,n}.$$

We have chosen the numbers $\tilde{A}$ and $\tilde{B}$ so that the conditions $n \geq \mathcal{N}(p)$ and $\deg(q) \geq D(n) - \deg(p)$ imply $n \leq \mathcal{N}(q)$. We apply Lemma 5.2 and estimate

$$\sum_{\substack{p\in\mathcal{B},n\in\mathbb{Z}:\\ n\geq\mathcal{N}(p)}} \phi_{p,n}\, n^{\tilde{\alpha}+1} \leq \sum_{\substack{p\in\mathcal{B},n\in\mathbb{Z}:\\ n\geq\mathcal{N}(p)}} n^{\tilde{\alpha}+1}\, C \deg(p)^{\beta} \cdot \sum_{\substack{q\in\mathcal{B}:\\ \deg(q)>D(n)-\deg(p)}} \sum_{\substack{m\in\mathbb{Z}:\\ m\geq\mathcal{N}(q))}} m^{\tilde{\alpha}}\, n^{\tilde{\alpha}}\, \phi_{q,m}.$$

Here, C is a new constant equal to the product of the constant C from Lemma 5.1 and the constant C from Lemma 5.2. We change the order of summation and arrive at

$$\sum_{\substack{p\in\mathcal{B},n\in\mathbb{Z}:\\ n\geq\mathcal{N}(p)}} \phi_{p,n}\, n^{\tilde{\alpha}+1} \leq \sum_{\substack{m\in\mathbb{Z},q\in\mathcal{B}:\\ m\geq\mathcal{N}(q))}} m^{\tilde{\alpha}}\, \phi_{q,m} \sum_{\substack{p\in\mathcal{B},n\in\mathbb{Z}:\\ n\geq\mathcal{N}(p)\\ D(n)-\deg(p)\leq\deg(q)}} C\, n^{2\tilde{\alpha}+1} \deg(p)^{\beta}.$$

By Lemma 5.3, the inner sum on the right-hand side of this inequality satisfies

$$\sum_{\substack{p\in\mathcal{B},n\in\mathbb{Z}:\\ n\geq\mathcal{N}(p)\\ D(n)-\deg(p)\leq\deg(q)}} C\, n^{2\tilde{\alpha}+1} \deg(p)^{\beta} < m.$$

This inequality is guaranteed by our choice of function D. Suppose that the nonnegative numbers $(\phi_{q,m})_{\substack{m\in\mathbb{Z},q\in\mathcal{B}\\ m\geq\mathcal{N}(q)}}$ are not all zero. Then

$$\sum_{\substack{p\in\mathcal{B},n\in\mathbb{Z}:\\ n\geq\mathcal{N}(p)}} \phi_{p,n}\, n^{\tilde{\alpha}+1} < \sum_{\substack{q\in\mathcal{B},m\in\mathbb{Z}:\\ m\geq\mathcal{N}(q)}} \phi_{q,m}\, m^{\tilde{\alpha}+1}.$$

This is a contradiction. Therefore, our assumptions on the Schwartz function f imply that $\phi_{q,m} = 0$ whenever $m \geq \mathcal{N}(q)$. Moreover, Lemma 5.2 implies that $\phi_{q,n} = 0$ for all $q \in \mathcal{B}$ and $n \in \mathbb{Z} \geq 0$. Finally, we deduce from Theorem 1.2 that for all harmonic polynomials p in the basis $\mathcal{B}$ the functions f_p in the decomposition (2.1) of the Schwartz function f vanish. Therefore, f is also identically zero. This finishes the proof of Theorem 1.4. ■

ACKNOWLEDGMENTS

I thank Martin Stoller and Joao Ramos for fruitful discussions and comments on the manuscript. I am grateful to Andriy Bondarenko for sharing his conjecture about the existence of the interpolation formula in dimension one.

FUNDING

The author is supported by the SNSF grant "Optimal configurations in multidimensional spaces" 200021184927.

REFERENCES

[1] A. Bondarenko, D. Radchenko, and M. Viazovska, Optimal asymptotic bounds for spherical designs. *Ann. of Math.* (2013), 443–452.

[2] I. S. Gradshteyn, I. M. Ryzhik, Y. V. Geronimus, M. Y. Tseytlin, and A. Jeffrey, 8.411.10. In *Table of Integrals, Series, and Products*, edited by Daniel Zwillinger and Victor Hugo Moll, Translated by Scripta Technica, Inc. (8 ed.), Academic Press, Inc., (2015) [October 2014]. ISBN 978-0-12-384933-5, LCCN 2014010276.

[3] H. Iwaniec, *Topics in classical authomorphic forms*. American Mathematical Society, Providence, 1997.

[4] D. Radchenko and M. Viazovska, Fourier interpolation on the real line. *Publ. Math. Inst. Hautes Études Sci.* **129** (2019), 51–81.

[5] M. Stoller, Fourier interpolation from spheres. *Trans. Amer. Math. Soc.* **374** (2021), no. 11, 8045–8079.

[6] M. Stoller and J. P. G. Ramos, Perturbed Fourier uniqueness and interpolation results in higher dimensions. *J. Funct. Anal.* **282** (2022), no. 12, 109448.

MARYNA VIAZOVSKA

Institute of Mathematics, Ecole Polytechnique Federale de Lausanne, Lausanne, 1015, Switzerland, maryna.viazovska@epfl.ch

COMMUNICATION AND INFORMATION COMPLEXITY

MARK BRAVERMAN

ABSTRACT

Communication complexity is an area of computational complexity theory that studies the amount of communication required to complete a computational task. Communication complexity gives us some of the most successful techniques for proving impossibility results for computational tasks.

Information complexity connects communication complexity with Shannon's classical information theory. It treats information revealed or transmitted as the resource to be conserved. On the one hand, information complexity leads to extensions of classical information and coding theory to interactive scenarios. On the other hand, it provides us with tools to answer open questions about communication complexity and related areas.

This note gives an overview of communication complexity and some recent developments in two-party information complexity and applications. The note is based on a talk given by the author at the International Congress of Mathematicians in 2022. It expands on some of the themes from the talk. It also provides references that were omitted during the talk.

MATHEMATICS SUBJECT CLASSIFICATION 2020

Primary 68Q11; Secondary 68P30, 94A15

KEYWORDS

Computational complexity, communication complexity, information theory, information complexity, interactive computation

INTERNATIONAL CONGRESS OF MATHEMATICIANS
2022 JULY 6–14

© 2022 International Mathematical Union

Proc. Int. Cong. Math. 2022, Vol. 1, pp. 284–320
DOI 10.4171/ICM2022/208

Published by EMS Press and licensed under a CC BY 4.0 license

1. COMPUTATIONAL COMPLEXITY THEORY

Computational complexity theory is concerned with modeling, understanding, and mapping out the computational resources needed to solve various problems involving manipulation of information. Below we give a brief nontechnical overview to set communication and information complexity in context. A principled and extensive treatment of the area can be found in texts such as [2, 59].

1.1. Upper and lower bounds

Mapping out the limits of computation involves a combination of upper and lower bounds on the amount of resources being studied.

An *upper bound* is typically an algorithm with some provable properties. The primary goal of such an algorithm is to place a problem in a complexity class. Sometimes such an algorithm is practically useful, or may inspire a practically useful version later on.

For example, the problem of sorting n elements can be solved using $O(n \log n)$ comparisons. This upper bound can be established via the *MergeSort* algorithm [92], which is fairly straightforward to analyze. In practice, the *QuickSort* algorithm often performs better, but it is harder to analyze for the purposes of establishing an upper bound.

Some upper bounds have desirable properties, but are clearly not the most "practical" algorithms for the problem. For example, using recursion, one can show that the problem of raising an $n \times n$ matrix $A \in \mathbb{F}_2^{n \times n}$ to the power n can be done using $O(\log^2 n)$ bits of working memory. But alas, the resulting algorithm would run in $n^{\Theta(\log n)}$ steps of computation, and would be impractical. Beyond its theoretical value, the upper bound placing Matrix Powering in $\mathsf{SPACE}(\log^2 n)$[1] gives us a hint that basic linear algebraic operations may be amenable to parallelization—a direction that has seen a lot of work in practice [29, 36].

A *lower bound* involves a proof that some computational task is impossible to accomplish within a given constraint on resources. Lower bounds are often harder to prove than upper bounds, since upper bounds are constructive (providing an algorithm), while to prove a lower bound one needs to rule out *all* possible algorithms for a given problem. Still, in many situations provable lower bounds are possible. As we shall see, in many more situations lower bounds can be proved assuming a plausible conjecture, such as $\mathsf{P} \neq \mathsf{NP}$.

In the sorting problem mentioned earlier, the algorithm may output one of $n!$ possible orderings. Each comparison rules out at most half of the orderings. Therefore, at least $\log_2 n! = \Omega(n \log n)$ comparisons are needed to sort n elements.[2] Thus, the upper bound given by *MergeSort* is asymptotically optimal, and sorting requires $\Theta(n \log n)$ comparisons.

1.2. Abstraction and complexity classes

Abstraction is one of the two core ideas underpinning much of complexity theory, it allows us to develop models of computation.

---

1 In fact, it is in the slightly smaller complexity class $\mathbf{NC}^2$ [34].

2 This simple argument shows that $\Omega(n \log n)$ comparisons are needed in the worst case, but it is not hard to show that the same bound also holds *on average*.

There are many different mathematical models that focus on different aspects of computation. Problems can then be grouped into *complexity classes* based on the amount of resources required to solve the problem in a given model. There are hundreds of complexity classes that have been studied explicitly.

The *Turing Machine* is an example of an early and very successful abstraction introduced in the 1930s **[99]**. It gave a mathematical definition of computation, which is still accepted today. In modern terms, a Turing Machine is equivalent to a standard computer with unlimited (but finite at any point during the computation) memory. The class R of problems corresponds to problems solvable by a Turing Machine. Problems inside R are said to be "computable" and problems outside R are "noncomputable."[3]

The taxonomy of computable vs. noncomputable is a very coarse one. For example, the tasks of adding two n-bit numbers, breaking an n-bit cryptographic cipher, or simulating the n-body problem for 2^n time steps are all computable tasks, yet clearly some are more "tractable" than others. Such observations were the starting point for defining more complexity classes based on the setting and resources being constrained.

One plausible (and robust) definition of tractability is given by the class P – the class of problems that are solvable by a Turing Machine in time polynomial in input size. For example, a problem on graphs $G = (V, E)$ with n vertices and m edges is in P if it can be solved by a Turing Machine running in time at most n^c for some constant c. The class P abstracts enough details that we do not need to be concerned with the exact model of the Turing Machine.[4] We also do not need to worry about dependence on the number of edges m, since $m < n^2$, and any bound polynomial in n is also polynomial in m.

The class P abstracts away many details, yet it still gives a very useful definition of tractability. It is especially useful in its negation—if a problem is (for a "typical" input) $\notin$ P, then it is likely intractable in practice except on very small inputs. In its positive direction, being in P does not guarantee that the problem is "easy." For example, checking whether a graph G contains a clique K_{100} with 100 nodes is easily seen to be in P, yet no general algorithm for the problem that runs substantially faster than checking all n^{100} possible vertex sets is known,[5] and is suspected to not exist **[33]**.

The class P can be further refined by restricting the running time of the Turing Machine. For example, the class DTIME(n) restricts the number of steps to be linear in input length. In the case of graphs, this would be linear in $n + m$ – the total number of vertices and edges. Note that here we need to be more careful about the memory access model – linear-time algorithms are typically allowed random memory access. The field of

---

3 Formally, the class R contains decision problems, that are called "recursive" or "decidable." To simplify our current discussion, we blur the distinction between decision problems and general computation tasks.

4 For example, whether data is stored on a tape and addressed sequentially or in a random-access memory array.

5 Here "substantially" means $n^{o(k)}$, where $k = 100$ is the size of the clique we are looking for.

fine-grained complexity aims to classify problems within P based on their required running time, using plausible assumptions [103].

It is possible to reduce the allotted time even further and talk about sublinear time algorithms. Those are particularly important in databases and other "big-data" applications, where one wants to maintain a large data-structure, and to answer queries about it without reading it in its entirety.

Computation time (given by the number of steps performed) is only one of many resources one could consider. Other resources commonly considered include memory used (to store the algorithm's data), parallelization (e.g., is there an algorithm that can be completed in a very short amount of time in a parallel computer), whether the algorithms uses randomness (and how much), and latency caused by communication if the computation is a distributed one.[6] Specific applications (such as data structures) feature additional parameters, as one needs to consider the cost of updating the data structure and the cost of querying it. In addition to "physical" resources used by an algorithm, there are sometimes additional desirable properties such as fault-tolerance or privacy-preservation. These additional requirements may interact with the resource constraints (typically by making them harder to satisfy).

Given the long (but still partial!) list of possible resources and resource combinations to consider, it should not be a surprise that there are so many complexity classes! In fact, it may be surprising that classifying algorithmic problems into complexity classes has been such a productive enterprise at all. One possible explanation of this is that reductions—which we will discuss next—allow us to "cull" classes by showing equivalences. These equivalences are often nontrivial and very surprising.

1.3. Reductions and conditional lower bounds

Reduction is the other core idea in computational complexity theory. As discussed earlier, it is easier to prove that a computational task is attainable within given constraints (by demonstrating an algorithm) than to prove that it is unattainable (need to rule out all algorithms). A reduction allows to turn algorithms into (conditional or unconditional) lower bounds.

Let C be a complexity class (i.e., a class of problems solvable within some given resource constraints). For two problems A and B, one can often use an *algorithmic* construction to prove a statement of the form

$$A \in \mathsf{C} \Rightarrow B \in \mathsf{C}. \tag{1.1}$$

For example, if $\mathsf{C} = \mathsf{P}$, all one needs to do is to construct a polynomial-time algorithm that uses a black box for solving A[7] in order to solve B.

Relationship (1.1) is often denoted by $B \leq_{\mathsf{C}} A$, where the reduction from A to B is done using an algorithm from class C. Thus, for example, if $A \in \mathsf{P}$ and $B \leq_{\mathsf{P}} A$, then $B \in \mathsf{P}$.

6 In practice, latency of communication between processing cores is significantly slower than computation within a core.

7 The box is also assumed to run in polynomial time.

Taking the contrapositive of (1.1), we get

$$B \notin \mathsf{C} \Rightarrow A \notin \mathsf{C}. \tag{1.2}$$

Thus a lower bound on B translates into a lower bound on A.

In terms of theory building, (1.1) and (1.2) allow one to consolidate problems into complexity classes. As it turns out, many natural complexity classes C have a complete problem P_{C} such that all $A \in \mathsf{C}$ are reducible to P_{C}. For example, Cook–Levin's theorem asserts that boolean circuit satisfiability SAT is complete for the class NP. This means that for any problem $B \in \mathsf{NP}$, $B \leq_{\mathsf{P}}$ SAT. Therefore, if $\mathsf{P} \neq \mathsf{NP}$, then for some $B \in \mathsf{NP}$, $B \notin \mathsf{P}$, and by (1.2) SAT $\notin \mathsf{P}$.

In practice, whenever the assumption $\mathsf{P} \neq \mathsf{NP}$ is made, what is actually used is the assumption that SAT $\notin \mathsf{P}$. Under this assumption, to show that $A \notin \mathsf{P}$ it is enough to show that SAT $\leq_{\mathsf{P}} A$. The latter is an algorithmic problem. It may be a simple algorithm taught in introductory classes—such as the reduction **[63]** showing that

$$\text{SAT} \leq_{\mathsf{P}} [\text{deciding whether a given graph } G \text{ is 3-colorable}].$$

Or, it could be the result of stacking extremely complex reductions, such as optimal inapproximability of 3-SAT – one of the crowning achievements of the Probabilistically Checkable Proofs (PCP) program **[3,4]**. In either case, the result is ultimately algorithmic—unspooling the reduction would yield an algorithm that, given a black-box access to the problem being proven to be hard, solves SAT in polynomial time.

Reductions are very useful in consolidating complexity classes. Suppose that C_1 and C_2 are two complexity classes with complete problems $P_1 \in \mathsf{C}_1$ and $P_2 \in \mathsf{C}_2$.[8] Then to show that $\mathsf{C}_1 = \mathsf{C}_2$, it suffices to show that $P_1 \in \mathsf{C}_2$ and that $P_2 \in \mathsf{C}_1$ – again, solving two algorithmic problems.

The logic of (1.2) is very powerful in practice, as it allows one to maintain a list of reasonable hardness assumptions, and to prove tight lower bounds modulo these assumptions. Proving those assumptions may be out of reach (proving $\mathsf{P} \neq \mathsf{NP}$ appears to be currently out of reach). The assumptions may even be false, but nonetheless they can be useful in practice! An example of such an assumption is the Strong Exponential Time Hypothesis (SETH) – asserting that certain flavors of SAT on n variables cannot be solved in $2^{0.999n}$ computation steps.[9] There is a fair chance that the SETH assumption is false, although it has been open for about two decades **[56,57]**. Still, if someone works on an applied data structure, and designing a faster-than-trivial solution for the problem leads to a violation of SETH, the practical implication is that the algorithm designers may assume that there is no better solution than the trivial one (at least for now) – and focus their efforts on other aspects of the design.

In summary, the majority of results in complexity theory—even the deepest and most important ones—are algorithmic reductions, establishing connections between problems and between complexity classes. Most hardness results are conditional ones, using

---

8 Technically, the complexity classes also need to be closed under appropriate reductions, but this is rarely a problem.

9 Note that a brute-force search over all possible assignments takes time $2^n \cdot n^{O(1)}$.

reasonable assumptions such as $\mathsf{P} \neq \mathsf{NP}$, or more ambitious assumptions, such as the SETH. From the perspective of most engineering applications (with the notable exception of cryptography) this is good enough—one can accept a reasonably-aged conjecture as evidence of computational hardness. Computational complexity theory has thus been spectacularly successful in classifying problems into hardness classes based on conjectures. Proving those conjectures is a different matter altogether—progress in attaining *unconditional* lower bounds, i.e., ones where we do not have the luxury of reductions—has been very slow. Devising new attack routes and advancing existing ones is therefore a major challenge in attaining mathematical understanding of computation.

1.4. Unconditional lower bounds: some attack routes

The most general technique for proving unconditional lower bounds on computation is through *diagonalization*. The very first result in the theory of computation [99] used diagonalization to show that the Halting Problem is noncomputable. The Halting Problem asks, given a computer program and an input,[10] to decide whether the program eventually terminates, or runs indefinitely. The proof of the noncomputability of the Halting Problem is straightforward (assuming one accepts that one can program a compiler that takes an encoding of a program and executes it). It is similar to Cantor's proof that there is an uncountable number of real numbers. Many (perhaps most) proofs of noncomputability results work through a reduction to the Halting Problem.

Diagonalization is useful not just for proving noncomputability, but for proving hierarchy theorems, stating that giving programs asymptotically more time strictly increases the set of problems that can be solved. For example,[11]

$$\mathsf{DTIME}(n^{2.3}) \subsetneq \mathsf{DTIME}(n^{2.4}).$$

Still, there are reasons (namely "relativization" [6]) to believe that diagonalization cannot unconditionally prove results such as $\mathsf{P} \neq \mathsf{NP}$, and other currently open unconditional lower bounds.

For unconditional lower bounds that are most likely true but do not follow from diagonalization (such as $\mathsf{P} \neq \mathsf{NP}$), one would need a different set of lower bounds strategies. One approach to take is an incremental one: design an hierarchy of results of increasing difficulty, and incrementally prove them—hopefully discovering and developing new techniques in the process. As an added benefit, even the partial results can be used independently. As discussed earlier, through the magic of reductions, one unconditional lower bound can be converted into many interesting results across multiple settings.

Historically, the most prominent such hierarchy has been that of circuit complexity classes—it is not the main topic of this note, and therefore we will only review it briefly. The circuit complexity program has the advantage that strong enough results under the program

10 A Turing Machine in the original formulation.

11 Perhaps not surprisingly, the task that is easy to perform is time $n^{2.4}$ and impossible to perform in time $n^{2.3}$ is simulating a Turing Machine for $n^{2.35}$ time steps.

will immediately lead to strong lower bounds for complexity classes. One drawback of the program is that progress in the last 30 years has been slow, and it is unclear at this point what tools would be needed to make further progress.

Other than the circuit complexity program, two additional programs of note are arithmetic circuit complexity [96], and communication complexity. Discussing arithmetic circuit complexity is beyond the scope of this note. Communication complexity is going to be our main focus, and will be discussed in some detail.

Boolean circuit complexity program. A circuit is a directed acyclic graph with edges carrying boolean signals 0 or 1. Nodes with no incoming edges correspond to input variables. Other nodes correspond to gates. A gate computes a boolean function of the values of edges incoming into the node, and places the result on the outgoing edges.

Gates may be of fan-in-2, or of unbounded fan-in.[12] Bounded fan-in gates are typically **OR**, **AND**, and **NOT**. Unbounded fain-in gates may be computing an **OR** or **AND** of their inputs, or a more complicated function. Two particularly important functions are summation modulo k: $\oplus_{\mathbf{k}}(x_{1..m}) = \mathbf{1}_{\sum x_i \equiv 0 \pmod{k}}$ and majority $\mathbf{MAJ}(x_{1..m}) := \mathbf{1}_{\sum x_i \geq m/2}$.

A function computable in polynomial time by a Turing Machine can also be computed by a polynomial-size circuit.[13] The class of polynomial-size circuits is denoted by $\mathsf{P/poly}$. As we have just noted, $\mathsf{P} \subset \mathsf{P/poly}$, proving that SAT $\notin \mathsf{P/poly}$ would imply $\mathsf{P} \neq \mathsf{NP}$.

Within circuit complexity, the most natural hierarchy within $\mathsf{P/poly}$ is based on *circuit depth*: the largest number of gates from an input to the output of the circuit. Note that if each gate takes 1 time unit to evaluate, then circuit depth corresponds to the (parallel) latency needed to evaluate the circuit.

When only fan-in-2 gates are allowed, the class NC^i denotes the set of functions that can be evaluated by a circuit of depth $O((\log n)^i)$ and size $n^{O(1)}$. When unbounded fan-in **OR** and **AND** gates are allowed, AC^i denotes the set of functions that can be evaluated by a circuit of depth $O((\log n)^i)$. When in addition $\oplus_{\mathbf{k}}$ gates are allowed for some constant k, we get the class denoted by $\mathsf{AC}^{\mathsf{i}}_{\oplus\mathsf{k}}$. When the majority gate **MAJ** is allowed, we get the class denoted by TC^i.

A majority gate with $n^{O(1)}$ many inputs can be computed by a depth-$O(\log n)$ boolean circuit with fan-in-2 gates. This gives us the following chain of inclusions:

$$\mathsf{NC}^0 \subseteq \mathsf{AC}^0 \subseteq \mathsf{AC}^0_{\oplus\mathsf{k}} \subseteq \mathsf{TC}^0 \subseteq \mathsf{NC}^1 \subseteq \mathsf{AC}^1 \subseteq \mathsf{NC}^2 \subseteq \mathsf{P/poly}. \tag{1.3}$$

Recall that the program was to progressively prove lower bounds against circuit classes in (1.3), eventually building up to SAT $\notin \mathsf{P/poly}$.

---

12 "Fan-in" here is the number of inputs a gate can take. A fan-in-2 **AND** gate takes two inputs $x_1, x_2 \in \{0, 1\}$ and outputs $x_1 \wedge x_2$. A fan-in-n **AND** gate takes n inputs $x_1, \ldots, x_n \in \{0, 1\}$ and outputs $x_1 \wedge x_2 \wedge \cdots \wedge x_n$. A fan-in-$n$ **AND** gate can be computed by a depth-$(\log n)$-binary tree of fan-in-2 **AND** gates.

13 Roughly, in the circuit, each wire corresponds to the state of one bit of memory at a particular point of time in the computation. This reduction is used in designing ASIC circuits that need to be particularly fast or energy-efficient in performing a particular calculation, such as for cryptographic attacks or for routing internet traffic.

The class NC^0 contains circuits where the depth is constant (since $O((\log n)^0) = O(1)$), and each gate's fan-in is 2. Therefore, the n-bit **AND**, $\mathbf{AND}_n$, cannot be computed in NC^0, making the first inclusion strict.

Making the second inclusion in (1.3) strict already requires significant effort. A progression of results in the 1980s showed that an AC^0 circuit computing the parity $\oplus_2$ of n variables has to be of size exponential in n [42, 58]. The proof is combinatorial in nature, using the fact that a random restriction of the parity function to a subset of its coordinates yields another parity function. This line of work led to important results in boolean function analysis—showing that functions computed by AC^0 circuits are approximated by low-degree polynomials in Fourier space [71]. Still, these techniques do not appear to lead to any lower bounds against $\mathsf{AC}^0_{\oplus 2}$ – the class of constant depth circuits with unbounded fan-in **OR**, **AND**, and $\oplus_2$ parity gates.

Lower bounds against $\mathsf{AC}^0_{\oplus 2}$ (or $\mathsf{AC}^0_{\oplus p}$ for an arbitrary constant prime p) – given by Razborov and Smolensky [89, 97] also in the 1980s – require yet another set of ideas, this time algebraic. It turns out that a function computable by a polynomial-size $\mathsf{AC}^0_{\oplus p}$ circuit can be approximated by a low-degree polynomial over the field $\mathbb{F}_p$ (note that $\oplus_\mathbf{p}$ become simple addition over $\mathbb{F}_p$). A dimensionality/counting argument then shows that computing $\oplus_\mathbf{q}$ for any other prime $q \neq p$ cannot be done in $\mathsf{AC}^0_{\oplus p}$. These results only hold for primes. In particular, it is strongly believed, but not known, that $\oplus_5$ cannot be computed by a polynomial sized circuit in $\mathsf{AC}^0_{\oplus 6}$.

As of late 1980s, diagram (1.3) appears as

$$\mathsf{NC}^0 \subsetneq \mathsf{AC}^0 \subsetneq \mathsf{AC}^0_{\oplus p} \subsetneq \mathsf{AC}^0_{\oplus k} \subseteq \mathsf{TC}^0 \subseteq \mathsf{NC}^1 \subseteq \mathsf{AC}^1 \subseteq \mathsf{NC}^2 \subseteq \mathsf{P/poly}. \tag{1.4}$$

Since then, there has been no progress in diagram (1.4). There are several possible explanations for this. One possible explanation is that we are underestimating the power of TC^0 circuits, and that some of the inclusions are in fact not strict (or, at the very least, lower bounds against TC^0 are not much easier than general circuit lower bounds). There is some indirect evidence for the power of TC^0. Within circuit complexity, one surprising result about TC^0 is that it is capable of computing the Chinese Remainder representation of n-bit integers, leading to additional surprising upper bounds [53]. More informally, TC^0-circuits are able to represent artificial neural nets, which have shown a surprising degree of expressiveness in practice, providing indirect evidence for the computational power of the class.

Another possible reason for the relative lack of progress of the circuit complexity program is that the techniques involved appear to be related to logic and combinatorics (diagonalization is a logic technique, while most existing lower bounds are combinatorial), and that new connections are needed to make progress on this programs (or to obtain unconditional lower bounds in another way). This is something that has been noted very early in the study of theoretical computer science (and what became complexity theory). The following is a quote from John von Neumann [101]:

> *"There exists today a very elaborate system of formal logic, and, specifically, of logic as applied to mathematics. This is a discipline with many good sides,*

but also with certain serious weaknesses. This is not the occasion to enlarge upon the good sides, which I have certainly no intention to belittle. About the inadequacies, however, this may be said: Everybody who has worked in formal logic will confirm that it is one of the technically most refractory parts of mathematics. The reason for this is that it deals with rigid, all-or-none concepts, and has very little contact with the continuous concept of the real or of the complex number, that is, with mathematical analysis. Yet analysis is the technically most successful and best-elaborated part of mathematics. Thus formal logic is, by the nature of its approach, cut off from the best cultivated portions of mathematics, and forced onto the most difficult part of the mathematical terrain, into combinatorics.

The theory of automata, of the digital, all-or-none type, as discussed up to now, is certainly a chapter in formal logic. It would, therefore, seem that it will have to share this unattractive property of formal logic. It will have to be, from the mathematical point of view, combinatorial rather than analytical."

In the 70+ years since this quote, analysis has played an increasing role in both lower and upper bounds. Boolean function analysis [79] is an example of a relatively new field that has played a critical role in lower-bound reductions (in Probabilistically Checkable Proofs and Unique Games), and in upper bounds (for example in learning theory). Ideas from convex optimization (some dating back to von Neumann and his colleagues) are now used extensively in upper bounds for such "discrete" problems as Max-Flow [75]. Still, more "analytic" concepts of complexity, particularly ones that tensorize[14] are always helpful in moving the field forward. Communication complexity, and especially its subarea of information complexity, fit well within this general thrust.

Communication complexity and unconditional lower bounds. Like many other concepts within computational complexity theory, communication complexity has been primarily developed as an abstraction of concrete computational problems. Within theoretical computer science, the model was introduced in 1979 by Yao in [105]. Communication complexity arises naturally when studying the complexity of distributed computing, where oftentimes the delay cost of communication between nodes dominates the computational cost within the nodes.

Communication complexity theory has been very successful at producing unconditional lower bound results. Early results used combinatorial methods, but more recently analytical and information-theoretic methods (which are also continuous and analytical in many ways) have shown some success. As with circuit complexity, it is possible to construct a hierarchy of various communication complexity classes, although the hierarchy requires more formalism to define, so we will omit it here. Specific parameters affecting a particular

14 In complexity theory, tensorization is known as direct sum and direct product properties, which we discuss later in this note.

model of communication include the number of players (2 or more), the number of rounds of back-and-forth communication, whether randomness and errors are allowed, etc.

It should be noted that some of the most promising approaches to unconditional circuit lower bounds go through communication complexity. A notable example is Karchmer-Wigderson games [61,62] – a particular type of two-party deterministic communication complexity models for which a lower bound would give a lower bound against NC^1 circuits from (1.4). Another example is an implication of a result of Beigel and Tauri [10] about $\mathsf{AC}^0_{\oplus k}$ circuits, that certain multiparty communication lower bounds imply lower bounds for such circuits. Of course, to achieve these lower bounds further technical progress is needed in communication complexity lower bounds. We will return to this briefly at the end of this note.

1.5. Shannon's information theory and one-way communication

Note. Large parts of this section (as well as the next two) were previously presented in the note [12] by the author accompanying the talk at ICM 2014 in Seoul.

We begin with a very high-level overview of Shannon's information theory. We do this for two reasons. The first reason is that we will need its formalism when defining information complexity in Section 3. The second reason is that one-way information and coding theory is an example of a successful theory that gives very precise answers to many natural questions about data transmission. It serves as a kind of inspiration for what a theory of communication complexity (or even computational complexity) could aspire to—even if it turns out that some core aspects of this program cannot be extended to interactive or multiparty settings.

Information and coding theory is an enormous field of study, with subareas dealing with questions ranging from foundations of probability and statistics to applied wireless transmission systems. We will focus only on some of the very basic foundational aspects, which were set forth by Shannon in the late 1940s, or shortly after.

While our overview of information and coding theory in this section focuses on fairly simple facts, we present those in some detail nonetheless, as they will be used as a scaffold for the interactive coding discussion. A thorough introduction into modern information theory is given in [35].

Noiseless coding. Classical information theory studies the setting where one terminal (Alice) wants to transmit information over a channel to another terminal (Bob). Two of the most important original contributions by Shannon are the *Noiseless Coding* (or Source Coding) Theorem and the *Noisy Coding* (or Channel Coding) Theorem. Here we will only focus on the noiseless part of the theory. The Source Coding Theorem asserts that the cost of Alice transmitting n i.i.d. copies of a discrete random variable X to Bob *over a noiseless* *binary channel*[15] scales as Shannon's entropy $H(X)$ as $n \to \infty$:[16]

$$H(X) = \sum_{x \in \text{supp}(X)} \Pr[X = x] \log \frac{1}{\Pr[X = x]}. \quad (1.5)$$

If we denote by X^n the concatenation of n independent samples from X, and by $C(Y)$ the (expected) number of bits needed for Alice to transmit a sample of random variable Y to Bob, then the Source Coding Theorem asserts that[17]

$$\lim_{n \to \infty} \frac{C(X^n)}{n} = H(X). \quad (1.6)$$

This fact can be viewed as the operational definition of entropy, i.e., one that is grounded in reality. Whereas definition (1.5) may appear artificial, (1.6) implies that it is the right one, since it connects to the "natural" quantity $C(X^n)$. Another indirect piece of evidence indicating that $H(X)$ is a natural quantity is its additivity property,

$$H(X^n) = n \cdot H(X), \quad (1.7)$$

and more generally, if XY is the concatenation of random variables X and Y, then

$$H(XY) = H(X) + H(Y) \quad (1.8)$$

whenever X and Y are independent. Note that it is not hard to see that (1.7) and (1.8) fail to hold for $C(X)$, making $H(X)$ a "nicer" quantity to deal with than $C(X)$. Huffman coding (1.9) below blurs the distinction between the two, as they only differ by at most one additive bit, but we will return to it later in the analogous distinction between communication complexity and information complexity.

For noiseless coding in the one-way regime, it turns out that while $H(X)$ does not exactly equal the expected number of bits $C(X)$ needed to transmit a *single* sample from X, it is very close to it. For example, the classical Huffman's coding **[55]** implies that

$$H(X) \le C(X) < H(X) + 1, \quad (1.9)$$

where the "hard" direction of (1.9) is the upper bound. The upper bound showing that $C(X) < H(X) + 1$ is a *compression result*, showing how to encode a message with low average information content (i.e., entropy) into a message with a low communication cost (i.e., number of bits in the transmission). Note that this result is much less "clean" than the limit result (1.6): in the amortized case the equality is exact, while in the one-shot case a gap is created. This gap is inevitable if only for integrality reasons, but as we will see later, it becomes crucial in the interactive case.

---

15 A noiseless binary channel allows the sender to transmit to a receiver a single bit without error at a unit cost.

16 All logs in this paper are base-2, with ln denoting the natural logarithm.

17 In fact, Shannon's Source Coding Theorem asserts that due to concentration the *worst case* communication cost scales as $H(X)$ as well, if we allow negligible error. We ignore this stronger statement at the present level of abstraction.

Beyond giving the exact answer to the source coding question (equation (1.6)), Shannon's theory has two important benefits. First, it turns "communication" into a continuous resource—much more analytical than combinatorial. This is even more pronounced in the *Noisy Channel Coding* theorem, which allows one to denominate the capacity of a communication channel in bits of information, and to separate the ability of the channel to carry communication from the content of that communication.

Second, it gives us a powerful formalism for talking about information relationships between random variables, which naturally translate informal statements into mathematical expressions. We will give a brief exposition here of notions that we will use in Section 3.

For a single random variable X, entropy $H(X)$ gives a way to quantify the inherent uncertainty in the draw of this variable. For a pair of random variables X and Y, the conditional entropy $H(X|Y)$ can be thought of as the amount of uncertainty remaining in X for someone who knows Y:

$$H(X|Y) := H(XY) - H(Y) = \mathbf{E}_{y\sim Y} H(X|Y = y). \tag{1.10}$$

In the extreme case where X and Y are independent, we have $H(X|Y) = H(X)$. In the other extreme, when $X = Y$, we have $H(X|X) = 0$. The *mutual information* $I(X;Y)$ between two variables X and Y measures the amount of information that revealing Y reveals about X, i.e., the reduction in X's entropy as a result of conditioning on Y. Thus

$$I(X;Y) := H(X) - H(X|Y) = H(X) + H(Y) - H(XY) = I(Y;X). \tag{1.11}$$

Conditional mutual information is defined similarly to conditional entropy,

$$I(X;Y|Z) := H(X|Z) - H(X|YZ) = I(Y;X|Z). \tag{1.12}$$

The expression $I(X;Y|Z)$ is translated into English as "the (expected) amount of information learning variable Y reveals about X to someone who already knows Z."

A very important property of conditional mutual information is the *chain rule*

$$I(XY;Z|W) = I(X;Z|W) + I(Y;Z|WX) = I(Y;Z|W) + I(X;Z|WY). \tag{1.13}$$

Again, an informal interpretation of (1.13) is that XY reveal about Z what X reveals about Z, plus what Y reveals about Z to someone who already knows X.

2. COMMUNICATION COMPLEXITY

For the majority of this discussion we will focus on 2-party computation, returning to the general case at the end of the note.

Communication complexity was introduced by Yao in [105], and is the subject of the texts [68,85]. It has found numerous applications for unconditional lower bounds in a variety of models of computation, including Turing machines, streaming, sketching, data structure lower bounds, and VLSI layout, to name a few. In the basic (two-party) setup, the two parties Alice and Bob are given inputs $X \in \mathcal{X}$ and $Y \in \mathcal{Y}$, respectively, and are required to compute a function $F(X,Y)$ of these inputs (i.e., both parties should know the answer at the end of the communication), while communicating over a noiseless binary channel (sending 0/1 bits to each other). The parties are computationally unbounded, and their only goal is to minimize the number of bits transmitted in the process of computing $F(X, Y)$.

In a typical setup, F is a function $F : \{0, 1\}^n \times \{0, 1\}^n \to \{0, 1\}$. Examples of functions commonly discussed and used include the Equality function

$$\mathrm{EQ}_n(X, Y) := \mathbf{1}_{X=Y}(X, Y) = \bigwedge_{i=1}^{n} \big((X_i \wedge Y_i) \vee (\neg X_i \wedge \neg Y_i)\big), \tag{2.1}$$

and the Disjointness function

$$\mathrm{Disj}_n(X, Y) := \bigwedge_{i=1}^{n} (\neg X_i \vee \neg Y_i). \tag{2.2}$$

The basic notion in communication complexity is the *communication protocol*. A communication protocol over a binary channel formalizes a conversation, where each message only depends on the input to the speaker and the conversation so far:

Definition 2.1. A (deterministic) protocol π for $F : \mathcal{X} \times \mathcal{Y} \to \{0, 1\}$ is defined as a finite rooted binary tree, whose nodes correspond to partial communication transcripts, such that the two edges coming out of each vertex are labeled with a 0 and 1. Each leaf ℓ is labeled by an output value $f_\ell \in \{0, 1\}$. Each internal node v is labeled by a player's name and either by a function $a_v : \mathcal{X} \to \{0, 1\}$, or $b_v : \mathcal{Y} \to \{0, 1\}$ corresponding to the next message of Alice or Bob, respectively.

The protocol $\pi(X, Y)$ is executed on a pair of inputs (X, Y) by starting from the root of the tree. At each internal node labeled by a_v the protocol follows the child $a_v(X)$ (corresponding to Alice sending a message), and similarly at each internal node labeled by b_v the protocol follows $b_v(Y)$. When a leaf ℓ is reached the protocol outputs f_ℓ.

By a slight abuse of notation, $\pi(X, Y)$ will denote both the transcript and the output of the protocol; which one it is will be clear from the context. The communication cost of a protocol is the depth of the corresponding protocol tree. A protocol *succeeds* on input (X, Y) if $\pi(X, Y) = F(X, Y)$. Its communication cost on this pair of inputs is the depth of the leaf reached by the execution. The *communication complexity* $\mathrm{CC}(F)$ of a function F is the lowest attainable communication cost of a protocol that successfully computes F. In the case of deterministic communication we require the protocol to succeed on all inputs.

A deterministic communication protocol π induces a partition of the input space $\mathcal{X} \times \mathcal{Y}$ into sets S_ℓ by the leaf ℓ that $\pi(X, Y)$ reaches. Since at each step the next move of the protocol depends only on either X or Y alone, each S_ℓ is a combinatorial rectangle of the form $S_\ell = S_\ell^{\mathcal{X}} \times S_\ell^{\mathcal{Y}}$. This key combinatorial property is at the heart of many combinatorial communication complexity lower bounds. To give an example of such a simple combinatorial proof, consider the *rank* bound. Let $N = |\mathcal{X}|$, $M = |\mathcal{Y}|$, and consider the $N \times M$ matrix M_F over $\mathbb{R}$ whose (X, Y)th entry is $F(X, Y)$. Each protocol π with leaf set $\mathcal{L}$ of size L, induces a partition of $\mathcal{X} \times \mathcal{Y}$ into combinatorial rectangles $\{S_\ell\}_{\ell \in \mathcal{L}}$. Let M_ℓ be the matrix whose entries are equal to $M_{X,Y}$ for $(X, Y) \in S_\ell$ and are 0 elsewhere. Since

$\{S_\ell\}_{\ell\in\mathcal{L}}$ is a partition of $\mathcal{X}\times\mathcal{Y}$, we have $M_F = \sum_{\ell\in\mathcal{L}} M_\ell$. Assuming π is always correct, each M_ℓ is *monochromatic*, i.e., either all-0, or all-1 on S_ℓ, depending on the value of f_ℓ. Thus, $\text{rank}(M_\ell) \le 1$, and

$$\text{rank}(M_F) \le \sum_{\ell\in\mathcal{L}} \text{rank}(M_\ell) \le L. \tag{2.3}$$

In fact, a stronger bound of $L-1$ holds unless M_F is the trivial all-1 matrix. Thus any protocol computing F must have a communication cost of at least $\log(\text{rank}(M_F)+1)$, and it follows that the communication complexity of F is at least $\log(\text{rank}(M_F)+1)$. As an example of an application, if $F = \text{EQ}_n$ is the Equality function, then $M_{\text{EQ}_n} = I_{2^n}$ is the identity matrix, and thus $\text{CC}(\text{EQ}_n) \ge n+1$. In other words, the trivial protocol where Alice sends Bob her input X (n bits), and Bob responds whether $X = Y$ (1 bit), is optimal.

As in many other areas of theoretical computer science, there is much to be gained from randomization. For example, in practice, the Equality function does not require linear communication as Alice and Bob can just hash their inputs and compare the hash keys. The shorter protocol may return a false positive, but it is correct with high probability, and reduces the communication complexity from $n+1$ to $O(\log n)$.

More generally, a randomized protocol is a protocol that tosses coins (i.e., accesses random bits), and produces the correct answer with high probability. The *distributional setting*, where there is a prior probability distribution μ on the inputs and the players need to output the correct answer with high probability with respect to μ is closely related to the randomized setting, as will be seen below. In the randomized setting there are two possible types of random coins. *Public coins* are generated at random and are accessible to both Alice and Bob at no communication cost. *Private coins* are coins generated privately by Alice and Bob, and are only accessible by the player who generated them. If Alice wants to share her coins with Bob, she needs to use the communication channel. In the context of communication complexity the pubic-coin model is clearly more powerful than the private coin one. Fortunately, the gap between the two is not very large [78], and can be mostly ignored. For convenience reasons, we will focus on the public-coin model.

The definition of a randomized public-coin communication protocol π_R is identical to Definition 2.1, except a public random string R is chosen at the beginning of the execution of the randomized π_R, and all functions at the nodes of π_R may depend on R in addition to the respective input X or Y. We still require the answer f_ℓ to be unequivocally determined by the leaf ℓ alone. The communication cost $|\pi_R|$ of π_R is still its worst-case communication cost (for historic reasons; an average-case notion would also have been meaningful to discuss here).

The randomized communication complexity of F with error $\varepsilon > 0$ is given by

$$R_\varepsilon(F) := \min_{\pi_R:\forall X,Y\ \Pr_R[\pi_R(X,Y)=F(X,Y)]\ge 1-\varepsilon} |\pi_R|. \tag{2.4}$$

For a distribution μ on $\mathcal{X}\times\mathcal{Y}$ the *distributional* communication complexity $D_{\mu,\varepsilon}(F)$ is defined as the cost of the best protocol that achieves expected error ε with respect to μ. Note that in this case fixing public randomness R to a uniformly random value does not change (on average) the expected success probability of π_R with respect to μ. Therefore, without loss of generality, we may require π to be deterministic,

$$D_{\mu,\varepsilon}(F) := \min_{\pi:\mu\{X,Y:\, \pi(X,Y)=F(X,Y)\}\geq 1-\varepsilon} |\pi|. \tag{2.5}$$

It is easy to see that for all μ, $D_{\mu,\varepsilon}(F) \leq R_\varepsilon(F)$. By an elegant minimax argument **[106]**, a partial converse is also true: for each F and ε, there is a distribution against which the distributional communication complexity is as high as the randomized,

$$R_\varepsilon(F) = \max_\mu D_{\mu,\varepsilon}(F). \tag{2.6}$$

For this reason, we will be able to discuss distributional and randomized communication complexity interchangeably.

How can one prove lower bounds for the randomized setting? This setting is much less restrictive than the deterministic one, making lower bounds more challenging. Given a function F, one can guess the hard distribution μ, and then try to lower bound the distributional communication complexity $D_{\mu,\varepsilon}(F)$ – that is, show that there is no low-communication protocol π that computes F with error $\leq \varepsilon$ with respect to μ. Such a protocol π of cost $k = |\pi|$ still induces a partition $\{S_\ell\}_{\ell\in\mathcal{L}}$ of the inputs according to the leaf they reach, with $L \leq 2^k$ and each S_ℓ a combinatorial rectangle. However, it is no longer the case that when we consider the corresponding submatrix M_ℓ of M_F it must be monochromatic—the output of π is allowed to be wrong on a fraction of S_ℓ, and thus for some inputs the output of π on S_ℓ may disagree with the value of F. Still, it should be true that for *most* leaves the value of F on S_ℓ is strongly biased one way or the other, since the contribution of S_ℓ to the error is

$$e(S_\ell) = \min\big(\mu(S_\ell \cap F^{-1}(0)), \mu(S_\ell \cap F^{-1}(1))\big). \tag{2.7}$$

In particular, a fruitful lower bound strategy is to show that all "large" rectangles with respect to μ have $e(S_\ell)/\mu(S_\ell) \gg \varepsilon$, and thus there must be many smaller rectangles—giving a lower bound on $L \leq 2^{|\pi|}$. One simple instantiation of this strategy is the *discrepancy* bound: for a distribution μ, the discrepancy $\mathrm{Disc}_\mu(F)$ of F with respect to μ is the maximum over all combinatorial rectangles R of

$$\mathrm{Disc}_\mu(R, F) := \big|\mu\big(F^{-1}(0) \cap R\big) - \mu\big(F^{-1}(1) \cap R\big)\big|.$$

In other words, if F has low discrepancy with respect to μ, then only very small rectangles (as measured by μ) can be unbalanced. With some calculations, it can be shown that for all $\varepsilon > 0$ (see **[68]** and references therein),

$$D_{\mu,\frac{1}{2}-\varepsilon}(F) \geq \log_2\big(2\varepsilon/\mathrm{Disc}_\mu(F)\big). \tag{2.8}$$

Note that (2.8) not only says that if the discrepancy is low then the communication complexity is high, but also that it remains high even if we are only trying to gain a tiny advantage over random guessing in computing F! An example of a natural function to which the discrepancy method can be applied is the n-bit Inner Product function $\mathrm{IP}_n(X, Y) = \langle X, Y\rangle \bmod 2$. This simple discrepancy method can be generalized to a richer family of *corruption bounds* that can be viewed as combinatorial generalizations of the discrepancy bound. More on this method can be found in the survey [70].

One of the early successes of applying combinatorial methods in communication complexity was the proof that the *randomized* communication complexity of the set disjointness problem (2.2) is linear, $R_{1/4}(\mathrm{Disj}_n) = \Theta(n)$. The first proof of this fact was given in the 1980s [60], and a much simpler proof was discovered soon after [88]. The proofs exhibit a specific distribution μ of inputs on which the distributional communication complexity $D_{\mu,1/4}(\mathrm{Disj}_n)$ is $\Omega(n)$. Note that the uniform distribution would not be a great fit, since uniformly drawn sets are nondisjoint with a very high probability. It turns out that the following family of distributions μ is hard: select each coordinate pair (X_i, Y_i) i.i.d. from a distribution on $\{(0,0), (0,1), (1,0)\}$ (e.g., uniformly). This generates a distribution on pairs of disjoint sets. Now, with probability $1/2$ choose a uniformly random coordinate $i \in_U [n]$ and set $(X_i, Y_i) = (1,1)$ (and with probability $1/2$ do nothing). Thus, under μ, X and Y are disjoint with probability $1/2$.

Treating communication complexity as a generalization of one-way communication and applying information-theoretic machinery to it is a very natural approach (perhaps the most natural, given the success of information theory in communication theory). Interestingly, however, this is not how the field has evolved. For example, a 2009 survey [70] was able to present the vast majority of communication complexity results up until then without dealing with information theory at all. It is hard to speculate why this might have been the case. One possible explanation is that the mathematical machinery needed to tackle the (much more complicated) interactive case from the information-theoretic angle was not available until the 1990s; another possible explanation is that linear algebra, linear programming duality, and combinatorics (the main tools in communication complexity lower bounds) are traditionally more central to theoretical computer science research and education than information theory.

A substantial amount of literature exists on communication complexity within the information theory community. See for example [81, 82] and references therein. The flavor of the results is usually different from the ones discussed above. In particular, there is much more focus on bounded-round communication, and significantly less focus on techniques for obtaining specific lower bounds on the communication complexity of specific functions such as the disjointness function. The most relevant work to our current discussion is a more recent line of work by Ishwar and Ma, which studied interactive amortized communication and obtained characterizations closely related to the ones discussed below [73, 74], building on earlier works of Wyner and Ziv [104] from the 1970s.

Within the theoretical computer science literature, in the context of communication complexity,[18] information theoretic tools were explicitly introduced in [31] in the early

18 As with many other concepts within theoretical computer science, it was introduced earlier in a more applied context, namely quantifying information-theoretic privacy of communication protocols [7, 90]. The two lines of work only converged later, after information complexity was developed in the context of direct sum in communication complexity.

2000s for the simultaneous message model (i.e., 2 noninteractive rounds of communication). Building on this work, [8] developed tools for applying information theoretic reasoning to fully interactive communication, in particular giving an alternative (arguably, more intuitive) proof for the $\Omega(n)$ lower bound on the communication complexity of Disj_n. The motivating questions for [31], as well as for subsequent works developing information complexity, were the *direct sum* [39] and *direct product* questions for (randomized) communication complexity.

The direct sum problem. In general, a direct sum theorem quantifies the cost of solving a problem F^n consisting of n subproblems in terms of n and the cost of each subproblem F. The value of such results to lower bounds is clear: a direct sum theorem, together with a lower bound on the (easier-to-reason-about) subproblem, yields a lower bound on the composite problem (a process also known as hardness amplification). For example, the Karchmer–Wigderson program for boolean formulae lower bounds can be completed via a (currently open) direct sum result for a certain communication model [62].

The direct sum property, while useful, is often untrue—sometimes in unexpected or profound ways. Consider the example of matrix–vector multiplication over $\mathbb{F}_2$. The matrix $A \in \mathbb{F}_2^{n\times n}$ is chosen at random and fixed. The input is $x \in \mathbb{F}_2^n$, and the n-bit output is Ax. The computational model is boolean circuits (as in P/poly discussed earlier). A simple counting argument shows that with high probability, for a randomly chosen A, computing Ax requires a circuit of size $\widetilde{\Omega}(n^2)$.[19] On the other hand, computing $Ax_1, \ldots, Ax_n$ for n vectors in parallel amounts to multiplying A by an $n \times n$ matrix. This can be done in time (and also circuit size) $n^\omega = O(n^{2.38}) \ll n \times n^2$, showing a violation of direct sum for this model. We will return to the direct sum problem for randomized communication complexity in the next section.

Direct product results further sharpen direct sum theorems by showing a "threshold phenomenon," where solving F^n with insufficient resources is shown to be impossible to achieve except with an exponentially small success probability. Classic results in complexity theory, such as Raz's Parallel Repetition Theorem [86] can be viewed as a direct product result. Direct product theorems are also important in the context of cryptography: by repeating a challenge n times, one hopes to boost the security of a system exponentially.

In the next section, we will formally introduce information complexity. We will first look at it as a generalization of Shannon's entropy to interactive tasks. We will then discuss its connections to the direct sum and product questions for randomized communication complexity, as well as other connections.

3. INFORMATION COMPLEXITY

Interactive information complexity. In this section we will work towards developing information complexity as the analogue of Shannon's entropy for interactive computation. It will sometimes be convenient to work with general *interactive two-party tasks* rather than just

19 Here $\widetilde{\Omega}(\cdot)$ hides factors polynomial in $\log n$.

functions. A task $T(X, Y)$ is any action on inputs (X, Y) that can be performed by a protocol. $T(X, Y)$ can be thought of as a set of distributions of outputs that are acceptable given an input (X, Y). Thus "computing $F(X, Y)$ correctly with probability $1 - \varepsilon$" is an example of a task, but there are examples of tasks that do not involve function or relation computation, for example "Alice and Bob need to sample strings A and B, respectively, distributed according to $(A, B) \sim \mu_{(X,Y)}$." For the purpose of our discussion, it suffices to think about T as the task of computing a function with some success probability. The communication complexity of a task T is then defined analogously to the communication complexity of functions. It is the least amount of communication needed to successfully perform the task $T(X, Y)$ by a communication protocol $\pi(X, Y)$.

The *information complexity* of a task T is defined as the least amount of information Alice and Bob need to exchange (i.e., reveal to each other) about their inputs to successfully perform T. This amount is expressed using mutual information (specifically, conditional mutual information (1.12)). We start by defining the *information cost* of a protocol π. Given a prior distribution μ on inputs (X, Y) the information cost is

$$\mathsf{IC}(\pi, \mu) := I(Y; \Pi | X) + I(X; \Pi | Y), \tag{3.1}$$

where Π is the random variable representing a realization of the protocol's transcript, including the *public* randomness it uses.[20] In other words, (3.1) represents the sum of the amount of information Alice learns about Y by participating in the protocol and the amount of information Bob learns about X by participating. Note that the prior distribution μ may drastically affect $\mathsf{IC}(\pi, \mu)$. For example, if μ is a singleton distribution supported on one input (x_0, y_0), then $\mathsf{IC}(\pi, \mu) = 0$ for all π, since X and Y are already known to Bob and Alice respectively under the prior distribution μ. Definition (3.1), which will be justified shortly, generalizes Shannon's entropy in the noninteractive regime. Indeed, in the transmission case, Bob has no input, thus $X \sim \mu$, $Y = \perp$, and Π allows Bob to reconstruct X, thus

$$\mathsf{IC}(\pi, \mu) = I(X; \Pi) = H(X) - H(X | \Pi) = H(X) - 0 = H(X).$$

The *information complexity* of a task T can now be defined similarly to communication complexity in (2.5),

$$\mathsf{IC}(T, \mu) := \inf_{\pi \text{ successfully performs } T} \mathsf{IC}(\pi, \mu). \tag{3.2}$$

One notable distinction between (2.5) and (3.2) is that the latter takes an infimum instead of a minimum. This is because while the number of communication protocols of a given communication cost is finite, this is not true about information cost. One can have a sequence $\pi_1, \pi_2, \ldots$ of protocols of ever-increasing communication cost, but whose information complexity $\mathsf{IC}(\pi_n, \mu)$ converges to $\mathsf{IC}(T, \mu)$ in the limit. Moreover, as we will discuss later, this

20 The protocol is also allowed to use private randomness, known to only one of the two parties, that is not automatically included in the transcript. Unlike the context of communication complexity, in information complexity private randomness is more useful than public randomness [27].

phenomenon is already observed in very simple tasks T, such as computing the conjunction of two bits.

Our discussion of information complexity will be focused on the slightly simpler to reason about *distributional* setting, where inputs are distributed according to some prior μ. In (3.2), if T is the task of computing a function F with error ε with respect to μ, the distribution μ is used twice: first in the definition of "success," and then in measuring the amount of information learned. It turns out that it is possible to define worst-case information complexity [13] as the information complexity with respect to the worst-possible prior distribution in the spirit of the minimax relationship (2.6). In particular, the direct sum property of information complexity which we will discuss below holds for prior-free information complexity as well.

3.1. Direct sum for information and amortized communication

Information complexity as defined here has been extensively studied (see, e.g., survey [102]). In particular, it is surprisingly simple to show that information complexity is additive for tasks over independent pairs of inputs. Let T_1 and T_2 be two tasks over pairs of inputs (X_1, Y_1), (X_2, Y_2), and let μ_1, μ_2 be distributions on pairs (X_1, Y_1) and (X_2, Y_2), respectively. Denote by $T_1 \otimes T_2$ the task composed of successfully performing both T_1 and T_2 on the respective inputs (X_1, Y_1) and (X_2, Y_2). Then information complexity is additive over these two tasks:

Theorem 3.1. $\mathsf{IC}(T_1 \otimes T_2, \mu_1 \times \mu_2) = \mathsf{IC}(T_1, \mu_1) + \mathsf{IC}(T_2, \mu_2)$.

Sketch; a complete proof of a slightly more general statement can be found in [13]. The "easy" direction of this theorem is the '$\leq$' direction. Take two protocols π_1 and π_2 that perform T_1 and T_2, respectively, and consider the concatenation $\pi = (\pi_1, \pi_2)$ (which clearly performs $T_1 \otimes T_2$). Consider what Alice learns from an execution of π with prior $\mu_1 \times \mu_2$. A straightforward calculation using, for example, repeated application of the chain rule (1.13) yields

$$I(Y_1 Y_2; \Pi_1 \Pi_2 | X_1 X_2) = I(Y_1; \Pi_1 | X_1) + I(Y_2; \Pi_2 | X_2),$$

and a similar statement is true about what Bob learns as well. Therefore $\mathsf{IC}(\pi, \mu_1 \times \mu_2) = \mathsf{IC}(\pi_1, \mu_1) + \mathsf{IC}(\pi_2, \mu_2)$. By passing to the limit as $\mathsf{IC}(\pi_1, \mu_1) \to \mathsf{IC}(T_1, \mu_1)$ and $\mathsf{IC}(\pi_2, \mu_2) \to \mathsf{IC}(T_2, \mu_2)$ we obtain the '$\leq$' direction.

The '$\geq$' direction is more interesting, even if the proof is not much more complicated. In this direction we are given a protocol π for solving $T_1 \otimes T_2$ with information cost $I = \mathsf{IC}(\pi, \mu_1 \times \mu_2)$, and we need to construct out of it two protocols for T_1 and T_2 of information costs I_1 and I_2 that add up to $I_1 + I_2 \leq I$. We describe the protocol $\pi_1(X_1, Y_1)$ below:

$\boldsymbol{\pi_1(X_1, Y_1)}$:

- Bob samples a pair $(X_2, Y_2) \sim \mu_2$, and sends X_2 to Alice;

- Alice and Bob execute $\pi((X_1, X_2), (Y_1, Y_2))$, and output the portion relevant to T_1 in the performance of $T_1 \otimes T_2$.

It is not hard to see that the tuple (X_1, Y_1, X_2, Y_2) is distributed according to $\mu_1 \times \mu_2$, and hence by the assumption on π, π_1 successfully performs T_1. Note that there is a slight asymmetry in π_1: X_2 is known to both Alice and Bob while Y_2 is only known to Bob. For the purpose of correctness, the protocol would have worked the same if Bob also sent Y_2 to Alice, but it is not hard to give an example where the information cost of π_1 in that case is too high. The information cost of π is thus given by the sum of what Bob learns about X_1 from π_1 and what Alice learns about Y_1 (note that (X_2, Y_2) are not part of the input),

$$I_1 = I(X_1; \Pi | X_2 Y_1 Y_2) + I(Y_1; \Pi | X_1 X_2).$$

The protocol $\pi_2(X_2, Y_2)$ is defined similarly to π_1 in a skew symmetric way:

$\boldsymbol{\pi_2(X_2, Y_2)}$:

- Alice samples a pair $(X_1, Y_1) \sim \mu_1$, and sends Y_1 to Bob;
- Alice and Bob execute $\pi((X_1, X_2), (Y_1, Y_2))$, and output the portion relevant to T_2 in the performance of $T_1 \otimes T_2$.

We get that π_2 again successfully performs T_2, and its information cost is

$$I_2 = I(X_2; \Pi | Y_1 Y_2) + I(Y_2; \Pi | X_1 X_2 Y_1).$$

Putting I_1 and I_2 together using the Chain Rule (1.13) we get

$$\begin{aligned} I_1 + I_2 &= I(X_1; \Pi | X_2 Y_1 Y_2) + I(Y_1; \Pi | X_1 X_2) + I(X_2; \Pi | Y_1 Y_2) + I(Y_2; \Pi | X_1 X_2 Y_1) \\ &= I(X_2; \Pi | Y_1 Y_2) + I(X_1; \Pi | X_2 Y_1 Y_2) + I(Y_1; \Pi | X_1 X_2) + I(Y_2; \Pi | X_1 X_2 Y_1) \\ &= I(X_1 X_2; \Pi | Y_1 Y_2) + I(Y_1 Y_2; \Pi | X_1 X_2) = I. \end{aligned}$$

Once again, passing to the limit, gives us the '$\geq$' direction, and completes the proof. ■

If we denote an n-time repetition of a task T by $T^{\otimes n}$, then repeatedly applying Theorem 3.1 yields

$$\mathsf{IC}(T^{\otimes n}, \mu^n) = n \cdot \mathsf{IC}(T, \mu). \tag{3.3}$$

Thus information complexity is additive and has the *direct sum property*: the cost of n copies of T scales as n times the cost of one copy. This fact can be viewed as an extension of the property $H(X^n) = n \cdot H(X)$ to interactive problems, but what does it teach us about communication complexity?

Information equals to amortized communication. Let us return to the communication complexity setting, fixing T to be the task of computing a function $F(X, Y)$ with some error at most $\varepsilon > 0$ over a distribution μ (the case $\varepsilon = 0$ seems to be different from $\varepsilon > 0$). We will denote by F_ε^n the task of computing n copies of F on independent inputs distributed according to μ^n, with error at most ε *on each copy* (note that computing F correctly with error at most ε on all copies simultaneously is a harder task).

It is an easy observation that the information cost of a protocol π is always bounded by its length $|\pi|$, and therefore information complexity is always bounded by communication complexity. Therefore, by (3.3),

$$\frac{1}{n} \cdot D_{\mu^n}\big(F_\varepsilon^n\big) \geq \frac{1}{n} \cdot \mathsf{IC}\big(F_\varepsilon^n, \mu^n\big) = \mathsf{IC}(F_\varepsilon, \mu). \tag{3.4}$$

It turns out that the converse is also true in the limit, as $n \to \infty$ **[22]**:[21]

$$\lim_{n\to\infty} \frac{1}{n} \cdot D_{\mu^n}\big(F_\varepsilon^n\big) = \mathsf{IC}(F_\varepsilon, \mu). \tag{3.5}$$

Equation (3.5) can be viewed as the interactive version of the Source Coding Theorem (1.6). In particular, it gives an operational characterization of information complexity exclusively in terms of communication complexity. The link given by (3.5) has been further refined in **[100]**, establishing the second-order term in the equation.

3.2. Direct sum and direct product for communication

Direct sum and interactive compression. Recall that the direct sum property asserts that solving n copies of a problem requires n times the resources it takes to solve one copy. It is one of the most generic tools one can deploy (or hope to deploy) in the quest for unconditional lower bounds.

Theorem 3.1 implies that the direct sum property holds exactly for information complexity. In addition, (3.5) immediately gives us a handle on the direct sum question for *communication complexity.*

The *direct sum* question for communication complexity asks whether

$$D_{\mu^n}\big(F_\varepsilon^n\big) = \Omega\big(n \cdot D_\mu(F_\varepsilon)\big)? \tag{3.6}$$

By (3.5), the question (3.6) is equivalent to

$$\mathsf{IC}(F_\varepsilon, \mu) = \Omega\big(D_\mu(F_\varepsilon)\big)? \tag{3.7}$$

Or, switching directions,

$$D_\mu(F_\varepsilon) = O\big(\mathsf{IC}(F_\varepsilon, \mu)\big)? \tag{3.8}$$

Note that the equivalence works on a per-problem basis, so whenever (3.8) holds for a given problem, direct sum for communication complexity holds for that problem. On the other hand, to show that direct sum for communication complexity fails in general, it suffices to give one example of a function where $D_\mu(F_\varepsilon) = \omega(\mathsf{IC}(F_\varepsilon, \mu))$.

One natural way to interpret (3.8) is through the lens of interactive compression—an interactive analogue of Huffman coding (1.9), where it does hold that $H(X) > C(X) - 1$. Huffman (one way) coding shows how to encode a low-entropy "uninformative" signal into

21 More precisely, the converse adds error that vanishes exponentially in n (and thus goes to 0 as $n \to \infty$). Such a statement would be false with no errors allowed **[77, 80]**. Therefore, (3.5) only holds when $\mathsf{IC}(F_\varepsilon, \mu)$ is continuous in ε as we approach from ε^+. In particular, this means that in many applications we need $\varepsilon > 0$ for it to hold, as there is often a discontinuity at $\varepsilon = 0$.

a short one. Its interactive version seeks to simulate a low information cost "uninformative" protocol π with a low communication protocol π'.

It turns out that such a compression scheme is impossible, disproving the direct sum conjecture through the information complexity route. In a series of breakthrough works, Ganor, Kol, and Raz [43–45] give an example of a function whose information complexity is exponentially smaller than its communication complexity. That is, in [44] – building on earlier works by the same authors – they present an F such that

$$D_\mu(F_\varepsilon) = 2^{\Omega(\mathsf{IC}(F_\varepsilon,\mu))} \gg \mathsf{IC}(F_\varepsilon, \mu). \tag{3.9}$$

In fact, the exponential gap is the largest possible, as it can be shown [13] for all F,

$$D_\mu(F_\varepsilon) = 2^{O(\mathsf{IC}(F_\varepsilon,\mu))}. \tag{3.10}$$

To prove the strongest possible direct sum theorem (3.6) we would have needed π' to be compressed all the way down to $O(I)$ bits of communication (the strongest possible interactive compression result). Even though such a compression is impossible, weaker interactive compression results lead to weaker (but still nontrivial) direct sum theorems. At present, the two strongest compression results, which partially resolve Problem 3.2, compress π to $\tilde{O}(\sqrt{C \cdot I})$ communication[22] [9] and $2^{O(I)}$ communication (3.10), respectively. Note that these results are incomparable since C can be much (e.g., double-exponentially) larger than I.

These result lead to direct sum theorems for randomized communication complexity. As the compression introduces an additional small amount of error, the first result implies for any constant $\rho > 0$,

$$D_{\mu^n}(F_\varepsilon^n) = \widetilde{\Omega}\big(\sqrt{n} \cdot D_\mu(F_{\varepsilon+\rho})\big), \tag{3.11}$$

and the second implies

$$D_{\mu^n}(F_\varepsilon^n) = \Omega\big(n \cdot \log\big(D_\mu(F_{\varepsilon+\rho})\big)\big). \tag{3.12}$$

In summary, we know that perfect compression a la Huffman is impossible in the two-party interactive setting. Mapping out the exact limits of interactive compression remains open:

Problem 3.2 (Interactive compression problem). Given a protocol π whose communication cost is C and whose information cost is I, what is the smallest amount of communication needed to (approximately) simulate π?

As noted above, we know that whenever $I \ll C$, the protocol can be compressed to $o(C)$ bits of communication. At the same time, it is unknown, for example, whether compression to $I^{O(1)} \cdot (\log C)^{O(1)}$ or even to $I^{O(1)} \cdot C^{o(1)}$ is possible. A candidate problem for such a lower bound on compression is presented in [14].

---

22 Here, the $\tilde{O}(\cdot)$ notation hides polylogarithmic factors.

Direct product for communication complexity. Next, we turn our attention to the more difficult *direct product* problem for communication complexity. The direct sum question talks about the amount of resources needed to achieve a certain probability of success on n copies of F. What if that amount of resources is not provided? For example, (3.4) implies that unless $n \cdot \mathsf{IC}(F_\varepsilon, \mu)$ bits of communication are allowed in the computation of F_ε^n, the computation of *some* copy of F will have $< 1 - \varepsilon$ success probability. What does it tell us about the success probability of *all* copies simultaneously? It only tells us that the probability of the protocol succeeding on all copies simultaneously is bounded by $1 - \varepsilon$. This is a very weak bound, since solving the n copies independently leads to a success probability of $(1 - \varepsilon)^n$, which is exponentially small for a constant ε. How can this gap be reconciled? In particular, can one show that Alice and Bob cannot "pool" the errors from all n copies onto the same instances, thus keeping the success probability for each coordinate, as well as the global success probability, close to $1 - \varepsilon$? The direct product problem addresses precisely this question. Let us denote by $\mathsf{suc}(F, \mu, C)$ the highest success probability (with respect to μ) in computing F that can be attained using communication $\leq C$. Thus $\mathsf{suc}(F, \mu, C) \geq 1 - \varepsilon$ is equivalent to $D_\mu(F_\varepsilon) \leq C$. Somewhat informally phrased, the direct product question asks whether

$$\mathsf{suc}\big(F^n, \mu^n, o(n \cdot C)\big) < \mathsf{suc}(F, \mu, C)^{\Omega(n)}? \tag{3.13}$$

The examples showing that (3.6) fails also show that direct product (3.13) for communication is false. The direct sum discussion already suggests that for $\mathsf{suc}(F, \mu, C) = 1 - \varepsilon$, the best scaling of the amount of communication one can hope for is $n \cdot I$, where $I = \mathsf{IC}(F_\varepsilon, \mu)$. This is because, as $n \to \infty$, the per-copy communication cost of computing F with error ε scales as $n \cdot I$.

Thus, the "right" question is whether the direct product property holds when communication scales as the *information complexity* of the problem. If we denote by $\mathsf{suc}^{\mathsf{i}}(F, \mu, I) \geq \mathsf{suc}(F, \mu, I)$ the best success probability one can attain solving F while incurring an *information cost* of at most I, the direct product question for information asks whether

$$\mathsf{suc}\big(F^n, \mu^n, o(n \cdot I)\big) < \mathsf{suc}^{\mathsf{i}}(F, \mu, I)^{\Omega(n)}? \tag{3.14}$$

Note that the success probability on the left-hand side is still with respect to communication. A statement such as this with respect to information cost is bound to be false: Information cost being an average-case quantity, one can attain an information-cost I_n protocol by doing nothing with probability $1 - \delta$, and incurring an information cost of $I_n/\delta \gg n \cdot I$ with probability δ that can be taken only *polynomially* (and not exponentially) small.

This latter version of the direct product theorem was shown to be true up to polylogarithmic factors for boolean functions in **[24, 25]**. To simplify parameters, suppose $\mathsf{suc}^{\mathsf{i}}(F, \mu, I) < 2/3$. Then there are constants c_1, c_2 such that

$$\text{if } T \log T < c_1 n \cdot I, \quad \text{then } \mathsf{suc}(F^n, \mu^n, T) < 2^{-c_2 n}. \tag{3.15}$$

The proof of (3.15) is quite involved and combines ideas from the proof of direct sum theorems and of parallel repetition theorems. The main idea is that an event that happens with probability $> 2^{-c_2 n}$ (namely, the event of succeeding on all coordinates) "confers" at most $\sim c_2$ bits of information onto each coordinate. If c_2 is a small constant, then this extra information is very small and can be ignored. The actual proof involves developing the right information-theoretic language to make this simple-sounding ideas rigorous.

We next turn our attention to an early application of information complexity: exact bounds on communication complexity. We briefly discuss additional applications in Section 3.4.

3.3. Exact communication complexity of set disjointness

One of the great successes of information theory as it applies to (classical, one-way) communication problems is its ability to give precise answers to fairly complicated asymptotic communication problems—those involving complicated dependencies between terminals or complicated channels. Using combinatorial techniques (in most cases) such precision is inaccessible in the two-party setting, since the techniques often lose constant factors by design. In contrast, information complexity extends the precision benefits of one-way information theory to the interactive setting.

We give one specific example of an exact communication complexity bound. Recall that the disjointness problem $\mathrm{Disj}_n(X, Y)$ takes two n-bit vectors X, Y and checks whether there is a location with $X_i = Y_i = 1$. Thus Disj_n is just a disjunction of n independent copies of the two bit $\mathrm{AND}(X_i, Y_i)$ function. Using techniques similar to the proof of Theorem 3.1, one can show that the communication complexity of disjointness is tightly linked with the information complexity of AND. Note that disjointness becomes trivial if many coordinates (X_j, Y_j) of the input are $(1, 1)$. However, any distribution of inputs where $\mu((X_j, Y_j) = (1, 1)) \sim 1/n \to 0$ will not be trivial. More formally, denote by 0^+ a function $f(n)$ of n such that $f(n) = o(1)$ and $f(n) \gg 2^{-O(n)}$. For example, one can take $f(n) = 1/n$. Denote by AND_0 the task of computing AND correctly on all four possible inputs. Then with some work one shows **[18]** that

$$R_{0^+}(\mathrm{Disj}_n) = \Big(\inf_{\mu:\mu(1,1)=0} \mathsf{IC}(\mathrm{AND}_0, \mu)\Big) \cdot n \pm o(n). \tag{3.16}$$

Thus, understanding the precise asymptotics of the communication complexity of Disj_n boils down to understanding the (0-error) information complexity of the two-bit AND function.[23]

The information-theoretically optimal protocol for the two-bit AND function (and for any other function) depends on the prior distribution of the inputs. The protocol attaining the optimal information complexity for the two-bit AND function for symmetric prior distributions (where $\mu(0, 1) = \mu(1, 0)$) is given in Figure 1.[24]

---

23 Note that even when $\mu(1, 1) = 0$ and thus $\mathrm{AND}(X, Y) = 0$ on $\mathrm{supp}(\mu)$, the task AND_0 requires the protocol to *always* be correct – even on the $(1, 1)$ input. Otherwise, $\mathsf{IC}(\mathrm{AND}_0, \mu)$ would trivially be 0.

24 The protocol for general μ is an extension of the protocol in Figure 1, and can be found in **[18]**.

Alice receives $x \in \{0,1\}$; Bob receives $y \in \{0,1\}$.
Goal: to compute $\mathrm{AND}(x,y) = x \wedge y$ *with no error.*

(1) *If $x = 0$ then Alice samples $N^A \in_R [0,1)$ uniformly at random.*
If $x = 1$ then Alice sets $N^A = 1$.

(2) *If $y = 0$ then Bob samples $N^B \in_R [0,1)$ uniformly at random.*
If $y = 1$ then Bob sets $N^B = 1$.

(3) *Alice and Bob monitor the clock C, which starts at value* 0.

(4) *The clock continuously increases to* 1. *If* $\min(N^A, N^B) < 1$, *when the clock reaches* $\min(N^A, N^B)$ *the corresponding player sends* 0 *to the other player, the protocol ends, the players output 0. If* $\min(N^A, N^B) = 1$, *once the clock reaches* 1, *Alice sends* 1 *to Bob, the protocol ends, and the players output* 1.

FIGURE 1
The information-theoretically optimal protocol for $\mathrm{AND}(x,y)$ under prior distribution μ with $\mu(0,1) = \mu(1,0)$

Observe that the "protocol" in Figure 1 is not an actual communication protocol: it involves a continuous-time clock, and not a finite sequence of discrete messages. The protocol can be approximated by a discrete protocol by sampling N^A and N^B from the discrete set $\{0, \frac{1}{r}, \frac{2}{r}, \ldots, \frac{r-1}{r}\}$ instead of $[0,1)$, and then having r iterations of the clock going over multiples of $\frac{1}{r}$.

Interestingly, even in the case of such a simple function as two-bit AND, the information complexity is not attained by any particular protocol, but rather by an infinite family of communication protocols! Moreover, if we denote by $\mathrm{IC}_r(\mathrm{AND}_0)$ the information complexity of AND_0 where the infimum in (3.2) is only taken over protocols of length r, then it turns out that $\mathrm{IC}_r(\mathrm{AND}_0) = \mathrm{IC}(\mathrm{AND}_0) + \Theta(1/r^2)$, implying that an asymptotically optimal protocol is only achieved with a super-constant number of rounds [18]. We do not yet know how general this $1/r^2$ gap phenomenon is, and which communication tasks admit a minimum in (3.2).

By calculating the information cost of the optimal protocol for AND, and maximizing it over all possible distributions μ with $\mu(1,1) = 0$, we obtain from (3.16) that

$$R_{0^+}(\mathrm{Disj}_n) = C_{\mathrm{DISJ}} \cdot n \pm o(n), \quad \text{where } C_{\mathrm{DISJ}} \approx 0.4827. \tag{3.17}$$

Small set Disjointness. An interesting special case of the set-disjointness problem is the *small set disjointness* case. In this setting, only at most $k \ll n$ of the X_i's are 1 and at most k of the Y_i's are 1. In other words, Alice and Bob each have a set of k elements over a universe of $n \gg k$ elements, and they wish to determine whether they have an element in common. Denote this problem by $\mathrm{Disj}_{n,k}$.

The naïve upper bound in this case is $O(k \log n)$, since it takes $O(\log n)$ bits to transmit a single element from the set $\{1, \ldots, n\}$.[25] Somewhat surprisingly, Håstad and Wigderson [51] showed that small set disjointness can be solved using communication linear in k,

$$R_{0^+}(\text{Disj}_{n,k}) = O(k). \tag{3.18}$$

Note that the $\Omega(n)$ lower bound for Disj_n immediately translates into an $\Omega(k)$ lower bound for $\text{Disj}_{n,k}$, leading to

$$R_{0^+}(\text{Disj}_{n,k}) = \Theta(k). \tag{3.19}$$

It turns out that the precise bound follows from the optimality of the protocol in Figure 1 almost immediately. The relevant distribution for the single AND instance is one where the probability of $X = 1$ is $\frac{k}{n}$. Calculating the information cost of the protocol with prior $\mu(1,0) = \mu(0,1) = \frac{k}{n}$, $\mu(0,0) = 1 - \frac{2k}{n}$ yields [18]

$$R_{0^+}(\text{Disj}_{n,k}) = \frac{2}{\ln 2} \cdot k \pm o(k). \tag{3.20}$$

3.4. Some other connections

Let us briefly mention some recent connections between information complexity and other subareas of theoretical computer science.

Streaming: do we need numbers to approximately count? Beyond answering questions such as the direct sum for randomized communication complexity, the main advantage of information complexity is that it allows us to phrase intuitive statements about computation and communication in a rigorous way. We will illustrate it with a sketch of a recent result about the streaming complexity of approximate majority [19].

In the *streaming* setup, inputs $X_1, \ldots, X_n$ arrive one-by-one, and the state of the computation is updated based on the input and the previous state. Thus, the computation can be represented as the following diagram:

$$M_0 \xrightarrow{X_1} M_1(M_0, X_1) \xrightarrow{X_2} M_2(M_1, X_2) \xrightarrow{X_3} \cdots \xrightarrow{X_n} M_n(M_{n-1}, X_n).$$

The answer is then computed from the final state M_n. Typically we are interested in either the average memory used by the algorithm $\bar{m} = \frac{1}{n} \sum_i |M_i|$, or the maximum amount of memory $m_{\max} := \max |M_i|$.[26]

Consider the following problem: *Given n i.i.d. coin tosses of $X_i \sim B_{1/2}$, compute* $\mathbf{MAJ}(X_1, \ldots, X_n)$ *while allowing a 1% error probability.*[27]

The simplest possible algorithm would just count the bits: set $M_0 = 0$ and

$$M_i(M_{i-1}, X_i) := M_{i-1} + X_i,$$

so that $M_n = \sum X_i$, from which one can compute $\mathbf{MAJ}(X_1, \ldots, X_n)$ *with no error*. This solution requires $\bar{m} \sim \log n$ memory. It is not hard to show that producing an exact count

25 The precise bound is $O(k \log(n/k))$, but this becomes $O(k \log n)$ whenever $n > k^{1+c}$.

26 Here $|M|$ is the length of M in bits.

27 That is, the algorithm needs to be correct at least 99% of the time.

requires this much memory. What about approximate counting? Can we avoid storing numbers if we only wish to count the numbers approximately? It turns out that the answer is 'no': indeed, $\bar{m} = \Omega(\log n)$ is necessary.

A key step of the construction is to correctly define the information cost of this streaming setup as[28]

$$\mathsf{IC}(M) := \sum_{i=1}^{n} \sum_{j=1}^{i} I(M_i; X_j | M_{j-1}). \tag{3.21}$$

Each term of the sum captures how much information the ith message still retains about input X_j that appeared earlier. As in many other cases, information here is a lower bound on $\sum |M_i|$. It turns out that for a typical pair we must have

$$I(M_i; X_j | M_{j-1}) \gtrsim \frac{1}{i - j + 1}, \tag{3.22}$$

and therefore

$$\bar{m} = \frac{1}{n} \cdot \sum_i |M_i| \geq \frac{1}{n} \cdot \mathsf{IC}(M) \gtrsim \frac{1}{n} \cdot \sum_{i=1}^{n} \sum_{j=1}^{i} \frac{1}{i - j + 1} = \Theta(\log n). \tag{3.23}$$

The main inequality (3.22) is proved by rephrasing the following intuition in information-theoretic terms. If we break the stream into $k = 2^r$ blocks $B_1, \dots, B_k$ of length n/k each, then at the end of each block B_i, the message $M_{i \cdot n/k}$ should contain at least 1 bit of information about the approximate count in the previous block. This translates into containing at least $\frac{k}{n}$ bits of information about a typical X_j in that block, leading to (3.22). This proof also gives intuition for the need to have $\Omega(\log n)$ bits of information in the streaming algorithm: ~ 1 bit of information needs to be dedicated to each of $\log n$ "scales" of the stream.

Distributed learning. All large-scale machine learning today is performed using a large number of processing cores. As a result, communication costs and delays often dominate the overall execution time. This motivates efforts to minimize communication between worker cores, and to understand the fundamental limits of communication needed to complete basic tasks—such as distributed parameter estimation **[108]**. Information complexity (and its ability to bring in tools from information theory, such as strong data processing inequalities) has led to tight results about problems such as distributed sparse parameter estimation **[17, 46]**.

Parallel repetition. Parallel repetition first appeared in the context of Probabilistically Checkable Proofs (PCP) and hardness amplification. Hardness amplification is accomplished here by taking a task T (e.g., a verification procedure that allows authorized provers to pass the test, while unauthorized provers pass with probability at most $1 - \varepsilon$), and creating a task T^n by taking n independent instances of T. It has been shown **[37, 54, 84, 86]** that as n grows,

---

28 An important benefit of an information-theoretic lower bound – as opposed to a combinatorial one – is that it can be used in the context of a direct sum theorem to lower bound the cost of doing multiple copies of a problem in parallel. Indeed, this is how it was used in **[19]**.

the success probability of unauthorized provers goes to 0. Unfortunately, it does not go to 0 as $(1-\varepsilon)^n$. Indeed, as shown by a counterexample constructed by Raz [87], the best rate one can hope for is $(1-\varepsilon^2)^n$. The reason for this, pointed out by an earlier example by Feige and Verbitsky [41], is that the answers can be arranged to align errors together, so that when the provers fail, they fail on a lot more than εn coordinates at the same time. This is possible when answers are allowed to be correlated.

It should not be surprising that the parallel repetition problem shares some similarities with the *direct product* problem in communication complexity. In both cases, the concern is that correlations between coordinates will lead to an unexpectedly high success probability—much higher than $(1-\varepsilon)^n$. Indeed, the proof of the direct product theorem for information complexity (3.15) can be combined with "standard" parallel repetition machinery to obtain the most general parallel-repetition theorem to-date [15].

In turn, parallel repetition has interesting connections to foams—low surface area tiling of $\mathbb{R}^n$ by $\mathbb{Z}^n$ [40], leading to new geometric constructions that implicitly use information theory [21,65].

Quantum information complexity. Information theory and its quantum extensions have been used to obtain key results in quantum communication complexity [58,66]. The basic notions of information complexity as discussed in the previous section can be adapted to the quantum setting [98]. Unlike classical information complexity, the quantum information complexity of the two-bit AND as in (3.16) actually vanishes as the number of rounds goes to ∞. This is consistent with the fact that the quantum communication complexity of disjointness is $O(\sqrt{n}) = o(n)$ [1,28], and an earlier result by Elitzur and Vaidman on quantum bomb testing [38]. Nonetheless, it is possible to use the information complexity machinery to get a near-tight bound on the information complexity of AND in terms of the number of rounds (the dependence is $\mathsf{IC}(\mathrm{AND}_0, \mu) = \widetilde{\Theta}(\frac{1}{r})$ for the best r-round protocol). This gives the tight bound of $\widetilde{\Omega}(\frac{n}{r} + r)$ on the r-round quantum communication complexity of Disj_n [16].

Interactive error-correcting codes. Most of the discussion so far focused on developing (two-party) information complexity as a tool for studying communication complexity and related models of computation. In other words, the motivation has been mostly complexity-theoretic.

The main aim of the original information theory project, starting from the work of Shannon in the 1940s was to further coding theory and practice. Coding theory is concerned with developing efficient codes that are robust to errors for data storage and transmission—information theory has become a tool for giving bounds (that are sometimes tight) on what codes are possibly attainable.

In the context of interactive communication one can view interactive information complexity (and even communication complexity) as one aspect of coding for interactive communication (one dealing with noiseless coding). Another important aspect of coding theory is dealing with *noisy communication*.

In the interactive setting, this gives rise to questions about interactive error-correcting codes: given a noisy channel,[29] encode the entire interactive computation in a way that is robust to noise. The problem was first studied by Schulman in the 1990s [91], who showed that it is possible to protect an interactive protocol against a small amount of adversarial noise. Note that "standard" techniques of encoding each message separately cannot work here, since in such an encoding an adversary would be able to derail an entire protocol by completely replacing one of the messages.

The area has seen a resurgence of activity since the work by the author with Rao [23], which showed that it is possible to encode an interactive protocol in a way that protects it against $\frac{1}{4} - \varepsilon$ adversarial error rate. Since then there has been much activity dealing with making the constructions more efficient, more error resilient, and apply in a broader set of regimes. A survey on the developments in the field as of 2017 can be found in [47].

In addition to developing interactive coding schemes, some of the fundamental questions such about interactive channel capacity (as the analogue of noninteractive Shannon's channel capacity) need to be revisited in the interactive setting [20, 67].

4. CHALLENGES AND NEXT STEPS

As we have seen in the last section, information complexity has been a useful tool (and the right "language") in a variety of settings involving communication. We have also briefly seen in Section 1.4 that there are several attack routes for obtaining strong (and currently apparently unreachable) separations between complexity classes using communication complexity. This raises the natural question of whether information complexity can be helpful with these communication complexity bounds.

There are several settings where information complexity (and, more generally, information-theoretic reasoning) appears to get "stuck". Specific examples include:

- Extending tight communication lower bounds to 3 or more parties in the number-on-the-forehead model (with $(\log n)^{O(1)}$ parties this would imply difficult circuit lower bounds [10]).
- Pătraşcu's multiphase conjecture [83] – a lower bound conjecture against a specific model of computation with 3 parties. The conjecture implies strong dynamic data structures lower bounds.
- The Arthur–Merlin model in communication complexity. This is a particularly challenging model for communication complexity lower bounds. In fact, it is the communication-complexity analogue of the corresponding Arthur–Merlin AM class in computational complexity [5]. We will not define it here, only mention that it can be thought of as a communication protocol with $2 + \varepsilon$ players; "ε" here is Merlin, who can provide Alice and Bob with an untrusted hint, but then cannot

29 As in the one-way communication case, the various models of noise include adversarial and various forms of random noise.

participate in the protocol. There is evidence that the Arthur–Merlin communication model is resistant to information complexity techniques [49].

- Extending parallel repetition results from the setting with two provers to settings with three or more provers. While tight bounds are known in the two-prover case, there is an exponential gap between the best upper and lower bounds even in some of the simplest settings with three provers [48].

There appears to be a common theme in terms of what makes these examples difficult—namely, the existence of *secure computation* in the relevant models.

Secure computation. Throughout most of this note information complexity was presented as the interactive extension of Shannon's entropy (emphasizing connections to amortized communication cost). Historically, the fist appearance of information complexity within theoretical computer science was in the context of privacy of communication protocols [7, 90].[30] Formula (3.2) for information complexity *exactly* quantifies the smallest possible information-theoretic privacy loss[31] that Alice and Bob can experience while successfully completing task T. It is important that the model here is *information-theoretic* security: it is possible to attain cryptographic security based on computational hardness assumptions [107].

In contrast with the cryptographic results, we now know that information-theoretic privacy in the honest-but-curious model is unattainable. Many of the communication complexity bounds, such as results (3.17) and (3.20) actually apply to information complexity as well, which means that for these problems there is (asymptotically) no gap between information and communication, and the shortest possible protocol is also the one that reveals the least information to the participants about the inputs. In other words, *information-theoretically secure two-party computation is impossible*.

Surprisingly, with three or more players information-theoretically secure computation becomes possible [11, 32]: if Alice, Bob and Charlie have inputs X, Y, Z, respectively, and have pairwise private channels,[32] then any function $F(X, Y, Z)$ can be computed in such a way that the only thing Alice learns about (Y, Z) is the value of $F(X, Y, Z)$ (and similarly for the other two players).

The result above means that while one can write the natural expression for 3-party information complexity, and even prove a direct sum result about it, the result will be vacuous: $n \times 0 = 0$, since the information complexity of any function in this model is zero.

This pattern repeats itself when one tries to prove Arthur–Merlin lower bounds using information complexity. Here, the relevant result about secure computation has to do with the channel used. Communication so far was defined over the binary channel where Alice and Bob send individual bits. A different kind of channel would take in input from both Alice and Bob, and then distribute an output to them. The simplest channel of this kind

30 And, more recently, in the context of differential privacy [76].

31 In the "honest-but-curious" model of privacy, where participants do not actively deviate from the protocol to learn information they are not supposed to learn.

32 That is, Alice can talk with Bob without Charlie listening.

is the Shannon–Blackwell Binary Multiplying channel [93]: Alice and Bob each send a bit $a \in \{0, 1\}$, $b \in \{0, 1\}$, respectively, into the channel, and the channel sends to both of them the value of $a \wedge b \in \{0, 1\}$. Note that in this channel, if Alice sends $a = 0$ into the channel, she does not learn anything about the value of b.

It turns out that over the Binary Multiplying channel (BMC) one can implement secure *two-party* computation [64]. Once again, one can write expressions for information complexity over the BMC, and obtain direct sum results similar to Theorem 3.1 above, but the result would be vacuous of the form $0 + 0 = 0$.

Analytic techniques to bypass the secure computation barrier? It remains to be seen whether the barrier to using information-theoretic techniques (or any techniques for that matter) for the problems discussed above is merely a technical one, or is related to something deeper.

It is worth noting that in the two-party case (for both communication and parallel repetition) it is possible to rephrase most proofs in analytic terms, in terms of values of relevant semidefinite programs on the function's value matrices [37,69,72,95]. In fact, in cases where both an analytic and an information-theoretic proof exists, the analytic proof often predated the information-theoretic one. A notable example of a problem for which we had a number of analytic proofs [30,94] before an information-theoretic one [52] is for the Gap Hamming Distance. In other cases, such as exact communication bounds (3.17) and (3.20), information complexity appears to be the right tool.

When moving from two to three or more parties, in the analytic setup, the main object of consideration becomes tensors instead of matrices (see, e.g., [26]). They are much more difficult to deal with, both because some of the nicer aspects of linear algebra are missing, and because the theory as a whole is much less developed. A promising strategy for pinning down the exact difficulty in the examples above would be to trace it to a statement about 3-dimensional tensors.

If that statement is true, the proof might be useful in communication and parallel repetition applications (as has been the case with the analytic tools in the two-party setting [37,69,72,95]). Moreover, using 2-party information complexity as a guiding map, it might lead to new "information-like" definitions that do not currently exist.

If that statement is false, or turns out to be very difficult to prove even in its analytic form, then we might have discovered a mathematical obstacle to computational complexity lower bounds that would guide future lower bound efforts.

In either case, we can look forward to exciting results on the quest towards unconditional lower bounds in various computation models.

ACKNOWLEDGMENTS

I would like to thank Ankit Garg, Sumegha Garg, and Omri Weinstein for their comments on earlier versions of this paper.

This note is based on many works with students, postdocs, and colleagues. Working with them is one of the biggest privileges of my career and I am deeply grateful for everything

I have learned from them and everything we have done and will continue to do in the future.

FUNDING

Work on this paper was supported in part by the NSF Alan T. Waterman Award, Grant No. 1933331, a Packard Fellowship in Science and Engineering, and the Simons Collaboration on Algorithms and Geometry. Work described here has been generously supported over the years by NSERC, the NSF, the Alfred P. Sloan Foundation, the University of Toronto, Princeton University, the Institute for Advanced Study, and the John Templeton Foundation, through a Turing Centenary Fellowship.

REFERENCES

[1] S. Aaronson and A. Ambainis, Quantum search of spatial regions. In *FOCS'03 proceedings*, pp. 200–209, IEEE, 2003.

[2] S. Arora and B. Barak, *Computational complexity: a modern approach*. Cambridge University Press, 2009.

[3] S. Arora, C. Lund, R. Motwani, M. Sudan, and M. Szegedy, Proof verification and the hardness of approximation problems. *J. ACM* **45** (1998), no. 3, 501–555.

[4] S. Arora and S. Safra, Probabilistic checking of proofs: a new characterization of np. *J. ACM* **45** (1998), no. 1, 70–122.

[5] L. Babai, P. Frankl, and J. Simon, Complexity classes in communication complexity theory. In *FOCS'86 proceedings*, pp. 337–347, IEEE, 1986.

[6] T. Baker, J. Gill, and R. Solovay, Relativizations of the $\mathcal{P} = ?\mathcal{NP}$ question. *SIAM J. Comput.* **4** (1975), no. 4, 431–442.

[7] R. Bar-Yehuda, B. Chor, E. Kushilevitz, and A. Orlitsky, Privacy, additional information and communication. *IEEE Trans. Inf. Theory* **39** (1993), no. 6, 1930–1943.

[8] Z. Bar-Yossef, T. S. Jayram, R. Kumar, and D. Sivakumar, An information statistics approach to data stream and communication complexity. *J. Comput. System Sci.* **68** (2004), no. 4, 702–732.

[9] B. Barak, M. Braverman, X. Chen, and A. Rao, How to compress interactive communication. *SIAM J. Comput.* **42** (2013), no. 3, 1327–1363.

[10] R. Beigel and J. Tarui, On ACC. *Comput. Complexity* **4** (1994), no. 4, 350–366.

[11] M. Ben-Or, S. Goldwasser, J. Kilian, and A. Wigderson, Multi-prover interactive proofs: how to remove intractability assumptions. In *STOC'88 proceedings*, pp. 113–131, ACM, 1988.

[12] M. Braverman, Interactive information and coding theory. In *Proceeding of the International Congress of Mathematicians, ICM'2014*, Kyung Moon Sa, 2014.

[13] M. Braverman, Interactive information complexity. *SIAM J. Comput.* **44** (2015), no. 6, 1698–1739.

[14] M. Braverman, A. Ganor, G. Kol, and R. Raz, A candidate for a strong separation of information and communication. In *ITCS'18 proceedings*, Schloss Dagstuhl-Leibniz-Zentrum für Informatik, 2018.

[15] M. Braverman and A. Garg, Small value parallel repetition for general games. In *STOC'15 proceedings*, pp. 335–340, ACM, 2015.

[16] M. Braverman, A. Garg, Y. K. Ko, J. Mao, and D. Touchette, Near-optimal bounds on the bounded-round quantum communication complexity of disjointness. *SIAM J. Comput.* **47** (2018), no. 6, 2277–2314.

[17] M. Braverman, A. Garg, T. Ma, H. L. Nguyen, and D. P. Woodruff, Communication lower bounds for statistical estimation problems via a distributed data processing inequality. In *STOC'16 proceedings*, pp. 1011–1020, ACM, 2016.

[18] M. Braverman, A. Garg, D. Pankratov, and O. Weinstein, From information to exact communication. In *STOC'13 proceedings*, pp. 151–160, ACM, 2013.

[19] M. Braverman, S. Garg, and D. P. Woodruff, The coin problem with applications to data streams. In *FOCS'20 proceedings*, pp. 318–329, IEEE, 2020.

[20] M. Braverman and J. Mao, Simulating noisy channel interaction. In *ITCS'15 proceedings*, pp. 21–30, ACM, 2015.

[21] M. Braverman and D. Minzer, Optimal tiling of the euclidean space using permutation-symmetric bodies. In *Complexity'21 proceedings*, Schloss Dagstuhl-Leibniz-Zentrum für Informatik, 2021.

[22] M. Braverman and A. Rao, Information equals amortized communication. In *FOCS'11 proceedings*, pp. 748–757, IEEE, 2011.

[23] M. Braverman and A. Rao, Towards coding for maximum errors in interactive communication. In *STOC'11 proceedings*, pp. 159–166, ACM, 2011.

[24] M. Braverman, A. Rao, O. Weinstein, and A. Yehudayoff, Direct products in communication complexity. In *FOCS'13 proceedings*, pp. 746–755, IEEE, 2013.

[25] M. Braverman and O. Weinstein, An interactive information odometer with applications. *Electron. Colloq. Comput. Complex.*, https://eccc.weizmann.ac.il/report/2014/047, 2014.

[26] J. Briët and T. Vidick, Explicit lower and upper bounds on the entangled value of multiplayer XOR games. *Comm. Math. Phys.* **321** (2013), no. 1, 181–207.

[27] J. Brody, H. Buhrman, M. Koucký, B. Loff, F. Speelman, and N. Vereshchagin, Towards a reverse Newman's theorem in interactive information complexity. *Algorithmica* **76** (2016), no. 3, 749–781.

[28] H. Buhrman, R. Cleve, and A. Wigderson, Quantum vs. classical communication and computation. In *STOC'98 proceedings*, pp. 63–68, ACM, 1998.

[29] A. Buttari, J. Langou, J. Kurzak, and J. Dongarra, A class of parallel tiled linear algebra algorithms for multicore architectures. *Parallel Comput.* **35** (2009), no. 1, 38–53.

[30] A. Chakrabarti and O. Regev, An optimal lower bound on the communication complexity of gap-hamming-distance. *SIAM J. Comput.* **41** (2012), no. 5, 1299–1317.

[31] A. Chakrabarti, Y. Shi, A. Wirth, and A. Yao, Informational complexity and the direct sum problem for simultaneous message complexity. In *FOCS'01 proceedings*, pp. 270–278, IEEE, 2001.

[32] D. Chaum, C. Crépeau, and I. Damgard, Multiparty unconditionally secure protocols. In *STOC'88 proceedings*, pp. 11–19, ACM, 1988.
[33] J. Chen, X. Huang, I. A. Kanj, and G. Xia, Linear FPT reductions and computational lower bounds. In *STOC'04 proceedings*, pp. 212–221, ACM, 2004.
[34] S. A. Cook, A taxonomy of problems with fast parallel algorithms. *Inf. Control* **64** (1985), no. 1–3, 2–22.
[35] T. M. Cover and J. A. Thomas, *Elements of information theory*. 2nd edn., J. Wiley and Sons, New York, 2006.
[36] J. W. Demmel, M. T. Heath, and H. A. Van der Vorst, Parallel numerical linear algebra. *Acta Numer.* **2** (1993), 111–197.
[37] I. Dinur and D. Steurer, Analytical approach to parallel repetition. In *STOC'14 proceedings*, pp. 624–633, ACM, 2014.
[38] A. C. Elitzur and L. Vaidman, Quantum mechanical interaction-free measurements. *Found. Phys.* **23** (1993), no. 7, 987–997.
[39] T. Feder, E. Kushilevitz, M. Naor, and N. Nisan, Amortized communication complexity. *SIAM J. Comput.* **24** (1995), no. 4, 736–750.
[40] U. Feige, G. Kindler, and R. O'Donnell, Understanding parallel repetition requires understanding foams. In *FOCS'07 proceedings*, pp. 179–192, IEEE, 2007.
[41] U. Feige and O. Verbitsky, Error reduction by parallel repetition – A negative result. *Combinatorica* **22** (2002), no. 4, 461–478.
[42] M. Furst, J. B. Saxe, and M. Sipser, Parity, circuits, and the polynomial-time hierarchy. *Math. Syst. Theory* **17** (1984), no. 1, 13–27.
[43] A. Ganor, G. Kol, and R. Raz, Exponential separation of information and communication. In *FOCS'14 proceedings*, pp. 176–185, IEEE, 2014.
[44] A. Ganor, G. Kol, and R. Raz, Exponential separation of information and communication for boolean functions. *J. ACM* **63** (2016), no. 5, 1–31.
[45] A. Ganor, G. Kol, and R. Raz, Exponential separation of communication and external information. *SIAM J. Comput.* **50** (2019), no. 3, STOC16–236.
[46] A. Garg, T. Ma, and H. Nguyen, On communication cost of distributed statistical estimation and dimensionality. *Adv. Neural Inf. Process. Syst.* **27** (2014).
[47] R. Gelles, Coding for interactive communication: a survey. *Found. Trends Theor. Comput. Sci.* **13** (2017), no. 1–2, 1–157.
[48] U. Girish, K. Mittal, R. Raz, and W. Zhan, Polynomial bounds on parallel repetition for all 3-player games with binary inputs. 2022, arXiv:2204.00858.
[49] M. Göös, T. Pitassi, and T. Watson, Zero-information protocols and unambiguity in arthur–merlin communication. *Algorithmica* **76** (2016), no. 3, 684–719.
[50] J. Håstad, Almost optimal lower bounds for small depth circuits. In *STOC'86 proceedings*, pp. 6–20, ACM, 1986.
[51] J. Håstad and A. Wigderson, The randomized communication complexity of set disjointness. *Theory Comput.* **3** (2007), no. 1, 211–219.
[52] U. Hadar, J. Liu, Y. Polyanskiy, and O. Shayevitz, Communication complexity of estimating correlations. In *STOC'19 proceedings*, pp. 792–803, ACM, 2019.

[53] W. Hesse, E. Allender, and D. A. M. Barrington, Uniform constant-depth threshold circuits for division and iterated multiplication. *J. Comput. System Sci.* **65** (2002), no. 4, 695–716.

[54] T. Holenstein, Parallel repetition: simplifications and the no-signaling case. In *STOC'07 proceedings*, pp. 411–419, ACM, 2007.

[55] D. A. Huffman, A method for the construction of minimum redundancy codes. *Proc. IRE* **40** (1952), no. 9, 1098–1101.

[56] R. Impagliazzo and R. Paturi, On the complexity of k-SAT. *J. Comput. System Sci.* **62** (2001), no. 2, 367–375.

[57] R. Impagliazzo, R. Paturi, and F. Zane, Which problems have strongly exponential complexity? *J. Comput. System Sci.* **63** (2001), no. 4, 512–530.

[58] R. Jain, J. Radhakrishnan, and P. Sen, A lower bound for the bounded round quantum communication complexity of set disjointness. In *FOCS'03 proceedings*, pp. 220–229, IEEE, 2003.

[59] S. Jukna, *Boolean function complexity: advances and frontiers. 5*. Springer, 2012.

[60] B. Kalyanasundaram and G. Schnitger, The probabilistic communication complexity of set intersection. *SIAM J. Discrete Math.* **5** (1992), no. 4, 545–557.

[61] M. Karchmer, *Communication complexity a new approach to circuit depth*. 1989.

[62] M. Karchmer, R. Raz, and A. Wigderson, Super-logarithmic depth lower bounds via the direct sum in communication complexity. *Comput. Complexity* **5** (1995), no. 3/4, 191–204.

[63] R. M. Karp, Reducibility among combinatorial problems. In *Complexity of computer computations*, pp. 85–103, Springer, 1972.

[64] J. Kilian, A general completeness theorem for two party games. In *STOC'91 proceedings*, pp. 553–560, ACM, 1991.

[65] G. Kindler, R. O'Donnell, A. Rao, and A. Wigderson, Spherical cubes and rounding in high dimensions. In *FOCS'08 proceedings*, pp. 189–198, IEEE, 2008.

[66] H. Klauck, A. Nayak, A. Ta-Shma, and D. Zuckerman, Interaction in quantum communication. *IEEE Trans. Inf. Theory* **53** (2007), no. 6, 1970–1982.

[67] G. Kol and R. Raz, Interactive channel capacity. In *STOC'13 proceedings*, pp. 715–724, ACM, 2013.

[68] E. Kushilevitz and N. Nisan, *Communication complexity*. Cambridge University Press, Cambridge, 1997.

[69] T. Lee and A. Shraibman, Disjointness is hard in the multiparty number-on-the-forehead model. *Comput. Complexity* **18** (2009), no. 2, 309–336.

[70] T. Lee and A. Shraibman, *Lower bounds in communication complexity*. Now Publishers Inc, 2009.

[71] N. Linial, Y. Mansour, and N. Nisan, Constant depth circuits, Fourier transform, and learnability. *J. ACM* **40** (1993), no. 3, 607–620.

[72] N. Linial, S. Mendelson, G. Schechtman, and A. Shraibman, Complexity measures of sign matrices. *Combinatorica* **27** (2007), no. 4, 439–463.

[73] N. Ma and P. Ishwar, Some results on distributed source coding for interactive function computation. *IEEE Trans. Inf. Theory* **57** (2011), no. 9, 6180–6195.

[74] N. Ma and P. Ishwar, The infinite-message limit of two-terminal interactive source coding. *IEEE Trans. Inf. Theory* **59** (2013), no. 7, 4071–4094.

[75] A. Madry, Continuous optimization: the 'right' language for graph algorithms? (invited talk). In *FSTTCS'16*, Schloss Dagstuhl-Leibniz-Zentrum fuer Informatik, 2016.

[76] A. McGregor, I. Mironov, T. Pitassi, O. Reingold, K. Talwar, and S. Vadhan, The limits of two-party differential privacy. In *FOCS'10 proceedings*, pp. 81–90, IEEE, 2010.

[77] M. Naor, A. Orlitsky, and P. W. Shor, Three results on interactive communication. *IEEE Trans. Inf. Theory* **39** (1993), no. 5, 1608–1615.

[78] I. Newman, Private vs. common random bits in communication complexity. *Inform. Process. Lett.* **39** (1991), no. 2, 67–71.

[79] R. O'Donnell, *Analysis of boolean functions*. Cambridge University Press, 2014.

[80] A. Orlitsky, Average-case interactive communication. *IEEE Trans. Inf. Theory* **38** (1992), no. 5, 1534–1547.

[81] A. Orlitsky and A. El Gamal, Communication complexity. In *Complexity in information theory*, pp. 16–61, Springer, 1988.

[82] A. Orlitsky and J. R. Roche, Coding for computing. In *FOCS'95 proceedings*, pp. 502–511, IEEE, 1995.

[83] M. Pătraşcu, Towards polynomial lower bounds for dynamic problems. In *STOC'10 proceedings*, pp. 603–610, ACM, 2010.

[84] A. Rao, Parallel repetition in projection games and a concentration bound. *SIAM J. Comput.* **40** (2011), no. 6, 1871–1891.

[85] A. Rao and A. Yehudayoff, *Communication complexity: and applications*. Cambridge University Press, 2020.

[86] R. Raz, A parallel repetition theorem. *SIAM J. Comput.* **27** (1998), no. 3, 763–803.

[87] R. Raz, A counterexample to strong parallel repetition. *SIAM J. Comput.* **40** (2011), no. 3, 771–777.

[88] A. Razborov, On the distributed complexity of disjointness. *Theoret. Comput. Sci.* **106** (1992).

[89] A. A. Razborov, Lower bounds for the size of circuits of bounded depth with basis $\vee, \oplus$. *Math. Notes Acad. Sci. USSR* **41** (1987), no. 4, 333–338.

[90] T. Satoh, K. Kurosawa, and S. Tsujii, Privacy for multi-party protocols. In *International workshop on the theory and application of cryptographic techniques*, pp. 252–260, Springer, 1992.

[91] L. J. Schulman, Coding for interactive communication. *IEEE Trans. Inf. Theory* **42** (1996), no. 6, 1745–1756.

[92] R. Sedgewick and K. Wayne, *Introduction to programming in Java: an interdisciplinary approach*. Addison-Wesley Professional, 2017.

[93] C. E. Shannon, Two-way communication channels. *Proc. 4th Berkeley Symp. Math. Stat. Prob* **1** (1961), no. 3, 611–644.

[94] A. A. Sherstov, The communication complexity of gap Hamming distance. *Theory Comput.* **8** (2012), no. 1, 197–208.

[95] A. A. Sherstov, The multiparty communication complexity of set disjointness. In *Proceedings of the forty-fourth annual ACM symposium on theory of computing*, pp. 525–548, ACM, 2012.

[96] A. Shpilka and A. Yehudayoff, Arithmetic circuits: a survey of recent results and open questions. *Found. Trends Theor. Comput. Sci.* **5** (2010), no. 3–4, 207–388.

[97] R. Smolensky, Algebraic methods in the theory of lower bounds for boolean circuit complexity. In *STOC'87 proceedings*, pp. 77–82, ACM, 1987.

[98] D. Touchette, Quantum information complexity. In *STOC'15 proceedings*, pp. 317–326, ACM, 2015.

[99] A. M. Turing, On computable numbers, with an application to the Entscheidungsproblem. *J. of Math.* **58** (1936), no. 345–363, 5.

[100] H. Tyagi, S. B. Venkatakrishnan, P. Viswanath, and S. Watanabe, Information complexity density and simulation of protocols. *IEEE Trans. Inf. Theory* **63** (2017), no. 11, 6979–7002.

[101] J. von Neumann, The general and logical theory of automata. In *John von Neumann, collected works*, chap. 9, pp. 288–328, Pergamon Press, 1951.

[102] O. Weinstein, Information complexity and the quest for interactive compression. *ACM SIGACT News* **46** (2015), no. 2, 41–64.

[103] V. V. Williams, On some fine-grained questions in algorithms and complexity. In *Proceeding of the International Congress of Mathematicians, ICM'2018*, pp. 3447–3487, World Scientific, 2018.

[104] A. Wyner and J. Ziv, The rate-distortion function for source coding with side information at the decoder. *IEEE Trans. Inf. Theory* **22** (1976), no. 1, 1–10.

[105] A. C. C. Yao, Some complexity questions related to distributive computing (preliminary report). In *STOC'79 proceedings*, pp. 209–213, ACM, 1979.

[106] A. C. C. Yao, Lower bounds by probabilistic arguments. In *Foundations of computer science, 1983, 24th annual symposium on*, pp. 420–428, IEEE, 1983.

[107] A. C. C. Yao, How to generate and exchange secrets. In *FOCS'86 proceedings*, pp. 162–167, IEEE, 1986.

[108] Y. Zhang, J. Duchi, M. I. Jordan, and M. J. Wainwright, Information-theoretic lower bounds for distributed statistical estimation with communication constraints. *Adv. Neural Inf. Process. Syst.* **26** (2013).

MARK BRAVERMAN

Department of Computer Science, Princeton University, Princeton, NJ 08540, USA, mbraverm@cs.princeton.edu

This Page Intentionally Left Blank

POPULARIZATION OF MATH: SKETCHES OF RUSSIAN PROJECTS AND TRADITIONS

NIKOLAI ANDREEV

ABSTRACT

We give a guided tour of unique Russian traditions of math popularization. While many have already become part of the worldwide practice, some are yet to follow suit. What unites them is the desire to not only amuse people with clever puzzle-solving, but explain the math behind it.

MATHEMATICS SUBJECT CLASSIFICATION 2020

Primary 00A09; Secondary 00A08, 97-01, 97U10

KEYWORDS

Popularization of mathematics, mathematical etudes

ICM INTERNATIONAL CONGRESS OF MATHEMATICIANS 2022 JULY 6–14

© 2022 International Mathematical Union
Proc. Int. Cong. Math. 2022, Vol. 1, pp. 322–330
DOI 10.4171/ICM2022/209

Published by EMS Press and licensed under a CC BY 4.0 license

Throughout the 20th century, Russia had accumulated a great wealth of unique traditions in mathematical education and popularization of mathematics. Many of those, such as mathematical circles or clubs and summer schools, have already become part of the worldwide practice of popularizing mathematics. Other ideas and projects, especially those that only emerged in the 21st century, are yet to follow suit. Below we try to give a guided tour of different traditions. What unites them is the desire to not only amuse people with clever puzzle-solving, but explain the math behind it.

The first printed mathematical textbook, Arithmetic by Leonty Magnitsky, was published in Russia in 1703. The textbook was written at the behest of tsar Peter the Great, for the Moscow School of Mathematics and Navigation that he had established. Arithmetic was followed by Geometry (The Methods of Compasses and Ruler, or the Selected Fundamentals in the Mathematical Arts) translated by the tsar's associate James Bruce. The year was 1709, the Northern War was far from over, and Peter edited the manuscript right at the front lines. A copy of Geometry with the tsar's edits is preserved in the archive. For one of the editions, Peter the Great personally wrote a chapter with precise geometric instructions on how to make an accurate sundial! At about the same time, in 1705, the first mathematical poster was published – a wall table *"A new method of arithmetic theoretic or visual,"* compiled by V. Kiprianov, engraved by F. Nikitin with M. Petrov.

ЗАНИМАТЕЛЬНЫЯ

и

УВЕСЕЛИТЕЛЬНЫЯ ЗАДАЧИ,

изданныя

ИВАНОМЪ БУТТЕРОМЪ.

МОСКВА.

въ типографіи Августа Семена,

при Императорской Медико-Хирург. Академіи.

1831.

Left: Geometry textbook with corrections by Peter the Great **Center:** The first Russian math poster
Right: Cover of the Amusing and Entertaining Problems and Riddles

In the next 170 years, mathematics in the Russian Empire became part of university and school curriculum. During the same period, Russian mathematics became associated with such names as L. Euler (1707–1783), N. I. Lobachevsky (1792–1856), N. D. Brashman (1796–1866), M. V. Ostrogradsky (1801–1861), V. Ya. Bunyakovsky (1804–1889), P. L. Chebyshev (1821–1894), and others. And while, at the time, there was no pressing need to popularize mathematics just yet, some steps were already taken in that direction.

The first Russian outreach book on arithmetic was called *"Arithmetic Guesswork for Fun and Pleasure"* (62 pages, 41 problems), by I. Krasnopolsky. This book (or rather, a small booklet by modern standards) was published in 1789. The year 1831 saw a remarkable book by I. Butter: *"Amusing and Entertaining Problems and Riddles."* Its second edition, with 56 pages, followed in 1844. From about the middle of the 19th century, amusing mathematical problems and publications about mathematicians and their achievements would even make an occasional appearance in mainstream newspapers and magazines. The first attempts at publishing specialized journals were made as well: the more specialized Educational Mathematical Journal (1833–1834, K. Ya. Kupfer) and the Bulletin of Mathematical Sciences (1861–1862, M. M. Gusev), with a target audience of both educators and anyone curious about math. One of the world's oldest mathematical societies, the Moscow Mathematical Society, began publishing its own journal, Mathematical Collection (or Sbornik in Russian), in 1866. During the first few years (from 1867 to 1882), the journal had not just the research section, but also a second section, designed for teachers of mathematics.

The appearance of the Bulletin of Experimental Physics and Elementary Mathematics https://vofem.ru/ in 1886 had a lasting influence on popularization of mathematics in Russia. In the first issue, editor-in-chief E. K. Shpachinsky wrote on behalf of the editorial board:

> *"Our journal is intended mainly, but not exclusively, for the young people studying at our educational institutions, and therefore it will, first of all, aim to satisfy, in the field of physical and mathematical sciences, the need to broaden one's mental horizons, which is especially strong in adolescence and is always found among the young students in the form of an irresistible urge to know more than what the official curriculum provides. [...] Furthermore, our journal is also intended for all teachers of physics and elementary mathematics in general, mainly for the purpose of uniting our educators, scattered as they are across Russia."*

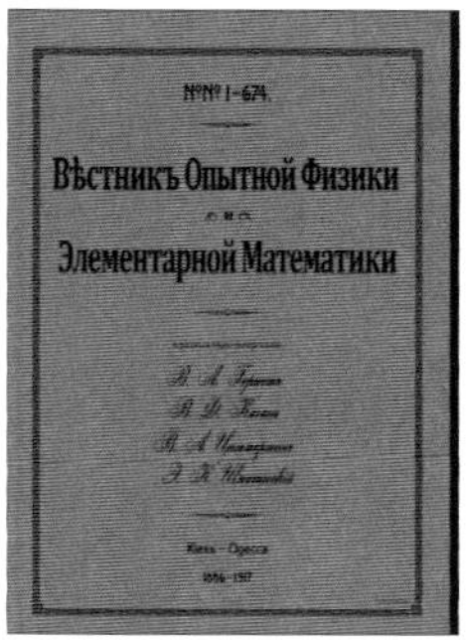

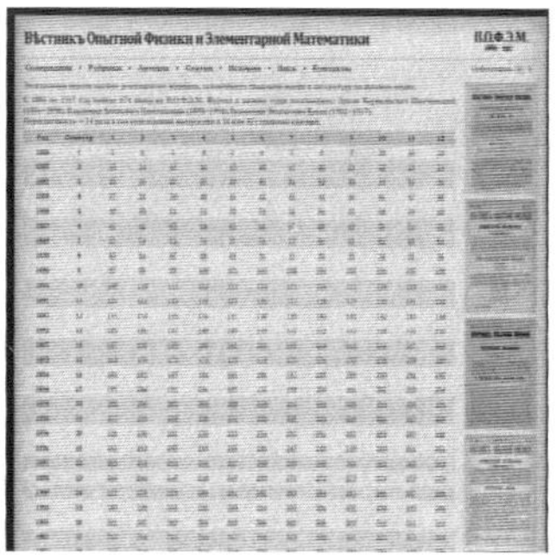

Left: The cover of the Bulletin of Experimental Physics and Elementary Mathematics
Center: Its web archive **Right:** Web archive of Mathesis

This Bulletin can be considered the first popular physics and mathematics journal in Russia. It was this journal that laid the foundations of this Russian-language popularization genre: fundamental articles, publication of new problems with solutions by the author, and updates on the events in the mathematical world, including reviews of new books.

Around the same time, different publishers started mass producing popular science literature. The most noteworthy books of that period include *In the Realm of Ingenuity*, by Ignatyev (a three-volume collection of entertaining problems and puzzles). Among the publishing houses, the Odesa-based Mathesis, a "publishing house of scientific and popular scientific works from the field of physical and mathematical sciences," stands out in particular https://www.mathesis.ru/. It published more than 150 books, many of which sold thousands of copies and are being republished to this day. Many of the books were translations, and in this field, Mathesis established a tradition of selecting, supplementing, and commenting on the source material (sometimes commentaries formed a significant part of the publication).

A truly large-scale popularization of mathematics in Russia began with the works of Y. I. Perelman (1882–1942). While still a student, he began to collaborate with popular science magazines and later became the author of many fascinating and accessible books on mathematics and physics that won recognition and popularity. In Perelman's books, math is given a modern twist: he both offers logic problems for young readers and popularizes the latest scientific advances in an animated and engaging way. Perelman's stories about the world around us become visual models for the reader's understanding of physics and mathematics. His books remain in print to this day, with many million copies printed in Russia alone!

In addition to being an author, Perelman was one of the main founders of the House of Amusing Science in 1935. This museum had over 500 large exhibits, along with many small ones, grouped by field: mathematics, physics (with an optics room), astronomy (with a meteorology section), and geography (with a geology section). Most of the exhibits were interactive, as Perelman believed that the visitor should be able to figure out the exhibit's structure and learn to work with it in a meaningful way. The House of Amusing Science resembled modern museums in many ways, including interactivity as much as the activities that took place there. It had more than 50 youth clubs, attended by students from different schools. Math and physics contests and debates were held regularly. Museum staff gave lectures at schools and factories, opened small exhibitions at the district young pioneer houses, and published miniature brochures on various fields of science and technology.

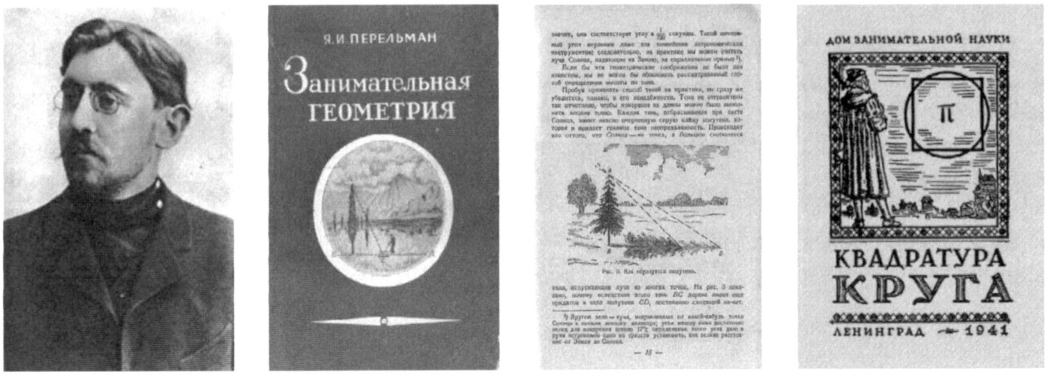

Yakov Perelman and two of his books

There is a wonderful historical anecdote about Perelman. In 1934, Herbert Wells, the father of modern science fiction, came to the Soviet Union. He was brought to a meeting with a group of Leningrad writers and popular scientists. During this meeting, Yakov Perelman took the floor and told Wells that the hero of his Invisible Man should have been blind. The reason is that if a person is truly, completely invisible—that is, transparent—his/her eye lens will also become transparent. And, therefore, it cannot refract light, making eyesight impossible.

The attitude toward the popularization of science in the Soviet Union was extremely serious even during the Second World War. On the 150th anniversary of N. I. Lobachevsky in 1942, the loudspeakers that would usually broadcast news from the front lines shared a popular science lecture about non-Euclidean geometry.

The middle of the 20th century saw the publication of a great deal of popular science literature. Books with the most serious educational content (both original ones and translations) often had circulation in hundreds of thousands. There were several very popular series of books: "Popular Lectures on Mathematics," "Math Club Library," and "Physics and Mathematics School Library," etc. Many of the best foreign books were translated.

In 1970, "Kvant" or "Quantum," a popular physics and mathematics magazine, saw the light for the first time https://kvant.ras.ru/. The publication was designed for schoolchildren, and its first editors-in-chief were prominent Russian scientists: the physicist I. I. Kikoin and the mathematician A. N. Kolmogorov. Articles for the magazine were written by teachers and prominent mathematicians. With a circulation of several hundred thousand, it reached many schoolchildren often becoming the gateway to serious science. It also had a monthly problem section, and schoolchildren could send in the solutions, which were checked and graded. An associated series of popular books "Kvant library" published many excellent popular introductions to science.

A back cover of Kvantik with a "picture-problem"

Quite possibly the most striking recent innovation has been the monthly "Kvantik" (or "Quantik", diminutive of Quantum) magazine https://kvantik.com/, which celebrates its 10th anniversary in 2022. The founder and permanent editor-in-chief is S. Dorichenko, chairman of the Tournament of Cities and member of the editorial board of the Kvant magazine. The "younger brother" of the Kvant journal is accessible to younger children and combines not only mathematics and physics, but also publishes meaningful, yet accessible and fascinating articles on linguistics, biology, and the history of science. This is a magazine for the curious schoolchildren, and not only them. The design of the magazine, which is created by a large team of artists with different visions, immediately attracts attention and draws children into reading. The journal has retained the tradition of monthly problem contests, the answers to which are sent by readers to the editorial office and the editorial office is in correspondence (often by paper letters!) with schoolchildren. And seeing their name published in their favorite journal among the contest winners is a big stimulus for children. School librarians in different regions of Russia note that Quantik has become the most popular demand among both schoolchildren and teachers. The back cover of the magazine is traditionally an interesting "picture-problem," containing almost no text. Each New Year, the magazine creates monthly calendars from such covers, which have become very popular.

One of the most important practices in the popularization of mathematics are the Olympiad competitions. The first modern style Mathematical Olympiad was organized in Leningrad by B. Delaunay in 1934. In the 1950s, the Olympiad competitions became a mass movement. The gained the now commonplace pyramid structure: the competitions were held in several stages, first within one school, then within the city, region, and so on. The first stage—the school stage—took place in most schools around the country. Olympiad participation encouraged the students to be passionate about mathematics, and prominent scientists delivered popular talks during the closing ceremonies of various stages. Besides being sport-like competition, Olympiads play an important role in attracting schoolchildren to mathematics with interesting problems, and their outreach is in the millions.

A. N. Kolmogorov with students of School 18 ("Kolomogorov's school"). Courtesy of A. N. Shiryaev

Such Olympiads became now universally popular around the world, but a few new practices were introduced in Russia. Among such new math Olympiads, Tournament of the Towns stands out https://www.turgor.ru/. Currently it takes place in more than 25 countries. A distinctive feature of the tournament is the scoring system. Participants are offered a list of tasks and are credited for making progress on three tasks, where they attained the maximum number of points. Those who submit the best solutions are invited to the Tournament's Summer Conference, where students attempt to attack research-grade problems.

Another nonstandard Olympiad is the Lomonosov Tournament. It stands out because it is a multisubject competition in math and mathematical games, physics, astronomy and earth sciences, chemistry, biology, history, linguistics, and literature. And it works! There have been cases where a school student, that was planning to tie his/her whole life to a specific science, suddenly found out at the tournament that other subjects are also interesting and became a professional in one of them. Both tournaments were created by N. Konstantinov.

A really singular Olympiad-style event is Matprazdnik (literally, "Mathematical Celebration"; https://olympiads.mccme.ru/matprazdnik/), a math Olympiad for children between 12 and 14 years of age, where they solve engaging math problems (tricky, but not requiring extensive knowledge) for two hours and then are entertained with lectures in popular science and games.

Finally, the Caucasus Mathematical Olympiad (CMO https://cmo.adygmath.ru/) is not simply a math competition, but also aims at building bridges of friendship and understanding between students of different countries of the Caucasus and the Black Sea region. For more details, see the article "Caucasus Mathematical Olympiad," (Russian Math. Surveys, 75:5 (2020), 991–993).

The Moscow Center for Continuous Mathematical Education was organized in 1995 (https://www.mccme.ru/) and quickly became a focal point for math popularization. Clubs, various olympiads, and tournaments, as well as seminars for teachers, are held within its walls and under its auspices. It is hosting the best publishing house of popular scientific literature on mathematics in Russia, both maintaining interest in old reprinted editions and releasing new books.

Math clubs or math circles (as they are called in Russian) are another excellent tradition. Once a week, school students come together to solve and discuss problems on various mathematical topics. Classes are supervised by both experienced teachers and university students. Most often, such clubs would function at universities, accepting students citywide. At some point, math clubs began to appear at specific schools with a strong math curriculum.

In addition to face-to-face clubs, there were long-distance correspondence schools. The most famous schools of this kind were the All-Union Correspondence Mathematical School at Moscow State University and the Correspondence Physics and Technology School at MIPT. Tasks and methodical brochures were mailed out to students. They, in turn, mailed back their solutions. Feedback was an important part of such schools. Many famous scientists participated in running those schools, starting with I. Gelfand.

Just like math clubs, Russia has a large number of summer schools or camps, which play a very important role. These are usually summer camps close to nature, where math classes are mingled with sports and recreation. Many math circles and mathematical schools organize them for many decades.

An informal seminar in Dubna

A very special annual Summer school "Modern Mathematics" has been operating on the Dubna and Volga riverside every summer since 2001 (https://mccme.ru/dubna/). During two weeks, about 100 participants – first-year college students and school students from the last two high school grades – take part in 100 lectures and seminars. This mix of both high school and university students creates a very special atmosphere. Apart from lectures, students can chat with prominent mathematicians in an informal and friendly setting. Brochures have been published on the materials of the classes and videos of most of the lectures have been made available online.

The summer school's traditions were established by V. Arnold (1968–2017). Its teachers were well-known mathematicians D. Anosov (1936–2014), L. Beklemishev, A. Bolibrukh (1950–2003), V. Buchstaber, V. A. Vassiliev, A. Gaifullin, A. Kirillov, A. Kuznetsov, S. Lando, Y. Matiyasevich, S. Novikov, A. Okounkov, D. O. Orlov, I. Panin, V. Protasov, A. Razborov, A. Raigorodskii, S. Smirnov, A. Sossinski, V. Tikhomirov, A. Shiryaev, and V. A. Uspensky (1930–2018), as well as recent graduates who have just started their career in science.

Proliferation of personal computers created new possibilities for math popularization. While it is difficult to surprise colleagues with computer databases of math problems, some stand out nevertheless. Moscow teacher R. K. Gordin has collected an outstanding database of geometric problems https://zadachi.mccme.ru/2012/jndex.html, with more than 15 thousand entries. Painstaking daily work for more than two decades has borne fruit – the database has been verified, cross-referenced, tagged, and beautiful and mathematically accurate illustrations have been drawn. Another well-known database that has absorbed math problems from most Russian textbooks is available at https://problems.ru/.

The addiction of today's children to smartphones opens up new opportunities to popularize mathematics. For example, the Euclidea app (https://www.euclidea.xyz/) presents a modern and unusual approach to studying geometry. Instead of memorizing theorems and ready-made recipes for geometric constructions, you discover the properties of figures and their relationships on your own: by trial and error, by applying logic and intuition, and by studying dynamic drawings. Euclidea requires you to solve problems using the smallest possible number of moves, which turns even standard problems into mindbenders.

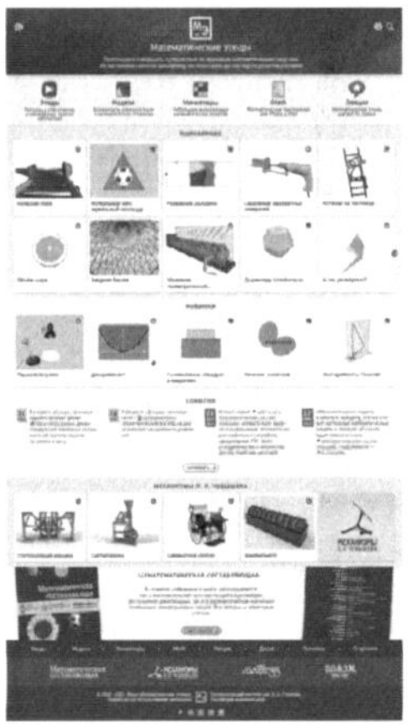

Left and center: Mechanisms of Tchebyshev DVD cover and web page **Right:** Mathematical Etudes webpage

As was mentioned, an exceptional feature of mathematical popularization in Russia has been the prominent part played by leading scientists. From its very beginning, the main role has always been played by leading scientists. Following this tradition, the Steklov Mathematical Institute launched in 2010 a specialized Laboratory of Popularization and Propaganda of Mathematics after a long preparatory work. This Laboratory became a pioneer in the promotion of mathematics in Russia, setting new standards in the popularization of mathematics, and stimulating the development of this field. Among the projects of Laboratory are

- films "Mathematical Etudes" (https://etudes.ru/). This is a series of more than 70 movies, made using modern 3D computer graphics, devoted to some solved and unsolved mathematical problems. The project includes an online encyclopedia of visual mathematical models;
- book "Mathematical Essence" (https://book.etudes.ru/), the authors of which are leading Russian mathematicians.The main purpose of this book is to unveil and emphasize the mathematical "essence" of some of the greatest achievements of our civilization, as well as to show the mathematical "content" inside regular everyday things;
- internet-museum "Mechanisms by Tchebyshev" (https://tcheb.ru/). This is a collection of movies and other materials on mechanisms suggested and constructed by the great Russian mathematician of the 19th century. Among other things, it includes animations and detailed explanations;
- mathematical park in the open air in the Republic of Adygea (https://math-park.ru/ru/). More details can be found in the article Mathematical Etudes: Evolution from Multimedia to a Book (EMS Newsletter December 2016, pp. 38–43).

NIKOLAI ANDREEV
Steklov Mathematical Institute of Russian Academy of Sciences, Moscow, 119991, Russia, andreev@etudes.ru

This Page Intentionally Left Blank

REPRESENTATIONS OF p-ADIC GROUPS OVER COMMUTATIVE RINGS

MARIE-FRANCE VIGNÉRAS

ABSTRACT

Motivated by the Langlands program in representation theory, number theory, and geometry, the theory of representations of a reductive p-adic group, originally in complex vector spaces, has been widely developed in modules over a commutative ring during the last two decades. This article surveys basic results obtained during this period, assuming some familiarity with the representation theory connected to the Langlands program. Addressed to a broader audience, the 2022 ICM Noether Lecture should be accessible without prerequisites and convey intuition on the most striking results.

MATHEMATICS SUBJECT CLASSIFICATION 2020

Primary 20G05; Secondary 11F70

KEYWORDS

p-adic groups, supercuspidal representations, blocks, pro-p Iwahori Hecke algebra

ICM INTERNATIONAL CONGRESS OF MATHEMATICIANS 2022 JULY 6–14

© 2022 International Mathematical Union

Proc. Int. Cong. Math. 2022, Vol. 1, pp. 332–374
DOI 10.4171/ICM2022/192

Published by EMS Press
and licensed under
a CC BY 4.0 license

1. INTRODUCTION

The theory of representations of a p-adic group G, for instance, $\mathrm{GL}(n, \mathbb{Q}_p)$, where $\mathbb{Q}_p$ is the p-adic completion of $\mathbb{Q}$ is an essential part of the Langlands program. At the beginning, it was a question of studying representations in a complex vector space. But the development of its links with number theory and geometry has required studying continuous representations in vector spaces defined over other fields than $\mathbb{C}$. There are many possibilities for such a generalization. It is easy to replace $\mathbb{C}$ by an algebraic closure $\mathbb{Q}_\ell^{\mathrm{ac}}$ of a local field $\mathbb{Q}_\ell$, where ℓ is a prime different from p. The choice of a field isomorphism $\mathbb{C} \simeq \mathbb{Q}_\ell^{\mathrm{ac}}$ identifies continuous complex representations of G and continuous ℓ-adic representations. A more difficult case is that of $\ell = p$ because the topology of a p-adic group and of $\mathbb{Q}_p$ is the same. One even considers representations with values not in a vector space, but in a module over some commutative ring like $\mathbb{Z}[1/p]$ or $\mathbb{Z}/p^i\mathbb{Z}, i \geq 1$. The representations over these different categories of coefficient rings are now essential in the theory of automorphic forms. Their theory has been widely developed since the beginning of the 21st century, and different versions of the local Langlands correspondence have emerged.

We review the main basic results for representations over coefficient rings[1] different from $\mathbb{C}$. In an attempt to make this paper accessible to readers with a wide range of backgrounds, we give fairly complete definitions of all terminology. Proofs are omitted, yet we give a short indication of the key points, we cite sources and provide examples. For the theory before 2002, the reader may consult our book[2] and our article in the proceedings of the Bejing ICM. The subject has remained confined in research articles since these last two decades, and we hope that this survey provides a navigable route to the literature.

2. NOTATION

We work with a triple (F, G, R) where F is the basic field, G the reductive p-adic group, and R the coefficient ring. We assume that F is a local non-archimedean field of ring of integers O_F, uniformizer p_F, and residue field k_F of characteristic p with q elements, G is the group $\underline{G}(F)$ of F-points of a connected reductive F-group $\underline{G}$, endowed with the topology generated by the open pro-p-subgroups[3] and R is a commutative ring.[4]

An R-representation V of G will always be *smooth* (continuous for the discrete topology on R). It is *admissible* if for all open compact subgroups K of G, the R-module V^K of vectors fixed by K is finitely generated.

The absolute Galois group Gal_E of a field E is the group of automorphisms of an algebraic closure E^{ac} fixing E. For a prime number r, $\mathbb{F}_r$ denotes a field with r elements,

---

1 That we are aware of, without geometry or derived functors.

2 Représentations ℓ-modulaires d'un groupe réductif p-adique avec ℓ différent de p, Birkhaüser, 1996.

3 Called a connected reductive p-adic group, but beware that some authors use this terminology only when F contains $\mathbb{Q}_p$.

4 A ring is supposed to have a unit.

$\mathbb{Z}_r$ denotes the ring of integers in the field $\mathbb{Q}_r$ of r-adic numbers, and $\mathbb{Z}_r^{ac}$ denotes the ring of integers of $\mathbb{Q}_r^{ac}$. We always denote by ℓ a prime number different from p.

The parabolic and parahoric subgroups play an essential role in the theory of representations of a reductive p-adic group.

The parabolic subgroups appear for the first time in the section on parabolic induction. We fix a maximal split[5] torus T of G of G-centralizer Z and a minimal parabolic subgroup $B = ZU$ of unipotent radical U and opposite $B^{op} = ZU^{op}$. A *standard parabolic subgroup* of G is a parabolic subgroup containing B, that is, $P = MN = MB$, with unipotent radical N contained in U and Levi subgroup M containing Z. The opposite parabolic subgroup $P^{op} = MN^{op} = MB^{op}$ is not standard.

The Weyl group W_G is equal to the quotient of the G-normalizer of T by Z. We denote by $Z^+ \subset Z$ the submonoid of elements contracting U by conjugation, Z^- those contracting U^{op}, $T^+ = T \cap Z^+$, $T^- = T \cap Z^-$. The group G is split if $T = Z$ and quasisplit if Z is a torus.[6]

The parahoric subgroups appear for the first time in the section on Hecke algebras. We fix a *special parahoric subgroup* K of G and a *pro-p-Iwahori subgroup* $\tilde{J}$ of G, as follows. We choose a special point x_0 of the apartment $\mathfrak{A}$ of T in the adjoint Bruhat–Tits building of G. The parahoric subgroup of G fixing the alcove in $\mathfrak{A}$ of vertex x_0 associated to B is an Iwahori subgroup J of G. Then K is the parahoric subgroup fixing x_0 and $\tilde{J}$ is the maximal open normal pro-p subgroup of J. For a standard parabolic subgroup $P = MN$ of G, $M^0 = M \cap K$ is a special parahoric subgroup of M and $\tilde{J}_M = \tilde{J} \cap M$ is a pro-p Iwahori subgroup of M. We denote $N^0 = K \cap N$.

The pro-p-Iwahori subgroups of G are all G-conjugate, but in general there are only finitely many G-conjugacy classes of special parahoric subgroups of G.

Examples. There are two conjugacy classes of special parahoric subgroups of $\mathrm{SL}(2, F)$.

The special parahoric subgroups of $\mathrm{GL}(n, F)$ are conjugate to $\mathrm{GL}(n, O_F)$.

The inverse image by the quotient map $\mathrm{GL}(n, O_F) \to \mathrm{GL}(n, k_F)$ of the (strictly) upper triangular group of $\mathrm{GL}(n, k_F)$ is a (pro-p) Iwahori subgroup of $\mathrm{GL}(n, F)$.

The split torus T has a unique parahoric subgroup, equal to the maximal compact subgroup $T^0 = T \cap K = T \cap J$, and the quotient T/T^0 is isomorphic to the group $X_*(T)$ of cocharacters of T via p_F. The compact mod center connected reductive group Z has a unique parahoric subgroup $Z^0 = Z \cap K = Z \cap J$, and the quotient Z/Z^0 is a commutative finitely generated group (Thomas Haines and Sean Rostami **[83]**).

5 This means, by a common abuse of notation, that $T = \underline{T}$ where $\underline{T}$ is a maximal F-split torus of $\underline{G}$.

6 When G is not quasisplit, Z is not commutative.

3. CHANGE OF BASIC FIELD

The basic field F is a finite extension of $\mathbb{Q}_p$ or of $\mathbb{F}_p((t))$. It is called a *p-adic field* in characteristic 0 and a *local function field* in characteristic p. Many geometric methods demand F to be a local function field. For example, the proof by Bao Chau Ngo[7] of the fundamental lemma, essential in the Langlands theory, which asserts an equality between certain linear combinations of integral orbitals over the Lie algebras of G and of endoscopic groups. On the other hand, when F is a local function field, the harmonic analysis is full of traps, there are inseparable semisimple elements, there is no exponential map to pass to the Lie algebra and G has no cocompact discrete subgroup (except for type A), G is not a p-adic Lie group.

But the basic field F appears only through the residual field in many constructions (endoscopy, buildings, Iwahori Hecke algebras). This is a key to transfer properties between basic fields of different characteristics. For instance, Jean-Loup Waldspurger **[201]** proved that the fundamental lemma for F of characteristic p implies the fundamental lemma for F of characteristic 0. There is another proof using the general transfer principle of Cluckers and Loeser in model theory and motivic integration **[31, 32]**. In the other direction, the fundamental lemma for the automorphic induction for $\mathrm{GL}(n, F)$ proved by Guy Henniart and Rebecca Herb for F of characteristic 0 was transferred to F of characteristic p by Henniart and Bertrand Lemaire **[102]** using *close local fields*. For a positive integer m, two non-archimedean local fields are m-closed, if their rings of integers modulo the mth power of their respective maximal ideals are isomorphic. The Deligne–Kazhdan philosophy can be loosely stated as follows: the representation theory of Galois groups (or of reductive groups) over m-close local fields is the same "up to level m". For instance, Radhika Ganapathy **[74]** proved that for two m-close local fields F, F' and $\underline{G}$ split, the category of complex representations of $\underline{G}(F)$ generated by their invariants by the m-filtration subgroup of an Iwahori subgroup is equivalent to the same category for representations of $\underline{G}(F')$. For $\underline{G}$ not split, she made sense of a natural connected reductive group $\underline{G}'$ over F' associated to $\underline{G}$, first when $\underline{G}$ is quasisplit (an F-form of a split group) and then when $\underline{G}$ is general (an inner form of a quasisplit group) **[75, 3.A AND 5.A]**.

The local field $\mathbb{Q}_p$ is a completion of $\mathbb{Q}$ and $\mathbb{Q}$ is a globalization of $\mathbb{Q}_p$. The local case is simpler than the global. The ring $\mathbb{Z}_p$ has only one prime ideal, namely $p\mathbb{Z}_p$, but the ring $\mathbb{Z}$ has infinitely many prime ideals. The absolute Galois group $\mathrm{Gal}_{\mathbb{Q}_p}$ of $\mathbb{Q}_p$ is simple compared to $\mathrm{Gal}_{\mathbb{Q}}$. In the same vein, the local field F is the completion of a (non-unique) global field[8] E and E is a globalization of F, the local group G is a localization of the group H of rational points of a connected reductive group over a global field, and H is a globalization of G.[9] An automorphic irreducible $\mathbb{C}$-representation V_A of the adelic group

---

7 Fields medal in 2020.

8 A global field is a finite extension of $\mathbb{Q}$ or of $\mathbb{F}_p(T)$.

9 For F of characteristic p, Wee-Teck Gan, Luis Lomeli **[72]**, for F of characteristic 0, Shahidi (A proof of Langland's Conjecture on Plancherel measures; Complementary Series of p-adic groups, The Annals of Math., Series 2, Vol. 132, 2 (1990), 273–330) when G is quasisplit, implying the general case as in **[72]**.

$H(A)$ gives by localization an irreducible $\mathbb{C}$-representation V of G and V_A is a globalization of V. The study of automorphic representations uses the theory of representations of local reductive groups. In the other direction, some theorems of representations of local groups are proved by embedding the local case into a global one.

The classical local Langlands correspondence introduced by Langlands in 1967–1970 is a generalization of local class field theory from abelian Galois groups to non-abelian Galois groups. The absolute Galois group Gal_{k_F} of the finite field k_F is topologically generated but the Frobenius $\mathrm{Frob}(x) = x^q$. The subgroup of elements in Gal_F with image an integral power of Frob in the natural quotient map $\mathrm{Gal}_F \to \mathrm{Gal}_{k_F}$ is the *Weil group* W_F of F.[10] The reciprocity map of local class field theory $F^* \tilde{\to} W_F^{ab}$ identifies the irreducible R-representations of $\mathrm{GL}(1, F)$ with the one-dimensional R-representations of W_F when R is an algebraically closed field. Langlands proposed a parametrization of the irreducible $\mathbb{C}$-representations of G in terms of $\mathbb{C}$-representations of W_F.

When $G = \mathrm{GL}(n, F)$, the complex local Langlands correspondence is a theorem which has been generalized to representations over $R = \mathbb{F}_\ell^{\mathrm{ac}}$, $\ell \neq p$.[11] The first proofs of local class field theory were global. Today the proofs of the local Langlands correspondence for $\mathrm{GL}(n, F)$ needs global arguments, except for $n = 2$ and $R = \mathbb{C}$, where there is a local proof (Colin Bushnell and Henniart **[21]**). When F has characteristic 0, Peter Scholze **[174]** gave a new local characterization of the complex local Langlands correspondence; a local Langlands correspondence over $R = \mathbb{F}_p^{\mathrm{ac}}$ is to-day a very active research area.[12]

The geometrization of a (semisimple) local Langlands correspondence for all F, G and $R = \mathbb{Z}_\ell$ for almost all $\ell \neq p$, obtained by Laurent Fargues and Scholze in 2021, is entirely local.

4. CHANGE OF COEFFICIENT RING

Many features of complex representations of G use harmonic analysis only apparently and can be generalized to representations over other coefficient rings. For instance,

(a) The theory of discrete series and tempered complex representations has an algebraic and combinatorial flavor.[13] It was extended by Dat **[38]** to an algebraically closed field R of characteristic different from p with a nontrivial valuation.

(b) The proof of the classification of the irreducible complex representations of an inner form of $\mathrm{GL}(n, F)$ by Tadic for $F \supset \mathbb{Q}_p$ uses harmonic analysis (the simple trace formula). Alberto Minguez and Vincent Sécherre **[139]** gave an

---

10 The kernel I_F of the quotient map is an extension of $\prod_{\ell \neq p} \mathbb{Z}_\ell$ by a pro-p group P_F.

11 Proved when $R = \mathbb{C}$ by Gérard Laumon, Michael Rapoport, and Ulrich Stuhler in 1993 if $F \supset \mathbb{F}_p((t))$, and if $F \supset \mathbb{Q}_p$ by Michael Harris and Richard Taylor in 2001 (Guy Henniart gave another proof), and extended by Vignéras in 2001 to $R = \mathbb{F}_\ell^{\mathrm{ac}}$, $\ell \neq p$.

12 There is nothing for F and R of characteristic p, to the best of my knowledge.

13 The asymptotic behavior of coefficients may be derived from the central exponents of the Jacquet modules.

algebraic proof for all F and all algebraic closed fields R of characteristic different from p.

A prime $\ell \neq p$ not dividing the order of a torsion element of G is called *banal*[14] for G. A field R is of *banal characteristic* for G if its characteristic is 0 or ℓ banal for G. A general principle is that the properties of complex representations of G described in purely algebraic terms transfer to representations of G over fields R of banal characteristic.

Example. The banal primes for $\mathrm{GL}(m, F)$ are those coprime with $q^i - 1$ for $1 \leq i \leq m$.

The R-representations of G form a locally small abelian Grothendieck category $\mathrm{Mod}_R(G)$ (Vignéras **[199]**). For a commutative ring S which is an R-algebra, the R-representations of G are related to the S-representations of G by the scalar extension[15]

$$S \otimes_R - : \mathrm{Mod}_R(G) \to \mathrm{Mod}_S(G)$$

and, by the restriction its right adjoint, an S-representation is considered as an R-representation. One says that an S-representation of G in the image of the scalar extension *descends* to R, or *is defined* on R.

When R is a field, many properties on admissible irreducible R-representations of G still assume R to be algebraically closed although this is not necessary. A good tool to show this is the bijection (Henniart–Vignéras **[106]**, **[107, SECTION 2]**)

$$V \mapsto BC(V)$$

- from the isomorphism classes of irreducible admissible R-representations V of G,
- onto the Galois orbits[16] $BC(V)$ of the isomorphism classes of the irreducible admissible R^{ac}-representations of G defined on a finite extension of R.

Here $BC(V)$ is the set of isomorphism classes of the irreducible subquotients V^{ac} of

$$R^{\mathrm{ac}} \otimes_R V \simeq \oplus^d \oplus_{V^{\mathrm{ac}} \in BC(V)} W(V^{\mathrm{ac}}),$$

where d is the reduced dimension of the division R-algebra $\mathrm{End}_{RG} V$ over its center E_V, the length of the R^{ac}-representation $R^{\mathrm{ac}} \otimes_R V$ of G is $d[E_V : R]$, the number of elements of $BC(V)$ is $[E_V^s : R]$ where E_V^s is the maximal separable subextension of E_V/R, and $W(V^{\mathrm{ac}})$ is an indecomposable R^{ac}-representation of G of irreducible subquotients isomorphic to V^{ac} and of length $[E_V : E_V^s]$. Any $V^{\mathrm{ac}} \in BC(V)$ is V-isotypic as an R-representation of G, and is defined on a maximal subfield of $\mathrm{End}_{RG} V$ (Justin Trias **[188]**).

Any irreducible admissible R^{ac}-representation of G is absolutely irreducible and has a central character by the Schur's lemma. If the characteristic of R is different from p,

---

14 See **[51]**, Lemma 5.22 and Corollary 5.23 for other characterizations.

15 Also called base change or induction.

16 An orbit under the group $\mathrm{Aut}_R(R^{\mathrm{ac}})$ of R-automorphisms of R^{ac}.

any irreducible R^{ac}-representation of G is admissible and defined on a finite extension of R [107].

As G is locally a pro-p group, there is no Haar measure on G with values in a commutative ring R where p is not invertible and the R-representations of G present new phenomena. To understand them is a crucial question.

For a field R of characteristic p, any irreducible R-representation V of G with $\dim_R V^K < \infty$ for some open pro-p subgroup K of G, is admissible (Vytautas Paskunas [155], a simple proof is given in Henniart [101]). For any open pro-p subgroup K of G, any nonzero representation of G has a nonzero vector invariant by K (like for finite groups).

Irreducible implies admissible when $G = \mathrm{GL}(2, \mathbb{Q}_p)$. Indeed, one reduces to $R = \mathbb{F}_p^{ac}$; in this case irreducible implies that the center acts by a character (Laurent Berger [14]) hence is admissible by Barthel-Livne and Breuil [16].

But, there exists an irreducible non-admissible $\mathbb{F}_p^{ac}$-representation of $\mathrm{GL}(2, F)$ for an unramified extension F of $\mathbb{Q}_p$ (Daniel Le [136]). One does not know if any infinite-dimensional irreducible non-admissible $\mathbb{F}_p^{ac}$-representation of G has a central character, because its dimension is equal to the cardinal of $\mathbb{F}_p^{ac}$ and the classical proof with the Schur's lemma does not apply.

It happens that a property of admissible irreducible representations of G over a field R transfers to representations of G over any coefficient field of the same characteristic. This is the case in the following examples:

(i) In characteristic different from p, for the classification of cuspidal irreducible R-representations of G by compact induction (Henniart–Vignéras [107]).

(ii) In characteristic p, for the classification of non-cuspidal[17] admissible irreducible R-representations of G, for the classification of non-supersingular simple modules of the pro-p-Iwahori Hecke R-algebra of G (Noriyuki Abe, Henniart, Florian Herzig, and Vignéras [8], Henniart–Vignéras [106]), for the existence of a supersingular admissible irreducible R-representation of G when $F \supset \mathbb{Q}_p$ (Herzig, Karol Koziol, and Vignéras [110]).

For a prime r,[18] an *r-adic representation* of G is a representation of G on a $\mathbb{Q}_r^{ac}$-vector space which is continuous for the r-adic topology on the vector space. A p-adic representation of G may be not smooth, but an ℓ-adic representation of G is smooth if $\ell \neq p$. In this article, an R-representation of G is supposed always to be smooth. A $\mathbb{Q}_r^{ac}$-representation of G is a smooth r-adic representation of G. The choice of an isomorphism

$$\mathbb{C} \simeq \mathbb{Q}_r^{ac}$$

identifies the complex representations of G and the $\mathbb{Q}_r^{ac}$-representations of G.

A *mod r representation* of G is a $\mathbb{F}_r^{ac}$-representation of G.

17 Cuspidal and supersingular will be defined later.

18 Letter ℓ is reserved for the primes different from p, think $r = \ell$ or p.

An admissible $\mathbb{Q}_r^{ac}$-representation V of G is called *integral* if V is defined on a finite extension $E/\mathbb{Q}_r$ and V contains an G-stable $\mathbb{Z}_r^{ac}$-lattice L,[19] admissible as a $\mathbb{Z}_r^{ac}$-representation of G and descending to O_E.[20] The mod r representation $\mathrm{red}_r(L) = L \otimes_{\mathbb{Z}_r^{ac}} \mathbb{F}_r^{ac}$ of G is called the reduction of L.

By the strong Brauer–Nesbitt theorem (Vignéras **[189]**), when $r = \ell \neq p$, the $\mathbb{Z}_\ell^{ac}[G]$-module L is finitely generated, of reduction $\mathrm{red}_\ell(L)$ of finite length, and the image of $\mathrm{red}_\ell(L)$ in the Grothendieck group of finite length $\mathbb{F}_\ell^{ac}$-representations of G does not depend on the choice of L; it is called the *reduction* of V. Two finite-length integral ℓ-adic representations of G are said to be *congruent modulo* ℓ when their reductions are isomorphic.

This does not hold true for $\mathbb{Q}_p^{ac}$-representations of G. For example, an irreducible $\mathbb{Q}_p^{ac}$-representation $V = \mathrm{ind}_K^G W$ of $G = \mathrm{PGL}(n, F)$ compactly induced from a representation W of $K = \mathrm{PGL}(n, O_F)$ contains an admissible G-stable $\mathbb{Z}_p^{ac}$-lattice L defined on some O_E as above, of infinite length reduction, and another one L' of finite length reduction. Take $L = \mathrm{ind}_K^G W_{\mathbb{Z}_p^{ac}}$ for a K-stable $\mathbb{Z}_p^{ac}$-lattice $W_{\mathbb{Z}_p^{ac}}$ of W and $L' = V \cap \mathrm{ind}_\Gamma^G 1_{\mathbb{Z}_p^{ac}}$ for a small enough discrete cocompact subgroup Γ of G.

5. PARABOLIC INDUCTION

For any triple (F, G, R) (as in the Notation section) and any parabolic subgroup P of G of Levi quotient M, the *parabolic induction*[21]

$$\mathrm{ind}_P^G : \mathrm{Mod}_R(M) \to \mathrm{Mod}_R(G)$$

allows constructing representations of G from representations of the smaller connected reductive p-adic group M. The parabolic induction has excellent properties, it commutes with small direct sums[22] (Vignéras **[199]**); for p nilpotent in R, it is fully faithful (Vignéras **[199]**); for a field R, the parabolic induction respects finite length representations with admissible subquotients (this depends on the classification of admissible irreducible representations if the characteristic of R is p).

The parabolic induction is exact and has a *left adjoint* L_P^G called the Jacquet functor, equal to the coinvariant functor $(-)_N$ with respect to the unipotent radical N of P, and a *right adjoint*[23] R_P^G (Vignéras **[199]**). By adjointness, L_P^G is right exact and R_P^G is left exact. The scalar extension commutes with the three parabolic functors (Henniart–Vignéras **[106]**).

For p invertible in R, the *second adjunction*

$$R_P^G = \delta_P L_{P^{\mathrm{op}}}^G,$$

---

19 A free $\mathbb{Z}_r^{ac}$-submodule of scalar extension V to $\mathbb{Q}_r^{ac}$.

20 The ring O_E is principal but not $\mathbb{Z}_r^{ac}$. The definition bypasses this difficulty.

21 $\mathrm{ind}_P^G(W)$ is the R-module of locally constant functions $f : G \to W$ such that $f(mng) = mf(g)$ for $m \in M$, $n \in N$, $g \in G$, where G acts by right translation.

22 When R is a field of characteristic p, ind_P^G commutes with direct products **[169]**.

23 By **[116, 8.3.27]**, as $\mathrm{Mod}_R(G)$ is a locally small abelian Grothendieck category and ind_P^G is right exact and commutes with small direct sums.

where δ_P is the modulus of P,[24] is a deep property proved this year by Dat, David Helm, Robert Kurinczuk, and Gilbert Moss **[52, COROLLARY 1.3]**, originally proved by Bernstein when $R = \mathbb{C}$. When R is noetherian, the parabolic induction ind_P^G respects admissibility, the second adjunction implies that $\mathrm{Mod}_R(G)$ is noetherian, that the parabolic induction respects projective (resp. finitely generated) R-representations **[52, COROLLARIES 1.4, 1.5]**, and that L_P^G respects admissibility. The functor L_P^G is exact, preserves infinite direct sums **[40]**, and when R is a field, L_P^G respects finite length because L_P^G respects the property of being finitely generated, and an admissible finitely generated R-representation of G has finite length (the proof uses the Moy–Prasad unrefined types when R is algebraically closed but being algebraically closed is not necessary).

For a field R of characteristic p, the adjoint functors L_P^G and R_P^G send an admissible irreducible R-representation of G to 0 or to an admissible irreducible R-representation of M. Irreducibility is necessary, an example of an admissible R-representation V of G with $L_P^G(V)$ not admissible is given in (Abe–Henniart–Vignéras **[10]**). But contrary to the complex case, the functors L_P^G and R_P^G fail to be exact (for R_P^G, see Emerton **[61]** and Koziol **[122]**), ind_P^G does not preserve finitely generated representations, R_P^G does not preserve infinite direct sums (Abe–Henniart–Vignéras **[10, SECTION 4.5]**).

When p is nilpotent in the commutative ring R, the right adjoint R_P^G respects admissibility (Abe–Henniart–Vignéras **[10]**); it is equal to the Emerton's functor $\mathrm{Ord}_{P^{\mathrm{op}}}^G$ of ordinary parts on admissible R-representations.[25] If, moreover, R is artinian, Matthew Emerton **[61]** extended the functor of ordinary parts to a δ-functor, expected to coincide with the derived functors when the characteristic of F is 0.

Example. When $G = \mathrm{SL}(2, \mathbb{Q}_p)$, Koziol **[122]** showed that the derived functors of R_B^G and Ord_B^G are equal on any absolutely irreducible $\mathbb{F}_p^{\mathrm{ac}}$-representations of G.

When the characteristic of F is p, surprisingly, R_P^G is exact on admissible $\mathbb{F}_p^{\mathrm{ac}}$-representations of G (Julien Hauseux **[88]**).

A representation of G over a field R is called *unramified* when it is trivial on the subgroup G^0 of G generated by its compact subgroups.[26] The group $\Psi_R(G)$ of unramified R-characters $\psi : G \to R^*$ of G is a torus. One says that (F, G, R) satisfies *generic irreducibility* property if for any parabolic subgroup P of G of Levi M and any irreducible R-representation W of M, the set of $\psi \in \Psi_R(M)$ such that $\mathrm{ind}_P^G(W \otimes \psi)$ is irreducible is Zariski-dense in $\Psi_R(M)$.

Generic irreducibility property is probably true for any F, G and any field R. It is known for R of characteristic p (Abe–Henniart–Vignéras **[10]**) or when $F \subset \mathbb{Q}_p$ and R algebraically closed of characteristic different from p (Dat **[38]**).

---

24 $\delta_P(m) = |\det \mathrm{Ad}_{\mathrm{Lie}\, N}(m)| \in q^{\mathbb{Z}}$.

25 There is no description of R_P^G on non-admissible representations.

26 This coincides with the classical definition (Henniart–Lemaire **[103, 2.12 REMARQUE 1]**).

Dat [38, THEOREM 3.11] extended the complex *Langlands quotient theorem* to any algebraically closed field R of characteristic different from p with a nontrivial valuation ν (for example, $\mathbb{Q}_\ell^{ac}$).

An admissible R-representation V of G is *ν-tempered* (Dat [38, DEFINITION 3.2]) if for any standard parabolic subgroup $P = MN$ such that $L_P^G(V) \neq 0$, any exponent χ in $L_P^G(V)$ satisfies $-\nu(\delta_P^{-1/2}\chi) \in \overline{{}^+\mathcal{A}_P^*}$.[27] It is called *discrete* if $-\nu(\delta_P^{-1/2}\chi) \in {}^+\mathcal{A}_P^*$. The *exponents* of $L_P^G(V)$ are the R-characters of the split component A_M of the center of M appearing in $L_P^G(V)$ seen as an R-representation of A_M.

Theorem 5.1 (Dat–Langlands quotient theorem). (i) *When $P = MN$ is a standard parabolic subgroup of G, W is a ν-tempered irreducible R-representation of M, and $\psi \in \Psi_R(M)$ satisfies $-\nu(\psi) \in (\mathcal{A}_P^*)^+$, then the R-representation* $\mathrm{ind}_P^G(W \otimes \psi)$ *has a unique irreducible quotient $J(M, W, \psi)$.*

(ii) *Any irreducible R-representation V of G is isomorphic to $J(M, W, \psi)$ for a unique triple (P, W, ψ).*

From the Dat's theory of ν-tempered representations, David Hansen, Tasho Kaletha, and Jared Weinstein deduced (see [84, C.2.2)]):

The Grothendieck group of finite length ℓ-adic representations of G is generated by representations of the form $\mathrm{ind}_P^G(W \otimes \psi)$, for a standard parabolic subgroup $P = MN$ of G, an integral irreducible ℓ-adic representation W of M and an unramified ℓ-adic character ψ of M.

6. ADMISSIBLE REPRESENTATIONS AND DUALITY

The classification of irreducible admissible R-representations of G is an objective of the local Langlands program. There are few finite-dimensional representations when G is not compact modulo the center, and admissibility is a crucial finiteness property.

When R is a noetherian commutative ring, a subrepresentation of an admissible R-representation of G is admissible. A quotient of an admissible R-representation of G is admissible [195] and the category $\mathrm{Mod}_R(G)^a$ of admissible R-representations of G is abelian if p is invertible in R, or if R is a finite field of characteristic p and $F \supset \mathbb{Q}_p$.[28]

Example. When $F \supset \mathbb{F}_p((T))$ and p is not invertible in R, there exists an admissible representation with a non-admissible quotient (Abe–Henniart–Vignéras [10]).

---

27 Let $\Delta(M)$ denote the set of simple roots of T in M, $\Delta(P)$ the set of simple roots in P of T_M, $\mathcal{A}^* = X \otimes_{\mathbb{Z}} \mathbb{R}$ where X is the lattice of rational characters of T, and $(\cdot,\cdot)$ a W_G-invariant scalar product on $\mathcal{A}^*$. Then $\overline{{}^+\mathcal{A}_P^*} = \sum_{\alpha\in\Delta(P)} \mathbb{R}_{\geq 0}\,\alpha$ and $(\mathcal{A}_P^*)^+$ is the cone $\{x \in \mathcal{A}^*, (x,\alpha) = 0 \text{ for } \alpha \in \Delta(M), (x,\alpha) > 0 \text{ for } \alpha \in \Delta(P)\}$.

28 The completed group algebra of $R[K]$ is noetherian when $F \supset \mathbb{Q}_p$ but not when $F \supset \mathbb{F}_p((T))$.

Let R be a field. The *smooth dual* $V^\vee$ of an R-representation V of G is the smooth part of the contragredient action of G on the linear dual $V^* = \mathrm{Hom}_R(V, R)$.[29]

For R of characteristic different from p, the smooth dual is an autoduality on $\mathrm{Mod}_R(G)^a$. In particular, $V^\vee$ is irreducible if and only if V is irreducible. The smooth dual and the parabolic induction and its left adjoint satisfy[30]:

$$(\mathrm{ind}_P^G W)^\vee \simeq \mathrm{ind}_P^G(W^\vee \delta_P), \quad L_P^G(V^\vee) \simeq \big(L_{P^{\mathrm{op}}}^G(V)\big)^\vee,$$

for any R-representation W of M and any admissible R-representation V of G.

For R of characteristic p, the smooth dual of any infinite dimensional admissible irreducible R-representation of G is zero! For F of characteristic 0, Jan Kohlhaase **[117]** developed a higher smooth duality theory on $\mathrm{Mod}_R(G)^a$. He studied the ith smooth duality functors $S^i : \mathrm{Mod}_R(G)^a \to \mathrm{Mod}_R(G)^a$ for $0 \leq i \leq d = \dim_{\mathbb{Q}_p} G$ under tensor product, inflation and induction and proved that for $V \in \mathrm{Mod}_R(G)^a$, the integer

$$d(V) = \max\{i \mid S^i(V) \neq 0)\}$$

satisfies

(i) $d(V) = 0$ if and only if V is finite dimensional,

(ii) $d(\mathrm{ind}_P^G W) = d(W) + \dim_{\mathbb{Q}_p} N$, for a parabolic subgroup $P = MN$ and $W \in \mathrm{Mod}_R(M)^a$,

(iii) $d(V) = 1$ and $S^1(V)$ coincides with the Colmez's contragredient introduced for the p-adic Langlands correspondence for $G = \mathrm{GL}(2, Q_p)$, $R = \mathbb{F}_p^{\mathrm{ac}}$, and V irreducible of infinite dimension; for the Steinberg representation St_G which is irreducible, $S^1(\mathrm{St}_G)$ is indecomposable of length 2!

For G unramified,[31] K a hyperspecial subgroup of G, $W \in \mathrm{Mod}_{\mathbb{F}_p^{\mathrm{ac}}}(K)$ and $i > \dim_{\mathbb{Q}_p} U$, we have $S^i(\mathrm{ind}_K^G W) = 0$ (Claus Sorensen **[185]**).

7. SUPERCUSPIDAL SUPPORT

An R-representation V of G is called *cuspidal* if it is killed by the left and right adjoints of the parabolic induction

$$L_P^G(V) = R_P^G(V) = 0,$$

for all parabolic subgroups $P \neq G$.

When p is invertible in R, the second adjunction implies that V is cuspidal if and only if $L_P^G(V) = 0$ for any proper parabolic subgroup P of G. Any irreducible R-

29 The smooth dual is the set of linear forms on V fixed by some open subgroup of G.

30 The normalized induction $\mathrm{ind}_P^G(W \otimes \delta_P^{1/2})$ commutes with the smooth dual, the second isomorphism is equivalent to the second adjunction.

31 G is quasisplit and splits over some unramified extension of F.

representation V of G is a subrepresentation of $\mathrm{ind}_P^G W$ for some cuspidal irreducible R-representation W. Assuming that R is an algebraically closed field,[32] the pair (M, W) is unique modulo G-conjugation; the G-conjugation class of (M, W) is called the *cuspidal support* of V. Twisting the cuspidal support by unramified characters, we get the *inertial cuspidal support* Ω of V. So, Ω is the set of (M', W') which are G-conjugate to $(M, W \otimes \Psi)$ for some $\psi \in \Psi_R(M)$. The subgroup of $w \in W_G$ fixing M acts on the R-representations of M. Let H be the group of $w \in W_G$ such that $W^w \simeq W \otimes \psi$ for some $\psi \in \Psi_R(M)$ and S the (finite) group of $\psi \in \Psi_R(M)$ such that $W \otimes \psi \simeq W$. Then, Ω is an algebraic variety with regular functions $\mathcal{O}(\Omega) = (R[M/M^0]^S)^H$.

When p is not invertible in R, one needs both L_P^G and R_P^G to define cuspidality. For a field R of characteristic p, the trivial representation 1_G of G and the Steinberg representation $\mathrm{St}_G = \mathrm{ind}_B^G(1_Z)/\sum_{P \supsetneq B} \mathrm{ind}_P^G(1_M)$ satisfy, for any parabolic subgroup P of Levi M,

$$L_P^G(1_G) = 1_M, \quad R_P^G(1_G) = 0, \quad L_P^G(\mathrm{St}_G) = 0, \quad R_P^G(\mathrm{St}_G) = \mathrm{St}_M .$$

The Steinberg representation is not a subrepresentation of $\mathrm{ind}_P^G W$ for any cuspidal admissible irreducible R-representation W. Any irreducible R-representation V of G is a subquotient of $\mathrm{ind}_B^G W$ for some R-representation W of Z (for R algebraically closed, see Abe–Henniart–Herzig–Vignéras [8, IV.1]). This is very different from the complex case!

An admissible irreducible R-representation of G which is not isomorphic to a subquotient of a proper parabolically induced representation $\mathrm{ind}_P^G W$ for all $P \neq G$, W an admissible irreducible R-representation of M, is called *supercuspidal*.[33]

A cuspidal irreducible admissible R-representation is always supercuspidal if R is a field of characteristic 0 or p, but not if the characteristic of R is $\ell \neq p$!

Example. When $G = \mathrm{GL}(2, \mathbb{Q}_p)$, $R = \mathbb{F}_\ell^{\mathrm{ac}}$, ℓ divides $p+1$, the unique infinite dimensional irreducible subquotient of the representation $\mathrm{ind}_B^G 1_Z$ indecomposable of length 3 is cuspidal and non-supercuspidal.

Any admissible irreducible R-representation V of G is a subquotient of $\mathrm{ind}_P^G W$ for some supercuspidal admissible irreducible R-representation W.

For a field R of characteristic p, (P, W) is unique modulo G-conjugation. This follows from the classification.

For a field R of characteristic different from p, the G-conjugation class of (M, W) is called a *supercuspidal support* of V. Contrary to the cuspidal support, the supercuspidal support is not always unique if the characteristic of R is $\ell \neq p$.

Examples. The supercuspidal support is not unique when $R = \mathbb{F}_\ell^{\mathrm{ac}}$, ℓ divides q^2+1 and G is the finite group $\mathrm{Sp}_8(\mathbb{F}_q)$ (Olivier Dudas [58]) or $\mathrm{Sp}_8(F)$ (Dat [49]).

---

32 Being algebraically closed is probably not necessary.

33 One does not need to suppose that W is irreducible when R is an algebraically closed field of characteristic different from p (Dat [49]).

The supercuspidal support is unique if R has characteristic 0, or G is an inner form of $\mathrm{GL}(n, F)$ (Minguez–Sécherre **[141]**), or G is the unramified unitary group $U(2, 1)$, $p \neq 2$ (Kurinczuk **[126]**), when R is algebraically closed (this is probably not necessary).

When R is algebraically closed, the twist by unramified characters of a supercuspidal support of V is called an *inertial supercuspidal support* of V; if all the irreducible R-representations of G have a unique supercuspidal support, the *Bernstein variety* $\mathcal{B}_R(G)$ is the disjoint union of the inertial supercuspidal supports of the irreducible R-representations of G.

An irreducible $\mathbb{Q}_\ell^{\mathrm{ac}}$-representation of G is integral if and only if its supercuspidal support is integral (Dat–Helm–Kurinczuk–Moss **[52, COROLLARY 1.6]**). Is any irreducible mod ℓ representation of G a subquotient of the reduction of an integral irreducible ℓ-adic representation?

For a field R of banal characteristic for G, any cuspidal irreducible R-representation of G is supercuspidal and projective in the category of R-representations of G with a given central character. The reduction of an integral cuspidal irreducible ℓ-adic representation of G is irreducible and cuspidal, and any cuspidal irreducible mod ℓ-representation of G lifts[34] to an integral cuspidal irreducible ℓ-adic representation of G (Dat–Helm–Kurinczuk–Moss, to appear). The reduction of an integral irreducible ℓ-adic representation of G may be reducible. Does any irreducible mod ℓ representation of G lift to an integral irreducible ℓ-adic representation of G?

8. HECKE ALGEBRAS

Hecke $\mathbb{Z}$-algebras appear everywhere in the theory of representations of G, giving algebraic proofs of properties proved earlier with harmonic analysis. An open subgroup K of G which is compact, or compact modulo the center of G, defines a *Hecke ring*

$$\mathcal{H}(G, K) = \mathrm{End}_{\mathbb{Z}[G]} \mathbb{Z}[K \backslash G],$$

naturally isomorphic to the opposite of $\mathbb{Z}[K\backslash G/K]$. For any commutative ring R, the Hecke R-algebra $\mathcal{H}_R(G, K) = \mathrm{End}_{R[G]} R[K\backslash G]$ is the scalar extension to R of $\mathcal{H}(G, K)$.

Finiteness property of $\mathcal{H}_R(G, K)$. The center $\mathcal{Z}_R(G, K)$ of $\mathcal{H}_R(G, K)$ is a finitely generated R-algebra and $\mathcal{H}_R(G, K)$ is a finitely generated $\mathcal{Z}_R(G, K)$-module, if R is a noetherian $\mathbb{Z}_\ell$-algebra.

This theorem of Dat–Helm–Kurinczuk–Moss **[52]** is the key of the proof of the second adjunction. It was proved by Deligne and Bernstein for complex Hecke algebras. It is equivalent to another statement, involving the endomorphism ring $\mathcal{Z}_R(G)$ of the identity functor of $\mathrm{Mod}_R(G)$, called the *Bernstein center*:

For R as above, any finitely generated R-representation V of G is $\mathcal{Z}_R(G)$-admissible and the natural image of $\mathcal{Z}_R(G) \to \mathrm{End}_{R[G]} V$ is a finitely generated R-algebra.

---

34 Is the reduction modulo ℓ of an integral cuspidal irreducible ℓ-adic representation of G.

The highly nontrivial proof uses the Fargues–Scholze local version of the Vincent Lafforgue's theory of excursion operators [65].

The finiteness theorem is true for the Iwahori and the pro-p Iwahori Hecke rings ($R = \mathbb{Z}$ and $K = J$ or $\tilde{J}$) (Vignéras [196]). Is it true for any Hecke ring?

The K-*invariant functor*

$$V \mapsto V^K \simeq \mathrm{Hom}_{R[G]}\big(R[K\backslash G], V\big) : \mathrm{Mod}_R(G) \to \mathrm{Mod}\, \mathcal{H}_R(G, K)$$

and its left adjoint $\mathcal{M} \to \mathcal{M} \otimes_{\mathcal{H}_R(G,K)} R[K\backslash G]$ relate the R-representations of G and the right $\mathcal{H}_R(G, K)$-modules. From now on, an $H_R(G, K)$-module will be a right module.

When R is a field and the order of any finite quotient of K is invertible in R, the K-invariant functor induces a bijection between the (isomorphism classes of) irreducible R-representations V of G with $V^K \neq 0$ and the (isomorphism classes of) simple $\mathcal{H}_R(G, K)$-modules.

If R is a field of characteristic different from p, an irreducible R-representation of G is admissible (Henniart–Vignéras [107, THEOREM 3.2]), any simple $\mathcal{H}_R(G, K)$-module has finite dimension. For any field R, a simple module of the Iwahori or pro-p Iwahori Hecke algebra has finite dimension.

Let $\mathrm{Mod}_R(G)(K)$ denote the category of R-representations of G generated by their K-invariant vectors. When any subrepresentation of any representation in $\mathrm{Mod}_R(G)(K)$ belongs to $\mathrm{Mod}_R(G)(K)$, the category $\mathrm{Mod}_R(G)(K)$ is abelian and equivalent by the K-invariant functor to

$$\mathrm{Mod}_R(G)(K) \xrightarrow{\sim} \mathrm{Mod}\, \mathcal{H}_R(G, K).$$

This is the case if $R = \mathbb{C}$ and K is an Iwahori subgroup J by a classical result of Borel, or a pro-p Iwahori subgroup $\tilde{J}$ (Vignéras [196]). The category $\mathrm{Mod}_{\mathbb{C}}(G)(J)$ is an indecomposable factor of $\mathrm{Mod}_{\mathbb{C}}(G)(\tilde{J})$, and $\mathrm{Mod}_{\mathbb{C}}(G)(\tilde{J})$ is a factor of $\mathrm{Mod}_{\mathbb{C}}\, G)$.

For R of characteristic p, the category $\mathrm{Mod}_{\mathbb{C}}(G)(\tilde{J})$ is not abelian in general. However, it is abelian if $R = \mathbb{F}_p^{\mathrm{ac}}$ and $G = \mathrm{GL}(2, \mathbb{Q}_p)$ or $\mathrm{SL}(2, \mathbb{Q}_p)$, $p \neq 2$ (Ollivier [146],[35] Koziol [119], Ollivier–Schneider [151]).

For a prime r, a $\mathbb{Q}_r$-representation V of G is called *locally integral* if for some finite extension $E/\mathbb{Q}_r$, V^K admits a $\mathcal{H}(G, K)$-stable O_E-lattice for all open compact subgroups K of G.

An integral irreducible $\mathbb{Q}_r^{\mathrm{ac}}$-representation is clearly locally integral. The converse is true if $r = \ell \neq p$ [38]. The equivalence between integral and locally integral for irreducible $\mathbb{Q}_p^{\mathrm{ac}}$-representations of G is an open question. It is the analogue of the Breuil–Schneider conjecture [20] restricted to smooth representations (Hu [113], Sorensen [183, 184]).

A finite length $\mathbb{Q}_p^{\mathrm{ac}}$-representation V of G is locally integral if and only if (Dat [41])

$$\nu(\delta_P^{-1/2}\chi) \in \rho_P - \overline{{}^{+}\mathcal{A}_P^*}$$

for any standard parabolic subgroup $P = MN$ of G with $L_P^G(V) \neq 0$, and any exponent χ of $L_P^G(V)$.[36]

---

35 Supposing that a uniformizer of F acts trivially.

36 ρ_P is half the sum of the roots of A_M in Lie P. The formula can be simplified!

9. REPRESENTATIONS OVER A FIELD OF CHARACTERISTIC DIFFERENT FROM p

For any commutative ring R, an R-representation W of an open subgroup K of G defines an R-representation $\mathrm{ind}_K^G W$ of G by *compact induction*.[37]

Example. $\mathrm{ind}_K^G 1_K = R[K\backslash G]$ for the trivial R-representation 1_K of K.

Assume that R is a field of characteristic different from p, until the end of this section.

All cuspidal irreducible R-representations of G are conjectured to be compactly induced from open subgroups of G compact modulo the center of G.

For R algebraically closed, the conjecture has been proved for the level 0[38] cuspidal representations of any G or when

- G has rank 1 (Martin Weissman [203]),
- G is an inner form of $\mathrm{GL}(n, F)$ (Minguez–Sécherre [141]), or of $\mathrm{SL}(n, F)$ (Peyi Cui [36, 37]),
- G is a classical group (Stevens [187], Stevens–Kurinczuk–Skodlerak [131]) or a quaternionic form of G (Skodlerak [181]), if $p \neq 2$.
- G splits on a moderately ramified extension of F and p does not divide the order of the absolute Weyl group (Fintzen [66]).

Being algebraically closed is not necessary and there is an explicit list $\mathcal{X}$ of pairs (K, W) of G where K is an open subgroup of G compact modulo the center and W an R-representation of K such that $\mathrm{ind}_K^G W$ is irreducible cuspidal satisfying (Henniart–Vignéras [107]):

(a) any cuspidal irreducible R-representation of G is isomorphic to $\mathrm{ind}_K^G W$ for some $(K, W) \in \mathcal{X}$ unique modulo G-conjugation,

(b) $\mathrm{ind}_K^G W$ and W have the same intertwining algebra

$$\mathrm{End}_{R[K]} W \simeq \mathrm{End}_{R[G]} \mathrm{ind}_K^G W,$$

(c) $\mathrm{ind}_K^G W$ is supercuspidal if and only if W is supercuspidal, for the "natural notion of supercuspidality" of W,[39]

(d) $\mathcal{X}$ is stable by automorphisms of R.

Until the end of this section, assume R algebraically closed and $G = \mathrm{GL}(m, D)$ where D is a central division algebra of dimension d^2 over F, $n = md$.

---

37 The R-module of functions $f : G \to W$ supported on finitely many cosets Kg, satisfying $f(kg) = \rho(k) f(g)$ for $k \in K$, $g \in G$ where G acts by right translation.

38 Definition in the section on Bernstein blocks.

39 Fintzen gave another proof when G is moderately ramified and p does not divide the order of the absolute Weyl group.

Minguez and Sécherre [140] classified the irreducible R-representations of G with a given supercuspidal support by "supercuspidal multisegments," and those with a given cuspidal support by "aperiodic cuspidal multisegments." This generalizes the Bernstein–Zelevinski classification of complex irreducible representations of $\mathrm{GL}(n, F)$. For R of characteristic ℓ, the proof uses the theory of ℓ-modular types (Minguez–Sécherre [141]) and deep results on affine Hecke algebras of type A at roots of unity.

Any irreducible ℓ-modular representation of G is a subquotient of the reduction of an integral irreducible ℓ-adic representation [140]. In the other direction, any irreducible ℓ-modular representation V of G lifts to an ℓ-adic representation when it is supercuspidal or "banal" or unramified[40] (Dat [38], Minguez–Sécherre [139,140,142]) or when it is cuspidal and $G = \mathrm{GL}(n, F)$. Contrary to the case $G = \mathrm{GL}(n, F)$, some irreducible cuspidal ℓ-modular representation of G may not lift and the reduction of a integral cuspidal irreducible ℓ-adic-representation of G may be reducible; but its irreducible components are cuspidal and in the same inertial class.

Example. When $q = 8$, $\ell = 3$, $d = 2$, any integral irreducible ℓ-adic representation of D^* containing an homomorphism $\chi : O_D^* \to (\mathbb{Q}_\ell^{ac})^*$ trivial on $1 + P_D$ such that $\chi \neq \chi^q$ has dimension 2 and its reduction is reducible. When $q = 4$, $\ell = 17$, $d = 2$, there exists an irreducible cuspidal ℓ-modular representation of $\mathrm{GL}(2, D)$ not lifting to $\mathbb{Q}_\ell^{ac}$ (Minguez–Sécherre [143]).

Let $\mathcal{D}_{\mathbb{C}}(G)$ denote the set of isomorphism classes of the essentially square integrable irreducible (or discrete series) complex representations of G. The complex *local Jacquet–Langlands correspondence*

$$\mathrm{JL}_{\mathbb{C}} : \mathcal{D}_{\mathbb{C}}\big(\mathrm{GL}(m, D)\big) \tilde{\to} \mathcal{D}_{\mathbb{C}}\big(\mathrm{GL}(n, F)\big)$$

is a bijection characterized by a character relation on matching elliptic regular conjugacy classes. Fixing an isomorphism $\mathbb{C} \simeq \mathbb{Q}_\ell^{ac}$, the complex local Jacquet–Langlands correspondence gives an ℓ-adic local Jacquet–Langlands correspondence

$$\mathrm{JL}_{\mathbb{Q}_\ell^{ac}} : \mathcal{D}_{\mathbb{Q}_\ell^{ac}}\big(\mathrm{GL}(m, D)\big) \tilde{\to} \mathcal{D}_{\mathbb{Q}_\ell^{ac}}\big(\mathrm{GL}(n, F)\big)$$

independent of the isomorphism $\mathbb{C} \simeq \mathbb{Q}_\ell^{ac}$, and respecting integrality. Minguez and Sécherre [143] proved that two integral representations of $\mathcal{D}_{\mathbb{Q}_\ell^{ac}}(\mathrm{GL}(m, D))$ are congruent modulo ℓ if and only if their transfers to $\mathrm{GL}(n, F)$ are congruent modulo ℓ. But there is no ℓ-modular local Jacquet–Langlands correspondence compatible with the ℓ-adic local Jacquet–Langlands correspondence by reduction, as, for example, when $d = 2$ and $q + 1 \equiv 0$ modulo ℓ, the trivial representation $1_{\mathbb{Q}_\ell^{ac}}$ of D^* corresponds to the Steinberg $\mathrm{St}_{\mathbb{Q}_\ell^{ac}}$ of $\mathrm{GL}(2, F)$ of reduction modulo ℓ of length 2 (Dat [43]). However, the Badulescu morphism [13]

$$\mathrm{LJ}_{\mathbb{Q}_\ell^{ac}} : \mathcal{G}r_{\mathbb{Q}_\ell^{ac}}\big(\mathrm{GL}(n, F)\big) \to \mathcal{G}r_{\mathbb{Q}_\ell^{ac}}\big(\mathrm{GL}(m, D)\big),$$

40 $V^{\mathrm{GL}(m, O_D)} \neq 0$, equivalent to V irreducibly parabolically induced from an unramified character of a Levi subgroup [142].

where $\mathcal{G}r_R(G)$ is the Grothendieck group of finite length R-representations of G, gives by reduction an ℓ-modular Badulescu morphism

$$\mathrm{LJ}_{\mathbb{F}_\ell^{\mathrm{ac}}} : \mathcal{G}r_{\mathbb{F}_\ell^{\mathrm{ac}}}\big(\mathrm{GL}(n,F)\big) \to \mathcal{G}r_{\mathbb{F}_\ell^{\mathrm{ac}}}\big(\mathrm{GL}(m,D)\big).$$

Sécherre and Stevens **[180]** introduced the interesting notions of mod ℓ inertial supercuspidal support and linkage for irreducible complex representations π, π' of G.

(a) Picking an isomorphism $\mathbb{C} \simeq \mathbb{Q}_\ell^{\mathrm{ac}}$ one supposes that π is an ℓ-adic representation of G. The inertial cuspidal support of π contains an integral cuspidal representation τ. The *mod ℓ inertial supercuspidal support* of π is the inertial supercuspidal support of any irreducible component of $r_\ell(\tau)$; it depends only on the isomorphism class of π.

(b) π, π' are *linked* if there are prime numbers $\ell_1, \dots, \ell_r$ different from p, and irreducible complex representations $\pi = \pi_0, \pi_1, \dots, \pi_r = \pi'$ such that, for each $i \in \{1, \dots, r\}$, the representations π_{i-1}, π_i have the same mod ℓ_i inertial supercuspidal support.

When π, π' are essentially square integrable, they are linked if and only if their images by the local Jacquet–Langlands correspondence $\mathrm{JL}_{\mathbb{C}}$ are linked if and only if (Dotto **[55]**) they have the same semisimple endoclass (a type invariant). When $G = \mathrm{GL}(n, F)$ and π, π' are cuspidal, they have the same *endoclass* if and only if the associated irreducible representations of Weil group W_F by the local Langlands correspondence share an irreducible component when restricted to the wild inertia group.

10. BERNSTEIN BLOCKS

For a commutative ring R, a nontrivial idempotent e in the Bernstein center $\mathcal{Z}_R(G)$ decomposes the abelian category

$$\mathrm{Mod}_R(G) = e\big(\mathrm{Mod}_R(G)\big) \times (1-e)\big(\mathrm{Mod}_R(G)\big)$$

into a direct product of two abelian full subcategories. When the idempotent $e \in \mathcal{Z}_R(G)$ is primitive, the subcategory $e(\mathrm{Mod}_R(G))$, where e acts by the identity, is indecomposable (no nontrivial factors) and called a *block*.

Bernstein and Deligne factorized $\mathrm{Mod}_{\mathbb{C}}(G)$ into blocks. Their arguments are valid for any algebraically closed field R of characteristic 0. The decomposition is based on the uniqueness of the supercuspidal support. We have

$$\mathrm{Mod}_R(G) = \prod_{\Omega \in \mathcal{B}_R(G)} \mathrm{Mod}_R(G)_\Omega$$

over the connected components Ω of the Bernstein variety $\mathcal{B}_R(G)$. The *Bernstein block* $\mathrm{Mod}_R(G)_\Omega$ consists of the R-representations of G all of whose irreducible subquotients have inertial supercuspidal support Ω. The center of the block $\mathrm{Mod}_R(G)_\Omega$ is the ring of regular functions on the variety Ω.

When G is an inner form of $\mathrm{GL}(n, F)$, two complex discrete series of G in the same block are inertially equivalent. The complex Jacquet–Langlands correspondence commutes with twisting by characters, and yields a bijection between the blocks containing discrete series. Andrea Dotto [55] parametrized these blocks by two algebraic invariants (one is the endo-class) and obtained a complete algebraic description of the Jacquet–Langlands correspondence at the level of inertial classes.

For an algebraically closed field R of characteristic different from p, the Deligne–Bernstein decomposition remains true (Sécherre and Stevens [179]). Bastien Drevon and Vincent Sécherre [57] described the block decomposition of the abelian category of finite length R-representations of G. Unlike the case of all R-representations of G, several non-isomorphic supercuspidal supports may correspond to the same block. A supercuspidal block is equivalent to the principal block of the multiplicative group of a suitable division algebra.

When R is an algebraically closed field of characteristic ℓ banal for G, it is expected that the Deligne–Bernstein decomposition remains true and that the reduction modulo ℓ gives a bijection between the blocks of ℓ-adic representations of G and the blocks of mod ℓ representations of G.

When $R = W(\mathbb{F}_\ell^{\mathrm{ac}})$ is the Witt ring of $\mathbb{F}_\ell^{\mathrm{ac}}$ and $G = \mathrm{GL}(n, F)$, Helm [96–98] showed that the block decomposition of $\mathrm{Mod}_{\mathbb{F}_\ell^{\mathrm{ac}}}(G)$ lifts to a block decomposition of $\mathrm{Mod}_{W(\mathbb{F}_\ell^{\mathrm{ac}})}(G)$,

$$\mathrm{Mod}_{W(\mathbb{F}_\ell^{\mathrm{ac}})}(G) = \prod_{\Omega \in \mathcal{B}_{\mathbb{F}_\ell^{\mathrm{ac}}}(G)} \mathrm{Mod}_{\mathbb{Z}_\ell^{\mathrm{ac}}}(G)_\Omega.$$

The block $\mathrm{Mod}_{W(\mathbb{F}_\ell^{\mathrm{ac}})}(G)_\Omega$ consists of the $W(\mathbb{F}_\ell^{\mathrm{ac}})$-representations of G such that any irreducible subquotient V

- has a supercuspidal support in Ω modulo isomorphism, if $\ell V = 0$,
- is such that the reduction modulo ℓ of an integral element in the inertial class of the supercuspidal support of V is in Ω modulo isomorphism, if $\ell V = V$.

The center of $\mathrm{Mod}_{W(\mathbb{F}_\ell^{\mathrm{ac}})}(G)$ is naturally isomorphic to the ring of endomorphisms of the Gelfand–Graev representation of G,[41] and the center of $\mathrm{Mod}_{W(\mathbb{F}_\ell^{\mathrm{ac}})}(G)_\Omega$ is a finitely generated, reduced, ℓ-torsion free $W(\mathbb{F}_\ell^{\mathrm{ac}})$-algebra.

The *principal block* of $\mathrm{Mod}_R(G)$ contains the trivial R-representation of G. When $R = \mathbb{C}$, the principal block is equivalent to the category of modules over the Iwahori Hecke $\mathbb{C}$-algebra. The blocks have been computed in a large number of examples with the theory of types. Many blocks are equivalent to the principal block of another group G'.

Example. For an algebraically closed field R of characteristic different from p and G an inner form of $\mathrm{GL}(n, F)$, each block of $\mathrm{Mod}_R(G)$ is equivalent to the principal block of a product of general linear groups [179].

---

41 $\mathrm{ind}_U^{\mathrm{GL}(n,F)} \psi$, where ψ is a generic $W(\mathbb{F}_\ell^{\mathrm{ac}})$-character of the unipotent radical U of a Borel subgroup of $\mathrm{GL}(n, F)$.

When $R = \mathbb{Q}_\ell^{ac}, \mathbb{Z}_\ell^{ac}$ or $\mathbb{F}_\ell^{ac}$, Dat explained the known coincidences between the blocks of $\mathrm{Mod}_R(G)$ and predicted many more by a functoriality principle involving dual groups **[47, 48]**.

For a commutative ring R where p is invertible, there is a decomposition of $\mathrm{Mod}_R(G)$ by the Moy–Prasad *depth* **[40, APPENDIX A]**.

An R-representation V of G has *depth* 0 if $V = \sum_x V^{\tilde{G}_x}$ is the sum of its invariants $V^{\tilde{G}_x}$ by the pro-p radicals $\tilde{G}_x$ of the subgroups of G fixing the vertices of the adjoint Bruhat–Tits building of G. The possible depths form a sequence of non-negative rational numbers $r_0 = 0 < r_1 < \cdots$. The category $\mathrm{Mod}_R(G)^{(r)}$ of R-representations of G of depth r is abelian with an explicit finitely generated projective generator but is generally not a block. We have

$$\mathrm{Mod}_R(G) = \prod_{n \in \mathbb{N}} \mathrm{Mod}_R(G)^{(r_n)}.$$

When $p = 0$ in R, the Bernstein center $\mathcal{Z}_R(G)$ of G is as small as possible, equal to the Bernstein center of the center $Z(G)$ of G (see Ardakov–Schneider **[12]** when R is a field, but their proofs are valid for a commutative ring, see also Dotto **[55]**)

$$\mathcal{Z}_R\big(Z(G)\big) = \varprojlim_K R\big[Z(G)/K\big], \quad K \subset Z(G) \text{ open compact subgroup.}$$

When $E/\mathbb{Q}_p$ is a finite extension of ring of integers O_E, the category of locally finite representations (equal to the union of their subrepresentations of finite length) of $\mathrm{GL}(2, \mathbb{Q}_p)$ on O_E-torsion modules with a central character decomposes as a product of blocks with a noetherian center (Paskunas and Shen-Nin Tung **[159]**).

11. SATAKE ISOMORPHISM

The structure of the Hecke ring of any special parahoric subgroup K of G is understood via the *Satake transform*

$$\mathrm{Sat} : \mathcal{H}(G, K) \to \mathcal{H}(Z, Z^0), \quad \mathrm{Sat}(f)(z) = \sum_{u \in U^0 \backslash U} f(uz) \text{ for } z \in Z.$$

It is an injective ring homomorphism, and as $\mathcal{H}(Z, Z^0) \simeq \mathbb{Z}[Z/Z^0]$ is commutative, it shows that the Hecke ring $\mathcal{H}(G, K)$ is commutative. A basis of the image of Sat is

$$S_\lambda = \sum_{\lambda' \in W(\lambda)} \delta^{1/2}(\lambda/\lambda') e_{\lambda'} \quad \text{for } \lambda \in Z^+/Z^0,$$

where $e_\lambda \in \mathcal{H}(Z, Z^0)$ corresponds to λ (Henniart–Vignéras **[105]**, **[104, PROPOSITION 2.3]**). This shows that modulo isomorphism, the commutative Hecke ring $\mathcal{H}(G, K)$ does not depend on the choice of K.

By scalar extension to a commutative ring R, the Satake transform extends to a map $\mathrm{Sat} : \mathcal{H}_R(G, K) \to \mathcal{H}_R(Z, Z^0)$. For $R = \mathbb{C}$, it is well known that $\delta_B^{1/2}$ Sat induces an isomorphism

$$\mathcal{H}_\mathbb{C}(G, K) \simeq \mathbb{C}\big[Z/Z^0\big]^{W_G}.$$

An all-important special case was singled out by Langlands, that is, where G is unramified and where K is a hyperspecial maximal compact subgroup of G. Langlands interpreted the Satake isomorphism as giving a parametrization of the isomorphism classes of complex irreducible representations of G with a nonzero K-fixed vector, by certain semisimple conjugacy classes in a complex group $\hat{G}$ "dual" to G.

For a field R of characteristic p, Sat induces an isomorphism (Henniart–Vignéras **[105]**)[42]

$$\mathcal{H}_R(G,K) \simeq R[Z^+/Z^0].$$

Instead of focusing on the trivial R-representation 1_K of K, one can consider two finitely generated R-representations W, W' of K and the Hecke R-bimodule

$$\mathcal{H}_R(G,K,W,W') \simeq \mathrm{Hom}_{R[G]}(\mathrm{ind}_K^G W, \mathrm{ind}_K^G W').$$

It is realized as a set of compactly supported functions $f : G \to \mathrm{Hom}_R(W, W')$ with a certain K-biinvariance. In the case $W = W'$, it is an algebra called an *Hecke algebra with weight* W that we rather write $\mathcal{H}_R(G,K,W)$; the Hecke algebra with trivial weight is the Hecke R-algebra $\mathcal{H}_R(G,K)$. For any standard parabolic subgroup $P = MN$, the Satake transform generalizes to an injective map

$$\begin{aligned} &\mathrm{Sat}_M : \mathcal{H}_R(G,K,W,W') \to \mathcal{H}_R(M, M^0, W_{N^0}, W'_{N^0}),\\ &\mathrm{Sat}_M(f)(m)(\overline{v}) = \sum_{n\in N^0\backslash N} \overline{f(nm)(v)} \end{aligned}$$

for $m \in M$, $v \in W$, where $v \to \overline{v}$ is the quotient map $W \to W_{N^0}$ (similarly for $W' \to W'_{N^0}$). The functional approach of Sat_M (Henniart–Vignéras **[104, SECTION 2]**) is a motivation to prefer it to another generalization considered in (see Herzig **[109]** when G is split, Henniart–Vignéras **[105]**)

$$\begin{aligned} &\mathrm{Sat}'_M : \mathcal{H}_R(G,K,W,W') \to \mathcal{H}_R(M, M^0, W^{N^0}, W'^{N^0}),\\ &\mathrm{Sat}'_M(f)(z)(v) = \sum_{u\in U^0\backslash U} f(uz)(v) \end{aligned}$$

for $v \in W^{N^0}$. The maps Sat'_M and Sat_M are related by taking duals **[104]**.

When R is an algebraically closed field of characteristic p and W, W' are irreducible, the generalized Satake transforms play a role in the modulo p and p-adic Langlands correspondence. In this situation W_{U^0}, W'_{U^0} have dimension 1, the Hecke bimodule $\mathcal{H}_R(G,K,W,W')$ is nonzero if and only if the R-characters of Z^0 on W_{U^0}, W'_{U^0} are Z-conjugate. For $M = Z$, there are explicit bases $(S_\lambda^{W,W'})$ of the image of Sat_Z, and $(T_\lambda^{W,W'})$ of $\mathcal{H}_R(G,K,W,W')$ such that

$$\mathrm{Sat}_Z(T_\lambda^{W,W'}) = S_\lambda^{W,W'}$$

for $\lambda \in Z^+(W,W')/Z^0$ where $Z^+(W,W')$ is a certain union of cosets of Z^0 in Z^+ (Abe–Herzig–Vignéras **[11]**). The proof relies on the theory the pro-p-Iwahori Hecke R-algebra.

42 With Z^-/Z^0 instead of Z^+/Z^0, but these monoids are isomorphic.

A simple consequence is the "change of weight"[43] which is an important step in the proof of the classification of admissible irreducible R-representations of G. There is also a change of weight the pro-p-Iwahori Hecke algebra giving another proof for the change of weight for G (Abe **[4]**). For an Hecke algebra $\mathcal{H}_R(G, K, W)$ with irreducible weight W, one gets an explicit inverse of the Satake isomorphism (Henniart–Vignéras **[104]**)[44]:

$$\mathrm{Sat}_Z : \mathcal{H}_R(G, K, W) \xrightarrow{\sim} \mathcal{H}_R(Z^+, Z^0, W_{U^0}).$$

For G quasisplit, $\mathcal{H}_R(Z^+, Z^0, W_{U^0}) \simeq R[Z^+/Z^0]$, hence $\mathcal{H}_R(G, K, W)$ is commutative and does not depend on the choice of (K, W) modulo isomorphism. For G general, the center of $\mathcal{H}_R(G, K, W)$ contains a finitely generated subalgebra $\mathcal{Z}_T$ isomorphic to $R[T^+/T^0]$, and $\mathcal{H}_R(G, K, W)$ is a finitely generated $\mathcal{Z}_T$-module.

One chooses an element s in the center of M which strictly contracts N by conjugation. There is a unique element $T_s \in \mathcal{H}_R(M, M^0, W_{N^0})$ with support $M^0 s$ such that $T_s(s)$ is the identity on W_{N^0}. The generalized Satake transform

$$\mathrm{Sat}_M : \mathcal{H}_R(G, K, W) \hookrightarrow \mathcal{H}_R(M, M^0, W_{N^0})$$

is a localization at T_s.[45] The natural intertwiner

$$I_V : \mathrm{ind}_K^G W \to \mathrm{ind}_P^G(\mathrm{ind}_{M^0}^M W_{N^0})$$

is injective and its localization at T_s is bijective when W satisfies a regularity assumption[46] (Herzig **[108]**, Abe **[3]**, Henniart–Vignéras **[104]**).

For a field R of characteristic p, the *supersingularity* of an admissible irreducible R-representation V of G is defined with the Satake homomorphism (Abe–Henniart–Herzig–Vignéras **[8]**). First, assuming R algebraically closed, an homomorphism $\mathcal{Z}_R(G, K, W) \to R$ from the center $\mathcal{Z}_R(G, K, W)$ of an Hecke algebra $\mathcal{H}_R(G, K, W)$ with irreducible weight is said to be supersingular if it does not extend to the center of $\mathcal{H}_R(M, M^0, W_{N^0})$ via the Satake homomorphism for any $P \neq G$. As V is admissible, there exists some irreducible representation W of K such that $\mathrm{Hom}_{R[G]}(\mathrm{ind}_K^G W, V) \neq 0$. If $\mathrm{Hom}_{R[G]}(\mathrm{ind}_K^G W, V)$ as a module over the center of $\mathcal{H}_R(G, K, W)$ contains an eigenvector with a supersingular eigenvalue, V is called supersingular. This does not depend on the choice of (K, W). For R not algebraically closed, V is called supersingular if some admissible irreducible R^{ac}-representation V^{ac} of G which is V-isotypic as an R-representation, is supersingular. This does not depend on the choice of V^{ac} (Henniart–Vignéras **[106]**).

For G unramified and K hyperspecial, using the geometric Satake equivalence, Xinwein Zhu **[209]** identified the Hecke ring $\mathcal{H}(G, K)$ with a ring associated to the Vinberg

---

43 The change of weight theorem is an isomorphism between two compactly induced representations.

44 This isomorphism for Sat′ is proved when G is split in Herzig **[109]**, and in general in Henniart–Vignéras **[105]**.

45 This means that the image of Sat_M contains T_s and that its localization at T_s is $\mathcal{H}_R(M, M^0, W_{N^0})$.

46 Meaning that the map $\mathcal{H}_R(M, M^0, V_{N^0}) \otimes_{\mathcal{H}_R(G,K,V)} \mathrm{ind}_K^G V \to \mathrm{ind}_P^G(\mathrm{ind}_{M^0}^M V_{N^0})$ is bijective, if the kernel of $V \to V_{N^0}$ contains $kV^{(N^{\mathrm{op}})^0}$ for all $k \in K \setminus P^0(P^{\mathrm{op}})^0$.

monoid of $\hat{G}$ and formulated a canonical Satake isomorphism. He proved that the commutative $\mathbb{Z}$-algebra $\mathcal{H}(G, K)$ is finitely generated. He extended his formulation to an Hecke algebra $\mathcal{H}_{O_E}(G, K, W)$ with weight a finite free O_E-module W arising from an irreducible algebraic representation $E \otimes_{O_E} W$ of G, where E/F is a finite extension.

For F of characteristic 0 and R a field of characteristic p, Claudius Heyer **[112, THEOREM 4.3.2]** defined a derived generalized Satake homomorphism.

For F of characteristic 0, G split, K hyperspecial and $R = \mathbb{Z}/p^a\mathbb{Z}, a \geq 1$, Niccolo Ronchetti **[167]** established a Satake homomorphism for the derived Hecke $\mathbb{Z}/p^a\mathbb{Z}$-algebra of (G, K) (a graded associative $\mathbb{Z}/p^a\mathbb{Z}$-algebra whose degree 0 subalgebra is $\mathcal{H}_{\mathbb{Z}/p^a\mathbb{Z}}(G, K)$). The relation with the Heyer derived Satake homomorphism is unclear.

12. PRO-p IWAHORI HECKE RING

The Iwahori Hecke ring $\mathcal{H}(G, J)$ and the pro-p Iwahori Hecke ring $\mathcal{H}(G, \tilde{J})$ modulo isomorphism depend only on G, because the Iwahori subgroups of G are conjugate, as well as the pro-p Iwahori subgroups.

They are both natural generalizations of affine Hecke $\mathbb{Z}$-algebras. We will focus on the pro-p Iwahori Hecke ring which is more involved, that we will denote by $\mathcal{H}(G)$, but all the results apply to Iwahori Hecke rings with some simplifications.

Our motivation to study the pro-p Iwahori Hecke ring instead of the Iwahori Hecke ring comes from the theory of mod p representations.[47] Any nonzero mod p representation of G has a nonzero $\tilde{J}$-fixed vector, and the pro-p radical of any parahoric subgroup of G is contained in some G-conjugate of $\tilde{J}$.

For any commutative ring R, the pro-p Iwahori Hecke R-algebra $\mathcal{H}_R(G) = R \otimes_{\mathbb{Z}} \mathcal{H}(G)$ is a specialization of the *generic pro-p Iwahori Hecke $R[\mathbf{q}_*]$-algebra $\mathcal{H}(G)(\mathbf{q}_*, c_*)$* of G, introduced by Nicolas Schmidt **[170, 171]** when G is split (Vignéras **[198]** in general). The $\mathbf{q}_*$ are finitely many indeterminates and the finitely many $c_* \in R[\mathbf{q}_*]$ satisfy simple conditions. The general principle is that one proves properties of the generic pro-p Iwahori Hecke $R[\mathbf{q}_*]$-algebra by specializing all $\mathbf{q}_*$ to 1, and then one transfers them to $\mathcal{H}_R(G)$ by specialization.

Example. The affine Yokonuma–Hecke algebra defined by Maria Chlouveraki and Loic Poulain d'Andecy is a generic pro-p Iwahori Hecke algebra (Chlouveraki and Sécherre **[30]**).

The main features[48] of affine Hecke R-algebras generalize to the generic pro-p Iwahori Hecke R-algebra, and by specialization to $\mathcal{H}_R(G)$. The $R[\mathbf{q}_*]$-module $\mathcal{H}(G)(\mathbf{q}_*, c_*)$ is free with an Iwahori–Matsumoto basis of elements satisfying braid relations and quadratic relations, with "alcove walk bases" associated to the Weyl chambers. There are product for-

---

47 Flicker **[69]** studied the pro-p Iwahori Hecke complex algebra when G is unramified.

48 The Iwahori Matsumoto presentation, the Bernstein basis, the Bernstein–Lusztig relations, the description of the center, and the geometric proofs of Görtz **[78]**.

mulas involving different alcove walk bases, and Bernstein–Lusztig relations from which one deduces an explicit canonical $R[\mathbf{q}_*]$-basis of the center [196].

Finiteness properties of the pro-p Iwahori ring $\mathcal{H}(G)$.

(i) The center $\mathcal{Z}(G)$ of $\mathcal{H}(G)$ is a finitely generated $\mathbb{Z}$-algebra and $\mathcal{H}(G)$ is a finitely generated $\mathcal{Z}(G)$-module.

(ii) $\mathcal{Z}(G)$ contains a canonical subring $\mathcal{Z}_T$ isomorphic to the affine semigroup $\mathbb{Z}$-algebra $\mathbb{Z}[T^+/T^0]$, and the $\mathcal{Z}_T$-modules $\mathcal{Z}$ and $\mathcal{H}$ are finitely generated.

(iii) The elements of the Iwahori–Matsumoto basis[49] of $\mathcal{H}(G)$ are invertible in $\mathbb{Z}[1/p] \otimes_{\mathbb{Z}} \mathcal{H}(G)$.

(iv) For any commutative ring R, the center of $\mathcal{H}_R(G)$ is $\mathcal{Z}_R(G) = R \otimes_{\mathbb{Z}} \mathcal{Z}(G)$.

For any field R, any simple $\mathcal{H}_R(G)$-module is finite dimensional by (i) and (iv) [101].

Xuhua He and Radhika Ganapathy [93] gave an Iwahori–Matsumoto presentation of the Hecke ring $\mathcal{H}(G, J_n)$ of the nth congruence subgroup J_n of J for any $n \in \mathbb{N}_{>0}$.

For a standard parabolic subgroup $P = MN$, although $\mathcal{H}_R(M)$ is not contained in $\mathcal{H}_R(G)$, there is a *parabolic induction*

$$\mathrm{ind}_{\mathcal{H}(M)}^{\mathcal{H}(G)} = - \otimes_{\mathcal{H}_R(M)} X_{G,P} : \mathrm{Mod}\,\mathcal{H}_R(M) \to \mathrm{Mod}\,\mathcal{H}_R(G), \quad X_{G,P} = \mathrm{ind}_P^G\big(R[\tilde{J}_M \backslash M]\big)$$

of right adjoint $\mathrm{Hom}_{\mathcal{H}_R(G)}(X_{G,P}, -)$ and of left adjoint a certain localization (hence the left adjoint is exact, a surprise when p is not invertible in R as the functor $(-)_N$ for representations is not exact). The parabolic induction and its right adjoint for the group and for the pro-p Iwahori Hecke algebra correspond to each other via the pro-p Iwahori invariant functors. The same holds true for the left adjoint functor if R is a field of characteristic different from p, but Abe gave a counterexample for $G = \mathrm{GL}(2, \mathbb{Q}_p)$ and R of characteristic p (Ollivier–Vignéras [154]). The parabolic induction is isomorphic to

$$\mathrm{ind}_{\mathcal{H}(P)}^{\mathcal{H}(G)} = - \otimes_{\mathcal{H}(P)} \mathcal{H}(G) : \mathrm{Mod}\,\mathcal{H}_R(M) \to \mathrm{Mod}\,\mathcal{H}_R(G),$$

where $\mathcal{H}(P) = \mathbb{Z}[(\tilde{J} \cap P)\backslash G/(\tilde{J} \cap P)]$ is the parabolic pro-p Iwahori Hecke ring of P for two ring homomorphisms $\mathcal{H}(M) \leftarrow \mathcal{H}(P) \to \mathcal{H}(G)$ (Heyer [111]).

For an algebraically closed field R of characteristic p and an irreducible R-representation W of a special parahoric subgroup K containing $\tilde{J}$, an inverse Satake-type isomorphism

$$f : \mathcal{H}_R(Z^-, Z^0, W^{U^0}) \tilde{\to} \mathcal{H}_R(G, K, W)$$

is obtained by composition of two natural algebra isomorphisms (Ollivier [149] when G is split, Vignéras [200] in general). The first isomorphism is associated to a "good" alcove walk basis

$$\mathcal{H}_R(Z^-, Z^0, W^{U^0}) \tilde{\to} \mathrm{End}_{\mathcal{H}_R(G)}\big(W^{\tilde{J}} \otimes_{\mathcal{H}_R(K,\tilde{J})} \mathcal{H}_R(G)\big).$$

49 The Iwahori–Matsumoto basis of $\mathcal{H}(G)$ is given by the characteristic functions of the double cosets of G modulo $\tilde{J}$.

The dimension of $W^{\tilde{J}}$ is 1. The second isomorphism

$$\mathrm{End}_{\mathcal{H}_R(G)}\big(W^{\tilde{J}} \otimes_{\mathcal{H}_R(K,\tilde{J})} \mathcal{H}_R(G)\big) \tilde{\to} \mathcal{H}_R(G, K, W)$$

is associated to a natural $H_R(G)$-module isomorphism $W^{\tilde{J}} \otimes_{\mathcal{H}_R(K,\tilde{J})} \mathcal{H}_R(G) \tilde{\to} (\mathrm{ind}_K^G W)^{\tilde{J}}$. When G is split, f is the inverse of the generalized Satake isomorphism Sat'_Z (Ollivier **[149]**).

13. MODULES OF PRO-p IWAHORI HECKE ALGEBRAS OVER A FIELD IN CHARACTERISTIC p

There is a numerical mod p local Langlands correspondence for the pro-p Iwahori Hecke algebra of $\mathrm{GL}(n, F)$ (Vignéras **[191]**). The following two sets have the same (finite) cardinality[50]:

(a) the isomorphism classes of the n-dimensional irreducible $\mathbb{F}_p^{\mathrm{ac}}$-representations of $\mathrm{Gal}(F^{\mathrm{ac}}/F)$ with a fixed value of the determinant of the action of a Frobenius;

(b) the isomorphism classes of the supersingular simple modules $\mathcal{H}_{\mathbb{F}_p^{\mathrm{ac}}}(\mathrm{GL}(n, F))$ with a fixed action of p_F embedded diagonally.

When $F \supset \mathbb{Q}_p$, this was significantly improved by Grosse-Kloenne **[80,81]**. He constructed an exact and fully faithful functor from the category of finite length supersingular $\mathcal{H}_{\mathbb{F}_{p^d}}(\mathrm{GL}(n, F))$-modules to the category of $\mathbb{F}_q^d$-representations of $\mathrm{Gal}(F^{\mathrm{ac}}/F)$, if $p^d \geq q$.[51]

We recall that the pro-p Iwahori Hecke ring $\mathcal{H}(G)$ of G is a finitely generated module over a central subring $\mathcal{Z}_T \simeq \mathbb{Z}[T^+/T^0]$. A nonzero (right) $\mathcal{H}_R(G)$-module $\mathcal{V}$ is called

ordinary if the action on $\mathcal{V}$ of any $z \in \mathcal{Z}_T$ corresponding to a non-invertible element of the semigroup T^+/T^0 is invertible;

supersingular if for any $v \in \mathcal{V}$ and any $z \in \mathcal{Z}_T$ corresponding to a non-invertible element of T^+/T^0, there exists $n \in \mathbb{N}$ such that $z^n v = 0$.

Let R be an algebraically closed field of characteristic p.

Classification of simple $\mathcal{H}_R(G)$-modules. The supersingular simple $\mathcal{H}_R(G)$-modules are classified (Vignéras **[200]**). The simple $\mathcal{H}_R(G)$-modules are classified in terms of the simple supersingular $\mathcal{H}_R(M)$-modules for the Levi subgroups M of the parabolic subgroups of G (Abe **[6]**; being algebraically closed is not necessary, see Henniart–Vignéras **[106]**):

For a standard parabolic subgroup $P = MN$ of G and a simple supersingular $\mathcal{H}_R(M)$-module $\mathcal{W}$, there is a notion of extension $e_{P'}(\mathcal{W})$ of $\mathcal{W}$ to $\mathcal{H}_R(M')$ for a parabolic

---

50 Equal to the number of irreducible unitary polynomials of degree n in $k_F[X]$.

51 $F^{\mathrm{sep}} = F^{\mathrm{ac}}$ as the characteristic of F is 0.

subgroup $P' = M'N'$ of G containing P. There is a maximal P' with this property, denoted by $P(\mathcal{W})$. For a parabolic subgroup Q with $P \subset Q \subset P(\mathcal{W})$, there is a generalized Steinberg $\mathcal{H}_R(M(\mathcal{W}))$-module

$$\mathrm{st}_Q^{P(\mathcal{W})}(\mathcal{W}) = \mathrm{ind}_{\mathcal{H}(Q)}^{\mathcal{H}(G)}\big(e_Q(\mathcal{W})\big) / \sum_{Q \subsetneq Q' \subset Q(\mathcal{W})} \mathrm{ind}_{\mathcal{H}(Q')}^{\mathcal{H}(G)}\big(e_{Q'}(\mathcal{W})\big).$$

The triple $(P, \mathcal{W}, Q)$ is called standard. The $\mathcal{H}_R(G)$-module

$$I_{\mathcal{H}(G)}(P, \mathcal{W}, Q) = \mathrm{ind}_{\mathcal{H}(P(\mathcal{W}))}^{\mathcal{H}(G)}\big(\mathrm{st}_Q^{P(\mathcal{W})}(\mathcal{W})\big)$$

is simple, and any simple $\mathcal{H}_R(G)$-module is isomorphic to $I_{\mathcal{H}(G)}(P, \mathcal{W}, Q)$ for some standard triple $(P, \mathcal{W}, Q)$ unique modulo G-conjugation. It is ordinary if and only if $P = B$.

Extensions. The extensions between simple $\mathcal{H}_R(G)$-modules

$$\mathrm{Ext}^i_{\mathcal{H}(G)}\big(I_{\mathcal{H}_R(G)}(P_1, \mathcal{W}_1, Q), I_{\mathcal{H}(G)}(P_2, \mathcal{W}_2, Q_2)\big), \quad i \geq 0,$$

are either 0, or extensions between supersingular simple modules of a specialization of a generic pro-p Iwahori Hecke algebra which is not of a pro-p Iwahori Hecke R-algebra (Abe **[2]**). In more details, considering the central characters, the extensions are 0 if $P_1 \neq P_2$. When $P = P_1 = P_2$, following the construction of the simple modules, we have

$$\mathrm{Ext}^i_{\mathcal{H}_R(G)}\big(I_{\mathcal{H}(G)}(P, \mathcal{W}_1, Q), I_{\mathcal{H}(G)}(P, \mathcal{W}_2, Q_2)\big) \simeq \mathrm{Ext}^i_{\mathcal{H}_R(M')}\big(\mathrm{st}_{Q_1'}^{P'}(\mathcal{W}_1), \mathrm{st}_{Q_2'}^{P'}(\mathcal{W}_2)\big)$$

for some P', Q_1', Q_2',

$$\mathrm{Ext}^i_{\mathcal{H}_R(G)}\big(\mathrm{st}_{Q_1}^G(\mathcal{W}_1), \mathrm{st}_{Q_2}^G(\mathcal{W}_2)\big) \simeq \mathrm{Ext}^{i-r}_{\mathcal{H}_R(G)}\big(e_G(\mathcal{W}_1), e_G(\mathcal{W}_2)\big)$$

for some explicit $r \in \mathbb{N}_{\geq 0}$, and using results of Ollivier–Schneider **[150]**,

$$\mathrm{Ext}^i_{\mathcal{H}_R(G)}\big(e_G(\mathcal{W}_1), e_G(\mathcal{W}_2)\big) \simeq \mathrm{Ext}^i_{\mathcal{H}_R(M)/I}(\mathcal{W}_1, \mathcal{W}_2)$$

for some ideal I of $\mathcal{H}_R(M)$ acting on $\mathcal{W}_1$, $\mathcal{W}_2$ by 0. Abe computed explicitly Ext^1 for two supersingular simple $\mathcal{H}_R(M)/I$-modules.

- When $G = \mathrm{GL}(2, F)$, Cédric Pépin and Tobias Schmidt proved:

  (i) The 2-dimensional supersingular simple $\mathcal{H}_{\mathbb{F}_p^{\mathrm{ac}}}(G)$-modules can be realized through the equivariant cohomology of the flag variety of the dual group $\hat{G}$ over $\mathbb{F}_p^{\mathrm{ac}}$ **[160]**.

  (ii) There is a version in families of the Breuil's semisimple mod p Langlands correspondence for $\mathrm{GL}_2(\mathbb{Q}_p)$ **[161]**.

  (iii) There is a Kazhdan–Lusztig theory for the generic pro-p Iwahori Hecke $\mathbb{Z}[\mathbf{q}]$-algebra of G, where the role of $\hat{G}$ is taken by the Vinberg monoid $V_{\hat{G}}$ and its flag variety; the monoid comes with a fibration $V_{\hat{G}} \to \mathbb{A}^1$ and the dual parametrization of $H_{\mathbb{F}_p^{\mathrm{ac}}}(G)$-modules is achieved by working over the 0-fiber. They introduce a generic pro-p antispherical module and a generic pro-p Satake homomorphism for a generic spherical Hecke $\mathbb{Z}[\mathbf{q}]$-algebra **[162]**.

14. REPRESENTATIONS OVER A FIELD OF CHARACTERISTIC p

In this section, R is a field of characteristic p. The admissible irreducible R-representations of G are classified in terms of the supersingular admissible irreducible R-representations of the Levi subgroups of G (Abe–Henniart–Herzig–Vignéras [8] for R algebraically closed, Henniart–Vignéras [106] for R not algebraically closed).

Classification. The representation $\operatorname{ind}_P^G W$ parabolically induced from an irreducible admissible supersingular R-representation W of a Levi subgroup M of a parabolic subgroup P of G, has multiplicity 1 and irreducible subquotients

$$I_G(P, W, Q) = \operatorname{ind}_{P(W)}^G\big(e(W) \otimes \operatorname{St}_Q^{P(W)}\big)$$

for the parabolic subgroups Q of G containing P and contained in the maximal parabolic subgroup $P(W)$ where the inflation of W to P extends to a representation $e(W)$, and

$$\operatorname{St}_Q^{P(W)} = (\operatorname{ind}_Q^{P(W)} 1_Q) / \sum_{Q \subsetneq Q' \subset P(W)} \operatorname{ind}_{Q'}^{P(W)} 1_{Q'}.$$

Any irreducible admissible R-representation V of G is isomorphic to $I_G(P, W, Q)$ for a unique triple (P, W, Q) modulo G-conjugation.

A similar classification holds true for the irreducible admissible genuine mod p representations of the metaplectic double cover of $\operatorname{Sp}_{2n}(F)$ (Koziol–Peskin [124]).

There is a complete description of $\operatorname{ind}_P^G W$ for any irreducible admissible R-representation W of M [106]. As a corollary, one obtains generic irreducibility and for any admissible irreducible R-representation V of G,

$$V \text{ supersingular} \Leftrightarrow V \text{ cuspidal} \Leftrightarrow V \text{ supercuspidal}.$$

When F has characteristic 0, the higher duals $(S^i(I_G(P, W, Q)))_{i \geq 0}$ are computed in terms of $(S^i(W))_{i \geq 0}$ in a few cases (Kohlhaase [117]).

The extensions between R-representations $\operatorname{ind}_P^G W$ of G, parabolically induced from supersingular absolutely irreducible R-representations W of Levi subgroups, are computed in many cases when G is split and R finite (Hauseux [86, 87, 89, 90]).

When $P = B$, the irreducible subquotients of $\operatorname{ind}_B^G W$ are called *ordinary*. An admissible R-representation of G with ordinary irreducible subquotients is called ordinary.

The $\tilde{J}$-invariant functor induces an equivalence between the category of finite length ordinary R-representations of G generated by their $\tilde{J}$-invariant vectors and the category of the finite length ordinary $\mathcal{H}_R(G)$-modules, assuming R algebraically closed (Abe [5]).

The pro-p Iwahori invariant $I_G(P, W, Q)^{\tilde{J}}$ is computed and depends only on the pro-p Iwahori invariant $W^{\tilde{J}_M}$ (Abe–Henniart–Vignéras [9, 10]).

The supersingular admissible irreducible R-representations V of G are not understood, this remains an open crucial question for two decades and a stumbling block for the search of a mod p or p-adic local Langlands correspondence if $G \neq \mathrm{GL}(2, \mathbb{Q}_p)$. The supersingularity can be seen on the pro-p Iwahori invariants (Ollivier–Vignéras [154] for R algebraically closed, but being algebraically closed is not necessary Henniart–Vignéras [106]):

V is supersingular $\Leftrightarrow V^{\tilde{J}}$ is supersingular $\Leftrightarrow$ some nonzero subquotient of $V^{\tilde{J}}$ is supersingular.

The classification of simple supersingular $\mathcal{H}_R(G)$-modules does not help because we do not have enough information on the pro-p Iwahori invariant functor.

When $G = \mathrm{GL}(2, \mathbb{Q}_p)$, Breuil **[16]** relying on the work of Barthel–Livne classified the supersingular admissible irreducible mod p representations. This was the starting point of the mod p local Langlands correspondence for $\mathrm{GL}_2(\mathbb{Q}_p)$. There are two main novel features. The mod p local Langlands correspondence involves reducible representations and extends to an exact functor from finite length representations of $\mathrm{GL}_2(\mathbb{Q}_p)$ to finite length representations of $\mathrm{Gal}(\mathbb{Q}_p^{\mathrm{ac}}/\mathbb{Q}_p)$.

When $G \neq \mathrm{GL}(2, \mathbb{Q}_p)$, supersingular admissible irreducible mod p representations are classified only for some groups close to $\mathrm{GL}(2, \mathbb{Q}_p)$: for $\mathrm{SL}(2, \mathbb{Q}_p)$ (Abdellatif **[1]**, Cheng **[27]**), and for the unramified unitary group $U(1,1)(\mathbb{Q}_p^2/\mathbb{Q}_p)$ in two variables (Koziol **[118]**). When $F \neq \mathbb{Q}_p$, there can be many more supersingular admissible irreducible mod p representations of $\mathrm{GL}(2, F)$ than 2-dimensional irreducible representations of $\mathrm{Gal}(F^{\mathrm{sep}}/F)$ (Breuil–Paskunas **[19]**); they cannot be described as quotients of a compact induction by a finite number of equations (Hu **[115]** if $F \supset \mathbb{F}_p((T))$, Schraen **[176]** if $F/\mathbb{Q}_p$ is quadratic, Wu **[204]** in general if $F \supsetneq \mathbb{Q}_p$).

When R is a field of characteristic p and $F \supset \mathbb{Q}_p$, Herzig–Koziol–Vignéras **[110]** proved that any G admits a supersingular admissible irreducible R-representation, using a local method of Paskunas **[155]** if the semisimple rank r_G of G is 1, and a global method if $r_G > 1$. The existence is not known if $F \supset \mathbb{F}_p((T))$.

There have been recent advances which strongly suggest that the study of mod p representations of G is best done on the derived level. When R is a field of characteristic p, Schneider **[172]** introduced the unbounded derived category $D_R(G)$ of R-representations of G. When $\tilde{J}$ is torsion free (this forces F to be of characteristic 0), $D_R(G)$ is equivalent to the derived category of differential graded modules over a differential graded version $\mathcal{H}_R(G)^\bullet$ of the pro-p Iwahori Hecke R-algebra of G, by the derived $\tilde{J}$-invariant functor.

The parabolic induction $\mathrm{ind}_P^G : \mathrm{Mod}_R(M) \to \mathrm{Mod}_R(G)$ being exact extends to an exact derived parabolic induction $R\,\mathrm{ind}_P^G : D_R(M) \to D_R(G)$ between the unbounded derived categories. The total derived functor of R_P^G is right adjoint to $R\,\mathrm{ind}_P^G$. The category $D_R(G)$ has arbitrary small direct products and $R\,\mathrm{ind}_P^G$ commutes with arbitrary small direct products (Heyer **[112]**), hence $R\,\mathrm{ind}_P^G$ has a left adjoint.[52] When $\tilde{J}$ is torsion free, the functor $R\,\mathrm{ind}_P^G$ corresponds to the derived parabolic induction functor on the dg Hecke algebra side, via the derived $\tilde{J}$-invariant functor (Sarah Scherotzke and Schneider **[169]**).

The Kohlhaase duality functors are related to the derived duality functor $\mathrm{R}\underline{\mathrm{Hom}}(-, R)$ (Schneider–Sorensen **[173]**).

The cohomology algebra $\mathrm{Ext}^\bullet_{\mathrm{Mod}_R(G)}(R[\tilde{J}\backslash G], R[\tilde{J}\backslash G])$ is simpler than of $\mathcal{H}_R(G)^\bullet$; when $G = \mathrm{SL}(2, \mathbb{Q}_p)$, $p \geq 5$, its structure is explicited by Ollivier and Schneider **[152, 153]**.

52 By Brown representability; Heyer **[112]** gave another proof.

15. LOCAL LANGLANDS CORRESPONDENCES FOR GL(n, F)

The complex local Langlands correspondence for $\mathrm{GL}(n, F)$ is a bijection between the isomorphism classes of irreducible complex representations of $\mathrm{GL}(n, F)$ and the isomorphism classes of n-dimensional Weil–Deligne complex representations, given by local class field theory when $n = 1$, and characterized by the requirement that the L and ε factors[53] attached to corresponding pairs of complex representations coincide (Henniart **[100]**). An n-dimensional Weil–Deligne complex representation is a pair (σ, N) where σ is an n-dimensional Frobenius semisimple complex representation of the Weil group W_F and $N \in \mathrm{End}_{\mathbb{C}}\,\sigma$ is a nilpotent endomorphism satisfying $\sigma(w)N\sigma(w)^{-1} = q^{|w|}N$ for all $w \in W_F$.[54] The supercuspidal irreducible $\mathbb{C}$-representations of $\mathrm{GL}(n, F)$ correspond to the n-dimensional irreducible $\mathbb{C}$-representations of W_F.[55]

A twist of the correspondence by an unramified complex character of $\mathrm{GL}(n, F)$ is compatible with the automorphisms of $\mathbb{C}$. For a prime r, an isomorphism $\mathbb{C} \simeq \mathbb{Q}_r^{\mathrm{ac}}$ transfers the twisted complex local Langlands correspondence to a local Langlands correspondence for $\mathbb{Q}_r^{\mathrm{ac}}$-representations of $\mathrm{GL}(n, F)$. For $r = \ell \neq p$, the nilpotent part is related to the action of the tame inertia group on an ℓ-adic representation of W_F. By reduction modulo ℓ of the ℓ-adic local Langlands correspondence composed with the Zelevinski involution on ℓ-adic representations of $\mathrm{GL}(n, F)$, one obtains a *ℓ-modular local Zelevinski correspondence*. The ℓ-modular local Zelevinski correspondence is a parametrization for ℓ-modular irreducible representations of $\mathrm{GL}(n, F)$ by n-dimensional Weil–Deligne ℓ-modular representations, defined as above with $\mathbb{F}_\ell^{\mathrm{ac}}$ instead of $\mathbb{C}$. The supercuspidal irreducible ℓ-modular representations of $\mathrm{GL}(n, F)$ correspond to the n-dimensional irreducible ℓ-modular representations of W_F. But the nilpotent part N of the Weil–Deligne ℓ-modular representation has no obvious Galois interpretation.

Dat **[43–45]** obtained a geometric realization of the ℓ-adic local Zelevinski correspondence and of the ℓ-modular local Zelevinski correspondence on the unipotent[56] irreducible $\mathbb{F}_\ell^{\mathrm{ac}}$-representations of $\mathrm{GL}(n, F)$ when the order of q in $\mathbb{F}_\ell^*$ is at least n **[42]**,[57] and when $q \equiv 1 \bmod \ell$ and $\ell > n$ **[46]**.[58]

Kurinczuk and Matringe **[127–130]**, extended to ℓ-modular representations the Rankin–Selberg local constants of Jacquet, Piatetski-Shapiro, and Shalika of pairs of complex generic representations of linear groups, and the Artin–Deligne local constants of pairs of complex Weil–Deligne representations. These local constants are preserved by the complex local Langlands correspondence, but not by the ℓ-modular local Zelevinski correspondence. Enlarging the space of ℓ-modular Weil–Deligne representations to representations with not necessarily nilpotent operators, they suggested a *ℓ-modular local Langlands cor-*

---

53 For a fixed nontrivial $\mathbb{C}$-character of F.

54 $|w|$ is the power of q to which w raises the elements of the residue field k_F.

55 $N = 0$.

56 = in the principal block = subquotients of some $\mathrm{Ind}_B^G(\chi)$ for χ an unramified character of a Borel subgroup B, this is not the definition of Lusztig.

57 The regular case.

58 The limit case.

respondence compatible with the formation of local constants and characterized by a list of natural properties. When R is a noetherian $W(\overline{F}_\ell^{ac})$-algebra, using the Rankin–Selberg functional equations, Matringe and Moss [138] proved that an R-representation of $GL(n, F)$ of Whittaker type admits a Kirillov model.

When the characteristic of F is 0, Breuil and Schneider [20] motivated by an hypothetical p-adic extension of the local Langlands correspondence, suggested a *modified local Langlands correspondence* where the complex representations of $GL(n, F)$ are no more irreducible. The Langlands quotient theorem realizes an irreducible $\mathbb{C}$-representation V of $GL(n, F)$ as a quotient of a certain parabolically induced representation $\operatorname{ind}_P^G W$. In the modified version, V is replaced by a twist of $\operatorname{ind}_P^G W$ by an unramified character of $GL(n, F)$.

When the characteristic of F is 0, Emerton and Helm [62] motivated by a local Langlands correspondence in families and by global contexts, introduced the *generic ℓ-adic local Langlands correspondence* which has useful applications to the cohomology of Shimura varieties. For any finite extension $E/\mathbb{Q}_\ell$, it is a map $\rho \mapsto \pi(\rho)$ from n-dimensional continuous E-representations of the Galois group $\mathrm{Gal}(F^{ac}/F)$ to finite length E-representations of $GL(n, F)$ with an absolutely irreducible generic socle and no other generic irreducible subquotients.[59] Each $\pi(\rho)$ contains a $GL(n, F)$-stable O_E-lattice $\pi(\rho)^o$ of reduction having an absolutely irreducible socle and no other generic subquotients, unique modulo homotethy.

The *generic mod ℓ local Langlands correspondence* (Emerton–Helm [62]) is compatible with the generic ℓ-adic local Langlands correspondence by reduction modulo ℓ. Irreducible representations of $GL(n, F)$ are no longer irrreducible, Weil–Deligne representations are now Galois representations, and the Zelevinski involution does not intervene. For a finite extension $R/\mathbb{F}_\ell$, it is the unique map $\overline{\rho} \mapsto \overline{\pi}(\overline{\rho})$ from n-dimensional R-representations of $\mathrm{Gal}(F^{ac}/F)$ to finite length R-representations of $GL(n, F)$ such that

(1) $\overline{\pi}(\overline{\rho})$ has an absolutely irreducible generic socle and no other generic irreducible subquotients,

(2) For all finite extensions $E/\mathbb{Q}_\ell$ of ring of integers O_E and residue field k_E containing R, and continuous representation $\rho : \mathrm{Gal}(F^{ac}/F) \to GL(n, O_E)$ lifting $\overline{\rho} \otimes_R k_E$, the reduction of $\pi(\rho)^o$[60] embeds in $\overline{\pi}(\overline{\rho}) \otimes_R k_E$.

(3) $\overline{\pi}(\overline{\rho})$ is minimal with respect to the above two conditions.

For $GL(2, F)$, the correspondence is fairly concrete and explicit when $\ell \neq 2$ (Helm [95]).

Emerton and Helm [99] introduced also a notion of a *local Langlands correspondence in families*.[61] For any suitable complete local noetherean algebra R with finite residue field k, it is the unique map $\rho \mapsto \pi(\rho)$ from the continuous representations $\rho : \mathrm{Gal}(F^{ac}/F) \to GL(n, R)$ to the admissible R-representations of $GL(n, F)$ that interpolates the generic local

59 It is a slight modification of the Breuil and Schneider correspondence transferred to ℓ-adic representations; the socle of V is the maximal semisimple subrepresentation of V.

60 ρ identifies with a representation $\mathrm{Gal}(F^{ac}/F) \to GL(n, E)$.

61 For an example of a local p-adic Langlands correspondence in families for $GL(2, \mathbb{Q}_p)$, see Ildar Gaisin and Joaquin Rodrigues Jacinto [70].

Langlands correspondences over the points of Spec R and satisfies certain technical hypotheses.

The existence of the map amounts to showing that whenever there is a congruence between two ℓ-adic representations of $\mathrm{Gal}(F^{\mathrm{ac}}/F)$, there is a corresponding congruence on the other side of the ℓ-adic generic local Langlands correspondence. The key idea of the proof is the introduction of the Bernstein center $\mathcal{Z}$ of $\mathrm{Mod}_{\mathbb{Z}_\ell^{\mathrm{ac}}}(\mathrm{GL}(n, F))$ (Helm **[96–98]**), which encodes deep information about congruences between integral $\mathbb{Q}_\ell^{\mathrm{ac}}$-representations of $\mathrm{GL}(n, F)$. For instance, if two integral irreducible $\mathbb{Q}_\ell^{\mathrm{ac}}$-representations of $\mathrm{GL}(n, F)$ become isomorphic modulo ℓ, then $\mathcal{Z}$ acts on these representations by scalars congruent modulo ℓ.

When G is quasisplit, motivated by a local Langlands correspondence in families, Dat, Helm, Kurinczuk, and Moss **[51]** studied the scheme of Langlands parameters of G with coefficient the smallest possible ring $R = \mathbb{Z}[1/p]$. In particular, this allows studying a chain of congruences of Langlands parameters modulo several different primes. In a work in progress, they extend the Emerton–Helm–Moss local Langlands correspondence in families to a conjecture which asserts the existence of isomorphisms between

(a) the center of $\mathrm{Mod}_{\mathbb{Z}[q^{-1/2}]}(G)$,

(b) the ring of functions on the moduli stack of Langlands parameters[62] for G over $\mathbb{Z}[q^{-1/2}]$,

(c) the descent to $\mathbb{Z}[q^{-1/2}]$ of the endomorphisms of a Gelfand–Graev representation of G.

They prove the conjecture when G is any classical p-adic group after inverting an integer. The conjecture should follow from a Fargues–Scholze conjecture **[65, I.10.2]**.[63]

The blocks in the category of $\mathbb{Z}^{\mathrm{ac}}[1/p]$-representations of G of depth 0 are in natural bijection with the connected components of the space of tamely ramified Langlands parameters for G over $\mathbb{Z}^{\mathrm{ac}}[1/p]$; there is only one block (the category is indecomposable) if G is semisimple and simply connected, or unramified (Dat–Lanard **[53]**).

When the characteristic of F is 0, the p-adic local Langlands correspondence for $\mathrm{GL}(n, F)$ is a hypothetical correspondence between continuous unitary E-Banach space representations of $\mathrm{GL}(n, F)$ and n-dimensional continuous E-representations of $\mathrm{Gal}(F^{\mathrm{ac}}/F)$, for any finite extension $E/\mathbb{Q}_p$, given by local class field theory when $n = 1$. Using global methods, Ana Caraiani, Matthew Emerton, Toby Gee, David Geraghty, Vytautas Paskunas, and Sug Woo Shin **[26]** constructed a candidate when p does not divide $2n$. For $F = \mathbb{Q}_p$ and $n = 2$, it coincides with the p-adic local correspondence envisioned by Breuil 20 years ago, constructed by Pierre Colmez **[33]**, and analyzed by Paskunas **[157]**, Colmez, Dospinescu, Paskunas **[35]**.

62 Constructions of moduli spaces of Langlands parameters have been also proposed by Fargues and Scholze (**[65]** over $\mathbb{Z}_\ell$, $\ell \neq p$ using the condensed mathematics of Clausen–Scholze) and by Xinwen Zhu **[208]**. The local Langlands correspondence is now conjectured to exist at a categorical level (Denis Gaitsgory **[71]**).

63 Private communication of Dat.

When $n \geq 2$ and D_n is the central division algebra over F of invariant $1/n$, Scholze **[175]** constructed a candidate for a p-adic and mod p Jacquet–Langlands correspondence from $\mathrm{GL}(n, F)$ to D_n^* in a purely geometric way, using the cohomology of the infinite-level Lubin–Tate space. The mod p Jacquet–Langlands correspondence is a canonical map from the admissible mod p representations of $\mathrm{GL}(n, F)$ to the admissible mod p representations of D_n^* having a continuous action of $\mathrm{Gal}(F^{\mathrm{ac}}/F)$. For $F = \mathbb{Q}_p$ and $n = 2$, it is studied by Dospinescu–Paskunas–Schraen **[54]**.

16. GELFAND–KIRILLOV DIMENSION

Let R be a field and V an irreducible admissible R-representation of G. For any decreasing sequence $(K_i)_{i\geq 1}$ of open compact subgroups of G with limit the trivial group, the dimensions $\dim_R V^{K_i}$ for $i \geq 1$ are finite. If V is finite dimensional, $\dim_R V^{K_i} = \dim_R V$ when i is large enough. Generally, the dimension of V is infinite and the increasing sequence $(\dim_R V^{K_i})_{i\geq 1}$ tends to infinity, but how?

When F has characteristic 0, one can choose an O_F-lattice $\mathfrak{L}$ of the Lie algebra $\mathcal{G}$ of G on which the exponential map exp is defined and such that $K = \exp(\mathfrak{L})$ is a group, and consider the decreasing sequence $(K_i = \exp(p_F^{2i}\mathfrak{L}))_{i\geq 1}$. When $R = \mathbb{C}$, the Harish-Chandra local character expansion of V implies that $\dim_R V^{K_i}$ eventually becomes polynomial[64]

$$\dim_R V^{K_i} = P_{\mathfrak{L},V}(q^i), \quad P_{\mathfrak{L},V}(X) \in \mathbb{Q}[X] \text{ for } i \text{ large enough.}$$

The degree d_V of the polynomial $P_{\mathfrak{L},V}[X]$ does not depend on the choice of $\mathfrak{L}$. It is half the dimension of a unipotent conjugacy class in G,

$$0 \leq d_V \leq \dim_F U,$$

and is 0 if and only if V is finite dimensional. The integer q^{d_V} measures the growth of $(\dim_R V^{K_i})_{i\geq 1}$ for any choice of $\mathfrak{L}$.

For F of either characteristic 0 or p, when $G = \mathrm{GL}(n, F)$, $K_i = 1 + p_F^{i+1} M(n, O_F)$ for $i \geq 1$, if $R = \mathbb{C}$, the Roger Howe local character expansion implies that the dimensions

$$\dim_R V^{K_i} = P_V(q^i), \quad P_V(X) \in \mathbb{Z}[X]$$

are polynomial when i is large, for a polynomial $P_V(X)$ with integral coefficients and degree $d_V \leq n(n-1)/2$. When V is cuspidal (or more generally, generic), $d_V = n(n-1)/2$.

Example. For $\mathrm{GL}(2, F)$, V is infinite dimensional if and only if $d_V = 1$.

Any cuspidal irreducible ℓ-modular representation V of $\mathrm{GL}(n, F)$ lifts to an irreducible cuspidal ℓ-adic representation, implying that the dimensions $\dim_R V^{K_i}$ satisfy the same properties. This is probably true for any irreducible representation of $\mathrm{GL}(n, F)$ over any field R of characteristic ℓ.[65]

---

64 Harish-Chandra, Notes by Stephen DeBacker and Paul J. Sally, Admissible invariant distributions on reductive p-adic group, University Lecture Series Vol. 16, 1999, 97 pp.

65 Article in preparation.

For a finite field R of characteristic p, $G = \mathrm{GL}(2, \mathbb{Q}_p)$ and V admissible absolutely irreducible, Stefano Morra [144] computed $\dim_R V^{K_i}$ for $i \geq 1$. The dimensions satisfy the above properties.

For F of characteristic 0, R a finite field of characteristic p, K a uniformly powerful open pro-p subgroup of G, K_i the closed subgroup of K generated by $\{k^{p^i}, k \in K\}$ for $i \geq 1$, and V an admissible R-representation of G, there is a positive integer δ_V not depending on the choice of K and positive real numbers $a \leq b$ such that (Calegari–Emerton [24], Emerton–Paskunas [64], Dospinescu–Paskunas–Schraen [54]):

$$ap^{i\delta_V} \leq \dim_R V^{K_i} \leq bp^{i\delta_V}.$$

The integer δ_V which is a sort of Iwasawa dimension of the dual of V, is called the *Gelfand–Kirillov dimension* of V. When $F/\mathbb{Q}_p$ is unramified, the admissible R-representations V of $\mathrm{GL}_2(F)$ studied by Breuil–Herzig–Hu–Morra–Schraen [18] in mod p cohomology satisfy $\delta_V = [F : \mathbb{Q}_p]$. If V is isomorphic to $I_G(P, W, Q)$, we have[66]

$$\delta_{I_G(P,W,Q)} = \delta_W + \dim_{\mathbb{Q}_p} N_Q,$$

where N_Q is the unipotent radical of the parabolic subgroup Q of G.

REFERENCES

[1] R. Abdellatif, Classification des représentations modulo p de $\mathrm{SL}(2, F)$. *Bull. Soc. Math. France* **142** (2014), no. 3, 537–589.

[2] N. Abe, Extension between simple modules of pro-p Iwahori Hecke algebras. *J. Inst. Math. Jussieu* (2022, to appear).

[3] N. Abe, On a classification of irreducible admissible modulo p representations of a p-adic split reductive group. *Compos. Math.* **149** (2013), no. 12, 2139–2168.

[4] N. Abe, In *Change of weight theorem for pro-p-Iwahori Hecke algebras, Around Langlands correspondences*, Contemp. Math. 691, pp. 1–13, Amer. Math. Soc., Providence, RI, 2017.

[5] N. Abe, A comparison between pro-p-Iwahori Hecke modules and mod p representations. *Algebra Number Theory* **13** (2019), no. 8, 1959–1981.

[6] N. Abe, Modulo p parabolic induction of pro-p-Iwahori Hecke algebra. *J. Reine Angew. Math.* **749** (2019), 1–64.

[7] N. Abe, Parabolic inductions for pro-p-Iwahori Hecke algebras. *Adv. Math.* **355** (2019).

[8] N. Abe, G. Henniart, F. Herzig, and M.-F. Vignéras, A classification of admissible irreducible modulo p representations of reductive p-adic groups. *J. Amer. Math. Soc.* **30** (2017), 495–559.

[9] N. Abe, G. Henniart, and M.-F. Vignéras, On pro-p-Iwahori invariants of R-representations of p-adic groups. *Represent. Theory* **22** (2018), 119–159.

66 Article in preparation.

[10] N. Abe, G. Henniart, and M.-F. Vignéras, Mod p representations of reductive p-adic groups: Functorial properties. *Trans. Amer. Math. Soc.* **371** (2019), 8297–8337.

[11] N. Abe, F. Herzig, and M.-F. Vignéras, Inverse Satake isomorphism and change of weight. *Represent. Theory* **26** (2022), 264–324.

[12] K. Ardakov and P. Schneider, The Bernstein center in natural characteristic. 2021, arXiv:2105.06128.

[13] A. I. Badulescu, Jacquet–Langlands et unitarisabilité. *J. Inst. Math. Jussieu* **6** (2007), no. 3, 349–379.

[14] L. Berger, Central characters for smooth irreducible modular representations of $\mathrm{GL}_2(Q_p)$. *Rend. Semin. Mat. Univ. Padova* **128** (2012), 1–6.

[15] N. Bourbaki, *Éléments de mathématiques. Algèbre, Chap. 8. Modules et anneaux semi-simples*. Springer, Berlin–Heidelberg, 2012.

[16] C. Breuil, Sur quelques représentations modulaires et p-adiques de $\mathrm{GL}_2(\mathbb{Q}_p)$ I. *Compos. Math.* **138** (2003), 165–188.

[17] C. Breuil, F. Herzig, Y. Hu, S. Morra, and B. Schraen, Conjectures and results on modular representations of $\mathrm{GL}_n(K)$ for a p-adic field K. 2021, arXiv:2102.06188v3.

[18] C. Breuil, F. Herzig, Y. Hu, S. Morra, and B. Schraen, Gelfand–Kirillov dimension and mod p cohomology for GL_2. 2021, arXiv:2009.03127v4.

[19] C. Breuil and V. Paskunas, *Towards a modulo p Langlands correspondence for* GL(2). Mem. Amer. Math. Soc. 216, 2012.

[20] C. Breuil and P. Schneider, First steps towards p-adic Langlands functoriality. *J. Reine Angew. Math.* **610** (2007), 149–180.

[21] P. Broussous, V. Sécherre, and S. Stevens, *The local Langlands conjecture for* GL(2). Grundlehren Math. Wiss. 335, Springer, 2006.

[22] P. Broussous, V. Sécherre, and S. Stevens, Smooth representations of $\mathrm{GL}_m(D)$ V: Endo-classes. *Doc. Math.* **17** (2012), 23–77.

[23] P. Broussous, V. Sécherre, and S. Stevens, Modular local Langlands correspondence for GL_n. *Int. Math. Res. Not. IMRN* **15** (2014), 4124–4145.

[24] F. Calegari and M. Emerton, In *Completed cohomology. A survey, Non-abelian Fundamental Groups and Iwasawa Theory*, London Math. Soc. Lecture Note Ser. 393, pp. 239–257, Cambridge University Press, Cambridge, 2011.

[25] A. Caraiani, M. Emerton, T. Gee, D. Geraghty, V. Paskunas, and S. W. Shin, Patching and the p-adic Langlands program for $\mathrm{GL}_2(\mathbb{Q}_p)$. *Compos. Math.* **154** (2018), no. 3, 503–548.

[26] A. Caraiani, M. Emerton, T. Gee, D. Geraghty, V. Paskunas, and S. Woo Shin, Patching and the p-adic local Langlands correspondence. *Cambridge J. Math.* **4** (2016), no. 2, 197–287.

[27] C. Cheng, Mod p representations of $\mathrm{SL}_2(Q_p)$. *J. Number Theory* **133** (2013), no. 4, 1312–1330.

[28] G. Chinello, Hecke algebra with respect to the pro-p-radical of a maximal compact open subgroup for $\mathrm{GL}(n, F)$ and its inner forms. *J. Algebra* **478** (2017), 296–317.

[29] G. Chinello, Blocks of the category of smooth ℓ-modular representations of $\mathrm{GL}_n(F)$ and its inner forms: reduction to level 0. *Algebra Number Theory* **12** (2018), no. 7, 1675–1713.

[30] M. Chlouveraki and V. Sécherre, The affine Yokonuma–Hecke algebra and the pro-p Iwahori–Hecke algebra. *Math. Res. Lett.* **23** (2016), no. 3, 707–718.

[31] R. Cluckers, J. Gordon, and I. Halupczok, Integrability of oscillatory functions on local fields: Transfer principles. *Duke Math. J.* **163** (2014), no. 8, 1549–1600.

[32] R. Cluckers, T. Hales, and F. Loeser, Transfer Principle for the Fundamental Lemma. In *Stabilisation de la formule des traces, variétés de Shimura, et applications arithmétiques*, edited by L. Clozel, M. Harris, J.-P. Labesse, and B.-C. Ngo, International Press of Boston, 2011.

[33] P. Colmez, Representations de $\mathrm{GL}_2(\mathbb{Q}_p)$ et (φ, Γ)-modules. *Astérisque* **330** (2010), 281–509.

[34] P. Colmez, G. Dospinescu, J. Hauseux, and W. Niziol, p-adic étale cohomology of period domains. *Math. Ann.* **381** (2021), no. 1–2, 105–180.

[35] P. Colmez, G. Dospinescu, and V. Paskunas, The p-adic local Langlands correspondence for $\mathrm{GL}_2(\mathbb{Q}_p)$. *Cambridge J. Math.* **2** (2014), 1–47.

[36] P. Cui, Modulo ℓ-representations of p-adic groups $\mathrm{SL}(N, F)$. 2019, arXiv:1912.13473.

[37] P. Cui, Modulo ℓ-representations of p-adic groups $\mathrm{SL}_n(F)$: Maximal simple k-types. 2020, arXiv:2012.07492.

[38] J.-F. Dat, ν-tempered representations of p-adic groups I: l-adic case. *Duke Math. J.* **126** (2005), no. 3, 397–469.

[39] J.-F. Dat, Espaces symétriques de Drinfeld et correspondance de Langlands locale. *Ann. Sci. Éc. Norm. Supér.* **39** (2006), no. 1, 1–74.

[40] J.-F. Dat, Finitude pour les représentations lisses des groupes p-adiques. *J. Inst. Math. Jussieu* **8** (2009), no. 1, 261–333.

[41] J.-F. Dat, Représentations lisses p-tempérées des groupes p-adiques. *Amer. J. Math.* **131** (2009), no. 1, 227–255.

[42] J.-F. Dat, Lefschetz operator and local Langlands mod ℓ: The regular case. *Nagoya Math. J.* **208** (2012), 1–38 (Hiroshi Saito memorial volume).

[43] J.-F. Dat, Opérateur de Lefschetz sur les tours de Drinfeld et de Lubin–Tate. *Compos. Math.* **148** (2012), no. 2, 507–530.

[44] J.-F. Dat, Théorie de Lubin–Tate non-abélienne ℓ-entière. *Duke Math. J.* **161** (2012), no. 6, 951–1010.

[45] J.-F. Dat, Un cas simple de correspondance de Jacquet–Langlands modulo ℓ. *Proc. Lond. Math. Soc.* **104** (2012), 690–727.

[46] J.-F. Dat, Lefschetz operator and local Langlands mod ℓ: The limit case. *Algebra Number Theory* **8** (2014), no. 3, 729–766.

[47] J.-F. Dat, A functoriality principle for blocks of p-adic linear groups. In *Around Langlands Correspondences*, pp. 103–131, Contemp. Math. 691, Amer. Math. Soc., Providence, RI, 2017.
[48] J.-F. Dat, Equivalence of tame blocks for p-adic general linear groups. *Math. Ann.* **371** (2018), 565–613.
[49] J.-F. Dat, Simple subquotients of big parabolically induced representations of p-adic groups. *J. Algebra* **510** (2018), 499–507.
[50] J.-F. Dat, Depth 0 representations over $Z[1/p]$. 2022, arXiv:2202.03982v1.
[51] J.-F. Dat, D. Helm, R. Kurinczuk, and G. Moss, *Moduli of Langlands parameters*. 2020.
[52] J.-F. Dat, D. Helm, R. Kurinczuk, and G. Moss, Finiteness for Hecke algebras of p-adic groups. 2022, arXiv:2203.04929v1.
[53] J.-F. Dat and T. Lanard, Depth zero representations over $\overline{Z}[1/p]$. 2022, arXiv:2202.03982v1.
[54] G. Dospinescu, V. Paskunas, and B. Schraen, Gelfand–Kirillov dimension and the p-adic correspondence. 2022, arXiv:2201.12922v1.
[55] A. Dotto, The inertial Jacquet–Langlands correspondence. 2021, arXiv:1707.00635v3.
[56] A. Dotto, Mod p Bernstein centres of p-adic groups. *Math. Res. Lett.* (2021, accepted).
[57] B. Drevon and V. Sécherre, Décomposition en blocs de la catégorie des représentations ℓ-modulaires lisses de longueur finie de $\mathrm{GL}_m(D)$. 2021, arXiv:2101.05898v1.
[58] O. Dudas, Non-uniqueness of supercuspidal support for finite reductive groups. *J. Algebra* **510** (2018), 508–512.
[59] N. Dupré and J. Kohlhaase, Model categories and pro-p Iwahori–Hecke modules. 2021, arXiv:2112.03150.
[60] M. Emerton, Ordinary parts of admissible representations of p-adic reductive groups I. Definition and first properties. *Astérisque* **331** (2010), 355–402.
[61] M. Emerton, Ordinary parts of admissible representations of reductive p-adic groups II. *Astérisque* **331** (2010), 383–438.
[62] M. Emerton and D. Helm, The local Langlands correspondence for GL_n in families. *Ann. Sci. Éc. Norm. Supér.* **47** (2014), no. 4, 655–722.
[63] M. Emerton and V. Paskunas, On the effeacibility of certain δ-functors. *Astérisque* **331** (2010), 461–469.
[64] M. Emerton and V. Paskunas, On the density of supercuspidal points of fixed weight in local deformation rings and global Hecke algebras. *J. Éc. Poly. Math.* **7** (2020), 337–371.
[65] L. Fargues and P. Scholze, Geometrization of the local Langlands correspondence. 2021, arXiv:2102.13459v1.
[66] J. Fintzen, Tame cuspidal representations in non-defining characteristics. 2019, arXiv:1905.06374.

[67] J. Fintzen, On the construction of tame cuspidal representations. *Compos. Math.* **157** (2021), no. 12, 2733–2746.

[68] J. Fintzen, Types for tame p-adic groups. *Ann. of Math.* **193** (2021), no. 1, 303–346.

[69] Y. Z. Flicker, The Tame Algebra. *J. Lie Theory* **21** (2011), no. 2, 469–489.

[70] I. Gaisin and J. Rodriguez, Arithmetic families of (φ, Γ)-modules and locally analytic representations of $\mathrm{GL}_2(\mathbb{Q}_p)$. 2017, arXiv:1703.01627v2.

[71] D. Gaitsgory, From geometric to function-theoretic Langlands (or how to invent shtukas). 2016, arXiv:1606.09608.

[72] W.-T. Gan and L. Lomeli, Globalizations of supercuspidal representations over function fields and applications. *J. Eur. Math. Soc. (JEMS)* **20** (2018), no. 11, 2813–2858.

[73] R. Ganapathy, *The Deligne–Kazhdan philosophy and the Langlands conjectures in positive characteristic*. Pro-Quest LLC, Ann Arbor, MI, 2012. Thesis.

[74] R. Ganapathy, The local Langlands correspondence for GSp_4 over local function fields. *Amer. J. Math.* **137** (2015), no. 6, 1441–1534.

[75] R. Ganapathy, Congruences of parahoric group schemes. *Algebra Number Theory* **13** (2019), no. 6, 1475–1499.

[76] R. Ganapathy and S. Varma, On the local Langlands correspondence for split classical groups over local function fields. *J. Inst. Math. Jussieu* **16** (2017), no. 5, 987–1074.

[77] A. Genestier and V. Lafforgue, Chtoucas restreints pour les groupes réductifs et paramétrisation de Langlands locale. 2018, arXiv:1709.00978v2.

[78] U. Görtz, Alcove Walks and Nearby Cycle on Affine Flag Manifolds. *J. Algebraic Combin.* **26** (2007), no. 4, 415–430.

[79] E. Grosse-Kloenne, On special representations of p-adic reductive groups. *Duke Math. J.* **163** (2014), no. 12, 2179–2216.

[80] E. Grosse-Kloenne, From pro-p Iwahori–Hecke modules to (φ, Γ)-modules II. *Int. Math. Res. Not.* **2018** (2016), no. 3, 865–906.

[81] E. Grosse-Kloenne, From pro-p-Iwahori Hecke modules to (φ, Γ)-modules I. *Duke Math. J.* **165** (2016), no. 8, 1529–1595.

[82] E. Grosse-Kloenne, Supersingular modules as Galois representations. *Algebra Number Theory* **14** (2020), no. 1, 67–118.

[83] T. Haines and S. Rostami, The Satake isomorphism for special maximal parahoric Hecke algebra. *Represent. Theory* **14** (2010), 264–284.

[84] D. Hansen, T. Kaletha, and J. Weinstein, On the Kottwitz conjecture for local shtuka spaces. 2022, arXiv:1709.06651v4.

[85] M. Harris, Speculations on the mod p representation theory of p-adic groups. *Ann. Fac. Sci. Toulouse Math.* **6** (2016), no. 25, 2–3.

[86] J. Hauseux, Extensions entre séries principales p-adiques et modulo p de $G(F)$. *J. Inst. Math. Jussieu* **15** (2016), no. 2, 225–270.

[87] J. Hauseux, Compléments sur les extensions entre séries principales p-adiques et modulo p de $G(F)$. *Bull. Soc. Math.* **145** (2017), no. 1, 161–192.

[88] J. Hauseux, On the exactness of ordinary parts over a local field of characteristic p. *Pacific J. Math.* **295** (2018), no. 1, 7–30.

[89] J. Hauseux, Parabolic induction and extensions. *Algebra Number Theory* **12** (2018), no. 4, 779–831.

[90] J. Hauseux, Sur une conjecture de Breuil–Herzig. *J. Reine Angew. Math.* (2019), no. 751, 91–119.

[91] J. Hauseux, T. Schmidt, and C. Sorensen, Deformation rings and parabolic induction. *J. Théor. Nombres Bordeaux* **30** (2018), no. 2, 695–727.

[92] J. Hauseux, T. Schmidt, and C. Sorensen, Functorial properties of generalised Steinberg representations. *J. Number Theory* **195** (2019), 312–329.

[93] X. He and R. Ganapathy, Tits groups of Iwahori–Weyl groups and presentations of Hecke algebras. 2021, arXiv:2107.01768v1.

[94] D. Helm, On ℓ-adic families of representations of $\mathrm{GL}_2(\mathbb{Q}_p)$. *Math. Res. Lett.* **17** (2010), 805–822.

[95] D. Helm, On the modified mod p local correspondence for $\mathrm{GL}_2(\mathbb{Q}_\ell)$. *Math. Res. Lett.* **20** (2013), 489–500.

[96] D. Helm, The Bernstein center of the category of smooth $W(k)[\mathrm{GL}_n(F)]$-modules. *Forum Math. Sigma* **4** (2016).

[97] D. Helm, Whittaker models and the integral Bernstein center for GL_n. *Duke Math. J.* **165** (2016), no. 9, 1597–1628.

[98] D. Helm, Curtis homomorphisms and the integral Bernstein center for GL_n. *Algebra Number Theory* **14** (2020), no. 10, 2607–2645.

[99] D. Helm and G. Moss, Converse theorems and the local Langlands correspondence in families. *Invent. Math.* **214** (2018), 999–1022.

[100] G. Henniart, Une caractérisation de la correspondance de Langlands locale pour $\mathrm{GL}(n)$. *Bull. Soc. Math.* **130** (2002), no. 4, 587–602.

[101] G. Henniart, Sur les représentations modulo p de groupes réductifs p-adiques. *Contemp. Math.* **489** (2009).

[102] G. Henniart and B. Lemaire, Intégrales orbitales tordues sur $\mathrm{GL}(n, F)$ et corps locaux proches: Applications. *Canad. J. Math.* **58** (2006), no. 6, 1229–1267.

[103] G. Henniart and B. Lemaire, Représentations des espaces tordus sur un groupe réductif connexe p-adique. *Astérisque* **387** (2017).

[104] G. Henniart and M.-F. Vignéras, Comparison of compact induction with parabolic induction. Special issue to the memory of J. Rogawski. *Pacific J. Math.* **260** (2012), no. 2, 457–495.

[105] G. Henniart and M.-F. Vignéras, The Satake isomorphism modulo p with weight. *J. Reine Angew. Math.* **701** (2015), 33–75.

[106] G. Henniart and M.-F. Vignéras, Representations of a p-adic group in characteristic p. For Joseph Bernstein. *Proc. Sympos. Pure Math.* **101** (2019), 171–210.

[107] G. Henniart and M.-F. Vignéras, Representations of a reductive p-adic group in characteristic distinct from p. *Tunis. J. Math.* (2022).

[108] F. Herzig, The classification of admissible irreducible modulo p representations of a p-adic GL_n. *Invent. Math.* **186** (2011), 373–434.

[109] F. Herzig, A Satake isomorphism in characteristic p. *Compos. Math.* **147** (2011), no. 1, 263–283.

[110] F. Herzig, K. Koziol, and M.-F. Vignéras, On the existence of admissible supersingular representations of p-adic reductive groups (with an appendix by Sug Woo Shin). *Forum Math. Sigma* **8** (2020).

[111] C. Heyer, Parabolic induction via the parabolic pro-p Iwahori Hecke algebra. 2021, arXiv:2010.08435v2.

[112] C. Heyer, The left adjoint of the derived parabolic induction. 2022, arXiv:2204.11581v1.

[113] Y. Hu, Normes invariantes et existence de filtrations admissibles. *J. Reine Angew. Math.* **634** (2009), 107–141.

[114] Y. Hu, Sur quelques représentations supersingulières de $\mathrm{GL}\,2(Q_{p^f})$. *J. Algebra* **324** (2010), 1577–1615.

[115] Y. Hu, Diagrammes canoniques et représentations modulo p de $\mathrm{GL}\,2(F)$. *J. Inst. Math. Jussieu* **11** (2012), 67–118.

[116] M. Kashiwara and P. Shapira, *Categories and Sheaves*. Grundlehren Math. Wiss. 332, Springer, Berlin–Heidelberg, 2006.

[117] J. Kohlhaase, Smooth duality in natural characteristic. *Adv. Math.* **317** (2017), 1–49.

[118] K. Koziol, A classification of the irreducible mod p representations of $U(1,1)(\mathbb{Q}_{p^2}/\mathbb{Q}_p)$. *Ann. Inst. Fourier (Grenoble)* **66** (2016), no. 4, 1545–1582.

[119] K. Koziol, Pro-p-Iwahori invariants for SL_2 and L-packets of Hecke modules. *Int. Math. Res. Not.* **4** (2016), 1090–1125.

[120] K. Koziol, Hecke module structure on first and top pro-p-Iwahori Cohomology. *Acta Arith.* **186** (2018), 349–376.

[121] K. Koziol, Homological dimension of simple pro-p-Iwahori–Hecke modules. *Math. Res. Lett.* **26** (2019), no. 3, 769–804.

[122] K. Koziol, Derived right adjoints of parabolic induction: an example. 2022, arXiv:2202.09915v1.

[123] K. Koziol, F. Herzig, M.-F. Vignéras (appendix, and S.-W. Shin, On the existence of admissible supersingular representations of p-adic reductive groups. *Forum Math. Sigma* **8**(e2) (2020), 73.

[124] K. Koziol and L. Peskin, Irreducible admissible mod-p representations of metaplectic groups. *Manuscripta Math.* **155** (2018), no. 3–4, 539–577.

[125] K. Koziol and P. Xu, Hecke modules and supersingular representations of $U(2,1)$. *Represent. Theory* **19** (2015), 56–93.

[126] R. J. Kurinczuk, ℓ-modular representations of unramified p-adic $U(2,1)$. *Algebra Number Theory* **8** (2014), no. 8, 1801–1838.

[127] R. Kurinczuk and N. Matringe, Rankin–Selberg local factors modulo ℓ. *Selecta Math. (N.S.)* **23** (2017), no. 1, 767–811.

[128] R. Kurinczuk and N. Matringe, The ℓ-modular local Langlands correspondence and local factors. *J. Inst. Math. Jussieu* **20** (2019), no. 5, 1585–1635.

[129] R. Kurinczuk and N. Matringe, Test vectors for local cuspidal Rankin–Selberg integrals of $\mathrm{GL}(n)$ and reduction modulo ℓ. *Nagoya Math. J.* **233** (2019), 170–192.

[130] R. Kurinczuk and N. Matringe, Characterization of the relation between two ℓ–modular correspondences. *C. R. Math.* **358** (2020), no. 2, 201–209.

[131] R. Kurinczuk, D. Skodlerack, and S. Stevens, Endo-parameters for p-adic classical groups. *Invent. Math.* **223** (2021), 597–723.

[132] R. Kurinczuk and S. Stevens, Cuspidal ℓ-modular representations of p-adic classical groups. *J. Reine Angew. Math.* **2020** (2020), no. 764, 23–69.

[133] T. Lanard, Sur les ℓ-blocs de niveau zéro des groupes p-adiques. *Compos. Math.* **154** (2018), no. 7, 1473–1507.

[134] T. Lanard, Sur les ℓ-blocs de niveau zéro des groupes p-adiques II. 2018, arXiv:1806.09543.

[135] T. Lanard, Unipotent ℓ-blocks for simply connected p-adic groups. 2020, arXiv:2011.01165v1.

[136] D. Le, On some nonadmissible smooth irreducible representations for GL_2. *Math. Res. Lett.* **26** (2019), no. 6, 1747–1758.

[137] F. Ly, Représentations de Steinberg modulo p pour un groupe réductif sur un corps local. *Pacific J. Math.* **277** (2015), no. 2, 425–462.

[138] N. Matringe and G. Moss, The Kirillov model in families. 2022, arXiv:2005.13484v3.

[139] A. Minguez and V. Sécherre, Représentations banales de $\mathrm{GL}(m, D)$. *Compos. Math.* **149** (2013), 679–704.

[140] A. Minguez and V. Sécherre, Représentations lisses modulo ℓ de $\mathrm{GL}_m(D)$. *Duke Math. J.* **163** (2014), 795–887.

[141] A. Minguez and V. Sécherre, Types modulo ℓ pour les formes intérieures de GL_n sur un corps local non archimédien, avec un appendice par V. Sécherre et S. Stevens. *Proc. Lond. Math. Soc.* **109** (2014), no. 4, 823–891.

[142] A. Minguez and V. Sécherre, Unramified ℓ-modular representations of $\mathrm{GL}(n, F)$ and its inner forms. *Int. Math. Res. Not.* **8** (2014), 2090–2118.

[143] A. Minguez and V. Sécherre, Correspondance de Jacquet–Langlands locale et congruences modulo ℓ. *Invent. Math.* **208** (2017), no. 2, 553–631.

[144] S. Morra, Invariant elements for p-modular representations of $\mathrm{GL}_2(\mathbb{Q}_p)$. *Trans. Amer. Math. Soc.* **365** (2013), no. 12, 6625–6667.

[145] B. C. Ngo, Le lemme fondamental pour les algèbres de Lie. *Publ. Math. Inst. Hautes Études Sci.* **111** (2010), 1–169.

[146] R. Ollivier, Le foncteur des invariants sous l'action du pro-p Iwahori de $\mathrm{GL}_2(\mathbb{Q}_p)$. *J. Reine Angew. Math.* **635** (2009), 149–185.

[147] R. Ollivier, Parabolic Induction and Hecke modules in characteristic p for p-adic GL_n. *Algebra Number Theory* **4** (2010), no. 6, 701–742.

[148] R. Ollivier, Compatibility between Satake and Bernstein isomorphisms in characteristic p. *Algebra Number Theory* **8** (2014), no. 5, 1071–1111.

[149] R. Ollivier, An inverse Satake isomorphism in characteristic p. *Selecta Math. (N.S.)* **21** (2015), no. 3, 727–761.

[150] R. Ollivier and P. Schneider, Pro-p Iwahori Hecke algebras are Gorenstein. *J. Inst. Math. Jussieu* **13** (2014), no. 4, 753–809.

[151] R. Ollivier and P. Schneider, A canonical torsion theory for pro-p Iwahori–Hecke modules. *Adv. Math.* **327** (2018), 52–127.

[152] R. Ollivier and P. Schneider, The modular pro-p Iwahori–Hecke Ext-algebra. *Proc. Sympos. Pure Math.* **101** (2019), 255–308.

[153] R. Ollivier and P. Schneider, On the pro-p Iwahori Hecke Ext-algebra of $\mathrm{SL}_2(Q_p)$. 2021, arXiv:2104.13422v1.

[154] R. Ollivier and M.-F. Vignéras, Parabolic induction modulo p. *Selecta Math. (N.S.)* **24** (2018), no. 5, 3973–4039.

[155] V. Paskunas, Coefficient systems and supersingular representations of $\mathrm{GL}_2(F)$. *Mém. Soc. Math. Fr. (N.S.)* **99** (2004).

[156] V. Paskunas, Extensions for supersingular representations of $\mathrm{GL}_2(\mathbb{Q}_p)$. *Astérisque* **331** (2010), 317–353.

[157] V. Paskunas, The image of Colmez's Montreal functor. *Publ. Math. Inst. Hautes Études Sci.* **188** (2013), 1–191.

[158] V. Paskunas, Blocks for mod p representations of $\mathrm{GL}_2(Q_p)$. In *Automorphic forms and Galois representations*, pp. 231–247, London Math. Soc. Lecture Note Ser. 415, Cambridge University Press, Cambridge, 2014.

[159] V. Paskunas and S.-N. Tung, Finiteness properties of the category of mod p representations of $\mathrm{GL}_2(Q_p)$. *Forum Math. Sigma* **e80** (2021).

[160] C. Pépin and T. Schmidt, Mod p Hecke algebras and dual equivariant cohomology I: The case of GL_2. 2019, arXiv:1907.08808.

[161] C. Pépin and T. Schmidt, A semisimple mod p Langlands correspondence in families for $\mathrm{GL}_2(Q_p)$. Preprint, 2020.

[162] C. Pépin and T. Schmidt, Generic and mod p Kazhdan–Lusztig theory for GL_2. 2021, arXiv:2007.01364.

[163] D. Renard, *Représentations des groupes réductifs p-adiques*, Cours Spéc. 17, S.M.F., 2010.

[164] A. Roche, Parabolic induction and the Bernstein decomposition. *Compos. Math.* **134** (2002), no. 2, 113–133.

[165] A. Roche, *Notes on the Bernstein decomposition*. 2004.

[166] A. Roche, The Bernstein decomposition and the Bernstein centre. In *Ottawa Lectures on Admissible Representations of Reductive p-Adic Groups*, pp. 3–52, Fields Inst. Monogr. 26, American Mathematical Society, Providence, RI, 2009.

[167] N. Ronchetti, A Satake homomorphism for the mod p derived Hecke algebra. 2019, arXiv:1808.06512.

[168] M. M. Schein, A family of irreducible supersingular representations of $\mathrm{GL}_2(F)$ for some ramified p-adic fields. 2022, arXiv:2109.15244v2.

[169] S. Scherotzke and P. Schneider, Derived parabolic induction. *Bull. Lond. Math. Soc.* **54** (2022), no. 1, 264–274.

[170] N. A. Schmidt, *Generische pro-p-algebren*, Diplomarbeit, 2009.

[171] N. A. Schmidt, Generic pro-p-Hecke algebras. 2017, arXiv:1801.00353v1.

[172] P. Schneider, Smooth representations and Hecke modules in characteristic p. *Pacific J. Math.* **279** (2015), 447–464.

[173] P. Schneider and C. Sorensen, Duals in natural characteristic. 2022, arXiv:2202.01800v1.

[174] P. Scholze, The Local Langlands Correspondence for GL_n over p-adic fields. *Invent. Math.* **192** (2013), no. 3, 663–715.

[175] P. Scholze, On the p-adic cohomology of the Lubin–Tate tower, with an appendix of Michael Rapoport. *Ann. Sci. Éc. Norm. Supér. (4)* **51** (2018), no. 4, 811–863.

[176] B. Schraen, Sur la présentation des représentations supersingulières de $\mathrm{GL}_2(F)$. *J. Reine Angew. Math.* **704** (2015), 187–208.

[177] V. Sécherre and S. Stevens, Représentations lisses de $\mathrm{GL}(m, D)$, IV, représentations supercuspidales. *J. Inst. Math. Jussieu* **7** (2008), no. 3, 527–574.

[178] V. Sécherre and S. Stevens, Smooth representations of $\mathrm{GL}(m, D)$, VI: Semisimple types. *Int. Math. Res. Not.* (2011).

[179] V. Sécherre and S. Stevens, Block decomposition of the category of ℓ-modular smooth representations of $\mathrm{GL}_n(F)$ and its inner forms. *Ann. Sci. Éc. Norm. Supér. (4)* **49** (2016), no. 3, 669–709.

[180] V. Sécherre and S. Stevens, Towards an explicit local Jacquet–Langlands correspondence beyond the cuspidal case. *Compos. Math.* **155** (2019), no. 10, 1853–1887.

[181] D. Skodlerack, Cuspidal irreducible complex or ℓ-modular representations of quaternionic forms of p-adic classical groups for odd p. 2019, arXiv:1907.02922v2.

[182] D. Skodlerack, Semisimple characters for inner forms II: Quaternionic inner forms of classical groups. *Represent. Theory* **24** (2020), no. 11, 323–359.

[183] C. M. Sorensen, A proof of the Breuil–Schneider conjecture in the indecomposable case. *Ann. of Math.* **177** (2013), 1–16.

[184] C. M. Sorensen, The Breuil–Schneider conjecture. A survey. In *Adv. in the Theory of Numbers. Proceedings of the CNTA XIII*, edited by A. Alaca, S. Alaca, and K. S. Williams, Fields Inst. Commun. 77, 2015.

[185] C. M. Sorensen, A vanishing result for higher smooth duals. *Algebra Number Theory* **13** (2019), no. 7, 1735–1763.

[186] S. Stevens, Semisimple characters for p-adic classical groups. *Duke Math. J.* **127** (2005), no. 1, 123–173.

[187] S. Stevens, The supercuspidal representations of p-adic classical groups. *Invent. Math.* **172** (2008), 289–352.

[188] J. Trias, Correspondance thêta locale ℓ-modulaire I: Groupe métaplectique, représentation de Weil et θ-lift. 2020, arXiv:2009.11561.

[189] M.-F. Vignéras, On highest Whittaker models and integral structures. In *Contributions to automorphic forms, geometry, and number theory*, pp. 773–816, John Hopkins University Press, 2004.

[190] M.-F. Vignéras, Representations modulo p of the p-adic group $\mathrm{GL}(2, F)$. *Compos. Math.* **140** (2004), 333–358.

[191] M.-F. Vignéras, Pro-p Iwahori Hecke ring and supersingular $\overline{\mathbb{F}}_p$-representations. *Math. Ann.* **331** (2005), no. 3, 523–556. Erratum: **333** (2005), no. 3, 699–701.

[192] M.-F. Vignéras, Algèbres de Hecke affines génériques. *Represent. Theory* **10** (2006), 1–20.

[193] M.-F. Vignéras, Représentations irréductibles de $\mathrm{GL}(2, F)$ modulo p. In *L-functions and Galois representations*, edited by B. N. Burns, LMS Lecture Notes 320, 2007.

[194] M.-F. Vignéras, Série principale modulo p de groupes réductifs p-adiques. *Geom. Funct. Anal.* **17** (2007), 2090–2112.

[195] M.-F. Vignéras, Représentations p-adiques de torsion admissibles (shorter version of Admissibilité des représentations p-adiques et lemme de Nakayama, janvier 2007 (pdf)). In: *Number Theory, Analysis and Geometry: In Memory of Serge Lang*, edited by D. Golfeld, P. Jones, D. Ramakrishnan, K. Ribet, and J. Tate, Springer, 2011.

[196] M.-F. Vignéras, The pro-p Iwahori Hecke algebra of a reductive p-adic group II, Muenster. *J. Math.* **7** (2014), 363–379.

[197] M.-F. Vignéras, The pro-p Iwahori Hecke algebra of a reductive p-adic group V (parabolic induction). *Pacific J. Math.* **279** (2015), 499–529.

[198] M.-F. Vignéras, The pro-p Iwahori Hecke algebra of a reductive p-adic group I. *Compos. Math.* **152** (2016), 693–753.

[199] M.-F. Vignéras, The right adjoint of the parabolic induction. In *Hirzebruch Volume Proceedings Arbeitstagung 2013*, pp. 405–424, Progr. Math. 319, Birkhäuser, 2016.

[200] M.-F. Vignéras, The pro-p-Iwahori Hecke algebra of a reductive p-adic group III (spherical Hecke algebras and supersingular modules). *J. Inst. Math. Jussieu* **16** (2017), no. 3, 571–608.

[201] J.-L. Waldspurger, Endoscopie et changement de caractéristique. *J. Inst. Math. Jussieu* **5** (2006), no. 3, 423–525.

[202] J.-L. Waldspurger, Endoscopie et changement de caractéristique: intégrales orbitales pondérées. *Ann. Inst. Fourier* **59** (2009), no. 5, 1753–1818.

[203] M. H. Weissman, An induction theorem for groups acting on trees. *Represent. Theory* **23** (2019), 205–212.

[204] Z. Wu, A note on presentations of supersingular representations of $\mathrm{GL}_2(F)$. *Manuscripta Math.* **165** (2021), 583–596.

[205] J. K. Yu, Construction of tame supercuspidal representations. *J. Amer. Math. Soc.* **14** (2001), 579–622.

[206] J. K. Yu, Bruhat–Tits theory and buildings, pp. 53–77, On the local Langlands correspondence for tori, pp. 177–183. In *Ottawa Lectures on Admissible Representations of Reductive p-Adic Groups*, Fields Inst. Monogr. 26, American Mathematical Society, Providence, RI, 2009.

[207] X. Zhu, The Geometric Satake Correspondence for Ramified Groups. *Ann. Sci. Éc. Norm. Supér.* **48** (2015), 409–451.

[208] X. Zhu, Coherent sheaves in the stack of Langlands parameters. 2021, arXiv:2008.02998v2.

[209] X. Zhu, A note on integral Satake isomorphisms. 2021, arXiv:2005.13056v3.

MARIE-FRANCE VIGNÉRAS

Université de Paris Cité, Institut de Mathématiques de Jussieu, 4 place Jussieu, Paris 75005, France, marie-france.vigneras@imj-prg.fr

POPULAR SCIENTIFIC EXPOSITIONS

THE ISING MODEL IN OUR DIMENSION AND OUR TIMES

ANDREI OKOUNKOV

ABSTRACT

While the author is a professional mathematician, he is by no means an expert in the subject area of these notes. The goal of these notes is to share the author's personal excitement about some results of Hugo Duminil-Copin with mathematics enthusiasts of all ages, using maximally accessible, yet precise mathematical language. No attempt has been made to present an overview of the current state field, its history, or to place this narrative in any kind of broader scientific or social context. See the references in Section 5 for both professional surveys and popular science accounts that will certainly give the reader a broader and deeper understanding of the material.

MATHEMATICS SUBJECT CLASSIFICATION 2020

Primary 82B20; Secondary 97

KEYWORDS

Ising model

ICM INTERNATIONAL CONGRESS OF MATHEMATICIANS 2022 JULY 6–14

© 2022 International Mathematical Union
Proc. Int. Cong. Math. 2022, Vol. 1, pp. 376–413
DOI 10.4171/ICM2022/200

Published by EMS Press
and licensed under a CC BY 4.0 license

Phase transitions are dramatic physical phenomena. A physical system undergoing a phase transition may exhibit large spatial fluctuations, a detailed understanding of which presents an important challenge to physicists and mathematicians alike. Thanks to Hugo Duminil-Copin and his collaborators, the recent years saw a great progress in our understanding of the phase transition in the 3-dimensional Ising model, perhaps the most famous model of mathematical statistical physics.

The goal of these notes is to explain an introductory portion of this progress to the broadest possible audience of mathematics enthusiasts. Before we get to say anything of substance about the new results, there is a certain amount of language to develop and background to review.

1. MATHEMATICS AND PHYSICS

Mathematics provides the universal language of science. While human languages have words that describe natural phenomena, they lag far, far behind the language of mathematics in their precision and predictive power. It is easy to fill a sizable volume with quotes to this effect from the most prominent scientists of all epochs.[1]

Wouldn't the task of writing these notes be really simple if mathematics were *only* a language? There would probably be usable automatic translation available at a click. In fact, it is a very common request to translate from mathematics to a natural language.[2] Richard Feynman, in particular, talks about it in the second of his 1964 lectures about the Character of Physical Law.[3]

While mathematics has its special words and symbols, as well as grammatical rules that govern what is the correct logical use of these symbols and what is not, the real treasure of mathematics is the much deeper level of understanding that this language empowers. By defining the boundaries of precise reasoning, and removing all other boundaries between ideas, mathematics allows humans to use the most inventive and unexpected mathematical constructions and arguments to discover deep truths about the world around us. Instead of being lost in the confusing woods of natural languages, mathematics makes it possible for our thought to fly *safely*.

These notes are about mathematical physics, the field where mathematics and physics come together. A mathematical physicist starts by *defining* her or his object of

1 Gibbs measures, named so in honor of J. Willard Gibbs (1839–1903), will play the central role in our narrative. Gibbs is remembered as very unsociable and the only words he ever said in the Yale faculty meeting were *Mathematics is a language*. See the biography **[30]** of Gibbs by Muriel Rukeyser. She also wrote a poem about Gibbs inspired by this quote.

2 In the narrator's personal experience, good progress in science often happens when trying to answer the opposite question, namely, *can you translate what you just said to mathematics?*

3 Feynman says, in particular, this about translating mathematics: *But I do not think it is possible, because mathematics is not just another language. Mathematics is a language plus reasoning; it is like a language plus logic. Mathematics is a tool for reasoning. It is in fact a big collection of the results of some person's careful thought and reasoning. By mathematics it is possible to connect one statement to another.*

study, introducing a mathematical object that captures some essential features of one or many physical phenomena. One calls it a *model*, which, unlike many other words used by mathematicians, is a term that stays fairly close to its meaning in natural languages. Having defined a model, a mathematical physicist is free to be arbitrarily creative in her or his choice of mathematical tools to study this model. This investigation is going through all the natural stages of research in mathematics: one asks precise questions, considers examples, formulates conjectures, obtains partial results, and, as a proof of having achieved a really good understanding of the model, one can prove mathematical theorems about it.

For example, in his Mathematical Principles of Natural Philosophy, Newton introduced differential equation as a mathematical language to describe the motion of celestial as well as terrestrial bodies. This gives him a *model* for motion of planets around the sun. In the approximation that ignores the mutual attraction of the planets, he then mathematically *proves* the planets follow Kepler's empirical laws of planetary motion.

The language and models evolve. Each chapter in that great book of the Universe to which Galileo refers[4] in Il Saggiatore is written in a new mathematical language that has to be discovered every time. Newtonian mechanics is an approximation that is good at modeling some phenomena but not others. Quantum mechanics had to be created to describe the behavior of molecules, atoms, and other tiny constituents of the universe. Statistical physics had to be created to describe phenomena in which the myriads of particles that form planets and other macroscopic objects do not just move as one, but instead create very complex patterns and materials through spatial interactions. The actual mathematics used in each case is very different. The Ising model, which will be our focus of attention in this narrative, is perhaps the most famous model of statistical physics.

A question often asked about mathematical physics is: where is the boundary between mathematics and physics in it? In the personal view of this narrator, there is no boundary.[5] It is a really joint endeavor between mathematics and physics, where each side contributes something extremely important. Among other things, physics provides invaluable *intuition*, rooted in laboratory and numerical experiments, as well as parallels and correspondences that extends across different branches of physics. These can guide mathematics at any of the research stages discussed above. For mathematical physicists, following the logic of the subject is much more important than departmental affiliation. For instance, the first truly amazing mathematical result about the Ising model was obtained by Lars Onsager, the winner of the 1968 Nobel Prize in chemistry. In the narrator's personal experience, physicists are very proud when they find a mathematical proof and mathematicians are very proud when they discover a good physical explanation.

---

4 Philosophy is written in this grand book, which stands continually open before our eyes (I say the 'Universe'), but cannot be understood without first learning to comprehend the language and know the characters as it is written. It is written in mathematical language, and its characters are triangles, circles and other geometric figures, without which it is impossible to humanly understand a word; without these one is wandering in a dark labyrinth.

5 If it really exists, the boundary is as diffuse as the boundary in (1).

It is fitting to end this quotation-filled section with a quote from the hero of these notes Hugo Duminil-Copin: "*Mathematical Physics gathers everything I always dreamt of as a researcher: it satisfies my curiosity to understand the physical world in which we live, and it rewards the mathematicians with beautiful and elegant rigorous proofs.*"

2. THE ISING MODEL

2.1. Stuff fluctuates in space

It is good to have a picture in mind as we discuss the definition of the Ising model. Here is one, a simulation by Stanislav Smirnov. Its meaning will be made clear gradually.

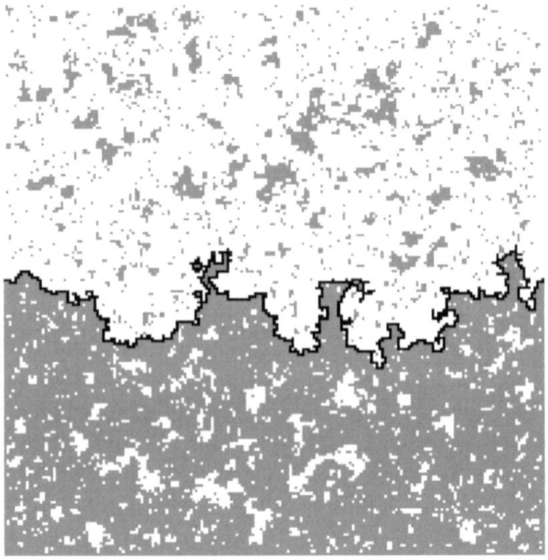

(1)

Clearly, this is something *random*. The language of statistical physics is based on randomness and probabilities.

Our world is fundamentally random. In statistical physics randomness is introduced from the very beginning.[6] From the very beginning, statistical physics talks about the *probabilities* for a physical system to be in such or such state.

Very importantly, the randomness in the figure of (1) happens in *space*, here a 2-dimensional space. In other words, in (1) we have a random spatial pattern. Note that while obviously complex, this pattern is not pure noise. We see a very diffuse boundary between black and white, with many islands or lakes of one color inside another. These have intricate shapes and may be nested, that is, there can be an island on a lake in the middle of a larger island on a larger lake, etc.

People are usually introduced to probability theory through coin tosses, rolls of dice, and similar random events that have a few possible outcomes and no spatial structure.

---

6 In quantum mechanics, randomness is also present from the very beginning. In principle, Newtonian mechanics makes exact predictions about the behavior of its models. However, for systems of large size and complexity, think Avogadro number, many billiard balls bouncing off each other, these predictions are so complicated as to be effectively random. This is the subject of ergodic theory, the development of which was very much stimulated by the quest to see statistical physics emerge from Newtonian mechanics.

Successive games of chance and similar data sets (think stock prices, air temperature, etc.) produce random time series like that in (2). These have a 1-dimensional structure to it. They are like beads threaded by the axis of time. In probability theory, these are known as random processes.

(2)

Statistical physics really starts in dimension 2 or larger, studying random objects fluctuating in the corresponding number of dimensions. Importantly, the behavior of the Ising model (and most models of statistical physics) very strongly depends on the dimension. The Ising model is dull in dimension 1, very interesting in dimensions 2 and 3, and regresses to the generic, and hence not as exciting,[7] Gaussian behavior in dimensions ≥ 4.

Many past glorious mathematical successes of statistical physics concern the 2-dimensional Ising model. While we will say a few words about it, our goal in this narrative is to report on the great recent progress in our physical dimension 3 achieved by Hugo Duminil-Copin and his collaborators. This will come in due course. We have not defined the Ising model, yet.

2.2. A lattice in space

While it is good to imagine that a random process like that in (2) happens in continuous time, the actual data (a stock is traded, air temperature is recorded, etc.) comes in discrete bits. It makes both mathematical and practical sense to similarly discretize the space in which the Ising model will live.

Mathematically, instead of having a random object defined for all points of the d-dimensional space $\mathbb{R}^d$, it will be defined only on the vertices of the d-dimensional cubic lattice Λ, like in the figure of (3).

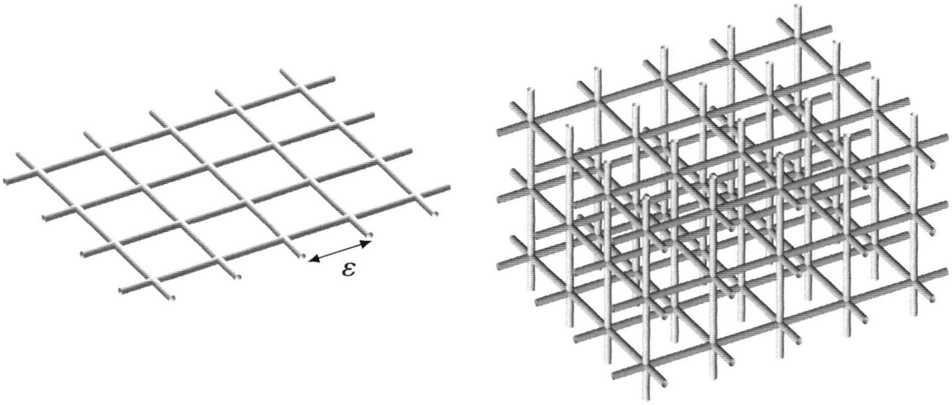

(3)

7 It still takes very exciting mathematics and lots of deep ideas to *prove* the behavior is Gaussian in dimensions $d \geq 4$.

Let ε denote the mesh size of this lattice. Then the vertices of the lattice are the points whose coordinates are integer multiples of ε, that is,

$$\text{vertices}(\Lambda) = \varepsilon\mathbb{Z}^3 = \{(\varepsilon n_1, \varepsilon n_2, \varepsilon n_3)\} \subset \mathbb{R}^3, \tag{4}$$

where $n_1, n_2, n_3 \in \mathbb{Z}$ are integers. We will use the words "lattice vertex" and "lattice point" interchangeably.

From the human scale point of view, we should imagine ε is vanishingly small, like the atomic scale. So, on the human scale, Λ is very dense. But from the atomic scale point of view, we can take $\varepsilon = 1$. For an infinite lattice, both points of view are mathematically completely equivalent.

2.3. Signs on a lattice

Now it is time to assign some fluctuating degrees of freedom to the vertices of the lattice. In the Ising model, one makes the simplest possible *binary* choice. That is, at every vertex $\mathsf{v} \in \Lambda$, there is a random variable $\sigma(\mathsf{v})$ that can take two possible values. The reader may choose any name she or he likes for these values: black/white, blue/red, 0/1, ± 1, etc. We will stick to the convention that

$$\sigma(\mathsf{v}) = \pm 1, \tag{5}$$

and we will call these variables' *signs*. For historical reasons, they are normally called *spins*, which may be rather confusing for those familiar with spins and not familiar with the history of the Ising model. One advantage of (5) over 0/1 and other choices is that it stresses the *symmetry* between two possibilities. This symmetry is very important in the Ising model.

Minimalistically, a fragment of a configuration of the 2-dimensional Ising model may be represented like this:

$$\begin{matrix} + & - & + & - & + \\ + & + & + & + & - \\ - & - & + & + & - \\ - & - & - & - & + \end{matrix} \tag{6}$$

Ising, and his adviser Lenz, created the model as a model of ferromagnetism, and they imagined a miniature magnet at every site of a lattice pointing in one of two possible directions. For them, (6) represented something like this:

(7)

As it typically happens in mathematics, physics, and elsewhere, once introduced, mathematical models live their own life and follow their own logic. In particular, they may be used to understand phenomena that are very different from the originally envisioned applications.

History often preserves the context where people first stumble upon an important discovery. For instance, the mineral bauxite, the world's source of aluminium, is named after the village of Les Baux where it was first described by P. Berthier in 1821. In the Earth's crust, it occurs mainly in places that are very far away from Provence and its current uses are probably very far from what Berthier could have envisioned.

2.4. Probabilities and energy

A common misconception about probability theory is that all possible outcomes of a random event are equally probable. While this is a good approximation for coin tosses and dice rolls, this would not be a very interesting assignment of probabilities in the Ising model. Indeed, if all signs were equally likely, the picture in (1) would be pure noise, indistinguishable from the noise introduced by the structure of the paper or the printing process. In particular, there would be no spatial structure to it, as it would be a collection of independent random bits, not affected by each other in any way. If they do not feel each other, they can be rearranged arbitrarily and hence there cannot be any significance to their particular spatial arrangement.

Ludwig Boltzmann and Willard Gibbs, the early architects of statistical physics, understood the connection between probabilities and *energy*. Energy is a central concept in physics which appears in the Newtonian, quantum, and statistical physics in slightly different, but compatible incarnations. Informally, it is supposed to be a universal equivalent that, just like ordinary human money, determines the intensity of any physical process. While people may have different attitudes towards money, energy is certainly making the physical world go round.

Without plunging into economic or metaphysical depths, mathematically energy is just a *function*

$$\{\text{configurations } \mathscr{C} \text{ of signs } \sigma\} \xrightarrow{\text{Energy}} \mathbb{R}, \tag{8}$$

that will be used to assign probabilities to configurations. Boltzmann and Gibbs understood that, in *equilibrium*, the probability of any configuration decays *exponentially* with its energy. To put these words into a formula, we have

$$\operatorname{Prob}(\mathscr{C}) = \frac{1}{Z(T)} \exp\left(-\frac{\operatorname{Energy}(\mathscr{C})}{T}\right). \tag{9}$$

In this formula, we have two proportionality coefficients T and $Z(T)$ that both deserve a comment. Let us start with T.

Only dimensionless numbers make sense inside the exponential, but energy has physical dimension, namely

$$[\text{energy}] = [\text{mass}][\text{length}]^2[\text{time}]^{-2}, \tag{10}$$

as exemplified by the familiar $\frac{1}{2}mv^2$ formula for the kinetic energy in the Newtonian mechanics. Therefore, we need a dimensional constant T to convert energy into dimensionless numbers. The constant T determines how fast probability decays with energy. Intuitively, it sets the scale of energy fluctuations. From

$$\frac{\mathrm{Prob}(\mathscr{C}_1)}{\mathrm{Prob}(\mathscr{C}_2)} = \exp\left(\frac{\mathrm{Energy}(\mathscr{C}_2) - \mathrm{Energy}(\mathscr{C}_1)}{T}\right), \tag{11}$$

we see that two configurations $\mathscr{C}_1$ and $\mathscr{C}_2$ make a comparable contribution only if the difference of their energies is not much larger than T.

From purely mathematical perspective, (9) may be taken as a *definition* of a statistical equilibrium, which depends on a constant $T \geq 0$ called the *temperature*.[8]

The coefficient $Z(T)$ is defined so that the probabilities of all possible configurations $\mathscr{C}$ sum to 1. This is an interesting function of T which, for historical reasons, is often called the *partition function*. Note that if we shift the energies of all configurations by the same constant E_0, then $Z(T)$ gets a factor of $e^{-E_0/T}$ and the probabilities do not change. In other words, only *energy differences* are important in (9). This is also clear from (11).

The case $T = 0$, interpreted using $e^{-\infty} = 0$, means the absolute zero temperature: only energy-minimizing configurations occur, and they are all equally likely.

2.5. Energy vs. entropy

The dramatic plot of statistical mechanics is the competition between energy and *entropy*. Formula (9) gives preference to energy-saving configurations. They may each have a relatively large probability, but there are typically not so many of them. Having a close-to-minimal energy is a special property that most configurations will fail. But the competitive advantage of most configurations is that there are many of them.

To make a mathematical question out of this, we can ask how is the energy distributed in the system described by (9)? The value of the energy in a random state of the system is a random variable, so it is a fair question. Anticipating the fact that in a system of large size the energy will scale linearly with a suitably defined volume V of the system, it is better to look at energy E per unit volume. One defines its entropy by

$$S(E) = \frac{1}{V} \ln\left(\text{number of states with } \frac{\text{Energy}}{V} = E\right). \tag{12}$$

Here, again, we normalize the logarithm by the volume V because we expect the counts of different possible states of the system to grow exponentially with the volume V. An equivalent of (12) is inscribed on Boltzmann's tombstone in Vienna's Zentralfriedhof.

---

8 While (9) is a definition, it is still worth explaining why T is called temperature. Imagine two systems in equilibrium at temperatures T_1 and T_2, respectively, which can exchange energy but otherwise do not interact. So, the configurations of the combined system are pairs $(\mathscr{C}_1, \mathscr{C}_2)$ and

$$\mathrm{Prob}\big((\mathscr{C}_1, \mathscr{C}_2)\big) = \mathrm{Prob}(\mathscr{C}_1)\ \mathrm{Prob}(\mathscr{C}_2), \quad \mathrm{Energy}\big((\mathscr{C}_1, \mathscr{C}_2)\big) = \mathrm{Energy}(\mathscr{C}_1) + \mathrm{Energy}(\mathscr{C}_2).$$

From (9) the combined system is in equilibrium if and only if $T_1 = T_2$. It thus suffices to check that (9) agrees with any other definition of a temperature for any one standard system, such as the ideal gas. Note that many thermometers work by putting some standard probe in contact and equilibrium with the system in question.

Formula (12) is a definition, like formula (9). From these definitions, we conclude

$$\operatorname{Prob}\left(\frac{\text{Energy}}{V} = E\right) = \frac{1}{Z(T)} \exp\left(\frac{V}{T} \underbrace{\left(-E + TS(E)\right)}_{\text{maximize}}\right), \tag{13}$$

where we hid the inessential proportionality factor in gray.

When V is very large, only those energies that minimize $E - TS(E)$, a quantity known as free energy, will be observed in the system, not those that simply minimize E. The character of this minimum depends on the temperature. For $T = 0$, only the energy counts, and we get strict energy minima. For $T = \infty$, energy means nothing and entropy decides. For other values of T, both energy and entropy count, in different proportions. We will see this principle in action in the Ising model.

2.6. Interactions in the Ising model

Now it is time to specify the energy function in the Ising model. Let $\mathscr{C}$ be a configuration of signs. We can write it as a function

$$\sigma : \Lambda \longrightarrow \{\pm 1\}, \tag{14}$$

assigning each vertex $\mathsf{v} \in \Lambda$ a sign. When mathematicians talk about a function σ, they write $\sigma(\mathsf{v})$ to denote its value at the argument v, and use the symbol σ to denote the "whole" function. A configuration in the Ising model is a function (14) and we do not need another symbol $\mathscr{C}$ to denote it. What we need is to assign a number to it that will be called Energy(σ).

The spatial structure of the lattice will be taken into account by declaring that only neighboring signs interact. That is,

$$\operatorname{Energy}(\sigma) = \sum_{\text{edges } \mathsf{v}-\mathsf{v}'} E\big(\sigma(\mathsf{v}), \sigma(\mathsf{v}')\big), \tag{15}$$

where the edges are the edges in the lattice (3), the vertices v and v' are the two endpoints of a given edge, and $E(\pm 1, \pm 1)$ is some interaction energy of the neighboring spins to be specified momentarily.

Note that all edges contribute equally to (15), no matter where in the lattice they occur and in which of the coordinate directions they are pointing. In other words, the interactions in (15) are as homogeneous and as isotropic as the presence of a lattice in space allows.

It remains to specify 4 numbers $E(\pm 1, \pm 1)$. Since we want plus and minus to be symmetric, we need to have

$$E(1,1) = E(-1,-1), \quad E(1,-1) = E(-1,1),$$

where the latter equality also follows from the symmetry of the interaction between two neighbors. Recall that the overall shift of energy changes nothing and note that the overall scale of energy is equivalent to rescaling the temperature. In the end, there are no meaningful free parameters left, and we can set

$$E\big(\sigma(\mathsf{v}), \sigma(\mathsf{v}')\big) = -\sigma(\mathsf{v})\sigma(\mathsf{v}'). \tag{16}$$

The minus sign here means that signs lower the energy by being equal.[9] In other words, the function (14) likes to be a constant function, or the spins in (7) like to be pointing in the same direction. This desire, however, is not expressed globally, but only through local interaction of each sign with its immediate neighbors.

A careful reader may have been worried for a long time by the sum in (15) being an infinite sum of ± 1's for the infinite cubic lattice Λ. This worry is well justified and related to some core mathematical and physical issues. We will devote many pages below to dealing with it carefully. For now, let us replace the infinite lattice Λ by any finite piece of it or any finite graph. Then (15) is a finite sum, the probabilities in (9) are well-defined, and we have defined the Ising model in finite volume.

2.7. Clusters and interfaces

Grouping together neighboring vertices of the same sign, we get *clusters* of pluses and minuses, as in the following figure:

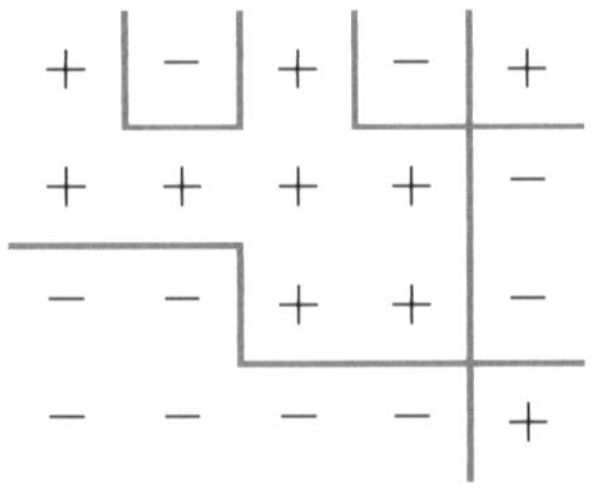

(17)

The boundary between the clusters is the *interface* between pluses and minuses. It is a $(d-1)$-dimensional object glued out of sides of a unit square/cube, so a path for $d = 2$, a surface for $d = 3$, etc. One component of the interface is highlighted in the figure of (1). For $d = 3$, the interface may look something like (48) in Section 3.3.3.

From (15) we have

$$\text{Energy}(\sigma) = \text{const} + 2\,\text{Area(interface)}, \tag{18}$$

where area (or length) is the $(d-1)$-dimensional lattice area, meaning that each side of the unit cube has area 1. This means that the Ising model can be interpreted as describing a fluctuating lattice interface, where the energy of the interface is its lattice area.

The lattice area has some peculiarities compared to usual area in $\mathbb{R}^d$. For instance, in $\mathbb{R}^2$, any two points are joined by a unique shortest path—a straight line segment, while the shortest path enclosing a given volume is a circle. For the 2-dimensional lattice distance, there are many shortest paths connecting two points. Indeed, any path that goes up/right from

---

9 The opposite choice of sign in (16) mathematically means negative temperature and, for other discretization of space, e.g., the triangular lattice, may correspond to a very different physics. For the cubic lattice, however, it can be reduced to the minus sign by flipping half of the signs in a checkerboard fashion.

the start to the finish in the following picture has a minimal length:

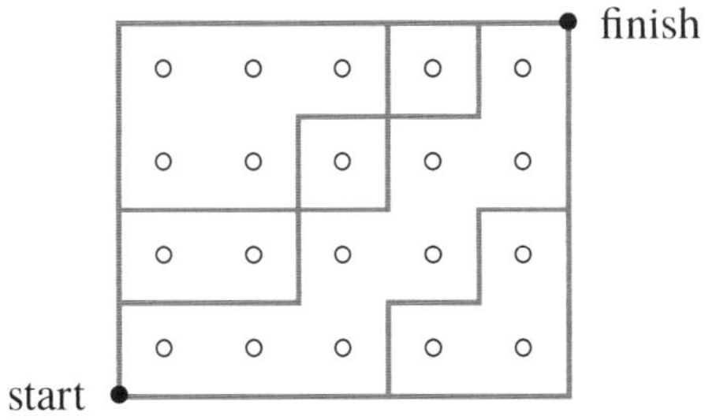

(19)

The shortest lattice paths enclosing a given volume are close to squares, not circles. Similar features persist in all dimensions.

In this narrative, we will be concerned with the *phase transition* in the Ising model that happens at a certain *critical temperature* T_c. These terms will be introduced properly below. For now we remark that for $T < T_c$, the Ising model reflects, due to the peculiarities of the lattice area, the behavior of materials that are similarly anisotropic, for instance, crystals. Indeed, for a crystal, a square or a cube is the shape one should expect to see, not the sphere.

At $T = T_c$ however, this anisotropy disappears, a rather remarkable phenomenon. In fact, at the critical temperature, not only rotation invariance is restored, but some further symmetries appear. This is an incredibly interesting topic, but since it it would be a side trip for our story, we refer the reader to [20, 32] for details.

This concludes our brief discussion of the Ising model in finite volume. It is time to make sense of the energy and probabilities for the whole infinite lattice. This will be our task in Section 3.

In the process of defining the Ising model, there were choices, and we always made the simplest possible nontrivial symmetric choice. A reader may get the impression we defined a little mathematical toy, a basic wooden block set, which may be good for play but seriously oversimplifies the nature. What is the place of the Ising model in the broader landscape of statistical physics? An interested reader will find an introductory discussion of this question in Appendix A. In short, mathematical physicists believe the Ising model provides a *universal* description of a very large class of phenomena in which a ± 1 symmetry becomes *broken* below a certain temperature.

3. GIBBS MEASURES

3.1. Definition

Our goal now is to define probabilities in the Ising model on the *infinite* cubic lattice. More precisely, we want to know what is the probability to see any particular pattern π of signs in any given finite subset Ω of Λ. For instance, for $d = 2$, we want to know

$$\operatorname{Prob}\left(\sigma\Big|_{\substack{\text{a fixed } 3\times 3\\ \text{square } \Omega}} = \boxed{\begin{matrix} + & - & + \\ + & + & - \\ - & - & + \end{matrix}} \right) = \,?\, \in [0,1]. \tag{20}$$

Mathematicians denote by $\sigma|_\Omega$ the restriction of a function σ to a subset Ω of arguments. We will sometimes call the subset Ω a *window*. If we have a finite window into an infinite system, it is reasonable to ask what is the probability to see some pattern π in it.

If $\Omega \subset \Omega'$, then the probabilities for the smaller window Ω are determined from the probabilities for the larger window Ω'. Therefore, it is enough to define the probabilities for larger and larger cubes

$$\Omega_L = [-L, \ldots, L]^d \subset \mathbb{Z}^d = \Lambda, \quad L = 1, 2, 3, \ldots, \tag{21}$$

because any finite subset of Λ is contained in some Ω_L. For $d = 2$, the square Ω_1 looks like the square in (20).

The main issue with formula (9) for the infinite lattice was that the energy (15) is infinite. Recall, however, that the important thing in physics is not the energy itself, but rather the *difference* in energies and note that energy difference

$$\Delta \,\text{Energy} = \text{Energy}(\sigma) - \text{Energy}(\sigma') \tag{22}$$

is well defined if σ and σ' differ only at finitely many vertices.

Let us look at the example in (23), where the difference in signs is circled:

$$\sigma_+ = \begin{matrix} & \vdots & \vdots & \vdots & \\ \cdots & + & - & + & \cdots \\ \cdots & + & \oplus & - & \cdots \\ \cdots & - & - & + & \cdots \\ & \vdots & \vdots & \vdots & \end{matrix}, \qquad \sigma_- = \begin{matrix} & \vdots & \vdots & \vdots & \\ \cdots & + & - & + & \cdots \\ \cdots & + & \ominus & - & \cdots \\ \cdots & - & - & + & \cdots \\ & \vdots & \vdots & \vdots & \end{matrix}. \tag{23}$$

Assuming this is the only difference, that is, assuming that the dots in (23) represent some choice of signs for σ_+ and an *identical* choice for σ_-, we can compute the energy difference as follows:

$$\text{Energy}(\sigma_+) - \text{Energy}(\sigma_-) = 4. \tag{24}$$

Indeed, only the edges incident to the circled vertices change their energy, and their energy is $3 - 1 = 2$ for σ_+ and $1 - 3 = -2$ for σ_-. In exactly the same way, we can determine the change of energy if we flip some signs in the *interior* of the window Ω_L for any L.

Denote

$$\pi_+ = \boxed{\begin{matrix} + & - & + \\ + & + & - \\ - & - & + \end{matrix}}, \qquad \pi_- = \boxed{\begin{matrix} + & - & + \\ + & - & - \\ - & - & + \end{matrix}}. \tag{25}$$

We may interpret formula (9) as saying that

$$\frac{\text{Prob}(\sigma|_{\Omega_1} = \pi_-)}{\text{Prob}(\sigma|_{\Omega_1} = \pi_+)} = \exp(4T^{-1}). \tag{26}$$

More generally, if π and π' are signs pattern in Ω_L that differ only in the interior, we may interpret formula (9) as saying that

$$e^{\text{Energy}(\pi)/T}\, \text{Prob}(\sigma|_{\Omega_L} = \pi) = e^{\text{Energy}(\pi')/T}\, \text{Prob}(\sigma|_{\Omega_L} = \pi'). \tag{27}$$

Note this formula operates with finite quantities. Moreover, (27) are *linear* equations on the probabilities.

Here comes the important moment. We say that an assignment of probabilities to the events that $\sigma|_{\Omega_L} = \pi$ is a *Gibbs measure*[10] if it satisfies (27) for all L and all π and π' that differ only in the interior of Ω_L. This key mathematical definition goes back to 1960s and the work of Roland Dobrushin, Oscar Lanford, and David Ruelle. And, no, Gibbs measures were not studied by Gibbs.

Note the change in perspective. Instead of saying that the formula (9) gives the probabilities, we have rewritten (9) as a system of equations that the probabilities have to satisfy. As with any equations, one naturally wonders: do they have a solution? If they do, how many solutions are there?

The existence of Gibbs measures for any temperature T is a very general and soft mathematical fact, see Section 4.4.4. The question of how many Gibbs measures there are for a given value of T is really the central question for us in these notes.

Formula (9) was meant to describe a statistical system in equilibrium at temperature T. But an infinite system can be in *many different* equilibria at given T, unable to fluctuate from one to another due to an infinite energetic cost. Concretely in the Ising model, each sign $\sigma(\mathsf{v})$ likes to be the same as that of its neighbors. Hence, a strong local preference for $+1$ or -1 may be self-reproducing in fluctuations. It could be a preference for either $+1$ or -1, and if there such a preference, then the system is stuck with it. The reader will probably have no difficulty thinking of real-life examples of this phenomenon.

Anticipating the fact that there may be many Gibbs measures at a given temperature, we will denote by $\boldsymbol{\mu}$ a Gibbs measure and write $\boldsymbol{\mu}(A)$ for the probability that $\boldsymbol{\mu}$ assigns to some event denoted by A. For example, A can say that $\sigma|_{\Omega_L} = \pi$.

In practice, it is convenient to use the averages

$$\big\langle \sigma(\mathsf{v}_1)\sigma(\mathsf{v}_2)\cdots\sigma(\mathsf{v}_n)\big\rangle_{\boldsymbol{\mu}} = \boldsymbol{\mu}(\text{this product equals 1}) - \boldsymbol{\mu}(\text{it equals } -1) \tag{28}$$

with respect to $\boldsymbol{\mu}$, where $\mathsf{v}_1, \dots, \mathsf{v}_n$ are some vertices of Λ. The averages (28) are called *correlation functions*, and when one wants to stress the number of different lattice points involved, one talks about n-point correlation functions.

In general, the averages (also known as expectations, or integrals) with respect to a Gibbs measure $\boldsymbol{\mu}$ are defined as follows. Let $f(\sigma)$ be a function that depends on finitely many signs $\sigma(\mathsf{v}_i)$, $\mathsf{v}_i \in \Lambda$. Then f takes finitely many values f_j, and we can define

$$\langle f \rangle = \sum_j f_j \boldsymbol{\mu}(f = f_j). \tag{29}$$

In measure theory, general integrals with respect to a measure $\boldsymbol{\mu}$ are defined by approximating the integrand f by functions taking finitely many values.

10 The word measure denotes a very important concept in mathematics, which we will leave without a proper discussion. The power of measure theory lies in being able to measure (meaning, assign some version of length, volume, probability, etc.) rather general sets. In our case, the probability is assigned to simple events of the form $\sigma|_{\Omega_L} = \pi$ and we hope the reader will have no difficulty thinking about this.

3.2. High temperature

3.2.1.

It is easiest to start the discussion of Gibbs measures at the infinite temperature $T = \infty$. Since $1/T = 0$, energy disappears from (27) and we conclude that all sign patterns are equally likely. In other words, each sign is an independent symmetric coin toss. This is the complete description of the unique Gibbs measure for $T = \infty$.

3.2.2.

For high enough temperatures, the unique Gibbs measure can be written as a series in the inverse temperature

$$\beta = \frac{1}{T}, \tag{30}$$

following a very general *perturbation theory* ideas, used everywhere in mathematical physics.

3.2.3.

Let $\boldsymbol{\mu}_0$ be a Gibbs measure at inverse temperature β_0, from which we want to construct a Gibbs measure $\boldsymbol{\mu}$ at inverse temperature $\beta \approx \beta_0$. Let us first consider a finite piece $\Omega \subset \Lambda$ of the infinite lattice. For a finite graph Ω, the unique Gibbs measure at inverse temperature β is defined by (9). We can transform this definition as follows:

$$\begin{aligned}
\langle f(\sigma)\rangle_{\Omega,\beta} &= \frac{\sum_\sigma e^{-\beta\,\mathrm{Energy}(\sigma)} f(\sigma)}{\sum_\sigma e^{-\beta\,\mathrm{Energy}(\sigma)}} && (31)\\
&= \frac{\sum_\sigma e^{-\beta_0\,\mathrm{Energy}(\sigma)} e^{(\beta_0-\beta)\,\mathrm{Energy}(\sigma)} f(\sigma)}{\sum_\sigma e^{-\beta_0\,\mathrm{Energy}(\sigma)} e^{(\beta_0-\beta)\,\mathrm{Energy}(\sigma)}} && (32)\\
&= \frac{\langle e^{(\beta_0-\beta)\,\mathrm{Energy}} f(\sigma)\rangle_{\Omega,\beta_0}}{\langle e^{(\beta_0-\beta)\,\mathrm{Energy}}\rangle_{\Omega,\beta_0}}, && (33)
\end{aligned}$$

where the summation in (31) and (32) ranges over all possible values of signs $\sigma(\mathbf{v})$ for $\mathbf{v} \in \Omega$.

Since the number $\Delta\beta = \beta - \beta_0$ is small, it may be useful to expand the exponentials in (33) in a series, using

$$e^x = 1 + x + \frac{x^2}{2} + \cdots + \frac{x^n}{n!} + \cdots. \tag{34}$$

3.2.4.

For the infinite lattice Λ, formula (33) will seemingly run into the old problem of energy being infinite for an infinite lattice. However, it may happen that the infinities cancel between the numerator and denominator in (33) in each term of the series expansion in powers of $\Delta\beta$.

For concreteness, let us examine the first order of the expansion of a 1-point correlation function $\langle\sigma(\mathrm{v}_1)\rangle_{\boldsymbol{\mu}}$. We have

$$e^{(\beta_0-\beta)\,\mathrm{Energy}} = 1 + \Delta\beta \sum_{\text{edges } \mathrm{v}_2 - \mathrm{v}_3} \sigma(\mathrm{v}_2)\sigma(\mathrm{v}_3) + \cdots, \tag{35}$$

where dots stand for terms of degree 2 or larger in $\Delta\beta$. Therefore,

$$\left\langle e^{(\beta_0-\beta)\,\mathrm{Energy}}\sigma(\mathsf{v}_1)\right\rangle_{\boldsymbol{\mu}_0} = \left\langle\sigma(\mathsf{v}_1)\right\rangle_{\boldsymbol{\mu}_0} + \Delta\beta \sum_{\text{edges } \mathsf{v}_2 - \mathsf{v}_3} \left\langle\sigma(\mathsf{v}_1)\sigma(\mathsf{v}_2)\sigma(\mathsf{v}_3)\right\rangle + \cdots . \tag{36}$$

Dividing (36) by the average of (35), we obtain

$$\begin{aligned}\left\langle\sigma(\mathsf{v}_1)\right\rangle_{\boldsymbol{\mu}} &= \left\langle\sigma(\mathsf{v}_1)\right\rangle_{\boldsymbol{\mu}_0} \\ &\quad + \Delta\beta \sum_{\text{edges } \mathsf{v}_2 - \mathsf{v}_3} \left(\left\langle\sigma(\mathsf{v}_1)\sigma(\mathsf{v}_2)\sigma(\mathsf{v}_3)\right\rangle_{\boldsymbol{\mu}_0} - \left\langle\sigma(\mathsf{v}_1)\right\rangle_{\boldsymbol{\mu}_0}\left\langle\sigma(\mathsf{v}_2)\sigma(\mathsf{v}_3)\right\rangle_{\boldsymbol{\mu}_0}\right) + \cdots .\end{aligned} \tag{37}$$

The sum over the edges in (37) is infinite. However, if the edge v_2–v_3 is far away from the vertex v_1, we expect the corresponding signs to be approximately independent random variables. Approximate independence means that

$$\left\langle\sigma(\mathsf{v}_1)\sigma(\mathsf{v}_2)\sigma(\mathsf{v}_3)\right\rangle_{\boldsymbol{\mu}_0} \approx \left\langle\sigma(\mathsf{v}_1)\right\rangle_{\boldsymbol{\mu}_0}\left\langle\sigma(\mathsf{v}_2)\sigma(\mathsf{v}_3)\right\rangle_{\boldsymbol{\mu}_0}. \tag{38}$$

So, the difference in (37) measures how close $\sigma(\mathsf{v}_1)$ and $\sigma(\mathsf{v}_2)\sigma(\mathsf{v}_3)$ are to being independent or, equivalently, how much they are *correlated*. If they decorrelate sufficiently fast with the distance between v_1 and v_2 then the sum in (37) will be convergent. Similar considerations apply to all other higher terms in the expansion (37).

3.2.5.

Given two random variables f_1 and f_2, the difference

$$\langle f_1 | f_2\rangle = \langle f_1 f_2\rangle - \langle f_1\rangle\langle f_2\rangle \tag{39}$$

is called their covariance. For example, the covariance of $f_1 = \sigma(\mathsf{v}_1)$ and $f_2 = \sigma(\mathsf{v}_2)\sigma(\mathsf{v}_3)$ with respect to $\boldsymbol{\mu}_0$ appears in (37).

In statistical physics, it is typical for covariance (39) to decay if f_1 and f_2 depend on spatially separated arguments, like in (37). If this decay is exponential in the spatial separation then we will say that the model has an exponential decay of correlations or is exponentially decorrelated.

3.2.6.

There are higher analogs of the covariance, involving three of more arguments. For instance, one defines

$$\left\langle f_1 | f_2 | f_3\right\rangle = \langle f_1 f_2 f_3\rangle - \langle f_1 f_2\rangle\langle f_3\rangle - \langle f_1 f_3\rangle\langle f_2\rangle - \langle f_2 f_3\rangle\langle f_1\rangle + 2\langle f_1\rangle\langle f_2\rangle\langle f_3\rangle. \tag{40}$$

These are called *cumulants* and are related to the combinatorial principle of inclusion–exclusion. They measure finer mutual dependencies between 3 or more random variables and appear naturally in perturbation series for the following reason.

The general formula for cumulants may be obtained from the identity

$$\ln\left\langle e^{f_1+f_2+\cdots}\right\rangle = \sum_n \frac{1}{n!} \sum_{i_1,\dots,i_n} \left\langle f_{i_1} | f_{i_2} | \dots | f_{i_n}\right\rangle, \tag{41}$$

in which one expands the exponential as in (34) and equates terms containing the same functions f_i. In particular, let us replace f_1 in (41) by tf_1 and compute the value of $\frac{\partial}{\partial t}$ at $t = 0$. We get

$$\frac{\langle f_1 e^{f_2+f_3+\cdots}\rangle}{\langle e^{f_2+f_3+\cdots}\rangle} = \sum_n \frac{1}{n!} \sum_{i_1,\ldots,i_n \geq 2} \langle f_1 | f_{i_1} | f_{i_2} | \ldots | f_{i_n} \rangle. \tag{42}$$

Voila, this is just what we need in (33), with $f_1 = \prod \sigma(\mathsf{v}_i)$ and

$$f_2 + f_3 + \cdots = \sum_{\text{edges } \mathsf{v}-\mathsf{v}'} \sigma(\mathsf{v})\sigma(\mathsf{v}').$$

Later in Section 4.3, we will meet random variables for which all cumulants with $n \geq 3$ vanish, meaning that that any $\langle f_1 \ldots f_n \rangle$ may be written entirely in terms of the expectations $\langle f_i \rangle$ and the covariances $\langle f_i | f_j \rangle$. Such random variables are called *Gaussian*. See Section A.3.4 for more on this. In a certain precise technical sense, nonzero cumulants with $n \geq 3$ measure the nonlinearity of the model.

3.2.7.

Going back to the special case $\beta_0 = 0$ and the unique Gibbs measure $\boldsymbol{\mu}_0$ at $T = \infty$, we observe that signs at different lattice sites are totally independent for $\boldsymbol{\mu}_0$. Thus the $\Delta\beta$ term in (38) is simply zero. In fact, great simplifications happen for $\boldsymbol{\mu}_0$ and a nice convergent combinatorial series can be written down for $\boldsymbol{\mu}$ provided the inverse temperature β is sufficiently small.[11] In this high-temperature range, the Gibbs measure remains unique.

Since $\boldsymbol{\mu}_0$ and the energy are invariant under flipping all signs, this property is inherited by the perturbation series. The invariance of $\boldsymbol{\mu}$ under flipping all signs also follows from its uniqueness. It follows that

$$\big\langle \text{product of odd number of } \sigma(\mathsf{v}_i) \big\rangle_{\text{high T}} = 0, \tag{43}$$

and in particular that

$$\big\langle \sigma(\mathsf{v}) \big\rangle_{\text{high T}} = 0, \tag{44}$$

for any v. The expected value of a single sign in (44) is the simplest measure of a possible ± 1 asymmetry of a Gibbs measure. It is a very important parameter of the Gibbs measure called *magnetization*.

Also note that the uniqueness of the high-temperature Gibbs measure implies it is invariant under *shifts* of the lattice Λ. This *translational invariance* is an important property for a Gibbs measure to have or not to have. For a translation-invariant Gibbs measure, the magnetizations at all vertices of the lattice are equal.

---

11 Since the energy (35) is a product of terms like

$$e^{\beta\sigma(\mathsf{v}_2)\sigma(\mathsf{v}_3)} = \cosh(\beta)\big(1 + \tanh(\beta)\sigma(\mathsf{v}_2)\sigma(\mathsf{v}_3)\big),$$

it is more convenient to write this series in powers of the hyperbolic tangent of β,

$$\tanh(\beta) = \beta - \frac{1}{3}\beta^3 + \frac{2}{15}\beta^5 - \cdots.$$

3.2.8.

One should stress that the perturbative series expansion for (33) has no guarantee of success in general. In particular, it will fail when either $\boldsymbol{\mu}_0$ or $\boldsymbol{\mu}$ have long-range correlations, meaning that the signs at distant vertices do not become decorrelated sufficiently fast. Needless to say, these are precisely the situations of maximal interest and significance!

3.3. Low temperature

3.3.1.

What about the opposite case $T = 0$? Equation (27) has the following meaning at $T = 0$:

$$\text{Energy}(\pi) > \text{Energy}(\pi') \quad \Rightarrow \quad \text{Prob}(\sigma|_{\Omega_L} = \pi) = 0. \tag{45}$$

In other words, if the energy of a configuration can be lowered by flipping finitely many signs, then its probability vanishes.

In terms of the interface between the pluses and minuses, formula (18) says that it should be *minimal*, meaning that its length/area cannot be made any smaller by finite modifications.[12]

What minimal interfaces can we think of? First, there is the empty interface. If the interface is empty then all signs are equal. A measure $\boldsymbol{\mu}_+$ that assigns probability 1 to the configuration in which all signs are $+$, and zero probability to all other configurations, is a Gibbs measure at $T = 0$. No randomness is a special case of randomness and it may happen that the probability of just one particular configuration equals 1.

Since with an empty interface all signs can be $+$ or all signs can be $-$, we have *two* Gibbs measures $\boldsymbol{\mu}_\pm$ already. But there is more. For instance, the plane $x_1 = \frac{1}{2}$, or any plane of the form $x_i = \frac{1}{2} + \text{integer}$, $i = 1, 2, 3$, defines a minimal interface, see the figure in (46).

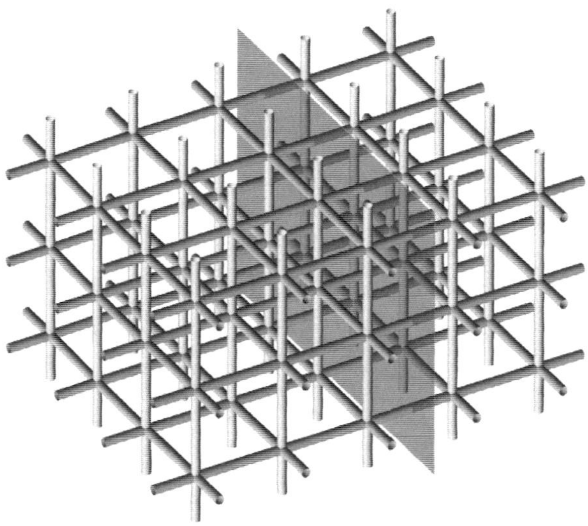

(46)

12 For $d = 1$, there is a difference between finite modifications of signs and of the interface. We will consider finite modifications of the interface.

We can put pluses on either side of such wall and this gives many more zero-temperature Gibbs measures, all of which we will denote by $\boldsymbol{\mu}_{\text{wall}}$. They are not translation-invariant and, in fact, they can be all taken to one another by a symmetry of the lattice Λ.

A curious reader may think about more Gibbs measures at $T = 0$, but it is already clear that there are plenty of them. They very visibly break the symmetries between ± 1 and also between different lattice points.

Recall that at $T = \infty$ we have a total *disorder*, which persisted to all high temperatures and manifested itself, in particular, by the vanishing magnetization. By contrast, the $T = 0$ measures exhibit a very strong spatial *order*.

3.3.2.

When there is more than one Gibbs measure, the following point should be kept in mind. Let $\boldsymbol{\mu}_1$ and $\boldsymbol{\mu}_2$ be two Gibbs measures. Then their mixture of the schematic form

$$\boldsymbol{\mu}_{\text{mix}} = 0.71\boldsymbol{\mu}_1 + 0.29\boldsymbol{\mu}_2, \tag{47}$$

where 0.71 can be replaced by any number between 0 and 1, is also a Gibbs measure. Indeed, it assigns probabilities in $[0, 1]$ to all events and satisfies the linear equations (27) from the definition of a Gibbs measure.

What does the equation (47) mean? Imagine there are two different labs in a physics department, labeled by $i = 1, 2$. In the lab number i, our physical system is kept in a state described by the measure μ_i. We go to a *random* lab and do the measurement. If our chance to go to the first lab is 0.71 then the outcome of our measurement will be described by (47).

Clearly, this is rather silly. It is quite unnatural and adds nothing to our understanding of the system. Therefore, when there is more than one Gibbs measure, people usually restrict their attention to those Gibbs measures that cannot be nontrivially written in the form (47). They are called *extremal* or *pure*.

3.3.3.

How will the $T = 0$ Gibbs measure perturb for small positive T? The interface between plus and minus no longer has to be minimal, but every time its area increases by 1 the probability decreases by $e^{-2/T}$. It is therefore natural to organize the expansion in powers of $e^{-2/T}$ which is a small parameter for T positive and small.

For example, we may perturb the wall in (46) as follows:

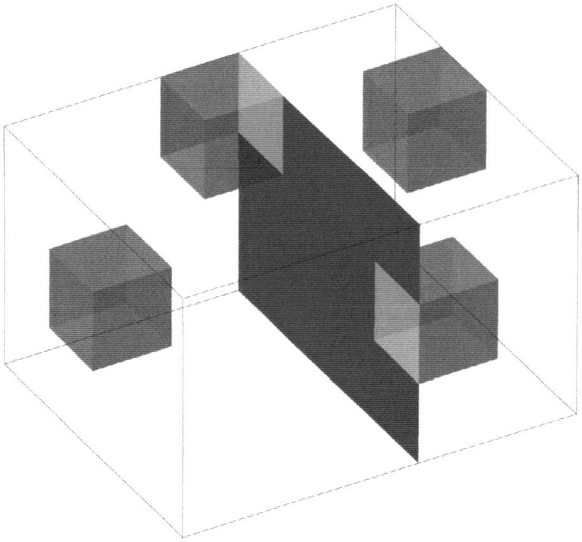

(48)

This perturbation increases the area by 20, so it counts with the weight $(e^{-2/T})^{20} = e^{-40/T}$.

The fate of the perturbation series for zero temperature Gibbs measures is *different* in different dimensions.

3.3.4.

For $d = 1$, there are only the measures $\boldsymbol{\mu}_\pm$ at $T = 0$. Their perturbation series breaks down at the very first term. Indeed, let $\boldsymbol{\mu}_+$ denote the hypothetical Gibbs measure for which the signs are positive at the positive infinity of the lattice $\mathbb{Z}$.

The $e^{-2/T}$ term in the perturbation series for $\boldsymbol{\mu}_+$ is then a sum over all configurations like this

$\cdots$ 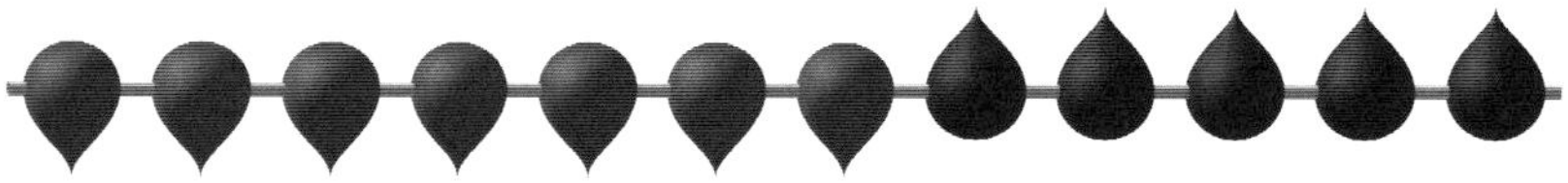 $\cdots$, (49)

where the switch of signs can occur at any place. This gives infinitely many equal terms that affect the sign at any lattice point.

In fact, for $d = 1$, the high temperature disordered behavior happens for all $T > 0$. No wonder we cannot access any positive temperature by a perturbation of the $T = 0$ description. Historically, Ernst Ising studied precisely the $d = 1$ case and reached this conclusion in his 1924 dissertation.

The absence of order for any $T > 0$ in $d = 1$ led to a certain temporary dip of interest in the Ising model. See [17] for a much more informative account of the many chapters of the Ising model history.

3.3.5.

For $d \geq 2$, the perturbation series for $\boldsymbol{\mu}_\pm$ converges! This was first noted, in essence, in 1936 by Rudolf Peierls, who observed that the number of relevant interfaces of given area is bounded by C^{Area} for some constant C. This makes the series converge as long as $e^{2/T} > C$

and proves that, for $d \geq 2$, the Ising model can exhibit both order and disorder, depending on the temperature.

3.3.6.

For $d = 2$, the perturbation series for $\boldsymbol{\mu}_{\text{wall}}$ breaks down at the first possible term when the length is allowed to increase by 2. We already discussed in Section 2.7 that the lattice length has just too many minimizers, and this is another consequence of this fact. In fact, in $d = 2$, the measures $\boldsymbol{\mu}_{\pm}$ can be shown to be the *only* pure Gibbs measures for $T > 0$.

By contrast, for $d \geq 3$, the series for $\boldsymbol{\mu}_{\text{wall}}$ converges. The corresponding measures were first studied by Dobrushin and bear his name [10, 11]. This means that at low temperature and in dimensions $d \geq 3$, the Ising model can *break* both the $\pm$ symmetry *and* the symmetries of the lattice. One says that some symmetry $\mathbf{g}$ of the system is broken by a Gibbs measure $\boldsymbol{\mu}$ if $\mathbf{g}$ takes $\boldsymbol{\mu}$ to another Gibbs measure, different from $\boldsymbol{\mu}$.

3.4. Critical temperature

We have talked about the behavior of the Ising model at high and low temperatures, respectively. This behavior differs strikingly. At high temperatures, we have a homogenous disorder. The system expresses no ± 1 preference and looks the same everywhere. At low temperatures, vertices prefer one sign over the other and this preference may change from vertex to vertex.

What happens for temperatures in the middle? This question must be on everybody's mind by now. Is there some intermediate range of temperatures for which yet another qualitatively different behavior is observed?

For the Ising model, and related models of statistical physics, there is exactly one *critical* value T_c of the temperature at which the balance between order and disorder, energy and entropy tips. A numerical simulation, done by Stanislav Smirnov, may help visualize this transition.

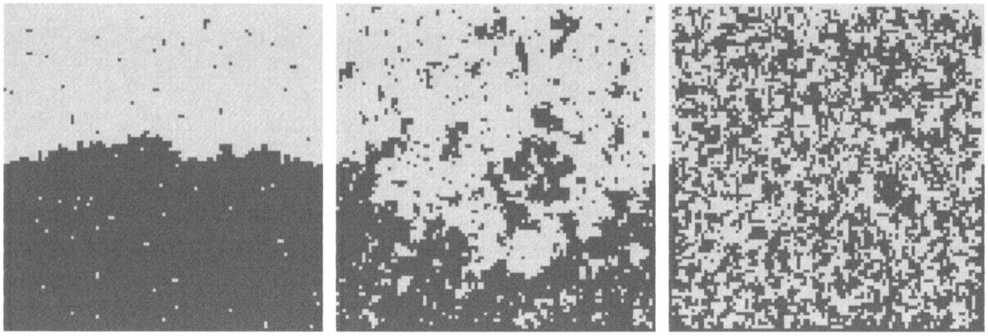

(50)

In (50) we see the $d = 2$ Ising model simulated on the 100×100 grid for $T < T_c$, $T = T_c$, and $T > T_c$, respectively. For a finite piece of the lattice, the notion of *boundary conditions* is important. Formula (27) tells us about the probabilities of different signs patterns inside the square, while signs along the boundary may be in principle assigned arbitrarily. In (50) and also in (1), we have $+1$ along one the lower half of the boundary and -1 along the upper half, see the figure in (51), as if we are trying to simulate the nonexistent Gibbs measure μ_{wall} in $d = 2$.

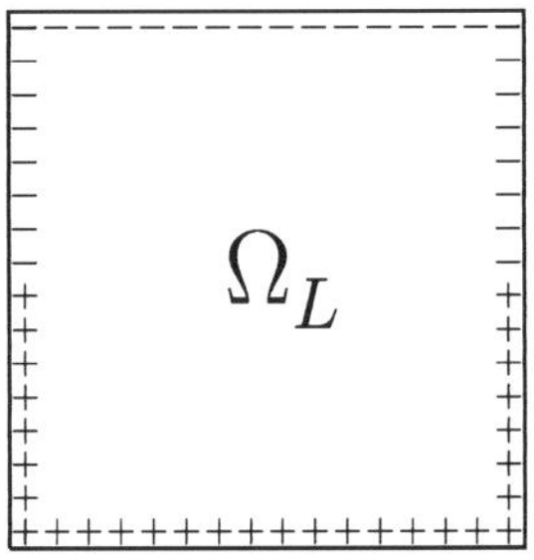

(51)

Such a boundary condition allows one to focus on the properties of the interface by singling out one special component of the interface that goes from one vertical side of the boundary to another. This interface is clearly visible for $T < T_c$, has essentially evaporated for $T > T_c$, and it is both visible and very strongly fluctuating at the critical point $T = T_c$.

All features of the $T > T_c$ are microscopic, everything happens on the lattice scale. The $T < T_c$ has one macroscopic feature—the interface, but its fluctuations are again small in size and would not be visible from far away.[13] What is common between the $T \neq T_c$ pictures is that the signs at different lattice points become independent *exponentially* fast with the lattice distance. The rate of this exponential decay sets the typical scale of observed features in both pictures.

By contrast, the $T = T_c$ picture has some features on *all scales*, enabled by the slow *polynomial* decorrelations. In fact, the lattice mesh $\varepsilon \to 0$ limit of the critical $d = 2$ Ising model is invariant not just under scaling, but under all *conformal* transformations. These are transformations that look like scaling and rotation in a very small neighborhood of any point.[14] For the proof of conformal invariance of the $d = 2$ critical Ising model, Stanislav Smirnov was awarded the 2010 Fields Medal, the only Fields Medal previously awarded for the study of the Ising model. While our focus in these notes is on $d = 3$, we hope a curious reader will open **[20, 32]** for more on conformal invariance. After the groundbreaking 1984 work of three Alexanders, Belavin, Polyakov, and Zamolodchikov, conformal invariance and the language of Conformal Field Theory (CFT) grew to be the most powerful tool for understanding 2-dimensional critical phenomena.

It is the job of a statistical physicist to predict macroscopic properties of materials from the macroscopically invisible fluctuations that take place on the atomic scale. It is a very, very interesting job, with its challenges and rewards, and the study of the Ising model at $T \neq T_c$ is no exception. But the statistical physicist's finest hour is when she or he gets to

---

13 It may be useful to explain, in terms of the figure in (50), why there is no Gibbs measure μ_{wall} in $d = 2$ for $0 < T < T_c$. If we fix any finite window at exactly the middle height, the interface will pass over it or under it with probability almost $\frac{1}{2}$. As a result, we will observe $\mu_{\pm}$ in our window with probabilities $\frac{1}{2}$ as the square in (51) grows to infinity. For $d = 3$ and $T < T_c$, the interfaces fluctuates less, and we will see it in the window as the measure μ_{wall}.

14 It is easier to define conformal transformation as the transformations that preserve all angles between curves. They can scale and rotate by different amounts in the vicinity of different points. The case $d = 2$ is special in that there is an abundance of such transformations.

describe a system that does exhibit macroscopic fluctuations. Phase transitions are these kind of phenomena. In particular, the order/disorder phase transition in the Ising model certainly packs more excitement, and is much more widely applicable, than what happens at $T \neq T_c$. So, what happens at $T = T_c$? This question calls for the start of a new section.

4. WHAT HAPPENS AT $T = T_c$?

4.1. Critical Gibbs measures

With our focus on Gibbs measures in this narrative, it is clear what our next question is going to be. Is there one or are there many Gibbs measures at $T = T_c$? This question may be phrased as *continuity* of the phase transition. Indeed, if there are many Gibbs measures at $T = T_c$, there will be one of them, say, $\boldsymbol{\mu}_c$, which is *not* the $T \downarrow T_c$ limit of the unique high-temperature measure $\boldsymbol{\mu}_{\text{high T}}$. Thus, for a system in state $\boldsymbol{\mu}_c$, a tiny increase in temperature will lead to a jump to $\boldsymbol{\mu}_{\text{high T}}$, meaning a jump in physical properties.

In a live or online class, it may be a good idea to take a poll on this question. Do you think it is going to be continuous? Or not? Phase transitions come in both flavors in nature. When the water melts or boils, its properties change discontinuously. At the pressure of 1 atmosphere, water boils at 100°C. Increasing the pressure increases the boiling point monotonically until, at the pressure of 217.7 atmospheres, we reach a very special point called the water critical point. After it, the difference between liquid and vapor disappears. When going through this point, the properties of the system remain continuous. Admittedly, this is a much more delicate example than simply boiling the water.

Another example of a continuous phase transition is the Curie critical point, the original motivation for the introduction of the Ising model. Magnets loose their magnetic properties when heated. For an iron magnet, this happens at 770°C, and the loss of magnetic properties is continuous. The Ising model is not a particularly convincing model of magnetism for several reasons, so we should be careful with drawing conclusions from this example.

4.2. The Potts model

A very important difference between a magnet and Ising model signs is that magnetization is a vector that can be rotated in all possible ways. These rotational symmetries are very different from the simple ± 1 symmetry of the binary degrees of freedom in the Ising model. Rotations form a continuous Lie group. Importantly, rotations can be arbitrarily small.

Closer to the Ising model are the models with a larger, but still finite symmetry group. The most important example is the Q-state Potts model, the Ising model being the $Q = 2$ case of the Potts model. In the Potts model, the function

$$\sigma : \Lambda \longrightarrow \{1, 2, \dots, Q\} \tag{52}$$

can take Q possible values, and the energy has the same form (15) with

$$E(a,b) = \begin{cases} E_=, & a = b, \\ E_{\neq}, & a \neq b. \end{cases} \tag{53}$$

This is invariant under all $Q!$ possible permutations of the values in (52). As with the Ising model, the exact values of constants in (53) are not important as long as $E_{\neq} > E_=$.

Simulations of the critical Q-state Potts model by V. Beffara for $d = 2$ and $Q = 2, 3, 4, 5, 6, 9$ may be seen in (54).

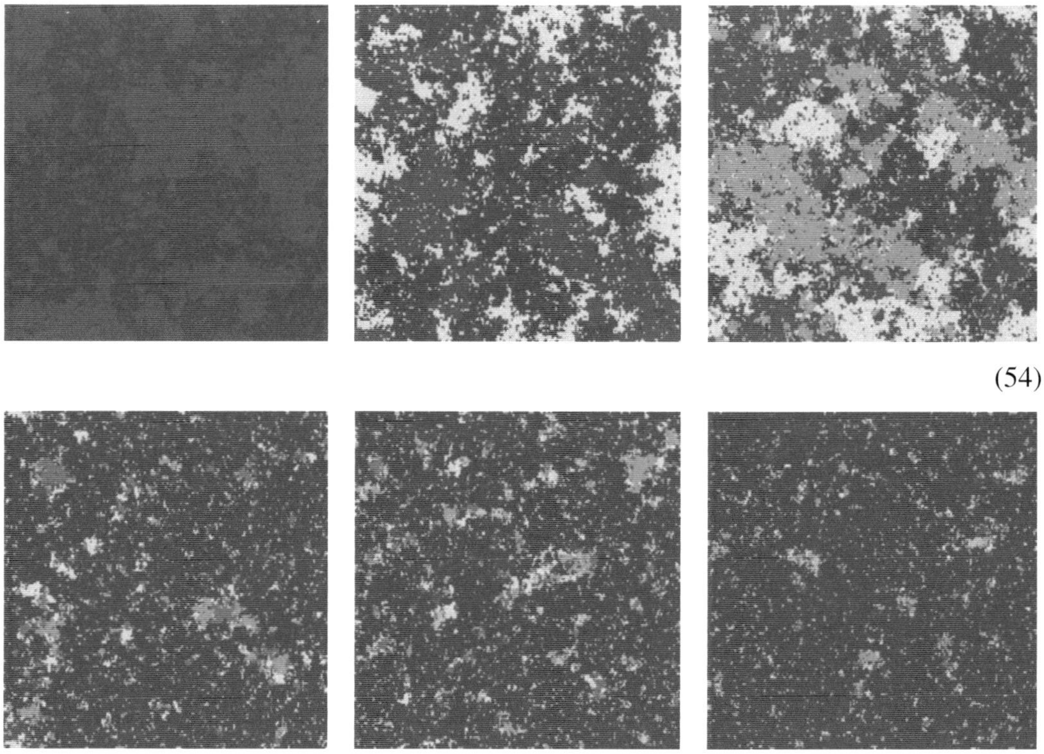

(54)

Different colors represent clusters of different values of Q. The reader will notice a clear difference in behavior between the top $Q \leq 4$ row and the bottom $Q \geq 5$ row.

4.3. Theorems

Our goal on the preceding pages was to ignite the reader's interest in what happens in the 3-dimensional Ising model at the critical temperature. To add to the suspense, we start with the following fundamental result of Hugo Duminil-Copin and his collaborators in the $d = 2$ case.

Theorem 1 ([18,19]). *The phase transition in the Q-state Potts model for $d = 2$ is continuous if and only if $Q \leq 4$.*

Theorem 1 is a logical conjunction of two different results proven using two different sets of tools. The continuity for $Q \leq 4$ is proven in the paper [19] by Hugo Duminil-Copin, Vladas Sidoravicius, and Vincent Tassion, building, in particular, on the earlier solo work

[12] of Hugo Duminil-Copin. The discontinuity for $Q > 4$ is proven in the paper [18] by Hugo Duminil-Copin, Maxime Gagnebin, Matan Harel, Ioan Manolescu, and Vincent Tassion.

After the depth of the continuity question has been underscored yet another time by Theorem 1, we can finally state the following result of Michael Aizenman, Hugo Duminil-Copin, and Vladas Sidoravicius.

Theorem 2 ([2]). *The phase transition in the $d = 3$ Ising model is continuous.*

It is expected that the phase transition for the $d = 3$ Potts model is discontinuous for $Q \geq 3$, see [16]. As to the higher dimension we have the following result. Recall we have met the Gaussian random fields in Section 3.2.5, see also Section A.3.4 in the Appendix. The scaling limit refers to the limit when we take correlations, suitably scaled, for larger and larger spatial separations. It describes what we would actually observe on our human scale.[15]

Theorem 3 ([1]). *The scaling limit of the critical Ising model in $d = 4$ is Gaussian.*

I hope the readers share the narrator's sense of awe at this absolutely amazing mathematics and join me in warmest congratulations on it being recognized by the Fields Medal. I also hope the readers got the sense that today's mathematics is not just extraordinarily powerful, but also concrete, understandable, and fun, once one finds the right idea and the right point of view. While finding that right point of view is not at all easy, my biggest hope is to have inspired my youngest readers to believe that mathematics can be beautiful and rewarding, both as a subject and as a profession. Maybe this is also a good place for me to thank Hugo Duminil-Copin, Stanislav Smirnov, and Martin Hairer for this special opportunity to be introduced to their wonderful subject.

4.4. Contours of proofs, seen in the distance

4.4.1.

We hope the reader agrees that the majestic view of Theorems 1, 2, and 3 was worth the uphill hike through the foothills of the Ising range. We also hope the reader will not be discouraged to learn that a much longer and steeper climb is needed to get a good view of the actual mathematics that goes into the proof of these theorems. As we stressed at the beginning, having a mathematical proof is a measure of our understanding of the model and, certainly, understanding is a great reward for any effort.

To help the reader master the subject, there are brilliant expositions available, in particular by Hugo Duminil-Copin himself. In [16], the reader will find a very fun, colorful, and engaging explanation of Theorems 1, 2, and many other results.

4.4.2.

Let us start with Theorem 2 and a discussion of the basic logic of how something like this could be proven. One logical point we should make from the very beginning is that we *do not* know the value of T_c for $d = 3$.

15 if we lived in a corresponding number of spatial dimensions

It is a gift of nature to mathematical physicists that many fascinating and highly nontrivial exact results can be obtained in the Ising, Potts, and related models when $d = 2$. In particular, it is known that for the square lattice Potts model we have

$$T_c = \frac{2}{\ln(1+\sqrt{Q})}, \quad d = 2. \tag{55}$$

This goes back to the 1941 work of Kramers and Wannier in the $Q = 2$ Ising case and is proven by Vincent Beffara and Hugo Duminil-Copin in [5], in general.

Formulas like this are extremely sensitive to the exact lattice formulation of the model, and other $d = 2$ models presumably converging to the same critical CFT at their (unknown!) critical point loose the magic. In addition to being a huge help in the study of the Ising and the Potts models proper, exact results very much contributed to how mathematical physicists think about their subject in general. We will say a few words about them below.

Nothing of the kind was ever discovered for $d = 3$, and there are many different strong hints that the physics, and the mathematics, in the plane and in the space are just different.

4.4.3.

Recall our discussion of the $T = 0$ Gibbs measures and note that, of all possible Gibbs measures, the measure $\boldsymbol{\mu}_+$ clearly has the most pluses, while the measure $\boldsymbol{\mu}_-$ has the least possible number of them. This basic comparison persists to all temperatures. All possible Gibbs measures are, in a certain precise mathematical sense, sandwiched between $\boldsymbol{\mu}_-$ and $\boldsymbol{\mu}_+$. Hence, the continuity question may be phrased as

$$\boldsymbol{\mu}_+ \stackrel{?}{=} \boldsymbol{\mu}_-, \quad T = T_c. \tag{56}$$

To see whether $\boldsymbol{\mu}_+ \stackrel{?}{=} \boldsymbol{\mu}_-$, one does not need to compute all correlation functions. Well-developed techniques in the subject reduce the question to the comparison of 1-point correlation functions, that is, magnetizations, at all vertices. Since $\boldsymbol{\mu}_\pm$ are both translation-invariant and differ by exactly the flip of all signs, the continuity question is equivalent to

$$\langle \sigma(\text{any one point}) \rangle_{\boldsymbol{\mu}_+} \stackrel{?}{=} 0, \quad \text{at } T = T_c. \tag{57}$$

This may sounds like we made good progress until we remind ourselves that we do not know the value of T_c, or any equation that determines this number. An approximate value of T_c is known from numerical experiments, but it is not useful for us now. The only thing we know about T_c is that

$$T_c = \inf\{T \text{ such that } \langle \sigma(\mathrm{v}) \rangle_{\boldsymbol{\mu}_+} = 0\}. \tag{58}$$

But since the $T \downarrow T_c$ continuity is precisely the crux of the matter, we did not progress much. We will be just going in circles until we can relate the question (57) to something which is either:

$$\begin{aligned} &\text{(A) true for ALL temperatures, or} \\ &\text{(C) is manifestly CONTINUOUS as } T \downarrow T_c. \end{aligned} \tag{59}$$

4.4.4.

In the (c) category in (59), one can actually describe the limit of the unique high-temperature Gibbs measure as $T \downarrow T_c$. One has

$$\lim_{T \downarrow T_c} \boldsymbol{\mu}_{\text{high T}} = \boldsymbol{\mu}_{\text{free}}|_{T=T_c}, \tag{60}$$

where $\boldsymbol{\mu}_{\text{free}}$ is the *free boundary* Gibbs measure that can be constructed as follows.

There is a universal way to produce Gibbs measures for any temperature. Recall the discussion of the boundary condition from Section 3.4. For all sufficiently large L, say $L \geq 100$, fix some configurations of signs

$$\sigma|_{\partial\Omega_L} = \tau_L, \quad L = 100, 101, 102, \dots, \tag{61}$$

along the boundary $\partial\Omega_L$ of the cube Ω_L from (21) as in the figure of (62). The values of τ_L can be all $+1$, can be all -1, can be like the Dobrushin's boundary conditions from (51), or can be anything at all. In particular, they do not have to be related to each other for different values of L.

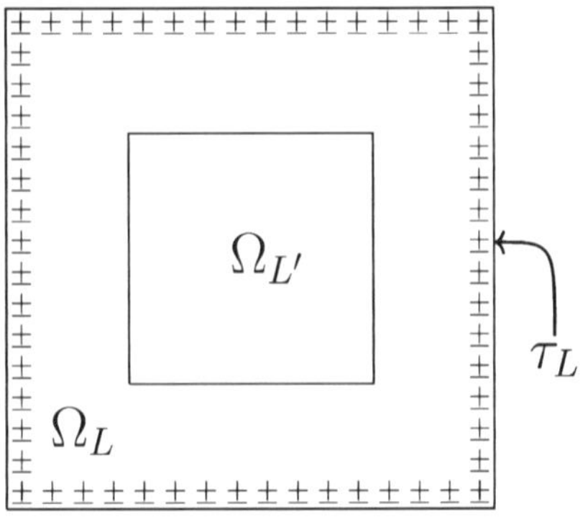

(62)

For all L, equation (27) defines a unique probability distribution $\boldsymbol{\mu}_L$ for signs in Ω_L and thus probability distributions for signs in any smaller cube $\Omega_{L'} \subset \Omega_L$. Possible values of $\sigma|_{\Omega_{L'}}$ form a finite set and probability distributions on a finite (or compact) set are compact.[16] Hence there is a subsequence of L for which the limit

$$\boldsymbol{\mu}(\sigma|_{\Omega_{L'}} = \pi) = \lim_{L \to \infty} \boldsymbol{\mu}_L(\sigma|_{\Omega_{L'}} = \pi) \tag{63}$$

exists for all L' and π. Since it is a limit of solutions of (27), it is a Gibbs measure. Incidentally, this proves that the set of Gibbs measures is nonempty for any T.

Any Gibbs measure may be obtained in this way. Indeed, if we make τ_L random and let it be distributed according to some Gibbs measure $\boldsymbol{\mu}$ then $\boldsymbol{\mu}_L = \boldsymbol{\mu}$.

The measure $\boldsymbol{\mu}_{\text{free}}$ is obtained when instead of fixing the boundary signs, we take all possible sign configuration in Ω_L, weighted according to the temperature and their energy. This means we really sum over all possible configurations of $\pm$ along the boundary in (62) with the corresponding weights. The measure $\boldsymbol{\mu}_{\text{free}}$ is not pure. In fact, it is known that **[6]**

$$\boldsymbol{\mu}_{\text{free}} = \frac{1}{2}\boldsymbol{\mu}_+ + \frac{1}{2}\boldsymbol{\mu}_-, \quad T < T_c. \tag{64}$$

---

16 The key property of a compact set that we need here is that any infinite sequence of elements of a compact set has a converging subsequence.

While $\boldsymbol{\mu}_\pm$ are at the two extremes of the set of the Gibbs measure, the measure $\boldsymbol{\mu}_{\text{free}}$ is its very center. In particular, it is ±-symmetric. It thus makes sense, and can be shown formally, that it is the $T \downarrow T_c$ limit of the unique high-temperature measure in (60).

4.4.5.

In the (A) category in (59), we would like a comparison between $\boldsymbol{\mu}_+$ and $\boldsymbol{\mu}_{\text{free}}$ which is valid for *all* temperatures.

From (64), we see that it is hopeless to compare magnetizations, as they will definitely differ below T_c. However, since $\boldsymbol{\mu}_+$ and $\boldsymbol{\mu}_-$ differ by a sign flip, their n-point correlation functions are equal for any *even* number n. This means that both above and below T_c we have

$$\left\langle \prod_{i=1}^{n} \sigma(\mathsf{v}_i) \right\rangle_{\boldsymbol{\mu}_+} = \left\langle \prod_{i=1}^{n} \sigma(\mathsf{v}_i) \right\rangle_{\boldsymbol{\mu}_{\text{free}}}, \quad T \neq T_c, \quad n \text{ is } \underline{\text{even}}, \tag{65}$$

and thus it is a reasonable hope to extend this to $T = T_c$.

In fact, if (65) can be extended to $T = T_c$ for $n = 2$, that would be the end of the proof because of the following argument.

On the one hand, as a very special case of a general FKG inequality published in 1971 by Cees Fortuin, Pieter Kasteleyn, and Jean Ginibre **[23]**, one has

$$\langle \sigma(\mathsf{v}) \rangle_{\boldsymbol{\mu}} \langle \sigma(\mathsf{v}') \rangle_{\boldsymbol{\mu}} \le \langle \sigma(\mathsf{v}) \sigma(\mathsf{v}') \rangle_{\boldsymbol{\mu}} \tag{66}$$

for any Gibbs measure $\boldsymbol{\mu}$. For a translation-invariant measure $\boldsymbol{\mu}$, it follows that[17]

$$\langle \sigma(\mathsf{v}) \rangle_{\boldsymbol{\mu}}^2 \le \lim_{\|\mathsf{v}-\mathsf{v}'\| \to \infty} \langle \sigma(\mathsf{v}) \sigma(\mathsf{v}') \rangle_{\boldsymbol{\mu}}. \tag{67}$$

For $\boldsymbol{\mu}_{\text{free}}$, the right-hand side of (67) can be seen to vanish at T_c as a consequence of the $T \downarrow T_c$ continuity. If the 2-point functions for $\boldsymbol{\mu}_{\text{free}}$ and $\boldsymbol{\mu}_+$ are the same then (67) implies $\langle \sigma(\mathsf{v}) \rangle_{\boldsymbol{\mu}_+} = 0$ at $T = T_c$, and we are done.

4.4.6.

What do we remember about the sign configurations if we forget all n-point correlations for n odd? It is easy to see that we remember precisely the clusters of equal signs which we talked about in Section 2.7, see the figure in (68).

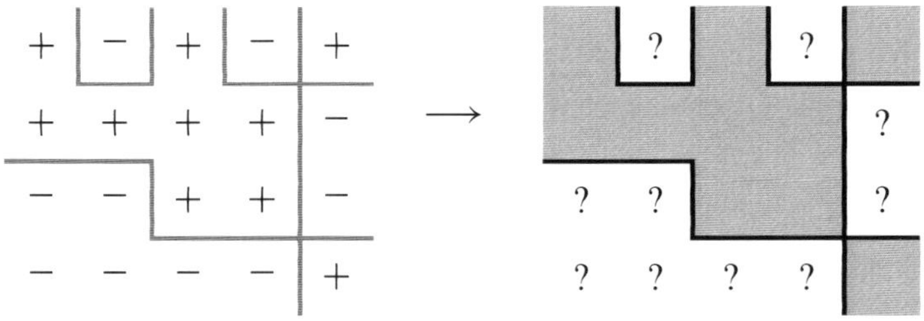

(68)

17 As a side remark, the inequality in (67) is, in fact, an equality for $\boldsymbol{\mu}_+$ and this is how Onsager's formula (71) for magnetization for $d = 2$ was originally derived.

The Ising model thus becomes a *random cluster model*, the random object in which is a random partition of lattice vertices into clusters.[18] Such random cluster models play a very important role in mathematical physics and are closely related to various *percolation* models, see **[16]**. In a percolation model, the edges of a lattice, or of a more general graph, are kept or erased with some probabilities and the connected pieces of what remains are called the percolation clusters.

The analog of a magnetization for a random cluster model is the probability that two *neighboring* vertices v–v′ belong to the same cluster. A closely related quantity is $\langle\sigma(\mathsf{v})\sigma(\mathsf{v}')\rangle_{\boldsymbol{\mu}}$, where v–v′ is an edge of the lattice. By an analysis reminiscent of how (56) is deduces from (57), the authors of Theorem 2 show:

$$\begin{matrix} \langle\sigma(\mathsf{v})\sigma(\mathsf{v}')\rangle_{\boldsymbol{\mu}_+} = \langle\sigma(\mathsf{v})\sigma(\mathsf{v}')\rangle_{\boldsymbol{\mu}_{\text{free}}} \\ \text{for an edge } \mathsf{v}\text{–}\mathsf{v}' \end{matrix} \quad \Rightarrow \quad \begin{matrix} \text{the random cluster models} \\ \text{for } \boldsymbol{\mu}_+ \text{ and } \boldsymbol{\mu}_{\text{free}} \text{ are equal.} \end{matrix} \tag{69}$$

Recall that the equality of the random cluster models implies the equality (65) for all T.

4.4.7.

It "only" remains to show that

$$\big\langle\sigma(\mathsf{v})\sigma(\mathsf{v}')\big\rangle_{\boldsymbol{\mu}_+} - \big\langle\sigma(\mathsf{v})\sigma(\mathsf{v}')\big\rangle_{\boldsymbol{\mu}_{\text{free}}} = 0 \tag{70}$$

for one edge v–v′. And this is where the *real* ascent or perhaps even flight begins and our excursion wraps up.

We will just say that the authors of **[2]** estimate the left-hand side in (70) using a certain auxiliary percolation model, the edges in which are kept or erased by a procedure that takes its input from the Ising model or, more precisely, from the *random current* representation of the Ising model. This random current representation may be compared and contrasted with the high-temperature expansion from Section 3.2.

Recall how we talked in the beginning about a mathematician's freedom do introduce and use any auxiliary mathematical structure that may shed new light on the question at hand. Just like a geometer is free to introduce any auxiliary construct, a mathematical physicist is free to introduce any auxiliary model, limited only by one's own imagination. While there is no physical percolation happening in the Ising model, one can learn a great deal about the Ising model from the percolation model studied in **[2]**.

4.4.8.

The proof of Theorem 1 is very different and is based on certain highly nontrivial *exact* results for the square lattice $d = 2$ Q-state Potts model. From the early days of the Ising model to the modern heights of Theorem 1, exact results played a very important role in the development of statistical physics, and mathematical physics in general.

---

18 In specialized literature, the term *random cluster model* often refers to a particular class of models that are related to the Ising clusters by a further random refinement, see **[16]**.

For instance, the continuity in the $d = 2$ Ising model case follows at once from the celebrated formula of Onsager (see, e.g., [3,4] for a historical account)

$$\langle\sigma(\mathrm{v})\rangle_{\mu_+} = \left(1 - \frac{1}{\sinh^4(2/T)}\right)^{\frac{1}{8}}, \quad T \le T_c, \tag{71}$$

where the formula (55) for the critical temperature is obtained solving the equation

$$\sinh(2/T_c) = 1.$$

The plot of this function can be seen in the figure of (72).

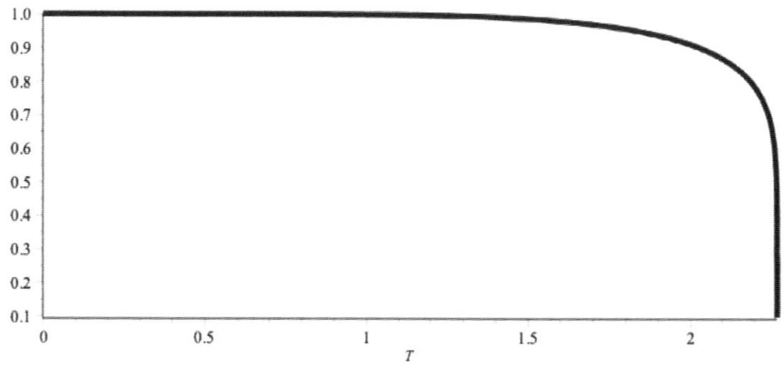

(72)

In addition to the continuity, we observe the remarkable fact that magnetization behaves like $(T_c - T)^{\frac{1}{8}}$, a result that found a deep explanation in conformal field theory.[19]

4.4.9.

Ultimately, the algebraic structure responsible for exact computations is a certain infinite-dimensional symmetry algebra present in the Q-state Potts model on the square lattice. It extends to the discrete lattice level the infinite-dimensional symmetries of the CFT limit—a most remarkable phenomenon. The algebra in question is a q-deformation of the Lie algebra of 2×2 matrices with entries in Laurent polynomials in one variable. The parameter q of this deformation is related to the parameter Q by

$$q + q^{-1} = -\sqrt{Q}.$$

Hence the difference between $Q > 4$ and $Q \le 4$ is the difference between q being negative real and q being a complex number on the unit circle $|q| = 1$.

While this is a strikingly beautiful story in mathematics, it does not really belong in our narrative, with our focus on the amazing $d = 3$ breakthrough achieved in a total absence of exact results. We do suggest, however, that the interested reader opens [8] for a general introduction to quantum group with a view towards their applications, [26,29] for classic treatments by some of the key figures in the development of the subject, and maybe also [25] for a representation-theoretic take on the origin of the structures used in the $Q \le 4$ part of

19 Numerical bootstrap computations [21] predict that in $d = 3$ the magnetization behaves like $(T_c - T)^{\beta}$ where $\beta = 0.326419\ldots$ This is a much more robust and universal number than the critical temperature, but a good mathematical understanding of it awaits future generations of mathematical physicists.

the proof. All of this, of course, in addition to the brilliant exposition of the actual proofs in [16].

Asked whether he likes exact results or estimates better, Hugo Duminil-Copin says: *"I prefer estimates. They usually offer a more robust approach to critical phenomena, and I am as much as possible trying to obtain proofs that are not based on exact formulae."*

"I certainly agree with Hugo, inequalities are more versatile. But I am not sure we would have advanced so far without having some exact identities first. It is a miracle that there are any equalities concerning the Ising model. Onsager's calculation was shocking at the time, as it provided an exact formula for a function exhibiting a phase transition. This and later miraculous equalities, together with inequalities, forged our understanding of the Ising model. Hugo is a master of both inequalities and equalities, and has really moved the frontier of statistical physics with many beautiful theorems with equally beautiful proofs. I congratulate Hugo from all my heart, bravo!", says Stanislav Smirnov.

"Hugo Duminil-Copin's work has brought unprecedented clarity to our mathematical understanding of phase transitions in statistical mechanics. The elegance of his proofs truly makes them seem come straight out of The Book", says Martin Hairer.

5. FURTHER READING

Popular accounts of these and related developments include [9, 34]. See especially the popular piece [15] written by Hugo Duminil-Copin for the Oberwolfach's snapshots of modern mathematics.

We quoted many times from [16] and an interested reader is certainly advised to continue her or his exploration of the subject following these lectures. Among other survey articles written by Hugo Duminil-Copin, one may list [13, 14, 17].

To anyone who can read French, Hugo Duminil-Copin wholeheartedly recommends the lectures [27] by Jean-François Le Gall and the book [33] by Wendelin Werner. Another very important book is the subject is [24] by Geoffrey Grimmett.

I hope the reader has a lot of fun studying these sources as well as the original articles including [2].

A. THE UNIVERSAL ATTRACTION OF THE ISING MODEL

A.1. Universality

It is hard to tell the atomic composition of a liquid by watching it evaporate or freeze. There is a good reason it took humans millenia to figure out the microscopic composition of macroscopic objects. Part of the reason is that a great many different microscopic systems have the same macroscopic behavior.

Molecules live on a nanometer (that is, 10^{-9}m) scale and there is incredibly many of them in a macroscopic piece of any material ($18 = 2 + 16$ grams of H_2O contain about $6 \cdot 10^{23}$, the Avogadro number, of molecules). It sounds completely impossible that their individual behavior could be observed by us. Instead, we observe only the combined, or averaged, effect of myriads of molecules.

For instance, if we have a container of gas, we can measure the density, the pressure, the temperature, etc. These measure the average number of molecules[20] per unit of volume, the average force per unit area exerted by the gas on the wall of the container, and the average kinetic energy of molecules, respectively. In principle, we could measure more quantities, but the equation of state (an important concept in statistical physics) tells that temperature and pressure are enough. Add to this the vector of the wind, and there is no further information about the gas that a weather station can provide.

Mathematically, what does it mean that there is no further information? Recall the concept of a Gibbs measure from Section 3. It assigns a probability to every event one can detect and hence an average, or expected value, to any observable quantity. If some finite number of these expectations already determine the whole Gibbs measure then they determine the outcome of *every* possible measurement in our system.

Going back to the gas, on a macroscopic scale, it is described by 2 scalars, temperature and pressure, and one vector, wind velocity. These vary in space and time if the gas is not in global equilibrium and are the ingredients in the mathematical models of motion of gases and fluids. In some sense, the job of a statistical physicist is to provide the arrow

$$\boxed{\text{description on the 1 nm scale}} \xrightarrow{\text{statistical physics}} \boxed{\text{description on the 1 m scale}}, \tag{73}$$

connecting two very different kind of physics, and two communities of mathematical physicists studying the corresponding phenomena using very different models and mathematical tools. Since the source and the target in (73) are so very different, it is impossible for the target to be a faithful image of the source. To reiterate, we cannot tell the atomic composition of air just by feeling a cool breeze. Put differently, an enormous amount of information is discarded by the arrow (73).

This loss of information is a win for a statistical physicist. It means there is no pressing need to study every possible scenario of microscopic interactions. People call it *universality*, meaning the macroscopic conclusions should hold universally and independently of most microscopic details. Within each universality class, it is thus reasonable to restrict our attention to the simplest possible microscopic model.

Universality is a very important ingredient in how statistical physicists think about their subject. To be clear, it is always an enormous mathematical challenge to prove any universality statement rigorously. However, there is an appealing heuristic description of the universality classes based on the *renormalization group* idea of Kenneth Wilson. We will say a word about it below.

---

20 If we have a mixture of several gases, there will be separate densities for each kind of molecules. These can be traded for the corresponding partial pressures.

A.2. Models like the Ising model

A.2.1.

How can stuff fluctuate in space? We should have some fluctuating degrees of freedom, which we may describe by an N-tuple of numbers

$$\phi(x) = \big(\phi_1(x), \phi_2(x), \dots, \phi_N(x)\big) \in \mathbb{R}^N.$$

Here, the argument x is a d-dimensional vector, which we will discretize to a lattice $\Lambda \subset \mathbb{R}^d$. It is good to visualize Λ as a fine mesh $\varepsilon\mathbb{Z}^d \subset \mathbb{R}^d$ approximating the space $\mathbb{R}^d$ in the *continuous* limit $\varepsilon \to 0$.

In parallel with (14), we can write $\phi(x)$ as a random function

$$\Lambda \xrightarrow{\ \phi\ } \mathbf{\Phi} \subset \mathbb{R}^N, \tag{74}$$

where $\mathbf{\Phi}$ is the range of the possible values of ϕ. In the Ising model, for instance, $N = 1$ and $\mathbf{\Phi} = \{\pm 1\}$.

A.2.2.

The interactions are described by an energy function, such as the energy (15) in the Ising model. In general, one imagines

$$\text{Energy} = \text{External potential} + \text{Pair potential} + \cdots, \tag{75}$$

where dots stand for other possible interactions. We will assume (75) is translation-invariant. Then the first term has the form

$$\text{External potential} = \sum_{x\in\Lambda} U_1\big(\phi(x)\big), \tag{76}$$

for some function U_1 on $\mathbf{\Phi}$ in (74). In the continuous $\varepsilon \to 0$ limit, the sum in (76) becomes the integral of $U_1(\phi)$.

In the Ising model, one can add such term. This is called Ising model in an *external field*. It breaks the ± 1 symmetry and destroys the critical point. It is very interesting, however, to study the response of the critical Ising model to a small external field.

If (76) is the only nonzero term in (75), then from (9) we conclude that the values of $\phi(x)$ are independent identically distributed N-dimensional random variables with probability density function proportional to $e^{-U_1(\phi)/T}$. In space, this is a complete noise, with some nontrivial distributions of values, hence not something of great interest to us now.

A.2.3.

A translation-invariant pair potential has the form

$$\text{Pair potential} = \sum_{x,y\in\Lambda} U_2\big(\phi(x), \phi(y), x - y\big). \tag{77}$$

This term puts spatial interactions in (75). We may assume each term in (77) is $x \leftrightarrow y$ symmetric.

For the cubic lattice Ising model, U_2 vanishes unless $x - y = \pm e_i$, where e_i are the coordinate vectors. In principle, one can allow next-nearest neighbors to interact, as well as lattice sites further away.[21] It is important, however, for the interaction to decay rapidly with the distance between x and y. Models in which everything interacts with everything behave like a crowd and are usually well-described by the crowd average $\overline{\phi}$ fluctuating in some potential $U_{\text{effective}}(\overline{\phi})$ derived from U_1 and U_2.

For a pair of neighbors v, v' in the Ising model, we can write

$$-\sigma(\mathsf{v})\sigma(\mathsf{v}') = -1 + \frac{1}{2}\frac{|\sigma(\mathsf{v}) - \sigma(\mathsf{v}')|^2}{\|\mathsf{v} - \mathsf{v}'\|^2} \tag{78}$$

because $\mathsf{v} - \mathsf{v}'$ is a unit vector and σ takes values ± 1. The fraction on the right in (78) is a lattice version of the square of the derivative of σ in the direction of $\mathsf{v} - \mathsf{v}'$. Since an overall shift of energy does nothing, we see that the pair energy in the Ising model can be written as a discretization of $\frac{1}{2}\|\nabla\sigma\|^2$, where ∇ denotes the gradient of the function.

In general, we may think of (74) as of discretization of a map $\mathbb{R}^d \to \mathbf{\Phi}$, like in the figure of (79).

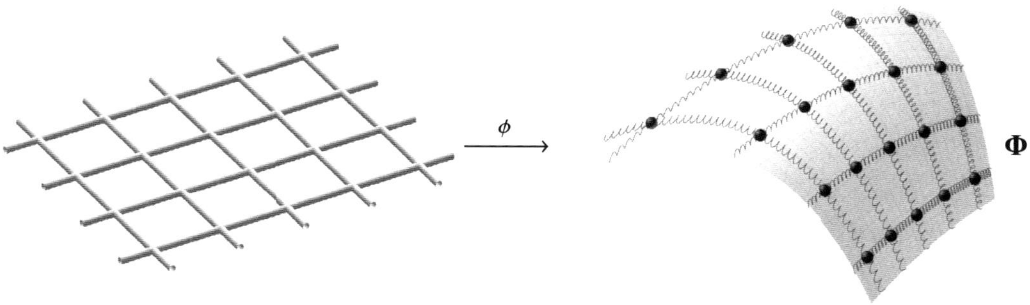

(79)

The role of the pair potential is to hold the values of this map together by putting an energy price on wild oscillations. One natural notion of energy for a continuous map is the Dirichlet energy $\frac{1}{2}\int \|\frac{\partial\phi}{\partial x}\|^2$, the construction of which in general requires a metric in the domain and target of ϕ. The anisotropic and anharmonic relatives of the Dirichlet energy are certainly possible and important in the description of materials with the corresponding properties.

A.3. Critical points

A.3.1.

Wilson's idea[22] was that the arrow in (73) can be presented as a composition of many similar arrows that each change the scale by modest factor, such as 2,

$$\begin{array}{ccccccccccc} 2^{-30}\,m & \longrightarrow & & & & & & & & & 1\,m \\ \| & & & & & & & & & & \| \\ 2^{-30}\,m & \longrightarrow & 2^{-29}\,m & \longrightarrow & 2^{-28}\,m & \longrightarrow & \cdots & \longrightarrow & \cdots & \longrightarrow & 1\,m. \end{array} \tag{80}$$

21 In fact, the authors of Theorem 2 prove it in much wider generality than described in these notes.

22 Like any fundamental idea in science, this one had many precursors in the work of many people. See, e.g., [7, 22, 28, 31, 35–37] for various perspectives.

Stepping off the firm mathematical grounds for the rest of this section, we may hypothesize that the change of scale by 2 corresponds to some *renormalization* transformation

$$(U_1, U_2, U_3, \dots) \xrightarrow{R} (U_1', U_2', U_3', \dots), \tag{81}$$

where U_3 corresponds to possible triple interactions in (75), etc. Since the arrow in (73) is the transformation R raised to some very large power, we should put two theories in the same universality basket if they become identical after many iterations of (81).

A.3.2.

While heuristic, this argument underscores the importance of *scale-invariant* models. If there really was a well-defined transformation (81), such theories will be its fixed points. It makes sense that the result in (73) is scale invariant, since we certainly expect the same macroscopic description to be valid at both the $1m$ and $2m$ scales.

The invariance here should be understood up to redefinition of the fields. Indeed, if $\phi_1(x)$ is measured in meters then R should act on it by $\phi_1(x) \mapsto 2^{-1}\phi_1(2x)$. In general, if

$$\phi_i(x) \xrightarrow{R} 2^{-\Delta_i}\phi_i(2x),$$

then the number Δ_i is called the scaling dimension of ϕ_i. For a lattice model, scale-invariance means scale-invariance of the mesh $\varepsilon \to 0$ limit, in which we rescale the fields ϕ_i by ε^{Δ_i}. It is this limit that we actually observe on the macroscopic scale. See [7] for a superb exposition of scaling and renormalization.

Near a fixed point of R, we have much better chance of understanding what R does. Many nearby theories will be attracted back to the fixed points by repeated applications of R. These should be put in the same universality class.

A.3.3.

The critical Ising model should be scale-invariant. For $T \neq T_c$, there is a microscopic scale in the model set by the scale at which the signs exponentially decorrelate. At $T = T_c$, this becomes infinite and scale-invariance should appear. Currently, there is no mathematical proof of this for $d = 3$ and it remains an important open problem. Numerical experiments [21] give

$$\Delta_{\sigma,3} = 0.518154\dots$$

as the scaling dimension of the spin field in $d = 3$.

A.3.4.

Importantly, for $d = 2$ and $d = 3$, this is not a Gaussian fixed point. A Gaussian random field is a generalization of a Gaussian process with d-dimensional time. For a Gaussian field, $\mathbf{\Phi} = \mathbb{R}^N$ and the functions U_1 and U_2 are quadratic. In suitable coordinates, the field thus becomes a superposition of many noninteracting Gaussian random variables. While certainly a very, very important part of probability theory and mathematical physics, Gaussian fields *do not* describe materials with nonlinear interactions.

By contrast, Theorem 3 implies that for $d \geq 4$, the critical Ising model is Gaussian and $\Delta_{\sigma,4} = 1$. It is a very difficult and important mathematical theorem to prove, and it underscores the crucial importance of dimensionality in statistical mechanics.

A.3.5.

Which other microscopic models will fall into the critical Ising fixed point? The crucial feature of the Ising model is the ± 1 symmetry between the two possible values of $\sigma(\mathsf{v})$. One should expect that any potential U_1 which has two symmetric minima, like the function in (82),

$U_1(\phi) =$ 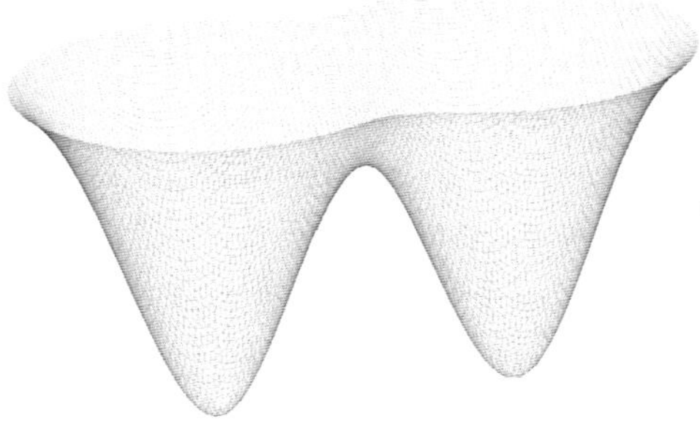 , (82)

has the same critical scaling limit. For $T < T_c$, there should be Gibbs measures that prefer to stay close to one of the minima in (82), thus breaking the symmetry. For $T > T_c$, one expects a unique symmetric Gibbs measure.

While the details of these measures, as well as the values of the critical temperature T_c, will depend U_1 and U_2, the large-scale fluctuations of the $T = T_c$ measure (as captured by the the mesh $\varepsilon \to 0$ limit of the lattice model) should probably be universal and the same as for the critical Ising model. In other words, the critical Ising model should give the universal description of the transition between the broken and unbroken ± 1 symmetry.

In light of Theorem 1, the reader may wish to contemplate what happens if the potential U_1 has 3 minima

$U_1(\phi) =$ 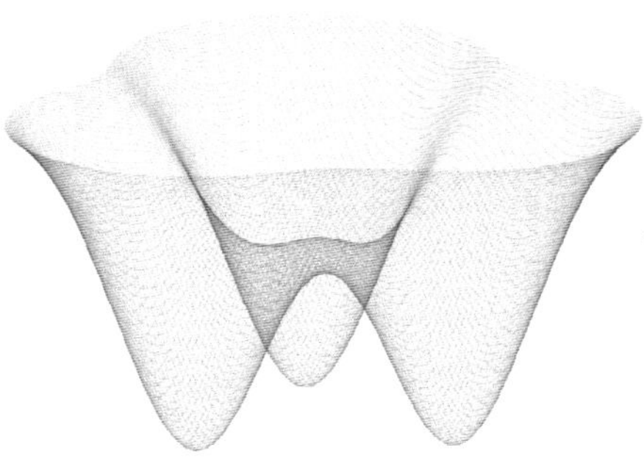 , (83)

which can be permuted in all possible ways by the symmetries of the theory.

REFERENCES

[1] M. Aizenman and H. Duminil-Copin, Marginal triviality of the scaling limits of critical 4D Ising and φ_4^4 models. *Ann. of Math. (2)* **194** (2021), no. 1, 163–235.

[2] M. Aizenman, H. Duminil-Copin, and V. Sidoravicius, Random currents and continuity of Ising model's spontaneous magnetization. *Comm. Math. Phys.* **334** (2015), no. 2, 719–742.

[3] R. J. Baxter, Onsager and Kaufman's calculation of the spontaneous magnetization of the Ising model. *J. Stat. Phys.* **145** (2011), no. 3, 518–548.

[4] R. J. Baxter, Onsager and Kaufman's calculation of the spontaneous magnetization of the Ising model: II. *J. Stat. Phys.* **149** (2012), no. 6, 1164–1167.

[5] V. Beffara and H. Duminil-Copin, The self-dual point of the two-dimensional random-cluster model is critical for $q \geq 1$. *Probab. Theory Related Fields* **153** (2012), no. 3–4, 511–542.

[6] T. Bodineau, Translation invariant Gibbs states for the Ising model. *Probab. Theory Related Fields* **135** (2006), no. 2, 153–168.

[7] J. Cardy, *Scaling and renormalization in statistical physics*. Camb. Lect. Notes Phys. 5, Cambridge University Press, Cambridge, 1996.

[8] V. Chari and A. Pressley, *A guide to quantum groups*. Cambridge University Press, Cambridge, 1995. Corrected reprint of the 1994 original.

[9] N. Curien, Hugo Duminil-Copin et les transitions de phase. *Images des Mathématiques, CNRS* (2017). https://images.math.cnrs.fr/Hugo-Duminil-Copin-et-les-transitions-de-phase.

[10] R. L. Dobrushin, An investigation of Gibbs states for three-dimensional lattice systems. *Teor. Veroyatn. Primen.* **18** (1973), 261–279 (Russian, with English summary).

[11] R. L. Dobrushin and S. B. Shlosman, The problem of translation invariance of Gibbs states at low temperatures. In *Mathematical physics reviews*, pp. 53–195, Sov. Sci. Rev. Sect. C: Math. Phys. Rev. 5, Harwood Academic Publ., Chur, 1985.

[12] H. Duminil-Copin, Divergence of the correlation length for critical planar FK percolation with $1 \leq q \leq 4$ via parafermionic observables. *J. Phys. A* **45** (2012), no. 49, 494013, 23.

[13] H. Duminil-Copin, Random currents expansion of the Ising model. In *European Congress of Mathematics*, pp. 869–889, Eur. Math. Soc., Zürich, 2018.

[14] H. Duminil-Copin, Sixty years of percolation. In *Proceedings of the International Congress of Mathematicians—Rio de Janeiro 2018. Vol. IV. Invited lectures*, pp. 2829–2856, World Sci. Publ., Hackensack, NJ, 2018.

[15] H. Duminil-Copin, Counting self-avoiding walks on the hexagonal lattice. In *Snapshots of modern mathematics from Oberwolfach*, 2019. https://www.imaginary.org/sites/default/files/snapshots/snapshots-2019-006.pdf.

[16] H. Duminil-Copin, Lectures on the Ising and Potts models on the hypercubic lattice, Random graphs, phase transitions, and the Gaussian free field. In *Springer Proc. Math. Stat.*, pp. 35–161, 304, Springer, Cham, 2020.

[17] H. Duminil-Copin, 100 Years of the (critical) Ising Model on the Hypercubic Lattice. In *Proceedings of the International Congress of Mathematicians, Vol. 1*, pp. 164–210, EMS Press, 2022.

[18] H. Duminil-Copin, M. Gagnebin, M. Harel, I. Manolescu, and V. Tassion, Discontinuity of the phase transition for the planar random-cluster and Potts models with $q > 4$. *Ann. Sci. Éc. Norm. Supér. (4)* **54** (2021), no. 6, 1363–1413 (English, with English and French summaries).

[19] H. Duminil-Copin, V. Sidoravicius, and V. Tassion, Continuity of the phase transition for planar random-cluster and Potts models with $1 \leq q \leq 4$. *Comm. Math. Phys.* **349** (2017), no. 1, 47–107.

[20] H. Duminil-Copin and S. Smirnov, Conformal invariance of lattice models. In *Probability and statistical physics in two and more dimensions*, pp. 213–276, Clay Math. Proc. 15, Amer. Math. Soc., Providence, RI, 2012.

[21] S. El-Showk, M. F. Paulos, D. Poland, S. Rychkov, D. Simmons-Duffin, and A. Vichi, Solving the 3d Ising model with the conformal bootstrap II. c-minimization and precise critical exponents. *J. Stat. Phys.* **157** (2014), no. 4–5, 869–914.

[22] M. Fisher, The renormalization group in the theory of critical behavior. *Rev. Modern Phys.* **46** (1974), no. 4, 597.

[23] C. M. Fortuin, P. W. Kasteleyn, and J. Ginibre, Correlation inequalities on some partially ordered sets. *Comm. Math. Phys.* **22** (1971), 89–103.

[24] G. Grimmett, *Percolation*. 2nd edn., Grundlehren Math. Wiss. 321, Springer, Berlin, 1999.

[25] Y. Ikhlef, R. Weston, M. Wheeler, and P. Zinn-Justin, Discrete holomorphicity and quantized affine algebras. *J. Phys. A* **46** (2013), no. 26.

[26] M. Jimbo and T. Miwa, *Algebraic analysis of solvable lattice models*. CBMS Reg. Conf. Ser. Math. 85, Published for the Conference Board of the Mathematical Sciences, Washington, DC; American Mathematical Society, Providence, RI, 1995.

[27] J.-F. Le Gall, *Intégration, Probabilités et Processus Aléatoires*. 2006. https://www.imo.universite-paris-saclay.fr/~jflegall/IPPA2.pdf.

[28] H. Maris and L. Kadanoff, Teaching the renormalization group. *Am. J. Phys.* **46** (1978), no. 6, 652–657.

[29] N. Reshetikhin, Lectures on the integrability of the six-vertex model. In *Exact methods in low-dimensional statistical physics and quantum computing*, pp. 197–266, Oxford Univ. Press, Oxford, 2010.

[30] M. Rukeyser, *Willard Gibbs*. Doubleday, Doran and Company, New York, 1942.

[31] D. Shirkov, Fifty years of the renormalization group. *CERN Cour.* **41** (2001), 19–22.

[32] S. Smirnov, Discrete complex analysis and probability. In *Proceedings of the International Congress of Mathematicians. Volume I*, Hindustan Book Agency, New Delhi, 2010.

[33] W. Werner, *Percolation et modèle d'Ising*. Cours Spéc. 16, Société Mathématique de France, Paris, 2009(French).

[34] A. Whitten, Mathematicians Prove Symmetry of Phase Transitions. *Quanta Mag.* (July 8 2021). https://www.quantamagazine.org/mathematicians-prove-symmetry-of-phase-transitions-20210708/.

[35] K. Wilson, Renormalization group and critical phenomena. I. Renormalization group and the Kadanoff scaling picture. *Phys. Rev. B* **4** (1971), no. 9, 3174.

[36] K. Wilson, Renormalization group and strong interactions. *Phys. Rev. D* **3** (1971), no. 8, 1818.

[37] J. Zinn-Justin, Renormalization and renormalization group: From the discovery of UV divergences to the concept of effective field theories. In *Quantum Field Theory: Perspective and Prospective*, pp. 375–388, Springer, 1999.

ANDREI OKOUNKOV

Department of Mathematics, University of California, Berkeley, 970 Evans Hall Berkeley, CA 94720–3840, USA, okounkov@math.columbia.edu

COMBINATORIAL GEOMETRY TAKES THE LEAD

ANDREI OKOUNKOV

ABSTRACT

While the author is a professional mathematician, he is by no means an expert in the subject area of these notes. The goal of these notes is to share the author's personal excitement about some results of June Huh with mathematics enthusiasts of all ages, using maximally accessible, yet precise mathematical language. No attempt has been made to present an overview of the current state field, its history, or to place this narrative in any kind of broader scientific or social context. See the references in Section 9 for both professional surveys and popular science accounts that will certainly give the reader a broader and deeper understanding of the material.

MATHEMATICS SUBJECT CLASSIFICATION 2020

Primary 05B35; Secondary 97

KEYWORDS

Matroids

INTERNATIONAL CONGRESS OF MATHEMATICIANS
2022 JULY 6–14

© 2022 International Mathematical Union

Proc. Int. Cong. Math. 2022, Vol. 1, pp. 414–458
DOI 10.4171/ICM2022/201

Published by EMS Press

and licensed under
a CC BY 4.0 license

1. POINTS, LINES, AND PLANES

Points and lines are the simplest geometric shapes and really primordial mathematical objects. Euclid opens his *Elements* by giving a definition of a point and a line, and his first postulate is that one can draw a straight line from any point to any point. While the need and standards for precise definitions in mathematics have only grown in the past $2.3 \cdot 10^3$ years, we imagine the reader has a good enough informal or formal grasp on lines and points to skip the definitions and focus on the basic geometric fact that two distinct points P_1 and P_2 determine a unique line through them. Somewhat unconventionally, we will denote this line $P_1 \vee P_2$.

While two points always lie on a line, three points P_1, P_2, P_3 may or may not be on a line. As we move the points around, the point P_3 is *typically* or *generically* not on the line $P_1 \vee P_2$, but in *special* cases it may be. The italicized words are important mathematical notions; we hope their meaning is intuitively clear.

Suppose we have $n = 3, 4, \ldots$ distinct points $P_1, \ldots, P_n$ in the plane, not all of them on same line. Generically, no three of these points will be on the same line, meaning that all lines $P_i \vee P_j$ will be distinct. Their number is thus the number of unordered pairs of numbers from $\{1, \ldots, n\}$, which can be computed as follows:

$$\# \text{ lines} = \frac{n(n-1)}{2} = 3, 6, 10, 15, \ldots, \quad n = 3, 4, 5, 6, \ldots$$

See the figure in (1) for an illustration for $n = 7$. In particular, $n \geq 3$ generic points in the plane always determine n or more lines.

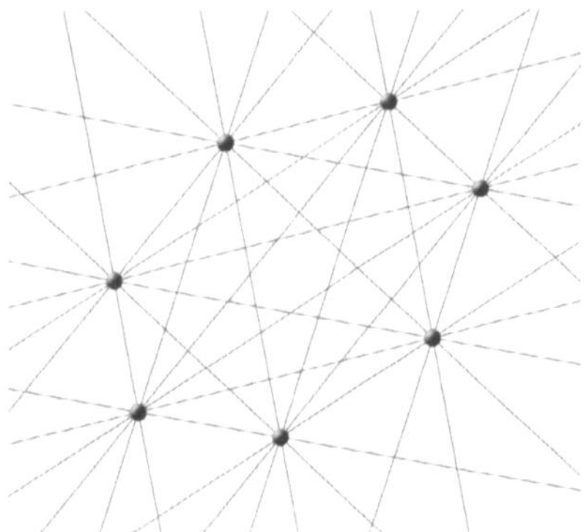

(1)

Let us see how the number of lines changes if we move the points into a special position. For example, let us put $n - 1$ of them on a line, as in the figure of (2). In this case, we get n lines, so again at least as many lines as points.

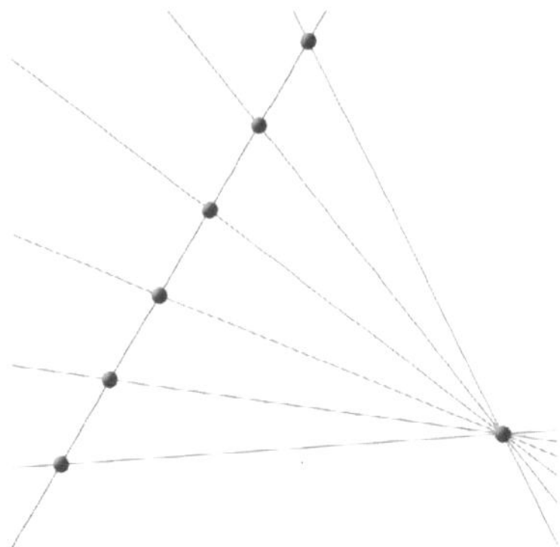

(2)

In general, it is classic result of de Brujin and Erdős [13] from 1948 that the number of lines determined by n points in the plane is at least n, unless all points lie on a single line.

Book 11 of the Elements opens with the definition of a solid, and Euclid proceeds with the development of the 3-dimensional geometry. Instead of *the* plane which previously contained the points $P_1, \dots, P_n$, in three dimensions there are many planes and any triple of points P_i, P_j, P_k, not contained in a line, determines a unique plane $P_i \vee P_j \vee P_k$ that meets them.

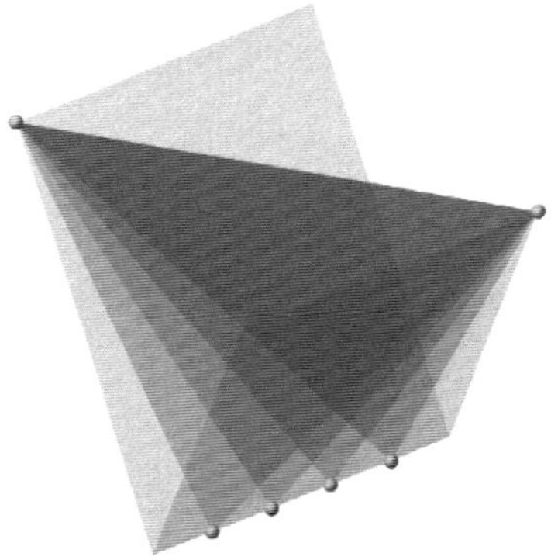

(3)

It is possible for n points to determine exactly n planes; see the figure in (3) in which all but two points lie on a line. Theodore Motzkin [36] showed in 1951 that this indeed is the minimal possible number of planes.[1]

2. POINTS, LINES, PLANES, ETC.

It took a long time since Euclid for mathematicians and scientists to realize that it is both natural and important to study d-dimensional geometry for general $d = 1, 2, 3, 4, 5, \dots$ A simple clear mathematical language of coordinates

$$\mathbb{R}^d = \{d\text{-tuples } (x_1, \dots, x_d) \text{ of real numbers}\} \tag{4}$$

to describe the real d-dimensional space $\mathbb{R}^d$ was introduced in the 17th century by Fermat and Descartes. Tuples of numbers, sometimes with very large d, are abundant in both theoretical and applied contexts. But it was not until much later, less than 200 years ago, that the necessity and advantages of thinking about such d-tuples geometrically was realized.

A plane in a 3-dimensional space $\mathbb{R}^3$ is described by a linear equation

$$a_0 + a_1 x_1 + a_2 x_2 + a_3 x_3 = 0, \tag{5}$$

in which at least one of the coefficients a_1, a_2, or a_3 is not zero. Two sets of coefficients (a_0, a_1, a_2, a_3) and (a'_0, a'_1, a'_2, a'_3) determine the same plane if and only if

$$\begin{aligned}(a'_0, a'_1, a'_2, a'_3) &= c(a_0, a_1, a_2, a_3) \\ &= (ca_0, ca_1, ca_2, ca_3),\end{aligned} \tag{6}$$

1 In fact, Motzkin first conjectured this in his 1936 PhD thesis [35]!

for some nonzero number $c \neq 0$. Of course, multiplying an equation by a nonzero number does not change its solutions.

A line in $\mathbb{R}^3$ is an intersection of two planes, thus the set of solution of a system of 2 linear equations. There are infinitely many planes containing a given line, and we can pick any two among them. In terms of the equations, this means that many transformations of a system of equations preserve their solutions. For instance, we can add to one of the equations any multiple of another equation.

Finally, a point P in $\mathbb{R}^3$ is a solution of 3 linear equations, which we can choose to have the confusingly simple form

$$x_i = \text{the } i\text{th coordinate of } P, \quad i = 1, 2, 3. \tag{7}$$

In exactly the same fashion, a linear equation in $\mathbb{R}^d$ is said to determine a *hyperplane*, and points, lines, and *flats* of all other dimensions are described as the intersection of the corresponding number of hyperplanes, that is, as solutions of systems of linear equations. There is hardly anything more basic and fundamental in mathematics, science, technology, data analysis, etc., than systems of linear equations. It is very likely that many, or most, readers of these notes have met them before. Those who would like a reminder or an explanation will find it in Appendix A.

The basic geometric facts like:

$$\begin{aligned}
&2 \text{ points } P_1, P_2, \text{ when distinct, lie on a unique line } P_1 \vee P_2,\\
&3 \text{ points } P_1, P_2, P_3, \text{ not contained in a line, lie in a unique plane } P_1 \vee P_2 \vee P_3,\\
&\dots\\
&r \text{ points } P_1, \dots, P_r, \text{ not contained in a } (r-2)\text{-dimensional flat,}\\
&\quad \text{lie in a unique } (r-1)\text{-dimensional flat } P_1 \vee P_2 \vee \dots \vee P_r
\end{aligned} \tag{8}$$

continue to hold in any dimension d. The minimal flat containing some points $P_1, \dots, P_k$ will be denoted $P_1 \vee P_2 \vee \dots \vee P_k$ and called the *span* of these points.

It is natural to ask how many flats of each dimension can n points in $\mathbb{R}^d$ determine. Since it takes r points to determine an $(r-1)$-dimensional flat, we will define the *rank* of such flat to equal r.

For instance, n generic points $P_1, P_2, \dots, P_n$ determine

$$\binom{n}{2} \text{ lines, } \binom{n}{3} \text{ planes, } \dots, \binom{n}{r} \text{ rank } r \text{ flats, } \dots \tag{9}$$

because we can choose so many r-element subsets from an n-element sets. Here

$$\binom{n}{r} = \frac{n!}{r!(n-r)!}, \quad n! = 1 \cdot 2 \cdot 3 \cdots (n-1) \cdot n,$$

denote the binomial coefficients $\binom{n}{r}$ and the *factorial* $n!$ of n, respectively. Factorials and binomial coefficients are as fundamental to mathematics and as ancient as points and lines, appearing in very old Indian, Persian, and Chinese texts long before becoming known in Europe in late Renaissance.

Many beautiful elementary properties of the binomial coefficients inspire combinatorialists to look for similar patterns in other, more complicated, sequences of numbers. In (9) we get some initial segment of the binomial coefficients, which look as follows for $d = 6$, $n = 10$ and $r = 1, \dots, d$:

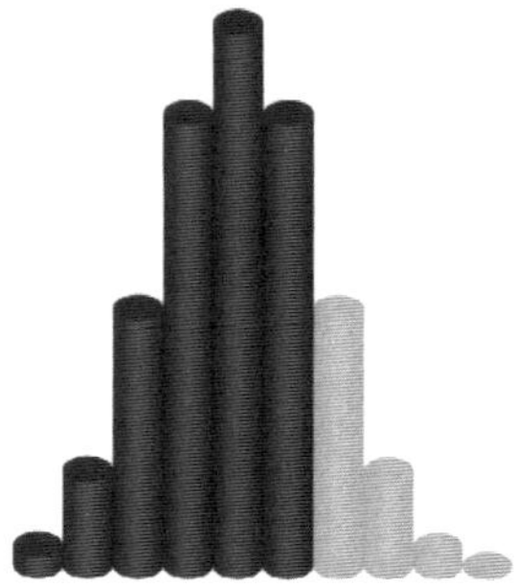

(10)

The transparent columns here represent the unused binomial coefficients with $r > d$. Two properties of this sequence of numbers are apparent. First, it is *unimodal*, that is, the numbers first increase and then decrease. Second, it is *top-heavy*, which can be quantified as

$$\binom{n}{r} \le \binom{n}{d-r}, \quad \text{provided } 2r \le d \le n. \tag{11}$$

So far, this was about n generic points in $\mathbb{R}^d$. Now let us allow the points P_1, P_2, $\dots$, P_n to be in some special position (and there are a great many ways in which a point configuration can be special for n large). Let $\mathscr{F}_r$ denote the set of rank r flats determined by the P_i's. In particular,

$$\mathscr{F}_1 = \{P_1, \dots, P_n\}, \tag{12}$$

while $\mathscr{F}_2$ are the lines in $\mathbb{R}^d$ containing at least two of the P_i's. Generalizing what we have seen for generic points, Rota [39] conjectured the *unimodality* of the sequence $|\mathscr{F}_r|$, where $|\mathscr{F}_r|$ denotes the numbers of elements, or *cardinality*, of $\mathscr{F}_r$. Dowling and Wilson [15,16] conjectured that the sequence $|\mathscr{F}_r|$ is top-heavy. These questions remained open for a very long time, but now the top-heavy conjecture and the increasing part of the unimodality conjectures are proven as a corollary of a theorem of June Huh and Botong Wang that will be discussed in the next section.

Why is the top-heavy conjecture so interesting? "*It indicates a deep hidden reciprocity!*", says Gil Kalai who presented June Huh's Fields Medal laudatio at ICM 2022. June Huh says he became interested in the top-heavy conjecture as a result of being intrigued by the "top-heavy phenomena" for lower Bruhat intervals in Coxeter groups that are proved using Elias–Williamson's combinatorial Hodge theory for Soergel bimodules. A curious reader will find out what this is about in the references [9,17,23,34].

3. MATCHING FLATS TO FLATS

Suppose we want to prove that one set, such as $\mathscr{F}_r$, has fewer elements than some other set, such as $\mathscr{F}_{r'}$. These sets may be complicated and the exact counts of elements in each of them may be hard to perform. However, we may be able to prove the inequality between $|\mathscr{F}_r|$ and $|\mathscr{F}_{r'}|$ without actually doing either count. It suffices to assign to each element $F \in \mathscr{F}_r$ an element $\iota(F) \in \mathscr{F}_{r'}$ so that distinct $F_1 \neq F_2$ are assigned distinct $\iota(F_1) \neq \iota(F_2)$.

Mathematicians have special words for any procedure ι that assigns an element $\iota(F)$ of some "target" set like $\mathscr{F}_{r'}$ to an element F of some "source" set like $\mathscr{F}_r$. We say that ι is a function or a map from $\mathscr{F}_r$ to $\mathscr{F}_{r'}$ and write

$$\iota : \mathscr{F}_r \to \mathscr{F}_{r'}. \tag{13}$$

When (13) takes distinct elements to distinct elements, we say that ι is *injective* or one-to-one. An injective map between two finite sets exists if and only if the cardinality of the source is less than or equal to the cardinality of the target.

Conversely, a map is called *surjective* or onto, if every element in the target is assigned to some element of the source. A surjective map implies the opposite inequality between the cardinalities of the two sets. A schematic example of an injective and a surjective set from a set of circles to a set of stars can be seen in (14).

(14)

For an injective map, every star is the target of ≤ 1 arrows; for a surjective map, every star is the target of ≥ 1 arrows.

Since the source and target in (13) have a geometric meaning, we can ask for the map ι to reflect this geometric meaning. It is nice to require that the flat $\iota(F)$ contains the flat F for all F. We will call such assignment a *matching*. In the figure of (15), the reader can see examples of a noninjective and an injective matching between points and lines from (2).

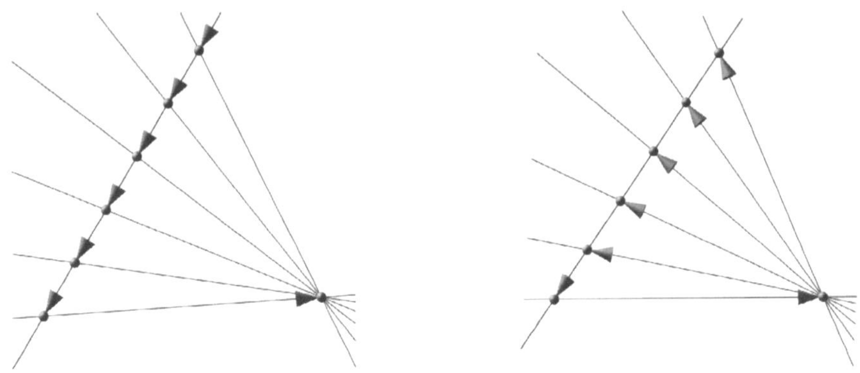

(15)

In (15), we give each point a conic tail in the direction of the matched line.

These notes are about some very general and powerful results proved by June Huh and his collaborators Tom Braden, Jacob P. Matherne, Nicholas Proudfoot, and Botong Wang. We will be stating and explaining them in ascending generality, starting with the following most basic version:

Theorem 1 ([24]). *For any n-tuple of points $P_1, \dots, P_n \in \mathbb{R}^d$ not contained in a hyperplane, there exists an injective matching* (13) *from rank r flats to rank r′ flats spanned by these points provided $r \le r'$ and $r + r' \le d + 1$.*

For instance, in the plane $d = 2$, the only interesting case is $r = 1$ and $r' = 2$. Theorem 1 then says that for every point one can chose a line containing it, in such a way that different points are assigned different lines. In $\mathbb{R}^3$ we can have $(r, r') = (1, 2)$ or $(r, r') = (1, 3)$, which means that for every point we can choose a line, and also a plane containing it. In every dimension, Greene showed that points can be matched to the hyperplanes they define [19].

It is not so clear at present what happens outside the $r + r' \le d + 1$ range, including the decreasing part of the unimodality conjecture. For instance, Dilworth and Greene constructed in [14] a configuration of 21 points in a 10-dimensional space such that there is no injective or surjective matching $\mathscr{F}_6 \to \mathscr{F}_7$.

4. RANK AND MATROIDS

Suppose that for some concrete collection $P_1, \dots, P_n$ we want a computer program to either construct or verify an injective matching described in Theorem 1. Or maybe we would like to experiment in the range $r + r' > d + 1$. Whatever our goals, we will need the program to manipulate the information about the position of the points $P_1, \dots, P_n$. It should be able to either determine or remember which subsets

$$S \subset \{P_1, \dots, P_n\}$$

of points lie on a line, in a plane, etc. So it reasonable to think that in our program there should be a procedure that either computes or looks up the function

$$\operatorname{rank}(S) = \dim \operatorname{span}(S) + 1. \tag{16}$$

It is easy to see that all other notions discussed so far can be easily expressed in terms of the function (16). For instance, a subset S corresponds to a flat of rank r if and only $\operatorname{rank}(S) = r$ and

$$\operatorname{rank}(S \cup P_i) = \operatorname{rank}(S) + 1, \quad \text{for any } P_i \notin S. \tag{17}$$

In other words, flats $F \in \mathscr{F}_r$ corresponds to subsets S of rank r that are maximal with respect to inclusion.

As we change the position of the points $P_1, \dots, P_n$, the corresponding rank functions will also change, but they will always satisfy the equalities:

$$\operatorname{rank}\big(\{P_i\}\big) = 1, \quad i = 1, \dots, n, \tag{18}$$

and the inequalities

$$\text{for any } S \subset S', \quad \operatorname{rank}(S) \le \operatorname{rank}(S'), \tag{19}$$

$$\text{for any } S_1, S_2, \quad \operatorname{rank}(S_1 \cup S_2) \le \operatorname{rank}(S_1) + \operatorname{rank}(S_2) - \operatorname{rank}(S_1 \cap S_2). \tag{20}$$

The intersection $S_1 \cap S_2$ in (20) may be the empty set $\varnothing$, and by definition

$$\operatorname{rank}(\varnothing) = 0. \tag{21}$$

The geometrically obvious inequality $\operatorname{rank}(S) \leq |S|$ is then a formal corollary of (20).

One may wonder whether, in fact, the above properties characterize all possible rank functions for points in a space of dimension d, where $d = \operatorname{rank}(\{P_1, \dots, P_n\}) - 1$? And maybe a proof of Theorem 1 may be found by exploring formal consequences of (19) and (20)? It turns out that the answers to these questions are emphatic "no" and "yes," respectively.

The above properties of the rank function give one of the many equivalent axiomatic definitions of a matroid.[2] Matroids were introduced by Hassler Whitney in 1935 as combinatorial generalization of incidence relations between flats of different dimensions, and have since found an abundance of applications across mathematics and computer science, both pure and applied. "*Matroid theory is a triumph in the pursuit of both abstraction and concrete simple examples,*" says Gil Kalai.

While we will discuss a few examples below, it should be made very clear now that matroids constitute a very rich and diverse universe, much larger than what we will explore in these notes. This makes the following result of Tom Braden, June Huh, Jacob Matherne, Nicholas Proudfoot, and Botong Wang extremely remarkable and powerful

Theorem 2. [10] *The injective matching*

$$\iota : \mathscr{F}_r \to \mathscr{F}_{r'}, \quad r \leq r',$$

as in Theorem 1, *exists for flats of any matroid M provided $r + r' \leq \operatorname{rank}(M)$.*

While the definition of a matroid is very short, the argument leading to the proof of Theorem 2 is very, very complex. The best we can hope to do in these introductory notes is to explain some earlier results and ideas in various areas of mathematics that may be listed among the precursors and inspirations for the fantastic achievement of [10].

At several points in our narrative, we will be coming back the following extraordinary feature of Theorem 2. In geometry, there is a constant dialog between the continuous and discrete. Of course, there is a fundamental unity in mathematics, and good mathematics is constantly transcending apparent boundaries between different subfields. Still, there is a clear difference between a matroid, which is combinatorial abstraction of a geometric configuration, and objects like a geodesic on a manifold, a minimal surface, or a harmonic form that require noncombinatorial methods to define and study.

One can compare and contrast the continuous and the discrete in many different ways, but one fundamental difference is the presence of *limits* in the continuous world. Of course, limits are absolutely central to mathematics and many crucial mathematical objects,

---

2 More precisely, the condition (18) means that here we focus on so-called loopless matroids. Given a rank function, one defines the flats of a matroid as in (17). Conversely, the rank function may be reconstructed from the data of the flats.

like the exponential function e^x, are transcendental in the sense that a limit is required to define or compute it. It is, however, an interesting question how much extra mileage one can get from using analytic tools to investigate combinatorial objects. As we will see, at the heart of Theorem 2, lies a certain *hard Lefschetz* property, which for many years was firmly associated with continuous, noncombinatorial geometry.

Theorem 2 applies to an *arbitrary* matroid, a purely combinatorial object. This is already very remarkable. But what is really amazing is the combinatorial and algebraic framework built in [10] to prove Theorem 2. It produces the required hard Lefschetz property from purely combinatorial, finite ingredients.

We will come back to these points later in the narrative. First, in the next section we want to discuss some examples of matroids beyond what we have seen so far. We warn the reader that these examples still cover a vanishing fraction of the universe of matroids.

5. SOME EXAMPLES OF MATROIDS

5.1. Points in $\mathbb{F}^d$, where $\mathbb{F}$ is a field

The first generalization concerns the coordinates in (4). There we took a d-tuple of real numbers, while instead we could have taken x_i to be rational numbers $x_i \in \mathbb{Q}$, or complex numbers $x_i \in \mathbb{C}$, or elements in an arbitrary field.

In mathematics, a field $\mathbb{F}$ is a set with special elements $0, 1 \in \mathbb{F}$ and binary operations $+, -, \times, /$ obeying all the usual laws of arithmetic for rational numbers $\mathbb{Q}$. An attentive reader will notice that the division is not really binary, since one cannot divide by zero. Instead, it is a function of the form

$$\mathbb{F} \times \big(\mathbb{F} \setminus \{0\}\big) \xrightarrow{(a,b)\mapsto a/b} \mathbb{F}.$$

For a dramatic example, we can take $\mathbb{F}_2 = \{0, 1\}$. Since $a + 0 = a$, $a \times 0 = 0$, and $a \times 1 = a$ in any field, most sums and products are already specified. The only interesting one is $1 + 1 = ?$ We invite the reader to check that

$$1 + 1 = 0 \tag{22}$$

is the only logical option and this indeed defines a field with two elements.

All constructions of Section 2 extend verbatim to any field. Note, however, that some configurations of points and lines can be realized using $\mathbb{F}_2$ and cannot be realized with real numbers, see (33) for an example.

Inspired by (22), we can ask when it is possible that

$$\underbrace{1 + 1 + \cdots + 1}_{p \text{ times}} = 0 \tag{23}$$

in a field $\mathbb{F}$. Minimal possible p with this property should be a *prime*, as we invite the reader to check. It is called the *characteristic* of the field $\mathbb{F}$. An example of a field of characteristic p is given by the residues

$$\mathbb{F}_p = \{0, 1, \dots, p - 1\}$$

modulo a prime p. The operations are defined by

$$(a, b) \xrightarrow{\;+\;} (a+b) \bmod p, \quad (a, b) \xrightarrow{\;\times\;} ab \bmod p, \tag{24}$$

where $a+b$ and ab are the usual addition and multiplication of integers. But since there is no division of integers, the existence of multiplicative inverse requires a minute of thought. For instance, one may notice that the multiplication by a map

$$\mathbb{F}_p \ni b \mapsto ab \in \mathbb{F}_p \tag{25}$$

is injective for $a \neq 0$, and therefore also surjective, as the source and target sets have the same cardinality. A map that is both injective and surjective is called *bijective*; those are the maps that have an inverse map. The inverse to multiplication by a is, by definition, the division by a. For fun, the reader may want to compute the inverses of elements in $\mathbb{F}_p$ for $p = 3, 5, 7$.

Mathematicians usually think about fields in terms of field *extensions*. Concretely, one describes new fields $\mathbb{F}'$ in terms of some previously understood subfield $\mathbb{F} \subset \mathbb{F}'$ and the new elements $x_1, x_2, \ldots \in \mathbb{F}'$ that have to be added to $\mathbb{F}$ to generate all elements of $\mathbb{F}'$ by arithmetic operations. One writes $\mathbb{F}' = \mathbb{F}(x_1, x_2, \ldots)$ to denote this situation. For example,

$$\mathbb{C} = \mathbb{R}(\sqrt{-1}).$$

All information about the field extension is contained in the polynomial equations satisfied by the elements $x_1, x_2, \ldots$ with coefficients in $\mathbb{F}$. For example, the element $i = \sqrt{-1}$ satisfies the equation

$$i^2 + 1 = 0. \tag{26}$$

Using this equation, we simplify powers i^k and, in particular, compute the product of two complex numbers as follows:

$$(a_1 + a_2 \cdot i)(b_1 + b_2 \cdot i) = (a_1 b_1 - a_2 b_2) + (a_1 b_2 + a_2 b_1) \cdot i. \tag{27}$$

In parallel to (25), we invite the reader to check that this is injective and surjective for $a_1^2 + a_2^2 \neq 0$ and compute $(a_1 + a_2 \cdot i)^{-1}$. After this exercise, the reader may want to find the formula for the inverse in $\mathbb{Q}(\sqrt{2})$.

To reiterate, while field theory is ultimately about solutions of polynomial equations, it is much more effective to *use* equations to learn about their solutions as opposed to "solving" the equations in the sense of looking for some complicated formulas giving the solutions in terms of the coefficients. We took this little detour into algebra now because later it will be crucial to *use* certain algebraic equations to deduce the information about flats in a matroid.

5.2. Projective spaces

While two points in a plane $\mathbb{R}^2$ always determine a line, two lines in a plane usually intersect in a single point, but not always. Sometimes lines can become parallel and then their point of intersection runs off to infinity. Projective geometry adds these points at infinity to the plane $\mathbb{R}^2$ so that two lines always intersect at a point. Many other geometric statements no longer require considering various cases either.[3]

The d-dimensional projective space over an arbitrary field $\mathbb{F}$ is easy to define using coordinates as follows:

$$\mathbb{P}^d(\mathbb{F}) = \big\{(d+1)\text{-tuples } (x_0, x_1, \dots, x_d) \text{ of elements of } \mathbb{F}, \\ \text{not all zero, up to proportionality } \sim\big\}, \tag{28}$$

where proportionality means

$$(x_0, x_1, \dots, x_d) \sim c(x_0, x_1, \dots, x_d), \quad c \in \mathbb{F} \setminus \{0\}. \tag{29}$$

Recall that we already met such an identification of proportional tuples in (6) when we talked about hyperplanes in $\mathbb{R}^d$.

The d-dimensional space $\mathbb{F}^d$ is naturally embedded in the projective space as the set where $x_0 \neq 0$. Indeed, when $x_0 \neq 0$, we can choose a unique c in (29) to make $x_0 = 1$, and so we get

$$\mathbb{F}^d = \big\{(1, x_1, \dots, x_d)\big\} \subset \mathbb{P}^d(\mathbb{F}). \tag{30}$$

The points with $x_0 = 0$ are the points "at infinity." They form a smaller projective space $\mathbb{P}^{d-1}(\mathbb{F})$.

By definition, a hyperplane in $\mathbb{P}^d(\mathbb{F})$ is defined by an equation of the form

$$a_0 x_0 + a_1 x_1 + \cdots + a_d x_d = 0, \tag{31}$$

in which some of the coefficients $a_i \in \mathbb{F}$ are not zero. In particular, the "infinity" is the hyperplane $x_0 = 0$.

Note that (31) is unchanged if we scale all variables x_i or all variables a_i by some constant $c \neq 0$. Thus the hyperplanes in $\mathbb{P}^d(\mathbb{F})$, as described by their coefficients $(a_0, \dots, a_d)$, form another projective space, called the *dual* projective space. This basic duality underlies many remarkable facts in geometry and combinatorics.

If the field $\mathbb{F}$ is finite, one can get very interesting matroids by taking *all* points of $\mathbb{P}^d(\mathbb{F})$ as the points $P_1, \dots, P_n$. For example, if $\mathbb{F} = \mathbb{F}_2$, there is no need to worry about proportionality (29), and so we get 7 points

$$\mathbb{P}^2(\mathbb{F}_2) = \big\{(1,0,0), (0,1,0), (0,0,1), (1,1,0), (0,1,1), (1,0,1), (1,1,1)\big\}. \tag{32}$$

By duality, there are 7 hyperplanes in $\mathbb{P}^2(\mathbb{F}_2)$. Each of them, in some coordinates, represents $\mathbb{P}^1(\mathbb{F}_2)$, and hence contains 3 points from (32). We invite the reader to check that the

---

3 For instance, the hyperbola, parabola, and ellipse are the same geometric shapes in projective geometry!

resulting configuration of points and lines looks as follows:

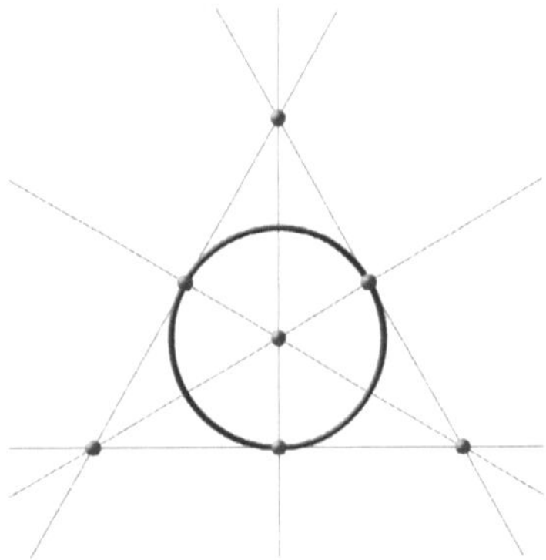

(33)

People who have read about the math behind the game of Dobble/Spot it! certainly recognize this picture. In (33), the line

$$x_1 + x_2 + x_3 = 0$$

is plotted as a circle precisely because this line does not meet the points (32) with real coefficients. With real coefficients, the three points on the circle define three different lines.

In fact, the matroid (33) can be realized in $\mathbb{P}^2(\mathbb{F})$ for some field $\mathbb{F}$ if and only if the characteristic of the field $\mathbb{F}$ equals 2, as noted already by Whitney. In general, whether a given matroid can be put into $\mathbb{P}^d(\mathbb{F})$, or as one says can be *linearly realized* over $\mathbb{F}$, is an interesting and important question.

Projective geometry is a very classical and very beautiful subject. Any result about incidence of points, lines, etc., in projective geometry is a potential obstruction to linear realizability of a given matroid over any field. For instance, the Pappus theorem says that if the three top points in (34) are collinear, and the three bottom points are also collinear, then so are the three middle points. The line, which Pappus proved exists, is highlighted in (34).

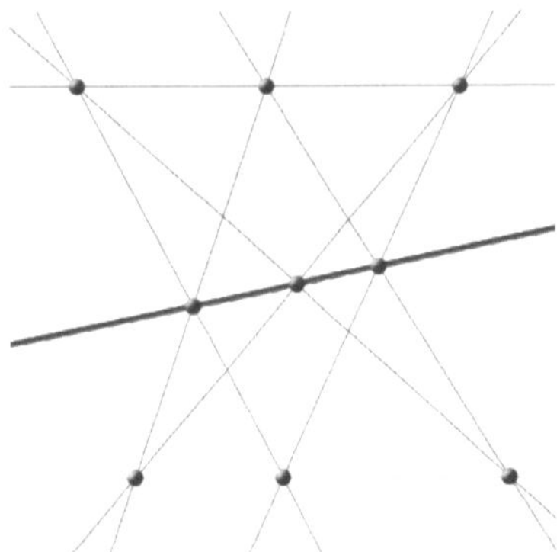

(34)

One can, however, declare the three middle points to be noncollinear without violating the matroid axioms. This gives a concrete example of a matroid that is not linearly realizable. In fact, the vast majority of matroids are such [37]. One way to think about them is to take a realizable matroid and add/remove some flats making sure the axioms are still obeyed, like we just did with with the non-Pappus matroid.

5.3. Field extensions

This class of examples is a little more advanced, so may be skipped on the first reading. Let $\mathbb{F}$ be a field and consider an extension

$$\mathbb{F}' = \mathbb{F}(y_1, \dots, y_n).$$

Note that we have n elements generating the extension, whereas earlier n was the number of points. This is intentional and we will define a rank function on subsets $S \subset \{1, \dots, n\}$ by

$$\operatorname{rank} S = \text{transcendence degree of } \mathbb{F}\big(\{y_i\}_{i \in S}\big) \text{ over } \mathbb{F}. \tag{35}$$

Here the transcendence degree of a field over a subfield $\mathbb{F}$ is the maximal number t of elements $f_1, f_2, \dots, f_t$ that satisfy no polynomial equation $P(f_1, \dots, f_t) = 0$ with coefficients in $\mathbb{F}$. The rank function (35) satisfies the matroid axioms.

Algebraic geometers associate a geometric image to this algebraic definition by thinking about

$$\mathbb{F}' = \text{rational functions on } Y, \quad Y \subset \mathbb{F}^n, \tag{36}$$

where $y_1, \dots, y_n$ are the coordinate functions on $\mathbb{F}^n$ and Y is an irreducible algebraic variety. By definition, an algebraic variety is defined by polynomial equations and it is irreducible if it is not a union of two other varieties. In (36) we identify two rational functions $f(y)$ and $f'(y)$ if they are equal on Y. With such reformulation

$$\operatorname{rank} S = \dim\big(\text{projection of } Y \text{ to the coordinates } \{y_i\}_{i \in S}\big). \tag{37}$$

Matroids that can be put in this form are called *algebraically realizable*. For instance, the non-Pappus matroid on 9 points is algebraically realizable over $\mathbb{F} = \mathbb{F}_2$; see **[32]**.

The reader should check that a linear realization yields an algebraic realization with $Y \subset \mathbb{F}^n$ being a linear subspace. In this case, it was shown in **[3]** that the matroid (35) controls many important properties of the closure

$$\overline{Y} \subset \big(\mathbb{P}^1(\mathbb{F})\big)^n \tag{38}$$

of the linear space Y in a product of projective lines over $\mathbb{F}$. We will come back to (38) in Section 8.3.

In another direction, Ingleton notes **[25]** that the tangent space to the generic points of Y provides a linear representation of an algebraic matroid, provided the characteristic of $\mathbb{F}$ is zero. This does not work in positive characteristic as the non-Pappus matroid demonstrates.[4]

Some simple explicit matroids can be shown to be algebraically nonrealizable; see **[26]**.

---

4 In characteristic p, one struggles with tangent spaces due to the fact that $(x^p)' = px^{p-1} = 0$.

5.4. Tropical realization of matroids

So far, we have looked at different classes of examples of matroids, always stressing the fact that these do not cover the great diversity of the world of matroids. Very remarkably, however, there is a class of examples that gives *all* matroids. This is the case for the *tropical* analog of the construction from Section 5.3, in which it is enough to take Y to be a tropical linear space.

This was discovered by Bernd Sturmfels in **[44]**. See Appendix C for a few introductory comments, **[27, 30, 33]** for a proper introduction to the subject, and **[1, 2, 6–8, 18]** for a sample of exciting recent advances in this direction. June Huh says: "*Mathematicians discovered tropical varieties by tropicalizing algebraic varieties, but only a tiny fraction of tropical varieties are tropicalizations of algebraic varieties. Tropical varieties are geometric objects that try to teach us a new kind of geometric intuition through their diversity.*"

In the spirit of this narrative, one should wish the best of success to all present and future combinatorial geometers in extending classical geometric results to this combinatorial setting. It is both very beautiful and important for applications.

6. GRADED MÖBIUS ALGEBRA

A certain algebraic language will be required the capture the essence of that is happening in Theorem 2. Most importantly, we will need explain one more meaning that the mathematicians attach to the word *algebra*.

6.1. Algebras

Consider a field extension $\mathbb{F}' = \mathbb{F}(x)$ generated by one element satisfying a polynomial equation of degree d with coefficients in $\mathbb{F}$. For instance, it can be $\mathbb{F}' = \mathbb{Q}(\sqrt[4]{-2})$, which means that the coefficients $\mathbb{F} = \mathbb{Q}$ are rational numbers and the new element x satisfies the equation

$$x^4 + 2 = 0. \tag{39}$$

We can think of $x^4 = -2$ as a substitution rule that we can instruct our computer to apply any time it sees a power x^k with $k > 3$. Using this substitution rule, we can describe $\mathbb{F}'$ by 4-tuples of rational numbers

$$\mathbb{F}' = \{a_0 + a_1 x + a_2 x^2 + a_3 x^3,\ a_i \in \mathbb{Q}\}, \tag{40}$$

and thus we can picture $\mathbb{F}'$ as a 4-dimensional linear space[5] $\mathbb{Q}^4$. To multiply two general elements of $\mathbb{F}'$,

$$(a_0 + a_1 x + a_2 x^2 + a_3 x^3)(b_0 + b_1 x + b_2 x^2 + b_3 x^3) = ?,$$

---

5 Readers who would like a bit more details about linear spaces will find them in Section A.3.

we have to expand out and use the substitution rule (39). This rule can phrased as follows:

$$\begin{aligned}&(a_0 + a_1x + a_2x^2 + a_3x^3)(b_0 + b_1x + b_2x^2 + b_3x^3)\\&\quad = \underbrace{(c_0 + c_1x + c_2x^2 + c_3x^3)}_{\text{product in } \mathbb{F}'} + \underbrace{\text{something} \cdot (x^4 + 2)}_{\text{discard}},\end{aligned} \tag{41}$$

where something refers to some polynomial in x. To see that multiplication by a nonzero element is invertible, it suffices to check that $\mathbb{F}'$ has no nontrivial divisors of zero. (We already used this logic when we were inverting (25).) But any $a_0 + a_1x + a_2x^2 + a_3x^3$ that divides zero in $\mathbb{F}'$ will have to divide $x^4 + 2$, whereas this polynomial cannot be nontrivially factored into polynomials with rational coefficients.[6] (Check this!)

What if we replaced 2 by 0 in (39), that is, what if we used a simpler equation $x^4 = 0$? The presentation (40) would still be valid and the multiplication would take a simpler form

$$\begin{aligned}&(a_0 + a_1x + a_2x^2 + a_3x^3)(b_0 + b_1x + b_2x^2 + b_3x^3)\\&\quad = \underbrace{(c_0 + c_1x + c_2x^2 + c_3x^3)}_{\text{product}} + \underbrace{\text{terms of degree} \geq 4 \text{ in } x}_{\text{discard}},\end{aligned} \tag{42}$$

meaning that

$$\begin{aligned}c_0 &= a_0b_0,\\c_1 &= a_1b_0 + a_0b_1,\\c_2 &= a_2b_0 + a_1b_1 + a_0b_2, \quad \text{etc.}\end{aligned}$$

Of course, this will no longer be a field, because multiplication by x is not invertible. But it is still a viable algebraic object that we will denote by a different letter

$$\begin{aligned}\mathbb{A} &= \{a_0 + a_1x + a_2x^2 + a_3x^3,\ a_i \in \mathbb{Q}\}\\&= \mathbb{Q}[x]/(x^4 = 0),\end{aligned} \tag{43}$$

lest somebody thinks it is a field.

We see that $\mathbb{A}$ is a linear space over a field $\mathbb{F}$ that has a product operation satisfying all the rules of arithmetic except those involving division. Mathematicians call such an object an *algebra*,[7] not to be confused with algebra as an area of mathematics that studies fields, algebras, and many other important structures. To distinguish between field and algebras, mathematicians use square brackets in (43) and they also like to to write the equations imposed on x as in (43). If there were no equations for x, we would simply get the algebra of polynomials in x with coefficients in $\mathbb{Q}$. That algebra is denoted $\mathbb{Q}[x]$.

The reader may wonder what could be the purpose of studying equations like $x^4 = 0$. Doesn't this just mean that $x = 0$? In fact, no, and there are very natural geometric situation

---

6 Mathematicians say (39) is *irreducible*. We have already used this term in exactly this meaning Section 5.3.

7 Or a *commutative algebra* to be more precise, since the product in $\mathbb{A}$ still obeys the commutative law of the arithmetic.

where relations like this appear. Let us look at the following table:

$$\begin{array}{c|c|c|c|c} 1 & x & x^2 & x^3 & x^4 = 0 \\ \hline \text{space} & \text{plane} & \text{line} & \text{point} & \varnothing \end{array}, \tag{44}$$

and note that the following parallels:

$$\begin{aligned} x \cdot x &= x^2 && \text{two general planes in space intersect in a line,} \\ x \cdot x^2 &= x^3 && \text{a plane and a general line in space intersect in a point,} \\ x \cdot x^3 &= 0 && \text{a plane and a general point in space intersect in an empty set.} \end{aligned}$$

Note that the 3-space here can be over an arbitrary field, which has nothing to do with the rational coefficients we had in (43). For a d-dimensional space, we should replace $x^4 = 0$ by $x^{d+1} = 0$. We will come back to these parallels in Section 8, but for now notice the potential for algebras to encode combinatorial information.

6.2. Graded algebras

This potential to encode combinatorial information gets amplified when we consider graded algebras. Let

$$\mathbb{A} = \mathbb{F}[x_1, \dots, x_N]/(\text{relations})$$

be an algebra generated by generators x_i subject to some relations. By definition, $\mathbb{A}$ is graded if every generator x_i is assigned a positive integer $\deg x_i = 1, 2, \dots$ so that all relations only involve monomials of the same total degree in $x_1, \dots, x_N$. For instance, $x^4 = 0$ is a good relation to have in a graded algebra, while $x^4 + x = 0$ is not. In a graded algebra, the subspaces

$$\mathbb{A}_k = \text{span of monomials in generators of total degree } k \tag{45}$$

intersect only in zero for different k. Mathematicians put a circle around the plus sign in

$$\mathbb{A} = \bigoplus_k \mathbb{A}_k \tag{46}$$

to stress this fact. One says that (46) is a *direct sum*.

Each $\mathbb{A}_k$ is a finite-dimensional linear space over $\mathbb{F}$ and its dimension $\dim_{\mathbb{F}} \mathbb{A}_k$ is a number which may be an interesting combinatorial function of k. For a combinatorial classic, consider the example

$$\mathbb{A}_{\text{W}} = \mathbb{Q}[x_1, x_2, x_3, x_4, x_5, x_6]/\big(x_1^{m_1+1} = 0, \dots, x_6^{m_6+1} = 0\big),$$

where the degrees of variables, which we write as a vector, are the denominations of the Korean won coins

$$\deg x = (1, 5, 10, 50, 100, 500).$$

The reader should check that

$$\begin{aligned} \dim \mathbb{A}_{\text{W},k} = {} & \text{number of ways to pay } k \text{ won} \\ & \text{using } \le m_i \text{ coins of each denomination.} \end{aligned}$$

For instance, if all $m_i = 1$, we get the sequence

$$\begin{array}{c|ccccccccccccc} k & 0 & 1 & 2 & 3 & 4 & 5 & 6 & 7 & 8 & 9 & 10 & 11 & 12 & \dots \\ \hline \dim \mathbb{A}_{₩,k} & 1 & 1 & 0 & 0 & 0 & 1 & 1 & 0 & 0 & 0 & 1 & 1 & 0 & \dots \end{array}, \tag{47}$$

while for all sufficiently large m_i one gets the sequence

$$\begin{array}{c|ccccccccccccc} k & 0 & 1 & 2 & 3 & 4 & 5 & 6 & 7 & 8 & 9 & 10 & 11 & 12 & \dots \\ \hline \dim \mathbb{A}_{₩,k} & 1 & 1 & 1 & 1 & 1 & 2 & 2 & 2 & 2 & 2 & 4 & 4 & 4 & \dots \end{array}. \tag{48}$$

The number 4 here comes from the fact that $10 = 5 + 5 = 5 + 1 + \cdots + 1 = 1 + \cdots + 1$ are all valid ways to pay 10 won.

The reader should further check that when all m_i are finite, the sequence $\dim \mathbb{A}_{₩,k}$ is always palindromic, that is, equals to itself read backwards. Equivalently,

$$\dim \mathbb{A}_{₩,k} = \dim \mathbb{A}_{₩,\text{topdeg}-k}, \quad \text{topdeg} = \sum_{i=1}^{6} m_i \deg x_i, \tag{49}$$

where the top degree is the total sum of money in our possession.[8] In particular, the sequence starts and ends with

$$\dim \mathbb{A}_{₩,0} = \dim \mathbb{A}_{₩,\text{topdeg}} = 1,$$

but in the middle can have many ups and down, as exemplified by (47). In particular, it is not in general a *unimodal* sequence, where unimodality is the concept we recall from the discussion following (10).

6.3. Hard Lefschetz property

Remarkably, in a closely related situation, one finds a sequence which is not just palindromic, but also unimodal. Every country in the euro area has its own euro cent coin, so there are many different coins, each worth €0.01. The number of ways to pay k cents, using at most m_i cents from the country number $i = 1, \dots, N$, is related to the algebra

$$\mathbb{A}_{€} = \mathbb{Q}[x_1, \dots, x_N] / \big(x_1^{m_1+1} = 0, \dots, x_N^{m_N+1} = 0\big),$$

with

$$\deg x = (1, \dots, 1).$$

If $m_i = 1$ for all i then $\dim \mathbb{A}_{€,k} = \binom{N}{k}$ is the binomial coefficient. If all $m_i \geq 2$ then

$$\dim \mathbb{A}_{€,k} = 1, N, \frac{N(N+1)}{2}, \dots, \frac{N(N+1)}{2}, N, 1,$$

where the last 1 occurs in the maximal degree $\sum m_i$. Moreover, for any m_i's, it will always be a unimodal sequence! (This is a combinatorial classic, for which the reader can try finding her or his own proof. It is probably the easiest to prove that the sequence of logarithms $\ln \dim \mathbb{A}_{€,k}$ is a *concave* function of k; see, for example, [11].)

---

8 Instead of paying k won, we can just give all our money and ask for topdeg $-k$ won in change.

There is certain algebraic property of $\mathbb{A}_\epsilon$ that is stronger than the symmetry (49) and the unimodality. It concerns multiplication by the element

$$\omega = \sum_{i=1}^{N} x_i \in \mathbb{A}_{\epsilon,1}.$$

The following diagram, plotted for $(m_1, m_2, \dots) = (3, 1, 0, 0, \dots)$ may help the reader visualize what this multiplication operator does:

$$\begin{array}{ccccccc} 1 & \longrightarrow & x_1 & \longrightarrow & x_1^2 & \longrightarrow & x_1^3 \\ \downarrow & & \downarrow & & \downarrow & & \downarrow \\ x_2 & \longrightarrow & x_1 x_2 & \longrightarrow & x_1^2 x_2 & \longrightarrow & x_1^3 x_2 \end{array} \tag{50}$$

In (50), we have listed all monomials that do not reduce to zero applying the rules $x_1^4 = x_2^2 = x_3 = \cdots = 0$. The graded pieces $\mathbb{A}_{\epsilon,k}$, $k = 0, 1, 2, 3, 4$ correspond to the diagonals in (50). The operations of multiplication by x_1 and x_2, when nonzero, are represented by the horizontal and vertical arrows, respectively. Thus, multiplication by ω is the sum of all outgoing arrows. We hope the reader has no problem visualizing the general case from this small example.

With these preparations, consider the multiplication map

$$\mathbb{A}_{\epsilon,i} \xrightarrow{\omega^{\text{topdeg}-2i}} \mathbb{A}_{\epsilon,\text{topdeg}-i}, \quad i < \frac{\text{topdeg}}{2}. \tag{51}$$

In the example (50), the possibilities for i are $i = 0, 1$. For $i = 0$, we get

$$\omega^4 \cdot 1 = 4x_1^3 x_2.$$

So, multiplication by ω^4 identifies $\mathbb{A}_{\epsilon,0} = \mathbb{Q}$ and $\mathbb{A}_{\epsilon,4} = \mathbb{Q} \cdot x_1^3 x_2$. Similarly,

$$\omega^2 \cdot (c_1 x_1 + c_2 x_2) = c_1 x_1^3 + (c_2 + 2c_1) x_1^2 x_2, \quad c_1, c_2 \in \mathbb{Q},$$

which is easily seen to be injective and surjective (one is enough, since this is linear map between spaces of the same dimension). Thus, in the example (50), we observe that the maps (51) are isomorphisms.[9]

In fact, the maps (51) are isomorphisms for the algebra $\mathbb{A}_\epsilon$ for any $(m_1, m_2, \dots)$. The reader who wants to prove this directly will probably find it a good challenge.

In general, one says that a graded algebra $\mathbb{A}$ satisfies the hard Lefschetz property (HLP) if the multiplication maps (51) are isomorphisms for some $\omega \in \mathbb{A}_1$. Among other things, HLP implies that the multiplication map

$$\mathbb{A}_{\epsilon,i} \xrightarrow{\omega} \mathbb{A}_{\epsilon,i+1}, \quad i < \frac{\text{topdeg}}{2}, \tag{52}$$

is injective, whence the unimodality of the sequence $\dim \mathbb{A}_{\epsilon,i}$.

---

9 Mathematicians call a bijective map between two sets an isomorphism when it preserves some further structures that these sets possess. Multiplication by ω preserves multiplication by $\mathbb{A}_0$ and addition.

Solomon Lefschetz was a very famous topologist, who proved[10] HLP for *cohomology* algebras of a certain class of *manifolds*; see Section 8. For the algebra $\mathbb{A}_{\in}$, the manifold in question is the product

$$\mathbb{P}^{m_1}(\mathbb{C}) \times \mathbb{P}^{m_2}(\mathbb{C}) \times \cdots \times \mathbb{P}^{m_N}(\mathbb{C})$$

of complex projective spaces.

The geometric story of HLP and its generalizations is an immensely beautiful subject, about which we will try to say a few words in Section 8. There are two reasons our contact with this subject will be only tangential. First, going any deeper into this story requires a level of mathematical sophistication that is well beyond the style of these notes. Second, and more importantly, the work of [10] completely *bypasses* the old HLP story, creating a totally new combinatorial alternative for it. It is true that the old HLP served as in important inspiration and, in fact, the original proof of Theorem 1 relied on it. But progress in mathematics also sometime includes letting go of very beautiful constructions that are no longer logically required.

6.4. The graded Möbius algebra, finally

Given a matroid M, its graded Möbius algebra $\mathbb{H}(M)$ is defined as follows. As a linear space over $\mathbb{Q}$, it has a basis y_F indexed by the flats F of M. It is graded by

$$\deg y_F = \operatorname{rank}(F).$$

Therefore,

$$|\mathscr{F}_r| = \dim \mathbb{H}(M)_r, \quad r = 0, 1, 2, \ldots$$

The multiplication is defined by

$$y_F y_{F'} = \begin{cases} y_{F \vee F'}, & \operatorname{rank}(F \vee F') = \operatorname{rank}(F) + \operatorname{rank}(F'), \\ 0, & \text{otherwise}, \end{cases} \tag{53}$$

where $F \vee F'$ is the minimal flat that contains F and F'. Recall we have used this notation to denote the line $P_1 \vee P_2$ spanned by points P_1 and P_2, etc.

In particular, $y_\varnothing \in \mathbb{H}(M)_0$ is the identity for this product. Also note that

$$y_F \cdot \mathbb{H}(M) \subset \operatorname{span}\big(\{y_{F'}\}_{\text{such that } F \subset F'}\big). \tag{54}$$

Theorem 2 easily follows from the following property of multiplication by the element:[11]

$$\omega = \sum_{F \in \mathscr{F}_1} y_F \tag{55}$$

in the algebra $\mathbb{H}(M)$.

---

10 Lefschetz's arguments were not entirely rigorous. Different correct proofs of HLP were given, in various geometric contexts, by Hodge, Chern, Deligne, and others. The influence of Lefschetz's work, however, was such that no one considers not naming this property after him.

11 More generally, one can replace each y_F in (55) by $c_F y_F$, where c_F is an arbitrary positive rational number.

Theorem 3 ([10]). *For $r \leq r'$ and $r + r' \leq \mathrm{rank}(M)$, the linear map*

$$\mathbb{H}(M)_r \xrightarrow{\text{multiplication by } \omega^{r'-r}} \mathbb{H}(M)_{r'} \tag{56}$$

is injective.

Here is how Theorem 3 implies Theorem 2. Let

$$\mathsf{A} = (a_{F',F})$$

be the matrix of multiplication by $\omega^{r'-r}$ in the bases $\{y_F\} \subset \mathbb{H}(M)_r$ and $\{y_{F'}\} \subset \mathbb{H}(M)_{r'}$. See Appendix A for a reminder about matrices and bases.

By (54), the matrix entry $a_{F',F}$ vanishes unless $F \subset F'$. Since A is injective, it has is a square invertible submatrix A' of size $|\mathscr{F}_r|$. Since A' is invertible, its determinant is not zero,

$$\det \mathsf{A}' \neq 0. \tag{57}$$

Since $\det \mathsf{A}' \neq 0$, there is at least one nonzero term in the formula (110). The corresponding permutation determines an injective matching of flats in $\mathscr{F}_r$ to flats in $\mathscr{F}_{r'}$. Quod erat demonstrandum.

Note how the logic of the proof goes from combinatorics to linear algebra and back. A linear map takes a basis vector to a linear combination of basis vectors, and this gives linear maps important extra flexibility. The argument above is about how one can go back, and obtain an injective map between bases from an injective linear map.

In broadest possible strokes, the strategy of the proof of Theorem 3 is the following. The authors of [10] construct a larger graded linear space

$$\mathbb{H}(M)_r \subset \mathbb{IH}(M)_r, \quad r = 0, 1, \ldots, \mathrm{rank}(M),$$

which is no longer an algebra, but still has multiplication by elements of $\mathbb{H}(M)$. Mathematicians say $\mathbb{IH}(M)$ is a *module* for the algebra $\mathbb{H}(M)$. Since it is a module, it makes sense to consider the map

$$\mathbb{IH}(M)_r \xrightarrow{\text{multiplication by } \omega^{\mathrm{rank}(M)-2r}} \mathbb{IH}(M)_{\mathrm{rank}(M)-r}, \tag{58}$$

for $r < \frac{1}{2}\mathrm{rank}(M)$. Evidently, (58) being an isomorphism implies that (64) is injective. It is this HLP for the map (58) that is really the key to Theorems 1, 2, and 3.

7. THE BIG INDUCTION

In a computer code, it is sometimes very convenient to allow a procedure to call another instance of itself. Of course, if done carelessly, this can easily lead to an infinite loop and failure. To make sure the code terminates, there have to be, first, some base cases, when the procedure returns the answer without calling itself, and, second, it should each time call itself on a smaller and simpler input, which gets closer and closer to a base case.

Imagine we already coded a data type `matroid` and we want to code, in some imaginary relative of the C programming language, a procedure

```
print_theorem_3 _proof(matroid M){
  if (rank M < 2){
  print("multiplication by ω^0 is an isomorphism")          (59)
      ... ,
```

where we have already indicated the base case. Definitely, a matroid M' is simpler than M if it has fewer points, so it is OK for this procedure to call `print_theorem_3_proof(M')` inside itself for such M'.

There are two important ways to construct a smaller matroid from M and a flat F of M. They are denoted M^F and M_F. The matroid M^F keeps only points P_i and flats F' contained in F. The matroid M_F keeps only those flats F' that contain F. The latter are determined by which points we should add to F to get F', hence M_F is a matroid on the points P_j that are *not* contained in F.

Of course, to have a mathematical proof of Theorem 3 it is not necessary to actually run the procedure. It is enough to know that a proof for M can be found if we have a proof for all smaller matroids M_F and M^F, and in the base case. Mathematicians call such proofs as proofs by *induction*.

A very important insight from **[10]** is that it is much more natural to prove a *stronger* theorem than Theorem 3. In an inductive proof, there is always a tension between trying to prove too much or too little. The logic of induction says that we can get from the result for M' to the result for M. So, assuming we can prove the statement for M', we can prove it for M. There is a certain climb between M' and M, and it becomes impossible if the starting point is too low or the goal is too high.

Analogies aside, what the authors of **[10]** actually prove is the whole *Kähler package* for the space $\mathbb{IH}(M)$. In addition to HLP, this package includes a nondegenerate bilinear form

$$(\cdot,\cdot) : \mathbb{IH}(M)_i \times \mathbb{IH}(M)_{\text{rank}-i} \to \mathbb{Q} \tag{60}$$

such that for $\alpha \in \mathbb{IH}(M)_i$ we have

$$\omega^{\text{rank}-2i+1}\alpha = 0 \quad \Rightarrow \quad (-1)^i(\omega^{\text{rank}-2i}\alpha, \alpha) > 0. \tag{61}$$

For realizable matroids, these properties have an interpretation and history in topology, at which we will hint in Section 8. Namely, (60) is the Poincaré duality and (61) are the Hodge–Riemann relations. But as we have already stressed at many points of this narrative, the amazing feature of Theorem 3 is that it works with no input from topology or algebraic geometry, and applies to absolutely all matroids, realizable or not.

The body of the procedure (59) is a marvel of combinatorics and combinatorial algebra, and it is way beyond the sophistication level of these notes to try to look any further in it. Let us just say it is not at all simple. There is a reason mathematics like this is recognized by the highest prizes in mathematics. In fact, it is miracle that some people can construct proofs like this.

An interested reader will find further reading suggestions in Section 9. We should also mention that Theorem 3 is not first time a combinatorial replacement of Hodge theory appears in mathematics.

June Huh says: "*Important precursors include the intersection cohomology* $\mathbb{IH}(P)$ *of a convex polytope* P [29] *and the Soergel bimodule* $\mathbb{IH}(w)$ *for a Coxeter group element* w [17]. *Both* $\mathbb{IH}(P)$ *and* $\mathbb{IH}(w)$ *satisfy Poincaré duality, the hard Lefschetz theorem, and the Hodge–Riemann relations, and these reveal fundamental properties of* P *and* w*: The generalized lower bound conjecture for the number of faces in the case of* P [29, 41] *and the nonnegativity conjecture for the coefficients of Kazhdan–Lusztig polynomials of Bruhat intervals in the case of* w [17, 31]. *Each of the known proofs of the three combinatorial Kähler packages involves numerous details that are unique to that specific case.*"

Speaking of Kazhdan–Lusztig polynomials, the authors of [10] prove, in fact, much more than we managed to explain in these notes. In particular, they prove the nonnegativity of KL polynomials for all matroids.

I hope the readers share the narrator's sense of awe at this absolutely amazing mathematics and join me in warmest congratulations on it being recognized by the Fields Medal. I also hope the readers got the sense that today's mathematics is not just extraordinarily powerful, but also concrete, understandable, and fun, once one finds the right idea and the right point of view. While finding that right point of view is not at all easy, my biggest hope is to have inspired my youngest readers to believe that mathematics can be beautiful and rewarding, both as a subject and as a profession. Maybe this is also a good place for me to thank June Huh and Gil Kalai for this special opportunity to be introduced to their wonderful subject.

8. INSPIRATIONS FROM TOPOLOGY

8.1. Cohomology

Consider a graph Γ drawn on a torus Σ, that is, on the surface of a bagel. By definition, a graph is a collection of *vertices* and *edges*. Since it is drawn on a surface Σ, it partitions Σ into some regions that we will call *faces*. The vertices, edges, and faces are the 0-, 1-, and 2-dimensional objects in (62).

We assume, and it is an important assumption, that every face is a polygon. For instance, the unique face in the figure of (62) is obtained by gluing the opposite sides of the hexagon in (65). For a very different example of a graph on the torus, the reader may take the fine square mesh representing the torus in (62).

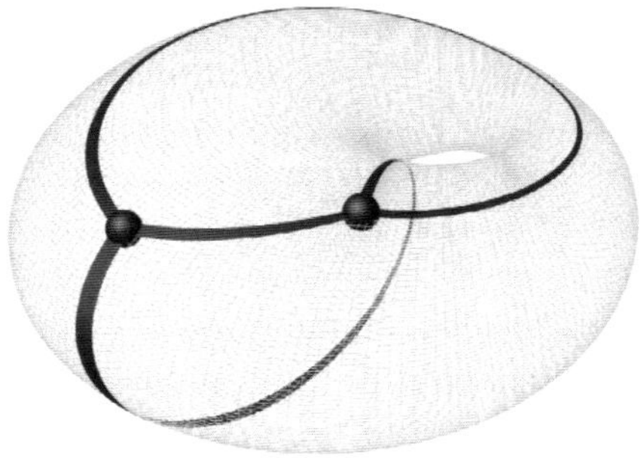

(62)

Let f_0 be a function defined on the vertices $\{V_i\}$ of Γ. We define its *gradient* df_0 as follows. This will be a function of an *oriented* edge E_{ij} of Γ. If E_{ij} goes from the vertex V_i to the vertex V_j, schematically

$$V_i \bullet \xrightarrow{E_{ij}} \bullet V_j,$$

then

$$df_0(E_{ij}) = f(V_j) - f(V_i). \tag{63}$$

For the opposite orientation of the edge, one gets the opposite sign,

$$df_0(\overleftarrow{E}) = -df_0(\overrightarrow{E}). \tag{64}$$

This elementary construction is found in abundance is both theoretical and applied situations. For instance, one may interpret df_0 as the current through edges of Γ generated by a potential function f_0 defined on its vertices.

Now let f_1 be a function on oriented edges satisfying the sign rule (64). We may interpret f_1 as a flow or a vector field going along the edges of Γ. Given on oriented face F, its *boundary*

$$\partial F = E_{ij} \cup E_{jk} \cup \cdots$$

is a collection of edges, which gets an orientation from the orientation of F; see the figure in (65).

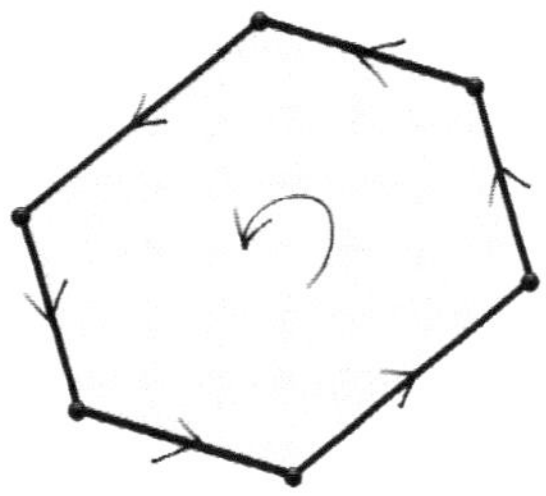

(65)

We define

$$df_1(F) = \sum_{E \in \partial F} f_1(E). \tag{66}$$

This has a natural interpretation as the circulation, or the curl, of the flow f_1 around the face F. This again changes sign upon switching the orientation of F. The reader may want to pause and write (63) in a form that resembles (66).

So far, we did not specify the range of the functions f_i. Let them take values in some field $\mathbb{F}$. Functions on vertices form a linear space over $\mathbb{F}$ that we will denote $\Omega^0(\Gamma)$, and similarly for functions $\Omega^i(\Gamma)$, $i = 1, 2$ on oriented edges and faces. We always assume these functions change sign as in (64) upon switching the orientation. To fix a basis in these spaces, we can fix some orientation of each edge and face arbitrarily.

Above, we have constructed two linear maps

$$\Omega^0(\Gamma) \xrightarrow{\quad d_0 \quad} \Omega^1(\Gamma) \xrightarrow{\quad d_1 \quad} \Omega^2(\Gamma), \tag{67}$$

where we marked the two maps d with lower indices for notational convenience. The key property of (67) is that the composition

$$d^2 = d_1 d_0 = 0 \tag{68}$$

is zero. This is known in may different guises, e.g., the circulation of a gradient vector field is zero, and reflects the geometric fact that $\partial^2 = 0$, meaning that a boundary has no boundary.

A classical question appearing in many branches of mathematics is: Does every vector field with zero curl come from a potential? In other words, is it true that the kernel $\operatorname{Ker} d_1$ equals the image $\operatorname{Im} d_0$? Or, using the language introduced in Section A.4, is the sequence (67) *exact* at the middle term?

More generally, one calls a sequence of maps composing to zero like (67) a *complex*, with the stress on the second syllable. From $d^2 = 0$, we see that $\operatorname{Im} d_{i-1} \subset \operatorname{Ker} d_i$ and one defines the *cohomology* groups of a complex by

$$H^i = \operatorname{Ker} d_i / \operatorname{Im} d_{i-1}.$$

In (67) and in general, we assume that the maps d_i not indicated are the zero maps. The image of a zero map is the zero subspace and the kernel of a zero map is the whole space.

Clearly,

$$\operatorname{Ker} d_0 = \text{constant functions} = \mathbb{F},$$

hence $\dim H^0 = 1$. It fun to check that

$$\operatorname{Im} d_1 = \operatorname{Ker}\Big(\Omega^2(\Gamma) \xrightarrow{\quad \int_\Sigma \quad} \mathbb{F}\Big)$$

where the integration map is defined by

$$\int_\Sigma f_2 = \sum_{\text{all faces } F} f_2(F),$$

with the orientation of each face induced by some chosen orientation of Σ. Therefore, $\dim H^2 = 1$. The remaining dimension $\dim H^1$ we can infer from

$$\begin{aligned} \dim H^0 - \dim H^1 + \dim H^2 &= \dim \Omega^0 - \dim \Omega^1 + \dim \Omega^2 \\ &= |\text{vertices}| - |\text{edges}| + |\text{faces}| \\ &= \text{Euler characteristic of } \Sigma \\ &= 0, \end{aligned} \tag{69}$$

where the first equality is a general property of all complexes which follows from (98) and (100), and where at the last step we used Euler's formula, one of the first topological results in mathematics.

Thus $\dim H^1 = 2$ and, in fact, a vector field f_1 is a gradient if and only if its circulation around each face *and* around the two blue loops as in (70) vanishes.

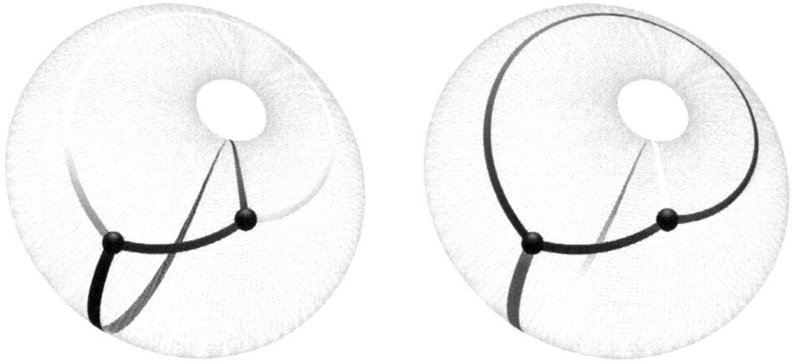

(70)

While we started with a graph on a torus, the eventual outcome of our computation is really about the torus itself and not about any particular graph drawn on it. One way to see it, is to take a a refinement Γ' of the graph Γ. This means that every edge of Γ

$$E_{ij} = E'_{ik_1} \cup E'_{k_1 k_2} \cup \cdots \cup E'_{k_l j}$$

is a union of edges of Γ'. It follows that every face of Γ is a union of faces of Γ'. A refinement of a face may look something like in the figure of (71).

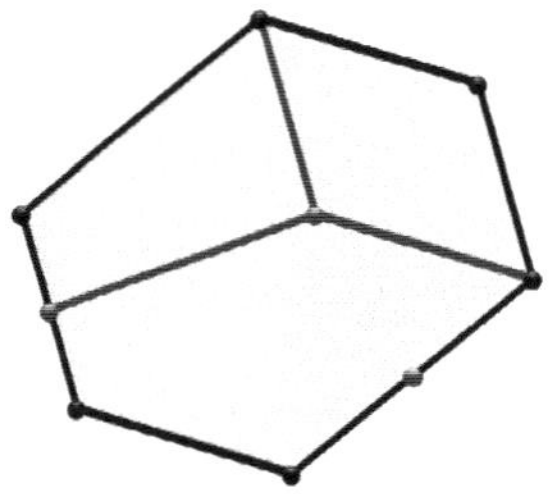

(71)

From the flows and circulations in the graph Γ', we can compute the flows and circulations in the graph Γ. This gives vertical maps in the diagram

$$\begin{array}{ccccc} \Omega^0(\Gamma) & \xrightarrow{d_0} & \Omega^1(\Gamma) & \xrightarrow{d_1} & \Omega^2(\Gamma) \\ \uparrow & & \uparrow & & \uparrow \\ \Omega^0(\Gamma') & \xrightarrow{d'_0} & \Omega^1(\Gamma') & \xrightarrow{d'_1} & \Omega^2(\Gamma'), \end{array} \tag{72}$$

which can be seen to identify the cohomology. Since any two graphs Γ_1 and Γ_2 have a common refinement Γ', the cohomology is really attached to the torus Σ and not to any particular decomposition of Σ into vertices, edges, and faces. What is really fundamental about the torus is that all possible loops in Σ, up to boundaries, span a two-dimensional space with a basis plotted in (70).

We hope it is easy for the reader to imagine the generalization of this story in which one replaces the torus Σ by any *topological space* that can be glued out of polytopes of different dimension. The elementary story told above was known in the 19th century, and since then topologists have developed really powerful tools to attach various topological invariants, including cohomology $H^i(X,\mathbb{F})$, to topological spaces X.

The torus is not only a topological space, it is also a *complex algebraic variety*. Namely, consider the solutions

$$\{P(x_0,x_1,x_2)=0\}\subset\mathbb{P}^2(\mathbb{C}), \tag{73}$$

where P is a homogeneous polynomial of degree 3, meaning that

$$P(tx_0,tx_1,tx_2)=t^3P(x_0,x_1,x_2), \quad \text{for any } t\in\mathbb{C}. \tag{74}$$

While the coordinates (x_0,x_1,x_2) on $\mathbb{P}^2$ are defined only up to proportionality, by (74) the zero set (73) is defined unambiguously. If the partial derivatives $\frac{\partial}{\partial x_i}P$ do not vanish simultaneously, then (73) is a torus. Not to be outdone by the topologists, algebraic geometers have defined equally powerful cohomology theories for algebraic varieties. These agree with the topological definitions over the field $\mathbb{C}$ of complex numbers.

It is a really inspiring lesson in the unity of mathematics that different cohomology theories, with very different starting points and emphasis on very different geometric objects, in the end all agree on their common domains of applicability.

8.2. Multiplication and Poincare duality

We were interested in cohomology because of the graded algebra structure on the direct sum

$$H^\bullet(X,\mathbb{F})=\bigoplus H^i(X,\mathbb{F}),$$

that is, because of the multiplication operation

$$H^i(X,\mathbb{F})\times H^j(X,\mathbb{F})\xrightarrow{\ \cup\ }H^{i+j}(X,\mathbb{F}). \tag{75}$$

The product (75) exists for *very* abstract reasons. For any topological space X, there is the diagonal map

$$X\to X\times X,$$

sending a point x to the pair (x,x). A map between topological spaces induces a map on cohomology the other way. Using the Künneth isomorphism

$$H^\bullet(X\times X,\mathbb{F})=H^\bullet(X,\mathbb{F})\otimes H^\bullet(X,\mathbb{F}), \tag{76}$$

where $\otimes$ denotes the tensor product, one obtains (75). A less general, but more intuitive description says that (75) is dual to *intersection*, and it goes as follows.

By definition, the complex of *dual* maps

$$\Omega^0(\Gamma)^\vee\xleftarrow{\ d_0^\vee\ }\Omega^1(\Gamma)^\vee\xleftarrow{\ d_1^\vee\ }\Omega^2(\Gamma)^\vee, \tag{77}$$

computes the *homology* groups $H_i(\Sigma,\mathbb{F})$. The geometric meaning of (77) is transparent. The bases in $\Omega^i(\Gamma)^\vee$, $i=0,1,2$, are indexed by vertices, edges, and faces, respectively, and the maps are the boundary maps ∂. Collectively, the vertices, edges, and faces are called *cells*, and $\mathbb{F}$-linear combinations of cells are called *chains*. Chains with no boundary are called *cycles*. Thus, homology describes cycles up to boundaries.

For the torus Σ, the homology complex (77) can be identified with the cohomology complex for the *dual graph* $\Gamma^\vee$, where the dual graph to (62) can be seen in (78).

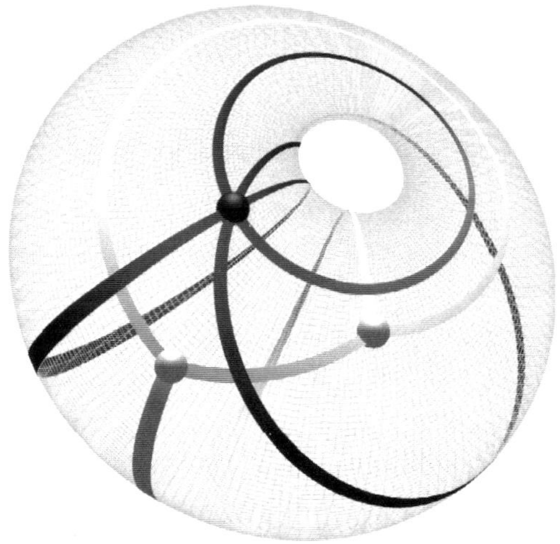

(78)

The vertices, edges, and faces of the dual graph correspond to the faces, edges, and the vertices of the original graph, respectively. Moreover, each cell intersects the dual cells in exactly one point. This gives the Poincaré duality isomorphism

$$H_i(\Sigma, \mathbb{F}) \cong H^{\dim \Sigma - i}(\Sigma, \mathbb{F}). \tag{79}$$

It works just the same for a doughnut with $g = 2, 3, \ldots$ holes and for any oriented closed (meaning, compact and without boundary) manifold M.

Manifolds are particularly nice topological spaces that, in a certain technical sense, look just like the linear space in the vicinity of every point. The linear space $\mathbb{R}^n$ is a manifold, but not a compact manifold. The n-dimensional sphere S^n and also the real and complex projective spaces are closed manifolds. Recall we insisted that $\frac{\partial}{\partial x_i} P \neq 0$ for some i at every point of (73). This was to make sure that (73) defines a manifold.

For cycles, one would like to define an intersection product

$$H_i(X, \mathbb{F}) \times H_j(X, \mathbb{F}) \xrightarrow{\cap} H_{i+j-\dim X}(X, \mathbb{F}), \tag{80}$$

that would turn into the $\cup$-product upon the identification (79). This does not really make sense for a general topological space, since it is not even clear what notion of dimension one should use in (80). But on a manifold, it works beautifully, especially if one intersects cycles defined using a graph Γ with the cycles defined using the *dual* graph $\Gamma^\vee$. (Recall that any two graphs Γ and Γ' define the *same* space of cycles up to boundaries.)

One very important detail here that one should count intersections with a sign according to orientation. This is crucial to make boundaries have zero intersection with any cycle. Let us look at the intersection of a cycle γ with a boundary ∂F in the figure of (81). If we keep track of the orientations, we can tell whether γ enters or exists F at a given point of intersection. Hence if we count intersections with signs then we get $\gamma \cap \partial F = 0$.

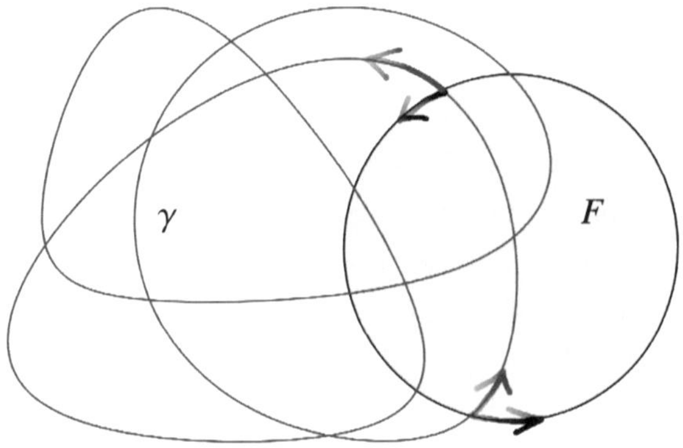

(81)

As a result, cohomology $H^\bullet(\Sigma, \mathbb{F})$ is not commutative but rather *supercommutative*,[12]

$$\alpha_1 \cup \alpha_2 = (-1)^{d_1 d_2} \alpha_2 \cup \alpha_1, \quad \alpha_i \in H^{d_i}. \tag{82}$$

If the reader has not seen it before, it is a good exercise to work out multiplication for $H^\bullet(\Sigma, \mathbb{F})$ and also for $H^\bullet(\Sigma_g, \mathbb{F})$, where Σ_g is the surface of the doughnut with g holes.

While the signs in (82) are an important fact of nature, they are of little concern for us here since we are interested in even cohomology, which is commutative. In particular, topological intersection of algebraic cycles, that is, those defined by polynomial equations, is commutative and agrees with its counterpart in algebraic geometry.

For cycles of complementary dimension, we can interpret interpret the isomorphism (79) as the Poincaré duality pairing:

$$(\cdot, \cdot) : H^i(X, \mathbb{F}) \times H^{\dim X - i}(X, \mathbb{F}) \to \mathbb{F}. \tag{83}$$

The purely combinatorial relative of this pairing appears in (60). One can also interpret (83) as the composition of the cup product with the isomorphism $H^{\dim X}(X, \mathbb{F}) \cong H_0(X, \mathbb{F}) = \mathbb{F}$.

It is an excellent student project to prove that

$$H^\bullet\big(\mathbb{P}^d(\mathbb{C}), \mathbb{F}\big) = \mathbb{F}[x]/(x^{d+1} = 0), \quad x \in H^2, \tag{84}$$

where x is dual to the class of the hyperplane in $H_{2d-2}(\mathbb{P}^d(\mathbb{C}))$. This formalizes the table (44). Similarly,

$$\mathbb{A}_{\epsilon,k} = H^{2k}\big(\mathbb{P}^{m_1}(\mathbb{C}) \times \cdots \times \mathbb{P}^{m_N}(\mathbb{C}), \mathbb{Q}\big). \tag{85}$$

8.3. The hard Lefschetz property

The title of this subsection prompts the question: For which even-dimensional manifolds X is there a class $\omega \in H^2(X, \mathbb{Q})$ such that the multiplication map

$$H^i(X, \mathbb{Q}) \xrightarrow{\omega^{\frac{1}{2}\dim_{\mathbb{R}} X - i}} H^{\dim_{\mathbb{R}} X - i}(X, \mathbb{Q}), \quad i < \frac{1}{2} \dim_{\mathbb{R}} X, \tag{86}$$

is an isomorphism? Here by the dimension of X we mean its real dimension, even though X may have been originally defined as a complex manifold. Thus, $\dim_{\mathbb{R}} \Sigma = 2$ for the torus (73).

12 The tensor product in Künneth theorem (76) should also be understood with signs.

It is not enough for X to be smooth: the even-dimensional spheres $X = S^{2k}$ have $H^2(S^{2k}) = 0$, and hence no chance of satisfying (86) for $k > 1$.

One classical answer is that a smooth projective $X \subset \mathbb{P}^N(\mathbb{C})$ satisfies the hard Lefschetz property (86) with ω dual to the class of a hyperplane section. Projective means that X is defined by polynomial equations just like (73), and smooth means it is a manifold. A more general class of *Kähler* manifold also satisfies the HLP.

If $X \subset \mathbb{P}^N(\mathbb{C})$ is not smooth then it is called *singular*. For a singular X, there is a more delicate cohomology theory that satisfies the HLP. It is called intersection cohomology and its development is one of the true highlights of geometry and topology, achieved in the 1970s and 1980s by Mark Goresky, Robert MacPherson, Pierre Deligne, Alexander Beilinson, Joseph Bernstein, and other amazing mathematicians.

For a matroid linearly realizable over $\mathbb{C}$, the Möbius algebra $\mathbb{H}(M)$ is the cohomology algebra of the variety $\overline{Y}$ associated to M in (38). Note that the generators $\mathbb{H}(M)$ commute and square to zero, so it is a quotient of $H^\bullet((\mathbb{P}^1)^n)$. The module $\mathbb{IH}(M)$ is the intersection cohomology of the same variety $\overline{Y}$. Thus, there is a topological proof the HLP for $\mathbb{IH}(M)$ for realizable matroids. This was used in the original proof of Theorem 1 given in **[24]**.

Hard Lefschetz property for cohomology and intersection cohomology has a history of very powerful application to combinatorial problems. One great example is Richard Stanley's proof of McMullen's conjectural characterization of f-vectors[13] of simplicial convex polytopes. (Stanley proved the necessity of McMullen's conditions, the sufficiency was proven about the same time by Billera and Lee.) See **[42]** for a discussion of this and other combinatorial applications of the HLP.

9. FURTHER READING

The *Quanta Magazine* has published popular accounts of these and related developments, see **[12,20]**.

Among surveys written by top experts in the field, one should mention **[5,28]**, including expositions by June Huh himself **[22,23]**.

Among textbook introductions to different areas mathematics discussed in our narrative, the reader will surely find something which suits her or his interests and style among **[4,21,38,40,43]**.

I hope the reader has a lot of fun studying these sources as well as the original articles **[10,24]**.

---

13 For a convex polytope, its f-vector records the number of faces of each dimension.

A. A RICE BOWL OF LINEAR ALGEBRA

A.1. Linear equations

A system of N linear equations for M unknowns $x_1, \ldots, x_M$,

$$\begin{cases} a_{11}x_1 + \cdots + a_{1M}x_M = c_1, \\ \qquad\qquad\qquad \cdots \\ a_{N1}x_1 + \cdots + a_{NM}x_M = c_N, \end{cases} \tag{87}$$

is conveniently written in matrix notation

$$\mathsf{Ax} = \mathsf{c}, \tag{88}$$

where

$$\mathsf{A} = \begin{bmatrix} a_{11} & \ldots & a_{1M} \\ \vdots & & \vdots \\ a_{N1} & \ldots & a_{NM} \end{bmatrix}, \quad \mathsf{x} = \begin{bmatrix} x_1 \\ \vdots \\ x_M \end{bmatrix}, \quad \mathsf{c} = \begin{bmatrix} c_1 \\ \vdots \\ c_N \end{bmatrix}. \tag{89}$$

Solutions of (87) are unchanged, if we multiply the ith equation, where $i = 1, \ldots, N$, by a nonzero number t, or add to the ith equation t times the jth equation. Such transformations are called *elementary*. In matrix form, they have the form

$$(\mathsf{A}, \mathsf{c}) \to \big(\mathsf{g_{ij}}(t)\mathsf{A}, \mathsf{g_{ij}}(t)\mathsf{c}\big),$$

where $\mathsf{g_{ij}}(t)$ is an *elementary* matrix, which means a matrix of the following form:

$$\underbrace{\begin{bmatrix} 1 & 0 & 0 & \ldots & 0 \\ 0 & 1 & 0 & \ldots & 0 \\ 0 & 0 & 1 & & \vdots \\ \vdots & \vdots & & \ddots & 0 \\ 0 & 0 & \ldots & 0 & 1 \end{bmatrix}}_{\text{identity matrix}} \xrightarrow{\text{put } t \neq 0 \text{ in the } i\text{th row and } j\text{th column}} \mathsf{g_{ij}}(t). \tag{90}$$

The rice in our rice bowl is the following statement, called row reduction, or Gaussian elimination. Any system of linear equations can be transformed by elementary operations[14] to a unique matrix of the schematic form

$$\mathsf{A}_{\text{rowred}} = \begin{bmatrix} 0 & 0 & 1 & * & 0 & 0 & * & 0 & * \\ 0 & 0 & 0 & 0 & 1 & 0 & * & 0 & * \\ 0 & 0 & 0 & 0 & 0 & 1 & * & 0 & * \\ 0 & 0 & 0 & 0 & 0 & 0 & 0 & 1 & * \\ 0 & 0 & 0 & 0 & 0 & 0 & 0 & 0 & 0 \end{bmatrix}, \tag{91}$$

where stars stand for some unspecified numbers. The 1's in (91) have to be in different rows and columns. A star $*$ can follow a 1 in a row, unless there is a 1 in the same column. The number of 1's in (91) is called the rank of A.

---

14 In practical implementations of row reduction, it is very convenient to permute equations. Abstractly, however, a permutation of two equations may be achieved by elementary transformations, as the reader will easily check.

We have

$$\text{solutions}(\mathsf{A}\mathsf{x} = \mathsf{c}) = \text{solutions}\big(\mathsf{A}_{\text{rowred}}\mathsf{x} = \mathsf{c}'\big), \tag{92}$$

where c' is the result of applying the sequence of elementary transformations $\mathsf{g}_{\mathsf{ij}}(t)$ to the vector c. For a reduced matrix, the solutions can be described very easily.

The zero rows in (91) lead the equations of the form $0 = c_i'$. These have either no solutions if $c_i' \neq 0$ or can be discarded if $c_i' = 0$. Thus (88) has solutions if and only if c satisfies $N - \text{rank}(\mathsf{A})$ linear equations given by $c_i' = 0$, $i = \text{rank}(\mathsf{A}) + 1, \ldots, N$.

After we have dealt with the zero rows in (91), the remaining equations may be solved uniquely for the variables x_j that have a 1 in their columns. All other variables are free parameters. Thus, when they exist, the solutions are parametrized by $M - \text{rank}(\mathsf{A})$ free parameters.

Row reduction is fundamental. Everything else in this section is a topping.

A.2. Linear maps

In Section A.1, we never specified the algebraic nature of the variables x_j or the coefficients a_{ij} and c_i. The reader may have assumed they are real or rational numbers. In fact, they can be taken to be elements in any field $\mathbb{F}$ without changing anything at all in the analysis.

Column vectors x of size M with entries in $\mathbb{F}$ form the coordinate linear space $\mathbb{F}^M$ of dimension M. It has operations of addition and multiplication by elements $t \in \mathbb{F}$, both defined coordinate by coordinate. A map

$$\mathsf{A} : \mathbb{F}^M \to \mathbb{F}^N \tag{93}$$

is said to be linear if it preserves these operations, that is,

$$\mathsf{A}(\mathsf{x} + \mathsf{x}') = \mathsf{A}(\mathsf{x}) + \mathsf{A}(\mathsf{x}'), \quad \mathsf{A}(t\mathsf{x}) = t\mathsf{A}(\mathsf{x}).$$

The reader should check that such maps are precisely those given by a matrix multiplication, and hence we can write $\mathsf{A}\mathsf{x}$ in place of $\mathsf{A}(\mathsf{x})$. From what we just learned about linear equations, it follows that:

- A is injective if and only if $\text{rank}(\mathsf{A}) = M$.
- A is surjective if and only if $\text{rank}(\mathsf{A}) = N$.
- A is bijective, or an isomorphism, or *invertible* if and only if $\text{rank}(\mathsf{A}) = N = M$. In particular, there is no isomorphism $\mathbb{F}^M \to \mathbb{F}^N$ if $M \neq N$.
- Any isomorphism $\mathsf{g} : \mathbb{F}^M \to \mathbb{F}^M$ is a product of elementary matrices $\mathsf{g}_{\mathsf{ij}}(t)$.

It is very important to remember that linear spaces have a lot of nontrivial isomorphisms $\mathsf{g} : \mathbb{F}^M \to \mathbb{F}^M$. These can be composed and inverted, thus form a group denoted $\mathrm{GL}(M, \mathbb{F})$. When we act by $\mathsf{g} \in \mathrm{GL}(M, \mathbb{F})$, we say that the we change the basis, or do a linear change of coordinates. While coordinates provide a very concrete and convenient description of geometric objects, a truly geometric construction should work equally well in any linear coordinates.

Formula (91) describes a standard form to which a matrix can be brought by post-composing with an isomorphism, that is, by a change of basis in the target of the map. We invite the reader to check that by an independent[15] change of basis in the source and the target, a matrix can be brought to a particularly simple form

$$\mathsf{A}_{\text{rowcolred}} = \begin{bmatrix} 1 & 0 & 0 & 0 & 0 & 0 & 0 & 0 & 0 \\ 0 & 1 & 0 & 0 & 0 & 0 & 0 & 0 & 0 \\ 0 & 0 & 1 & 0 & 0 & 0 & 0 & 0 & 0 \\ 0 & 0 & 0 & 1 & 0 & 0 & 0 & 0 & 0 \\ 0 & 0 & 0 & 0 & 0 & 0 & 0 & 0 & 0 \end{bmatrix}. \tag{94}$$

The number of 1's in (94) is still the rank of A.

This means that, in some coordinates, a linear map from one linear space to another just forgets some coordinates, and pads the remaining ones by zeros. Thus a linear map from one linear space to another may be pictured as follows:

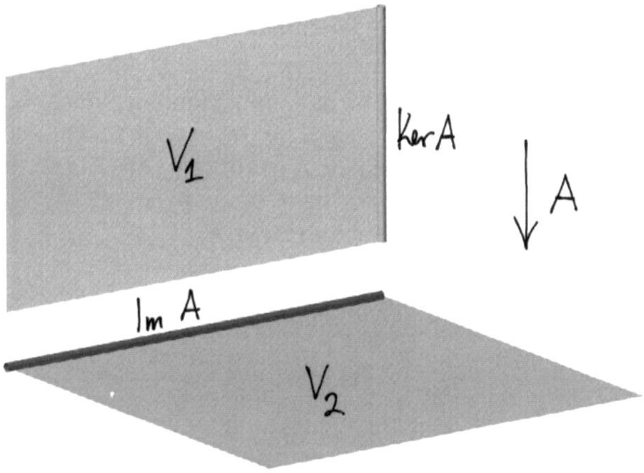

(95)

The symbols $\operatorname{Ker}\mathsf{A}$ and $\operatorname{Im}\mathsf{A}$ will be explained in Section A.4.

A.3. Abstract linear spaces

A set V is called a linear space over a field $\mathbb{F}$ if it has a special element $0 \in \mathsf{V}$, an operation of addition, and an operations of multiplication by $t \in \mathbb{F}$, satisfying the same rules as the corresponding operations in $\mathbb{F}^M$.

Any collection of vectors $\mathsf{v}_1, \ldots, \mathsf{v}_M$ determines a linear map

$$\mathbb{F}^M \to \mathsf{V}$$

given by

$$\begin{bmatrix} x_1 \\ \vdots \\ x_M \end{bmatrix} \mapsto \sum x_i \mathsf{v}_i. \tag{96}$$

15 Something much more interesting happens if the source and target are the *same* space and we have to use the *same* change of variables in both.

The linear space is said to have *finite dimension* if for some collection $v_1, \dots, v_M$ the map (96) is surjective. By a version of row reduction, there is then a subset of v_i's for which the map (96) becomes an isomorphism. Such a set of vectors is called a *basis* for V, and its cardinality is denoted $\dim \mathsf{V}$. From what we already know, V has many bases, which can be all taken to each other by the group $\mathrm{GL}(\mathsf{V}) = \mathrm{GL}(\dim \mathsf{V}, \mathbb{F})$.

A.4. Kernel, image, and quotient

Given a linear map

$$\mathsf{A} : \mathsf{V}_1 \to \mathsf{V}_2,$$

one defines its *kernel* by

$$\operatorname{Ker} \mathsf{A} = \{\mathsf{v} \in \mathsf{V}_1,\ \mathsf{Av} = 0\}. \tag{97}$$

This is a *linear subspace* of the source space of A of dimension

$$\dim \operatorname{Ker} \mathsf{A} = \dim \mathsf{V}_1 - \operatorname{rank} \mathsf{A}. \tag{98}$$

The projection in (95) is the projection *along* the kernel of A. One defines the image of A as the image of this projection, or more formally

$$\operatorname{Im} \mathsf{A} = \left\{\mathsf{v}' \in \mathsf{V}_2,\ \mathsf{v}' = \mathsf{Av} \text{ for some } \mathsf{v} \in \mathsf{V}_1\right\}. \tag{99}$$

This is a linear subspace in the target space of A of dimension

$$\dim \operatorname{Im} \mathsf{A} = \operatorname{rank} \mathsf{A}. \tag{100}$$

Thus, A may be factored as a projection and embedding

$$\underbrace{\mathsf{V}_1 \xrightarrow{\text{projection}} \operatorname{Im} \mathsf{A} \xrightarrow{\text{embedding}} \mathsf{V}_2}_{\mathsf{A}} \ . \tag{101}$$

Mathematicians call a sequence of maps of the form

$$0 \to \operatorname{Ker} \mathsf{A} \xrightarrow{\text{embedding}} \mathsf{V}_1 \xrightarrow{\text{projection}} \operatorname{Im} \mathsf{A} \to 0 \tag{102}$$

a *short exact sequence*. This is an important word to remember and use. It is *exact* because the the kernel of each arrow in (102) is the image of the preceding map. It is short because it contains only 3 nontrivial terms. One also says that a projection is the *quotient* of V_1 by its kernel.

A.5. Dual vector spaces

For an $\mathbb{F}$-linear space V, the space

$$\mathsf{V}^\vee = \{\text{linear functions } \mathsf{V} \xrightarrow{\ \xi\ } \mathbb{F}\}$$

is also an $\mathbb{F}$-linear space, since we can add linear functions and multiply them by numbers. The dual space to $\mathbb{F}^M$ is best visualized as the space of row vectors

$$(\mathbb{F}^M)^\vee = \left\{[\xi_1, \dots, \xi_M]\right\}, \quad \xi_1, \dots, \xi_M \in \mathbb{F}, \tag{103}$$

with

$$\boldsymbol{\xi}(\mathsf{x}) = [\xi_1, \ldots, \xi_M] \begin{bmatrix} x_1 \\ \vdots \\ x_M \end{bmatrix} = \sum \xi_i x_i \stackrel{\text{def}}{=} \langle \boldsymbol{\xi}, \mathsf{x} \rangle. \tag{104}$$

Here the notation $\langle \boldsymbol{\xi}, \mathsf{x} \rangle$ is introduced to stress the symmetry between $\boldsymbol{\xi}$ and x. Mathematicians like to stress that, while V and $\mathsf{V}^\vee$ are vector spaces of the same dimension, there is no *natural*, that is, coordinate-independent identification between them. By contrast, the symmetry in (104) shows that $(\mathsf{V}^\vee)^\vee = \mathsf{V}$ in a coordinate-free way for any finite-dimensional vector space.

For any linear subspace $\mathsf{V}' \subset \mathsf{V}$, there is the *annihilator* subspace

$$(\mathsf{V}')^\perp = \left\{ \boldsymbol{\xi} \text{ such that } \left\langle \boldsymbol{\xi}, \mathsf{v}' \right\rangle = 0 \text{ for all } \mathsf{v}' \in \mathsf{V}' \right\}.$$

This is a subspace of $\mathsf{V}^\vee$ of dimension

$$\dim(\mathsf{V}')^\perp = \dim \mathsf{V} - \dim \mathsf{V}', \tag{105}$$

satisfying $((\mathsf{V}')^\perp)^\perp = \mathsf{V}'$.

A linear map $\mathsf{A} : \mathsf{V}_1 \to \mathsf{V}_2$ induces the dual map

$$\mathsf{A}^\vee : \mathsf{V}_2^\vee \to \mathsf{V}_1^\vee$$

by precomposing a function with A. In other words,

$$\left\langle \mathsf{A}^\vee \boldsymbol{\xi}, \mathsf{x} \right\rangle = \langle \boldsymbol{\xi}, \mathsf{A}\mathsf{x} \rangle.$$

For row vectors as in (103), this is just *left* multiplication $\boldsymbol{\xi} \mapsto \boldsymbol{\xi}\mathsf{A}$. It is important that duality *reverses* the order of the composition

$$(\mathsf{A}_1\mathsf{A}_2)^\vee = \mathsf{A}_2^\vee \mathsf{A}_1^\vee.$$

We have

$$\operatorname{Ker} \mathsf{A}^\vee = (\operatorname{Im} \mathsf{A})^\perp, \quad \operatorname{Im} \mathsf{A}^\vee = (\operatorname{Ker} \mathsf{A})^\perp,$$

and in particular

$$\operatorname{rank} \mathsf{A}^\vee = \operatorname{rank} \mathsf{A}.$$

If we insist on identifying the row vectors with column vectors by switching the rows and columns then $\mathsf{A}^\vee$ becomes the *transposed matrix*

$$(a_{ij})^{\mathrm{T}} = (a_{ji}). \tag{106}$$

A.6. Rank and rank

Let $P_1, \ldots, P_N$ be a collection of points in $\mathbb{P}^d(\mathbb{F})$, where the projective space is defined in Section 5.2. If

$$\mathbf{P} = N \times (d+1) \text{ matrix with rows } P_i,$$

then the equation

$$\mathbf{Pa} = 0, \quad \mathbf{a} = \begin{bmatrix} a_0 \\ \vdots \\ a_d \end{bmatrix} \tag{107}$$

describes the hyperplanes containing the points $P_1, \dots, P_N$. Thus

$$\begin{aligned} \operatorname{rank}(\{P_1, \dots, P_N\}) &= \dim \operatorname{span}(\{P_1, \dots, P_N\}) + 1 \\ &= (d+1) - \dim \text{solutions of } (107) \\ &= \operatorname{rank} \mathbf{P}. \end{aligned} \tag{108}$$

More generally, the rank of any subset of $\{P_1, \dots, P_N\}$ is the rank of the corresponding submatrix in $\mathbf{P}$.

For a practical computation of the rank, it is enough to bring a matrix by row operations to the row echelon form

$$\mathsf{A}_{\text{row echelon}} = \begin{bmatrix} 0 & 0 & \star & * & * & * & * & * & * \\ 0 & 0 & 0 & 0 & \star & * & * & * & * \\ 0 & 0 & 0 & 0 & 0 & \star & * & * & * \\ 0 & 0 & 0 & 0 & 0 & 0 & 0 & \star & * \\ 0 & 0 & 0 & 0 & 0 & 0 & 0 & 0 & 0 \end{bmatrix}, \tag{109}$$

where $\star$ stands for some nonzero element of $\mathbb{F}$.

A theoretical formula for the rank may given using determinants; see Appendix B. Namely, the rank is the maximal size of a square submatrix with nonzero determinant.

B. DETERMINANT

B.1. Formula

A matrix $\mathsf{A} : \mathbb{F}^N \to \mathbb{F}^N$ is invertible if and only if $\det \mathsf{A} \neq 0$, where $\det \mathsf{A}$ is a certain polynomial in matrix elements a_{ij}. An explicit formula for this polynomial was needed in Section 6.4 to deduce the existence of a matching from equation (57).

This formula is a sum over *permutations* σ of $\{1, \dots, N\}$. It reads

$$\det A = \sum_{\text{permutations } \sigma} \operatorname{sgn}(\sigma) a_{1,\sigma(1)} a_{2,\sigma(2)} \cdots a_{N,\sigma(N)}, \tag{110}$$

where permutations and their signs are defined as follows. See further below for one possible explanation of the formula (110).

B.2. Permutations

By definition, a permutation σ of a finite set is a bijective map from a set to itself, like the following example:

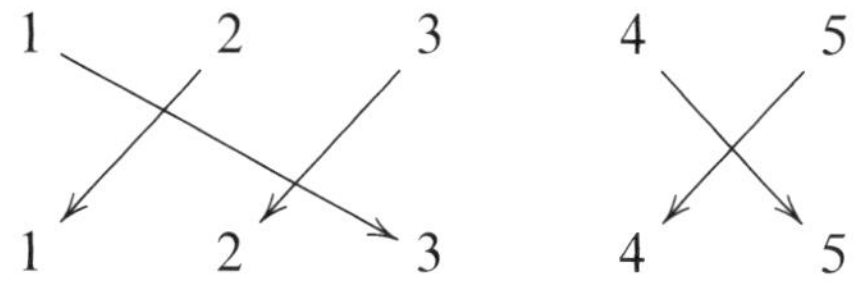

(111)

for $N = 5$. A permutation has a sign defined by

$$\operatorname{sgn}(\sigma) = (-1)^{|\text{crossings in (111)}|} = (-1)^{|\text{inversions}|}, \tag{112}$$

where an inversion of σ is a pair $i < j$ such that $\sigma(i) > \sigma(j)$. For example, $(1,2)$, $(1,3)$, and $(4,5)$ are the inversions for σ in (111) and hence $\operatorname{sgn}(\sigma) = -1$ for this particular permutation.

It is a nice exercise in the spirit of the figure in (81) to check that

$$\operatorname{sgn}(\sigma_1\sigma_2) = \operatorname{sgn}(\sigma_1)\operatorname{sgn}(\sigma_2),$$

where $\sigma_1\sigma_2$ denotes the composition of two permutations. In particular, the sign does not depend of how we order an N-element set. Compare the sign in (111) and below:

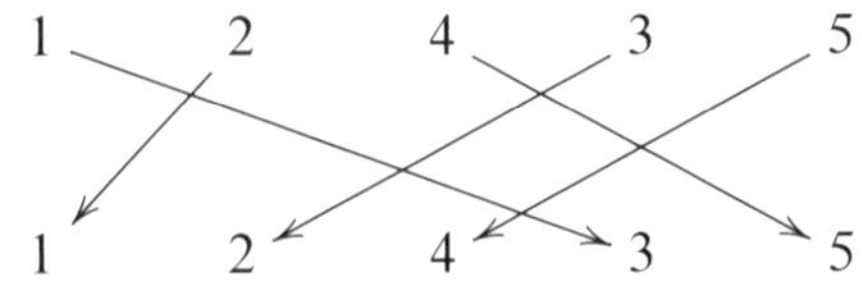

B.3. The $N = 2$ case and the cohomology of the torus

Let

$$\mathsf{A} = \begin{bmatrix} a & b \\ c & d \end{bmatrix}$$

be a 2×2 matrix. Then

$$\det \mathsf{A} = ad - bc \tag{113}$$

and

$$\mathsf{A}^{-1} = \frac{1}{\det \mathsf{A}} \begin{bmatrix} d & -b \\ -c & a \end{bmatrix}, \tag{114}$$

over any field $\mathbb{F}$, as can be checked directly. Thus indeed we see that $\det \mathsf{A} \neq 0$ is equivalent to invertibility of A.

Let us see what the cohomology of the torus $H^\bullet(\Sigma, \mathbb{F})$ can tell us about the formula (114). We hope the reader did the exercise suggested in Section 8.2 and remembers that

$$H^0(\Sigma, \mathbb{F}) = \mathbb{F}, \quad H^1(\Sigma, \mathbb{F}) = \mathbb{F}\gamma_1 \oplus \mathbb{F}\gamma_2, \quad H^2(\Sigma, \mathbb{F}) = \mathbb{F}\gamma_1 \cup \gamma_2, \tag{115}$$

for some basis $\{\gamma_1, \gamma_2\}$ of the 2-dimensional space $H^1(\Sigma, \mathbb{F})$. The description (115) means that $H^\bullet(\Sigma, \mathbb{F})$ is generated by γ_1 and γ_2 using the cup product, and the relations that these generators satisfy are

$$\gamma_1 \cup \gamma_1 = \gamma_2 \cup \gamma_2 = \gamma_1 \cup \gamma_2 + \gamma_2 \cup \gamma_1 = 0. \tag{116}$$

These can be written more compactly as follows:

$$\text{for all } \gamma \in H^1, \quad \gamma \cup \gamma = 0. \tag{117}$$

Now, H^1 is 2-dimensional linear space over $\mathbb{F}$ with a basis, hence we can act by the matrix A in it. This action preserves the relation (117), and so induces the action on H^2. Since

$$(a\gamma_1 + c\gamma_2) \cup (b\gamma_1 + d\gamma_2) = (ab - cd)\gamma_1 \cup \gamma_2,$$

we conclude that A acts on H^2 as follows:

$$H^2(\Sigma, \mathbb{F}) \xrightarrow{\text{multiplication by } \det \mathsf{A}} H^2(\Sigma, \mathbb{F}).$$

Writing the cup product as the Poincaré duality pairing (83), we conclude

$$(\mathsf{A}\gamma, \mathsf{A}\gamma') = (\gamma, \gamma') \det \mathsf{A}$$

for any $\gamma, \gamma' \in H^1$. Introducing a new variable $\gamma'' = \mathsf{A}\gamma$, we see that

$$(\mathsf{A}^{-1}\gamma'', \gamma') = \frac{1}{\det \mathsf{A}}(\gamma'', \mathsf{A}\gamma'),$$

which is equivalent to (114).

B.4. The general case

The torus $\Sigma = S^1 \times S^1$ is the product of two circles. We have

$$H^0(S^1, \mathbb{F}) = \mathbb{F}, \quad H^1(S^1, \mathbb{F}) = \mathbb{F}\gamma, \quad \gamma \cup \gamma = 0.$$

The relation

$$\gamma_1 \cup \gamma_2 = -\gamma_2 \cup \gamma_1$$

that we have in

$$H^\bullet(\Sigma, \mathbb{F}) = H^\bullet(S^1, \mathbb{F}) \otimes H^\bullet(S^1, \mathbb{F})$$

is an illustration of how one is supposed to put signs in the Künneth theorem (76) for odd cohomology classes.

Now we can take

$$H^\bullet\big((S^1)^N, \mathbb{F}\big) = \mathbb{F}\langle \gamma_1, \dots, \gamma_n \rangle / \big(\gamma_i^2 = 0, \gamma_i\gamma_j + \gamma_j\gamma_i = 0\big), \tag{118}$$

where angle brackets means we do not assume that γ_i and γ_j commute. Indeed, they anti-commute in the algebra (118). Note that the dimensions $\dim H^i((S^1)^N, \mathbb{F}) = \binom{N}{i}$ are the binomial coefficients, and hence the symmetry of the binomial coefficients may be interpreted as an instance of Poincaré duality.

When we act by A in the basis $\{\gamma_1, \dots, \gamma_n\}$ of H^1, we get, unraveling the definitions,

$$H^N\big((S^1)^N, \mathbb{F}\big) \xrightarrow{\text{multiplication by (110)}} H^N\big((S^1)^N, \mathbb{F}\big).$$

The Poincaré duality between H^1 and H^{N-1} gives the Cramer's formula for A^{-1}.

C. TROPICAL LINES, PLANES, ETC.

C.1.

Consider the line

$$x_2 = ax_1 + b \subset \mathbb{C}^2, \quad a, b \neq 0, \tag{119}$$

equivalently, the graph of the function $x_1 \mapsto ax_1 + b$. What does it look like when x_1 and x_2 are exponentially large or small? The question being a little vague, let us start by describing the set of possible values of $v_i = \ln|x_i|$ for (x_1, x_2) satisfying (119).

The complex numbers $\{-x_2, ax_1, b\} \subset \mathbb{C}$ sum to zero, hence we can form a triangle in the complex planes with these vectors as sides. The triangle inequality says that a triangle with side lengths L_1, L_2, L_3 exists if an only if each L_i is less than or equal to the sum of the other two numbers. This means

$$\begin{aligned} |x_2| &\leq |ax_1| + |b|, \\ |ax_1| &\leq |x_2| + |b|, \\ |b| &\leq |ax_1| + |x_2|, \end{aligned} \tag{120}$$

which is equivalent to

$$\ln\big(\pm|a|e^{\nu_1} \mp |b|\big) \leq \nu_2 \leq \ln\big(|a|e^{\nu_1} + |b|\big). \tag{121}$$

See the plot on the left in (122) for $|a| = 2$, $|b| = 3$. Note that this graph dips to $\nu_2 = -\infty$ precisely at the value of ν_1 that corresponds to the unique root $x_1 = -b/a$ of $ax_1 + b = 0$.

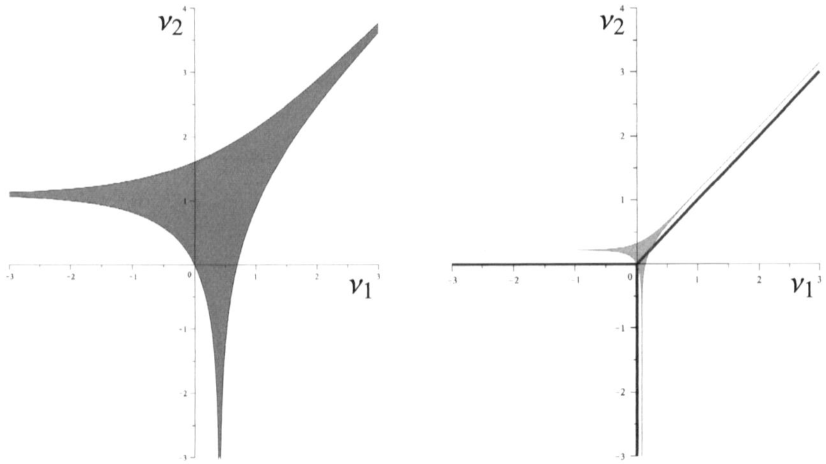

(122)

What does the plot on the left in the figure of (122) look like at a very large scale? We should rescale it by $1/T$, where T some large number, and take the limit when $T \to +\infty$. For $T = 5$, we get the shape in gray on the right in the figure of (122), and as $T \to \infty$ these shapes converge to the union of 3 rays plotted in blue. This union of three rays is called the *tropicalization* of the line (119).

C.2.

Here is an alternative way to talk about x_i being exponentially large or small. Above, we had a parameter $T \gg 0$, and after rescaling ν_i by T, the absolute values $|x_i|$ were of the order $e^{\nu_i T}$. To introduce objects like $e^{\nu_i T}$, where T is a parameter, into the framework of linear algebra, there should be some extension of the field $\mathbb{C}$ that contains elements $e^{\xi T}$ for all $\xi \in \mathbb{R}$. Their multiplication is clear

$$e^{\xi_1 T} \cdot e^{\xi_2 T} = e^{(\xi_1 + \xi_2)T},$$

but how should we compute inverses like $(e^{\xi_1 T} + e^{\xi_2 T})^{-1}$?

Since we are interested in the $T \to +\infty$ limit, we should focus on the larger of the two exponents ξ_i in $e^{\xi_1 T} + e^{\xi_2 T}$. Suppose $\xi_1 > \xi_2$. Then we can write

$$\frac{1}{e^{\xi_1 T} + e^{\xi_2 T}} = \frac{1}{e^{\xi_1 T}} \frac{1}{1 + e^{(\xi_2 - \xi_1)T}} = e^{-\xi_1 T} \sum_{n=0}^{\infty} (-1)^n e^{n(\xi_2 - \xi_1)T}. \tag{123}$$

The series in (123) is a geometric series which converges in the usual sense of calculus if T is a positive real number. For our purposes, a much weaker notion of convergence will be sufficient.

By definition, an absolute value on an algebra $\mathbb{A}$ is a function

$$\mathbb{A} \xrightarrow{\|\cdot\|} \mathbb{R}_{\geq 0}$$

that satisfies

$$\|x\| = 0 \quad \Leftrightarrow \quad x = 0,$$

$$\|xy\| = \|x\| \|y\|, \tag{124}$$

$$\|x + y\| \leq \|x\| + \|y\|. \tag{125}$$

For example, the usual absolute value on $\mathbb{C}$ used in (120) satisfies the above conditions, and we have used the triangle inequality (125) in the derivation of (120).

Another example of an absolute value is

$$\left\| \sum c_i e^{\xi_i T} \right\|_{\sim} = e^{\xi_{\max}}, \quad \xi_{\max} = \max_{c_i \neq 0} \xi_i, \tag{126}$$

where is subscript is chosen to remind us that the absolute value (126) records the leading *asymptotics* in the $T \to +\infty$ limit. Instead of (125), this absolute value satisfies a stronger property

$$\|x + y\|_{\sim} \leq \max\big(\|x\|_{\sim}, \|y\|_{\sim}\big), \quad \text{and, moreover,} \tag{127}$$

$$\|x + y\|_{\sim} = \max\big(\|x\|_{\sim}, \|y\|_{\sim}\big) \quad \text{if } \|x\|_{\sim} \neq \|y\|_{\sim}. \tag{128}$$

Such absolute values are called *nonarchimedian*.

The series in (123) converges with respect to the absolute value (126) in the sense that

$$\left\| \sum_{n=N}^{\infty} e^{n(\xi_2 - \xi_1)} \right\|_{\sim} \to 0, \quad N \to \infty. \tag{129}$$

More generally, all series of the following form converge:

$$\mathbb{F}_{\sim} = \left\{ \sum c_i e^{\xi_i T}, \text{ where } c_i \in \mathbb{C} \text{ and } \lim \xi_i = -\infty \right\}. \tag{130}$$

The reader should check that (130) is a field. The formula (126) defines an absolute value on this field. To save on notation, one can denote $t = e^{-T}$. The series (130) are then a series in ascending real powers of t.

The unifying feature in the figure of (122) is that in both cases we have the image of the line (119) under the map

$$(x_1, x_2) \mapsto \big(\ln \|x_1\|, \ln \|x_2\|\big).$$

Mathematicians call such images amoebas because they will look a little bit like an amoeba if we replace the line by a plane curve defined by an equation of degree ≥ 3. In other words, the tropical line is a *nonarchimedian amoeba* of a line.

C.3.

Given an absolute value $\|\cdot\|$, we define

$$\nu(x) = \ln \|x\| \in \mathbb{R} \cup \{-\infty\}. \tag{131}$$

For a nonarchimedian absolute value $\|\cdot\|$, this satisfies

$$\begin{aligned} \nu(x) &= -\infty \quad \Leftrightarrow \quad x = 0, \\ \nu(xy) &= \nu(x) + \nu(y), \qquad (132) \\ \nu(x+y) &\in \max\nolimits_{\curlyvee}\big(\nu(x), \nu(y)\big), \qquad (133) \end{aligned}$$

where (133) combines the two cases (127) and (128) into one formula using a multivalued function

$$\max\nolimits_{\curlyvee}(\nu_1, \ldots, \nu_n) = \begin{cases} \max \nu_i, & \text{if this maximum is unique,} \\ [-\infty, \max \nu_i], & \text{otherwise.} \end{cases} \tag{134}$$

The subscript in (134) is to remind us what the graph of this function looks like. Indeed, the graph on the right in (122) is the plot of the multivalued function

$$\max\nolimits_{\curlyvee}(0, \xi_1) = \text{possible values of } \nu(ax_1 + b),$$

where $\nu((a, b, x_1)) = (0, 0, \xi_1)$.

C.4.

Now we are ready to generalize this discussion to a hyperplane

$$x_{n+1} = \sum_{i=1}^{n} a_i x_i + a_0, \quad (x_1, \ldots, x_{n+1}) \in \mathbb{C}^{n+1}, \tag{135}$$

where all coefficients a_i are nonzero complex numbers. All arguments above generalize verbatim and give

$$\xi_{n+1} = \max\nolimits_{\curlyvee}(0, \xi_1, \ldots, \xi_n) \tag{136}$$

as the tropicalization of (135). This is what the plot of this function looks like for $n = 2$. This is a tropical hyperplane in 3-space.

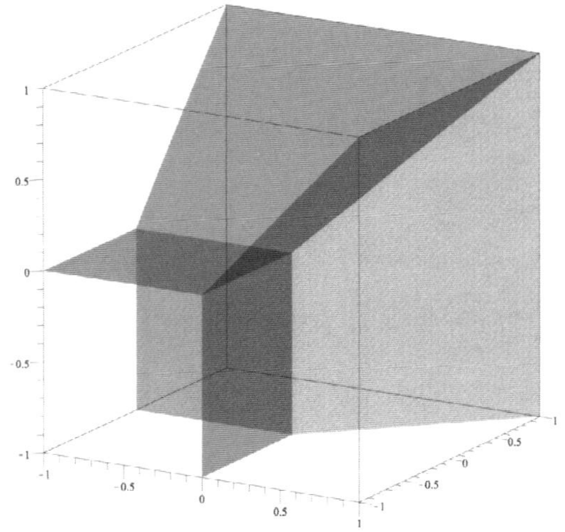

(137)

Note that we have

$$\max_{\curlyvee}(0, \xi_1, \xi_2) = -\infty = \nu(0), \tag{138}$$

precisely for (ξ_1, ξ_2) forming a tropical line. This is a tropical analog of the obvious fact the the intersection of $x_{n+1} = 0$ with another hyperplane is a hyperplane in $\mathbb{C}^n$.

C.5.

In place of a linear polynomial in (135), we could have taken an arbitrary polynomial

$$x_{n+1} = P(x_1, \dots, x_n) = \sum_{\boldsymbol{\beta} \in \mathbb{Z}^n} a_{\boldsymbol{\beta}} x^{\boldsymbol{\beta}}, \quad a_{\boldsymbol{\beta}} \in \mathbb{F}_{\sim}, \tag{139}$$

where $\boldsymbol{\beta} = [\beta_1, \dots, \beta_n] \in \mathbb{Z}^n$ and

$$x^{\boldsymbol{\beta}} = \prod_{i=1}^{n} x_i^{\beta_i}.$$

In (139) we assume that only finitely many coefficients $a_{\boldsymbol{\beta}}$ are nonzero. Note that here we allow $a_{\boldsymbol{\beta}}$ to be any elements of $F_{\sim}$. In other words, we allow the coefficients to be exponentially large or small.

Arguing as above, we get

$$\xi_{n+1} = \max_{\curlyvee}\left\{\langle \boldsymbol{\beta}, \boldsymbol{\xi} \rangle + \nu(a_{\boldsymbol{\beta}})\right\}_{a_{\boldsymbol{\beta}} \neq 0} \tag{140}$$

as the tropicalization of (139). Here $\boldsymbol{\xi} = [\xi_1, \dots, \xi_n]^{\mathrm{T}}$ and the angle brackets were defined in (104).

The left-hand side in (140) is called a tropical polynomial.[16] The set (140) in $\mathbb{R}^{n+1}$ is the graph of this polynomial. The intersection of the graph with $\xi_{n+1} = -\infty$ is the tropicalization of the hypersurface $P(x_1, \dots, x_n) = 0$. Here is an example of a graph of a tropical

16 Real numbers $\mathbb{R}$ with operations $\{\max, +\}$ form a semiring called tropical semiring.

polynomial of degree 3 in two variables:

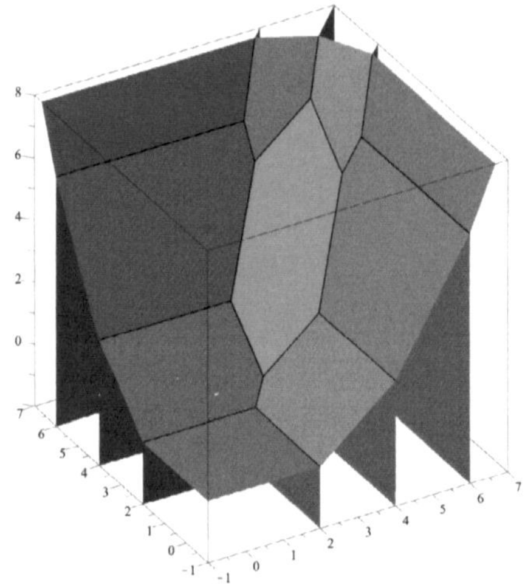

(141)

The facets in (141) have slopes $\boldsymbol{\beta} \in \{[0,0],[1,0],\ldots,[3,0],[2,1],[1,2],[0,3]\}$ and they move up and down as $\nu(a_{\boldsymbol{\beta}})$ change, leading to changes in the combinatorics. The reader should experiment to get some feeling for how this works. Graphs of linear polynomials from (122) and (137) change only by an overall translation if we take the coefficients from $\mathbb{F}_{\sim}$ instead of $\mathbb{C}$.

C.6.

Tropical varieties of codimension more than 1 are certain piecewise linear objects that are defined axiomatically; see [27, 33]. In particular, they *do not* need to be nonarchimedian amoebas of any an algebraic variety over $\mathbb{F}_{\sim}$ or some other field with a nonarchimedian norm.

In fact, this already happens for linear spaces. There is a tropical linear space for any matroid including the nonrealizable ones! To see how it works, let us go back to the settings of Section 5.3.

Let us first consider the case when the $Y \subset \mathbb{F}^n$ is a linear subspace, where $\mathbb{F}$ is a field. Given a subset $S \subset \{1,\ldots,n\}$, let $\mathbb{F}^S$ denote the quotient space of $\mathbb{F}^n$ with the coordinates y_i, where $i \in S$. We have

$$|S| = \operatorname{rank} S \quad \Leftrightarrow \quad \begin{array}{c}\text{the map } Y \to \mathbb{F}^S \\ \text{is surjective.}\end{array} \tag{142}$$

For any matroid, the subsets S such that $|S| = \operatorname{rank} S$ are called *independent*. Minimal dependent subsets are called *circuits*. One can reconstruct the matroid completely from the knowledge of either the independent sets or the circuits.

For a circuit S, the image of $Y \to \mathbb{F}^S$ is a hyperplane with an equation that involves all variables y_i with $i \in S$. We already know how to tropicalize it. One should take the tropical hyperplane

$$\max{}_{\curlyvee}\big(\{y_i\}_{i \in \text{circuit } S}\big) = -\infty. \tag{143}$$

Very remarkably, for an *arbitrary* matroid M, the equations (143), taken for all circuits S of M, describe a tropical linear space. This linear space provides a tropical realization of the matroid.

REFERENCES

[1] O. Amini and M. Piquerez, Hodge theory for tropical varieties. arXiv:2007.07826.

[2] O. Amini and M. Piquerez, Tropical Clemens–Schmid sequence and existence of tropical cycles with a given cohomology class. arXiv:2012.13142.

[3] F. Ardila and A. Boocher, The closure of a linear space in a product of lines. *J. Algebraic Combin.* **43** (2016), no. 1, 199–235.

[4] S. Axler, *Linear algebra done right*. 3rd edn., Undergrad. Texts Math., Springer, Cham, 2015.

[5] M. Baker, Hodge theory in combinatorics. *Bull. Amer. Math. Soc. (N.S.)* **55** (2018), no. 1, 57–80.

[6] M. Baker and N. Bowler, Matroids over partial hyperstructures. *Adv. Math.* **343** (2019), 821–863.

[7] M. Baker and O. Lorscheid, The moduli space of matroids. *Adv. Math.* **390** (2021), Paper No. 107883, 118.

[8] M. Baker and O. Lorscheid, Foundations of matroids I: Matroids without large uniform minors. arXiv:2008.00014.

[9] A. Björner and T. Ekedahl, On the shape of Bruhat intervals. *Ann. of Math. (2)* **170** (2009), no. 2, 799–817.

[10] T. Braden, J. Huh, J. Matherne, N. Proudfoot, and B. Wang, Singular Hodge theory for combinatorial geometries. arXiv:2010.06088.

[11] F. Brenti, Unimodal, log-concave and Pólya frequency sequences in y combinatorics. *Mem. Amer. Math. Soc.* **81** (1989), no. 413.

[12] J. Cepelewicz, He Dropped Out to Become a Poet. Now He's Won a Fields Medal, *Quanta Magazine*, July 5, 2022.

[13] N. G. de Bruijn and P. Erdős, On a combinatorial problem. *Ned. Akad. Wet., Proc.* **51** (1948), 1277–1279 = Indagationes Math. 10, 421–423 (1948).

[14] R. P. Dilworth and C. Greene, A counterexample to the generalization of Sperner's theorem. *J. Combin. Theory Ser. A* **10** (1971), 18–21.

[15] T. A. Dowling and R. M. Wilson, The slimmest geometric lattices. *Trans. Amer. Math. Soc.* **196** (1974), 203–215.

[16] T. A. Dowling and R. M. Wilson, Whitney number inequalities for geometric lattices. *Proc. Amer. Math. Soc.* **47** (1975), 504–512.

[17] B. Elias and G. Williamson, The Hodge theory of Soergel bimodules. *Ann. of Math. (2)* **180** (2014), no. 3, 1089–1136.

[18] A. Fink, Tropical cycles and Chow polytopes. *Beitr. Algebra Geom.* **54** (2013), no. 1, 13–40.

[19] C. Greene, A rank inequality for finite geometric lattices. *J. Combin. Theory* **9** (1970), 357–364.

[20] K. Hartnett, A path less taken to the peak of the math world. *Quanta Mag.* (June 27, 2017). https://www.quantamagazine.org/a-path-less-taken-to-the-peak-of-the-math-world-20170627/.

[21] A. Hatcher, *Algebraic topology*. Cambridge University Press, Cambridge, 2002.

[22] J. Huh, Combinatorial applications of the Hodge–Riemann relations. In *Proceedings of the International Congress of Mathematicians—Rio de Janeiro 2018. Vol. IV. Invited lectures*, pp. 3093–3111, World Sci. Publ., Hackensack, NJ, 2018.

[23] J. Huh, Combinatorics and Hodge theory. In *Proceedings of the International Congress of Mathematicians, Vol. 1*, pp. 212–239, EMS Press, 2022.

[24] J. Huh and B. Wang, Enumeration of points, lines, planes, etc. *Acta Math.* **218** (2017), no. 2, 297–317.

[25] A. W. Ingleton, Representation of matroids. In *Combinatorial Mathematics and its Applications (Proc. Conf., Oxford, 1969)*, pp. 149–167, Academic Press, London, 1971.

[26] A. W. Ingleton and R. A. Main, Non-algebraic matroids exist. *Bull. Lond. Math. Soc.* **7** (1975), 144–146.

[27] I. Itenberg, G. Mikhalkin, and E. Shustin, *Tropical algebraic geometry*. 2nd edn., Oberwolfach Semin. 35, Birkhäuser, Basel, 2009.

[28] G. Kalai, The Work of June Huh. In *Proceedings of the International Congress of Mathematicians, Vol. 1*, pp. 50–65, EMS Press, 2022.

[29] K. Karu, Hard Lefschetz theorem for nonrational polytopes. *Invent. Math.* **157** (2004), no. 2, 419–447.

[30] E. Katz, What is …tropical geometry? *Notices Amer. Math. Soc.* **64** (2017), no. 4, 380–382.

[31] D. Kazhdan and G. Lusztig, Schubert varieties and Poincaré duality. In *Geometry of the Laplace operator (Proc. Sympos. Pure Math., Univ. Hawaii, Honolulu, Hawaii, 1979)*, pp. 185–203 Proc. Sympos. Pure Math., XXXVI, Amer. Math. Soc., Providence, RI, 1980.

[32] B. Lindström, The non-Pappus matroid is algebraic. *Ars Combin.* **16** (1983), no. B, 95–96.

[33] D. Maclagan and B. Sturmfels, *Introduction to tropical geometry*. Grad. Stud. Math. 161, American Mathematical Society, Providence, RI, 2015.

[34] G. Melvin and W. Slofstra, Soergel bimodules and the shape of Bruhat intervals. 2020, preprint.

[35] Th. Motzkin. *Beiträge zur Theorie der linearen Ungleichungen*. Dissertation Thesis, University of Basel, Jerusalem, 1936.

[36] Th. Motzkin, The lines and planes connecting the points of a finite set. *Trans. Amer. Math. Soc.* **70** (1951), 451–464.

[37] P. Nelson, Almost all matroids are nonrepresentable. *Bull. Lond. Math. Soc.* **50** (2018), no. 2, 245–248.

[38] J. G. Oxley, *Matroid theory*. Oxford Sci. Publ., The Clarendon Press, Oxford University Press, New York, 1992.

[39] G.-C. Rota, Combinatorial theory, old and new, In *Actes du Congrès International des Mathématiciens (Nice, 1970)*, pp. 229–233, Gauthier-Villars, Paris, 1971.

[40] I. R. Shafarevich, *Basic algebraic geometry. 1*, Translated from the 2007 third Russian edition. Springer, Heidelberg, 2013.

[41] R. P. Stanley, The number of faces of a simplicial convex polytope. *Adv. Math.* **35** (1980), no. 3, 236–238.

[42] R. P. Stanley, Combinatorial applications of the hard Lefschetz theorem. In *Proceedings of the International Congress of Mathematicians, Vol. 1, 2 (Warsaw, 1983)*, pp. 447–453, PWN, Warsaw, 1984.

[43] R. P. Stanley, *Algebraic combinatorics. Walks, trees, tableaux, and more*. 2nd edn. of [MR3097651], Undergrad. Texts Math., Springer, Cham, 2018.

[44] B. Sturmfels, *Solving systems of polynomial equations*. CBMS Reg. Conf. Ser. Math. 97, Published for the Conference Board of the Mathematical Sciences, Washington, DC; American Mathematical Society, Providence, RI, 2002.

ANDREI OKOUNKOV

Department of Mathematics, University of California, Berkeley, 970 Evans Hall Berkeley, CA 94720–3840, USA, okounkov@math.columbia.edu

This Page Intentionally Left Blank

RHYMES IN PRIMES

ANDREI OKOUNKOV

ABSTRACT

While the author is a professional mathematician, he is by no means an expert in the subject area of these notes. The goal of these notes is to share the author's personal excitement about some results of James Maynard with mathematics enthusiasts of all ages, using maximally accessible, yet precise mathematical language. No attempt has been made to present an overview of the current state field, its history, or to place this narrative in any kind of broader scientific or social context. See the references in Section 11 for both professional surveys and popular science accounts that will certainly give the reader a broader and deeper understanding of the material.

MATHEMATICS SUBJECT CLASSIFICATION 2020

Primary 11N05; Secondary 97

KEYWORDS

Gaps in primes

INTERNATIONAL CONGRESS OF MATHEMATICIANS 2022 JULY 6–14

© 2022 International Mathematical Union
Proc. Int. Cong. Math. 2022, Vol. 1, pp. 460–490
DOI 10.4171/ICM2022/202

Published by EMS Press and licensed under a CC BY 4.0 license

1. THE ANCIENT SIEVE

It is hard to imagine a more fundamental arithmetic object than the multiplication table

$$\begin{array}{cccccccc} \mathbf{1} & 2 & 3 & 4 & 5 & 6 & 7 & \cdots \\ 2 & 4 & 6 & 8 & 10 & 12 & 14 & \cdots \\ 3 & 6 & 9 & 12 & 15 & 18 & 21 & \cdots \\ 4 & 8 & 12 & 16 & 20 & 24 & 28 & \cdots \\ 5 & 10 & 15 & 20 & 25 & 30 & 35 & \cdots \\ 6 & 12 & 18 & 24 & 30 & 36 & 42 & \cdots \\ 7 & 14 & 21 & 28 & 35 & 42 & 49 & \cdots \\ \vdots & \vdots & \vdots & \vdots & \vdots & \vdots & \vdots & \end{array} \qquad (1)$$

where the dots indicate that we imagine this table has infinitely many rows and columns. The numbers n that appear in the shaded area are called *composite* numbers. They can be written in the form $n = ab$ where both $a \neq 1$ and $b \neq 1$ are positive integers.

Numbers that are not 1 and not composite are called *prime*. For instance, 2, 3, 5, and 7 are prime, as one sees from (1). Indeed, every composite number ab appears in the multiplication table in the column a and row b, which are both *less* than the number ab. So, 2, 3, 5, 7 will never appear in the shaded part.

It is a fundamental arithmetic fact that every positive integer $n > 1$ can be factored as a product of primes, and this factorization is unique up to the order of the prime factors. One can compare and contrast factorization into primes with how molecules are built from atoms. One clear difference is that the order of prime factors does not matter, unlike the positions of the atoms in a molecule.

Primes form an infinite sequence which has mesmerized and puzzled mathematicians for millenia. Many mathematicians were first attracted to mathematics by the magic of prime numbers and remained true to their first mathematical love — number theory.

"It is the fact that primes are so fundamental (being the building blocks of whole numbers), but still so mysterious and poorly understood which makes them so fascinating to me," says James Maynard, the hero of these notes. Kannan Soundararajan, the presenter of Maynard's Fields Medal laudatio at ICM 2022, agrees: "*Like many others, I was drawn in by the extreme simplicity of problems involving primes, and the remarkable difficulty of proving anything about them. Twin primes and Goldbach in particular were especially fascinating problems. It's been amazing to witness such spectacular progress as the Green–Tao theorem and bounded gaps between primes over the last twenty years.*"

The following method for tabulating the primes goes at least far back as Eratosthenes (276–195/194 BC). To remove the composite numbers from the list of all numbers, we can successively cross out or punch trough all numbers from the grey columns in the multiplication table (1), that is, remove all nontrivial multiples of 2, of 3, of 5, etc. For instance, the list of natural numbers with 1 and multiples of 2 and 3 removed will look like this:

○	2	3	~~4~~	5	~~6~~	7	~~8~~	~~9~~	~~10~~
11	~~12~~	13	~~14~~	~~15~~	~~16~~	17	~~18~~	19	~~20~~
~~21~~	~~22~~	23	~~24~~	25	~~26~~	~~27~~	~~28~~	29	~~30~~
31	~~32~~	~~33~~	~~34~~	35	~~36~~	37	~~38~~	~~39~~	~~40~~
41	~~42~~	43	~~44~~	~~45~~	~~46~~	47	~~48~~	49	~~50~~
~~51~~	~~52~~	53	~~54~~	55	~~56~~	~~57~~	~~58~~	59	~~60~~
61	~~62~~	~~63~~	~~64~~	65	~~66~~	67	~~68~~	~~69~~	~~70~~
71	~~72~~	73	~~74~~	~~75~~	~~76~~	77	~~78~~	79	~~80~~
~~81~~	~~82~~	83	~~84~~	85	~~86~~	~~87~~	~~88~~	89	~~90~~
91	~~92~~	~~93~~	~~94~~	95	~~96~~	97	~~98~~	~~99~~	~~100~~
⋮	⋮	⋮	⋮	⋮	⋮	⋮	⋮	⋮	⋮

(2)

where dots indicate that this table has infinitely many rows. The reader may notice there is no need to worry about multiples of 4, 6, or any other composite number.

Once we remove all composite numbers from numbers up to a 100, the result will look like this (the colors will be explained momentarily):

○	2	**3**	○	**5**	○	**7**	○	○	○
11	○	**13**	○	○	○	**17**	○	**19**	○
○	○	23	○	○	○	○	○	**29**	○
31	○	○	○	○	○	37	○	○	○
41	○	**43**	○	○	○	47	○	○	○
○	○	53	○	○	○	○	○	**59**	○
61	○	○	○	○	○	67	○	○	○
71	○	**73**	○	○	○	○	○	79	○
○	○	83	○	○	○	○	○	89	○
○	○	○	○	○	○	97	○	○	○
⋮	⋮	⋮	⋮	⋮	⋮	⋮	⋮	⋮	⋮

(3)

This table has a lot of holes, just like a sieve. For this reason, the methods that produce an interesting set (e.g., primes) from a less interesting set (e.g., integers) by successively sifting out the unwanted elements are referred to as *sieve* methods.

The primes shown in green are the *twin primes*, that is, primes p such that $p+2$ or $p-2$ are also prime.[1] Twin primes are the simplest rhymes in the mysterious poem of primes. While it is very easy to see that there are infinitely many primes,[2] the infinitude of twin primes is a very old conjecture, still open today. However, the recent years saw an incredible progress in our understanding of various patterns in primes, recognized, in particular, by the Fields medal, the highest honor in mathematics, awarded in 2022 to James Maynard.

---

1 Can you prove that $p+2$ and $p-2$ cannot both be prime, except for $p=5$? Questions like this will be clarified when we talk about *admissible* patterns.

2 Every divisor of the number $n!+1$, where $n! = 1 \cdot 2 \cdot 3 \cdots n$, has to be larger than n. Since n is arbitrary, there are infinitely many primes.

In these notes, we will try to give a very basic introduction to this area of number theory and some of the results of Maynard and his predecessors. A more experienced reader can probably skip many sections of this narrative. All newcomers we wish some patience working through these notes, and very much hope this patience will be rewarded by the sense of awe that this mathematics inspires.

2. LAST DIGITS OF PRIMES

It is very noticeable in (3) that some columns have very few (in fact, zero or one) prime numbers in them. Given a number n, its column number in (3) is determined by the last digit of n in its decimal notation or, equivalently, by the remainder in the division of n by 10. Mathematicians have a special notation for the remainder, namely

$$89 \bmod 10 = 9.$$

One also says that the *residue* of 89 modulo 10 is 9. More generally, we write

$$a_1 = a_2 \bmod b$$

to mean that $a_1 - a_2$ is divisible by b. We say that a_1 and a_2 are equal mod b, or that they are in the same *residue class* modulo b.

If $n \bmod 10 = 8$ then n is even and not equal to 2, hence n cannot possibly be prime. Therefore, the 8th column in (3) is empty. Similar reasoning applies to the 2nd, 4th, 5th, 6th, and 10th columns. In due time we will see that prime numbers are approximately evenly distributed among the remaining 4 columns of table (3). Whether the column corresponding to a residue a modulo 10 has many or very few primes is determined by the greatest common divisor $\gcd(a, 10)$. The columns with $\gcd(a, 10) > 1$ contain at most one prime.

The base 10 of the decimal expansion can be replaced by any other base $b > 1$. For instance, $b = 2$ means binary expansions, as exemplified by

$$23 = 10111_{\text{binary}} = 1 \cdot 2^4 + 0 \cdot 2^3 + 1 \cdot 2^2 + 1 \cdot 2^1 + 1 \cdot 2^0. \tag{4}$$

Clearly, for all primes $p \neq 2$, we have $p = 1 \bmod 2$.

Generalizing what we have seen for $b = 10$ and $b = 2$, for any base b, primes are approximately evenly distributed among residue classes a modulo b such that $\gcd(a, b) = 1$. The residue classes with $\gcd(a, b) > 1$ contain at most one prime each.

For example, if we replace base $b = 10$ in (3), by $b = 211$, which is a prime number, we will get the following distribution of primes $p \leq 211^2$ (shown by blue or green squares, colors mean the same as in (3)).

 (5)

Primes indeed seem to be roughly evenly distributed among all columns,[3] except the very last one, which contains the multiples of 211. Of course, what catches the eye in this picture are the diagonal stripes. We invite the reader to explain them using the equality

$$211i + j = i + j \mod 210$$

and the factorization $210 = 2 \cdot 3 \cdot 5 \cdot 7$.

3. THE CHINESE REMAINDER THEOREM

One can add and multiply residue classes modulo b in the same way that one can tell the last digit of a sum $n_1 + n_2$ or a product $n_1 n_2$ from the last digits of n_1 and n_2. Such considerations of are both very basic and central to number theory. They can be simplified using the Chinese remainder theorem (CRT), which is a result nearly as ancient as the Eratosthenes sieve, appearing in Sunzi Suanjing treatise from the 3rd century CE.

CRT applies to residues modulo $b = b_1 b_2$, where b_1 and b_2 are *coprime*, meaning that $\gcd(b_1, b_2) = 1$. For example, $10 = 2 \cdot 5$ and $\gcd(2, 5) = 1$. Given a residue a modulo b, we can associate to it two numbers

$$a \longrightarrow (a_1, a_2) = (a \bmod b_1, a \bmod b_2). \tag{6}$$

---

3 Actually, the number of primes in any given column in (5) varies between 14 and 31, but it all evens out as we go further and further down the list of primes. It is a fact of life that it takes a while for primes p to equidistribute mod any fixed prime like $q = 211$. It is a very subtle business to find out how long exactly this while can be, for either some fixed q and or averaged over q. This is, in fact, one of the key technical questions in this part of number theory.

For instance, for $b = 10 = 2 \cdot 5$, consider the following table. The rows and columns of this table are indexed by residues mod 2 and 5, respectively, and we place each residue mod 10 in the corresponding row and column:

	1	2	3	4	0	mod 5
1 mod 2	1	7	3	9	5	.
0 mod 2	6	2	8	4	0	

(7)

We observe the remarkable fact that each residue $a = 0, 1, \ldots, 9$ mod 10 finds a unique place in this table, filling the table completely. In general, CRT says that the map (6) gives a one-to-one correspondence

$$\{\text{residues mod } b_1 b_2\} = \{\text{residues mod } b_1\} \times \{\text{residues mod } b_2\} \tag{8}$$

that preserves arithmetic operations. We invite the reader to prove the CRT and to generalize its statement to the case $b = b_1 b_2 \cdots b_r$.

Let us revisit table (3) from the point of view of CRT. Shading the residue classes that contain ≤ 1 primes, we get

	1	2	3	4	0	mod 5
1 mod 2	1	7	3	9	5	
0 mod 2	6	2	8	4	0	

(9)

which illustrates two key points:

- a is coprime to 10 if and only if a is coprime to 2 and 5,
- being coprime to 2 and 5 are *independent events*.

Here we think of residue classes a modulo 10 as all equally likely and we call two events $\mathscr{E}_1$ and $\mathscr{E}_2$ independent if

$$\text{Prob}(\mathscr{E}_1 \& \mathscr{E}_2) = \text{Prob}(\mathscr{E}_1)\,\text{Prob}(\mathscr{E}_2).$$

While primes are truly special and not random at all, after centuries of looking into patterns in primes most mathematicians would probably agree that primes behave as if they were completely random, subject to, first, all possible constraints imposed by the considerations of residues and, second, density constraints imposed by the unique factorization of integers into primes. It is therefore very useful to inject, following Cramér, some probabilistic terminology and intuition into our discussion.

4. INFINITY AND LIMITS

There is mystery and challenge in primes because there are infinitely many of them. Any list or plot of primes that we can examine, however long, contains only 0% of all primes, hence always at the best provides a warm-up for the real question. Which is: what happens for all sufficiently large primes?

In mathematics, there are lot of questions for which one is free to discard an arbitrary finite part of some infinite data set. As an example, let us take the concept of a *limit*, which is very important when talking about primes. In the discussion that follows, we will very often have a sequence of real numbers

$$(a_n) = (a_1, a_2, a_3, \dots),$$

that tends to a limit

$$a = \lim_{n\to\infty} a_n \tag{10}$$

as n goes to infinity. Slightly incorrectly, this means that every digit in the decimal expansions of a_n's equals to that of a, except for finitely many values of n. Any person trained in calculus will be quick to point out some problems with this definition, namely

$$a_n = 10^n \nrightarrow 0,$$

even though every digit of a_n is zero except for one value of n, while

$$a_n = 0.\underbrace{999\dots9}_{n \text{ times}} \to 1.0000\dots,$$

despite the fact that all displayed digits are different. Readers who are not sure how to fix these issues and feel they could use a more rigorous discussion, can find it in Appendix A.

With the notion of a limit, one can define infinite sums and products by

$$\sum_{n=1}^{\infty} a_n = \lim_{N\to\infty} \sum_{n=1}^{N} a_n, \quad \prod_{n=1}^{\infty} a_n = \lim_{N\to\infty} \prod_{n=1}^{N} a_n,$$

when these limits exist. For example, for any number $|x| < 1$, we have

$$x^\infty = \lim_{n\to\infty} x^n = 0, \tag{11}$$

and also

$$\sum_{n=0}^{\infty} x^n = \frac{1}{1-x}, \tag{12}$$

which we invite the reader to deduce from (11).

Limits are needed not only for talking about infinite sets, but also as a way to define some very important functions[4]

$$e^x = 1 + x + \frac{x^2}{2} + \frac{x^3}{2\cdot 3} + \frac{x^4}{2\cdot 3\cdot 4} + \cdots = \sum_{n=0}^{\infty} \frac{x^n}{n!}, \tag{13}$$

where $e = 2.71828\dots$ is a famous transcendental number that can be computed by substituting $x = 1$ in the above series. Another important constant that we will meet below is

---

4 The primary reason the exponential e^x and the natural logarithm $\ln y$ are so important in mathematics is because they solve the simplest *differential* equations, namely $(e^x)' = e^x$ and $(\ln y)' = 1/y$. The reader can check this using the series (13), (15), and the rule $(x^n)' = nx^{n-1}$.

the Euler constant

$$\gamma = \lim_{N\to\infty}\left(\ln N - \sum_{1}^{N}\frac{1}{n}\right) = 0.57721\ldots \tag{14}$$

Here and below $\ln y$ denotes the function inverse to (13), which means that by definition

$$\ln e^x = x.$$

It is called the *natural logarithm*, and for arguments in $(0, 2)$ it can be computed using the series

$$\ln(1+y) = y - \frac{y^2}{2} + \frac{y^3}{3} - \frac{y^4}{4} + \cdots = \sum_{n=1}^{\infty}(-1)^{n-1}\frac{y^n}{n}, \quad |y| < 1. \tag{15}$$

Readers unfamiliar with these functions will discover that the exponential e^x grows very quickly with x, making the inverse function $\ln y$ grow very slowly. Notice that the sum in (14), with its minus sign, is the partial sum for $y = -1$ in (15). No wonder it goes to $\ln 0 = -\infty$ as N grows.

While Zeno of Elea (c. 495–c. 430 BC) made a career out of being confused by the $x = 1/2$ case of (12), we want to stress there are no logical problems whatsoever in thinking about the infinity of primes and about limits. We encourage the reader to embrace these notions as something more true and fundamental than any finite approximations to it.

5. THE DENSITY OF PRIMES

If $\mathbb{N} = \{1, 2, \ldots\}$ is the set of natural numbers and $\mathscr{A} \subset \mathbb{N}$ is a subset of it, we define

$$\operatorname{density}(\mathscr{A}) = \lim_{N\to\infty}\frac{|\mathscr{A}\cap\{1,\ldots,N\}|}{N}, \tag{16}$$

assuming this limit exists. When the limit (16) exists, we will also say that this is the probability that a random natural number is in $\mathscr{A}$.

From table (9) it is clear that

$$\operatorname{density}\big(\{\text{coprime to } 10\}\big) = \frac{4}{10} = \frac{1}{2}\times\frac{4}{5}. \tag{17}$$

Similarly, if $p_1, p_2, \ldots, p_r$ are prime then

$$\operatorname{density}\big(\{\text{coprime to } p_1 p_2 \cdots p_r\}\big) = \prod_{i=1}^{r}\left(1 - \frac{1}{p_i}\right). \tag{18}$$

The equality (18) makes one wonder whether

$$\operatorname{density}\big(\{\text{primes}\}\big) \stackrel{?}{=} \prod_{\text{all primes } p}\left(1 - \frac{1}{p}\right). \tag{19}$$

This is indeed true, but with the clarification that

$$\prod_{\text{all primes } p}\left(1 - \frac{1}{p}\right) \stackrel{!}{=} 0, \tag{20}$$

as we will see momentarily. Let us look at the reciprocal of the product (18). We have the $x = 1/p$ special case of (12)

$$\frac{1}{1-\frac{1}{p}} = 1 + \frac{1}{p} + \frac{1}{p^2} + \frac{1}{p^3} + \cdots = \sum_{m \geq 0} \frac{1}{p^m},$$

and multiplying those out for different primes p_i, we get

$$\prod_{i=1}^{r} \left(1 - \frac{1}{p_i}\right)^{-1} = \sum_{m_1,\ldots,m_r \geq 0} \frac{1}{p_1^{m_1} p_2^{m_2} \cdots p_r^{m_r}}. \tag{21}$$

If the set $\{p_i\}$ contains all primes that are $\leq N$, then the sum on the right in (21) contains, in particular, the reciprocals of all natural numbers $\leq N$. Therefore, by the existence of the prime factorization, we conclude

$$\begin{aligned} \prod_{\text{all primes } p \leq N} \left(1 - \frac{1}{p}\right)^{-1} &= 1 + \frac{1}{2} + \cdots + \frac{1}{N} + \text{more terms} \\ &\geq 1 + \frac{1}{2} + \cdots + \frac{1}{N} \\ &= \ln N + \gamma + o(1), \end{aligned} \tag{22}$$

where γ is the Euler constant from (14) and $o(1)$ denotes a quantity that goes to 0 as $N \to \infty$. This shows that the rightmost term in

$$0 \leq \text{density}(\{\text{primes}\}) \leq \text{density}(\{\text{coprime to } N!\}) = \prod_{\text{all primes } p \leq N} \left(1 - \frac{1}{p}\right) \tag{23}$$

goes to 0 as $N \to \infty$ and completes the proof of (19).

It is curious to notice that taking logarithms in (22) and using that (15) says that $-\ln(1 - p^{-1}) \approx p^{-1}$ for large p, we get

$$\sum_{\text{primes } p} \frac{1}{p} = +\infty. \tag{24}$$

This means that the same computation (22) proves that the density of primes is zero and yet there are sufficiently many primes for the series (24) to diverge, as first noted by Euler.

While we may be disappointed in the fact that the number (19) vanishes, very similar considerations often lead to positive results. For instance, let us consider square-free numbers n, that is, numbers not divisible by m^2 for any $m > 1$. This means

$$n \bmod p^2 \neq 0,$$

for any prime p. Referring back to (4), this means that the *two* last digits of n in the expansion base p do not vanish simultaneously. Since this pair of digits is free to take any of the p^2 possible values, one can conclude

$$\text{density}(\{\text{squarefree}\}) = \prod_{\text{primes } p} \left(1 - \frac{1}{p^2}\right) = \zeta(2)^{-1} = \frac{6}{\pi^2} \approx 0.6. \tag{25}$$

Here we meet the infinitely famous Riemann ζ-function

$$\zeta(s) = \sum_{n=1}^{\infty} \frac{1}{n^s} = \prod_{\text{primes } p} \left(1 - \frac{1}{p^s}\right)^{-1}, \quad s > 1, \tag{26}$$

and its value $\zeta(2)$ first computed by Euler in 1735. Our earlier computation (20) means that $\zeta(1) = \infty$.

6. THE PRIME NUMBER THEOREM

For a set $\mathscr{A}$ of zero density, the numbers (16) go to 0 as $N \to \infty$. A finer measurement of the density is then the rate at which the limit 0 as approached. For prime numbers, the answer is given by the prime number theorem, which says that the density of primes around some large number N is about $1/\ln(N)$.

A mathematically precise way to phrase it uses the function

$$\pi(x) = \text{number of primes } p \text{ such that } p \leq x \tag{27}$$

and states that[5]

$$\pi(x) \sim \mathrm{Li}(x) \stackrel{\text{def}}{=} \int_2^x \frac{dy}{\ln y} \sim \frac{x}{\ln(x)}, \tag{28}$$

where $f_1(x) \sim f_2(x)$ means that $\frac{f_1(x)}{f_2(x)} \to 1$ as $x \to \infty$. The reader may find the following data, taken from the Online encyclopedia of integer sequences, convincing:

x	$\pi(x)$	$\mathrm{Li}(x)/\pi(x) - 1$
10	4	0.25
10^2	25	0.16
10^3	168	0.054
10^4	1229	0.013
10^5	9592	0.0039
10^6	78498	0.0016
10^7	664579	0.00051
10^8	5761455	0.00013
10^9	50847534	0.000033
10^{10}	455052511	0.0000068
10^{11}	4118054813	0.0000028
10^{12}	37607912018	0.0000010
10^{13}	346065536839	0.00000031
10^{14}	3204941750802	0.000000098
10^{15}	29844570422669	0.000000035
10^{16}	279238341033925	0.000000012
10^{17}	2623557157654233	0.0000000030
10^{18}	24739954287740860	0.00000000089
10^{19}	234057667276344607	0.00000000043
10^{20}	2220819602560918840	0.00000000010
10^{21}	21127269486018731928	0.000000000028
10^{22}	201467286689315906290	0.0000000000096

(29)

5 A limit procedure is part of the definition of such everyday notions as areas and volumes. The integral of a univariate or multivariate function f is the signed area or volume between the graph of f and the graph of the zero function. It is a continuous limit of summing the values of f over a finer and finer mesh.

Lest the reader concludes that the last column is always positive, it is known that, in fact, the function $\mathrm{Li}(x) - \pi(x)$ changes sign infinitely many times. Also, while all 3 functions in (28) grow at the same rate, the logarithmic integral $\mathrm{Li}(x)$ gives a much better approximation to $\pi(x)$ than the ratio $\frac{x}{\ln(x)}$.

The prime number theorem was first shown by Hadamard and de la Vallée Poussin in 1896, so more than 2000 years after Eratosthenes. Certainly, many additional ideas were required, and are still required today to prove (28). Therefore, we will say very little about the proof. The reader interested in a heuristic derivation of the $1/\ln(N)$ density from unique factorization can find it here (requires familiarity with integrals).

To extract the distribution of primes from (93), Hadamard and de la Vallée Poussin had to use some properties of $\zeta(s)$ for complex values of s. What happens with $\zeta(s)$ for complex s involves some of deepest problems in all of mathematics, including the infinitely famous Riemann hypothesis (RH), still completely open today. The RH says that all solutions of $\zeta(s) = 0$ are either the so-called trivial zeros $s = -2, -4, -6, \ldots$, or have real part $\Re s = \frac{1}{2}$.

The remarkable $\frac{1}{2}$ from the Riemann Hypothesis can be in fact seen in the table (29) if one notices that the number of 0's in the second column is about half the number of digits of $\pi(x)$, meaning that the difference $\pi(x) - \mathrm{Li}(x)$ is of the order $x^{1/2}$, give or take some logarithmic factors. If there was a zero with $\Re s = c > \frac{1}{2}$, the error $\pi(x) - \mathrm{Li}(x)$ would be at least of size x^c, and the argument of Hadamard and de la Vallée Poussin was really about proving that $\Re s < 1$ for all zeros of the ζ-function.

While this is an incredibly interesting topic, the plot of our narrative follows a different path. Asked about the RH, James Maynard says: "*The Riemann Hypothesis suggests that there is a deep hidden structure within the prime numbers. This must occur for a good reason – we just do not know what the reason is, yet.*"

7. INCLUSION–EXCLUSION

Let $\mathscr{A}$ be a set of integers, or even of objects of arbitrary nature. A very, very abstract formulation of a sieve involves some subsets $\mathscr{A}_p \subset \mathscr{A}$, labeled by p in some index set $p \in \mathscr{P}$, which we wish to remove or sift out from the set $\mathscr{A}$. In other words, our goal is to understand the complement $\mathscr{A} \setminus \bigcup_{p\in\mathscr{P}} \mathscr{A}_p$ of all sets $\mathscr{A}_p$ in $\mathscr{A}$.

In its most basic form, the principle of inclusion–exclusion refers to the following elementary observation. Assuming the number of elements $|\mathscr{A}|$ is finite, we have

$$\begin{aligned}
\Big|\mathscr{A} \setminus \bigcup_{p\in\mathscr{P}} \mathscr{A}_p\Big| &= |\mathscr{A}| && \text{count all elements of } \mathscr{A} \\
&- \sum_{p} |\mathscr{A}_p| && \text{subtract } |\mathscr{A}_p| \text{ for each } p \\
&+ \sum_{p_1<p_2} |\mathscr{A}_{p_1} \cap \mathscr{A}_{p_2}| && \text{correct for subtracting twice} \\
&- \sum_{p_1<p_2<p_3} |\mathscr{A}_{p_1} \cap \mathscr{A}_{p_2} \cap \mathscr{A}_{p_3}| + \cdots && \text{etc.}
\end{aligned} \tag{30}$$

For example, referring back to table (9), we may take

$$\mathscr{A} = \{\text{residues modulo } 10\},$$
$$\mathscr{A}_p = \{\text{multiples of } p\}, \qquad p \in \mathscr{P} = \{2, 5\},$$

in which case (30) gives

$$\left| \mathscr{A} \setminus \bigcup_{p=2,5} \mathscr{A}_p \right| = \left| \{\text{residues coprime to } 10\} \right| = 10 - 5 - 2 + 1.$$

In other words, subtracting 5 multiples of 2 and 2 multiples of 5, we subtract the zero residue twice, as the shading in table (9) illustrates. Hence we have to put it back.

If the subsets $\mathscr{A}_p \subset \mathscr{A}$ correspond to *independent* events, meaning that

$$\frac{|\mathscr{A}_{p_1} \cap \mathscr{A}_{p_2} \cap \cdots \cap \mathscr{A}_{p_r}|}{|\mathscr{A}|} = \prod_{i=1}^{r} \frac{|\mathscr{A}_{p_i}|}{|\mathscr{A}|}, \tag{31}$$

then formula (30) factors very nicely

$$\frac{|\mathscr{A} \setminus \bigcup_{p \in \mathscr{P}} \mathscr{A}_p|}{|\mathscr{A}|} = \prod_{p \in \mathscr{P}} \left(1 - \frac{|\mathscr{A}_p|}{|\mathscr{A}|}\right), \tag{32}$$

special instances of which we have observed in (17), (18), and (25).

For us, $\mathscr{A}$ will always be some set of integers or residue classes and $\mathscr{A}_d \subset \mathscr{A}$ will denote those divisible by a some number d. In this case, all possible intersections in (30) can be described very concretely

$$\mathscr{A}_{p_1} \cap \mathscr{A}_{p_2} \cap \cdots \cap \mathscr{A}_{p_r} = \mathscr{A}_{p_1 p_2 \ldots p_r}, \tag{33}$$

as is illustrated for $\mathscr{P} = \{2, 3, 5\}$ in the following diagram. In (34), we visualize a composite number as a kind of molecule formed by its factors. The primes in $\mathscr{P}$ are assigned three different colors.

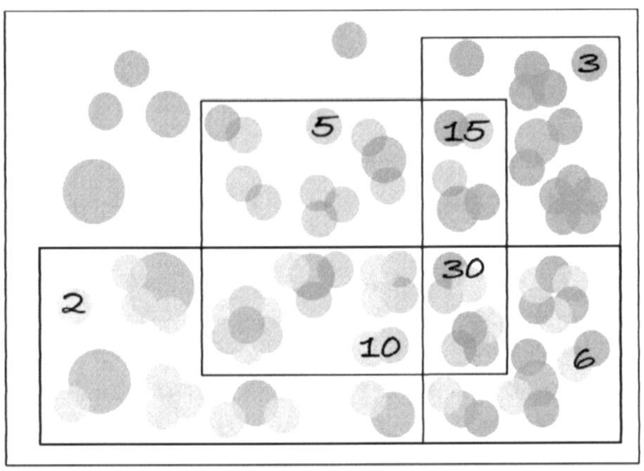

(34)

If (33) is the case, the terms in formula (30) correspond to square-free integers n all prime factors of which belong to $\mathscr{P}$. Thus (30) may be written more compactly

$$\left| \mathscr{A} \setminus \bigcup_{p \in \mathscr{P}} \mathscr{A}_p \right| = \sum_{d=1}^{\infty} \mu_{\mathscr{P}}(d) |\mathscr{A}_d|, \tag{35}$$

using a variant of the Möbius function

$$\mu_{\mathscr{P}}(d) = \begin{cases} (-1)^r, & d \text{ is a product of } r \text{ distict primes in } \mathscr{P}, \\ 0, & \text{otherwise.} \end{cases} \tag{36}$$

A more flexible language for the inclusion–exclusion principle uses the notion of *characteristic functions*. For any subset $S \subset \mathscr{A}$, we define its characteristic function δ_S by

$$\delta_S(n) = \begin{cases} 1, & n \in S, \\ 0, & n \notin S. \end{cases} \tag{37}$$

Then (35) can be refined to

$$\delta_{\mathscr{A} \setminus \bigcup_{p \in \mathscr{P}} \mathscr{A}_p} = \sum_{d=1}^{\infty} \mu_{\mathscr{P}}(d) \delta_{\mathscr{A}_d}. \tag{38}$$

Since

$$|S| = \sum_n \delta_S(n), \tag{39}$$

summing the values in (38) gives (35).

Formulas (35) and (39) require no assumption of independence like (31). This is very good because (31) is satisfied only approximately in the vast majority of sieve problems. Independence being only approximate is, in fact, a serious problem, to which we will come back below.

Another difficulty one encounters in real number-theoretic applications is that the set $\mathscr{A}$ is typically infinite. For example, we can have $\mathscr{A} = \mathbb{N}$, where $\mathbb{N} = \{1, 2, \dots\}$ is the set of natural numbers. The solution to this problem is to count elements $n \in \mathscr{A}$ not with weight 1 as in (39), but with some weight $\rho(n)$ such that the count converges. Schematically

$$|\mathscr{A}| = \sum_{n \in \mathscr{A}} 1 \xrightarrow{\text{generalize}} \rho(\mathscr{A}) = \sum_{n \in \mathscr{A}} \rho(n).$$

An example of such weight function is

$$\rho_\zeta(n) = n^{-s}, \quad s > 1, \tag{40}$$

used in the construction of the ζ-function. Multiplicativity of ρ, namely

$$\rho(n_1 n_2) = \rho(n_1)\rho(n_2), \tag{41}$$

satisfied by (40) and some other choices of ρ, implies an analog of independence (31) for weighted counts. For example, for $\mathscr{A} = \mathbb{N}$, $\mathscr{A}_p = p\mathbb{N}$, and a function ρ satisfying (41), formula (32) transforms into

$$\frac{\sum_{n \text{ coprime to } \mathscr{P}} \rho(n)}{\sum_{n \in \mathbb{N}} \rho(n)} = \prod_{p \in \mathscr{P}} \big(1 - \rho(p)\big). \tag{42}$$

We invite the reader to generalize formula (42) for functions ρ satisfying a weaker property

$$\gcd(n_1, n_2) = 1 \quad \Rightarrow \quad \rho(n_1 n_2) = \rho(n_1)\rho(n_2). \tag{43}$$

Other than (40), what other interesting functions satisfy (41)? For every N, the set

$$(\mathbb{Z}/N\mathbb{Z})^\times = \{\text{residue classes } a \bmod N \text{ such that } \gcd(a, N) = 1\} \tag{44}$$

is a finite abelian group with respect to multiplication. We take a character of χ of the group (44) that is, a complex-valued multiplicative function with $\chi(1) = 1$, and extend it by zero to all residues mod N. Examples of such functions are

$$\chi_3(n) = \begin{cases} 1, & n = 1 \bmod 3, \\ -1, & n = -1 \bmod 3, \\ 0, & n = 0 \bmod 3, \end{cases} \quad \chi_5(n) = \begin{cases} i^m, & n = 2^m \bmod 5, \\ 0, & n = 0 \bmod 5, \end{cases} \tag{45}$$

where the complex number $i = \sqrt{-1} \in \mathbb{C}$ is the imaginary unit. The weight

$$\rho_{N,\chi,s}(n) = \frac{\chi(n \bmod N)}{n^s}, \quad s > 1, \tag{46}$$

satisfies (41) and the corresponding analog of the ζ-function

$$L(\chi, s) = \sum_{n=1}^{\infty} \frac{\chi(n \bmod N)}{n^s}, \quad s > 1, \tag{47}$$

is called the Dirichlet L-function. Its properties are entirely parallel to the ζ-function with one crucial difference. Namely, if χ is *nontrivial*, that is, takes values other than 0 and 1, then, in contrast to the ζ-function having a singularity at $s = 1$ as in (93), the L-function has a *finite nonzero* value at $s = 1$. This allowed Dirichlet to show that primes are equally distributed among the residue classes (44).

8. THE FIRST CHALLENGE FOR SIEVES

As already emphasized above, the main difficulty with sieves is the fact that the independence (31) is only approximate and not exact. Here is an example. Take some large number x and consider the sets

$$\begin{aligned} \mathscr{A} &= \{\text{integers } n \text{ such that } \sqrt{x} < n \leq x\}, \\ \mathscr{P} &= \{\text{primes } p \text{ such that } p \leq \sqrt{x}\}. \end{aligned} \tag{48}$$

After sifting out $\mathscr{P}$, we get precisely the primes in the range $(\sqrt{x}, x]$, hence

$$\left| \mathscr{A} \setminus \bigcup_{p \in \mathscr{P}} \mathscr{A}_p \right| = \pi(x) - \pi(\sqrt{x}) \sim \frac{x}{\ln x},$$

by the prime number theorem. Let us see if, conversely, we can recover the prime number theorem from the sieve (48).

For fixed $p_1, \ldots, p_r$, the equality (31) is satisfied in the limit $x \to \infty$. However, the error terms present for finite x render the following reasoning *incorrect*. To warn the readers, will use $\overset{???}{=}$ to denote an incorrect equality. If we could just apply (32) to the $x \to \infty$ asymptotics, we would get

$$\frac{\pi(x) - \pi(\sqrt{x})}{x - \sqrt{x}} \sim \frac{\pi(x)}{x} \overset{???}{\sim} \prod_{\text{primes } p \leq \sqrt{x}} \left(1 - \frac{1}{p}\right), \quad x \to \infty. \tag{49}$$

Having seen products of this general shape before, the reader should not be surprised by the following exact result of F. Mertens:

$$\prod_{\text{primes } p \leq \sqrt{x}} \left(1 - \frac{1}{p}\right) \sim \frac{2e^{-\gamma}}{\ln x}, \tag{50}$$

where γ is the number from (22) and (93). Since $2e^{-\gamma} \approx 1.123$, this is somewhat close to the right answer and, in particular, gives the correct logarithmic dependence on x, but little else can be said in defence of a wrong formula.

This example is meant to illustrate that it is not easy to construct a good sieve, and not to discourage the reader from reading on! See also the references in Section 11, and in particular **[7]**.

9. PATTERNS IN PRIMES

So far, we have looked at primes individually, meaning that we studied expressions like

$$\pi(x) = \sum_{\text{primes } p} \delta_{[1,x]}(p), \quad \text{where} \quad \delta_{[1,x]}(y) = \begin{cases} 1, & y \in [1,x], \\ 0, & \text{otherwise}, \end{cases}$$

$$\ln \zeta(s) = - \sum_{\text{primes } p} \ln\left(1 - \frac{1}{p^s}\right),$$

given by summing some natural function $f(p)$ over the set of all primes. To a general science audience, we can say that we have been learning about 1-point correlations in the set of primes.

Recall we expect the primes to be as "random" as the constraints imposed by residues and density allow. To really put these ideas to the test, one should study multi-point correlations, that is, events or patterns that involve pairs, triples, etc., of primes.

To start with a concrete example, what is the probability that n and $n+1$ are both prime? The answer is clearly 0 because one of these numbers will have to be even, and so $n = 2$ is the only solution. What about n and $n+2$ being simultaneously prime? Such pairs are called *twin primes* and we saw many such pairs (green) in the Eratosthenes' sieve (3). Similarly, in the plot (5), twin primes are shown in green, all other primes in blue.

Twin primes provide an excellent test of our probabilistic intuition based on density and mod p considerations. From density alone, we should expect that the density of twin primes around N should be about $(\ln N)^{-2}$. However, this needs to be corrected from mod p considerations. Indeed, if n and $n+2$ were truly independent, the probability of both of them to be coprime to p would be $(1 - 1/p)^2$, while in reality it is $1/2$ for $p = 2$ and $(1 - 2/p)$ for $p > 2$. Whence the following constant in the 1923 conjecture of Hardy and Littlewood:

$$\pi_2(x) = \left|\{p \leq x \text{ such that } p+2 \text{ is prime}\}\right| \overset{?}{\sim} C_2 \int_2^x \frac{dy}{(\ln y)^2}, \quad x \to \infty, \tag{51}$$

where

$$C_2 = 2 \prod_{\text{primes } p > 2} \frac{1 - \frac{2}{p}}{(1 - \frac{1}{p})^2} = 1.32\ldots \tag{52}$$

In exactly the same fashion, the probability that n and $n + 2m$ are both coprime to p equals $(1 - 1/p)$ if p divides $2m$ and $(1 - 2/p)$ otherwise. Therefore, for any fixed m, one can conjecture that

$$\left|\{p \le x \text{ such that } p + 2m \text{ is prime}\}\right| \overset{?}{\sim} C_{2m} \int_2^x \frac{dy}{(\ln y)^2}, \quad x \to \infty, \tag{53}$$

where

$$\frac{C_{2m}}{C_2} = \prod_{p|m,\, p\neq 2} \frac{p-1}{p-2} \ge 1. \tag{54}$$

From this, it is clear that products of consecutive odd primes like $1155 = 3 \cdot 5 \cdot 7 \cdot 11$ should be particularly likely to occur as distances $p_2 - p_1$ between primes, while powers of 2 are the least likely values of $p_2 - p_1$. In (55) the function (54) is plotted in the ranges $m \in [1 \ldots 105]$ and $m \in [1 \ldots 1155]$, respectively.[6]

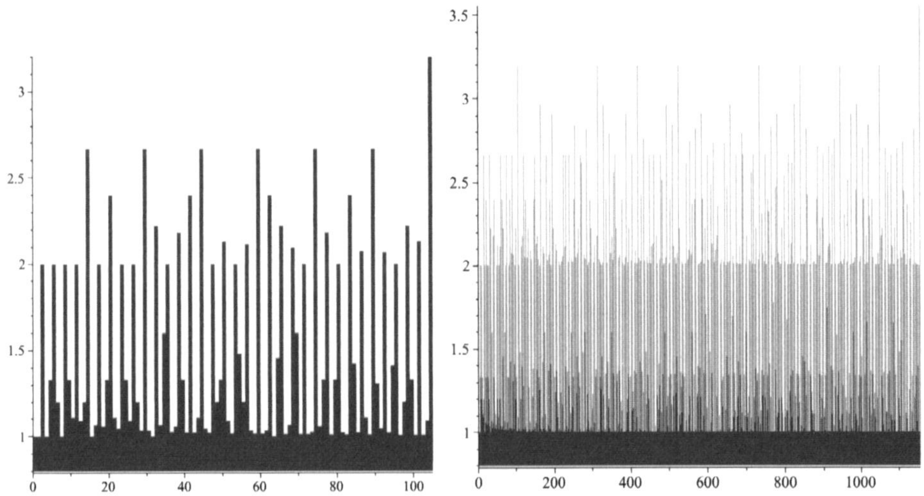

(55)

The conjecture (53) is in excellent agreement with data, especially if one considers the relative frequencies of distances. The following plot (56) compares the function C_{2m} with the actual distribution of the distances among first 10^6 odd primes:

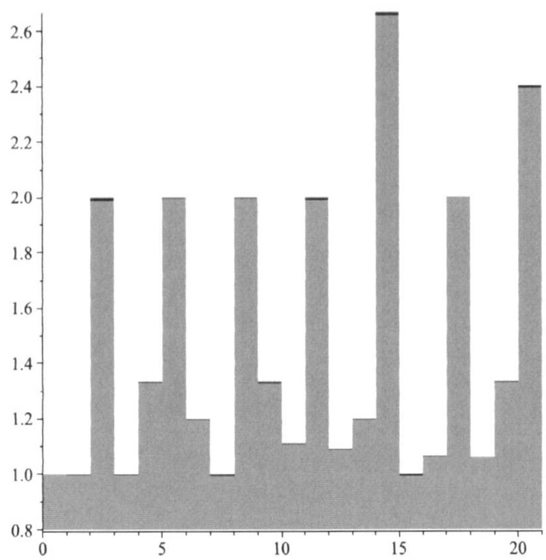

(56)

6 The reader may have to adjust the size/resolution of the graph to see the peak at 1155.

In (56) we have plotted the relative frequencies, normalized to exactly 1 for $m = 1$. The numerical data is in light blue and the theoretical prediction is in dark blue. The latter overshoots (with the exception of $m = 18$) the former by less than 1%, so it is just barely visible in the plot. Had we gone any deeper in the list of primes, the difference in graphs woud have become undetectable.

We note that the above discussion is for distances between primes, while a *prime gap* of length $2m$ means there are no other primes between p and $p + 2m$. However, since primes become sparser and sparser, finding another prime in an interval of fixed length becomes less and less probable as $p \to \infty$.

The exact same heuristic can be applied to any finite set of jumps

$$J = \{j_1 < j_2 < \cdots < j_l\} \subset \mathbb{N} \tag{57}$$

that we would like to find between primes. We denote by $n + J = \{n + j_1 < \cdots < n + j_l\}$ the shift of J by $n \in \mathbb{N}$ and by $n + J \subset \mathscr{P}$ the event that all numbers $n + j_i$ are prime. In parallel to (53), it is natural to expect that

$$\left|\{n \leq x \text{ such that } n + J \subset \mathscr{P}\}\right| \overset{?}{\sim} C_J \int_2^x \frac{dy}{(\ln y)^{|J|}}, \quad x \to \infty, \tag{58}$$

where

$$C_J = \prod_p \frac{1 - \frac{|J \bmod p|}{p}}{(1 - \frac{1}{p})^{|J|}}. \tag{59}$$

Here $|J \bmod p|$ is the number of distinct residue classes mod p in J. Since, for fixed J, this equals $|J|$ for all sufficiently large p, the contribution of all such p to (59) is $1 + O(\frac{1}{p^2})$. Therefore, the product (59) converges.

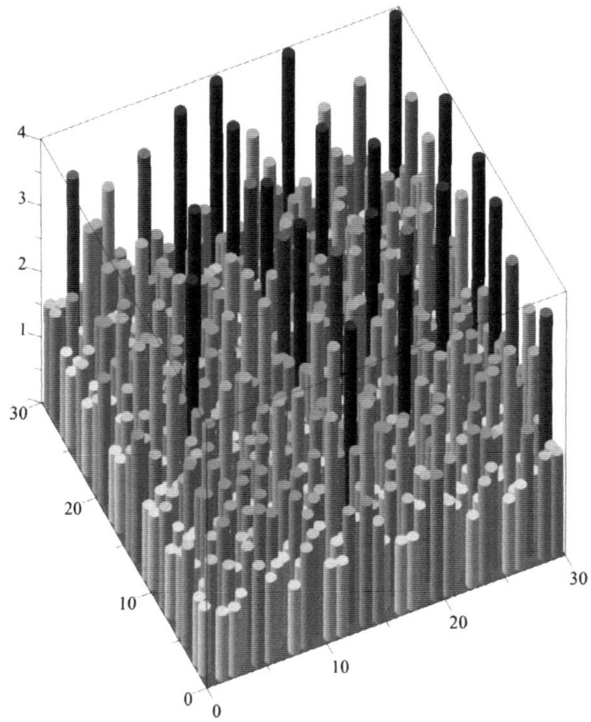

It is clear from (59) that the pattern in primes favor those J that contain a small fraction of residues modulo some prime p and prohibit those J for which $|J \bmod p| = p$. It is also clear from definitions that it suffices to consider the case $j_1 = 0$. The graph of the function $C_{\{0,2i,2(i+j)\}}/C_{\{0,2,6\}}$ is plotted on the left. It vanishes unless $ij(i+j) = 0 \bmod 3$, which explains the missing columns in the plot.

10. CLOSING THE GAP

Let us call a pattern J as in (57) *admissible* if $C_J \neq 0$, that is, if has a nonzero chance to occur in prime numbers. As a very, very special case of the above heuristic reasoning, one expects that any admissible pattern J will occur as a sequence of prime gaps *infinitely many* times.[7] In particular, one expects the set of twin primes to be infinite. This is known as the *twin prime conjecture*, and it is still open today. However, in contrast to the Riemann Hypothesis, there has been a truly dramatic progress in the recent years on such infinitude questions. This progress has been so dramatic that it inspires us to say that these conjectures are "almost" proven. It is quite incredible to see humans actually reach for the stars.

James Maynard does not quite agree with the narrator here. He says: "*Despite all the recent progress, it seems we are still missing an important idea to prove the Twin Prime Conjecture. But perhaps it is only one big idea.*"

Of course, the actual mathematics involved in proofs compares to what we have discussed so far like a modern airplane compares to a paper airplane. But if the reader tried to think about the issues discussed in Section 8, then she or he may begin to appreciate the amazing creativity and technical mastery required to design sieving arguments leading to the proofs of the breakthrough results below.

It is clear from the prime number theorem that for any constant $c > 1$ there are infinitely many pairs of primes p_1 and p_2 such that

$$p_1 < p_2 < p_1 + c \ln p_1. \tag{60}$$

Proving the same statement for some value $c < 1$ is not easy. Many brilliant mathematicians worked on this, finding proofs for smaller and smaller values of c, until Goldston, Pintz, and Yıldırım have shown that for *any* constant $c > 0$ there are infinitely many pairs of primes satisfying (60).

The new important ideas introduced by Goldston, Pintz, and Yıldırım opened the race to replace $c \ln p_1$ in (60) by some fixed constant B, that is, to prove the infinitude of pairs of primes that are within a fixed finite distance

$$p_1 < p_2 \leq p_1 + B \tag{61}$$

from each other. This race was won in a very dramatic fashion in April 2013 by Yitang Zhang.

---

7 This specific statement is known as the Dickson conjecture, made in 1904.

Even much more modest results in mathematics today require finding a new way through a real maze of possible ideas, techniques, and logical constructions, and hence moments of extraordinary concentration and clarity of mind. This is not unlike the need to be in a really, really top form for an athlete to set a world record. Research mathematicians (who do have time to do research as part of their job description, in addition to teaching, advising, and other professional duties) cherish these precious moments. Most athletes and mathematicians will surely agree that these special moments tend to be spaced further than $\ln N$ apart once we are past our prime. Zhang's proof is therefore particularly incredible and inspiring, since he had to find his way not just through the mathematical maze, but also through the many turns of his difficult career outside of academia, not giving up despite the big success finally coming to him only at the age of 55. His achievement was widely celebrated by the community, earning him a number of prestigious prizes including the 2013 Ostrowski Prize, the 2014 Cole Prize in Number Theory, and the 2014 Rolf Schock Prize. In the same year 2014, the Cole Prize in Number theory was also awarded to Goldston, Pintz, and Yıldırım for their influential work mentioned above.

We hope the reader will turn to [8,13,17,19,31,32] to learn more about these developments, and turn to the main hero of these popular notes, the winner of many awards including the 2022 Fields Medal. In the same eventful year 2013, James Maynard realized he can make the sieve a lot more effective, eclipsing Zhang's result in two key dimensions: getting a much stronger result by an easier method.

Speaking about the influences and inspirations that have lead to this result, James Maynard says: "*I was trying to understand the sieve intuition behind the groundbreaking work of Goldston–Pintz–Yıldırım, but in studying this I realised that it might be possible to modify their ideas to go further.*"

It is commonly said that great minds think alike, and the same sometimes happens to the greatest minds, also. In the suspenseful race to close the prime gap, Terry Tao arrived at the same results independently at the same time as James Maynard. "*I was a bit shocked when I first heard the news, but fortunately Tao was very generous and understanding. Simultaneous discovery happens more often than you'd imagine!*", says James Maynard.

To explain Maynard's and Tao's main result on small gaps in primes, it is important to make a certain change of perspective. In Section 9, we were interested in the event when *all* numbers

$$n + J = (n + j_1, n + j_2, \ldots, n + j_l) \tag{62}$$

are prime. But if one asks for less, one can prove more! Let us instead fix some $m < l$ and ask that at least m of the numbers (62) are prime for infinitely many values of n. We will not know which ones among (62) are prime, but we will know, for instance, that there are infinitely many primes within distance $j_l - j_1$ from each other.

The following is a special case of the spectacular main result of [20], which Kannan Soundararajan compares with "*sun amidst the stars*" in his Fields Medal laudatio.

Theorem 1. *For any m, for all sufficiently long admissible patterns J, at least m of the numbers* (62) *are prime for infinitely many n.*

In fact, for any given m, the required size of J in Theorem 1 can be made explicit. For $m = 2$, $|J| = 50$ suffices, and the following set being admissible

$$J = \{0, 4, 6, 16, 30, 34, 36, 46, 48, 58, 60, 64, 70, 78, 84, 88, 90, 94, 100, 106, 108, 114, 118, 126, 130, 136, 144, 148, 150, 156, 160, 168, 174, 178, 184, 190, 196, 198, 204, 210, 214, 216, 220, 226, 228, 234, 238, 240, 244, 246\} \quad (63)$$

shows there are infinitely many primes at most 246 apart.

For $m = 3$, $|J| = 35410$ suffices, and one can take,[8] for instance, the first 35410 primes larger than 35410

$$J = \{35419, 35423, \ldots, 469411, 469397\}.$$

Therefore, there are infinitely many triples of primes within 433992 of each other. In general, the best estimate for required length of J currently stands at $ce^{3.815m}$, see [1].

The more general result proven in [28] guarantees there are at least m primes among the numbers $a_1 n + j_1, \ldots, a_l n + j_l$ provided these are distinct and admissible. This stronger version of Theorem 1 leads to many further interesting conclusions about patterns in primes. For example, one can deduce that there are arbitrarily large sets of primes where any pair in the set differs in only 2 decimal places! Indeed, if we take

$$a_i = l!\,10^{l+2}, \quad j_i = 10^{i+1} + 1, \quad (64)$$

then all digits of $a_i n + j_i$, $i = 1, \ldots, l$ are the same, except the position of the 1 in the $(i+1)$st decimal place, which is changing its position within the string of l zeros.

I hope the readers share the narrator's sense of awe at this absolutely amazing mathematics and join me in warmest congratulations on it being recognized by the Fields Medal. I also hope the readers got the sense that today's mathematics is not just extraordinarily powerful, but also concrete, understandable, and fun, once one finds the right idea and the right point of view. While finding that right point of view is not at all easy, my biggest hope is to have inspired my youngest readers to believe that mathematics can be beautiful and rewarding, both as a subject and as a profession. Maybe this is also a good place for me to thank James Maynard and Kannan Soundararajan for this special opportunity to be introduced to their wonderful subject.

11. FURTHER READING

The *Quanta Magazine* has published several popular accounts of these and related developments, see [11, 13–15, 19].

Among surveys written by top experts in the field, one should mention [6, 8, 17, 27], including expositions by James Maynard himself [21–23].

---

8 As an exercise, the reader may check than any l-tuples of primes larger than l is admissible.

Among textbooks of different level, the reader will surely find something which suits her or his level and style among [3, 9, 16, 28, 29] or the more advanced [4, 12]. There is even a graphic detective novel [10]!

I hope the reader has a lot of fun studying these sources as well as the original articles [5, 20, 25, 26, 31].

12. A GLIMPSE INTO THE ARGUMENT

To help the reader make a transition to further popular and research reading, we will indicate some initial logical steps in the argument leading to the proof of Theorem 1. There is a certain distance that we can fly even on our paper airplane.

12.1. Being prime on average

We need to prove that at least m of the numbers (62) are prime for infinitely many n. It suffices to show that for any given integer N this is true for some $n \geq N$. Let $\mathscr{P}$ denote the set of all primes. Instead of trying to find a specific n for which the intersection $\{n+J\} \cap \mathscr{P}$ has at least m elements, we can ask about the average size of the intersection $|\{n+J\} \cap \mathscr{P}|$ with respect to some density $\rho(n) \geq 0$ on $[N, \ldots, 2N]$. This density ρ is something we are *bringing* into the argument, not something given to us in advance.

Clearly,

$$\operatorname{average}\big(\big|\{n+J\} \cap \mathscr{P}\big|\big) = \frac{\sum \rho(n)|\{n+J\} \cap \mathscr{P}|}{\sum \rho(n)} \leq \max\big(\big|\{n+J\} \cap \mathscr{P}\big|\big), \tag{65}$$

and so if we can bound the average in (65) below by m then we win. Now, since the numbers $j_k \in J$ are all distinct, we have

$$\frac{1}{\sum \rho(n)} \sum_{n=N}^{2N} \rho(n)\big|\{n+J\} \cap \mathscr{P}\big| = \sum_{k=1}^{l} \frac{\sum\limits_{n+j_k \text{ is prime}} \rho(n)}{\sum\limits_{N \leq n \leq 2N} \rho(n)}. \tag{66}$$

Hence, our strategy is to invent a function $\rho(n)$ for which each of the l ratios on the right-hand side of (66) can be shown to be large.

12.2. Looking for ρ, part I

A naive strategy would be to take

$$\rho_0(n) = \begin{cases} 1, & n+J \subset \mathscr{P}, \\ 0, & \text{otherwise}. \end{cases} \tag{67}$$

This makes the numerator and denominator in (66) equal, and so naively each fraction equals 1. What this overlooks is that $\frac{0}{0}$ is no good in (66), and that our original goal is precisely equivalent to showing that ρ_0 takes some nonzero values.

This underscores the point that we have not really advanced on the problem, yet, just put in a slightly more flexible framework by introducing the density ρ. Those who can design a good ρ are the great masters of the sieve.

Functions that only take values 0 or 1 are called *characteristic* functions as we recall from (37). These are also the functions that are equal to their own square. From the definitions,

$$\rho_0(n) = \delta_{[N,\dots,2N]}(n) \prod_{k=1}^{l} \delta_{\mathscr{P}}(n + j_k). \tag{68}$$

The next natural idea is to find a working replacement $\tilde{\delta}$ for $\delta_{\mathscr{P}}$ and get ρ by multiplying them together.

Plots of the function $\delta_{\mathscr{P}}$ look like barcodes, and here is an example

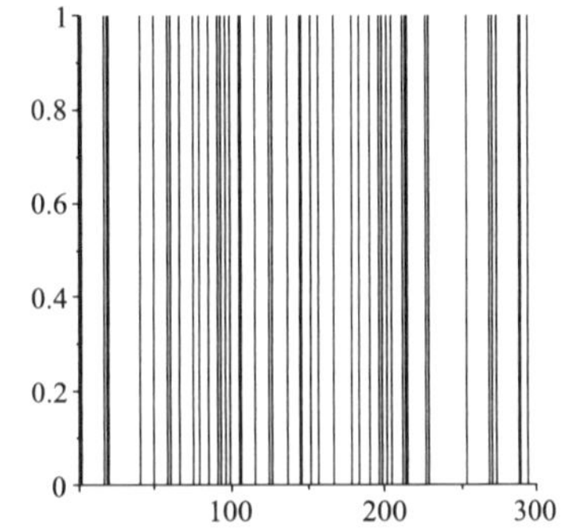

(69)

in which n takes odd values from $10^6 + 1$ to $10^6 + 599$. In principle, (38) gives a formula for $\delta_{\mathscr{P}}$, and we can approach the goal of finding a replacement $\delta_{\mathscr{P}}$ by tinkering with the formula (38). For instance, we just truncate summation over d to some maximal value D. That is, we define

$$\tilde{\delta}_0(n) = \Big(\sum_{d|n,\, d \le D} \mu(d) \Big)^2, \tag{70}$$

where we square the sum to make the result nonnegative. Since this equals 1 if n has no nontrivial divisors $d \le D$, it is natural to compare this function to the characteristic function $\delta_{\le D}$ of numbers without prime factors $p \le D$.

It is easy to plot the function $\tilde{\delta}_0 - \delta_{\le D}$ and the result

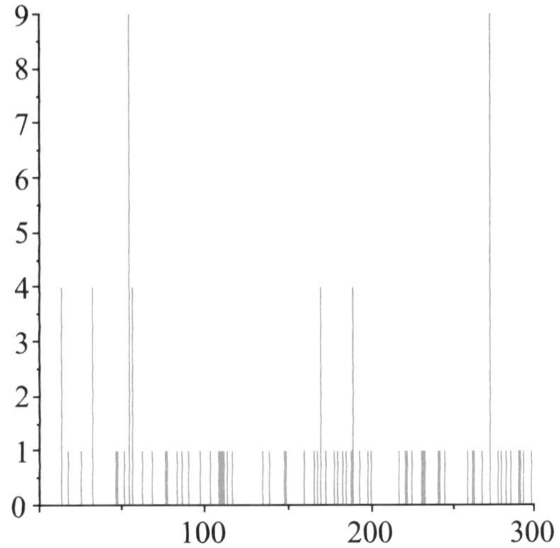

(71)

for $D = 100$ is not really satisfying. The two peaks in the graph correspond to the numbers

$$1000109 = 11 \cdot 23 \cdot 59 \cdot 67, \quad 1000545 = 3 \cdot 5 \cdot 7 \cdot 13 \cdot 733,$$

and, in general, the function (70) becomes large not because n is prime, but because there is a significant disbalance between its divisors $d \leq D$ with different parity of the number of prime factors. In other words, $\tilde{\delta}_0(n)$ is much more sensitive to the artificial cutoff introduced by us at $d \leq D$ than to what we set out to measure in the first place.

To get rid of this effect, it makes sense to replace the hard cutoff at $d \leq D$ by a more gentle one, through some weight function of d that gives 1 for prime numbers and vanishes at $d = D$. Let us try

$$\tilde{\delta}_k(n) = \frac{1}{(\ln D)^{2k}} \left(\sum_{d|n,\, d \leq D} \mu(d) \left(\ln \frac{D}{d} \right)^k \right)^2, \tag{72}$$

and this works much, much better for $k \geq 1$. For $D = 100$, the function $\tilde{\delta}_1 - \delta_{\leq D}$ looks like this:

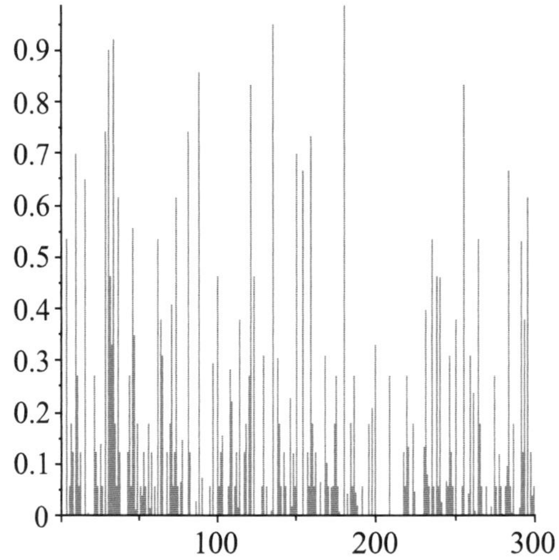

(73)

Not only does it take values in $[0, 1)$ in this plot, it also peaks at numbers with prime factors p of size close to D. Since the weight $\ln \frac{D}{p}$ gets small for such p, we certainly expect such numbers to contribute on par with the prime numbers.

12.3. Looking for ρ, part II

Functions (72) played an important role in the work of Goldston, Pintz, and Yıldırım. However magical, by themselves they are not enough to get to the Maynard–Tao theorem. If we just multiply them as in (68), then we loose the crucial synergy between different elements of the list J. Recall that the logic of Theorem 1 is such that the longer the list J gets, the easier it is to find many prime numbers in it. For this, there should be some nontrivial interaction between different j_k.

One key new ingredient in the Maynard–Tao method is to consider functions of the form

$$\rho(n) = \delta_{[N,\ldots,2N]} \left(\sum_{\substack{d_1|n+j_1,\ldots,d_l|n+j_l,\\ d_1 d_2 \cdots d_l \leq D}} \mu(d_1 d_2 \cdots d_l) F\left(\frac{\ln d_1}{\ln D}, \ldots, \frac{\ln d_l}{\ln D} \right) \right)^2, \tag{74}$$

where F is a multivariate function to be specified later. As before, we want F to be small if the arguments sum to 1 (meaning that $d_1 d_2 \cdots d_l = D$) to soften the effect of the summation cutoff introduced in (74).

By allowing F to depend on each divisor d_i, the Maynard–Tao method activates a very powerful principle of measure concentration in high-dimensional geometry. At the risk of being repetitive, one may note that there is really a lot of space in a space of a large dimension N. There is so much space that no probability distribution can cover all of it evenly as $N \to \infty$, and one could put this vague principle in a mathematically precise form; see, for instance, [18].

To make a negative statement positive, one can say that any high-dimensional probability density has to concentrate on some small portion of the whole space. For example, a probability measure ν on the line $\mathbb{R}$ is another name for a random variable x, and a product measure $\nu^{\otimes N} = \nu \times \cdots \times \nu$ on $\mathbb{R}^N$ is another name for a sequence of independent, identically distributed (i.i.d.) random variables $x_1, \ldots, x_N$. We know from basic probability theory that, with minimal assumptions about ν, the average $\frac{1}{N}\sum x_i$, and many other functions of i.i.d. random variables $x_1, \ldots, x_N$, will sharply peak, or concentrate, around their expected value as $N \to \infty$.

A reader not familiar with these notions, may experiment by working out the example in which ν is the uniform density on $[0, 1]$ and $\nu^{\otimes N}$ is a uniform density on an N-dimensional cube $[0, 1]^N$. Taking the sum $\sum x_i$ means projecting the cube onto the $(1, 1, \ldots, 1)$ axis, and the reader may enjoy actually plotting these densities for different values of N. It is also fun to compute the projection of a uniform measure on a high-dimensional sphere onto any axis.

It is by harnessing these concentration of measure phenomena that the density (74) can significantly improve upon (72).

12.4. Primes in arithmetic progressions, on average

Now let us plug the formula (74) into the numerator in (66), expand out the square, and do summation over the variable n first. We get a sum of the form

$$\sum_{n + j_k \text{ is prime}} \rho(n) = \sum_{\vec{d},\vec{d}'} \mu\mu FF \sum_{\text{certain } n} 1, \tag{75}$$

where the outer sum is over two sets of integers

$$\vec{d} = (d_1, \ldots, d_l) \quad \text{and} \quad \vec{d}' = \left(d_1', \ldots, d_l'\right),$$

there is a weight of the form

$$\mu\mu FF = \mu(\Pi d_i)\mu\left(\Pi d_i'\right) F\left(\frac{\ln \vec{d}}{\ln D}\right) F\left(\frac{\ln \vec{d}'}{\ln D}\right),$$

and the inner sum runs over n such that

$$n + j_i \quad = 0 \bmod \operatorname{lcm}\left(d_i, d_j'\right), \quad i = 1, \ldots, l, \tag{76}$$

$$n + j_k \quad \text{is prime}, \tag{77}$$

where $\mathrm{lcm}(d_i, d_j')$ denotes the least common multiple.

It is clear from this that we must have $d_k = d_k' = 1$. Since the remaining congruence conditions can be put into a single congruence condition using the Chinese Remainder Theorem, the sum over n thus counts primes in an arithmetic progression.

Time and time again in these notes we have stressed the technical importance of being able to accurately count primes in arithmetic progression in analytic number theory, also stressing that this may be very delicate if the progression is not much longer than its common difference.

The counting function (27) may be refined to count primes in a given residue class modulo b,

$$\pi(x, b, a) = \text{number of primes } p \text{ such that } p \leq x \text{ and } p = a \bmod b. \tag{78}$$

The Dirichlet theorem mentioned in Section 7 says that

$$\frac{\pi(x, b, a)}{\pi(x)} \to \begin{cases} \phi(b)^{-1}, & \gcd(a, b) = 1, \\ 0, & \text{otherwise}, \end{cases} \tag{79}$$

as $x \to \infty$, where $\phi(b)$ is the number of residue classes coprime to b. For fixed x, however, the function

$$(b, a) \mapsto \phi(b)\frac{\pi(x, b, a)}{\pi(x)} - 1 \tag{80}$$

behaves in a very irregular manner. This is illustrated in the following plot for $a < b \leq 100$:

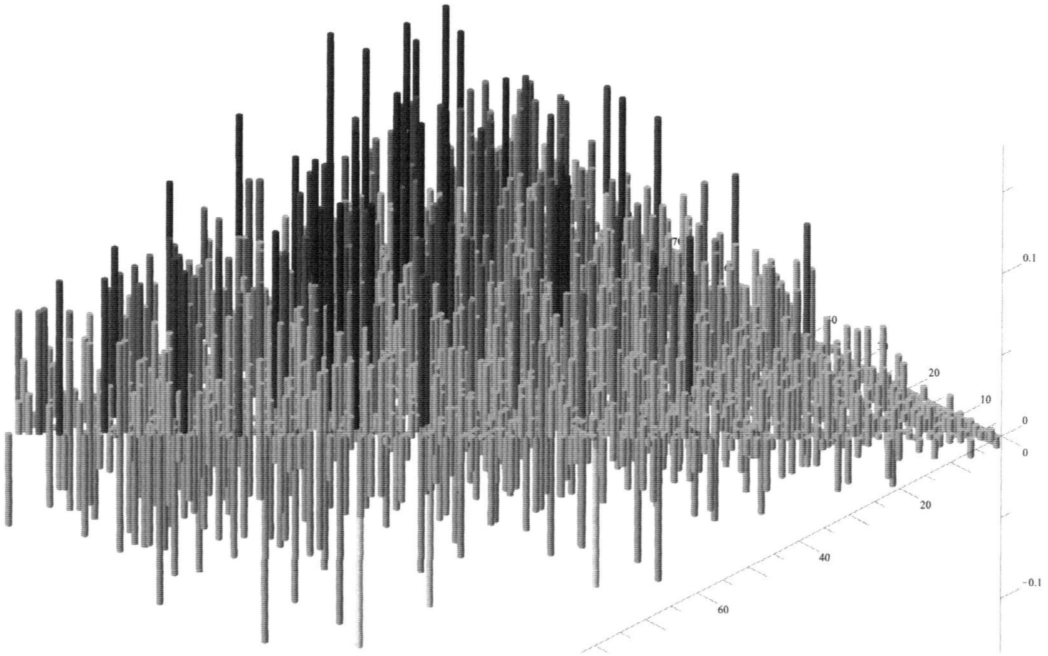

and the first 5000 primes, which means $x^{1/2} \approx 220$.

Very fortunately, in (75), we do not have to face the full complexity of this function. Since there is an outside summation over $\vec{d}$ and $\vec{d}'$, we only need to know its *average* over b.

Recall that the Riemann hypothesis implies error of size about $x^{1/2}$ in the prime number theorem. The conjectural extension of the Riemann hypothesis to Dirichlet L-functions (47) would give a similar error bound for $\pi(x, b, a)$. If one sums these errors for $b < x^{1/2}$, one thus expects to get something of order x. Remarkably, a slight weakening of this statement, known as the Bombieri–Vinogradov theorem, *has* been proven **[2, 30]**. In other words, the Riemann hypothesis for L-functions is a complete mystery, but its main consequence for the distributions of primes in arithmetic progression can be rigorously proven *on average*. The actual estimate one needs here has the form

$$\sum_{b<x^{1/2-\varepsilon}} \max_{\substack{a \\ \gcd(a,b)=1}} \left| \pi(x, b, a) - \frac{\pi(x)}{\phi(b)} \right| \le C(A, \varepsilon) \frac{x}{(\ln x)^A}, \tag{81}$$

which holds for any $A > 0$ and $\varepsilon > 0$ with some positive constant $C(A, \varepsilon)$ that depends on A and ε. In our example, the maxima over a in (81) and their running average over b can be seen in the following plot:

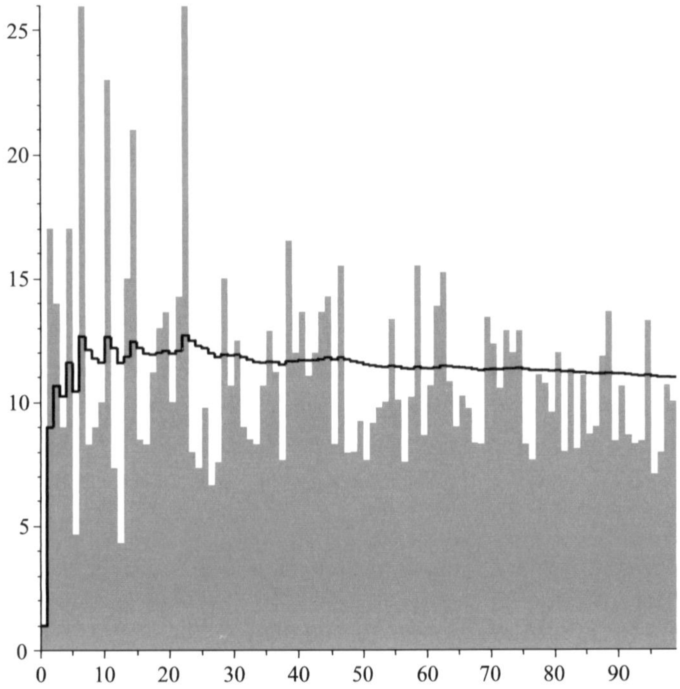

Averaging really does make the behavior a lot more regular and, hence, manageable.

We have discussed some of the key ingredient that go into the proof of the amazing result of Maynard and Tao. Perhaps, this discussion has given the reader the motivation and confidence to open more advanced literature written by the experts in the field, including the papers listed in Section 11. In any case, we hope to have communicated to the reader our own sense of awe at the beauty of mathematics.

A. LIMITS

Limits are defined not just for numerical sequences $(a_1, a_2, \dots)$ but for objects of arbitrary nature for which there is a notion of *neighborhoods*. Namely, a is the limit of the above sequence, if *every* neighborhood of a contains all elements a_n except maybe finitely many. The reader may find it useful to picture this as follows:

(82)

where the bin represents a neighborhood of a and spheres represent the elements a_n. Of course, since the sequence is infinite, any neighborhood of the limit point contains not just many, but infinitely many of the a_n's.

For real numbers, or any other set with the notion of distance, we may take the open balls of arbitrary positive radius $r > 0$,

$$B(a, r) = \{\text{all } x \text{ such that distance}(x, a) < r\},$$

as standard neighborhoods. The reader may check her or his understanding of the definition by proving (11) and (12), constructing a sequence or real numbers that does not have a limit, and proving that the limit of a sequence of real numbers is unique when it exists.

The slight issue with defining the limits digit by digit is that the set of all real numbers whose decimal expansion is fixed up to a certain point is a half-open interval, for instance,

$$\{\text{all } x \text{ such that } x = 2.71\dots\} = [2.71, 2.72).$$

To define limits for real numbers correctly, one should take *open* intervals, that is, those without both endpoints as neighborhoods. Back to the main text.

B. MELLIN TRANSFORM AND THE DENSITY OF PRIMES

Consider a simplified model, in which we forget about integrality and talk about real numbers $x > 1$. Let $\rho_1(x)$ be a certain density function on $[1, \infty)$. It will model the density of prime numbers. What should then correspond to the density $\rho_r(y)$ of the numbers y that have exactly r prime factors?

We have, by definition, $y = x_1 x_2 \dots x_r$, where x_i are distributed in the set

$$\{1 \le x_1 \le x_2 \le \cdots \le x_r\}$$

with density $\rho_1(x_1)\cdots\rho_1(x_r)$. Thus for *any* function $f(y)$, we have

$$\int f(y)\rho_r(y)\,dy = \int_{1\le x_1\le x_2\le\cdots\le x_r} f(x_1\cdots x_r)\prod \rho_1(x_i)\,dx_i. \tag{83}$$

Which functions $f(y)$ should we consider?

In mathematics, the success often depends on choosing the right point of view. If one has the right point of view, then one is able to see clearly where one is going.

A very nice choice here is to take $f(y) = y^{-s}$, where $s > 1$ is parameter. This is called *Mellin transform*, and it is a transform because it takes a function $\rho_r(y)$ of one variable y to another function $\rho_r^{\text{Mellin}}(s)$ of the parameter s. Thus one trades a function of one variable $\rho_r(y)$ for another function of one variable $\rho_r^{\text{Mellin}}(s)$, which seems like a fair exchange. In fact, one can reconstruct $\rho_r(y)$ from $\rho_r^{\text{Mellin}}(s)$, so no information is lost.

The Mellin transform is a close relative of the Fourier transform[9] and what makes the following computation work is the basic identity

$$(x_1x_2)^s = x_1^s x_2^s.$$

Because of this, the function $f(x_1\cdots x_r)$ in (83) factors as $f(x_1)\cdots f(x_r)$ and we can eventually reduce an r-fold integral in (83) to a product of r integrals.

We compute

$$\rho_r^{\text{Mellin}}(s) \overset{\text{def}}{=} \int_1^\infty y^{-s}\rho_r(y)\,dy \tag{84}$$

$$= \int_{1\le x_1\le x_2\le\cdots\le x_r} (x_1\cdots x_r)^{-s}\prod \rho_1(x_i)\,dx_i \tag{85}$$

$$= \frac{1}{r!}\int_{[1,\infty)^r} \prod x_i^{-s}\rho_1(x_i)\,dx_i \tag{86}$$

$$= \frac{1}{r!}\rho_1^{\text{Mellin}}(s)^r, \tag{87}$$

where, in going from (85) to (86), we used the fact that

$$[1,\infty)^r = \bigcup_{\substack{\text{permutations}\\ w:\{1,\dots,r\}\to\{1,\dots,r\}}} \{1\le x_{w(1)}\le x_{w(2)}\le\cdots\le x_{w(r)}\} \tag{88}$$

and that the integration over any of the $r!$ sets on the right-hand side of (88) gives the same result as (85).

If $\rho_\bullet$ is the density of numbers y having an arbitrary number of factors r, including the case when $r = 0$ and $y = 1$, then summing (87) over $r = 0, 1, 2, \ldots$ gives

$$\rho_\bullet^{\text{Mellin}}(s) = \exp\bigl(\rho_1^{\text{Mellin}}(s)\bigr), \tag{89}$$

where $\exp(x)$ is another notation for the function e^x from (13). The appearance of the exponential function here is typical in many inclusion–exclusion situations.

To model unique factorization, we want to take $\rho_\bullet = 1$ on $[1,\infty)$ which means

$$\rho_\bullet^{\text{Mellin}}(s) = \int_1^\infty x^{-s}\,dx = \frac{1}{s-1}, \quad s > 1. \tag{90}$$

---

9 Some readers may find the explanation of Fourier transform in **[24]** usable.

Thus, we expect

$$\int_1^\infty x^{-s}\rho_1(x)\,dx \overset{?}{=} \ln\frac{1}{s-1}, \quad s>1, \tag{91}$$

which is both good and bad news for the following reasons.

On the one hand, $\ln\frac{1}{s-1}$ is not a Mellin transform of any density ρ_1 on $[1,\infty)$ simply because it does not have a limit as $s\to+\infty$. The $s\to+\infty$ limit in (91) probes $\rho_1(x)$ for x very close to 1 because x^{-s} becomes very small on the whole interval $(1+\delta,\infty)$ as $s\to\infty$, for any fixed $\delta>0$. In particular, the Mellin transform of a bounded density function $\rho_1(x)$ on $[1,\infty)$ has to go to zero as $s\to+\infty$.

This means that we cannot accurately model prime numbers with real numbers and continuous densities. Of course, it was certainly silly to be asking for the density of small primes to begin with. However, our interest is precisely the opposite, as we want to know the behavior of $\rho_1(x)$ for large x. This region is probed by $s\to 1$ limit of the Mellin transform. In fact,

$$f(x)=f_0+O(x^{-c})\Rightarrow\int_1^\infty f(x)x^{-s}dx=\frac{f_0}{s-1}+\cdots, \tag{92}$$

where $O(x^{-c})$ means that $|\frac{f(x)-f_0}{x^{-c}}|$ remains bounded as $x\to\infty$, the double arrow $\Rightarrow$ denotes implication, and dots stand for a function which is analytic for $s>1-c$. (And also analytic for complex values of s such that $\Re s>1-c$.) In the $s\to 1$ limit, we may write

$$\int_1^\infty x^{-s}\rho_1(x)\ln(x)dx=-\frac{d}{ds}\int_1^\infty x^{-s}\rho_1(x)dx\sim-\frac{d}{ds}\ln\frac{1}{s-1}=\frac{1}{s-1},$$

which strongly suggests $\rho_1(x)\sim 1/\ln(x)$ for $x\to\infty$.

In place of continuous approximations, the proof of Hadamard and de la Vallée Poussin uses properties of the ζ-function (26), which, in the spirit of (84), can be interpreted as the average value of n^{-s} with respect to the measure that gives every positive integer n weight 1. The equality between the sum and product in (26) is the correct discrete version of the relation (89). It looks different because in the discrete situation we need to account for the nonzero chance of having two equal prime factors, the possibility of which was ignored in going from (85) to (86). The exact analog of (90) is the the following description:

$$\zeta(s)=\frac{1}{s-1}+\gamma+o(1), \quad s\to 1, \tag{93}$$

of the $s\to 1$ behavior of the ζ-function, where γ is the constant from (14) and (22). Back to the main text.

REFERENCES

[1] R. C. Baker and A. J. Irving, Bounded intervals containing many primes. *Math. Z.* **286** (2017), no. 3–4, 821–841.

[2] E. Bombieri, On the large sieve. *Mathematika* **12** (1965), 201–225.

[3] H. Davenport, *Multiplicative number theory*. 3rd edn., Grad. Texts in Math. 74, Springer, New York, 2000. Revised and with a preface by H. L. Montgomery.

[4] J. Friedlander and H. Iwaniec, *Opera de cribro*. Amer. Math. Soc. Colloq. Publ. 57, American Mathematical Society, Providence, RI, 2010.

[5] D. A. Goldston, J. Pintz, and C. Y. Yıldırım, Primes in tuples. I. *Ann. of Math. (2)* **170** (2009), no. 2, 819–862.

[6] D. A. Goldston, J. Pintz, and C. Y. Yıldırım, Small gaps between primes. In *Proceedings of the International Congress of Mathematicians—Seoul 2014. Vol. II*, pp. 419–441, Kyung Moon Sa, Seoul, 2014.

[7] A. Granville, Unexpected irregularities in the distribution of prime numbers. In *Proceedings of the International Congress of Mathematicians, Vol. 1, 2 (Zürich, 1994)*, pp. 388–399, Birkhäuser, Basel, 1995.

[8] A. Granville, Primes in intervals of bounded length. *Bull. Amer. Math. Soc. (N.S.)* **52** (2015), no. 2, 171–222.

[9] A. Granville, *Number theory revealed: a masterclass*. American Mathematical Society, Providence, RI, 2019.

[10] A. Granville and J. Granville, *Prime suspects. The anatomy of integers and permutations*. Princeton University Press, Princeton, NJ, 2019. Illustrated by R. J. Lewis.

[11] K. Hartnett, New proof settles how to approximate numbers like pi. *Quanta Mag.* (August 14, 2019). https://www.quantamagazine.org/new-proof-settles-how-to-approximate-numbers-like-pi-20190814/.

[12] H. Iwaniec and E. Kowalski, *Analytic number theory*. Amer. Math. Soc. Colloq. Publ. 53, American Mathematical Society, Providence, RI, 2004.

[13] E. Klarreich, Unheralded mathematician bridges the prime gap. *Quanta Mag.* (May 19, 2013). https://www.quantamagazine.org/yitang-zhang-proves-landmark-theorem-in-distribution-of-prime-numbers-20130519.

[14] E. Klarreich, Together and alone, closing the prime gap. *Quanta Mag.* (November 19, 2013). https://www.quantamagazine.org/mathematicians-team-up-on-twin-primes-conjecture-20131119/.

[15] E. Klarreich, Prime gap grows after decades-long lull. *Quanta Mag.* (December 10, 2014). https://www.quantamagazine.org/mathematicians-prove-conjecture-on-big-prime-number-gaps-20141210/.

[16] D. Koukoulopoulos, *The distribution of prime numbers*. Grad. Stud. Math. 203, American Mathematical Society, Providence, RI, 2019.

[17] E. Kowalski, Gaps between prime numbers and primes in arithmetic progressions [after Y. Zhang and J. Maynard]. *Astérisque* **367–368** (2015), Exp. No. 1084, ix, 327–366.

[18] M. Ledoux, *The concentration of measure phenomenon*. Math. Surveys Monogr. 89, American Mathematical Society, Providence, RI, 2001.

[19] T. Lin, After prime proof, an unlikely star rises. *Quanta Mag.* (April 2, 2015). https://www.quantamagazine.org/yitang-zhang-and-the-mystery-of-numbers-20150402.

[20] J. Maynard, Small gaps between primes. *Ann. of Math. (2)* **181** (2015), no. 1, 383–413.

[21] J. Maynard, Digits of primes. In *European Congress of Mathematics*, pp. 641–661, Eur. Math. Soc., Zürich, 2018.

[22] J. Maynard, Gaps between primes. In *Proceedings of the International Congress of Mathematicians—Rio de Janeiro 2018. Vol. II. Invited lectures*, pp. 345–361, World Sci. Publ., Hackensack, NJ, 2018.

[23] J. Maynard, The twin prime conjecture. *Jpn. J. Math.* **14** (2019), no. 2, 175–206.

[24] A. Okounkov, The magic of 8 and 24. In *Proc. Int. Cong. Math. 2022, Vol. 1*, pp. 492–545, EMS Press, 2022.

[25] D. H. J. Polymath, New equidistribution estimates of Zhang type. *Algebra Number Theory* **8** (2014), no. 9, 2067–2199.

[26] D. H. J. Polymath, Variants of the Selberg sieve, and bounded intervals containing many primes. *Res. Math. Sci.* **1** (2014), 83.

[27] K. Soundararajan, Small gaps between prime numbers: the work of Goldston–Pintz–Yıldırım. *Bull. Amer. Math. Soc. (N.S.)* **44** (2007), no. 1, 1–18.

[28] G. Tenenbaum and M. Mendès France, *The prime numbers and their distribution*. Stud. Math. Libr. 6, American Mathematical Society, Providence, RI, 2000. Translated from the 1997 French original by P. G. Spain.

[29] G. Tenenbaum, *Introduction to analytic and probabilistic number theory*. 3rd edn., Grad. Stud. Math. 163, American Mathematical Society, Providence, RI, 2015. Translated from the 2008 French edition by P. D. F. Ion.

[30] A. I. Vinogradov, The density hypothesis for Dirichet L-series. *Izv. Akad. Nauk SSSR Ser. Mat.* **29** (1965), 903–934.

[31] Y. Zhang, Bounded gaps between primes. *Ann. of Math. (2)* **179** (2014), no. 3, 1121–1174.

[32] Y. Zhang, Small gaps between primes and primes in arithmetic progressions to large moduli. In *Proceedings of the International Congress of Mathematicians—Seoul 2014. Vol. II*, pp. 557–567, Kyung Moon Sa, Seoul, 2014.

ANDREI OKOUNKOV

Department of Mathematics, University of California, Berkeley, 970 Evans Hall Berkeley, CA 94720–3840, USA, okounkov@math.columbia.edu

This Page Intentionally Left Blank

THE MAGIC OF 8 AND 24

ANDREI OKOUNKOV

ABSTRACT

While the author is a professional mathematician, he is by no means an expert in the subject area of these notes. The goal of these notes is to share the author's personal excitement about some results of Maryna Viazovska with mathematics enthusiasts of all ages, using maximally accessible, yet precise mathematical language. No attempt has been made to present an overview of the current state field, its history, or to place this narrative in any kind of broader scientific or social context. See the references in Section 5 for both professional surveys and popular science accounts that will certainly give the reader a broader and deeper understanding of the material.

MATHEMATICS SUBJECT CLASSIFICATION 2020

Primary 11H31; Secondary 97

KEYWORDS

Sphere packings

INTERNATIONAL CONGRESS OF MATHEMATICIANS 2022 JULY 6–14

© 2022 International Mathematical Union
Proc. Int. Cong. Math. 2022, Vol. 1, pp. 492–545
DOI 10.4171/ICM2022/203

Published by EMS Press and licensed under a CC BY 4.0 license

1. SPHERES KEEP THEIR DISTANCE

1.1. Spheres in a d-dimensional space

High-dimensional spaces really exist. A photo of a 3-dimensional object taken by our phone may seem to be a 2-dimensional representation of the original, but ... As we capture, process, store, transmit, or display photos, real manipulations are happening with long list of numbers $(x_1, \dots, x_d)$. Not just inside the phone, but also inside our brain, the image is processed out of many millions of electric potential readouts from the cone and rod cells.

We will call a list of numbers $\boldsymbol{x} = (x_1, \dots, x_d)$ a vector. Possible values of each x_i may be different in different contexts. It could be just a bit, meaning x_i equals either 0 or 1. It could take values from 0 to 255, as in many popular color specifications. If x_i records a value of the electric potential then, in principle, it is a real number that can take any value, arbitrarily small or large. While these different contexts all influence and enrich each other, our focus in this narrative will be on real vectors. Mathematicians denote real numbers by $\mathbb{R}$ and d-tuples of real numbers by $\mathbb{R}^d$. The number d is called the dimension.

To $\mathbb{R}^2$ and $\mathbb{R}^3$, we can attach a familiar geometric image. Via Cartesian coordinates, a point $\boldsymbol{x} = (x_1, x_2) \in \mathbb{R}^2$ corresponds to a point in the plane, whereas $\boldsymbol{x} = (x_1, x_2, x_3) \in \mathbb{R}^3$ corresponds to a point in the our native 3-dimensional space. While $\mathbb{R}^d$ may not be as familiar, it exists and it is important. With diverse uses and applications in mind, mathematicians, scientists, and engineers are all learning to wrap their 3-dimensional heads around the d-dimensional spaces.

A key geometric quantity in $\mathbb{R}^3$ is the distance between two points

$$\|\boldsymbol{x} - \boldsymbol{y}\| = \sqrt{\sum_{i=1}^{d} (x_i - y_i)^2}, \tag{1}$$

where $d = 3$. For $d = 2$, this is the distance between two points in the plane. For any d, this is the most natural way to define the distance between two points in $\mathbb{R}^d$. It is an important and useful notion in countless contexts, for instance, in statistical analysis.

For example, suppose we measured the values $\boldsymbol{x}' = (x'_1, \dots, x'_d)$ where we expected to see $\boldsymbol{x} = (x_1, \dots, x_d)$. Should we attribute the discrepancy to a small unavoidable random noise? Or have we observed something unexpected? The distance $\|\boldsymbol{x}' - \boldsymbol{x}\|$ is the principal measure of how well our measurements fit our predictions.

In this and other situations, it becomes important to separate the points $\boldsymbol{x}'$ whose distance from $\boldsymbol{x}$ is larger than some fixed threshold. One thus defines the ball and sphere in $\mathbb{R}^d$ with center $\boldsymbol{x}$ and radius r, respectively, by

$$B(\boldsymbol{x}, r) = \{\boldsymbol{x}' \text{ such that } \|\boldsymbol{x}' - \boldsymbol{x}\| \leq r\}, \tag{2}$$

$$S(\boldsymbol{x}, r) = \{\boldsymbol{x}' \text{ such that } \|\boldsymbol{x}' - \boldsymbol{x}\| = r\}. \tag{3}$$

We will use this terminology in all dimensions, even though for $d = 2$ this is usually called a disc and circle,[1] respectively.

1 And for $d = 1$, (2) is a segment and (3) are its endpoints.

The principal question for us in this narrative is how densely can one pack the spheres of a fixed radius in the d-dimensional space. One may compare and contrast a sphere with a cube

$$\mathrm{Cube}(\boldsymbol{x}, r) = \left\{\boldsymbol{x}' \text{ such that } \max_i \left|x_i' - x_i\right| \le r\right\}, \tag{4}$$

with center $\boldsymbol{x}$ and size $(2r) \times \cdots \times (2r)$. The maximum in (4) is an alternative measure of proximity of two vectors $\boldsymbol{x}$ and $\boldsymbol{y}$, and it is useful in different contexts. Spheres are exceptionally symmetric, preserved by all possible rotations around their center. Compared with spheres, cubes look heavy and boxy. Stacked side to side, cubes fill the whole space, leaving no voids between them. Two spheres can only touch at a point, and there will be voids left no matter how cleverly we try to pack them. However, what is the densest packing that can be achieved?

We will see that different dimensions vary significantly when it comes to sphere packings. In particular, in $\mathbb{R}^8$ and $\mathbb{R}^{24}$ there exist very special arrangements of spheres, denoted E_8 and Λ_{24}. They have been conjectured to be the densest possible in these dimensions.

Recently, this conjecture was proven in an absolutely stunning fashion by Maryna Viazovska in a solo work **[54]** for E_8 and by Viazovska and collaborators Henry Cohn, Abhinav Kumar, Stephen D. Miller, and Danylo Radchenko for Λ_{24} in **[11]**. For these and other phenomenal results, Maryna Viazovska was awarded the Fields Medal, the highest honor in mathematics, in 2022. Our modest goal in these notes is to share our personal excitement about the amazing math that goes into both the statement and the proof of these theorems with the broadest possible audience of mathematics enthusiasts.

1.2. Sphere packings in $\mathbb{R}^2$

The problem of sphere packing in two dimensions is familiar to anyone who tried to cut circular pieces from a rolled dough while preparing any of the delicious variations on the same universal theme, from vareniki to empanadas **[56]**.

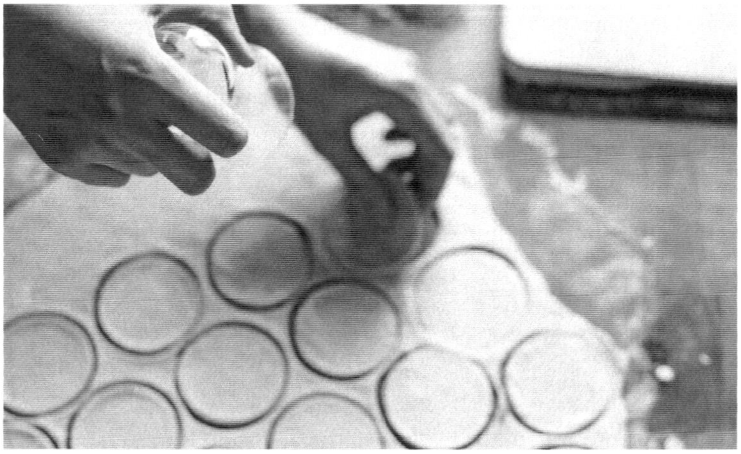

(5)

Naturally, one would like to minimize the fraction of the dough that gets discarded. If the size of the dough is much larger than the radius r of the cutter then this fraction is much more sensitive to the arrangement of circles than to the value of r. As a mathematical abstraction, one can consider an infinite piece of dough, and compute the fraction of the dough used (that is, the density of the sphere packing) as a *limit*[2] over larger and larger squares like in the figure of (6). We will compute the densities in (6) momentarily.

Note that for the infinite plane, a simple rescaling shows that the packing density does not depend on the radius r. This is true for sphere packing in all dimensions. In the analysis, one may leave r as a variable, or set it to any convenient value.

Let us look at the figure in (6) more closely:

(6)

On the left, we have stacked the circles just like squares. Hence, within each square, it is the inscribed circle that is used, and the rest discarded. Therefore, the packing density is

$$\frac{\text{area of inscribed circle}}{\text{area of a square}} = \frac{\pi}{4} = 0.785\ldots \tag{7}$$

If we slant the packing we can improve this. The other two arrangements in (6) are slanted at the angle of $\frac{5\pi}{12} = 75°$ and $\frac{\pi}{3} = 60°$, respectively. Therefore the distance between the horizontal rows of circles has decreased, and namely by a factor of

$$\sin\frac{5\pi}{12} = 0.965\ldots, \quad \sin\frac{\pi}{3} = \frac{\sqrt{3}}{2} = 0.866\ldots$$

As the horizontal rows get closer, the density increases by the reciprocals of these numbers. At $\frac{\pi}{3}$ this improvement has to stop, because each circle now touches not 4 but 6 other circles, and we cannot slant the figure any further.

Arguably, the hexagonal arrangement on the right in (6), with its 6-fold symmetry, is even more symmetric than the square arrangement on the left. It has a special name in mathematics, namely A_2. Here 2 stands for the dimension and the letter A will be discussed a bit later. We have

$$\text{density}(A_2) = \frac{\pi}{2\sqrt{3}} = 0.906\ldots \tag{8}$$

This is the densest the spheres can be packed in two dimensions. It is not simple to give a rigorous mathematical proof of this fact, but mathematicians succeeded a long time ago; see **[22, 23, 53]**. The person cutting the dough in the photograph (5) is evidently aware of this.

---

2 Readers unfamiliar with limits may probably find their discussion in **[43]** useful. To avoid worrying about the existence of the limit, it is a good idea to replace the limit by limit superior in this definition.

1.3. Contact number in $\mathbb{R}^3$

Let us see how well the life in three dimensions has prepared us for the analysis of the sphere packings in $\mathbb{R}^3$. As a warm-up, one can consider a local version of the packing problem, known as the contact number problem. It can be asked in any dimension and asks for the maximal number $\tau(d)$ of spheres of radius r in $\mathbb{R}^d$ that can be brought in contact with a given sphere of the same radius.

It is quite clear and will be revisited below that $\tau(2) = 6$, realized by the A_2 arrangement. In $\mathbb{R}^3$, the problem has a long history, the origin of which legend attributes to the notes taken by David Gregory during his conversations with Isaac Newton in 1694; see **[5]** for a critical analysis of this legend.[3]

Newton and Gregory apparently talked about celestial bodies, in which context it is natural to ask which percentage of the sky on one body, say the Earth, is occupied by image of another body, say the Moon, like in the figure of (9):

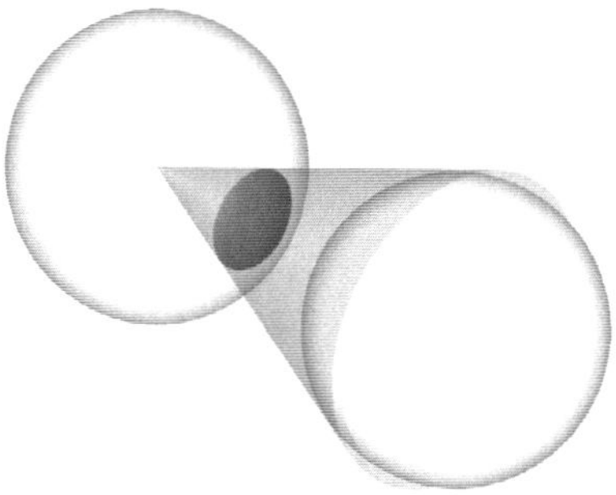

(9)

This percentage depends on the ratio of the distance to the Moon and Moon's radius. Suppose we have two touching spheres of the same radius r like in the figure of (10):

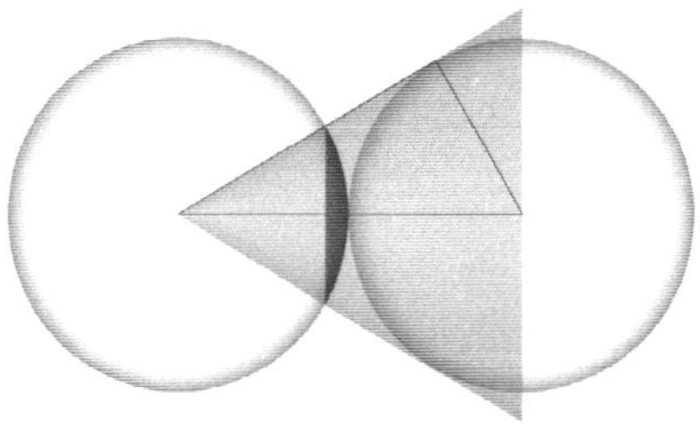

(10)

The right triangle in (10) has hypotenuse $2r$ and short side r. Therefore, the opposing angle equals

$$\arcsin \frac{1}{2} = \frac{\pi}{6}, \tag{11}$$

regardless of the dimension. Note that for $\mathbb{R}^2$ this already suffices to conclude $\tau(2) = 6$.

For $d = 3$, which fraction of the sky is occupied by the spheres in (10) in each other's sky? Consider the sphere

$$S(0, r) = \left\{(x_1, x_2, x_3) : x_1^2 + x_2^2 + x_3^2 = r^2\right\} \subset \mathbb{R}^3 \tag{12}$$

---

3 It is a problem in mathematics and human life in general that, lacking the time and resources to research every single topic, we mostly just repeat what we have been told. Not being able to break with this tradition, the narrator cannot do better than repeat what he read in **[5]**.

with center at the origin and radius r. The points with $x_3 \geq h$, where h is some fixed number between $-r$ and r, form what is called a *spherical cap*. The images in the sky in (9) and (10) are spherical caps.[4]

It was known already to Archimedes, and is commemorated as the comparison of the sphere with the cylinder on the back of the Fields Medal, that the area of a spherical cap is *proportional* to its height $r - h$.

Since the cap vanishes for $h = r$ and is the whole sphere for $h = -r$, we conclude

$$\frac{\text{area of the cap}}{\text{area of the sphere}} = \frac{r-h}{2r} = \frac{1-h/r}{2}. \tag{13}$$

For the cap in the figure of (10),

$$\frac{h}{r} = \cos\frac{\pi}{6} = \frac{\sqrt{3}}{2} \quad \Rightarrow \quad \frac{2r}{r-h} = \frac{4}{2-\sqrt{3}} = 14.92\ldots \tag{14}$$

Since each cap occupies more that $1/15$ of the surface area, 15 caps cannot fit without overlap, so the contact number in $\mathbb{R}^3$ is at most 14. It is easy to see that it is at least 12. Can it be equal 13? In the legend, Gregory thought yes, while Newton thought no.

There is an objective difficulty here, and it has to do with the fact that there are many different possible configurations of 12 spheres. One of them, realized for the densest packing, can be seen on the left in the figure of (27) below. But another possibility is to put the spheres in the 12 vertices of a regular icosahedron, like in (15), which also shows the plot of the spherical caps.

(15)

From the spherical caps, we see that the 12 spheres are not touching each other, hence can be moved around without losing the contact with the central sphere.[5] Perhaps we can make room

4 Of course, one gets the same geometric shape if one takes $x_1 \geq h$ or $x_2 \geq h$ instead of $x_3 \geq h$. Since caps are normally worn on the top of the head, we made the conventional choice of the vertical x_3-axis.

5 It is a fun fact to prove, see the appendix to Chapter 1 in **[15]**, that an *arbitrary* permutation of the 12 spheres may be achieved by rolling them around the central sphere.

for the 13th sphere? The problem of fitting caps into a sphere is a close spherical relative of the sphere packing problem[6] and belongs to a broad class of problems known as spherical codes and spherical designs [2,18].

Popular descriptions of the contact number problem often contain a suggestion for the reader to imagine a billiard ball hanging in mid-air, and 12 further billiard balls rolling around it. I envy those readers who have enough spatial intuition to imagine something like this. Even in our native $\mathbb{R}^3$, our geometric intuition often asks for help. Help may come from building models or from doing computations.

Geometers of all times have liked building models, using whatever materials the technology of the time made available. They would be surely thrilled to see the computer models that we can build today. There is a wonderful animated popular account of the contact number problem at Mathematical Etudes website. I am sure many readers will find it fascinating.

But the destination of our story being sphere packings in $\mathbb{R}^8$, it may be safe to expect the computations to overtake models in such high dimension. And indeed, at the heart of Viazovska's proof in [54] lies a brilliant inspired computation. It puts a big exclamation mark in a certain long line of argument. This line of argument was first born in the work of Philippe Delsarte in the discrete setting of coding theory [17] and was later adapted to spherical codes to compute

$$\tau(4) = 24, \quad \tau(8) = 240, \quad \tau(24) = 196560, \tag{16}$$

see [38,40,42]. It may be also used to show that $\tau(3) = 12$, see [1,40], but many other proofs of this fact were found earlier [5,47]. The Newton character from the legend was right.

Delsarte-type bounds, also known as linear programming bounds, were put to work in the sphere packings situation by H. Cohn and N. Elkies in [10]. We will talk about them in Section 3. They require a certain magic function to complete the proof. It is this elusive magic function that was discovered by Viazovska is her astonishing work [54].

1.4. The densest packings in $\mathbb{R}^3$

Let

$$\boldsymbol{v}_1, \ldots, \boldsymbol{v}_d \in \mathbb{R}^d$$

be a basis of $\mathbb{R}^d$, equivalently a set of linearly independent vectors.[7] By definition, the *lattice* Λ spanned by the vectors $\boldsymbol{v}_1, \ldots, \boldsymbol{v}_d$ is formed by all vectors

$$\Lambda = \mathbb{Z}\boldsymbol{v}_1 + \mathbb{Z}\boldsymbol{v}_2 + \cdots + \mathbb{Z}\boldsymbol{v}_d \subset \mathbb{R}^d \tag{17}$$

6 Strictly speaking, taking into account the spherical shape of the Earth, the person in the figure of (5) may be solving the spherical cap packing problem. In contrast to the sphere packing problem in $\mathbb{R}^d$, the radius of the spherical cap, or equivalently its angular size, cannot be scaled away and remains an important parameter in the problem. In the limit of very small caps, the problem reduces to sphere packing in the flat space $\mathbb{R}^d$. Of course, a person making vareniki is not taking the radius of the Earth into account!

7 Some readers may find the explanation of these notions given in [44] useful.

that can be obtained from the $\boldsymbol{v}_i$'s using addition and subtraction. The linear space $\mathbb{R}^d$ is a group under addition and lattices are special kinds of subgroups in it.

A sphere packing is called a *lattice packing* if the centers of the spheres form a lattice. For instance, the packings in (6) are lattice packings. There, we can take $\boldsymbol{v}_1 = (2r, 0)$ is all three cases, while

$$\boldsymbol{v}_2 = (2r\cos\phi, 2r\sin\phi), \quad \text{where } \phi = \frac{\pi}{2}, \frac{5\pi}{12}, \frac{\pi}{3},$$

depending on the slant angle ϕ.

Both the hexagonal packing and the corresponding hexagonal lattice are denoted by the symbol A_2. There is a cool way to realize this lattice inside $\mathbb{R}^3$ as the set

$$A_2 = \left\{(x_1, x_2, x_3), \sum x_i = 0\right\} \subset \mathbb{Z}^3 \tag{18}$$

of integer points with sum zero; see the figure in (19).

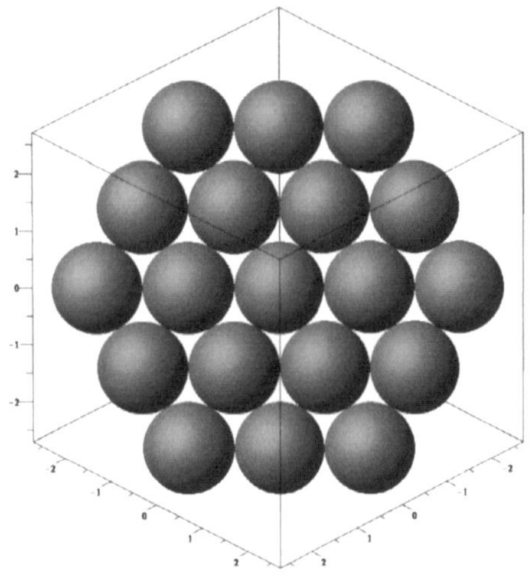

(19)

In (19), the sphere at $(0, 0, 0)$ is surrounded by 6 spheres with centers at all possible permutations of $(1, -1, 0)$. These are at distance $\sqrt{2}$ from the origin, and hence $r = \frac{1}{\sqrt{2}}$.

In a second we will need to talk about holes in the A_2 packing, shown in the figure of (21). These come in two different flavors according to the sign in

$$\text{hole center} = \pm\left(\frac{2}{3}, -\frac{1}{3}, -\frac{1}{3}\right) + \text{integer vector}. \tag{20}$$

The two kinds of holes are color-coded in (21). They are permuted by symmetries of A_2.

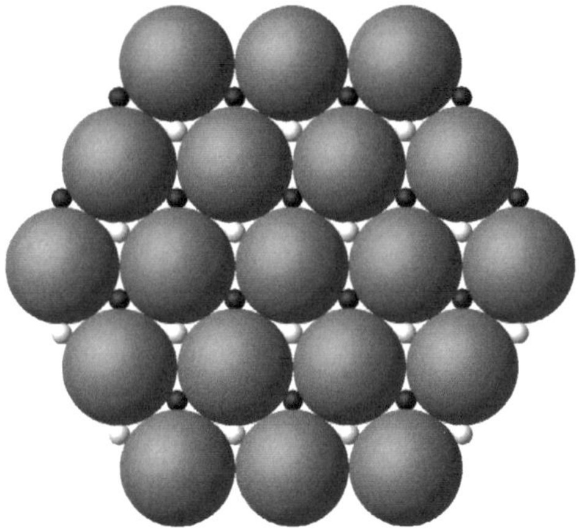

(21)

The densest sphere packing in $\mathbb{R}^3$ may be constructed by adding new layers of spheres (19) as in the figure of (22). Each new layer is a copy of (19) shifted so that the new spheres fit over the holes of the previous layer

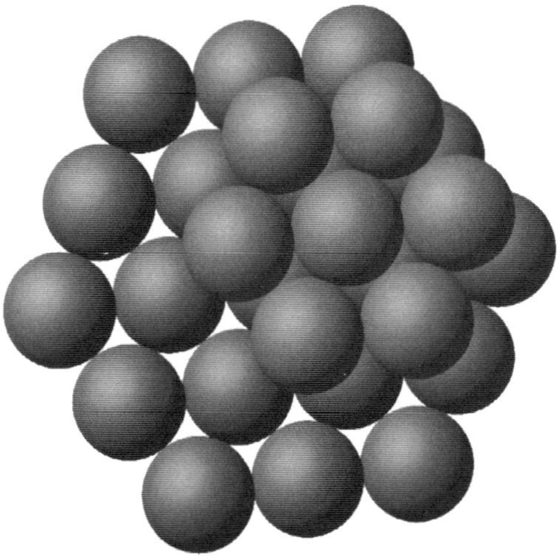

(22)

Since at every step we have 2 possible choice of the holes in (20), this gives 2^∞ different choice of packings with the same density! However, if we want it to be a lattice packing then there is only one choice up to an overall rotation or reflection. We can take

$$\begin{aligned} \boldsymbol{v}_1 &= (1,-1,\ \ 0), \\ \boldsymbol{v}_2 &= (0,\ \ 1,-1), \\ \boldsymbol{v}_3 &= (0,\ \ 1,\ \ 1), \end{aligned} \tag{23}$$

and these generate the lattice

$$D_3 = \left\{(x_1, x_2, x_3),\ \sum x_i \text{ is even}\right\} \subset \mathbb{Z}^3. \tag{24}$$

In general, one defines

$$A_d = \left\{(x_1, \dots, x_{d+1}),\ \sum x_i = 0\right\} \subset \mathbb{Z}^{d+1}, \tag{25}$$

$$D_d = \left\{(x_1, \dots, x_d),\ \sum x_i \text{ is even}\right\} \subset \mathbb{Z}^d. \tag{26}$$

For $d > 3$, these define different lattices and different sphere packings, but it is mathematical fact that A_3 is the same as D_3. Check this! Henry Cohn suggests the following exercise for when the reader visits the grocery store next. Find some fruit stacked as A_3 and some stacked as D_3. Then rotate your head until you are convinced that they are the same packing!

The proof of the fact that D_3 is the densest sphere packing in $\mathbb{R}^3$ is a monumental achievement of T. Hales and an inspiring story of computers helping humans to finish very complex proofs. See **[27–29,37]** for more about this. It is not known, but conjectured, that D_4 and D_5 are the densest sphere packings in $\mathbb{R}^4$ and $\mathbb{R}^5$, respectively.

It should be stressed emphatically that the optimality discussed in these notes concerns optimality among all sphere packings, not just lattice packings. Within the class of lattice packings, the optimality of D_3 was shown by Gauss in 1831 **[24]**, while the optimality of the D_4 and D_5 lattices was proven by Korkine and Zolotareff **[34,35]** in the 1870s.

Following our discussion of the contact number $\tau(3)$, it is fun to examine the arrangement of neighbors in the $A_3 = D_3$ packing. The sphere at $(0, 0, 0)$ has 12 neighbors with centers at the vectors $(\pm 1, \pm 1, 0)$ and their permutations. These are the 12 vectors $\boldsymbol{x}$ of length $\|x\| = \sqrt{2}$ in the lattice $A_3 = D_3$. The corresponding spherical caps can be seen in the figure of (27), together with 24 caps for the spheres with centers at points $\|x\|^2 = 6$ and 48 caps for the spheres at the distance $\|x\|^2 = 14$.

(27)

In this fashion, one can obtain very interesting collections of points on spheres from dense lattice packings in any dimension.

2. BEYOND THE 3-SPACE

2.1. 4, 5, 6, 7, 8, ...

The narrator of these notes is a complete novice in the field of sphere packing trying to share his first impressions of the striking beauty of the field with other mathematics enthusiasts. Among the mathematicians of older generations, I imagine I am not alone feeling like a schoolboy again, exploring spellbound the treasures described, in particular, in the treatise [15] by John Conway, Neil Sloane, and collaborators. The story starts deceptively simple but quickly leads to the highest heights and deepest depths of mathematics.

To continue the parallel with one's student years, each dimension d in the sphere packing problem feels like a new year of math classes. While it builds on and connects with the material form the previous years, many new phenomena and ideas appear each time.

In even further parallel to how mathematics courses change as we go trough high-school, college, graduate school, and so on, hopefully never stopping learning, the sphere packing problems seem to come in certain groups of dimensions. In dimensions up to 8, the densest packing are known or conjectured to be the lattice packings

$$\underline{A_1, A_2, A_3}, D_4, D_5, E_6, E_7, \underline{E_8}. \tag{28}$$

Here the known cases are underlined, the 8-dimensional case being Viazovska's breakthrough. This certainly feels like a story about exceptional Lie groups, which ends in dimension 8 with the largest exceptional Lie group E_8. The optimality of E_6, E_7, E_8 among *lattice* packings was shown by Blichfeldt [4] back in 1935.

Next come dimensions 9 through 24. Looking at the iconic picture in (29), reproduced here with permission from [49], we see that these dimensions start out as valley leading to an ascent to the sharp peak of the Leech lattice Λ_{24}. The Leech lattice is now proven to give the densest sphere packing in dimension 24 by Viazovska and collaborators [11].

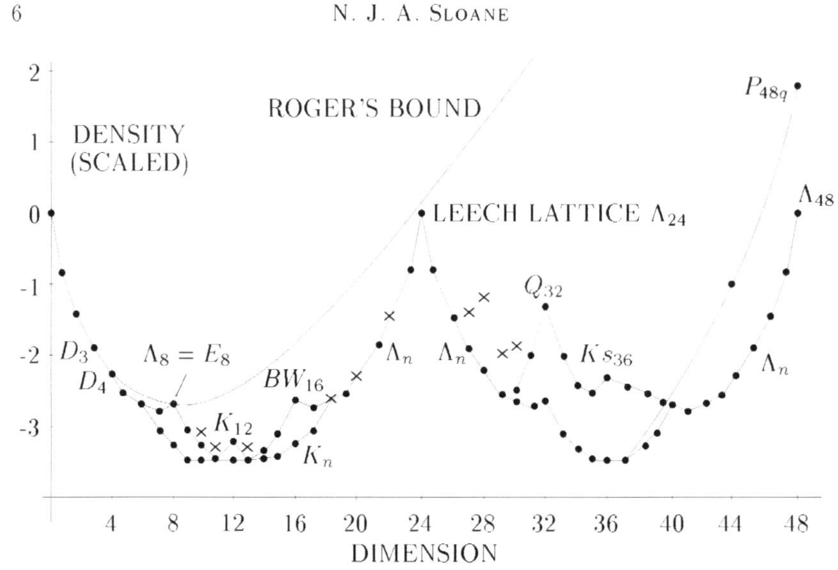

(29)

The Leech lattice, with its deep connections to the exceptional, or sporadic, finite simple groups including the Monster group of Bernd Fischer and Robert Griess, is the defining feature of the 9 to 24 valley. Again, the Monster being the largest exceptional group, the storyline has to change after 24.

What is next? We hope the reader's curiosity will lead her or him to explore, guided by [7,15]. See also the tables [9,41] of the densest sphere packings currently known in different dimensions.

2.2. Fluid diamond in $d = 9$

Here is one among the countless marvels of high-dimensional sphere packing. Recall the lattice D_d from (25) formed by integer vectors with even coordinate sum. The nearest neighbors in D_d are $\sqrt{2}$ away, so we can pack spheres of radius $r = \frac{1}{\sqrt{2}}$ using points of D_d as centers.

Consider the vector

$$\boldsymbol{\gamma} = \left(\frac{1}{2}, \frac{1}{2}, \frac{1}{2}, \dots, \frac{1}{2}\right). \tag{30}$$

What is the distance between $\boldsymbol{x}$ and the nearest point $\boldsymbol{v} \in D_d$? Since $\boldsymbol{v}$ has integral coordinates, we have

$$\|\boldsymbol{\gamma} - \boldsymbol{v}\|^2 \geq \underbrace{\frac{1}{2^2} + \cdots + \frac{1}{2^2}}_{d \text{ times}} = \frac{d}{4}.$$

Therefore, if $d = 8$, we can fit *two* copies of D_8 into $\mathbb{R}^8$ with a shift by $\boldsymbol{\gamma}$. The resulting lattice is nothing else than the magic E_8 lattice

$$E_8 = D_8 \cup (D_8 + \boldsymbol{\gamma}), \tag{31}$$

about which we will talk more in Section 2.3 below.

If $d = 9$, we can take the vector

$$\boldsymbol{\gamma}_{i,t} = \boldsymbol{\gamma} + t\boldsymbol{e}_i, \quad i = 1, \ldots, 9, \tag{32}$$

where $t \in \mathbb{R}$ is an arbitrary number and

$$\boldsymbol{e}_i = (0 \ldots, 0, \underset{i}{1}, 0, \ldots, 0) \tag{33}$$

is the ith coordinate vector. By the same argument as before,

$$\|\boldsymbol{\gamma}_t - \boldsymbol{v}\|^2 \geq 2, \quad \text{for} \quad \boldsymbol{v} \in D_9,$$

so we can pack the spheres of radius $r = \frac{1}{\sqrt{2}}$ using points of

$$\text{fluid diamond packing} = D_9 \cup (D_9 + \boldsymbol{\gamma}_{i,t}) \tag{34}$$

as centers. Note that since both t and i are arbitrary, half of the spheres in (34) can be shifted arbitrarily in one of the coordinate direction without running into the other half of the spheres—a rather fluid packing! And yet, its density matches, for any t, the highest known density in dimension 9. It is not so easy to imagine this possible based on our low-dimensional geometric intuition.

2.3. Stars align in E_8

The exceptionally dense and symmetric E_8 lattice packing which we met in (31) certainly merits a much longer discussion. One can start this discussion from many different angles, emphasizing different areas of mathematics where the E_8 lattice naturally appears.

2.3.1. Roots

The lattices A_d and D_d from (25) have the property that $\|\boldsymbol{v}\|^2$ is an even integer for any $\boldsymbol{v} \in D_d$. Such lattices are called *even*. How can we tell if a lattice Λ as in (17) is even? Using the concept of the *inner product*, recalled in Appendix A, it suffices to check that $(\boldsymbol{v}_i, \boldsymbol{v}_j) \in \mathbb{Z}$ and $(\boldsymbol{v}_i, \boldsymbol{v}_i) \in 2\mathbb{Z}$ for any basis of Λ. Since

$$(\boldsymbol{\gamma}, \boldsymbol{\gamma}) = 2 \quad \text{and} \quad (\boldsymbol{\gamma}, \boldsymbol{v}) \in \mathbb{Z},$$

for any $\boldsymbol{v} \in D_8$ and $\boldsymbol{\gamma}$ as in (30), we see that E_8 is an even lattice.

Given an even lattice Λ, vectors $\boldsymbol{\alpha} \in \Lambda$ of the minimal nonzero norm $\|\boldsymbol{\alpha}\|^2 = 2$ are call *roots*.[8] These are the centers of the spheres touching the central sphere. For example, the vectors

$$\boldsymbol{\alpha} = \pm\boldsymbol{e}_i \pm \boldsymbol{e}_j \in D_d, \quad i \neq j, \tag{35}$$

are roots. For E_8, we also have the root $\boldsymbol{\gamma}$, as well as

$$\boldsymbol{\alpha} = \left(\pm\frac{1}{2}, \pm\frac{1}{2}, \pm\frac{1}{2}, \ldots, \pm\frac{1}{2}\right) \quad \text{such that the sum is even,} \tag{36}$$

8 For lattices in which the squared norm takes both even and odd integer values, vectors of norm 1 should also count as roots. These have an important role to play in Lie theory and many other branches of mathematics, but not in our narrative.

of which there are $\frac{1}{2}2^8 = 128$ many. We invite the reader to check there are no other roots for E_8 and verify that the number of roots equals $\tau(8) = 240$. Thus the roots of E_8 give the solution of the contact number problem in $d = 8$, and in fact this solution is unique, very much unlike the $d = 3$ case discussed in Section 1.3.

2.3.2. Reflections

Every root $\boldsymbol{\alpha} \in \Lambda$ in an even lattice Λ generates a special symmetry of the lattice Λ, namely the orthogonal reflection $r_{\boldsymbol{\alpha}}$ in the hyperplane orthogonal to $\boldsymbol{\alpha}$. It sends $\boldsymbol{\alpha}$ to $-\boldsymbol{\alpha}$ and fixes all vectors $\boldsymbol{v}_\perp$ that are orthogonal to $\boldsymbol{\alpha}$.

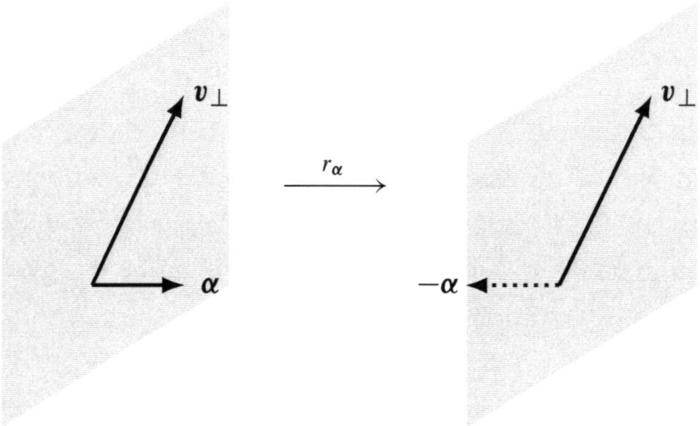

(37)

Explicitly, it is given by the formula

$$r_{\boldsymbol{\alpha}}(\boldsymbol{v}) = \boldsymbol{v} - (\boldsymbol{v}, \boldsymbol{\alpha})\boldsymbol{\alpha}, \tag{38}$$

which manifestly preserves the lattice Λ. Indeed, (101) shows the inner product takes integer values in an even lattice. For more on (38), see Section A.3.

For example, the roots $\boldsymbol{\alpha}$ of the A_2 lattice (18) are the permutations of the vector $(1, -1, 0)$. The reflection $r_{(1,-1,0)}$ swaps the first two coordinates. Similarly, for $A_d \subset \mathbb{R}^{d+1}$, each reflection $r_{\boldsymbol{\alpha}}$ swaps two coordinates of $\mathbb{R}^{d+1}$.

Orthogonal symmetries of a lattice Λ always form a finite group; see the brief introduction to this concept in Appendix B. In particular, its subgroup generated by the reflections $r_{\boldsymbol{\alpha}}$ is a finite group W generated by reflections. Such groups have been fully classified and studied in great detail due to their crucial importance in Lie theory, singularity theory, and many other branches of mathematics. As a corollary of this classification, we know that all even lattices spanned by roots are orthogonal direct sums of lattices of the form A_d, D_d, or E_6, E_7, E_8.

2.3.3. ADE classification

How does this classification work? In an even lattice Λ spanned by roots, one can always choose the basis of roots so that

$$(\boldsymbol{\alpha}_i, \boldsymbol{\alpha}_j) = 0 \text{ or } -1, \quad i \neq j. \tag{39}$$

For example, for E_8, we can take

$$\boldsymbol{\alpha}_i = \boldsymbol{e}_i - \boldsymbol{e}_{i+1}, \quad i = 1, \ldots, 6, \tag{40}$$

together with

$$\boldsymbol{\alpha}_7 = \boldsymbol{e}_6 + \boldsymbol{e}_7, \quad \boldsymbol{\alpha}_8 = -\boldsymbol{\gamma}. \tag{41}$$

We see that $(\boldsymbol{\alpha}_i, \boldsymbol{\alpha}_j) = 0$ for most pairs i, j, and that $(\boldsymbol{\alpha}_i, \boldsymbol{\alpha}_j) = -1$ precisely for the pairs connected by an edge in the following graph:

$E_8 =$

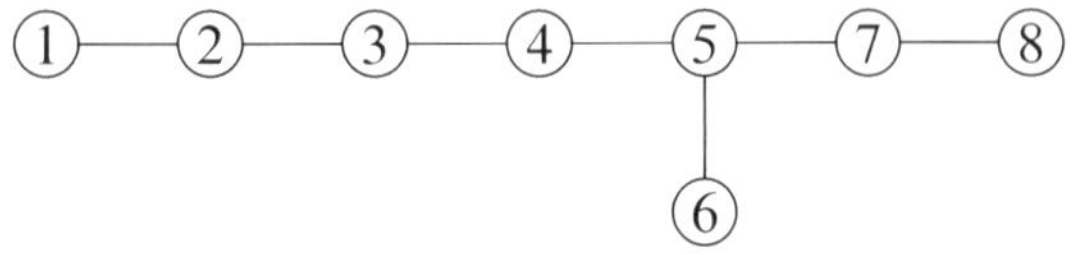

(42)

The graph (42) is a very convenient graphical way to represent the Gram matrix

$$B_\Lambda = \big((\boldsymbol{\alpha}_i, \boldsymbol{\alpha}_j)\big).$$

It is called the Dynkin diagram or the Coxeter diagram.

The ADE classification is really the classification of all possible diagrams like (42) for which the corresponding Gram matrix B_Λ is positive definite (a concept which will be explained and used in Section 3.1 below). This is not as difficult as it sounds, and revolves around the fact any subgraph of a positive definite diagram is a positive definite diagram.

For instance, if we erase nodes 1, 2, 3, 8, 7, 6, 5 from (42) in that order, we get diagrams for lattices E_7, E_6, D_5, D_4, $D_3 = A_3$, A_2, A_1, that are known or conjectured to give the densest packings in the corresponding dimensions.

In the opposite direction, if we try to invent the lattice E_9 with the diagram

"E_9" $=$

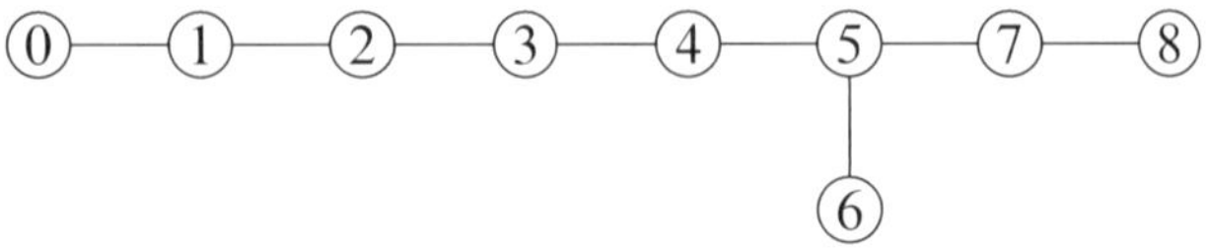

(43)

we see that this does not work because the determinant $\det B_{E_9}$ of the corresponding Gram matrix vanishes.[9]

2.3.4. Discriminant

In general, the determinant or the discriminant of an even lattice Λ,

$$\Delta_\Lambda = \det B_\Lambda, \tag{44}$$

is a very important quantity that enters the following formula for the density of the corresponding sphere packing.

The Λ-translates of the parallelepiped

$$\Pi_\Lambda = \left\{\boldsymbol{x} = \sum x_i \boldsymbol{\alpha}_i,\ \max |x_i| \le \frac{1}{2}\right\} \tag{45}$$

tile the whole space, each tile containing exactly one lattice point as its center. This just the story about the cube (4) in different coordinates. We have

$$\operatorname{Vol} \Pi_\Lambda = \sqrt{\Delta_\Lambda}. \tag{46}$$

9 In fact, the diagram (43) is the Dynkin diagram of the infinite affine reflection group of type $\hat{E}_8$ in $\mathbb{R}^8$.

Therefore, if Λ is even and the roots are the vectors of minimal length then

$$\text{density of the sphere packing} = \frac{\operatorname{Vol} B(0, \frac{1}{\sqrt{2}})}{\sqrt{\Delta_\Lambda}}. \tag{47}$$

The smaller the discriminant, the larger the density. Since the discriminant is an integer, 1 is the smallest it can be, and in fact

$$\Delta_{E_8} = 1. \tag{48}$$

Lattices with $\Delta_\Lambda = 1$ are called *unimodular*. Even unimodular lattices exist only in dimensions that are multiples of 8, and E_8 is the unique even unimodular lattice in $\mathbb{R}^8$.

In dimension 24, there exist 24 even unimodular lattices, and the superamazing Leech lattice is distinguished among them by having no roots! In other words, one can pack spheres of radius $r = 1$ with centers in Leech lattice instead of $r = \frac{1}{\sqrt{2}}$. While a meaningful discussion of the Leech lattice transcends the introductory nature of these notes, we hope that the reader's curiosity will be satisfied by the accounts in [15, 19, 52].

2.3.5. Codes

Recall how at the very beginning, in Section 1.1, we talked about the possible values of the entries x_i in a vector $\boldsymbol{x} = (x_1, \dots, x_d)$. While everywhere else in this narrative we consider the case of real entries x_i, let us turn our attention to the case $x_i \in \{0, 1\}$ for a brief moment. In other words, let us talk about binary vectors.

The natural distance between binary vectors is the Hamming distance

$$\|\boldsymbol{x} - \boldsymbol{x}'\|_{\text{Hamming}} = \sum |x_i - x_i'|. \tag{49}$$

It measures the number of entries in which $\boldsymbol{x}$ and $\boldsymbol{x}'$ differ, and it is very natural for error correction and other applications. If $\boldsymbol{x}$ and $\boldsymbol{x}'$ represent the input and output of a transmission through a binary communication channels, then (49) is the number of errors that occurred during the transmission. If we can pack nonintersecting Hamming balls of radius r in $\{0, 1\}^d$ then the centers $\mathscr{C} \subset \{0, 1\}^d$ of these balls give binary code words of length d that corrects up to r errors. A related concept is the minimal distance δ between the code words from $\mathscr{C}$. Evidently, $\delta > 2r$.

Given a code $\mathscr{C}$, we define $\widehat{\mathscr{C}} \subset \mathbb{Z}^d$ as the set of integer vectors that have the same parity as some code word from $\mathscr{C}$. Clearly, if $\boldsymbol{v} \neq \boldsymbol{v}'$ are two distinct points of $\widehat{\mathscr{C}}$ then

$$\|\boldsymbol{v} - \boldsymbol{v}'\| \geq \min(2, \sqrt{\delta}),$$

and hence we can pack sphere of half that radius with centers at $\widehat{\mathscr{C}}$.

For $d = 8$, there exists a remarkable code $\mathscr{C}$ with $\delta = 4$. It is obtained by adding the parity bit to Hamming $(7, 4)$-code, see [1, 15]. The corresponding packing $\widehat{\mathscr{C}}$ is isomorphic to the E_8 packing. Similarly, the Leech lattice can be obtained from the Golay code.

2.3.6. The Coxeter plane

There is the following cool way to visualize roots for any finite reflection group (requires familiarity with eigenvalues and also with complex numbers, see Section A.5). The material in this section may feel a bit advanced and it could be a good idea to come back to it after reading the material in the Appendix.

Recall the basis $\boldsymbol{\alpha}_i$ from Section 2.3.3 and consider the corresponding reflections $r_{\boldsymbol{\alpha}_i}$. Consider the product C of all these reflections taken in some order. The reflections do not commute, so C depends on the order. Remarkably, however, the conjugacy class of C is independent of the order. A a particularly nice choice is

$$\text{Coxeter element } \mathsf{C} = \underbrace{r_{\boldsymbol{\alpha}_1} r_{\boldsymbol{\alpha}_3} r_{\boldsymbol{\alpha}_5} r_{\boldsymbol{\alpha}_8}}_{\text{commute}} \underbrace{r_{\boldsymbol{\alpha}_2} r_{\boldsymbol{\alpha}_4} r_{\boldsymbol{\alpha}_6} r_{\boldsymbol{\alpha}_7}}_{\text{commute}}, \tag{50}$$

which presents C as a product of two involutions, that is, two elements that each square to $\mathbb{1}$, where $\mathbb{1}$ is the identity matrix.

The order and eigenvalues of a Coxeter element can be computed abstractly. For E_8, we have $\mathsf{C}^{30} = \mathbb{1}$ and the eigenvalues of C are exactly the primitive roots of unity of order 30 or, equivalently, the roots of the cyclotomic polynomial

$$z^8 + z^7 - z^5 - z^4 - z^3 + z + 1 = 0. \tag{51}$$

We can take any one of them and project the roots onto corresponding eigenspace $\mathbb{C} \subset \mathbb{C}^8$, called the Coxeter plane. The resulting collection of points are the centers of the circles in the figure of (52). The radii in that figure have no exact mathematical meaning and are simply adjusted to resemble a sphere packing. The colors will be explained below.

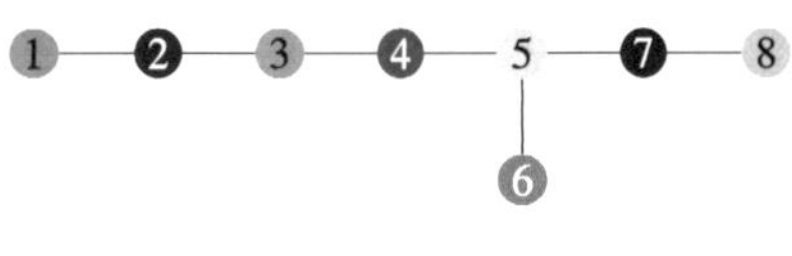

(52)

Consider the involutions

$$\mathsf{R}_{\text{green}} = \underbrace{r_{\boldsymbol{\alpha}_1} r_{\boldsymbol{\alpha}_3} r_{\boldsymbol{\alpha}_5} r_{\boldsymbol{\alpha}_8}}_{\text{commute}}, \quad \mathsf{R}_{\text{blue}} = \underbrace{r_{\boldsymbol{\alpha}_2} r_{\boldsymbol{\alpha}_4} r_{\boldsymbol{\alpha}_6} r_{\boldsymbol{\alpha}_7}}_{\text{commute}}, \tag{53}$$

where the colors refer to the color-coding in the Dynkin diagram in (52). By construction, the involutions (53) satisfy

$$\mathsf{R}_{\text{green}}\mathsf{C} = \mathsf{R}_{\text{blue}} = \mathsf{C}^{-1}\mathsf{R}_{\text{green}}, \tag{54}$$

and hence generate, together with C, the symmetry group of a regular 30-gon in the Coxeter plane. We have

$$\mathsf{R}_{\mathrm{green}}\, \boldsymbol{\alpha}_i = -\boldsymbol{\alpha}_i\,, \quad i \in \{1,3,5,8\}. \tag{55}$$

Therefore, all green vertices land on a line in the Coxeter plane—the line perpendicular to the line fixed by $\mathsf{R}_{\mathrm{green}}$. An identical argument works for $\mathsf{R}_{\mathrm{blue}}$.

We can use the figure in (52) to illustrate the following important concepts related to roots. First, the roots can be partitioned into positive and negative by a generic hyperplane in $\mathbb{R}^8$, which we can take to be the preimage of a line in the Coxeter plane. In (52), the negative roots are in gray, while the positive roots are colored.

Second, one can choose the roots $\boldsymbol{\alpha}_i$ from Section 2.3.3 as the *simple positive roots*. These are the positive roots that cannot be written nontrivially as a sum of positive roots. They are monochromatic in (52) where the colors correspond to the coloring of the Dynkin diagram as before.

Third, all positive roots are nonnegative integer linear combinations of simple roots. The proportions in which the simple roots combine to produce a given positive root are plotted as pie charts in the figure of (52). In particular, the dichromatic roots in (52) correspond to the roots

$$\boldsymbol{\alpha}_i + \boldsymbol{\alpha}_j = r_{\boldsymbol{\alpha}_i}(\boldsymbol{\alpha}_j) = r_{\boldsymbol{\alpha}_j}(\boldsymbol{\alpha}_i) \quad \text{when} \quad (\boldsymbol{\alpha}_i, \boldsymbol{\alpha}_j) = -1, \tag{56}$$

which exist for any pair of neighbors in the Dynkin diagram.

See Appendix F for more on connections between E_8 and regular polygons.

2.4. Very large dimensions

Our discussion of sphere packings in arbitrarily large dimensions will be very brief due to both objective lack of information about them and limits of the present narrative.

Let us call a sphere packing in $\mathbb{R}^d$ saturated if no additional sphere of the same radius r can be inserted into it. Remarkably, the density of a a saturated packing is at least 2^{-d}. We invite the reader to pause for a second and try to prove this. Maryna Viazovska says this is one of her favorite entry-level problems about sphere packings.

One way to prove this is to note that, for a saturated packing, balls of twice the radius with the same centers have to cover the whole $\mathbb{R}^d$. Otherwise, there would a point where we can insert another sphere of radius r. From

$$\operatorname{Vol} B(0,r) = 2^{-d} \operatorname{Vol} B(0,2r),$$

we get the sought lower bound for the density of a saturated packing.

As simple as this sounds, this bound is remarkable. As we review in Appendix E, the volume of $B(0,r)$ decays superexponentially with dimension d for any r. Hence a packing achieving a 2^{-d} density must have superexponentially many spheres in any cube $[0,L]^d \subset \mathbb{R}^d$ as $d \to \infty$. Also, the best known improvements to the 2^{-d} lower bound are only basically linear in d.

As to the upper bounds on density, that by Kabatiansky and Levenshtein [31] has been holding the world record at $2^{-0.5990\ldots d}$ since 1978, although the methods described below allowed Cohn and Zhao [13] to achieve a constant factor improvement.

3. UPPER BOUNDS ON PACKING DENSITY

3.1. Positive definite forms and functions

3.1.1.

Let us start with with simplest possible inequality. For any real number x, $x^2 \geq 0$. As trivial as this sounds, this proves, for instance, that

$$x_1^2 - 2x_1x_2 + 2x_2^2 - 2x_2x_3 + x_3^2 = (x_1 - x_2)^2 + (x_2 - x_3)^2 \geq 0. \tag{57}$$

3.1.2.

An expression of the form

$$B(\boldsymbol{x}) = \sum_{i,j=1}^{n} b_{ij} x_i x_j, \tag{58}$$

where $b_{ij} \in \mathbb{R}$ are coefficients, is called a *quadratic form* in the variables $\boldsymbol{x} = (x_1, \ldots, x_n)$. In the sum (58), we may and will assume that $b_{ij} = b_{ji}$. The symmetric array of numbers (b_{ij}) is called the matrix of the quadratic form (58).

A quadratic form is called *positive semidefinite* if it takes only nonnegative values, like that in (57). One writes $B \geq 0$. Forms that take positive values for nonzero arguments are called *positive definite*. For instance, (57) is positive semidefinite but not definite, since it vanishes for $\boldsymbol{x} = (1, 1, 1)$.

3.1.3.

If $B_1, B_2 \geq 0$ then

$$c_1 B_1 + c_2 B_2 \geq 0 \tag{59}$$

for all coefficients $c_1, c_2 \geq 0$. Mathematicians say that the set of positive semidefinite forms is a convex cone.

For example, for $n = 2$, we have

$$b_{11}x_1^2 + 2b_{12}x_1x_2 + b_{22}x_2^2 \geq 0 \quad \Leftrightarrow \quad \begin{matrix} b_{11} \geq 0, \ b_{22} \geq 0, \\ b_{12}^2 \leq b_{11}b_{22}. \end{matrix} \tag{60}$$

In the 3-space with coordinates (b_{11}, b_{22}, b_{12}), the set of the positive semidefinite forms is the familiar cone plotted in the figure of (61) with its vertex at the origin $(b_{11}, b_{22}, b_{12}) = (0, 0, 0)$.

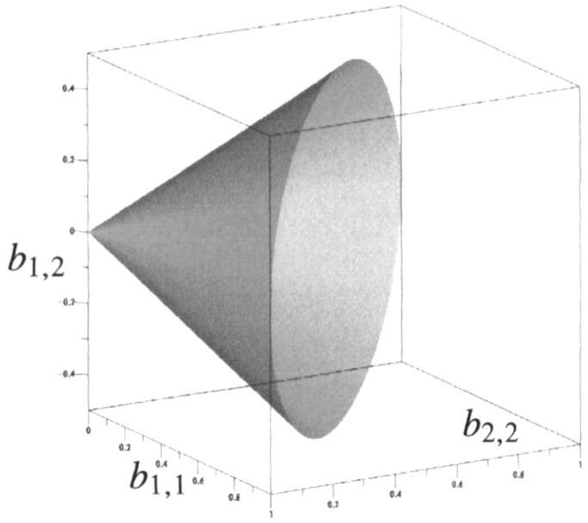

(61)

The interior of this cone corresponds to positive definite forms.

3.1.4.

Here is an example of an interesting positive semidefinite form. Fix some angles $\phi_1, \ldots, \phi_n$ and let

$$\boldsymbol{\varepsilon}_i = (\cos\phi_i, \sin\phi_i) \in \mathbb{R}^2, \quad i = 1, \ldots, n,$$

be unit vectors in $\mathbb{R}^2$ having the angle ϕ_i with the horizontal axis. Let us add them with coefficients $x_1, \ldots, x_n$; see the figure in (62).

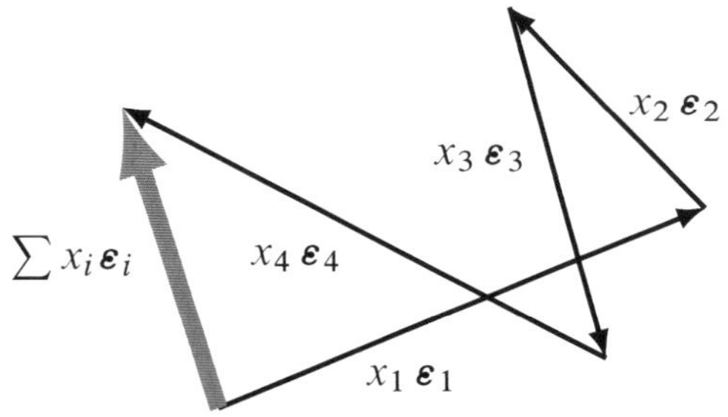

(62)

We define the form $B_{\cos}(\boldsymbol{x})$ as the squared length of this sum. Using inner products, see Appendix A, we compute

$$\begin{aligned} B_{\cos}(\boldsymbol{x}) &= \left\| \sum x_i \boldsymbol{\varepsilon}_i \right\|^2 = \left(\sum x_i \boldsymbol{\varepsilon}_i, \sum x_j \boldsymbol{\varepsilon}_l \right) \\ &= \sum_{ij} (\boldsymbol{\varepsilon}_i, \boldsymbol{\varepsilon}_l) x_i x_j = \sum_{ij} \cos(\phi_i - \phi_j) x_i x_j \geq 0. \end{aligned} \tag{63}$$

This is not obviously nonnegative based on coefficients, but we know it must be nonnegative as a squared length.

3.1.5.

In general, let $f(t)$ be an even function; that is, let $f(t)$ satisfy

$$f(t) = f(-t).$$

We say that f is *positive definite* if the quadratic form

$$B_f(\boldsymbol{x}) = \sum_{i,j=1}^{n} f(t_i - t_j) x_i x_j \geq 0 \tag{64}$$

is positive semidefinite for *any* choice of $t_1, \ldots, t_n$. It follows from (63) that $f(t) = \cos t$ is positive definite. Similarly, $f(t) = \cos\omega t$ is positive definite for any frequency ω.

To get a better feeling for positive-definite functions, the reader may want to deduce from (60) that $f(0) > 0$ for any positive-definite function $f(t)$ that is not identically zero.

3.1.6.

It follows from (59) that the function

$$f(t) = \sum_k c_k \cos(\omega_k t), \quad c_k \geq 0, \tag{65}$$

is positive definite for any frequencies ω_k as long as the coefficients c_k are nonnegative.

The coefficients c_k with which the different frequencies contribute to the function $f(t)$ will be very important in what follows and the generic notation c_k will not be adequate for them. We need some notation that incorporates the name of the function f, frequency ω_k, and the fact that we expand f in cosines and not some other periodic functions, specifically not in sines. A popular choice, satisfying all of these criteria is to replace c_k by $\hat{f}^{c}(\omega_k)$. So, we write

$$f(t) = \sum \hat{f}^{c}(\omega_k)\cos(\omega_k t), \quad \hat{f}^{c}(\omega_k) \geq 0. \tag{66}$$

A classical theorem of Bochner says that, conversely, every positive definite function is a limit of functions of the form (66). See Appendix B for more on this.

3.1.7.

In equation (64), it is perfectly OK to make the argument of f be a vector $\boldsymbol{t} \in \mathbb{R}^d$. We say that a function $f(\boldsymbol{t})$ is even if

$$f(\boldsymbol{t}) = f(-\boldsymbol{t}),$$

and we say it is *positive definite* if

$$B_f(\boldsymbol{x}) = \sum_{i,j=1}^{n} f(\boldsymbol{t}_i - \boldsymbol{t}_j)x_i x_j \geq 0 \tag{67}$$

for any $\boldsymbol{x}$ and any $\boldsymbol{t}_1, \ldots, \boldsymbol{t}_n$.

The only modification required in the formula (66) is that the frequencies also become vectors $\boldsymbol{\omega}_k$ and we replace the product $\omega_k t$ by the inner product $(\boldsymbol{\omega}_k, \boldsymbol{t})$. In sum, the function

$$f(t) = \sum \hat{f}^{c}(\boldsymbol{\omega}_k)\cos\big((\boldsymbol{\omega}_k, \boldsymbol{t})\big), \quad \hat{f}^{c}(\boldsymbol{\omega}_k) \geq 0, \tag{68}$$

is positive definite and every positive definite function of $\boldsymbol{t} \in \mathbb{R}^d$ is a limit of functions of the form (68).

3.1.8.

To feed the reader's curiosity, we note briefly that the differences $\boldsymbol{t}_i - \boldsymbol{t}_j$ in the definition of a positive definite function may be replaced by the ratios $\boldsymbol{t}_i \boldsymbol{t}_j^{-1}$ of elements $\boldsymbol{t}_i$ of an arbitrary group G. For the additive group of $\mathbb{R}^d$, we get the positive definite functions as discussed above.

Bochner's theorem is then interpreted as saying that f is a diagonal matrix element of an orthogonal representation of G. Viewed from the correct angle, this is very close to a tautology, as noted by Gelfand and Naimark [26] and Segal [48]. See Appendix B for more on this.

3.2. The fundamental bound

3.2.1.

We will now explain how positive definite functions may be used to bound the density of sphere packings following H. Cohn and N. Elkies [10]. Related considerations, in which translations are replaced by rotations, were used to bound the contact numbers; see [1, 38–40, 42]. As we already mentioned bounds of this type originated in coding theory [17].

3.2.2.

Consider a packing of spheres of radius r in $\mathbb{R}^d$ and suppose it is *periodic* in each coordinate direction with period L. For instance, consider the figure in (6) and let the square in (6) be the square $[0, L]^2 \in \mathbb{R}^2$. Then the leftmost packing in it is periodic, and the other two can be made periodic if we erase the spheres intersecting the boundary of the square $[0, L]^2$. In (69) one can see the result for the middle packing in (6).

For large L, the number of spheres intersecting the boundary of $[0, L]^2$ can be bounded from above by a constant multiple of L. Therefore, erasing these spheres changes the density by at most a constant multiple of L^{-1}. We conclude that we can come arbitrarily close to the optimal packing density using periodic packings.

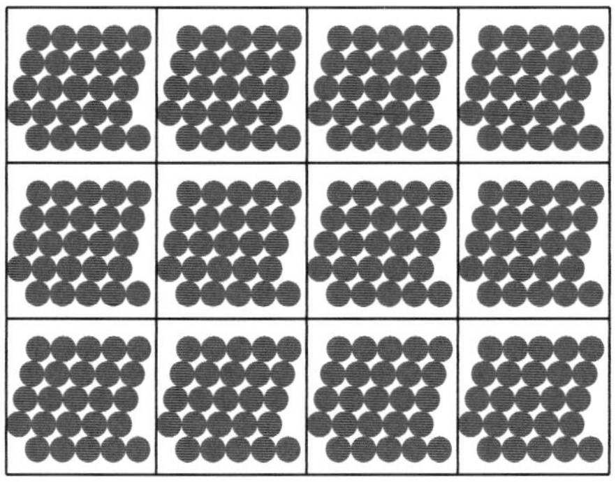

(69)

In dimension d, we may need to erase at most a constant multiple of L^{d-1} many spheres, and again this changes the density by at most a constant multiple of L^{-1}. We conclude that any upper bound on the density of periodic packings with an arbitrarily large period L gives an upper bound on the density of all sphere packings.

3.2.3.

So, returning to our periodic packing, suppose it has n spheres with centers in $[0, L]^d$ and let

$$\boldsymbol{t}_1, \ldots, \boldsymbol{t}_n \in [0, L]^d \tag{70}$$

be the centers of these spheres.

The translates of the cube $[0, L]^d$ tile the whole space $\mathbb{R}^d$ and any periodic sphere packing is just repeated in all these translates. Mathematicians call the basic tile $[0, L]^d$ the *fundamental domain*. So, the number n is the number of spheres per fundamental domain. It directly measures the density of packing by

$$\text{packing density} = \frac{n}{L^d} \operatorname{volume}\big(B(0, r)\big) \tag{71}$$

because the sphere packing is periodic. So, our goal is to bound the ratio n/L^d.

3.2.4.

To bound n/L^d, we will use a certain positive definite function $f(\boldsymbol{t})$, which will be similarly periodic with period L in all coordinates.

To make (68) periodic, the frequencies $\boldsymbol{\omega}$'s should be integer multiples of $\frac{2\pi}{L}$. We define

$$\boldsymbol{\omega}_{\boldsymbol{k}} = \frac{2\pi}{L}\boldsymbol{k}, \tag{72}$$

where $\boldsymbol{k} = (k_1, \ldots, k_d) \in \mathbb{Z}^d$ is a vector with integer entries, and consider a function of the form

$$f(\boldsymbol{t}) = \sum_{\boldsymbol{k}=(k_1,\ldots,k_d)\in\mathbb{Z}^d} \hat{f}^{c}(\boldsymbol{k}) \cos\left(\frac{2\pi}{L}(\boldsymbol{k},\boldsymbol{t})\right), \quad \hat{f}^{c}(\boldsymbol{k}) \geq 0. \tag{73}$$

We have written (73) as an infinite sum over all possible frequencies that produce functions with period L. Readers who are not comfortable with infinite sums yet may assume that only finitely many of the coefficients $\hat{f}^{c}(\boldsymbol{k})$ are nonvanishing in (73). Readers who have seen infinite series, should assume that (73) converges for those values of the argument that will be used below.

The series (73) is a *Fourier series*; see Appendix C for more on this.

3.2.5.

Recall that $f(0) > 0$ for any nonzero positive definite function. For other values of the argument, $f(\boldsymbol{t})$ may be positive or negative, as exemplified by $\cos \omega t$.

By periodicity, f is positive at any point whose coordinates are integer multiples of L. We will denote the set of all such points by $L\mathbb{Z}^d$. Imagine that we managed to arrange f so that

$$\text{distance}(\boldsymbol{t}, L\mathbb{Z}^d) \geq 2r \quad \Rightarrow \quad f(\boldsymbol{t}) \leq 0. \tag{74}$$

In other words, we would like the function $f(\boldsymbol{t})$ to look like function in the figure of (75), namely, positive near the points in $L\mathbb{Z}^d$ and negative away from them:

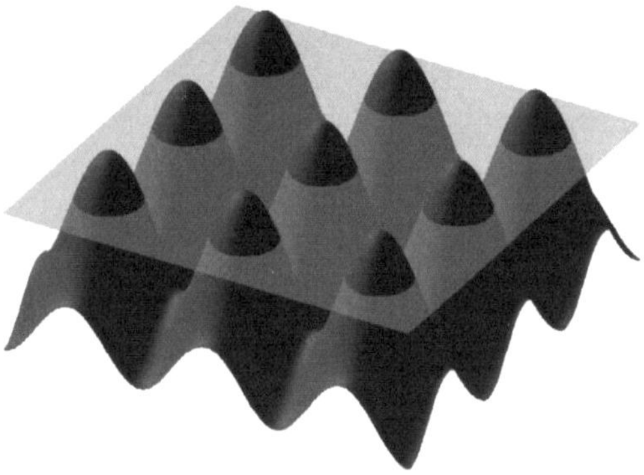

(75)

At this point, this is just a wish. There is absolutely no guarantee that we can find a suitable function. We are only saying that *any* positive definite function satisfying (74) will give some upper bound on the packing density. Whether this bound will be good or bad remains to be seen.

Stressing this logical point is important because the incredible brilliance of Viazovska's paper **[54]** is precisely in finding a certain magic positive definite function that makes everything work.

3.2.6.

Because the points (70) are the centers of a periodic sphere packing, we have

$$\operatorname{distance}\big(\boldsymbol{t}_i - \boldsymbol{t}_j, L\mathbb{Z}^d\big) \geq 2r, \quad i \neq j,$$

and hence $f(\boldsymbol{t}_i - \boldsymbol{t}_j) \leq 0$ for $i \neq j$. Therefore for the value of (67) at the point $\boldsymbol{x} = (1, \dots, 1)$, we obtain

$$B_f\big((1, 1, \dots, 1)\big) = \sum_{i,j} f(\boldsymbol{t}_i - \boldsymbol{t}_j) \leq \sum_i f(\boldsymbol{t}_i - \boldsymbol{t}_i) = nf(0). \tag{76}$$

This will be one side of the eventual inequality involving the number n of spheres in the packing.

3.2.7.

For the other side of the inequality, we note from (73) that the matrices of these quadratic forms satisfy

$$B_f = \sum_{\boldsymbol{k}} \hat{f}^{\mathrm{c}}(\boldsymbol{k}) B_{\cos(\frac{2\pi}{L}(\boldsymbol{k},\boldsymbol{t}))} \tag{77}$$

and that all terms in this sum are positive definite. Therefore the sum is at least as large as the $\boldsymbol{k} = 0$, that is, the B_1 term

$$B_f\big((1, 1, \dots, 1)\big) \geq \hat{f}(0) B_1\big((1, 1, \dots, 1)\big) = \hat{f}(0) n^2. \tag{78}$$

Here we dropped the superscript from $\hat{f}^{\mathrm{c}}(0)$ because the zero frequency means a constant function and there is no choice between the cosine and sine for $\boldsymbol{k} = 0$. Comparing (78) with (76), we deduce

$$n \leq \frac{f(0)}{\hat{f}(0)}. \tag{79}$$

This is the sought upper bound on the number n.

3.2.8.

To denominator $\hat{f}(0)$ in (79) may be interpreted as the *average* of the function f over the fundamental domain $[0, L]^d$. Indeed, the function $\cos(\frac{2\pi}{L}(\boldsymbol{k}, \boldsymbol{t}))$ with $\boldsymbol{k} \neq 0$ changes sign when shifted by $\frac{1}{2}$ of its minimal period due to

$$\cos(x + \pi) = -\cos(x).$$

Therefore, its average over the whole period vanishes. Thus

$$\begin{aligned} \hat{f}(0) &= \text{average of } f \text{ over } [0, L]^d \\ &= \frac{1}{L^d} \int_{[0,L]^d} f(\boldsymbol{t})\, d\boldsymbol{t}, \end{aligned} \tag{80}$$

where the second line is for readers familiar with integrals. Readers unfamiliar with integrals may want to consider (80) as the definition of the integral in terms of the average values of the function.

3.2.9.

Putting (79) and (80) together, we get

$$\frac{n}{L^d} = \frac{\text{number of spheres in } [0, L]^d}{\text{volume of } [0, L]^d} \leq \frac{f(0)}{\int_{[0,L]^d} f(\boldsymbol{t})\, d\boldsymbol{t}}. \tag{81}$$

Here $f(\boldsymbol{t})$ is a periodic function with period L in each coordinate, which is positive definite and satisfies (74).

3.2.10.

As we discussed before, to go from periodic packings to all packings, the period L in (81) should get arbitrarily large. Remarkably, there is a way to make *one* function f work for all periods L as follows.

We consider a function $f(\boldsymbol{t})$ such that

(i) $f(\boldsymbol{t})$ is positive definite,

(ii) $f(\boldsymbol{t}) \leq 0$ if $\|\boldsymbol{t}\| \geq 2r$,

(iii) $|f(\boldsymbol{t})|$ decays sufficiently fast as $\boldsymbol{t} \to \infty$.

As before, let (70) be the centers of the spheres in $[0, L]^d$. This means that all centers of the spheres have the coordinates $\{\boldsymbol{t}_i + L\mathbb{Z}^d\}$. If $|f(\boldsymbol{t})|$ decays sufficiently fast as $\boldsymbol{t} \to \infty$ then the series

$$\overline{f}(\boldsymbol{t}) = \sum_{\boldsymbol{v} \in L\mathbb{Z}^d} f(\boldsymbol{t} + \boldsymbol{v}) \tag{82}$$

converges, is periodic in $\boldsymbol{t}$, and is also positive definite. Evidently,

$$\int_{[0,L]^d} \overline{f}(\boldsymbol{t})\, d\boldsymbol{t} = \int_{\mathbb{R}^d} f(\boldsymbol{t})\, d\boldsymbol{t}. \tag{83}$$

As before, we have

$$f(0) \geq \frac{1}{n} \sum_{\boldsymbol{v} \in L\mathbb{Z}^d} \sum_{i,j} f(\boldsymbol{v} + \boldsymbol{t}_i - \boldsymbol{t}_j) = \frac{1}{n} \sum_{i,j} \overline{f}(\boldsymbol{v} + \boldsymbol{t}_i - \boldsymbol{t}_j) \geq \frac{n}{L^d} \int_{\mathbb{R}^d} f(\boldsymbol{t})\, d\boldsymbol{t}. \tag{84}$$

The first inequality here relies on the fact that $\boldsymbol{v} + \boldsymbol{t}_i - \boldsymbol{t}_j$ is a difference between two sphere centers, and hence has the norm at least $2r$ when nonzero. The second inequality is the inequality (78) applied to the periodic function $\overline{f}$.

We conclude

$$\boxed{\begin{matrix}\text{density of} \\ \text{sphere centers}\end{matrix} \leq \min_f \frac{f(0)}{\int_{\mathbb{R}^d} f(\boldsymbol{t})\, d\boldsymbol{t}},} \tag{85}$$

where the minimum is over all nonzero functions satisfying the properties (i)–(iii) above.

Bounds of this type are often called *linear programming* bounds because they ask for a minimum of a ratio of two linear functions on a convex set defined by conditions (i)–(iii). Instead of minimizing the ratio, we can consider positive definite functions normalized by $f(0) = 1$, which is an affine linear equation, and maximize the linear function $\int_{\mathbb{R}^d} f(\boldsymbol{t})\, d\boldsymbol{t}$ on the resulting convex set.

3.2.11.

Note that both the sets and functions to be extremized are invariant under rotations of $\mathbb{R}^d$, which is a compact group; see Section B.8. Compactness implies there is a well-defined average over all rotations of f, which is also a minimizer in (85). This average is a rotation-invariant function, that is, it depends on $\|\boldsymbol{t}\|$ only. Such functions are often called *radial*. To summarize, all we need is a function of one (radial) variable, not d variables.

4. VIAZOVSKA'S MAGIC FUNCTION

4.1. Lattice packings that saturate the bound

4.1.1.

Suppose there are a lattice Λ and a function f such that the corresponding packing saturates the bound (85). This implies at once that this packing is the densest possible, but also implies certain very special properties of the function f.

Indeed, the inequality (84) was obtained by discarding some nonpositive and non-negative terms, respectively. If the resulting inequality is an equality then this means all discarded terms vanish.

The first inequality in (84) is an equality if and only if

$$f(\boldsymbol{v}) = 0, \quad \text{for all } \boldsymbol{v} \in \Lambda \setminus \{0\}. \tag{86}$$

If f is radial then it vanishes for all vectors that have the same length as a nonzero vector from Λ. For E_8 this is the set $\sqrt{2n}$, for $n = 1, 2, \ldots$

4.1.2.

There is a very nice space of functions on $\mathbb{R}^d$ formed by functions that rapidly decay at infinity together with all their derivatives. It is called the Schwartz space. For functions f in the Schwarz space, the Fourier transform formulas (161) and (162) from Appendix C become nicely convergent integrals. The function $\hat{f}(\boldsymbol{k}) \geq 0$ in

$$f(\boldsymbol{t}) = \int_{\mathbb{R}^d} \hat{f}(\boldsymbol{k}) e^{2\pi i (\boldsymbol{k},\boldsymbol{t})}\, d\boldsymbol{k} \tag{87}$$

is also in Schwarz space and is even/radial if and only if $f(\boldsymbol{t})$ is even/radial. It is nonnegative because the function $f(\boldsymbol{t})$ is positive definite by our assumption.

As we will see momentarily, the second inequality in (84) becomes an equality precisely when the Fourier transform vanishes

$$\hat{f}(\boldsymbol{k}) = 0 \quad \text{for all } \boldsymbol{k} \in \Lambda^\vee \setminus \{0\} \tag{88}$$

for all nonzero vectors in the *dual* lattice, see Appendix C.7. Note the symmetry between (86) and (88). The symmetry is particularly pronounced for E_8 because $E_8^\vee = E_8$. For $\Lambda = E_8$ and a radial function f, this means the vanishing of the Fourier transform $\hat{f}(\boldsymbol{k})$ for all vectors $\boldsymbol{k}$ of length $\sqrt{2n}$, where $n = 1, 2, \ldots$

4.1.3.

To see (88), let us replace $[0, L]^d$ in the derivation of (84) by the fundamental parallelepiped for Λ. We redefine

$$\overline{f}(\boldsymbol{t}) = \sum_{\boldsymbol{v} \in \Lambda} f(\boldsymbol{t} + \boldsymbol{v}). \tag{89}$$

Since it is Λ-periodic, we have

$$\overline{f}(\boldsymbol{t}) = \frac{1}{\sqrt{\Delta_\Lambda}} \sum_{\boldsymbol{k} \in \Lambda^\vee} \hat{f}(\boldsymbol{k}) \exp\big(2\pi i (\boldsymbol{k}, \boldsymbol{t})\big), \quad \hat{f}(\boldsymbol{k}) \geq 0, \tag{90}$$

where the coefficients are found from (160) and (162) using

$$\int_{\mathbb{R}^d/\Lambda} \overline{f}(\boldsymbol{t}) e^{-2\pi i (\boldsymbol{k}, \boldsymbol{t})}\, d\boldsymbol{t} = \int_{\mathbb{R}^d} f(\boldsymbol{t}) e^{-2\pi i (\boldsymbol{k}, \boldsymbol{t})}\, d\boldsymbol{t}. \tag{91}$$

Because there is only *one* sphere in the fundamental parallelepiped, the inequality in (84) becomes

$$\overline{f}(0) = \frac{1}{\sqrt{\Delta_\Lambda}} \sum_{\boldsymbol{k} \in \Lambda^\vee} \hat{f}(\boldsymbol{k}) \geq \frac{1}{\sqrt{\Delta_\Lambda}} \hat{f}(0), \tag{92}$$

where $\frac{1}{\sqrt{\Delta_\Lambda}}$ is the density of the sphere centers, see (46). Clearly, (92) is an equality if and only if (88) holds.

4.1.4.

We conclude that to "finish" the proof of optimality of the E_8 lattice, one needs to find a function $f(x)$ of one variable satisfying the following constraints. We interpret $f(x)$ as a radial function on $\mathbb{R}^8$ and define the Fourier-transformed radial function $\hat{f}(x)$ by

$$\hat{f}(x) = \int_{\mathbb{R}^8} f\big(\|\boldsymbol{t}\|\big) e^{-2\pi i t_1 x}\, d\boldsymbol{t}, \tag{93}$$

where t_1 is the first coordinate of the vector $\boldsymbol{t}$. We need the function f and $\hat{f}$ to look like the functions in the figure of (94).

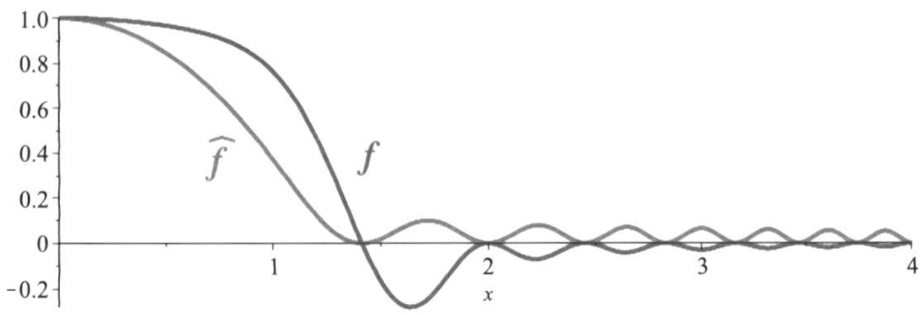

(94)

Namely, function $\hat{f}(x) \geq 0$ is nonnegative for all x while $f(x) \leq 0$ for $x \geq \sqrt{2}$. Further, both functions vanish for $x = \sqrt{2n}$, $n = 1, 2, \ldots$ Finally, since $\Delta_{E_8} = 1$, we may assume that $f(0) = \hat{f}(0) = 1$.

4.2. The wait is over

In his Fields medal laudatio [8] for Maryna Viazovska, Henry Cohn talks about his attempts to complete this last step, that is, to find the magic function f. In particular, he says:

> *"When Elkies and I proposed this method in 1999, Viazovska was still in secondary school. Without realizing how profoundly difficult the remaining step was, I imagined that we had almost solved the sphere packing problem in eight and twenty-four dimensions, and our inability to find the magic functions was extremely frustrating. At first, I worried that someone else would find an easy solution and leave me feeling foolish for not doing it myself. Over time I became convinced that obtaining these functions was in fact difficult, and others also reached the same conclusion. For example, Thomas Hales has said that* I felt that it would take a Ramanujan to find it [32]. *Eventually, instead of worrying that someone else would solve it, I began to fear that nobody would solve it, and that I would someday die without knowing the outcome. I am grateful that Viazovska found such a satisfying and beautiful solution, and that she introduced wonderful new ideas for the mathematical community to explore."*

Viazovska's solution is truly striking. She gives an extremely nontrivial explicit formula for the magic functions in terms of *modular forms*; see Appendix D. There is no way to tell if Ramanujan could have found it, but I would guess that seeing the solution would have made Ramanujan extremely, extremely happy.

Henry Cohn's laudatio [8] contains a very detailed masterfully written account of Viazovska's construction. I do hope the reader feels sufficiently prepared to work through it. While certainly not easy, it is very rewarding. There is a good reason computations like this are recognized by the highest honor in all of mathematics. I also hope the reader opens the interview [55] in which Maryna Viazovska talks, in particular, about her search for the elusive magic function.

4.3. Interpolation

4.3.1.

During her search for the magic function, Viazovska conjectured the following systematic way to construct functions like those in (94). Namely, she conjectured that a radial Schwartz function on $\mathbb{R}^8$ is uniquely specified by the values of f, f', $\hat{f}$, and $\hat{f}'$ at the points

$$x = \sqrt{2n}, \quad n = 1, 2, 3, \ldots$$

In other words, there exists an interpolation basis $a_n, b_n, \hat{a}_n, \hat{b}_n$ of the Schwartz space such that for every f we have

$$\begin{aligned} f(x) = \sum_{n=1}^{\infty} f(\sqrt{2n}) a_n(x) + \sum_{n=1}^{\infty} f'(\sqrt{2n}) b_n(x) \\ + \sum_{n=1}^{\infty} \hat{f}(\sqrt{2n}) \hat{a}_n(x) + \sum_{n=1}^{\infty} \hat{f}'(\sqrt{2n}) \hat{b}_n(x). \end{aligned} \tag{95}$$

In particular, the magic function f has to be proportional to $b_1(x)$ because all other coefficients vanish for it.

4.3.2.

This conjecture of Viazovska was proven in her work [12] with Henry Cohn, Abhinav Kumar, Stephen D. Miller, and Danylo Radchenko. They reformulate (95) as a certain functional equation for the following generating series $F(\tau, x)$ and $\hat{F}(\tau, x)$. By definition,

$$F(\tau, x) = \sum_{n=1}^{\infty} a_n(x) e^{2\pi i n\tau} + 2\pi i \tau \sum_{n=1}^{\infty} \sqrt{2n}\, b_n(x) e^{2\pi i n\tau}, \tag{96}$$

and similarly for $\hat{F}(\tau, x)$ with hats everywhere. Note that

$$F(\tau + 2, x) - 2F(\tau + 1, x) + F(\tau, x) = 0, \tag{97}$$

and similarly for $\hat{F}(\tau, x)$.

The radial function $f_\tau(\boldsymbol{t}) = e^{\pi i \tau \|\boldsymbol{t}\|^2}$, where $\boldsymbol{t} \in \mathbb{R}^8$, has Fourier transform

$$\hat{f}_\tau(\boldsymbol{t}) = \tau^{-4} e^{-\pi i \|\boldsymbol{t}\|^2/\tau}.$$

Therefore, for $f = f_\tau$, equation (95) reads

$$e^{\pi i \tau \|\boldsymbol{t}\|^2} = F(\tau, x) + \tau^{-4} \hat{F}(-1/\tau, x). \tag{98}$$

The authors of [12] solve equations (98) and (97) in terms of modular forms and deduce formulas for the interpolation basis in (96). In particular, this yields a formula for b_1, and hence for the E_8 magic function.

The appearance of τ and $-1/\tau$ in equation (98) is certainly a hint that modular forms have a role to play; compare with Section D.4. Note, however, that $F(\tau, x)$ is *not* periodic in τ, instead (97) says that $(T - 1)^2$ annihilates $F(\tau, x)$, where T shifts τ by 1. (This is a fancy way to say that $F(\tau + n, x)$ is linear in n for $n \in \mathbb{Z}$.) Ultimately, this is linked to the appearance of the modular functions for the subgroup $\Gamma(2)$ and also of the quasimodular Eisenstein series $\mathscr{E}_2$.

Like Viazovska's original construction, proving the interpolation formula (95) requires a certain cooperation between math and humans. Math has to make sure there is a miracle to be discovered. Humans have to send their brightest minds on the voyage to discover it.

4.3.3.

Similar results are also obtained in [12] for the Leech lattice. These stronger results imply the optimality of E_8 and Λ_{24} not just for sphere packing but also for certain more general geometric optimizations problems.

4.3.4.

I hope the readers share the narrator's sense of awe at this absolutely amazing mathematics and join me in warmest congratulations on it being recognized by the Fields Medal.

I also hope the readers got the sense that today's mathematics is not just extraordinarily powerful, but also concrete, understandable, and fun, once one finds the right idea and the right point of view. While finding that right point of view is not at all easy, my biggest hope is to have inspired my youngest readers to believe that mathematics can be beautiful and rewarding, both as a subject and as a profession. Maybe this is also a good place for me to thank Maryna Viazovska and Henry Cohn for this special opportunity to be introduced to their wonderful subject.

5. FURTHER READING

The *Quanta Magazine* has published several popular accounts of these and related developments, see [30, 32, 33].

Among introductory or survey articles written by top experts in the field, one could mention [7, 14, 20, 21, 49]. These were written prior to Viazovska's breakthrough. See [6, 8, 16] for expositions of Viazovska's breakthrough.

The reader will surely enjoy reading the textbooks [19, 52] and the comprehensive reference book [15]. A very interesting physics perspective on sphere packings may be found in [45].

I hope the reader has a lot of fun studying these sources as well as the original articles [10–12, 54].

A. INNER PRODUCTS

A.1.

In the following discussion we assume that the reader is familiar with basic linear algebra, in particular with the notion of a vector space such as $\mathbb{R}^d$. There exist many beautiful engaging professional expositions of the subject; see, for instance, [3, 36, 51]. Some readers may find the brief introduction in [44] usable.

A distance function like (1) is an *extra* structure on the linear space $\mathbb{R}^d$, meaning it is not part of the definition of a linear space. But is interacts very nicely with the linear space structures.

First, it is invariant under the translations. So it enough to specify the distance $\|\boldsymbol{x}\|$ to the point $\boldsymbol{x}$ from the origin $0 \in \mathbb{R}^d$. This is also called the *norm* of the vector $\boldsymbol{x}$. The formula

$$\|\boldsymbol{x}\|^2 = \sum x_i^2 \tag{99}$$

is valid in the coordinates with respect to the standard basis $\boldsymbol{e}_1, \dots, \boldsymbol{e}_d$ of $\mathbb{R}^d$, but will not remain valid in a different basis $\boldsymbol{e}'_1, \dots, \boldsymbol{e}'_d$. To describe the effect of a linear change of variables, and for many other computations, it is very convenient to introduce the *inner product* associated to (99). By definition,

$$(\boldsymbol{x}, \boldsymbol{y}) = \sum_i x_i y_i. \tag{100}$$

The norm (99) and the inner product (100) determine each other by $\|\boldsymbol{x}\|^2 = (\boldsymbol{x}, \boldsymbol{x})$ and

$$2(\boldsymbol{x}, \boldsymbol{y}) = \|\boldsymbol{x} + \boldsymbol{y}\|^2 - \|\boldsymbol{x}\|^2 - \|\boldsymbol{y}\|^2. \tag{101}$$

If

$$\boldsymbol{x} = \sum x_i \boldsymbol{e}_i = \sum x'_i \boldsymbol{e}'_i \tag{102}$$

is the expansion of $\boldsymbol{x}$ in two different bases, then

$$\|\boldsymbol{x}\|^2 = \sum x_i^2 = \sum_{ij} (\boldsymbol{e}'_i, \boldsymbol{e}'_j) x'_i x'_j. \tag{103}$$

We see that the squared norm is given by a *quadratic form* as in Section 3.1.2. Further, this quadratic form is positive definite because $\|\boldsymbol{x}\|^2 > 0$ for any $\boldsymbol{x} \neq 0$.

A.2.

Given any positive definite quadratic form $B(\boldsymbol{x})$, we can define a new norm by

$$\|\boldsymbol{x}\|_B^2 = B(\boldsymbol{x}). \tag{104}$$

Using a version of row reduction for the matrix (b_{ij}) called the Gram–Schmidt orthogonalization, we can always find a new basis $\boldsymbol{e}'_i$ in which

$$\|\boldsymbol{x}\|_B^2 = \sum (x'_i)^2.$$

Such a basis is called an *orthonormal* basis for the form (104).

Linear transformations g that preserve $\|x\|^2$ are called *orthogonal*. Linear isometries is another word for orthogonal transformations. They take orthonormal bases to orthonormal bases. Writing this condition in terms of matrix entries of $\mathsf{g} = (\mathsf{g}_{ij})$, we see it is equivalent to $\mathsf{g}^{\mathsf{T}}\mathsf{g} = \mathbb{1}$, where $\mathsf{g}^{\mathsf{T}} = (\mathsf{g}_{ji})$ is the transposed matrix and $\mathbb{1}$ denotes the identity matrix. Equivalently,

$$\mathsf{g}^{-1} = \mathsf{g}^{\mathsf{T}}, \tag{105}$$

where g^{-1} is the inverse matrix.

To summarize, invertible linear transformations g take the standard norm $\|x\|^2$ to all possible positive definite norms (104), and g takes $\|x\|^2$ to itself if and only if g is orthogonal. This means that all possible positive definite quadratic forms are the same as invertible linear transformations considered up to precomposing with an orthogonal transformation. See Appendix D for more on this.

A.3.

If $\boldsymbol{e}_1, \ldots, \boldsymbol{e}_d$ is a basis such that $(\boldsymbol{e}_i, \boldsymbol{e}_j) = 0$ for $i \neq j$ then the expansion $\boldsymbol{x} = \sum x_i \boldsymbol{e}_i$ can be written as

$$\boldsymbol{x} = \sum_i \frac{(x, \boldsymbol{e}_i)}{(\boldsymbol{e}_i, \boldsymbol{e}_i)} \boldsymbol{e}_i. \tag{106}$$

We will find it convenient in Section C.4 below.

Also note the link with the formula for the reflection in the hyperplane orthogonal to the vector $\boldsymbol{e}_1$,

$$r_{\boldsymbol{e}_1}(x) = x - 2\frac{(x, \boldsymbol{e}_1)}{(\boldsymbol{e}_1, \boldsymbol{e}_1)}\boldsymbol{e}_1. \tag{107}$$

Indeed, the transformation (107) changes the sign of the $\boldsymbol{e}_1$-coefficient in (106) and leaves all other coefficients unchanged. If $(\boldsymbol{e}_1, \boldsymbol{e}_1) = 2$, in particular, if $\boldsymbol{e}_1$ is a root in an even lattice, then (107) simplifies to (38).

A.4.

Let $\Lambda \subset \mathbb{R}^d$ be a lattice generated by vectors $\boldsymbol{v}_1, \ldots, \boldsymbol{v}_d$ as in Section 1.4. The matrix

$$B_\Lambda = \big((\boldsymbol{v}_i, \boldsymbol{v}_j)\big) \tag{108}$$

is called the Gram matrix. This is a positive definite symmetric matrix and any two collections of vectors $\boldsymbol{v}_1, \ldots, \boldsymbol{v}_d$ and $\boldsymbol{v}'_1, \ldots, \boldsymbol{v}'_d$ with the same Gram matrices (108) can be taken one to another by an orthogonal transformation of $\mathbb{R}^d$. Therefore, in the context of lattice sphere packings, we only care about the Gram matrices of lattices.

A.5.

Complex numbers are expressions of the form[10]

$$z = a + bi, \tag{109}$$

where a and b are real numbers and i is a symbol satisfying $i^2 = -1$. One can add and multiply complex numbers using this rule. We denote the set of complex numbers by $\mathbb{C}$.

The complex conjugate number is defined by

$$\bar{z} = a - bi. \tag{110}$$

Importantly,

$$z\bar{z} = a^2 + b^2, \tag{111}$$

which is a nonzero real number for $z \neq 0$. This means, in particular, that $z^{-1} = \frac{1}{a^2+b^2}\bar{z}$, which defines division by a nonzero complex number. In other words, complex numbers form a field.

Tuples $\boldsymbol{z} = (z_1, \ldots, z_d)$ of complex numbers form a linear space $\mathbb{C}^d$, in which one defines

$$\|\boldsymbol{z}\|^2 = \sum z_i \bar{z}_i, \quad (\boldsymbol{z}, \boldsymbol{z}') = \sum z_i \bar{z}'_i. \tag{112}$$

These norms and inner products are called *Hermitian,* and linear transformations that preserve them are called *unitary*.

---

10 The numbers a and b are called the real and imaginary part of the complex number z. Terminology notwithstanding, complex numbers really exist. For example, the imaginary unit i is the very first symbol in the Schrödinger equation, one of the fundamental equations describing our really complex world.

B. GROUPS AND POSITIVE DEFINITE FUNCTIONS

B.1.

In $\mathbb{R}^3$, consider rotations g around the origin. For a vector $\boldsymbol{v} \in \mathbb{R}^3$, we will denote by $\mathsf{g}\boldsymbol{v} \in \mathbb{R}^3$ the result of applying the rotation g to $\boldsymbol{v}$.

Remarkably, if we perform two rotations g_2 and g_1 in succession, the result is another rotation (can you prove this?) which we will denote by $\mathsf{g}_1\mathsf{g}_2$. It is called the composition or the *product* of two rotations. By construction,

$$(\mathsf{g}_1\mathsf{g}_2)\boldsymbol{v} = \mathsf{g}_1(\mathsf{g}_2\boldsymbol{v}) \tag{113}$$

for every $\boldsymbol{v}$. Note the order in which we write the product. It is important. In general, $\mathsf{g}_1\mathsf{g}_2 \neq \mathsf{g}_2\mathsf{g}_1$, as we invite the reader to check this in examples. From (113) it follows that

$$(\mathsf{g}_1\mathsf{g}_2)\mathsf{g}_3 = \mathsf{g}_1(\mathsf{g}_2\mathsf{g}_3), \tag{114}$$

so we do not need the brackets when we write the products.

There is a special identity rotation $\mathbb{1}$ that does nothing and satisfies

$$\mathbb{1}\mathsf{g} = \mathsf{g}\mathbb{1} = \mathsf{g} \tag{115}$$

for every g. Finally, for every rotation g there is the inverse rotation g^{-1} such that

$$\mathsf{g}^{-1}\mathsf{g} = \mathsf{g}\mathsf{g}^{-1} = \mathbb{1}. \tag{116}$$

B.2.

In mathematics, any set G with a special element $\mathbb{1} \in \mathsf{G}$, a binary product operation

$$(\mathsf{g}_1, \mathsf{g}_2) \xrightarrow{\text{product}} \mathsf{g}_1\mathsf{g}_2,$$

and a unary inverse operation

$$\mathsf{g} \xrightarrow{\text{inverse}} \mathsf{g}^{-1},$$

satisfying (114), (115), and (116) is called a group. A subset $\mathsf{G}' \subset \mathsf{G}$ closed under product and inverse is called a *subgroup*.

This is a very important notion, some examples of which are

$$\mathrm{GL}(n, \mathbb{R}) = \text{the group of all invertible } n \times n \text{ real matrices}, \tag{117}$$

$$O(n, \mathbb{R}) = \text{the subgroup of } n \times n \text{ orthogonal matrices}, \tag{118}$$

$$\mathrm{SO}(n, \mathbb{R}) = \text{orthogonal matrices with } \det \mathsf{g} = 1, \tag{119}$$

$$S(n) = \text{permutations of an } n\text{-element set}, \tag{120}$$

and so on. Orthogonal matrices were discussed in Section A.2. By construction, we have

$$\mathrm{SO}(n, \mathbb{R}) \subset O(n, \mathbb{R}) \subset \mathrm{GL}(n, \mathbb{R}) \tag{121}$$

and we can also embed

$$S(n) \subset O(n, \mathbb{R}) \tag{122}$$

by making $S(n)$ permute the basis vectors. The group $SO(3, \mathbb{R})$ is the group of rotations of $\mathbb{R}^3$ around the origin discussed above.

For a much more basic example of a group, one can take the group of real numbers $\mathbb{R}$ with the operation of addition. The zero $0 \in \mathbb{R}$ is the identity element for this operation. Similarly, $\mathbb{R}^d$ is a group with respect to addition. The group $\mathbb{R}^d$ is simpler that the groups in (117)–(120) in one important aspect. The operation in $\mathbb{R}^d$ is *commutative*, meaning that $g_1 g_2 = g_2 g_1$ for any g_1 and g_2.

In all examples above, $\mathbb{R}$ can be replaced by an arbitrary field. The field $\mathbb{C}$ of complex numbers and the group $U(n)$ of $n \times n$ unitary matrices are particularly important in mathematics.

B.3.

One can also use a ring with unit in place of $\mathbb{R}$ above, for instance, the ring $\mathbb{Z}$ of integers. In defining $GL(n, \mathbb{Z})$, one needs to make sure that the inverse g^{-1} of an integral matrix $g \in GL(n, \mathbb{Z})$ is also integral. For a commutative ring like $\mathbb{Z}$, it is enough to require that the determinant $\deg g \in \mathbb{Z}$ is an invertible element, meaning that $\deg g = \pm 1$.

The subgroup $GL(n, \mathbb{Z}) \subset GL(n, \mathbb{R})$ consists of matrices that preserve the standard lattice $\mathbb{Z}^n \subset \mathbb{R}^n$. Similarly, matrices preserving an arbitrary lattice $\Lambda \subset \mathbb{R}^n$ form a subgroup that becomes $GL(n, \mathbb{Z})$ in a suitable basis. This is an infinite group. By contrast, orthogonal matrices preserving a given lattice always form a finite group.[11] This group is the boring $\{\pm 1\}$ for a generic lattice, but can be very interesting for lattices like E_8 and Λ_{24}.

B.4.

A map between groups

$$G \to G'$$

preserving the group structure is called a group homomorphism. A special kind of homomorphism

$$\rho : G \to O(n, \mathbb{R}) \tag{123}$$

is called an orthogonal representation of G of dimension n. It represents every $g \in G$ by an orthogonal matrix $\rho(g)$ and we have

$$\rho(g_1 g_2) = \rho(g_1)\rho(g_2).$$

For example, (122) is an orthogonal representation. If the target group $O(n, \mathbb{R})$ is replaced by GL or the unitary group, one talks about linear or unitary representation.

Let an orthogonal representation as in (123) be given and and let $\boldsymbol{v} \in \mathbb{R}^n$ be a vector with $\|v\| = 1$. It defines a function on G by

$$f_{\rho, \boldsymbol{v}}(g) = \big(\rho(g)\boldsymbol{v}, \boldsymbol{v}\big). \tag{124}$$

---

11 Can you prove it? Note that we consider transformations that preserve the origin of Λ.

Such functions are called diagonal matrix elements. If $\boldsymbol{v}$ is the first basis vector in some basis of $\mathbb{R}^n$ then $f_{\rho,\boldsymbol{v}}(\mathsf{g})$ is the matrix element $\rho(\mathsf{g})_{1,1}$.

Since $\rho(\mathsf{g}^{-1}) = \rho(\mathsf{g})^{-1} = \rho(\mathsf{g})^{\mathrm{T}}$, we conclude that (124) is symmetric,

$$f_{\rho,\boldsymbol{v}}(\mathsf{g}^{-1}) = f_{\rho,\boldsymbol{v}}(\mathsf{g}). \tag{125}$$

If $\mathsf{g}_1, \ldots, \mathsf{g}_d \in \mathsf{G}$ are arbitrary group elements and $\boldsymbol{x} = (x_1, \ldots, x_n)$ is arbitrary then

$$\|\sum_i x_i \rho(\mathsf{g}_i)\boldsymbol{v}\|^2 = \sum_{i,j} f_{\rho,\boldsymbol{v}}(\mathsf{g}_i \mathsf{g}_j^{-1}) x_i x_j. \tag{126}$$

Clearly, the quadratic form in (126) is positive semidefinite.

Functions $f(\mathsf{g})$ that are symmetric, $f(\mathsf{g}^{-1}) = f(\mathsf{g})$, and produce positive semidefinite forms $\sum_{i,j} f(\mathsf{g}_i \mathsf{g}_j^{-1}) x_i x_j$ are called positive definite functions on G. If $f \neq 0$ then we can normalize it by $f(\mathbb{1}) = 1$.

For the additive group $\mathbb{R}^d$, this is the definition from Section 3.1.5. The analog of Bochner's theorem for G says that any positive definite function is a diagonal matrix element of an orthogonal representation. This representation could be infinite-dimensional, hence the need for limits in Bochner's theorem. A solid amount of mathematical care is required to work with infinite-dimensional representations, much beyond the introductory style of these notes. We will therefore consider the case of a finite group G, which already contains many key features of the general story.

B.5.

The simplest finite group is the group $\mathbb{Z}/m\mathbb{Z}$ of integers modulo m, with the addition operation. It is generated by one element 1, not to be confused with the identity $\mathbb{1}$. In this group, $\mathbb{1}$ is the zero element. For brevity, we denote it by $\mathbb{Z}/m$ in what follows.

The study of representations of groups is a generalization of the theory of eigenvalues and eigenvectors of a matrix. From the eigenvectors of the generator $\rho(1)$, one can conclude that any orthogonal representation of $\mathbb{Z}/m$, in a suitable basis, is the sum of 2×2 matrix blocks

$$\rho(j) = \begin{pmatrix} \cos \frac{2\pi k j}{m} & -\sin \frac{2\pi k j}{m} \\ \sin \frac{2\pi k j}{m} & \cos \frac{2\pi k j}{m} \end{pmatrix}, \quad k = 1, \ldots, m-1,$$

the trivial representation

$$\rho(j) = 1,$$

and the sign representation $\rho(j) = (-1)^j$, which exists for even m. In all cases, the diagonal matrix elements for $\mathbb{Z}/m$ are nonnegative combinations of the functions

$$f_k(j) = \cos \frac{2\pi k j}{m}, \quad k = 0, \ldots, m-1.$$

Our next goal is to show that these exhaust all positive definite functions f on $\mathbb{Z}/m$.

B.6.

To this end, we consider the representation ρ_{reg} of $\mathbb{Z}/m$ on an m-dimensional vector space with basis $\boldsymbol{\delta}_0, \ldots, \boldsymbol{\delta}_{m-1}$ given by

$$\rho_{\text{reg}}(j)\boldsymbol{\delta}_i = \boldsymbol{\delta}_{i+j \bmod m}.$$

This is called the regular representation. We introduce an inner product on it by

$$(\boldsymbol{\delta}_a, \boldsymbol{\delta}_b) = f(b-a).$$

This is symmetric because f is symmetric, positive semidefinite because f is positive definite, and preserved by the action of $\mathbb{Z}/m$. The vectors of zero norm form a linear subspace that is preserved by $\mathbb{Z}/m$, and the representation of $\mathbb{Z}/m$ on the quotient by this subspace is orthogonal. Finally,

$$f(j) = \left(\rho_{\text{reg}}(j)\boldsymbol{\delta}_0, \boldsymbol{\delta}_0\right),$$

and this finishes the proof.

B.7.

The above discussion may be further simplified if one uses complex numbers and unitary representations. For a unitary representation $\rho(\mathsf{g})$, we have

$$f_{\rho,\boldsymbol{v}}(\mathsf{g}^{-1}) = \overline{f_{\rho,\boldsymbol{v}}(\mathsf{g})}, \tag{127}$$

and (126) turns into a positive definite Hermitian form. A complex-valued function on a group is called positive definite if it satisfies these two properties.

For a commutative group like $\mathbb{Z}/m$, unitary representations are sums of 1-dimensional representations

$$\chi_k(j) = \exp\left(\frac{2\pi i j k}{m}\right), \quad k = 0, \ldots, m-1, \tag{128}$$

and the argument given above proves that positive definite functions on $\mathbb{Z}/m$ are nonnegative linear combinations of the functions (128). One-dimensional representations are called *characters* and often denoted by the letter χ.

In (128), the letter i denotes the imaginary unit, and the exponential of an imaginary number may be defined by

$$\begin{aligned} e^{it} &= 1 + it + \frac{(it)^2}{2} + \frac{(it)^3}{3!} + \cdots \\ &= \left(1 - \frac{t^2}{2} + \frac{t^4}{4!} - \cdots\right) + i\left(t - \frac{t^3}{3!} + \frac{t^5}{5!} - \cdots\right) \\ &= \cos(t) + i\sin(t), \end{aligned} \tag{129}$$

as discovered by Leonhard Euler around 1740. Note the famous special case

$$e^{\pi i} = -1. \tag{130}$$

B.8.

We started this section with a discussion of rotations of $\mathbb{R}^3$, which, in addition to forming a group, have two further important properties. First, rotations form a manifold, namely the real projective 3-space. Groups forming a manifold are called Lie groups. Second, this manifold is compact.

Compact Lie groups are very important in mathematics and they have been completely classified. This classification includes the classification of compact connected Lie groups and of finite simple groups. In both cases, there are certain well-understood infinite series as well as finitely many exceptional cases, surrounded by a much denser air of mystery. The classification of compact connected Lie groups is very close to the ADE classification[12] from Section 2.3.3, and it ends in the largest exceptional group E_8. Among the finite groups, there is the largest sporadic group called the Monster, which is very closely connected to the Leech lattice Λ_{24}.

C. FOURIER SERIES

C.1.

Let us revisit the regular representation of $\mathbb{Z}/m$ from Section B.6. Every group G acts on the linear space of functions $f : \mathsf{G} \to \mathbb{C}$ by the following rule:

$$\left[\rho_{\text{reg}}(\mathsf{g}) f\right](\mathsf{g}') = f(\mathsf{g}'\mathsf{g}). \tag{131}$$

In (131) we have the result of evaluation of the new function $\rho_{\text{reg}}(\mathsf{g}) f$ at a group element g'. The square brackets in (131) are put around $\rho_{\text{reg}}(\mathsf{g}) f$ just to stress that this is a new function, obtained by the action of g from the original function f.

The basis $\boldsymbol{\delta}_0, \ldots, \boldsymbol{\delta}_{m-1}$ from Section B.6 corresponds to the functions

$$\boldsymbol{\delta}_k(j) = \delta_{kj}, \quad \text{where } \delta_{kj} = \begin{cases} 1, & k = j, \\ 0, & \text{otherwise.} \end{cases} \tag{132}$$

C.2.

The following Hermitian product,

$$(f_1, f_2)_{\text{reg}} = \sum_{j \in \mathbb{Z}/m} f_1(j) \overline{f_2(j)}, \tag{133}$$

makes the regular representation of $\mathbb{Z}/m$ unitary. The basis (132) satisfies

$$(\boldsymbol{\delta}_k, \boldsymbol{\delta}_{k'})_{\text{reg}} = \delta_{kk'}. \tag{134}$$

---

12 In this classification, the lattices A_n correspond to special unitary groups $\mathrm{SU}(n+1)$, while the lattices D_n correspond to even orthogonal groups $\mathrm{SO}(2n)$.

C.3.

Now consider the functions (128) as elements of the regular representation. We compute

$$(\chi_k, \chi_{k'})_{\text{reg}} = \sum_{j \in \mathbb{Z}/m} \exp\left(\frac{2\pi i j(k-k')}{m}\right) = |\mathsf{G}|\delta_{k,k'}. \tag{135}$$

Indeed, the sum in (135) is a sum of a geometric progression and it vanishes if $k \neq k'$. We put the cardinality $|\mathsf{G}|$ in (135) instead of m because (135) is the simplest case of a very general relations known as orthogonality of characters (and other matrix elements of irreducible representation). In the case of a finite group, the cardinality $|\mathsf{G}|$ is the correct factor to put into these orthogonality relations.

C.4.

We have found two orthogonal bases $\{\boldsymbol{\delta}_k\}$ and $\{\chi_k\}$ in the space of complex-valued functions on $\mathsf{G} = \mathbb{Z}/m$. Let us expand a general function in these bases using (106).

The expansion in the basis $\{\boldsymbol{\delta}_k\}$ amounts to a tautology,

$$f = \sum_{k \in \mathbb{Z}/m} f(k)\boldsymbol{\delta}_k. \tag{136}$$

The expansion in the basis χ_k, by contrast, amounts to something very nontrivial. By (106), the coefficients $\hat{f}(k)$ in the expansion

$$f = \sum_{k \in \mathbb{Z}/m} \hat{f}(k)\chi_k \tag{137}$$

are given by

$$\hat{f}(k) = \frac{(f, \chi_k)_{\text{reg}}}{(\chi_k, \chi_k)_{\text{reg}}} = \frac{1}{m}\sum_{j=0}^{m-1} f(j)\exp\left(-\frac{2\pi i jk}{m}\right). \tag{138}$$

The expansion (137), written out, takes a very similar form

$$f(j) = \sum_{k=0}^{m-1} \hat{f}(k)\exp\left(\frac{2\pi i jk}{m}\right). \tag{139}$$

Formulas (138) and (139) describe the *Fourier transform* on the commutative group $\mathsf{G} = \mathbb{Z}/m$. By (139), every function $f(j)$ can be written as a combination of characters χ_k. The coefficients $\hat{f}(k)$ in (139) are the average values of $f\overline{\chi_k}$.

A similar Fourier transform on groups exists very generally. For noncommutative groups, one should take matrix elements of unitary representations instead of characters. This will remain entirely outside of our narrative.

C.5.

After talking about Fourier transform for a finite commutative group $\mathbb{Z}/m$, let us consider the simplest commutative connected Lie group SO(2) of rotations in $\mathbb{R}^2$ around the origin. A rotation is specified by the angle $\phi \in [0, 2\pi]$, with the endpoints 0 and 2π

representing the same identity element in SO(2). The group operation is the addition of angles ϕ, taken modulo 2π. Thus, we may think of SO(2) as the quotient

$$\mathrm{SO}(2) = \mathbb{R}/2\pi\mathbb{Z}$$

of a linear group by a lattice subgroup.

For any m, we have the subgroup

$$\mathbb{Z}/m = \left\{\phi = \frac{2\pi j}{m},\ j = 0, \ldots, m-1\right\} \subset \mathrm{SO}(2) \tag{140}$$

formed by rotations that preserve a regular m-gon. As m gets large, these become denser and denser. Let us *formally* take the $m \to \infty$ limit in the formulas (138) and (139), and see if we get the formulas for the Fourier transform on SO(2).

Let us rewrite the formulas (138) and (139) using the variable $\phi = \frac{2\pi j}{m}$. We get

$$\hat{f}(k) = \frac{1}{m} \sum_{\phi \in \mathbb{Z}/m} f(\phi) e^{-ik\phi}, \tag{141}$$

$$f(\phi) = \sum_{k \in \mathbb{Z}/m} \hat{f}(k) e^{ik\phi}. \tag{142}$$

In (141), we interpret $\mathbb{Z}/m$ as the subgroup (140), while in (142) we have a summation of a periodic function of k over any period of length m in $\mathbb{Z}$. As $m \to \infty$, the sum in (141) approximates the integral over the group SO(2), while the sum (142) becomes the sum over all integers k. Thus, we get

$$\hat{f}(k) = \frac{1}{2\pi} \int_0^{2\pi} f(\phi) e^{-ik\phi}\, d\phi, \tag{143}$$

$$f(\phi) = \sum_{k \in \mathbb{Z}} \hat{f}(k) e^{ik\phi}. \tag{144}$$

We stress that our derivation of these formulas was just by a formal analogy with the case of a finite group and much more serious work is required to both interpret these formulas correctly and prove them. Questions like these belong to the field of harmonic analysis, which is a very deep and important part of mathematics. We leave the reader by the entrance to this glorious edifice, referring to [46,50] for possible further reading. But to stimulate the reader's curiosity, we will do one example.

C.6.

Fix some angle ϕ_0 and consider the function

$$f(\phi) = \begin{cases} 1, & \cos(\phi) \geq \cos(\phi_0), \\ 0, & \text{otherwise}. \end{cases}$$

In other words, this function equal 1 on the "spherical cap" $[-\phi_0, \phi_0]$ and vanishes outside of it. From (143), we compute

$$\hat{f}(k) = \frac{1}{2\pi} \int_{-\phi_0}^{\phi_0} e^{-ik\phi}\, d\phi = \begin{cases} \frac{\phi_0}{\pi}, & k = 0, \\ \frac{\sin(k\phi_0)}{k\pi}, & k \neq 0, \end{cases} \tag{145}$$

where we used the relation (129) between the complex exponential and the trigonometric functions. Using (129) again, we can write (144) as follows:

$$f(\phi) \stackrel{?}{=} \frac{\phi_0}{\pi} + 2\sum_{k=1}^{\infty} \frac{\sin(k\phi_0)\cos(k\phi)}{k\pi}, \tag{146}$$

where the question mark indicates that the exact interpretation of this equality is beyond the scope of these notes.

A picture being worth a thousand words, we just plot the partial sums of the series above for $\phi_0 = \frac{\pi}{3}$ and k up to 3, 5, 10, and 50, respectively.

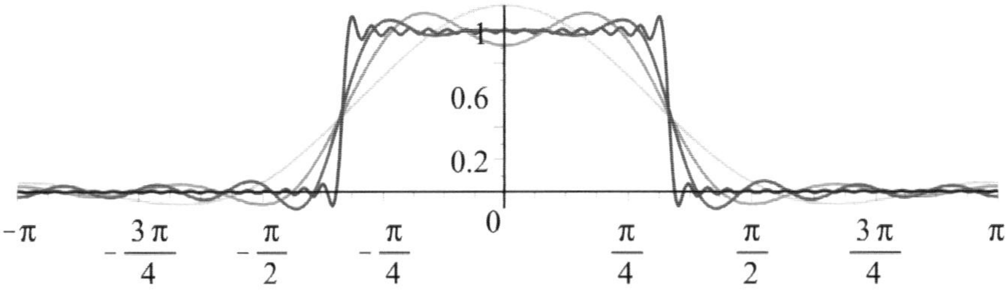

(147)

One salient feature of (147) are the very strong oscillations of the Fourier series near the point of discontinuity of the function, known as the Gibbs phenomenon.

Experimenting with Fourier series is a lot of fun, and we invite the reader to do more experiments! The functions that are actually needed in Viazovska's proof are infinitely differentiable and their Fourier expansions converge to them very nicely.

C.7.

The following common generalization of (138), (139), (143), (144) is valid for any commutative Lie group[13] G. It describes the expansion of a function f on G in terms of the characters of G.

Unitary characters of G, that is, continuous homomorphisms

$$\chi : \mathsf{G} \to U(1) = \{z \in \mathbb{C}, |z| = 1\}, \tag{148}$$

form a commutative group G^ with respect to pointwise multiplication of characters. The trivial character $\chi = 1$ is the identity of this group. If the group G is compact then G^ is discrete, and visa versa. The group G^ is called the Pontryagin dual group, or the dual group for short.

Mathematicians write

$$1 \to \mathsf{G}_1 \to \mathsf{G} \to \mathsf{G}_2 \to 1 \tag{149}$$

to indicate that G_1 is a Lie subgroup of G with quotient G_2. They call a sequence of the form (149) a short exact sequence. Duality reverses short exact sequences:

$$1 \to \mathsf{G}_2^{\wedge} \to \mathsf{G}^{\wedge} \to \mathsf{G}_1^{\wedge} \to 1, \tag{150}$$

13 as well as for more general locally compact commutative groups which means that characters of G_2 are the characters of G that are trivial when restricted to G_1, and vice versa. One replaces the 1's by 0's in short exact sequences when the group operation is written as addition.

For example, any inner product $(\cdot, \cdot)$ on $\mathbb{R}^d$ gives the identification $(\mathbb{R}^d)^\wedge \cong \mathbb{R}^d$ by

$$\chi_{\boldsymbol{k}}(\boldsymbol{t}) = \exp\big(2\pi i (\boldsymbol{k}, \boldsymbol{t})\big). \tag{151}$$

If $\Lambda \subset \mathbb{R}^d$ is a lattice then the quotient group in

$$0 \to \Lambda \to \mathbb{R}^d \to \mathbb{R}^d/\Lambda \to 0 \tag{152}$$

is a group abstractly isomorphic to $\mathrm{SO}(2)^d$ which can be realized concretely by gluing the opposite sides of the fundamental parallelepiped for Λ. Mathematicians call such group a *torus*. Using (151), we get the identifications

$$(\mathbb{R}^d/\Lambda)^\wedge = \Lambda^\vee, \tag{153}$$

$$\Lambda^\wedge = \mathbb{R}^d/\Lambda^\vee, \tag{154}$$

where $\Lambda^\vee$ is the *dual lattice*

$$\Lambda^\vee = \big\{\boldsymbol{k} \text{ such that } (\boldsymbol{k}, \boldsymbol{v}) \in \mathbb{Z} \text{ for all } \boldsymbol{v} \in \Lambda\big\}. \tag{155}$$

C.8.

While as abstract groups, all lattices and tori of the same dimension are isomorphic, they are all very different in the context of sphere packing and other problems involving distances and inner products. It is, therefore, important to distinguish clearly between a lattice Λ and the dual lattice $\Lambda^\vee$.

In general, $\Lambda^\vee$ is very different from Λ. For example, if we scale Λ by a factor then $\Lambda^\vee$ scales by the reciprocal factor. However,

$$(\mathbb{Z}^d)^\vee = \mathbb{Z}^d, \quad E_8^\vee = E_8, \tag{156}$$

and, in general, if a lattice is *integral*, which means that $(\boldsymbol{v}_1, \boldsymbol{v}_2) \in \mathbb{Z}$ for all $\boldsymbol{v}_1, \boldsymbol{v}_2 \in \Lambda$, and *unimodular*, which means that $\Delta_\Lambda = 1$ then[14] $\Lambda^\vee = \Lambda$.

C.9.

The Fourier transform on a general *compact commutative* group G takes the form

$$f(\mathsf{g}) = \sum_{\boldsymbol{k} \in \mathsf{G}^\wedge} \hat{f}(\boldsymbol{k}) \chi_{\boldsymbol{k}}(\mathsf{g}), \tag{157}$$

$$\hat{f}(\boldsymbol{k}) = \int_{\mathsf{G}} f(\mathsf{g}) \overline{\chi_{\boldsymbol{k}}(\mathsf{g})}\, d_{\mathrm{prob}}\mathsf{g}, \tag{158}$$

where the integration is with respect to the invariant measure on the group G of total volume 1. Measures of total volume 1 are often called *probability* measures, hence the subscript in (158).

14 Indeed, integrality implies that $\Lambda \subset \Lambda^\vee$, while Δ_Λ is the order of the group $\Lambda^\vee/\Lambda$.

For tori, the integral in (158) is just the usual integral over the fundamental domain, normalized so that the volume of the fundamental domain equals 1. Recall that this volume equals $\sqrt{\Delta_\Lambda}$ with respect to usual volume form $d\boldsymbol{t}$. Therefore, for $\mathsf{G} = \mathbb{R}^d/\Lambda$, the Fourier transform takes the form

$$f(\boldsymbol{t}) = \sum_{\boldsymbol{k}\in\Lambda^\vee} \hat{f}(\boldsymbol{k}) e^{2\pi i(\boldsymbol{k},\boldsymbol{t})}, \tag{159}$$

$$\hat{f}(\boldsymbol{k}) = \frac{1}{\sqrt{\Delta_\Lambda}} \int_{\mathbb{R}^d/\Lambda} f(\boldsymbol{t}) e^{-2\pi i(\boldsymbol{k},\boldsymbol{t})}\, d\boldsymbol{t}. \tag{160}$$

C.10.

We remind the reader that we glide over all the deep analytic issues involved in the Fourier transform on continuous groups. For our narrative, this is justified by the fact that the actual functions that come up in Viazovska's proof have very nice analytic properties.

While for noncompact groups G the Fourier transform presents further analytic difficulties, one can formally take the limit of a very large lattice Λ in (159) and (160) and obtain

$$f(\boldsymbol{t}) = \int_{\mathbb{R}^d} \hat{f}(\boldsymbol{k}) e^{2\pi i(\boldsymbol{k},\boldsymbol{t})}\, d\boldsymbol{k}, \tag{161}$$

$$\hat{f}(\boldsymbol{k}) = \int_{\mathbb{R}^d} f(\boldsymbol{t}) e^{-2\pi i(\boldsymbol{k},\boldsymbol{t})}\, d\boldsymbol{t}. \tag{162}$$

As Λ becomes very large, the dual lattice $\Lambda^\vee$ becomes very dense and the sum in (159) becomes the integral in (161).

D. MODULAR FORMS

D.1. The space of lattices

The many special lattices we met in these notes may be interpreted as some very special points in a space that parametrizes all possible lattices $\Lambda \subset \mathbb{R}^d$. How should think about this space?

An arbitrary lattice $\Lambda \subset \mathbb{R}^d$ may be obtained from the standard lattice $\mathbb{Z}^d \subset \mathbb{R}^d$ by a change of basis or, equivalently, as a result of linear transformation

$$\Lambda = \mathsf{g}\mathbb{Z}^d, \quad \mathsf{g} \in \mathrm{GL}(d,\mathbb{R}). \tag{163}$$

Further, $\mathsf{g}\mathbb{Z}^d = \mathbb{Z}^d$ if and only if $\mathsf{g} \in \mathrm{GL}(d,\mathbb{Z})$. Thus

$$\left\{\text{lattices in } \mathbb{R}^d\right\} = \mathrm{GL}(d,\mathbb{R})/\,\mathrm{GL}(d,\mathbb{Z}), \tag{164}$$

where the quotient sign means that we identify g_1 and g_2 if $\mathsf{g}_1^{-1}\mathsf{g}_2 \in \mathrm{GL}(d,\mathbb{Z})$.

For sphere packing and many other problems, we do not want to distinguish between isometric lattices, that is, lattices that differ by postcomposing g with an orthogonal transformation. Thus we consider

$$\left\{\begin{matrix}\text{lattices in } \mathbb{R}^d \\ \text{up to isometry}\end{matrix}\right\} = O(d,\mathbb{R})\backslash\,\mathrm{GL}(d,\mathbb{R})/\,\mathrm{GL}(d,\mathbb{Z}). \tag{165}$$

Finally, for the sphere packing problem, we can rescale the lattice arbitrarily, while simultaneously rescaling the radius of the spheres. Thus, one may want to consider

$$\left\{ \begin{array}{c} \text{lattices in } \mathbb{R}^d \text{ up to} \\ \text{scale and isometry} \end{array} \right\} = \left(\mathbb{R}_{>0} O(d,\mathbb{R})\right) \backslash \operatorname{GL}(d,\mathbb{R}) / \operatorname{GL}(d,\mathbb{Z}), \tag{166}$$

where $\mathbb{R}_{>0}$ is the subgroup of $\operatorname{GL}(d,\mathbb{R})$ consisting of positive multiples of the identity matrix.

D.2.

Let us see what the space (166) looks like for $d = 2$. Let Λ be a lattice and let $\boldsymbol{v} \in \Lambda$ be a vector of minimal length. We will complete $\boldsymbol{v}$ to a basis $\{\boldsymbol{v}, \boldsymbol{v}'\}$ of the lattice Λ by choosing a shortest vector $\boldsymbol{v}'$ among those not proportional to $\boldsymbol{v}$. Note, however, that $-\boldsymbol{v}'$ is another vector with the same properties as $\boldsymbol{v}'$.

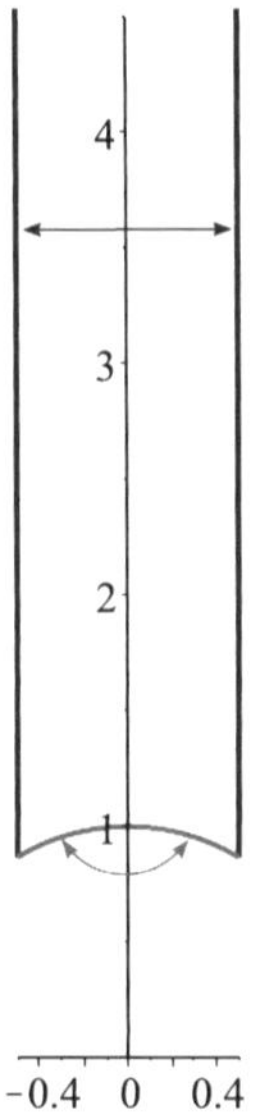

Since we take lattices up to scale and isometry, we may arrange so that $\boldsymbol{v} = \boldsymbol{e}_1 = (1,0)$ is the standard basis vector. What are the possibilities for $\boldsymbol{v}' = (v_1', v_2')$? First, we need to have

$$\|\boldsymbol{v}'\|^2 = {v_1'}^2 + {v_2'}^2 \geq 1. \tag{167}$$

Second,

$$\|\boldsymbol{v}' \pm \boldsymbol{e}_1\|^2 = \|\boldsymbol{v}'\|^2 \pm 2v_1' + 1, \tag{168}$$

which means that $\boldsymbol{v}'$ can be made shorter by adding or subtracting $\boldsymbol{e}_1$, unless

$$|v_1'| \leq \frac{1}{2}. \tag{169}$$

The combination of (167) and (169) describes the domain shown in the figure on the left plus the symmetric domain below the horizontal axis. Using the symmetry between $\boldsymbol{v}'$ and $-\boldsymbol{v}'$, we may restrict our attention to the figure on the left. Every point in this domain corresponds to a certain lattice in $\mathbb{R}^2$, but some pairs of points on the boundary correspond to the same lattices. Indeed, the vertical boundaries differ by a shift by $\boldsymbol{e}_1$. Since this shift does not change the lattice, they have to be glued as indicated. Points on the round boundary correspond to lattices generated by two vectors $\boldsymbol{v}, \boldsymbol{v}'$ of equal length. Since we can declare either one of them to be equal to $\boldsymbol{e}_1$, we may demand additionally that the angle between $\boldsymbol{v}$ and $\boldsymbol{v}'$ is not larger than $\pi/2$. This leads to gluing the round boundary to itself as indicated.

Once we do this gluing, we get a surface of the shape shown in the figure of (170).

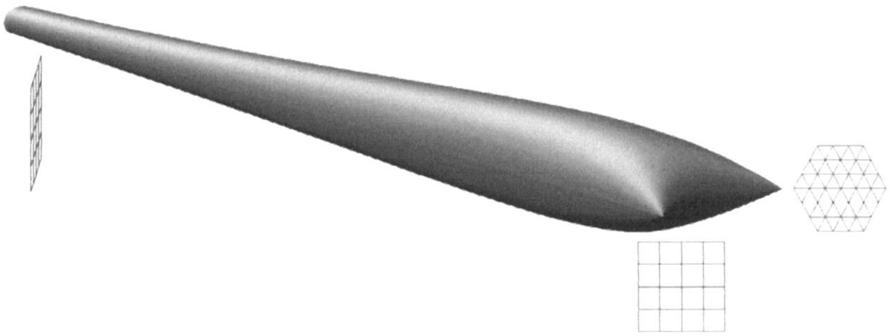

(170)

The two cone points of this surface correspond to the special lattices—the square lattice and the hexagonal latices. They are special because they are preserved by some nontrivial transformations in $O(2, \mathbb{R})$. The surface in (170) has an infinitely long neck, which is sometimes call the *cusp*. A lattice Λ runs off to infinity in this neck if $\frac{\|\boldsymbol{v}'\|}{\|\boldsymbol{v}\|} \to \infty$.

D.3.

Functions on quotient spaces like (164), (165), and (166) are called *automorphic functions*. They are objects of extreme beauty, complexity, and importance for mathematics. It suffices to say that they played a very essential role in Andrew Wiles' proof of Fermat's last theorem.

In these notes, we will limit ourselves to the discussion of one basic class of such functions for $d = 2$, called the Eisenstein series.[15] Let $\Lambda \subset \mathbb{R}^2$ be a lattice. We can identify $\mathbb{R}^2$ with the complex numbers $\mathbb{C}$ from Section A.5 and then Λ becomes a subset of $\mathbb{C}$. We define

$$\mathscr{E}_k(\Lambda) = \frac{1}{2} \sum_{z \in \Lambda_{\text{primitive}}} \frac{1}{z^k}, \tag{171}$$

where primitive means that z is not a positive multiple of another vector, in particular this means that $z \neq 0$. The series (171) converges absolutely for $k > 2$ and vanishes for k odd because the contributions of z and $-z$ cancel. For even k, z and $-z$ make the same contribution, hence the $\frac{1}{2}$ factor in front.

If we multiply the lattice Λ by a complex number w then

$$\mathscr{E}_k(w\Lambda) = w^{-k} \mathscr{E}_k(\Lambda). \tag{172}$$

Note that multiplication by w combines rotations and scaling of Λ. As a result, Eisenstein series are not exactly invariant under rotation and scaling, but rather transform in the way described by (172) under rotation and scaling. Using (172), we may assume that

$$\boldsymbol{v} = 1, \quad \boldsymbol{v}' = \tau, \tag{173}$$

where τ is a complex number in the upper half-plane. The series (171) being a sum of the terms $(n + m\tau)^{-k}$, where (n, m) runs over coprime pairs of integers, the Eisenstein series $\mathscr{E}_k$ is a holomorphic function of the parameter τ.

D.4.

Exchanging the roles of $\boldsymbol{v}$ and $\boldsymbol{v}'$ in (173) leads to the transformation

$$\tau \mapsto -1/\tau, \tag{174}$$

15 or, more precisely, holomorphic Eisenstein series

which takes the complex upper half-plane to itself. Here is what this transformation does to the Cartesian coordinates on the upper half-plane:

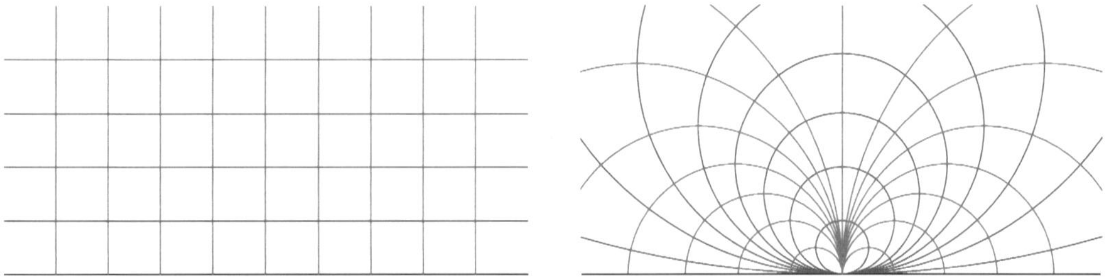

D.5.

The following beautiful formulas for the series $\mathscr{E}_k$ may be derived in terms of the variable $q = e^{2\pi i \tau}$. For τ in the upper half-plane, the corresponding q lies in the unit circle $|q| < 1$. We have

$$\mathscr{E}_k = 1 - \frac{2k}{B_k} \sum_{n=1}^{\infty} \Big(\sum_{d|n} d^{k-1} \Big) q^n, \tag{175}$$

where B_k are the Bernoulli numbers, and the coefficient of q^n is determined by summing over all divisors d of the number n. In particular, we have

$$\mathscr{E}_2 = 1-24q-72q^2-96q^3-168q^4-144q^5-288q^6-192q^7-360q^8-\cdots, \tag{176}$$

$$\mathscr{E}_4 = 1+240q+2160q^2+6720q^3+17520q^4+30240q^5+60480q^6+82560q^7+140400q^8+\cdots, \tag{177}$$

$$\mathscr{E}_6 = 1-504q-16632q^2-122976q^3-532728q^4-1575504q^5-4058208q^6-8471232q^7-17047800q^8-\cdots, \tag{178}$$

where we have added the series $\mathscr{E}_2$. Not being absolutely convergent, the series $\mathscr{E}_2$ may be summed with some further choices, and (176) is the result. The readers who feel they have already seen the number 240 somewhere recently are not mistaken.

D.6.

A holomorphic function $f(\Lambda)$ of a lattice $\Lambda \subset \mathbb{C}$ which satisfies (172) and remains bounded as long as the shortest vector $\boldsymbol{v} \in \Lambda$ is bounded away from 0 is called a modular form of weight k. We can multiply modular forms of different weights, and the weights add under multiplication. Thus modular forms form an algebra, and it is a classical theorem that

$$\text{Modular forms} = \mathbb{C}[\mathscr{E}_4, \mathscr{E}_6]. \tag{179}$$

The square brackets mean that $\mathscr{E}_4$ and $\mathscr{E}_6$ generate the algebra of modular forms freely, meaning, they do not satisfy any polynomial equation in two variables.

One often adds the series $\mathscr{E}_2$, whose converges requires some regularization, making its transformation law a bit more complicated. With this addition, the algebra (179) becomes

$$\text{Quasimodular forms} = \mathbb{C}[\mathscr{E}_2, \mathscr{E}_4, \mathscr{E}_6]. \tag{180}$$

D.7.

There are countless applications of modular forms to the study of lattices. A very major one is the subject of these introductory notes—Viazovska's gigantic breakthrough. For a much more basic one, consider the following situation.

Let $\Lambda \subset \mathbb{R}^d$ be a lattice. We can associate to it its theta series

$$\Theta_\Lambda(q) = \sum_{\boldsymbol{v} \in \Lambda} q^{\frac{1}{2}\|\boldsymbol{v}\|^2}. \tag{181}$$

This converges for $|q| < 1$ for any lattice Λ. If Λ is even then this is a series in q. And if Λ is additionally unimodular then this is a modular form of weight $d/2$.

In particular, for $\Lambda = E_8$ we should get a modular form of weight 4, and from (179) we see that it can only be a multiple of $\mathscr{E}_4$. Comparing the coefficients of q^0, we conclude

$$\Theta_{E_8} = \mathscr{E}_4. \tag{182}$$

Thus the coefficients in (177) count the vectors of a length $\sqrt{2n}$ in the lattice E_8. In particular, 240 is the number of roots.

E. THE VOLUME OF A d-DIMENSIONAL BALL

E.1.

Let $B(0, r)$ be the d-dimensional ball (2) of radius r. Its volume is proportional to r^d, namely

$$\operatorname{Vol} B(0, r) = \mathsf{v}_d r^d, \tag{183}$$

with some proportionality constant v_d. Our goal in this section is to compute this constant. As we will see, it is given in terms of a certain special function (184).

E.2.

The Gamma function is defined by the following integral:

$$\Gamma(s) = \int_0^\infty e^{-x} x^{s-1} dx, \tag{184}$$

which converges when $s > 0$. For complex s, the integral (184) converges when the real part $\Re s > 0$. Integration by parts gives

$$\Gamma(s+1) = s\Gamma(s), \tag{185}$$

and, using this formula, one can extend the definition of $\Gamma(s)$ to all values of s, except $s = 0, -1, -2, \ldots$

From (185) and the base case $\Gamma(1) = 1$, we conclude

$$\Gamma(n) = 1 \cdot 2 \cdot 3 \cdots (n-1) = (n-1)!, \quad n = 1, 2, 3, \ldots \tag{186}$$

The Gamma functions is, in a certain technical sense, the most natural extension of the factorial $(n-1)!$ to a function of a complex variable.

E.3.

Consider the plot (thick curve) of the logarithm of integrand in (184) for $s = 1000$.

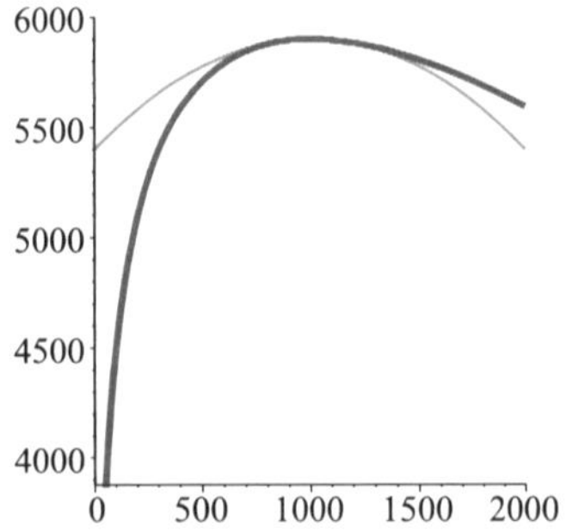

(187)

We will come back to the meaning of the thin curve later. In (187), we have the logarithm, meaning the integrand itself takes very large values. Their maximum is at the solution of

$$(e^{-x}x^{s-1})' = \left(-1 + \frac{s-1}{x}\right)e^{-x}x^{s-1} = 0 \quad \Rightarrow \quad x = s - 1. \tag{188}$$

Hence $\Gamma(1000)$ should be something of the order $(\frac{999}{e})^{999}$. Refining this argument, one can deduce a more precise asymptotic relation

$$\Gamma(s+1) \sim \sqrt{2\pi s}\left(\frac{s}{e}\right)^s \tag{189}$$

known as the Stirling formula. It is often used to approximate factorials.

E.4.

Let us put $x = y^2$ in (184). Since $dx = 2y\,dy$, we get

$$\Gamma(s) = 2\int_0^\infty e^{-y^2} y^{2s-1} dy. \tag{190}$$

In particular, we get the famous Gaussian integral for $s = \frac{1}{2}$,

$$\int_{-\infty}^\infty e^{-y^2} dy = \Gamma\left(\frac{1}{2}\right). \tag{191}$$

Let us multiply d copies of (191). We get

$$\begin{aligned}\Gamma\left(\frac{1}{2}\right)^d &= \int_{\mathbb{R}^d} e^{-(y_1^2+\cdots+y_d^2)}\, dy_1 \cdots dy_d \\ &= \int_{\mathbb{R}^d} e^{-\|\boldsymbol{y}\|^2}\, d\boldsymbol{y}.\end{aligned} \tag{192}$$

We observe the remarkable fact that the integrand in (192) depends only on the norm of the vector $\boldsymbol{y}$.

While in this section we have to assume that the reader has some familiarity with integrals, it may be worth recalling how Lebesgue integral is defined. One approximates the integrand by a function taking a discrete set of values and weighs each value by the volume of the set where this value is taken.

In particular, we can approximate the function $\|\boldsymbol{y}\|$ by the functions

$$\varepsilon\left\lceil \frac{\|\boldsymbol{y}\|}{\varepsilon} \right\rceil \to \|\boldsymbol{y}\|, \quad \varepsilon \to 0, \tag{193}$$

that take the value $r = k\varepsilon, k = 0, 1, 2, \ldots$, on the spherical shell formed by the difference of $B(0, r)$ and the smaller ball $B(0, r - \varepsilon)$. From (183), we conclude

$$\operatorname{Vol} B(0, r) - \operatorname{Vol} B(0, r - \varepsilon) \approx d\, \mathsf{v}_d\, r^{d-1} \varepsilon. \tag{194}$$

Therefore

$$\int_{\mathbb{R}^d} e^{-\|y\|^2}\, dy = d\, \mathsf{v}_d \int_0^\infty e^{-r^2} r^{d-1}\, dr = \frac{d}{2} \mathsf{v}_d \Gamma\left(\frac{d}{2}\right) = \mathsf{v}_d \Gamma\left(\frac{d}{2} + 1\right), \tag{195}$$

where we have used equalities (190) and (185).

Putting (192) and (195) together, we conclude

$$\mathsf{v}_d = \frac{\Gamma(\frac{1}{2})^d}{\Gamma(\frac{d}{2} + 1)}. \tag{196}$$

E.5.

To simplify (196), we note that the πr^2 formula for the area of circle computes the Gaussian integral! Indeed,

$$\mathsf{v}_2 = \pi \quad \Rightarrow \quad \Gamma\left(\frac{1}{2}\right) = \sqrt{\pi}. \tag{197}$$

In fact, the $\sqrt{2\pi s}$ prefactor in the Stirling formula (189) comes from approximating the integral (184) by a Gaussian integral peaked at $x = s - 1$ for $s \to \infty$. The Gaussian approximation for the integrand means the quadratic approximation for its logarithm, and the latter is plotted thin in the figure of (187).

This connects all the different appearances of the number π in this section.

E.6.

Therefore, we have the following great mnemonic formula:

$$\mathsf{v}_d = \frac{\pi^{d/2}}{(d/2)!}. \tag{198}$$

For odd dimensions, we should define the factorial using the Gamma functions, and concretely,

$$(d/2)! = \Gamma\left(\frac{d}{2} + 1\right) = \frac{d}{2} \frac{d-2}{2} \frac{d-4}{2} \cdots \frac{1}{2} \Gamma\left(\frac{1}{2}\right) = 2^{-(d+1)/2} d!! \sqrt{\pi}, \quad \text{for } d \text{ odd.} \tag{199}$$

Here $d!!$ means the double factorial of an integer d, that is, the product of odd (respectively, even) integers in $\{1, \ldots, d\}$.

E.7.

Note that from the Stirling formula, we have

$$\operatorname{Vol} B(0, r) \sim \frac{1}{\sqrt{\pi d}} \left(\frac{2\pi e\, r^2}{d}\right)^{d/2},$$

which leads to a very remarkable conclusion: the volume of a ball of *arbitrarily large* fixed radius r goes to 0 as $d \to \infty$ superexponentially fast!

F. MORE ON E_8 AND REGULAR m-GONS

F.1.

In addition to Coxeter elements C, which have order 30, the group $W(E_8)$ generated by reflections in the roots of the E_8 lattice has distinguished conjugacy classes of elements of order 24 and 20; see [25]. In this section, we will denote a representative of these conjugacy classes by C_{30}, C_{24}, and C_{20}. The eigenvalues of each C_m are the primitive mth roots of unity. There are exactly 8 of those in each case.

Projecting the roots on any of the eigenspaces, one gets the following patterns:

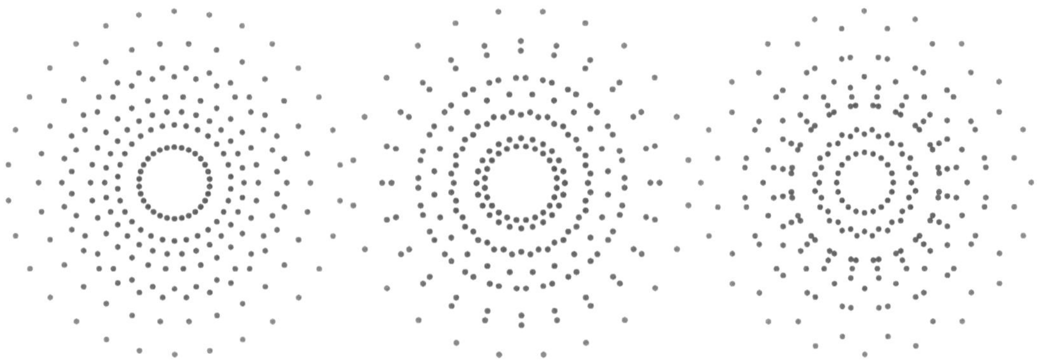

F.2.

From a slightly different angle, the relation between E_8 and regular polygons may be seen as follows. For any m, the mth roots of unity $\{1, \zeta, \zeta^2, \dots, \zeta^{m-1}\}$, where $\zeta = \exp(\frac{2\pi i}{m})$, are the images of the group $\mathsf{G} = \mathbb{Z}/m$ in a 1-dimensional representation. For instance, for $m = 6$ these are the vertices of a regular hexagon

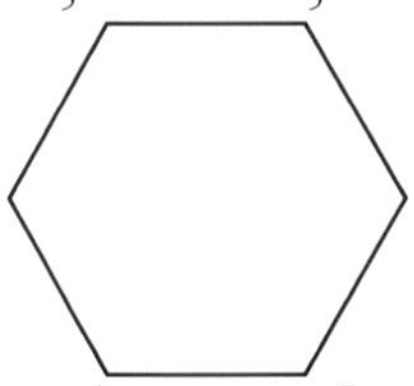

(200)

We can also consider the image of the group ring $\mathbb{Z}\mathsf{G}$, that is, the subring

$$\Lambda = \mathbb{Z}[\zeta] \subset \mathbb{C}.$$

This is the set of points that can be obtained by adding and subtracting the vertices of a regular m-gon. For instance, for $m = 4, 6$, this will be the square lattice $A_1 \oplus A_1$ and the hexagonal lattice A_2, respectively.

The powers $1, \zeta, \zeta^2, \dots$ are linearly independent over $\mathbb{Q}$ until we get to $\zeta^{\phi(m)}$, where $\phi(m)$ is number of residues modulo m that are coprime to m, also known as Euler's totient. The number $\zeta^{\phi(m)}$ is an integral linear combination of the numbers $1, \zeta, \dots, \zeta^{\phi(m)-1}$, given by the coefficients of the cyclotomic polynomial

$$\Psi_m(x) = \prod_{\gcd(i,m)=1} (x - \zeta^i) = x^{\phi(m)} + \cdots \in \mathbb{Z}[x]. \tag{201}$$

See (51) for the explicit form of Ψ_{30}. Thus,

$$\Lambda \cong \mathbb{Z}^{\phi(m)} \subset \mathbb{C} \tag{202}$$

as a group under addition. Since we want to construct E_8, we focus our attention on the case

$$\phi(m) = 8 \quad \Rightarrow \quad m \in \{15, 20, 24, 30\}. \tag{203}$$

The number 15 here corresponds to the element C_{30}^2. We skip it, since the 15-gon and the 30-gon generate the same Λ.

Lest the reader imagine $\Lambda \subset \mathbb{C}$ as a lattice, we plot the points $\sum_{i=0}^{29} c_i \zeta^i$, where $\zeta = \exp(\frac{\pi i}{15})$ and $c_i \in \{0, 1, 2, 3\}$; Λ is a free abelian subgroup of $\mathbb{C}$, but it is not a lattice.

(204)

F.3.

Let $\boldsymbol{v} \in E_8$ be a vector such that the vectors $\mathsf{C}_m^i \boldsymbol{v}$ span the lattice E_8, and consider the diagonal matrix element

$$\eta_{\boldsymbol{v}}(\zeta^i) = \left(\mathsf{C}_m^i \boldsymbol{v}, \boldsymbol{v}\right)_{E_8}. \tag{205}$$

Since $\Psi_m(\mathsf{C}_m) = 0$, this gives a well-defined linear function on Λ. Put slightly differently, since the eigenvalues of C_m are the primitive roots of unity, only those Fourier coefficients of $\eta_{\boldsymbol{v}}$, viewed as a function on G, do not vanish. This makes it well-defined as a function on Λ. Furthermore, the function (58) being positive definite, these Fourier coefficients are positive.

For example, let us take the particular Coxeter element constructed in (50) and the following vectors:

$$\boldsymbol{v} = \boldsymbol{\alpha}_5, \quad \boldsymbol{v}' = 2\boldsymbol{e}_8, \tag{206}$$

that is, the triple node in the Dynkin diagram and twice the last coordinate vector. From the explicit expression

$$\mathsf{C}_{30} = \frac{1}{4}\begin{bmatrix} -1 & -1 & 3 & -1 & -1 & 1 & 1 & -1 \\ 3 & -1 & -1 & -1 & -1 & 1 & 1 & -1 \\ -1 & -1 & -1 & -1 & 3 & 1 & 1 & -1 \\ -1 & 3 & -1 & -1 & -1 & 1 & 1 & -1 \\ -1 & -1 & -1 & -1 & -1 & -3 & 1 & -1 \\ -1 & -1 & -1 & 3 & -1 & 1 & 1 & -1 \\ -1 & -1 & -1 & -1 & -1 & 1 & -3 & -1 \\ -1 & -1 & -1 & -1 & -1 & 1 & 1 & 3 \end{bmatrix}, \tag{207}$$

one can check that the corresponding functions η and η' are given by

$$\eta(\zeta^i) = \left\lfloor\!\left\lfloor 2\cos\left(\frac{\pi i}{15}\right) \right\rfloor\!\right\rfloor, \quad \eta'(\zeta^i) = \left\lfloor\!\left\lfloor 4\cos\left(\frac{\pi i}{15}\right) \right\rfloor\!\right\rfloor, \tag{208}$$

where $\lfloor\!\lfloor x \rfloor\!\rfloor$ denotes the integer between 0 and x that is closest to x. These formulas may be illustrated as follows:

$$\eta(\zeta^i) = \tag{209}$$

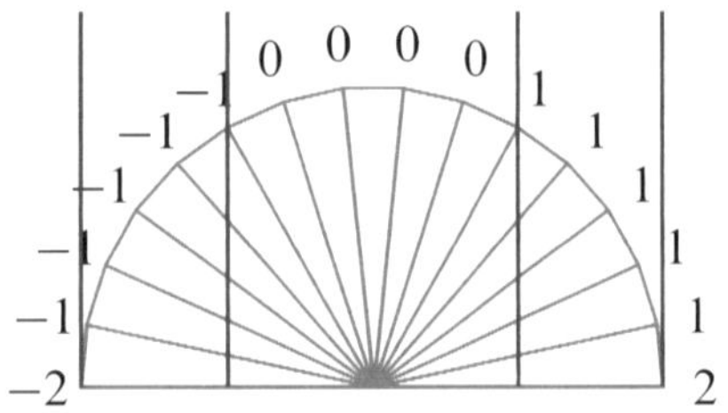

and

$$\eta'(\zeta^i) = \tag{210}$$

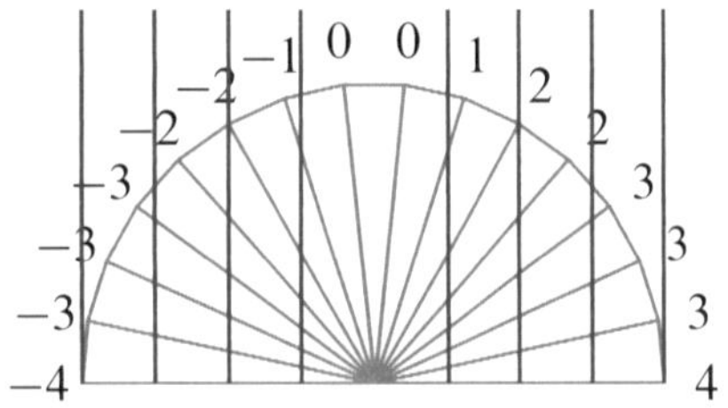

The proximity of these functions to the cosine function can be interpreted as the proximity of the vectors $\boldsymbol{v}$ and $\boldsymbol{v}'$ to the plane in which C_{30} acts as a rotation by $\pi/15$.

F.4.

In the style of Appendix B.6, one can turn this construction around as follows. One can check directly that the Fourier coefficients satisfy

$$\hat{\eta}(j) \text{ is } \begin{cases} > 0, & \gcd(j, m) = 1, \\ = 0, & \gcd(j, m) > 1, \end{cases} \tag{211}$$

and similarly for η'. We can then *define* the E_8 lattice as the group Λ with the quadratic form

$$(\boldsymbol{v}, \boldsymbol{v}')_{E_8} = \eta(\boldsymbol{v}\overline{\boldsymbol{v}'}), \quad \boldsymbol{v}, \boldsymbol{v}' \in \Lambda. \tag{212}$$

In fact, the functions

$$\eta_m(\zeta^i) = \left\lfloor\!\left\lfloor 2\cos\left(\frac{2\pi i}{m}\right) \right\rfloor\!\right\rfloor, \quad \eta'_m(\zeta^i) = \left\lfloor\!\left\lfloor 4\cos\left(\frac{2\pi i}{m}\right) \right\rfloor\!\right\rfloor, \quad m \in \{20, 24, 30\}, \tag{213}$$

all work and exhibit the E_8 lattice as a lattice with an isometry C_m of the corresponding order. In this realization, the isometry is given by multiplication by ζ. For specific m, the numbers 2 and 4 in (213) can be replaced by other even integers.

REFERENCES

[1] A. V. Akopyan, G. A. Kabatiansky, and O. R. Musin, Contact numbers, codes, and spherical polynomials. *Mat. Pros., Ser.* **16** (2012), 57–74 (Russian). Available from http://www.mathnet.ru/php/archive.phtml?wshow=paper&jrnid=mp&paperid=289&option_lang=eng.

[2] N. N. Andreev and V. A. Yudin, Extreme placements of points on a sphere. *Mat. Pros., Ser.* **3** (1997), no. 1, 115–125 (Russian).

[3] S. Axler, *Linear algebra done right*. 3rd edn., Undergrad. Texts Math., Springer, Cham, 2015.

[4] H. F. Blichfeldt, The minimum values of positive quadratic forms in six, seven and eight variables. *Math. Z.* **39** (1935), no. 1, 1–15.

[5] B. Casselman, The difficulties of kissing in three dimensions. *Notices Amer. Math. Soc.* **51** (2004), no. 8, 884–885.

[6] H. Cohn, A conceptual breakthrough in sphere packing. *Notices Amer. Math. Soc.* **64** (2017), no. 2, 102–115.

[7] H. Cohn, Packing, coding, and ground states. In *Mathematics and materials*, IAS/Park City Math. Ser. 23, pp. 45–102, Amer. Math. Soc., Providence, RI, 2017.

[8] H. Cohn, The work of Maryna Viazovska. In *Proceedings of ICM 2022, Vol. 1*, pp. 82–105, EMS Press, 2022.

[9] H. Cohn, *Sphere packing*. https://cohn.mit.edu/sphere-packing.

[10] H. Cohn and N. Elkies, New upper bounds on sphere packings. I. *Ann. of Math. (2)* **157** (2003), no. 2, 689–714.

[11] H. Cohn, A. Kumar, S. D. Miller, D. Radchenko, and M. Viazovska, The sphere packing problem in dimension 24. *Ann. of Math. (2)* **185** (2017), no. 3, 1017–1033.

[12] H. Cohn, A. Kumar, S. D. Miller, D. Radchenko, and M. Viazovska, Universal optimality of the E_8 and Leech lattices and interpolation formulas. arXiv:1902.05438.

[13] H. Cohn and Y. Zhao, Sphere packing bounds via spherical codes. *Duke Math. J.* **163** (2014), no. 10, 1965–2002.

[14] J. H. Conway and N. J. A. Sloane, What are all the best sphere packings in low dimensions? *Discrete Comput. Geom.* **13** (1995), no. 3–4, 383–403.

[15] J. H. Conway and N. J. A. Sloane, *Sphere packings, lattices and groups*. 3rd edn., Grundlehren Math. Wiss. 290, Springer, New York, 1999. With additional contributions by E. Bannai, R. E. Borcherds, J. Leech, S. P. Norton, A. M. Odlyzko, R. A. Parker, L. Queen and B. B. Venkov.

[16] D. de Laat and F. Vallentin, A breakthrough in sphere packing: the search for magic functions. *Nieuw Arch. Wiskd. (5)* **17** (2016), no. 3, 184–192. Includes an interview with Henry Cohn, Abhinav Kumar, Stephen D. Miller and Maryna Viazovska.

[17] P. Delsarte, Bounds for unrestricted codes, by linear programming. *Philips Res. Rep.* **27** (1972), 272–289.

[18] P. Delsarte, J. M. Goethals, and J. J. Seidel, Spherical codes and designs. *Geom. Dedicata* **6** (1977), no. 3, 363–388.

[19] W. Ebeling, *Lattices and codes*. 3rd edn., Adv. Lectures Math., Springer, Wiesbaden, 2013. A course partially based on lectures by Friedrich Hirzebruch.

[20] N. D. Elkies, Lattices, linear codes, and invariants. I. *Notices Amer. Math. Soc.* **47** (2000), no. 10, 1238–1245.

[21] N. D. Elkies, Lattices, linear codes, and invariants. II. *Notices Amer. Math. Soc.* **47** (2000), no. 11, 1382–1391.

[22] L. Fejes, Über einen geometrischen Satz. *Math. Z.* **46** (1940), 83–85 (German).

[23] L. Fejes Tóth, *Lagerungen in der Ebene auf der Kugel und im Raum.* 2nd edn., Grundlehren Math. Wiss. 65, Springer, Berlin–New York, 1972 (German).

[24] C. F. Gauss, Besprechung des Buchs von L. A. Seeber: Untersuchungen iiber die Eigenschaften der positiven terniiren quadratischen Formen usw. *Göttingsche Gelehrte Anzeigen* (1831, July 9) = Werke, II (1876), 188–196.

[25] M. Geck and G. Pfeiffer, *Characters of finite Coxeter groups and Iwahori–Hecke algebras.* London Math. Soc. Monogr. New Ser. 21, The Clarendon Press, Oxford University Press, New York, 2000.

[26] I. Gelfand and M. Neumark, On the imbedding of normed rings into the ring of operators in Hilbert space. *Rec. Math. [Mat. Sb.] N.S.* **12(54)** (1943), 197–213 (English, with Russian summary).

[27] T. C. Hales, Cannonballs and honeycombs. *Notices Amer. Math. Soc.* **47** (2000), no. 4, 440–449.

[28] T. C. Hales, Historical overview of the Kepler conjecture. *Discrete Comput. Geom.* **36** (2006), no. 1, 5–20.

[29] T. C. Hales, *Dense sphere packings.* London Math. Soc. Lecture Note Ser. 400, Cambridge University Press, Cambridge, 2012. A blueprint for formal proofs.

[30] P. Honner, The math of social distancing is a lesson in geometry. *Quanta Mag.* (March 30, 2016). https://www.quantamagazine.org/the-math-of-social-distancing-is-a-lesson-in-geometry-20200713/.

[31] G. A. Kabatjanskiĭ and V. I. Levenšteĭn, Bounds for packings on the sphere and in space. *Problemy Peredachi Informatsii* **14** (1978), no. 1, 3–25 (Russian).

[32] E. Klarreich, Sphere packing solved in higher dimensions. *Quanta Mag.* (March 30, 2016). https://www.quantamagazine.org/sphere-packing-solved-in-higher-dimensions-20160330/.

[33] E. Klarreich, Out of a magic math function, one solution to rule them all. *Quanta Mag.* (May 13, 2019). https://www.quantamagazine.org/universal-math-solutions-in-dimensions-8-and-24-20190513/.

[34] A. Korkine and G. Zolotareff, Sur les formes quadratiques positives quaternaires. *Math. Ann.* **5** (1872), no. 4, 581–583(French).

[35] A. Korkinge and G. Zolotareff, Sur les formes quadratiques positives. *Math. Ann.* **11** (1877), no. 2, 242–292 (French).

[36] A. I. Kostrikin and Y. I. Manin, *Linear algebra and geometry*, Revised reprint of the 1989 English edition, Algebra, Logic and Applications, vol. 1, Gordon and Breach Science Publishers, Amsterdam, 1997. Translated from the second Russian (1986) edition by M. E. Alferieff.

[37] J. C. Lagarias, Bounds for local density of sphere packings and the Kepler conjecture. *Discrete Comput. Geom.* **27** (2002), no. 2, 165–193.

[38] V. I. Levenšteĭn, Boundaries for packings in n-dimensional Euclidean space. *Dokl. Akad. Nauk SSSR* **245** (1979), no. 6, 1299–1303 (Russian).

[39] O. R. Musin, The kissing problem in three dimensions. *Discrete Comput. Geom.* **35** (2006), no. 3, 375–384.

[40] O. R. Musin, The kissing number in four dimensions. *Ann. of Math. (2)* **168** (2008), no. 1, 1–32.

[41] G. Nebe and N. Sloane, Table of densest packings presently known. http://www.math.rwth-aachen.de/~Gabriele.Nebe/LATTICES/density.html.

[42] A. M. Odlyzko and N. J. A. Sloane, New bounds on the number of unit spheres that can touch a unit sphere in n dimensions. *J. Combin. Theory Ser. A* **26** (1979), no. 2, 210–214.

[43] A. Okounkov, Rhymes in primes. In *Proc. Int. Cong. Math. 2022, Vol. 1*, pp. 460–490, EMS Press, 2022.

[44] A. Okounkov, Combinatorial geometry takes the lead. In *Proc. Int. Cong. Math. 2022, Vol. 1*, pp. 414–458, EMS Press, 2022.

[45] G. Parisi, F. Zamponi, and P. Urbani, *Theory of simple glasses: exact solutions in infinite dimensions*. Cambridge University Press, 2020.

[46] M. Reed and B. Simon, *Methods of modern mathematical physics. II. Fourier analysis, self-adjointness*. Academic Press [Harcourt Brace Jovanovich, Publishers], New York–London, 1975.

[47] K. Schütte and B. L. van der Waerden, Das Problem der dreizehn Kugeln. *Math. Ann.* **125** (1953), 325–334 (German).

[48] I. E. Segal, Irreducible representations of operator algebras. *Bull. Amer. Math. Soc.* **53** (1947), 73–88.

[49] N. J. A. Sloane, The sphere packing problem. In *Proceedings of the International Congress of Mathematicians, Vol. III (Berlin, 1998)*, Doc. Math., pp. 387–396, 1998.

[50] E. M. Stein and R. Shakarchi, *Fourier analysis. An introduction*. Princeton Lect. Anal. 1, Princeton University Press, Princeton, NJ, 2003.

[51] G. Strang, *Linear algebra and its applications*. 2nd edn., Academic Press [Harcourt Brace Jovanovich, Publishers], New York–London, 1980.

[52] T. M. Thompson, *From error-correcting codes through sphere packings to simple groups*. Carus Math. Monogr. 21, Mathematical Association of America, Washington, DC, 1983.

[53] A. Thue, Über die dichteste Zusammenstellung von kongruenten Kreisen in einer Ebene, Christinia Vid. Selsk. Skr. 1 (1910), 1–9. Universitetsforlaget, Oslo, 1910. Reprinted in *Selected mathematical papers*, With an introduction by Carl Ludwig Siegel and a biography by Viggo Brun; Edited by Trygve Nagell, Atle Selberg, Sigmund Selberg, and Knut Thalberg.

[54] M. S. Viazovska, The sphere packing problem in dimension 8. *Ann. of Math. (2)* **185** (2017), no. 3, 991–1015.

[55] M. S. Viazovska. Interview. IMU webpage.

[56] Image credit: https://www.pexels.com/photo/person-cutting-dough-with-clear-glass-5947603/.

ANDREI OKOUNKOV

Department of Mathematics, University of California, Berkeley, 970 Evans Hall Berkeley, CA 94720–3840, USA, okounkov@math.columbia.edu

SUMMARIES OF PRIZE WINNERS' WORK

2022 ABACUS MEDAL: MARK BRAVERMAN

ALLYN JACKSON

ABSTRACT

This article describes the work of Mark Braverman, winner of the 2022 Abacus Medal (formerly known as the Rolf Nevanlinna Prize), which was presented by the International Mathematical Union in conjunction with ICM2022. The Abacus Medal honors outstanding contributions in mathematical aspects of information sciences, including theoretical computer science.

MATHEMATICS SUBJECT CLASSIFICATION 2020

Primary 68-03; Secondary 01A70

KEYWORDS

Theoretical computer science, Mark Braverman

ICM INTERNATIONAL CONGRESS OF MATHEMATICIANS 2022 JULY 6–14

© 2023 International Mathematical Union
Proc. Int. Cong. Math. 2022, Vol. 1, pp. 548–553
DOI 10.4171/ICM2022/218

Published by EMS Press and licensed under a CC BY 4.0 license

In the past decade, Mark Braverman has emerged as a major leader in theoretical computer science. He has an uncommon versatility and fearlessness that has allowed him not only to tackle specific outstanding problems, but also to work on deep theoretical questions. As a researcher, he exhibited exceptional maturity at a young age, producing results that brought new insights and stimulated new research.

Of the many subfields of computer science, the one known as theoretical computer science is the closest to mathematics. It draws on, as well as develops, abstract mathematical notions in order to address questions inspired by concrete problems in computation, communication, information transmission, and related areas. A major goal of theoretical computer science is to establish precise, mathematically rigorous results about how quickly and efficiently problems can be solved. Emblematic of this goal is the famous P versus NP problem, a major unsolved question in both theoretical computer science and mathematics.

At the age of 38, Braverman already has a publication list of more than 100 papers written with a total of more than 85 collaborators. Because his oeuvre is extensive and diverse, we focus here on three areas to which he has contributed results that exemplify the depth and power of his work.

COMPUTING JULIA SETS

Even in his earliest work Braverman took on fundamental questions. One of them centered on investigations of what becomes possible – and impossible – when one changes the theoretical basis for computing.

Modern computers are based on the model for computing formulated by Alan Turing in the 1930s and are essentially discrete systems: Any task a computer carries out boils down to manipulating 0s and 1s. By contrast systems in nature – the swinging of a pendulum, the development of a weather pattern, the fractal geometry of a coastline – are continuous rather than discrete. Mathematicians and computer scientists have therefore investigated alternative computational models that are continuous rather than discrete. Braverman has worked with one model that represents computations not as 0s and 1s but rather as real numbers (the set of real numbers contains all numbers, including ones like π that have infinite, nonrepeating decimal expansions).

The bedrock of mathematics shifted when Turing's model for computing revealed the concept of uncomputability: There exist numbers that can be described in a perfectly clear and precise way but that cannot be computed explicitly. What kinds of uncomputability phenomena arise with continuous computational models? This is the question that Braverman explored starting already with his master's thesis and in subsequent work, much of it with Michael Yampolsky.

Braverman focused on a continuous computing model that, intuitively, is based on the idea that a set is computable if it can be drawn pixel-by-pixel on a computer screen. Among the mathematical objects drawn in this way are Julia sets, which are fractals originally discovered by Gaston Julia in the early 20th century and popularized by Benoit Mandelbrot in the 1980s. The beauty and intricacy of Julia sets have made them the subject not only of art exhibitions but also of intensive study within the theory of dynamical systems, the branch of mathematics treating systems that evolve over time.

Although Julia sets can exhibit highly complicated behavior, each is characterized by a single parameter. Braverman and Yampolsky identified values of this parameter such that the associated Julia set is uncomputable in the continuous computing model they used. These parameter values are few and far between; you are not likely to hit upon one when entering a parameter in one of the many web programs that draw Julia sets. This points to a kind of instability of uncomputable structures and might offer hints about why they are rarely encountered in real-world problems.

In 2009 Braverman and Yampolsky published a book *Computability of Julia Sets*, which provides an overview of this area. Braverman has made further contributions to computability of other phenomena in dynamical systems. For example, in a 2007 paper with Ilia Binder and Yampolsky, Braverman investigated the computability of the Riemann mapping, a fundamental mathematical notion from the area of complex analysis. And in 2015 he published a paper with Jonathan Schneider and Cristóbal Rojas that presents a refinement of a pillar of computing theory, the Church–Turing Thesis.

Starting around 2010, Braverman's attention turned to information complexity, which we will discuss next. Here too the theme of discrete versus continuous arises in his work.

INFORMATION COMPLEXITY

In 1948, Claude Shannon published a paper that provided a comprehensive theory governing the transmission of information. He showed that, even when information is represented as discrete bits – that is, as strings of 0s and 1s – it can be modeled as a continuous quantity using probability theory. One can then define the notion of the *entropy* of a transmitted message, which intuitively speaking is the amount of information it contains. For example, suppose Alice sends Bob a message giving a year's worth of data about two daily events: Whether the sun rose that day, and whether it rained that day. Although she has 365 bits of sun-rising data, that information could be compressed into one bit; its entropy is very small. By contrast, the 365 bits of rain data could not be compressed so much; its entropy is higher. Shannon's theory shows that the entropy establishes a natural limit on how much a message can be compressed without losing information.

Now suppose that instead of the communication being one-way, it's two-way: Alice and Bob each have some information and send bits back and forth. Their goal might be, for example, to understand how their knowledge differs. It could happen that what Alice knows differs by only one bit from what Bob knows, but establishing that fact requires sending many bits back and forth. In such a case, the communication cost, which is the number of bits exchanged, is large, but the information cost, which is the amount of *information* exchanged, is low. Does information theory shed light on how to make such an exchange more efficient?

Starting around 1980, the area of communication complexity grew up around such questions. A good deal of progress was made addressing specific problems, and this had a big impact in applications to tasks like streaming algorithms and data structures. However, the theoretical underpinning remained somewhat undeveloped, partly because the necessary mathematical machinery was lacking. When Braverman came on the scene starting around 2010, he revived and expanded the field through a series of striking results that supplied new and more precise theoretical foundations.

Basic to this area is the direct sum question, which asks the following. Suppose C is the cost of Alice and Bob interacting to carry out a certain task once. If they carry out k independent repetitions of the task, is the final cost always equal to k times C? In the case of information cost, Braverman proved that the answer is yes. For communication cost the answer is generally no and depends on the nature of the task.

But in 2010, Braverman, together with Boaz Barak, Xi Chen, and Anup Rao, showed that the cost of k repetitions is at least $\sqrt{k}$ times the cost of doing the task once. The following year, Braverman and Rao proved an "amortization" result, establishing that in the limit as k gets very large, the average communication cost of one repetition approaches the information cost. This result has had a wide impact by providing a natural line of attack for proving results about efficiency for specific communication tasks. Braverman has also had important results in regimes where the communication can be corrupted by transmission errors or sabotage.

In the realm of computation, the dominant theoretical problem is P versus NP. Work on this problem has to a large extent remained in the discrete realm and has not benefited much from continuous tools from analysis, which is one of the most sophisticated and highly developed areas of mathematics. The work of Braverman and his collaborators on efficiency in communication might provide a glimmer of hope that continuous tools could one day have a larger impact on P versus NP than they have so far.

MECHANISM DESIGN

As algorithmic economics has provided the foundation for much of online commerce, mechanism design has grown into one of its most active subfields. Here too Braverman has made several significant contributions.

Right after his doctorate, Braverman held a research position at the Microsoft Research New England laboratory, where he worked with the lab's health care group. There he investigated machine-learning tools for studying factors leading to patient rehospitalization. This experience led him to realize that such questions are often more economic and game-theoretic than they are computational and sparked his interest in algorithmic economics.

An algorithm takes an input, carries out a step by step procedure, and produces an output. The algorithm carries out the same procedure regardless of what the input is; one might say that the input doesn't care what the algorithm is. But in many economic procedures, for example in auctions, the inputs are provided by agents who do care what the algorithm is and are seeking, by their inputs, to influence the output. This is the setting for mechanism design, which aims to construct protocols that take inputs elicited from agents having a stake in the output and that also drives the agents towards inputs that result in desirable output.

A well known example of a mechanism is the Vickrey auction. Bidders submit secret bids, and the person submitting the highest bid is allowed to buy the item, but at a price equal to the *second-highest* bid. This system drives bidders to be honest about what the item is actually worth to them: Underbidding cannot reduce the purchase price and could cause them to lose the opportunity to buy the item. In the more general Vickrey–Clarke–Groves (VCG) mechanism, multiple items are distributed among bidders, and each bidder must pay for the "harm" that buying an item causes to the others, thereby achieving a socially optimal solution. VCG produces excellent theoretical results, but its practical implementation is marred by instabilities and other problems.

In today's world of cheap computing and interconnectedness, algorithms are increasingly manipulated by strategic agents. A major goal is therefore to find ways to convert algorithms to mechanisms, and this is the focus of Braverman's latest research. In particular, he has been looking at how to incorporate the VCG mechanism into algorithms that are based on many implementations of local optimization. Such an algorithm contains many sub-algorithms, each of which uses local optimization on just one small chunk of the problem and takes incremental steps towards optimal solutions within that chunk. Those locally optimal solutions are then combined by the main algorithm to solve the problem. Braverman's idea is to bring the VCG mechanism into the algorithm at the level of the local optimization, where the larger problems of VCG can be effectively controlled. Because local optimization is used in many systems, including in machine learning, Braverman's approach has potential for wide impact in applications. It has already borne fruit in the realm of theory; in 2021 Braverman used it to strengthen an economics result from more than 40 years ago.

PROBLEM-SOLVING PROWESS AND THEORETICAL INSIGHT

This brief account of Braverman's work shows how he is able to make progress on difficult questions that require sustained focus and development over time. But he has also worked in a different mode, solving isolated and highly abstract open questions that called on his problem-solving prowess. An example of this is his 2010 proof of the Linial–Nisan conjecture. Too technical to describe here, this conjecture arose in the area of pseudorandomness and had stumped researchers since it was first proposed in 1990. Braverman's strikingly original solution stunned experts and was especially surprising because the problem lay so far from the areas in which he had been working.

Mark Braverman's combination of potent problem-solving ability and deep theoretical insight has produced results of exceptional impact. His work embodies the spirit of theoretical computer science, with its emphasis on marrying the power of abstract mathematics to the real-world struggle for speed and efficiency. His influence on the field, already large for such a young researcher, will no doubt continue to grow.

ALLYN JACKSON
Allyn Jackson is a freelance writer specializing in mathematics and theoretical computer science. For further information visit allynjackson.com.

2022 CHERN MEDAL: BARRY MAZUR

ALLYN JACKSON

ABSTRACT

This article describes the work of Barry Mazur, winner of the 2022 Chern Medal, which was presented by the International Mathematical Union in conjunction with ICM2022. The Chern Medal honors an individual of any age or vocation whose accomplishments warrant the highest level of recognition for outstanding achievements in the field of mathematics.

MATHEMATICS SUBJECT CLASSIFICATION 2020

Primary 11-03; Secondary 01A70

KEYWORDS

Number theory, Barry Mazur

INTERNATIONAL CONGRESS OF MATHEMATICIANS 2022 JULY 6–14

© 2023 International Mathematical Union

Proc. Int. Cong. Math. 2022, Vol. 1, pp. 554–558
DOI 10.4171/ICM2022/219

Published by EMS Press
and licensed under a CC BY 4.0 license

Barry Mazur is a singular figure in the international mathematical community. His research achievements cover several areas of mathematics, from topology to algebraic geometry to number theory, and assure his position as one of the greatest mathematicians of our time. His ability to move from one area to another is already unusual; what is extraordinary is his perception of deep analogies between them. These analogies have not only brought solutions to outstanding problems but also sparked the development of new research areas.

Mazur is in many ways a very concrete mathematician, taking on and solving specific problems. He also has the ability to shift effortlessly to higher levels of generality and a big-picture, abstract viewpoint. He can therefore discuss mathematics on many different levels, making him an uncommonly effective interlocutor. This trait, combined with his buoyant zeal and the uncommon generosity with which he shares ideas, has proven to be a magnet for students, postdocs, and colleagues, amplifying his influence on the field. Moreover his charm and friendliness have made him a truly beloved member of the mathematical community.

And yet his influence goes beyond this community. Mazur's tireless intellect does not stop at the borders of mathematics but ranges into literature, law, philosophy, and physics. His many nontechnical writings have explored new genres for discussing mathematical ideas. He has also crossed academic barriers to teach courses in collaboration with colleagues in other fields.

To give a flavor of Mazur's rich and diverse mathematical oeuvre, we consider a few highlights.

THE "MAZUR SWINDLE"

As a doctoral student in the 1950s, Mazur formulated a deep question about the fundamental nature of space. Only after solving it did he find out it was a major open question in topology known as the Schoenflies problem.

A closed curve divides the plane into two regions: the region inside the curve and the region outside. What's more, no matter how complicated and undulating your curve is, you can stretch out the bumps to morph the interior into a disk; in mathematical terms, one says the interior is homeomorphic to a disk. The Schoenflies problem asks whether the analogous phenomena occur in higher dimensions. One need only add one dimension to encounter an obstacle: it is possible to create a surface so complicated that its interior is *not* homeomorphic to a three-dimensional ball. The Alexander horned sphere is a celebrated example.

Mazur came up with a mild restriction to rule out wild examples like the horned sphere. Then he answered the Schoenflies problem in the affirmative – for *all* dimensions. To do this, he created a technique, now called the "Mazur swindle", that eliminates difficulties by pushing them off to infinity. Seemingly magical but perfectly rigorous, the Mazur swindle was a powerful insight. As Valentin Poénaru wrote: "Barry's work, handling all the dimensions at once, came like a thunderbolt and was also a psychological revolution that, together with other developments, paved the way for what came next in high-dimensional topology."

Topology is the study of shapes, called manifolds, that can expand, bend, and move while retaining their most basic characteristics, like how many holes they have. Mazur's early work centered on topology; there is even a type of manifold called the Mazur-Poénaru manifold (the two discovered it independently around 1960). Then, like many other mathematicians in the 1960s, Mazur came under the influence of Alexander Grothendieck, who envisioned unity between the fluid, continuous world of shapes and the more rigid, discrete world of numbers. In realizing this vision, Grothendieck reworked the foundations of algebraic geometry, a branch of mathematics that uses geometric and topological ideas to study number-theoretic objects like curves that represent the solutions of polynomial equations.

THE LURE OF ALGEBRAIC GEOMETRY

One of the signs of the evolution of Mazur's interests is a remarkable paper he wrote in the mid-1960s describing an analogy (which he credited to David Mumford) between knots and prime numbers. Though the paper went unpublished, the ideas it set forth continued to blossom, forming the basis for a new area called arithmetic topology.

When Mazur received the Steele Prize of the American Mathematical Society in 2000, he looked back on the early 1960s, when Grothendieck posed to him an inspiring question. The question raised the intriguing possibility, Mazur said, "that different topologies might be 'unified' by virtue of the fact that they arose as different avatars of the same algebraic geometry." Lured in this way into algebraic geometry, Mazur launched a collaboration with Michael Artin, which he called "one of the most important mathematical experiences for me and ... enormous fun."

The Steele Prize honored one of Mazur's most influential papers, "Modular curves and the Eisenstein ideal," published in 1977. This paper represented the first time that the full power of the Grothendieck revolution in algebraic geometry was brought to bear on a purely number-theoretic problem – in fact on an important problem that had gone unsolved for more than 70 years. While the paper was immediately hailed as a significant advance, its real impact became apparent only with the passage of time, as other researchers used it as a springboard for new advances.

The paper continued mathematicians' millenia-long conversation about Diophantine equations, which are polynomial equations with whole-number coefficients. The solutions to a specific collection of Diophantine equations – those having two variables where the highest degree of the variables is three – form objects known as an elliptic curves. For example, solutions to the equation $y^2 - x^3 = 2$ form a curve in the plane, which looks a bit like the silhouette of a fish with a round body trailed by an infinitely long, infinitely widening tail.

Elliptic curves have some intriguing features. If you draw a line connecting two points on an elliptic curve, the line generally hits a third point on the curve, which can be thought of as the "sum" of the first two points. Miraculously, this summing operation makes the points into a *group*. Pervasive across mathematics and the sciences, the concept of a group organizes the myriad structures arising when a set is endowed with an operation that can combine pairs of elements and that can also be reversed to "uncombine" them. Groups usually arise through considering the collection of symmetries of an object; for example, the symmetries of a molecule with the operation of rotations is a group.

Among the mathematicians beguiled by the beauty of elliptic curves were the highly ingenious Italian algebraic geometers working in the early 20th century. They explored the group structure of the rational points on elliptic curves – that is, those points whose coordinates are rational solutions of the equation governing the curve. They observed that the groups – more precisely, the *torsion subgroups* – that arose were very limited in type. Why did so few types arise? And exactly which ones?

In a paper with John Tate in the mid-1970s, Mazur honed his intuition about elliptic curves by studying in detail some particular examples. That intuition formed the basis for Mazur's prize-winning 1977 paper, which completely answered the questions the Italians had wondered about, by describing the exact structure of all the possible torsion subgroups that could occur.

Beyond providing a definitive solution to a venerable problem, Mazur's paper opened new avenues of research through its many insightful asides and open questions, which other researchers took up to make further advances. The paper laid the foundation for many of the most important results in arithmetic algebraic geometry over the last 50 years, and its long echo is still felt at the frontier of research today. The paper also played a major role in reviving interest in the study of elliptic curves, which remains a central topic in number theory.

DEFORMING GALOIS REPRESENTATIONS

Such a result might have been the crowning achievement of an outstanding career in mathematics. But Mazur went on to do further seminal work. One example is his introduction of what are now known as "deformations of Galois representations." We can give only a very rough picture of this sophisticated notion.

The pioneering work of Evariste Galois, whose short life ended in the first part of the 19th century, teaches us that a certain group, now called the Galois group, is key to understanding solutions to polynomial equations. One way to get information about the Galois group is to study how it acts on other mathematical objects, most importantly vector spaces over finite fields and p-adic fields.

Mazur discovered a method for lifting a Galois representation over a finite field to a collection of deformations over a p-adic field. The reason it is useful to consider individual deformations is that they encode arithmetic information about concrete geometric objects like elliptic curves. Mazur's method endows the collection of all deformations with extra mathematical structures that are of great interest in their own right and remain part of a lively area of investigation.

Mazur's 1989 paper introducing this discovery did not solve a specific problem. Rather, it launched a new theory, the theory of deformation of Galois representations, which unveiled an entirely new viewpoint and which over the ensuing decades other researchers have used to make new advances. One of these is the application of sophisticated counting arguments to the set of Galois deformations satisfying certain conditions.

The first spectacular argument of this sort came in Andrew Wiles's epoch-making proof of Fermat's Last Theorem in 1993. This proof completed a grand edifice of which several of Mazur's ideas, including those arising in his 1977 paper on elliptic curves, are important cornerstones. The theory of deformation of Galois representations has also been the basis for advances in the Langlands Program, which offers a unifying view of mathematics by suggesting deep relations among geometry, algebra, number theory, and analysis.

BEYOND MATHEMATICS

As outstanding as Mazur's mathematical accomplishments are, they do not tell the whole story of his impact on the field. His students and colleagues speak of his unfailing graciousness and the generosity with which he shares ideas. A gifted communicator, he is unusually perceptive in his ability to pitch explanations at the right level for his listeners. In advising PhD students – he's had close to 60 in all – he guides and motivates without imposing his own views of what directions they should take. Mazur is surely a leader, but he's also an inspirer, a facilitator, a kind of intellectual midwife whose sensitive radar helps others give birth to their own creativity.

Mazur's passion for ideas has had an impact beyond mathematics. He has written several expository works that attempt to give those outside of the field an authentic sense of its depth and beauty. One example is his 2003 book *Imagining Numbers (particularly the square root of minus fifteen)*, in which the protagonist is the concept of imaginary numbers. Tracing the life story of this concept, Mazur calls on his wide knowledge of literature, philosophy, and history to explore the nature of mathematical imagination as a collective pursuit by human beings across millennia.

Mazur holds a cross-disciplinary appointment at Harvard University, the Gerhard Gade University Professorship, which allows him to teach in various academic areas. He has collaborated with colleagues in the law school to teach courses on the nature of evidence, and with those in the history of science to teach courses on ancient geometry. When in 2018 students and colleagues held a conference at Harvard to honor Mazur in his 80th year, the proceedings ran for five days and included a stellar lineup of mathematical lectures together with panels on the history of science, on literature and poetry, and on law, philosophy, and physics.

Mazur's work has shown us that these fields are not isolated entities. The ideas that populate them are organically connected in the fabric of human knowledge. By illuminating the warp and weft of mathematics within this fabric, Mazur has enriched us all.

ALLYN JACKSON

Allyn Jackson is a freelance writer specializing in mathematics and theoretical computer science. For further information visit allynjackson.com.

This Page Intentionally Left Blank

2022 GAUSS PRIZE: ELLIOTT H. LIEB

ALLYN JACKSON

ABSTRACT

This article describes the work of Elliott H. Lieb, winner of the 2022 Carl Friedrich Gauss Prize, which was presented by the International Mathematical Union in conjunction with ICM2022. The Gauss Prize honors scientists whose mathematical research has had an impact outside mathematics, whether in technology, in business, or simply in people's everyday lives.

MATHEMATICS SUBJECT CLASSIFICATION 2020

Primary 82-03; Secondary 01A70

KEYWORDS

Statistical mechanics, Elliott H. Lieb

ICM INTERNATIONAL CONGRESS OF MATHEMATICIANS 2022 JULY 6–14

© 2023 International Mathematical Union
Proc. Int. Cong. Math. 2022, Vol. 1, pp. 560–564
DOI 10.4171/ICM2022/220

Published by EMS Press and licensed under a CC BY 4.0 license

For more than six decades Elliott Lieb has been among the most influential figures in mathematical physics. From his first work in the late 1950s through research that continues to the present day, he has displayed an uncanny ability to perceive the mathematical structures that lie at the heart of physical systems. In elucidating these structures, he has enriched both mathematics and physics.

DIFFERENT FIELDS, DIFFERENT GOALS

The two fields have always had a symbiotic relationship: Mathematics supplies a rigorous basis for expressing physical intuitions, and physics supplies rich inspiration for new mathematics. Nevertheless the two fields are very different in their goals, outlook, and culture. Lieb is nearly unique in having repeatedly made profound and ground-breaking contributions to both fields. Both have awarded him top honors; in this year alone, he receives not only the Gauss Prize, but also the 2022 APS Medal for Exceptional Achievement in Research, the highest honor of the American Physical Society.

Lieb is very much a mathematician in the way he applies the utmost rigor to problems from physics. He has produced mathematical results about classical questions that, at the time he addressed them, were not fashionable in physics but that later turned out to have an impact in that field. One example is Lieb's 1973 work with Mary Beth Ruskai, which proved a key result about relations among certain characteristics of quantum mechanical systems. That result, known as "strong subadditivity of the entropy," is today one of the cornerstones of the burgeoning field of quantum information theory.

At the same time, Lieb works like a physicist in that his main aim is to understand physical reality. His physical intuition has identified many ideas in physics that subsequently had a significant impact in mathematics. For example, in 1976 Lieb and Herm Jan Brascamp were led by their work in statistical mechanics to develop a new tool now called the Brascamp–Lieb inequalities. Thirty years later, these inequalities had a major impact in the branch of mathematics known as harmonic analysis, and they and their relatives appear in some of the work that earned Terence Tao a Fields Medal in 2006. Even more recently, the Brascamp–Lieb inequalities have had an impact in theoretical computer science.

Comprising over 400 publications across a variety of subjects, Lieb's opus is impossible to summarize in a short space. Instead we provide here a closer look at three examples of his work that convey a sense of his taste in problems and his approach to solving them.

SQUARE ICE

In the late 1950s, mathematical physics was concerned largely with classical mechanics and dynamics. Lieb and others forged a completely new line of research by using tools from mathematical analysis to attack problems in quantum and statistical mechanics. A signal example of this is Lieb's 1967 solution to the "square ice" problem from physical chemistry.

In landmark experiments in the 1930s, researchers were able to bring ice to extremely low temperatures and measure its "residual entropy." This quantity captures the amount of entropy, or disorder, that remains despite the low temperature and that cannot be accounted for by vibrations within the crystalline lattice of the water molecules.

Abstractly, one can picture frozen H_2O as a three-dimensional lattice, in which the oxygen atoms lie on the nodes of the lattice and the hydrogen atoms lie on lines connecting the nodes. A 1935 paper by Linus Pauling proposed what came to be known as the "ice rule." In the abstract lattice, the bonds between H and O atoms can be represented by arrows pointing inward towards the O atoms. The ice rule says that each node in the lattice has exactly two inward-pointing arrows.

The number of possible lattice configurations abiding by the ice rule grows enormously as the size of the lattice grows. It is this proliferation of configurations that produces the disorder, and thus the residual entropy, in ice. The two quantities – the number of configurations and the residual entropy – ought to be related by a simple mathematical expression. So if one knew the number of configurations and plugged it into that expression, would it match the residual entropy measured in the experiments?

This was the question Pauling asked. An exact calculation of the number of configurations was out of reach. Instead, Pauling made a careful estimate and found that it accorded very well with the experimental value. This has been hailed as one of the most successful confirmations of the validity of statistical mechanics.

But because the result relied on an estimate, its potential was unfulfilled. In the mid-1960s Lieb took up the two-dimensional version of the ice problem, which is called "square ice." In the square ice model, one has a two-dimensional lattice where the nodes in the lattice are connected by arrows that obey the ice rule: Each node has exactly two incoming arrows.

In 1967, Lieb used insights from mathematical combinatorics, together with concepts imported from a different part of physics, to calculate the exact number of configurations of square ice. This "magic number," as Freeman Dyson once called it, also aligned closely to the experimental value and confirmed the validity of the ice rule.

Immediately recognized as a turning point, this result ushered in the flourishing field of what is now known as exactly soluble models, which lies at the border of mathematics and physics. Lieb continued to make decisive contributions to this field, some of which subsequently had wide impact within mathematics. One example is a construct known as the Temperley–Lieb algebra, which Lieb invented with Neville Temperley and which played a key role in the revolutionary work in knot theory that earned Vaughan Jones a Fields Medal in 1990.

STABILITY OF MATTER

Lieb's square-ice result exemplifies a theme that has pervaded his work ever since: the quest to understand matter in the lowest energy states. It is in such states that one can hope to perceive the most fundamental structures of matter and investigate them mathematically. This was the motivation behind another facet of Lieb's work that we will now consider, his work on the stability of matter.

By the mid-1960s, the 40-year-old theory of quantum mechanics had been widely confirmed. But at its heart lay a basic unanswered question: Why is matter stable? Quantum mechanics says that the basic components of matter are electrons and positively charged nuclei. These oppositely charged particles ought to simply implode and collapse. But they don't. Instead, all matter around us – rocks, people, trees – remains stable. Can quantum mechanics account for this?

The first proof that the answer is yes came in 1967–68, in long papers by Freeman Dyson and Andrew Lenard. The goal is to show that the minimal energy of N particles scales not like N^2 – that is, the number of interactions among the particles – but rather like N. Dyson and Lenard reached this goal, showing that the minimal energy is less than a constant times N. However, due to an accumulation of inefficient estimates, that constant was so huge, on the order of -10^{15}, that it was physically meaningless.

Together with Walter Thirring, Lieb came up with a completely new and greatly improved proof of the stability of matter. Just four pages long, their 1976 paper was not only far simpler mathematically but also shed new light on the physics. In particular, they greatly sharpened the constant that Dyson and Lenard had groped for. This epitomizes a major theme in Lieb's work, which is to optimize constants to elucidate their physical meaning.

Together with Thirring and others, Lieb went on to investigate in a mathematically precise way how stability of matter is governed by two basic tenets of quantum mechanics, the Pauli exclusion principle and the Heisenberg uncertainty principle. They showed how both principles can most usefully be captured in what became known as the Lieb–Thirring inequality, which is a vast generalization of the classic mathematical result called the Sobolev inequality. The Lieb–Thirring inequality has also found applications beyond the problem of stability of matter.

This work fed back into mathematics, as Lieb and his collaborators worked on generalizing and sharpening related inequalities, such as the Hardy–Littlewood–Sobolev inequality. In the process, they uncovered symmetries that brought new meaning and usefulness to these tools. This work has had a major impact within mathematics, especially in the fields of analysis and geometry.

BOSE-EINSTEIN CONDENSATE

Our third example from Lieb's work concerns a state of matter called the Bose–Einstein condensate, a state that can be reached only at extremely low temperatures close to absolute zero. In this extraordinary state, quantum mechanical effects, which normally operate only at the microscopic level, emerge at the macroscopic level. Many of the properties of this state come from quantum mechanical dynamics having no classical analog.

The phenomenon was predicted in the mid-1920s by Albert Einstein, following ideas of Satyendra Nath Bose. However, the technical capability of bringing matter to such low temperatures took another 70 years to develop. The physicists who produced the first Bose-Einstein condensate in 1995 received the Nobel Prize for their achievement. That landmark work set off a burst of new research.

It was in the early 1960s that Lieb first took up this problem. Earlier work had resulted in a formula for the ground-state energy in a Bose–Einstein condensate. While correct physically, the formula lacked a rigorous mathematical basis. Lieb hoped to supply that basis by proving the validity of the formula. In 1963 he managed to re-derive the formula in a new way, providing additional confirmation of its basic correctness. However, he was not able to prove its validity.

In a tour de force that exemplifies Lieb's persistence and long-term view, he finally produced the proof 40 years later, in a 1998 paper with Jakob Yngvason. Coming on the heels of the 1995 experiments, the Lieb–Yngvason paper added to the surge of interest in Bose–Einstein. The topic has since become one of the most active areas of research in mathematical physics.

In related work, Lieb, together with Ian Affleck, Tom Kennedy, and Hal Tasaki, invented and solved what is now known as the AKLT quantum spin system. Carried out in 1987, this work provides an early example of a system exhibiting what is today referred to as a topological state of matter, a subject of great current interest.

SHAPING DECADES OF RESEARCH

Over his long career, Lieb has had more than 100 co-authors. Many of these collaborations have had an intense, exhilarating quality, due to Lieb's prodigious intellectual energy, immense powers of concentration, and exacting work ethic. These traits have also marked his interactions with young researchers, including his ten doctoral students, all of whom have gone on to flourishing careers of their own. Some of them appeared on the stellar list of speakers for a conference honoring Lieb's 90th birthday, held 30 July to 1 August, 2022.

Lieb has also made notable contributions to support the professions of mathematics and of physics. He twice served as president of the International Association of Mathematical Physics (1982–1984 and 1997–1999). During 1992–1995, he served as a Member-at-Large of the Council of the American Mathematical Society. His exceptional probity and integrity led in 1994 to his appointment to a committee that formulated the Society's first-ever ethical guidelines.

In shaping decades of research in mathematics and in physics, Elliott Lieb has reached to the very roots of these twin trees of human knowledge. He stands out as one of the great thinkers of our time.

ALLYN JACKSON

Allyn Jackson is a freelance writer specializing in mathematics and theoretical computer science. For further information visit allynjackson.com.

This Page Intentionally Left Blank

2022 LEELAVATI PRIZE: NIKOLAI ANDREEV

ALLYN JACKSON

ABSTRACT

This article describes the work of Nikolai Andreev, winner of the 2022 Leelavati Prize, which was presented by the International Mathematical Union in conjunction with ICM2022. The Leelavati Prize honors outstanding contributions to increasing public awareness of mathematics as an intellectual discipline and of the crucial role it plays in diverse human endeavors.

MATHEMATICS SUBJECT CLASSIFICATION 2020

Primary 00A09; Secondary 01A70

KEYWORDS

Mathematics popularization, Nikolai Andreev

ICM INTERNATIONAL CONGRESS OF MATHEMATICIANS 2022 JULY 6–14

© 2023 International Mathematical Union

Proc. Int. Cong. Math. 2022, Vol. 1, pp. 566–570
DOI 10.4171/ICM2022/221

Published by EMS Press and licensed under a CC BY 4.0 license

On the screen is a schematic diagram showing three rods – one long, one medium, one short. Their arrangement looks a bit like the Greek letter lambda, λ. The background is blank, revealing nothing. Touches of color give hints: red for fixed ends of the rods, blue for free ends, and gray for joints.

As you start to wonder "What is this thing?" it all starts to move. The free end of the short rod traces out a circle. That motion drives the long rod to trace out a closed curve shaped like a mushroom cap, flat on the bottom, domed on the top. Nothing goes by too fast, so you absorb without effort the equality of the phase of the circular and mushroom-cap motions.

Wooden model of Tchebyshev's walking machine

Now a new rod takes its place, one end attached so as to trace out the mushroom curve, the other sporting a flat block looking vaguely like a foot. A mirror-image duplicate of the ensemble joins the first, linked to coordinate the phases. Now there are two feet, and it dawns on you that this thing is *walking*.

It ambles out of the picture, leaving the screen momentarily blank. That was the skeleton. Enter now the *real* walking machine, physically rendered in wood. As it executes its precise gait, the circle and the mushroom curve hover like ghosts in your mind. The wooden creature – it somehow has life to it – turns away and continues its journey off into the distance. The impression is quietly electrifying.

The movie just described is one of the many works of Nikolai Andreev, head of the Laboratory of Popularization and Promotion of Mathematics at the Steklov Mathematical Institute. This particular movie is based on the original mechanism designed and built by the mathematician Pafnuty Tchebyshev (1821–1894) for the 1878 World Fair in Paris. When the movie was posted on the web in 2007, it garnered more than one hundred thousand views on its first day. Since then has been shared and reproduced countless times.

AN UNUSUAL APPROACH

Trained as a mathematician, Andreev completed his PhD degree in 2000 in the Faculty of Mechanics and Mathematics of Moscow State University. That same year he joined the staff at the Steklov Institute and began to work on popularization. He became head of the laboratory when it was founded in 2010, and today he has a team of three: Roman Koksharov (illustrator and web-developer), Alexander Leshchinsky (woodworker and creator of mathematical models), and Nikita Panyunin (mathematics researcher). Together they are inventing new forms through which to bring the rich and distinctive Eastern European culture of mathematics to a wide audience.

The team's approach to popularization is unusual. They are not trying to explain the latest mathematics, nor even to focus mainly on topics important in the history of the

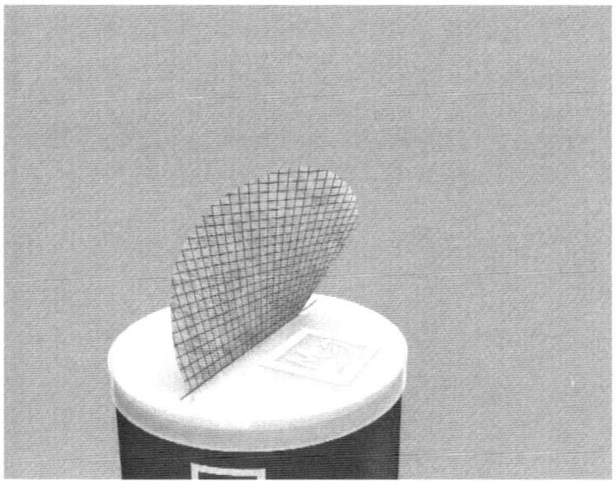

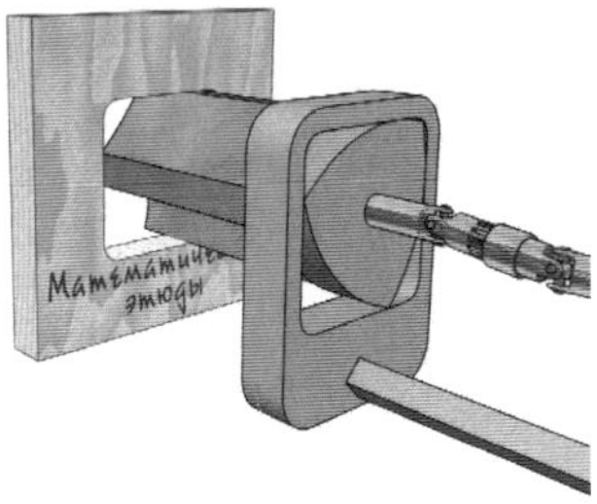

Left: A hyperbolic paraboloid passes through a slit in the lid of a potato chip can **Right:** The cross-section of the orange piece is a Reuleaux triangle, which allows one to drill a square holes.

subject. They do not rely on the usefulness of mathematics as a lure. Their materials are not primarily didactic, though they have been used in education. There is no exciting packaging, no attempt to dazzle or divert. In shedding preconceptions about what makes popularization successful, Andreev's group allows the mathematics to shine.

Their main activity is production of animations, which now number close to 200 and all of which are freely available on the internet. In today's video-saturated world, where expositions about mathematics often feature complicated graphics like pictures of fractals or even cute cartoon characters, Andreev's movies stand out for their minimalist approach.

Without background sound, the mathematics unfolds in serenity. Three-dimensional graphics are state-of-the-art and purposely spare. Exactly which details are included, the proportion of the figures, the choice of the colors, the pacing of the action – everything conspires to evoke the mathematics, so much so that one often has the sense of seeing mathematics in its native land. Nevertheless the movies are not somber affairs. They strike a contemplative tone but with a touch of whimsy. One senses the smile of the artist behind each one.

A POTATO CHIP, A SAUSAGE, A SHEET OF PAPER

If you cut a straight slit in the lid of a potato chip can, could you slide a chip through? You could if the chip belongs to the class of surfaces known as ruled surfaces, as shown in the animation "Chips as Hyperbolic Paraboloid." How can you create a perfect sine wave? In "Sine wave: cylinder net," a knife slices a sausage crosswise at an angle of 45 degrees, then slices the casing along the length of the sausage. The sausage disappears, leaving just the casing, which unrolls to reveal that one of its edges is now a sine wave. Viewers can use the animations as a basis to craft physical models of their own.

Folding of paper is a running theme in the animations. "Piecewise Linear Embedding of a Polyhedron" creates out of a flat piece of paper a three-dimensional shape that is curved everywhere and nowhere flat. In another video, a paper-folding technique illustrates a proof that the infinite sum of reciprocals of powers of two equals 1. Another theme in the animations is the Reuleaux triangle, a two-dimensional figure that rolls as smoothly as a circle and can perform all sorts of amazing feats, like drilling a square hole in wood.

A few of the videos touch on unsolved problems. For example, one video invites viewers to enter sequences of digits and locates those sequences in the infinite, nonrepeating decimal expansion of π. This leads to the question: Is π a *normal* number? That is, do all finite sequences of digits have equal likelihood of appearing in the decimal expansion of π? Although the question has been around for more than a century, no one knows the answer.

One group of animations, coming under the rubric "Tchebyshev's Mechanisms," provide on-screen elucidation of the ingenious mechanical devices, like the walking machine discussed above, that Tchebyshev designed and had built. Andreev is the curator for some of Tchebyshev's original wooden models, while others are housed in various museums. Those models served as source material for the "Tchebyshev's Mechanisms" animations.

FROM MULTIMEDIA TO PRINT

Andreev's lab has a few other historical models; one is a model of inscribed Platonic solids that the mathematician V. I. Arnold (1937–2010) made as a child from pieces of wooden ski poles. In addition, the lab has continued the tradition of creating wooden models that "do math" by crafting many of its own. Team member Alexander Leshchinsky is a master woodworker who has developed deep intuition for creating wooden models that are both artistic and mathematically precise.

In his two decades as a mathematics popularizer, Andreev has made around a thousand presentations: talks for schoolchildren, seminars for schoolteachers, conference lectures, and master classes. In these presentations he often uses physical models to illustrate the ideas, and he has worked closely with schoolteachers to explain how the models are designed and produced. The teachers then help their students to build their own models. If the students successfully present a model in their school, they can earn a small prize from Andreev's lab.

In English, the book's title might be "A Mathematical Take on Things"

The prize is a book, which stands as yet another distinctive work of the lab. After spending a decade on multimedia materials, the members of the lab returned to old roots in the written word. There were several reasons for this, one being that reading a book often leads more easily to deep contemplation than does staring at a computer screen. The members of the lab served as editors for the book and solicited contributions by thirty-two leading mathematicians, including three Fields Medalists.

The book, written in Russian, has a title that is difficult to translate adequately into English; possibilities include *Mathematical Component*, *Mathematical Essence*, or perhaps even *A Mathematical Take on Things*. The book is close in spirit to *Kvant*, the legendary mathematics magazine that began in the Soviet Union in 1970 and remained highly popular until it ceased publication in 2011. The aim of the book is to evoke the mathematics that is present all around us, in the great achievements of human civilization as well as in the more modest setting of everyday life.

In addition to the thirty-two solicited essays, the lab team wrote a few dozen of their own. The essays run from two to perhaps five or six pages and cover topics from cryptography to mechanics, from language to pattern formation. As with the animations, the choice of topics is highly original; nothing is trite. The first edition appeared in 2015 and is one-third the size of the second edition, which came out in 2019. The book has been very popular and won two awards for literature about science aimed at a general audience. Hopes are high for translations into other languages.

REACHING ACROSS BARRIERS

The work of Nikolai Andreev and his team constitutes an outstanding artistic and scientific achievement. Reaching across barriers of geography, language and culture, it brings the delight of mathematics to people of all ages. Perhaps more importantly, it inculcates respect for truth and rational thought. Their work is a positive force for unity the world over.

ALLYN JACKSON

Allyn Jackson is a freelance writer specializing in mathematics and theoretical computer science. For further information visit allynjackson.com.

LIST OF CONTRIBUTORS

European Mathematical Society – EMS – Publishing House GmbH
Institut für Mathematik
Technische Universität Berlin
Straße des 17. Juni 136
10623 Berlin, Germany

ISBN 978-1-7138-9868-9
9 781713 898689

CURRAN ASSOCIATES INC.
proceedings.com

International Conferences Interfaces and Human Computer Interaction (IHCI2020); and Game and Entertainment Technologies (GET2020)

Held at MCCSIS 2020 (Held online due to COVID-19)

Lisbon, Portugal
21 - 25 July 2020

iadis